AF513822

A Textbook on Hydrogen Energy

Pratibha Sharma

A Textbook on Hydrogen Energy

Production, Storage, Transport and Utilization

Pratibha Sharma
Department of Energy Science
and Engineering
Indian Institute of Technology Bombay
Mumbai, Maharashtra, India

ISBN 978-981-95-4600-8 ISBN 978-981-95-4601-5 (eBook)
https://doi.org/10.1007/978-981-95-4601-5

This Springer imprint is published by the registered company Springer Nature Singapore Pte Ltd.
The registered company address is: 152 Beach Road, #21-01/04 Gateway East, Singapore 189721, Singapore

Competing Interests The author has no competing interests to declare that are relevant to the content of this manuscript.

Contents

About the Author

Prof. Pratibha Sharma is currently the Associate Dean (Academic Programmes) and a Professor in the Department of Energy Science and Engineering at the Indian Institute of Technology Bombay, India. She also holds the Cummins Chair Professorship at the Institute. Prof. Sharma is a Fellow of the Indian National Academy of Engineering and a recipient of several recognitions, including the Outstanding Service Award from the International Association of Hydrogen Energy. She has contributed as an expert to various inter-ministerial committees and is on the evaluation panels of various National funding agencies in the field of hydrogen research and development. She is a member of the advisory group to India's National Green Hydrogen Mission. Her research focuses on hydrogen storage materials, metal hydride based systems, hydrogen infrastructure development, solid-state storage technologies for various applications including both stationary and vehicular. She has several patents, multiple peer reviewed Journal publications, thesis supervisions, prototypes and technology demonstrations to her credit. She has also led several national and international collaborative projects with academic, governmental, and industrial partners, with several outcomes contributing to national hydrogen initiatives. Along with her research contributions, she has been engaged in capacity-building efforts through national and international workshops, professional training programs, short term courses and a widely attended NPTEL course on "Hydrogen Energy: Production, Storage, Transportation and Safety".

Chapter 1
Properties of Hydrogen

Introduction

Energy is one of the most debated issues in the current global scenario, despite the availability of a wide range of fuel sources and technologies are available. The major economies in the world are expanding, and the energy requirements of most of the nations are growing at a faster rate. The need is to sustain economic growth while managing greenhouse gas emissions and ensuring energy security. In fact, the two major global concerns are: insecurity towards the availability of fossil fuels due to current geopolitical situations and climate change. Among all the sectors, such as industrial, residential, transportation, and heating, the sector that will get most affected is transportation. Today's energy and transportation system is mainly based on fossil fuels. At present, oil, with a share of 33% in the global primary energy mix, remains the largest primary fuel, accounting for more than 90% of the energy demand in the transport sector, which in turn accounts for two-thirds of the global oil use. Unfortunately, the combustion of hydrocarbon fuels for transportation and heating contributes over 16% of all greenhouse gas emissions and a significant fraction of air pollutant emissions. The global energy requirements (coal: 27.4%, oil: 33%, and natural gas: 27.6% as per the Statistical Review of World Energy) are primarily met by fossil fuels, and the energy-related carbon emissions have reached 40.8 Gt of CO_2. Thus, the modern world faces an urgent need to develop alternative fuels. Among the varioust alternatives available, hydrogen as a fuel offers substantial benefits due to its diverse production pathways, high energy density and reduced emissions of pollutants and greenhouse gases. The key criteria for any fuel are its inexhaustibility, cleanliness, convenience, and independence from imports/control. Hydrogen possesses all these properties and is well-suited for almost all sectoral applications.

P. Sharma, *A Textbook on Hydrogen Energy*,
https://doi.org/10.1007/978-981-95-4601-5_1

Hydrogen as a Fuel

Hydrogen is a promising energy carrier due to several advantages; it is a versatile fuel with a wide range of applications, can be produced from a wide variety of sources, has the highest energy density per unit weight, is the most abundant element, does not release carbon-based emissions at the point of use, is the lightest and simplest molecule, and can contribute to energy security. The use of hydrogen can not only address the issues associated with climate change which are of major concern globally, but also reduce dependence on fossil fuel imports, as hydrogen can be produced from any local feedstock available. The use of hydrogen is possible in a variety of sectors, including industries, domestic, heating, cooling, and transportation. Some of the sectors may have other competing technologies and options; however, there are certain hard-to-abate sectors where hydrogen proves to be the most promising alternative to decarbonize, and other alternatives may not be as feasible. The use of hydrogen can provide a lot of flexibility to the energy system through options like gas to X, where X can be gas, liquid, fuel, or power among others. Industries, especially fertilizer and refineries sectors, have been utlizing hydrogen for various processes and product synthesis for several decades. While the transition to using cleaner fuel or feedstock hydrogen cannot happen abruptly, a swift transition will allow for the use of existing assets and skill sets to the new fuel arena, wherein blending can provide a short-term or medium-term option. With the scaling up of technologies, economies of scale, improvements in efficiency, cost reductions, and progress along learning curve, the use of hydrogen will become more feasible and economically viable.

Properties of Hydrogen

Hydrogen was first discovered in sixteenth century when Paracelsus accidentally poured sulfuric acid on iron fillings and noticed a gas being released, which was flammable. However, Henry Cavendish in 1766 for the first time identified that the gas released was a separate element. It was in 1780s that Antoine Lavoisier named the gas hydrogen. The term hydrogen originates from the Greek words hydro and genes, where hydro means water and genes means to produce; thus, hydrogen means "water producer". Hydrogen is the most abundant element in the universe, accounting for approximately 75% by mass of the universe and 90% by volume. However, in the earth's atmosphere, the quantity of hydrogen is very small, due to its high diffusivity. Hydrogen doesn't exist in its atomic or molecular form; it is found in nature mostly combined with various other elements like oxygen, and carbon, forming hydrocarbons, water, and other organic molecules.

Hydrogen does not exist on earth in its pure form (except at a few parts per million), and energy is required to produce it from hydrogen-containing compounds. Therefore, hydrogen is called an energy carrier, not an energy source.

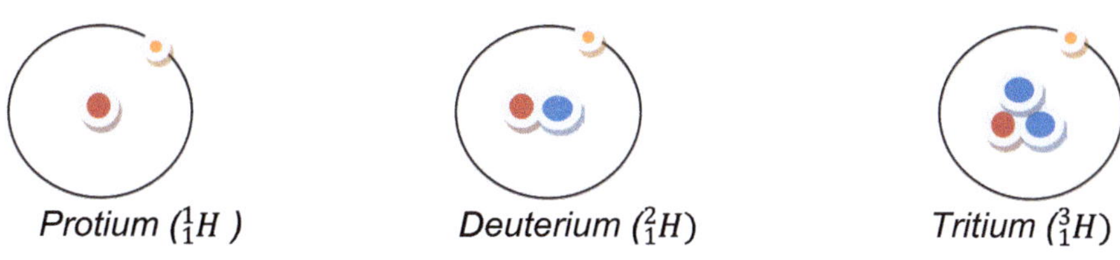

Fig. 1.1 Different isotopes of hydrogen

Hydrogen is the first element in the periodic table, and although it is non-metallic in nature, it is kept with alkali metals in period 1. It has one proton in the nucleus and one electron and has an atomic mass of 1.008 g/mol. Hydrogen forms a diatomic molecule and has a very high dissociation energy of 435 kJ/mol; i.e., this much energy is required to break the hydrogen molecule into atomic hydrogen. In its atomic form, it is highly reactive and can combine with various other elements, forming either a covalent bond (enabling an octet of other atoms, as hydronium ion) or an ionic bond (in the form of hydride, which combines with cations).

Hydrogen is, in fact, the lightest of all molecules (molecular weight = 2.016), the first element of the periodic table with atomic number 1 and configuration $1s^1$. Hydrogen has three isotopes. *Protium* ($^1_1$H) contains one proton and one electron, has an atomic mass of 1.008 g/mol, and a relative abundance of 99.98%. D*euterium* ($^2_1$H) contains one proton, one electron, and one neutron, with an atomic mass of 2.014 g/mol, and an abundance of 0.02%. T*ritium* ($^3_1$H) contains one proton, one electron, and two neutrons, has an atomic mass of 3.016 g/mol and negligible natural abundance, and is formed during nuclear reactions (Fig. 1.1).

Two forms of hydrogen exist, named as ortho hydrogen and para hydrogen, with similar chemical properties but slightly different physical properties. The primary difference between the two states is the nuclear spin, with parallel nuclear spins in ortho hydrogen and antiparallel spins in para hydrogen. Under ambient conditions, normal hydrogen consists of 75% ortho and 25% para hydrogen. However, the liquefaction of hydrogen requires the conversion of ortho hydrogen into para hydrogen because at lower temperatures (< 80 K) equilibrium favors the existence of para hydrogen, which is the lower energy state. The conversion of one form to another occurs under certain conditions. The conversion process is exothermic, and the heat of conversion is temperature dependent. It's a very slow process and requires catalysts like activated charcoal, magnetic impurities, or a small concentration of oxygen. The heat of conversion for the o-p transition at 300 K is 270 kJ/kg, at 77 K is 519 kJ/kg, at temperatures below 77 K is 523 kJ/kg, and at 20 K is 703 kJ/kg. The conversion enthalpy at 21 K is –1.063 kJ/mol, which is even higher than the evaporation enthalpy of + 0.904 kJ/mol. Since the process of ortho-to-para hydrogen conversion is a very slow process, thus under a non-equilibrium situation, there remains a certain amount of ortho hydrogen, which on conversion to para will evaporate the hydrogen, leading to its loss (can be 18% at 20 K within 24 h). At room temperature, 75% ortho and 25% para hydrogen exist. At 80 K both exist in equal amounts, i.e., 50% each. At 20 K, 99.821% of para and 0.179% of ortho hydrogen exist, while at 0 K, only para hydrogen exists. The physical properties of the two forms of hydrogen, ortho and para hydrogen, are slightly different; for example, their boiling and melting points differ by 0.1 K. However, the heat capacity and thermal conductivity are different and are higher for para hydrogen, and they differ by a factor of 1.4 and 1.3, respectively, in the temperature range of 150–200 K. The basic difference arises due to the large energy difference between the ortho and para forms. The normal boiling point of hydrogen is 20.3 K, and its melting point is 14.1 K.

The phase diagram for hydrogen is shown in Fig. 1.2. The triple point exists at $T = 13.8$ K (−259 °C) and $P = 7.2$ kPa. Since the pressure involved is low, any air ingress under these conditions could be of concern. The critical point for hydrogen is at $T = 33.2$ K (−240 °C) and$P = 1.3$ MPa, and this is the highest temperature, till which hydrogen gas can be liquefied under mentioned pressure. Beyond the critical temperature, liquefaction is not possible by increasing pressure. The latent heat of vaporization at NBP is 445.6 kJ/kg, while the heat released on ortho-to-para conversion at a temperature of 20 K is 670 kJ/kg, which thus makes it essential to remove this heat released while liquefying hydrogen. Heat of melting (fusion) at the melting (freezing) point is 58.8 kJ/kg. Heat of sublimation is 379.6 kJ/kg.

In the event of a liquid hydrogen spill, since the density of saturated hydrogen vapor is high, a cloud is formed that will either move horizontally or move downward. There are other safety challenges associated with liquid hydrogen due to its extremely low temperature. In the event that air enters the system, it can lead to

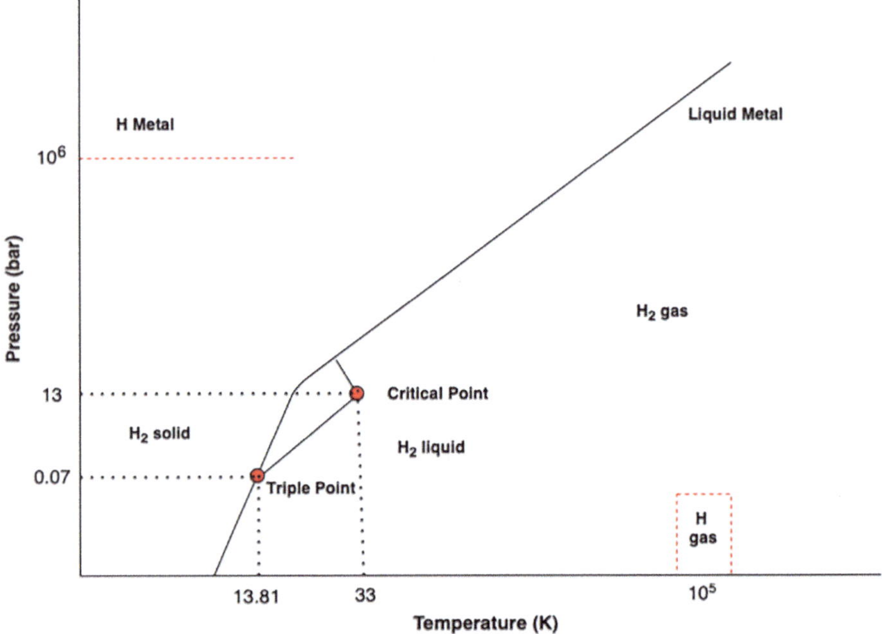

Fig. 1.2 Phase diagram of hydrogen

the solidification of air in the pipes or valves, resulting in blockages. Besides the ingress of unwanted gases into the liquid storage system will lead to their condensation, thereby reducing their volume, which could further create a vacuum, resulting in more air entering the system. This solidification of air can create problems when there is a heat inflow resulting in evaporated hydrogen or when it is intentionally heated from liquid hydrogen to gaseous hydrogen. The mixture can form a combustible mixture and also lead to very high pressures inside the system involved. Thus, as a safety check, containers with liquid hydrogen should not have oxygen enrichment, which can lead to increased flammability. Therefore, such vessels should be periodically warmed and purged to keep the oxygen content below 2% (IS/TR 15,916:2004).

Density of Hydrogen

Hydrogen is the lightest element, with a density of only 0.0899 kg m^{-3} at normal temperature (273.15 K) and pressure (101.325 kPa), i.e., 7% of the density of air. Liquid hydrogen also has a low density of 70.8 kgm^{-3} (7% of that of water). The liquid hydrogen has a very low boiling point (20.3 K) and thus requires fairly sophisticated technology to produce and maintain in this state. By virtue of its exceptionally low density, hydrogen has the best energy-to-weight ratio (heating value) of any fuel, but its energy-to-volume ratio is poor. The higher heating value of hydrogen is 141.78 MJ kg^{-1} H_2, and its lower heating value is 120 MJ kg^{-1} H_2. A comparison of properties of hydrogen against other fuels is shown in Table 1.1. Hydrogen has a low vapor density, which makes it the most buoyant gas; thus it is safe in case of unwanted leakage.

Diffusivity and Buoyancy

The diffusivity of hydrogen is highest, being the smallest molecule. The rate of diffusion of hydrogen in air is, in fact, four times that of rate of diffusion of air within air. Diffusivity of hydrogen is 0.63 cm^2s^{-1} as compared to 0.08 cm^2s^{-1} for gasoline. Hydrogen has both high diffusivity and high buoyancy, and thus it mixes with air at a faster rate. With its high diffusivity and buoyancy, it rises quickly and becomes diluted in the event of leaks, resulting in a non-flammable concentration very rapidly. The small molecular size and low viscosity of hydrogen result in a high flow rate in the event of a leak. It can leak from joints and welds, and even from, places that would remain intact for other gases. The low viscosity of hydrogen, its small size, high diffusivity, and high buoyancy are assets in the event of a leakage, which can lead to its faster dispersion and result in dilution in air forming concentrations below the lower flammability limit, in an unconfined space. The release in a confined space can lead to its accumulation underneath the roof. Thus, its mixing rate with air is

Table 1.1 Comparison of the properties of hydrogen with other fuels [1–5]

Properties	Units	Hydrogen	Methane	Propane	Methanol	Ethanol	Gasoline	Ammonia
Chemical formula		H_2	CH_4	C_3H_8	CH_3OH	C_2H_5OH	C_xH_y (x = 4–12)	NH_3
Molecular weight	g/mol	2.02	16.04	44.1	32.04	46.07	100−105	17.03
Density (NTP)	kg/m^3	0.0838	0.668	1.87	791	789	751	0.769
Normal boiling point	°C	−253	−162	−42.1	64.5	78.5	27–225	−33.35
Flammability range in air	Vol%	4.0–75.0	5.0–15.0	2.1–10.1	6.7–36.0	4.3–19	1.4–7.6	15–28
Autoignition temperature in air	°C	585	540	490	385	423	230−480	650
Liquid density	kg/m^3	70.8	425	507	797	789	846	771
Ignition energy	mJ	0.02	0.3	0.25	0.14		0.25	380–680

determined by both buoyancy and diffusivity. The high diffusivity and buoyancy have both advantage as well as disadvantage. The advantage is that if there is a hydrogen leak, it can mix with air to dilute very quickly, and any hazardous situation is excluded. However, its main disadvantage is that it mixes rapidly with air and can form a flammable mixture. If a source of ignition is present nearby, and can lead to accidents. Thus, when hydrogen release occurs in unconfined regions, high buoyancy is a safety asset. However, in the case of a hydrogen leak occuring in confined spaces, hydrogen can accumulate under the roof. When compared to other fuels, such as heavier hydrocarbons, which forms a combustible cloud on release and can cause fire or explosion in the presence of an ignition source, the high buoyancy and diffusivity of hydrogen make it safer than these heavier hydrocarbons. Hydrogen, being the smallest element with the lowest molecular weight, and its low viscosity, has a propensity toward leakage at higher molecular flow rates than any of the other gases. The addition of an odourant or colorant for the detection of hydrogen in the case of a leak is difficult, unlike the other flammable gases. At 20 °C and 1 bar, the viscosity of gaseous hydrogen is 0.88 (10^{-5} Pa-s), that of air is 1.82 (10^{-5} Pa-s), and that of methane is 1.10 (10^{-5} Pa-s).

The heat capacity and thermal conductivity of hydrogen are higher than the other gases. Another important property related to liquid hydrogen is tremendous volume increase that occurs upon its phase change to gaseous hydrogen. From liquid hydrogen to gaseous hydrogen at NTP, the volume expansion is 847, meaning that it expands 847 times its initial volume. If the liquid hydrogen was initially at 1 bar pressure, this much volume rise can result in overpressurization in the storage vessels and increase the pressure in such case, from 1 bar to 177 MPa, making it essential to have pressure relief valves in liquid-state storage facilities. Compared to the expansion ratio (final volume to initial volume) of liquid hydrogen, the corresponding gaseous hydrogen expansion ratio is small, and value is 240 for expansion from 250 bar to 1 bar.

Joule–Thomson Effect

The Joule–Thomson coefficient is defined as

$$\mu_{JT} = \left(\frac{\partial T}{\partial P}\right)_H = -\frac{V(1 - \beta T)}{C_p}$$

where $\beta = \left(\frac{\partial V}{\partial T}\right)_P$.

A negative Joule–Thomson coefficient means that the gas cools on expansion, while a positive coefficient indicates heating of the gas on expansion. For ideal gases, the Joule–Thomson coefficient is zero, while for real gases, when different temperature–pressure curves are drawn at constant enthalpy, a trend as shown in Fig. 1.3 is observed.

The curve passes through a maximum called the inversion point (the point where the JT coefficient changes sign), and the curve is called the inversion curve. The Joule–Thomson coefficient at the inversion point ($T = 1/\beta$) is zero. For hydrogen, the inversion temperature is 202 K; below this temperature, the Joule–Thomson coefficient is negative, and it cools on expansion. In contrast, for temperatures above the inversion temperature, the Joule–Thomson coefficient is positive, and it warms on expansion.

Ideal Gas and Real Gas Behavior

The ideal gas equation is applicable for hydrogen at moderate pressures and a wide temperature range. However, at high pressures, i.e., in the case of compressed hydrogen, the non-ideal behavior needs to be included, and this can be considered by defining the compressibility factor Z, with

$$Z = 1 + \frac{aPM}{\mathrm{RT}}$$

where, a is a constant with value 7.96 X 10^{-3} m^3/kg.

P is pressure.

M is the molecular mass of hydrogen, 2.016 kg/kmol.

R is the universal gas constant 8.3145 J/mol/K.

At a temperature of 293.15K, the compressibility factor Z has the values: 1.01 at 15.7 bar, 1.1 at 157 bar, and 1.5 at 786 bar.

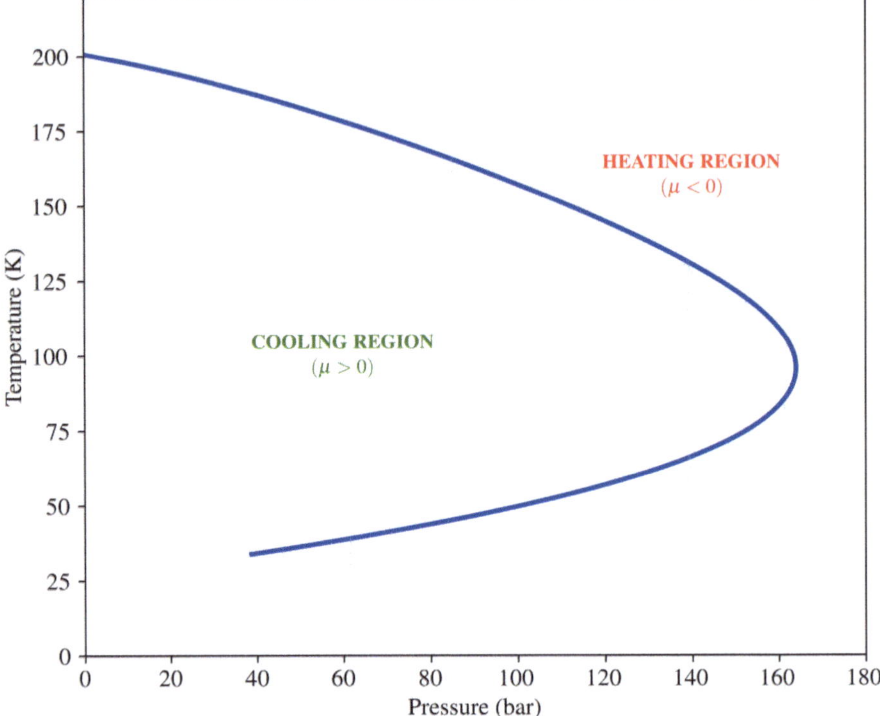

Fig. 1.3 Inversion temperature for a gas

Properties of Hydrogen as a Fuel

Hydrogen is colorless, tasteless, odorless, flammable, non-corrosive, non-toxic, physiologically not harmful but can lead to asphyxiation, i.e., displaces oxygen. Hydrogen burns cleanly in air, and water vapor is the only product, as such it is clean energy carrier at the point of use. There are various properties of hydrogen that are essential to be known before studying in detail the various aspects of hydrogen. Some of these properties are advantageous; however, some can lead to safety challenges in case of accidents.

Flammability Limit

Flammability limit is defined as the range of concentration of hydrogen at a specific temperature and pressure condition, which can result in fire or explosion in the presence of an ignition source and oxidant. It can also be defined as the concentration range of the combustible mixture that supports a self-propagating flame in the event of ignition. The two limits of the range are known as lower and upper flammability limits (LFL and UFL, respectively), where LFL is the lowest concentration that can support a self-propagating flame in the presence of an ignition source and UFL is the highest concentration that can support a self-propagating flame in the case of ignition. Below LFL, the mixture is too lean in hydrogen, and above UFL, the concentration of hydrogen is too high or the oxidant concentration is too low to sustain combustion.

For hydrogen, the flammability limit is 4–75 vol% at NTP, which is very wide compared to other fuels, as mentioned in Table 1.1; for example, for gasoline it is 1–7.6%. Thus, a leaner or a concentrated mixture of hydrogen with an oxidant can ignite in the presence of an ignition source.

The flammability range is a function of temperature, and at high temperatures, it becomes more wider. Forexample, at 400 °C, the LFL is 1.5 vol%, and UFL is 87.5 vol%. It's also a function of pressure; the LFL increases from 4 vol% to 5.6 vol% when the pressure increases from 1 to 50 bar. However, the change in UFL with pressure does not show a trend; it varies differently in different pressure ranges. Hydrogen detonates when the concentration is in the range of 18.3–59% by volume; this range is the explosive limit for hydrogen. It is defined as the range of concentration that will cause an explosion when mixed with air and in case gets ignited. However, there is a difference between fire and explosion; for an explosion, containment is required. So, when a combustible mixture is contained, is within the explosive limits, and gets ignited, then the temperature and pressure within the containment increase in such a manner that it gains sufficient energy to violently release from confinement and lead to an explosion.

The wide flammability limit of hydrogen has both advantage as well as disadvantage. The advantage is that when hydrogen is used as a fuel in hydrogen IC engines, even dilute mixtures can undergo combustion. This results in better combustion characteristics: even leaner and more concentrated mixtures can burn, have easier start of combustion, and engine operation is more stable with reduced emissions. At the same time this is a disadvantage; if there is a leak, even a very dilute mixture or a very concentrated mixture, both can ignite and result in a fire or explosion, posing a safety concern.

Stoichiometric Mixture

A stoichiometric mixture of hydrogen and oxidant is a mixture in which both fuel and oxidiser are entirely consumed in the process, i.e., complete combustion occurs to form the combustion product(s).

Stoichiometric concentration of hydrogen in air mixture: 29.59 vol. %.

Assuming air consists of 21 vol. % of oxygen (O_2) and 79 vol. % nitrogen (N_2):

$$2H_2 + 3.76N_2 + O_2 \rightarrow 2H_2O + 3.76N_2$$

Stoichiometric concentration of hydrogen in oxygen mixture: 66.66 vol. %.

Equivalence Ratio

Equivalence ratio is defined as the actual fuel-to-air (or oxidant) ratio divided by the stoichiometric fuel-to-air (or oxidant) ratio. For hydrogen, the range is 0.1–7.1, while for gasoline, the range is 0.7–4. Thus, for hydrogen this is quite wide.

The equivalence ratio is represented as:

$$\varphi = \frac{(\text{fuel}/\text{oxidant})_{\text{actual}}}{(\text{fuel}/\text{oxidant})_{\text{stoichiometric}}}.$$

$\varphi = 1$ for stoichiometric, < 1 for lean,and > 1 for concentrated mixture.

Ignition Energy

Ignition energy is the minimum energy required to ignite a hydrogen and oxidant mixture under a given set of conditions, i.e., the minimum energy required to initiate the combustion of a combustible mixture of fuel and oxidant. For hydrogen, the value of this energy is very small, i.e., 0.02 mJ, which is an order of magnitude less than that of gasoline (for gasoline, it is 0.24 mJ). Thus, a very small energy input even, a static charge, is enough to initiate combustion within the flammability limit. Any ignition source, such as a hotspot, static charge, open flame, electrical short circuit, or spark, can lead to the ignition of a combustible mixture with hydrogen. This raises a safety concern at the place of production, storage, transportation, and utilization of hydrogen. The ignition energy depends on the concentration of hydrogen, temperature, and pressure and varies based on the values of these parameters. Ignition energy decreases with the increase in temperature, pressure, and oxidant concentration.

Thus, any flow or agitation in the case of gaseous or liquid hydrogen, leading to friction or resulting in electrostatic charges, can initiate ignition. As such, wherever hydrogen is contained, those vessels or storage tanks are usually grounded. Now, this small ignition energy again has an advantage, as it leads to prompt ignition when used in IC engines; even a leaner mixture can cause ignition.

However, at the same time, the major challenge that arises from this small ignition energy is that even hotspots, hot gases, or hot surfaces can also serve as a means for ignition, resulting in premature ignition and flashback.

Autoignition Temperature

Autoignition temperature is defined as the minimum temperature, which is required to initiate self-sustained combustion in a combustible fuel oxidant mixture, in the absence of any external ignition source. The autoignition temperature of hydrogen is higher than that of other fuels; it is 585 °C for hydrogen, while for gasoline, it lies in the range of 240–460 °C. A high autoignition temperature is an advantage when hydrogen is used as a fuel in internal combustion engines, as it allows for a higher compression ratio, resulting in higher efficiency. A higher autoignition temperature makes hydrogen–air mixture difficult to ignite using heat alone, meaning that an additional ignition source is required to initiate combustion. Thus, there is a requirement for a spark to initiate combustion in the engine; therefore, hydrogen is suitable for spark ignition engines.

Hydrogen Embrittlement

Hydrogen gas can dissolve in various materials and liquids and can permeate into the container materials. In the presence of hydrogen, metals can lose their ductility and become brittle. Consequently, the diffusion of hydrogen can result in cracking, or existing cracks can propagate due to the diffusion of hydrogen. The material failures occur due to cracking caused by the stresses being created. However, the degree of embrittlement in a material depends on the amount of hydrogen absorbed and material's microstructre. In order to avoid embrittlement at various stages of the hydrogen pathway, appropriate materials need to be selected, e.g., special alloy steels, regions of stress concentrations (like joints and welding) need to be avoided, and time-to-time monitoring of the devices/containers needs to be carried out. The parameters on which the extent of embrittlement depends include the temperature and pressure of the vessel or pipeline or joints, ambient conditions, concentration and purity of hydrogen, exposure time, stress rate, material properties such as physical and mechanical, microstructure, and surface properties. For liquid hydrogen, the thermal expansion, yield, tensile strength, ductility, and stability subjected to repeated thermal cycling are important parameters that impact hydrogen embrittlement.

Thermal Conductivity and Specific Heat

Thermal conductivity of gaseous hydrogen at NTP is 0.187 W/m–K, while that of liquid hydrogen is 0.0989 W/m-K. The specific heat at constant pressure of gaseous hydrogen is 14.85 kJ/kg. K and that of liquid hydrogen is 9.66 kJ/kg. K. The heat capacity of hydrogen and its thermal conductivity are higher than other gases.

Flame Speed and Quenching Gap

The maximum flame propagation speed for a hydrogen–air stoichiometric mixture is 975 m/s. Flame speed differs from burning velocity, which refers to the speed at which a plane, smooth combustion wave can travel through a stationary flammable mixture of hydrogen and oxidant. The burning velocity depends on temperature, pressure, and concentration. Its value for a hydrogen–air mixture at stoichiometric ratio and under ambient conditions is 2.55 m/s, while in oxygen it is 11.5 m/s, which is significantly higher than that of other hydrocarbon fuels. Due to this higher burning velocity, the chances of deflagration to denotation transition (DDT) are higher. The concentration range in which it can detonate is 18–59 vol% in air, while in oxygen, it can be 15–90%. This range also depends on various parameters, including system size. The quenching distance, defined as the minimum distance between two surfaces through which a flame can propagate, is 0.51 mm for hydrogen. This makes extinguishing hydrogen flames very difficult. This quenching distance, however, is dependent on the hydrogen concentration, temperature, and pressure among other factors.

Hydrogen burns with a colorless and smokeless flame, until the flame burns other materials present in the vicinity, which could result in smoke or volatiles. During the combustion or explosion of a flammable hydrogen–air mixture, a pressure wave results, which depends on the ignition mode. The resulting damage is determined by the overpressure created by this pressure wave, which can be safe up to 10 kPa but becomes dangerous beyond 240 kPa. Due to the invisible flame, the only way hydrogen flame can be detected is through the radiated thermal energy, which corresponds to its higher heating value.

Hydrogen burns with a colorless flame, and the flame velocity of hydrogen is much higher than that of conventional gaseous fuels. When mixed with pure oxygen in a 2:1 molecular ratio of hydrogen to oxygen and ignited, hydrogen detonates violently since there is no inert gas or nitrogen to slow down the wave front. The energy required to ignite a hydrogen–air mixture is very low it is 0.02 mJ and is about one-fourteenth of the energy needed to ignite natural gas. The combination of all these physical properties of hydrogen (such as low density, low boiling point, wide flammability range, low ignition energy, high diffusivity in air, and high flame velocity) is unique among all fuels, and this combination has significant safety implications for the use of hydrogen in bulk. But the low density and high diffusivity in air together, in case of any leakage make hydrogen to disperses off very fast and high buoyancy makes it rise faster, leading to safer diffusion in case of leak. However, the heavier hydrocarbons are capable of forming large combustible clouds, as had happened in the past, in the cases of disastrous explosions at Flixborough in 1974 and Buncefield in England in 2005.

Safety, while in the hydrogen energy value chain, is important, but it has been observed that under many practical situations, hydrocarbons may pose stronger fire and explosion hazards than hydrogen. Hydrogen's high buoyancy significantly affects its dispersion, more so than its high diffusivity. Hydrogen diffusivity is greater than that of helium and approximately three times that of nitrogen in air at ambient conditions. Gaseous H_2 also readily diffuses through solids. For example, the effective diffusion coefficient of hydrogen through gypsum panels is 1.4×10^{-5} m^2/s at room temperature.

Solved Problem 1. Consider that a vehicle consumes around 40 L of petrol, overall, to travel from Mumbai to Ahmedabad, when it is fitted with an ICE having an efficiency of 20%. We now replace the ICE with a hydrogen-powered fuel cell having overall efficiency of 45%, and that, hydrogen is stored in the compressed state. Identify the amount of hydrogen required to complete the same journey. If the hydrogen is compressed at 150 bar (type-II cylinders), what will be the number of cylinders required if each cylinder has a volume of 45 L? What would be the number of cylinders required if it was charged at 700 bar and each cylinder had a volume of 20 L? ($T = 300$ K; LHV of petrol as 44.4 MJ kg^{-1}; density as 0.77 kg/L; LHV of Hydrogen: 120 MJ kg^{-1}).

Solution:

Step 1: Calculate Energy Used by the Car with ICE

- The vehicle consumes **40 L of petrol** to travel from Mumbai to Ahmedabad.
- Efficiency of ICE = 20%.
- Energy content of petrol $= \left(44.4\ \text{MJ kg}^{-1} \times 0.77\ \text{kg L}^{-1}\right) = 34.2$ MJ/L.

Total energy input from fuel:

$$E_{\text{input}} = 40 \times 34.2 = 1368\,\text{MJ}.$$

Useful energy output (i.e., work done):
Since only 20% is converted to usable work:

$$E_{\text{useful}} = 0.20 \times 1368 = 273.6\,\text{MJ}$$

Step 2: Calculate Hydrogen Required

- Considering a hydrogen fuel cell with 45% efficiency.
- LHV of H_2 = 120 MJ/kg.

To do the same amount of work (273.6 MJ), we need:
Hydrogen energy input $= \frac{273.6}{0.45} = 608$ MJ.
This corresponds to the mass of hydrogen to be $m_{H_2} = \frac{608\ \text{MJ}}{120\ \text{MJ kg}^{-1}} = 5.07\text{kg}$..
Step 3: Volume of Compressed Hydrogen Gas
We now calculate the volume required to store 5.07 kg of hydrogen at two different pressures.

Properties of Hydrogen Gas

- Molar mass of $H_2 = 2$ g mol^{-1}.
- Number of moles $n = \frac{5.07\ \text{kg}}{2\text{g}} = 2.535\text{kmol}$.

- Consider the pressure of 150 bar for the initial calculation.
- Using ideal gas equation:

$$PV = nRT$$

$$\left(150 \times 10^5\right)V = (2535\text{mol})(8.314)(300)$$

$$V = 0.4215\text{m}^3 = 421.5\text{L}.$$

Hence, the number of type-II cylinders needed would be 421.5/45 = 9.36 ~ **10 cylinders**.

At a pressure of 700 bar, the volume of hydrogen required is

$$150 \times 421.5 = 700 \times V;$$

$$V = 90.32L$$

The number of type-IV cylinders required will be:

$$N = \frac{90.32}{20} = 4.52 \sim \mathbf{5\,cylinders}.$$

Solved Problem 2. Consider the given information

Gas	Diffusivity (cm^2/s)	Molar mass (g/mol)	Lower flammability limit (LFL)
Hydrogen	0.63	2	4% or 1.63 mol/m^3
Methane	0.20	16	5% or 2.03 mol/m^3
Gasoline vapors	0.08	100	1.2% or 0.49 mol/m^3

In an event of a small leak in a poorly ventilated indoor environment, which one of the given gases reaches its LFL in a duration of 30 min of leakage? (Assume that the gases escape from a point source such that the radius of the gas front is given by $r = \sqrt{2\,\text{Dt}}$, that they form a hemispherical gas front, and that the flow rate of escape is constant with time; at T = 298 K and P = 1 atm.)

Solution:

We assume the gases to be ideal and the gas front to be hemispherical.

Using the equation of diffusion, the radius of the gas front is obtained as

$$r_{H_2} = \sqrt{2\,\text{Dt}} = \sqrt{2 \times 0.63 \times (30 \times 60)} = 47.6\,\text{cm}$$

$$r_{CH_4} = \sqrt{2\text{Dt}} = \sqrt{2 \times 0.2 \times (30 \times 60)} = 26.83\,\text{cm}$$

$$r_{\text{Gasoline}} = \sqrt{2\text{Dt}} = \sqrt{2 \times 0.08 \times (30 \times 60)} = 16.97\text{cm}.$$

Accordingly, the volume of the gas front hemisphere, calculated by $V = \frac{2}{3}\pi r^3$, becomes

$$V_{\text{H}_2} = 0.226\text{m}^3$$

$$V_{\text{CH}_4} = 0.0404\text{m}^3$$

$$V_{\text{Gasoline}} = 0.0102\text{m}^3.$$

Consider the concentration of 1 mole of each gas in the given situations:

$$c_{\text{H}_2} = \frac{1}{0.226} = 4.42\text{molm}^{-3}$$

$$c_{\text{CH}_4} = \frac{1}{0.0404} = 24.75\text{molm}^{-3}$$

$$c_{\text{Gasoline}} = \frac{1}{0.0102} = 98.04\text{molm}^{-3}.$$

Thus, all the three gases would be above the flammability limit after 30 min of leakage.

Solved Problem 3. In the above question, identify the amount of time taken by each gas to reach its lower flammability limits.

Solution:

- Diffusion radius:

$$r = \sqrt{2\text{Dt}} \Rightarrow t = \frac{r^2}{2\text{D}}.$$

- Volume of a hemisphere:

$$V = \frac{2}{3}\pi r^3.$$

Hence, we can obtain the radius in terms of concentration as:

$$C = \frac{n}{V} = \frac{1}{\frac{2}{3}\pi r^3} \Rightarrow r = \left(\frac{3}{2\pi C}\right)^{\frac{1}{3}}.$$

Putting the values of lower flammability limits in place of C, we obtain

$$r_{H_2} = 0.66\text{m}$$

$$r_{CH_4} = 0.62\text{m}$$

$$r_{\text{Gasoline}} = 0.99\text{m}.$$

Hence, we have: $r = \sqrt{2\text{Dt}} \rightarrow t = \frac{r^2}{2D}$.
Hence, we get:

$$t_{H_2} = 57.6\text{min}$$

$$t_{CH_4} = 160\text{min}$$

$$t_{\text{Gasoline}} = 1021\text{min}.$$

Solved Problem 4. Consider the flammability and ignition behavior of hydrogen and gasoline vapors released in a **10 m³closed room** at standard conditions (25 °C, 1 atm). What is the minimum amount of fuel needed:

(i) to reach a stoichiometric mixture with air,
(ii) comment on whether that mixure is ignitable using a 0.1 mJ spark system.

Density of air: 1.225 kg/m³.
Use the given information:

Parameter	Hydrogen	Gasoline vapors
Molecular weight (g/mol)	2.02	114.23
Stoichiometric air–fuel ratio (AFR)	34:1	14.7:1
Lower flammability equivalence ratio $f_{\min}$	0.1	0.7
Ignition energy (minimum)	0.02 mJ	0.24 mJ
Molar volume at STP	22.4 L/mol	–
Density of gasoline vapors (approx.)	–	3.5 kg/m³

Solution:
Consider the mass of air in the given room:

$$m_A = 1.225 \times 10 = 12.25\text{kg}.$$

The flammability equivalence ratio is given by

$$f = \frac{\left(\frac{\text{Fuel Mass}}{\text{Air Mass}}\right)_{\text{Actual}}}{\left(\frac{\text{Fuel Mass}}{\text{Air Mass}}\right)_{\text{stoichiometric}}}.$$

At a flammability equivalence ratio of f=0.1, hydrogen fuel mass required is:

$$\text{Fuel mass actual} = f\frac{\text{Air Mass actual } (m_A)}{(\text{AFR})_{\text{stoichiometric}}} = 0.1\frac{(12.25)}{34} = 0.036\text{kg} = 36\text{ g}.$$

Hence, fuel concentration needed for ignition:

$$\frac{36}{10} = 3.6\text{g m}^{-3}.$$

Similarly, for gasoline, the fuel concentration will be 58.3 g/m^3 at $f = 0.7$.

Given that the ignition energy is 0.02 mJ for hydrogen and that available energy is 0.1 mJ, hydrogen will ignite.

Also, since required ignition energy is 0.24 mJ, whereas available energy is 0.1 mJ, gasoline vapors will not ignite.

References

1. NIST chemistry WebBook. http://webbook.nist.gov/chemistry/
2. Perry's Chemical Engineers' Handbook (7th Edition), 1997, McGraw-Hill
3. Alternatives to traditional transportation fuels: An overview. DOE/EIA-0585/O. Energy information administration. U.S. Department of Energy. Washington, DC. June 1994
4. Hydrogen fuel cell engines and related technologies. Module 1: Hydrogen properties. U.S. DOE. 2001, http://www.eere.energy.gov/hydrogenandfuelcells/tech_validation/pdfs/fcm01r0.pdf
5. US DOE website https://h2tools.org/hyarc/hydrogen-data/comparative-properties-hydrogen-and-other-fuels

Chapter 2
Hydrogen Production from Hydrocarbons

Energy is essential for meeting the basic needs of human beings and for human development. But the energy usage pattern has led to serious environmental implications. There are serious global challenges wherein the energy systems can play a crucial role, like sustainable economic and human development, climate protection, health issues, energy security, etc. Since the beginning of the industrial revolution, the rise in the dependence on fossil fuels has been explosive, where the fuel for day-to-day needs and muscular power has been replaced mostly by fossil fuel-based systems. There are two clear changes being observed since 1800, wherein the first transition was from the use of biomass to coal, which became the prime mover in the first half of the twentieth century, and the later transition to oil around 1970, thereafter natural gas also took center stage.

The biggest challenges in the current energy scenario are the depleting fossil fuels, fluctuating prices of fossil fuels, and energy security due to the changing geopolitical situations, and on top of that, the climate change-related issues being observed globally. Hydrogen is a potential energy carrier which has an answer to address these challenges and can decarbonize the different sectors.

Hydrogen can be produced from any resource which contains hydrogen; it can be produced either in a distributed or centralized manner. For near term, due to the higher feasibility, along with the availability of local feedstock and least infrastructure to be laid down, distributed hydrogen production onsite at refueling sites or end-use places seems to be more economical option. However, with due economies of scale, centralized production may become more favorable for the production of low-cost hydrogen, but will need low-cost transport and delivery infrastructure and measures to manage the GHG emissions like carbon capture and sequestration (Fig. 2.1).

Hydrogen can be produced by several methods from a variety of resources depending on the available feedstock, output capacity requirements, costs involved (both feed as well as processing steps), purity needs, the possible value-added by-products, and the supply of energy sources available. Hydrogen can be produced from hydrocarbons through different routes. Current commercial hydrogen production is mainly from fossil fuels globally, and the fossil fuels not only act as a

P. Sharma, *A Textbook on Hydrogen Energy*,
https://doi.org/10.1007/978-981-95-4601-5_2

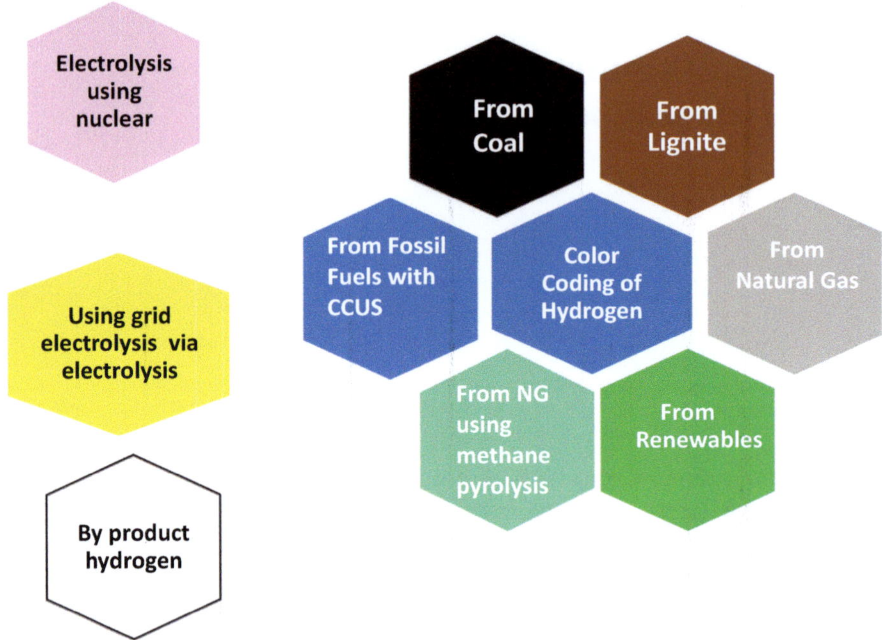

Fig. 2.1 Different colors of hydrogen based on source of production

source of hydrogen (as they have in their chemical formula) but also serve to provide the required energy for the production processes. Hydrogen is classified by color based on the resource and method used for its production. Green hydrogen is produced from renewables; grey from natural gas; blue from hydrocarbons combined with carbon capture, utilization, and sequestration (CCUS); turquoise from methane decomposition; brown and black hydrogen from coal; yellow and pink hydrogen from electrolysis using grid and nuclear electricity, respectively; and white hydrogen occurs naturally in geological formations.

A broad classification of the technologies in use for hydrogen production on the basis of feedstock is shown in Fig. 2.2.

1. Hydrogen Production from Steam Reforming

In this chapter, a detailed description on the methods of hydrogen production from different hydrocarbons excluding coal is included. The processes like reforming or gasification are endothermic, and the heat required to carry the process can be either supplied from outside the reactor, i.e., allothermically, or from within the reactor by combustion of the part of the feedstock, i.e., autothermically. The choice of the process - whether allothermic or autothermic - depends on the volatility of the feedstock. For example, natural gas and naphtha are highly volatile, so allothermic heating is used, whereas coal and oil are less volatile and the reaction can be provided

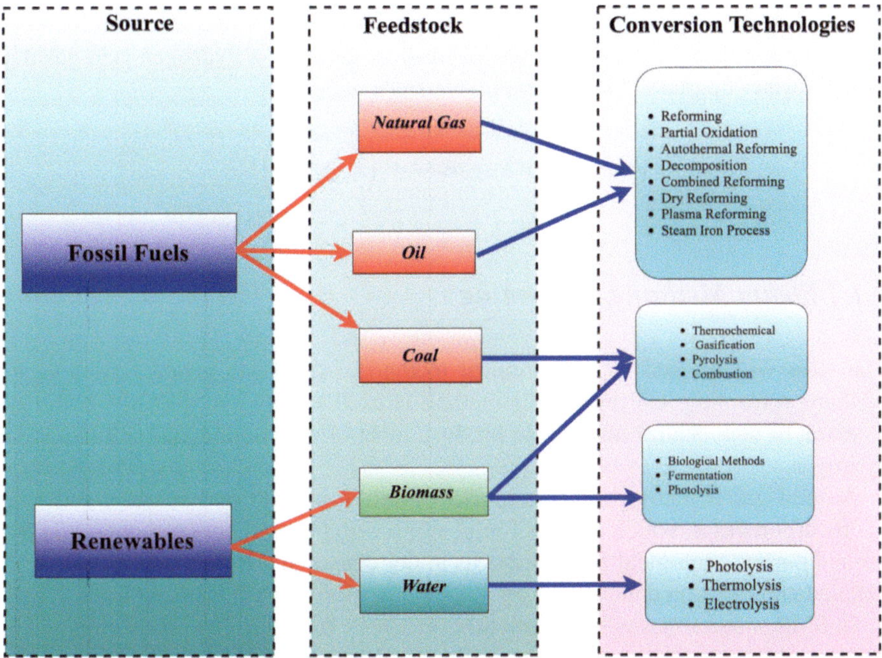

Fig. 2.2 Hydrogen production technologies from different energy sources

autothermically. Hydrocarbons can be converted into hydrogen either in the presence of an oxidant or in the absence of oxidant. The conversion in the presence of oxidants like steam, oxygen, both steam and oxygen, carbon dioxide, etc., constitutes the different oxidative processes wherein the major conversion technology is reforming. However, the conversion of hydrocarbons in the absence of oxidant is called non-oxidative process wherein the major conversion technology is decomposition. Reforming takes place when either a hydrocarbon or alcohol along with the oxidant is passed over catalyst bed. In oxidative processes, the oxidant can be steam, oxygen, a combination of steam and oxygen, or CO_2. Based on the oxidant used, the process is known as steam methane reforming (SMR), partial oxidation (POx), autothermal reforming (ATR), or dry reforming, respectively.

1.1 Steam Reforming of Lighter Hydrocarbons

Reforming is the process where the reactants which include the hydrocarbons, are getting reformed to give the desired products. Catalytic steam reforming involves removal of hydrogen from the fuel (which can be hydrocarbon or alcohol) and water in the presence of a catalyst. Hydrogen can be produced by steam reforming of various hydrocarbons from lighter to the heavier fractions like natural gas, naphtha, ethane, methanol, dimethyl ether, propane, butane, isooctane, diesel, gasoline, etc.

Catalytic steam reforming is the main route to reduce the tars and other hydrocarbons in the product gas from gasifiers and thereby to generate a hydrogen-rich gas that can be further upgraded to other energy carriers.

$$C_nH_m + nH_2O \rightleftharpoons nCO + \left(n + \frac{1}{2}m\right)H_2.$$

1.1.1 Steam Methane Reforming

The most widely used and economical method of hydrogen production is steam methane reforming. Currently, this method is used widely and globally, either on a central or semicentral manner, to produce most of the commercial and industrial hydrogen.

Natural gas is a colorless and odorless mixture of both hydrocarbon and non-hydrocarbon gases, which often is found along with the coal or petroleum in underground porous geological formations. Its major constituent is methane; however, its detailed composition depends on the source or region of its existence (Table 2.1).

In SMR, natural gas is used both as a feedstock for hydrogen production and as fuel, supplying the energy required for the highly endothermic reforming reactions. The major processes involved in the steam reforming of natural gas from supplying feed to getting the high-purity hydrogen involve pretreatment of feedstock, reforming to synthesis gas, conversion to a hydrogen-rich gas, and finally purification to get the desired hydrogen product as per the required specifications for utilization in an

Table 2.1 Composition of natural gas

Component	Typical analysis (mole %)	Range (mole %)
Methane	95.0	87.0–97.0
Ethane	3.2	1.5–7.0
Propane	0.2	0.1–1.5
Iso-butane	0.03	0.01–0.3
Normal-butane	0.03	0.01–0.3
Iso-pentane	0.01	Trace–0.04
Normal-pentane	0.01	Trace–0.04
Hexanes plus	0.01	Trace–0.06
Nitrogen	1.0	0.2–5.5
Carbon dioxide	0.5	0.1–1.0
Oxygen	0.02	0.01–0.1
Hydrogen	Trace	Trace–0.02
Specific gravity	0.58	0.57–0.62
Gross heating value (MJ/m^3), dry basis[a]	38.0	36.0–40.2

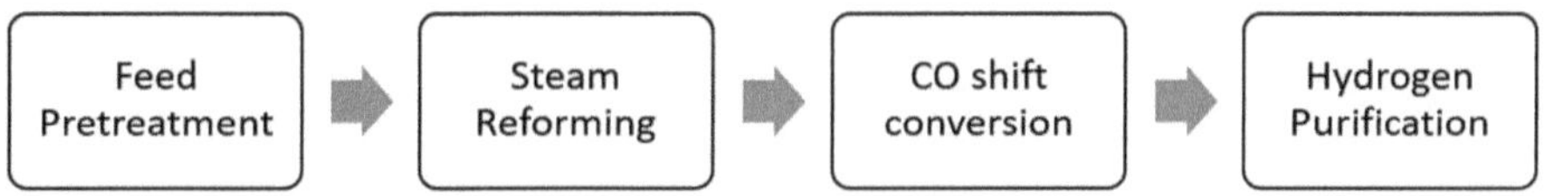

Fig. 2.3 Process flow diagram of steam reforming method of hydrogen production

energy conversion process or device. Based on the purification mechanism used, the final stage in the process flow can be different.

In brief, SMR is a process which involves introducing preheated natural gas along with steam over a suitable catalyst to get a gaseous product output with a composition of roughly 75–80% hydrogen (including SMR and WGS steps). Thus, the two major constituents in a steam methane reforming process are hydrogen and carbon monoxide; thereafter, purification to obtain pure hydrogen can be carried out using either pressure swing adsorption (PSA); or with solvent-based removal of CO_2 and a final methanator or purification using MEA scrubber. After the reformation step, the hydrogen content is about 50% on dry basis, and CO makes up 40–45% of the gas. This CO is further converted into CO_2 and hydrogen in the two stages of the WGS reactions. High-temperature WGS takes advantage of the higher reaction rate, while low-temperature WGS further reduces the CO content to as low as 1%. In general, the process flow diagram is shown in Fig. 2.3 and Fig. 2.5.

The description of each of the steps involved in SMR is included in the following sections.

1.1.1.1 Feedstock Pretreatment

Feedstock pretreatment is the first step in steam reforming process. It removes sulfur, chlorides, olefins, and other impurities that can poison the different catalysts used in the different downstream processes like reforming or water–gas shift reactions and can also lead to the corrosion of the equipment involved. If the starting feed material is a heavier hydrocarbon, the feedstock pretreatment involves several processes. These include hydrogenation, converting of olefins to paraffins; hydrodesulfurization, converting sulfur-containing compounds to hydrogen sulfid;, and hydrodechlorination, converting to alkyl chlorides to hydrogen chloride.

If the feedstock contains a small amount of sulfur, it is converted to H_2S, which can then be removed using ZnO bed, forming ZnS. For feedstocks with higher sulfur content in the feedstock, solvent-based scrubbing is employed to absorb H2S efficiently. In case of organic sulfur compounds, hydrodesulfurization is required to convert them to H2S..

However, in case of steam methane reforming with natural gas as the feedstock, the major process involving pretreatment is hydrodesulfurization. Sulfur-containing impurities present in natural gas can poison the catalysts used for reforming or water–gas shift processes. For example, in reformer, the presence of sulfur deactivates the catalyst, and if this situation is not controlled, it can lead to excessive tube wall temperature, resulting into reduced tube life and even rupture of the catalyst-filled tubes. Thus, pretreatment in the form of desulfurization can increase catalyst as well as tube life. During hydrodesulfurization, under medium pressure and on a sulfur-resistant catalyst, the different sulfur compounds are converted into hydrogen sulfide. For example, the sulfur organic compounds like mercaptans and thiols are converted to H_2S by catalytic hydrogenation reaction at temperatures 290–370 °C in the presence of Co–Mo catalyst. Hydrogen sulfide can then be removed using zinc oxide, which reacts with sulfur to form zinc sulfide, and the reaction takes place at 340–390 °C. The reaction can remove all the sulfur to less than 0.1 ppmv. The desulfurization unit typically has two vessels piped in a lead/lag configuration, allowing the removal of the desulfurization catalyst even at the time of planned operation.

$$H_2S + ZnO \rightarrow ZnS + H_2O \tag{2.1}$$

Chlorides if present don't affect reforming catalyst much but have a significant poisoning effect on the catalysts used for low-temperature water–gas shift reaction. Thus, an alumina guard bed is used before reforming to convert chlorides to hydrochloride.

1.1.1.2 Pre-reforming

If higher hydrocarbons (C2+) are present, depending upon their contents in the feedstock, an additional pre-reforming step is added after the feedstock pretreatment step and before the primary reformer. Also, the presence of olefins in the hydrocarbons can result in much more carbon deposition on catalyst surface than even the higher hydrocarbons; in such case, addition of pre-reformer can take care of these compounds. Since higher hydrocarbons are more reactive than methane, they can decompose to form coke leading to deactivation of the catalyst. In the pre-reforming step, an additional split stream of product hydrogen is used to convert these higher hydrocarbons into C1 species like methane, CO, CO_2, and H_2 in a fixed bed adiabatic reactor at temperature of 300–525 °C with Ni catalyst on Al_2O_3 support along with promoters. The pre-reformed gas is then preheated to 650 °C before it enters into the tubular steam reformer. The major reactions occurring in pre-reformer are reforming, cracking, and gasification.

The advantage of adding this pre-reforming unit prior to primary reforming is that it provides fuel flexibility and a variety of feedstocks can be used which can be converted into lower hydrocarbons or methane, before entering into the steam reformer unit. A pre-reformer will allow the processing of heavy feed including naphtha at low steam-to-carbon ratios (S/C) in primary reformer. The addition of a pre-reformer offers several advantages. It reduces the overall energy requirement and allows the process to operate at lower temperature. It also reduces coking of the catalysts used in hydrogen production, increases hydrogen capacity, and lowers pressure drop and coke deposition in the primary reformer, thereby extending its lifetime. Additionally, thepre- reformer enables the SMR to operate at 8–10% higher ratesthroughput at the same heat load by reducing the required reforming duty. The process of pre-reforming is similar to the primary reforming with the difference that it occurs at relatively lower temperature. Use of pre-reformer allows higher-temperature input to the primary reformer, which is the exit from pre-reformer. Also it increases the efficiency of the process by removing higher hydrocarbons thereby reducing the steam requirement in the primary reformer and increasing the plant efficiency as the amount of energy required to produce superheated steam will decrease. Advantages of inclusion of pre-reformer are:

- reduces coke deposition on catalyst,
- the steam reformer can be operated at lower S/C ratio resulting in better efficiency,
- can enable feedstock flexibility,
- can reduce size of main reformer,
- allow use of less expensive catalysts in the main reformer by eliminating coke formation reactions.

1.1.1.3 Steam Reforming of Methane

After initial feed pretreatment, the natural gas enters into the catalytic reforming unit, where it is mixed with steam and passed over catalyst to produce synthesis gas. The pretreated and preheated natural gas is mixed with steam and enters into the steam reformer (pre-heating results in higher inlet temperature to the reformer, making it easier to attain the desired reaction temperature inside the reformer, resulting in improved kinetics and improved efficiency). In the catalytic reforming reactor, there are several reactor tubes filled with Ni catalyst and kept in a furnace where these are externally heated to high temperatures. When the pretreated NG along with steam is passed through these catalyst-filled reformer tubes, it is converted into synthesis gas (i.e., CO and H_2) at temperatures of 850–900 °C. The reaction takes place as per the equation:

$$CH_4 + H_2O \rightarrow CO + 3H_2 \quad \Delta H^0 = +206 \text{ kJ/mol}. \tag{2.2}$$

The reaction is highly endothermic as the products are in higher energy state than the reactants; both reactants are being inert. Thus, energy is required to activate those reactants to react or provide the necessary activation energy to break the bonds C–C and C–H.

In Eq. (2.2), two molecules of reactants react to give four molecules of product; there is an increase in the net number of product molecules compared to the reactants, and thus by Le Chatelier's principle forward reaction is favored at low pressures. However, most of the applications require hydrogen at pressures of at least 2 MPa. Operating at higher pressure also allows for a more compact reactor design and high throughput, and it can eliminate the need for additional compression of the product.

Thus, mostly the reformers are operated at pressures in the range of 2–2.6 MPa. This increased pressure will further require higher temperatures which will be limited by the highest allowable tube wall temperature, i.e., metallurgical choice becomes important. The desired conditions are a uniform heat input and the flow of process gas inside the reformer. Although Eq. (2.2) suggests that one mole of steam is required for one mole of natural gas (NG) to result into the desired products, but in actual practice the reaction is performed using high steam-to-carbon ratio (S/C), commonly this ratio lies between 2.5 and 3, and this is required in order to prevent the catalyst deactivation by deposition of carbon (coke) on the catalyst surface. Thus, the presence of excess of steam suppresses carbon deposition and finally can avoid plant shutdown resulting from catalyst deactivation. The most commonly used catalyst in industrial processes includes Ni supported on alumina. Ni is favorable catalyst due to its low cost and sufficient activity, while alumina is suitable as support material because it is a refractory and stable material. However, the major disadvantage with Ni catalyst is deactivation due to coke formation and sintering of metallic Ni. Worldwide research is going on to address these issues by considering different methods of catalyst preparation, use of promoter, and other support materials. Coke deposition can lead to heat transfer problems and pressure drop across the reactor. The coke formation can lead to catalyst deactivation, and this can result in excessive tube wall temperature and finally lead to the rupture of tubes. The failure of tubes can be taken care of, by reduced deactivation of catalyst, thereby controlling the tube wall temperature and internal pressure.

The deposition of carbon on the catalyst surface occurs mainly due to the following three reactions:

(1) Decomposition of methane:

$$CH_4 \leftrightarrow C + 2H_2 \quad \Delta H^0 = +74.6 \text{ kJ/mol} \tag{2.3}$$

(2) Disproportionation of CO: (Boudouard's Reaction)

$$2CO \leftrightarrow C + CO_2 \quad \Delta H^0 = -172 \text{ kJ/mol} \tag{2.4}$$

(3) Heterogeneous water–gas reaction (reverse of gasification reaction) or CO reduction:

$$CO + H_2 \leftrightarrow H_2O + C \quad \Delta H^0 = -131 \text{ kJ/mol} \tag{2.5}$$

Equations (2.4) and (2.5) are favored at lower temperatures, while Eq. (2.3) is favored at higher temperatures. The temperature range in which reforming reaction occurs leads to less deactivation because (2.4) and (2.5) are not favorable, and the operating temperature is not sufficient enough to promote (2.3).

These reactions for carbon formation are reversible, and the carbon formed can be removed by gasification reaction. However, some carbon can polymerize and become less reactive. This carbon can diffuse into the catalyst, reaching the interface between the catalyst and support, which may lift the catalyst from the support surface and reduce its activity. Additionally, the formation of carbon whiskers can cause physical breakage of the catalyst. The majority of carbon deposition occur in the first half of the reformer tubes; this is because the concentration of the hydrocarbons is high and the equilibrium is low. This equilibrium also depends upon temperature, and up to 600 °C, and under the conditions of low steam-to-carbon ration and high-pressure, the equilibrium favors carbon formation. At the top of the reformer tube, the reaction rate is slow and the heat flux is high (in top fired configuration), however toward the bottom of the tube is the reverse, i.e., low heat flux while higher reaction rate. Besides that the tube wall temperatures are higher than the core temperatures, so the reaction occurs faster near the side walls then the core, as such the unreacted methane in the core region in the reformer tube can lead to fouling of the catalyst.

If the reforming process involves heavy hydrocarbons as feedstock instead of methane, then carbon deposition can occur both by decomposition as well as cracking; however, the gasification of carbon can further lead to the formation of methane, CO, CO_2, H_2. Also the decomposition of heavier feedstocks can form intermediates which can further undergo reforming to produce the products like CO, CO_2, H_2, and CH_4. The use of potash can help toward increasing the basicity of the support and also increasing the gasification of carbon. The major effects of carbon formation include catalyst poisoning, caused by the deposition of carbon on active sites, and physical breakage of the catalyst breaking due to lifting up and movement within the catalyst bed, which can increase pressure. Carbon can also fill voids inthe refformer tube, leading to partial or complete blockages, uneven flow distribution and overheating. These issues create hotspots, resulting in rise in the tube wall temperature.

However, an excess of steam shifts the equilibrium of the reforming reaction away from the formation of carbon.

The gas stream obtained after reforming can have a composition of around 70% H_2, 16.5% CO, 6% CO_2, 7.5% CH_4 on dry basis (the gas compositions vary based on feedstock compositions and process conditions). The outlet stream temperature from reformer is 700–900 °C, and this has to be within the temperature limitation of the reformer tube metallurgy. The outlet gas is obtained at a pressure lying between 15 and 30 bar, and methane conversion achieved in reforming is about 90–92%. This high conversion is achieved at high temperature, which is the thermodynamic limitation of the process.

SMR plants in industries like ammonia synthesis facilities are large in capacity, which can be even of capacity around 1000 ton/day. Steam methane reformers can be either top/bottom fired or wall fired depending on the position of the burners placed to heat and provide the required temperature of reaction to the hundreds of catalyst-filled tubes as shown in Fig. 2.4. These catalyst-filled tubes in reformer are arranged in rows, typically there are 400–600 such tubes, the tubes have inner diameter ranging between 7.5 and 12.5 cm, length 9–15 m, and thickness of tubes between 1 and 2 cm. Due to the presence of catalyst, a pressure drop is observed inside these tubes, which depends upon the gas flow rate and could be somewhere ranging between 1.5 and 5.0 bar; however, it is required that the pressure drop should be as low as possible. The reformer tubes undergo stringent reaction conditions and thus need to be metallurgically stable. These tubes are usually made up of Cr–Ni alloy steel (32–35 Ni/23–27 Cr stabilized with about 1.5% Nb or Cr 24–28%, Ni 18–22%; C 0.35–0.45%; Mn 2%; Si 2%; P and S 0.05%) to sustain high temperature of around 1150^{0}C and pressure of around 40 bar or more. The melting temperature of the alloy is around 1370 °C. Some of the more expensive materials that can be used include Pyrotherm G24/24 Nb and Manaurite 36X.

The parameters to be considered while designing the reformer include: the temperature, the pressure, rupture strength of the material used as well as the efficient heat transfer consideration. The heat transfer calculations include radiation from the furnace walls and the tubes, convection by the gases from the burner to the tubes, conduction across the tubes, convection from the tube wall to the catalyst inside the tubes and to the reactants. Besides these design parameters, the chemical reactions occurring inside the tubes, the reaction kinetics, interaction of catalyst with the reactants, heat and mass transfer inside the tubes, all of these are very important parameters to be known and can provide an understanding of the carbon formation inside the tubes. Well-designed reformers can operate for very long periods without failure. The critical challenge is not the operation under stable steady conditions, but transient events, such as startup and shutdown, which can have the greatest impact.

1.1.1.4 Water–Gas Shift Reaction

In order to extract more hydrogen from the carbon monoxide by-product of the reforming reaction, the next step is water–gas shift, as given by Eq. (2.6). Removal of CO_2 during hydrogen purification can be done by either absorption or adsorption to obtain pure hydrogen, but removal of CO is relatively more difficult; as such it is converted to CO_2 in this WGS step. WGS is an exothermic process, as such results in the rise in temperature. The water–gas shift reaction can be performed in two steps including a high-temperature water–gas shift (HTWGS) and a low-temperature water–gas shift (LTWGS); however, use of one or both depends on the purification process employed at the end of the process flow. The HTWGS is employed to have faster kinetics, while an additional LTWGS is incorporated to get further conversion and to obtain as much hydrogen as possible in the reforming process. The choice of catalysts can be made based on the process conditions. The exit stream from reformer is first cooled to 350 °C (Fe–Cr-based catalyst are active above 300 °C) as the reaction (2.5) is an exothermic reaction and is favored at lower temperatures. Following the SMR step, the excess heat, i.e., SMR outlet gas temperature which lies in the range of 750–900 °C while that for HTWGS required temperature is 350 °C, the heat removed in the process to cool to HTWGS temperature can be used for generating steam or pre-heating natural gas. The outlet gas from the reformer is passed into the HTWGS reactor, where the high-temperature water–gas shift reaction is carried out, the catalysts used are oxides of iron and chromium, and this reaction reduces the carbon monoxide level to 2–5%. The high-temperature water–gas shift (HTWGS), wherein the carbon monoxide is converted into carbon dioxide and water according to the following reaction:

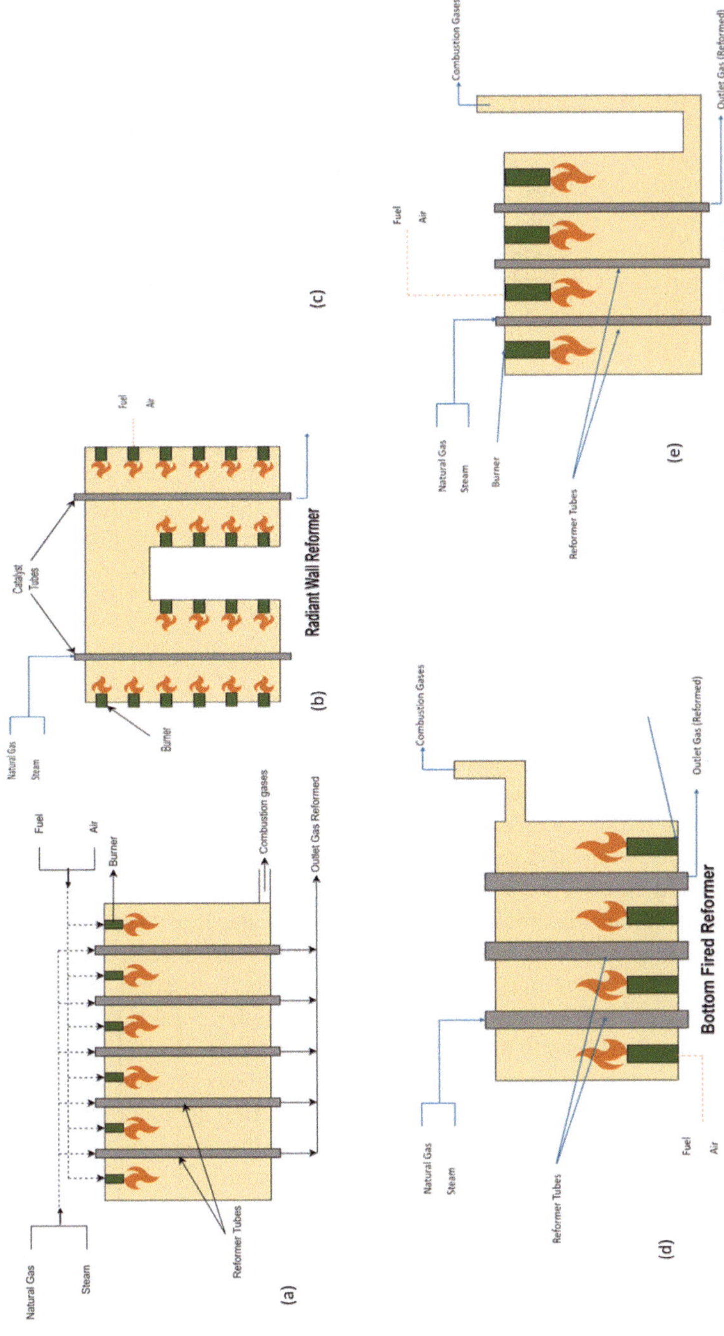

Fig. 2.4 **a** Schematic of a typical steam reformer reactor with catalyst-filled tubes, **b** radiant fired, **c** terrace fired, **d** bottom fired, **e** top wall fired

$$CO + H_2O \rightarrow CO_2 + H_2 \quad \Delta H^0 = -41 \text{ kJ/mol.} \tag{2.6}$$

Depending on the purification method used, the exit gas from HTWGS has to be cooled to 220 °C and passed through LTWGS, where copper, zinc oxide, and alumina are used as catalysts (these catalysts are active at lower temperature like 200 °C but are more sensitive to sulfur poisoning) and the LTWGS reaction reduces the level of CO to around < 0.5% [1]. The catalyst used for LTWGS needs to have higher activity because the temperature of the reaction is further reduced compared to HTWGS. These two steps (HTWGS and LTWGS) are carried out in separate adiabatic fixed bed reactor with intermediate cooling to reduce the temperatures. The Cu–Zn-based catalysts used for LTWGS are highly sensitive to the presence of sulfur; as such the sulfur content is removed to less than 1 ppm level, during the desulfurization step. Water–gas shift reaction equilibrium is not affected by pressure, as there is no volume change involved, as per reaction 2.6. However, reduced temperature favors forward reaction; this is due to the exothermic nature of the reaction. Due to the thermodynamic constraints with the WGS reactions, the complete conversion of CO to CO_2 doesn't occur; still 2–3% of CO is left in the outlet stream. After shift reaction, the exiting gas has roughly a composition 70–75% H_2, 0.2–1% CO, 15–25% CO_2, 1-3% CH_4 (mol% and on dry basis) (the gas compositions vary based on feedstock compositions and process conditions); this is finally purified to obtain pure hydrogen.

1.1.1.5 Purification Step

After the water–gas shift step, the remaining CO is converted to CO_2 by preferential oxidation process using noble metal catalysts. Thereafter the final gas cleaning process to get pure hydrogen can be carried out using either solvent-based removal method or pressure swing adsorption (PSA) or cryogenic separation or membrane-based separation. In the PSA process, by the change of pressure called pressure swing, the various impurities are adsorbed and removed; finally, the adsorbent bed is regenerated. The product gas after shift reaction is passed through an adsorption column, where impurities are preferentially adsorbed. After a fixed duration, the feed gas to the reactor column is diverted to another parallel column. At this stage, the first column is depressurized, which allows the bed to regenerate. It is a cyclic process but with the use of multiple beds is made to be a continuous process; there can be two, four, or more beds which can be alternately pressurized and depressurized.

For commercial hydrogen production, usually the purification step in order to get pure hydrogen is carried out using pressure swing adsorption (PSA); when PSA is used, then LTWGS is not required.

However, there are other methods of hydrogen purification like solvent-based removal systems, in this method low-temperature water–gas shift follows the HTWGS, followed by carbon dioxide scrubbing (using monoethanolamine or hot potash) and finally preferential oxidation and methanation steps are carried out as mentioned in equations 2.7 and 2.8:

$$CO + 0.5O_2 \rightarrow CO_2 \quad \Delta H^0 = -283 \text{ kJ/mol} \tag{2.7}$$

$$\begin{aligned} CO + 3H_2 &\rightarrow CH_4 + H_2O \quad \Delta H^0 = -205.9 \text{ kJ/mol} \\ CO_2 + 3H_2 &\rightarrow CH_4 + 2H_2O \quad \Delta H^0 = -164 \text{ kJ/mol} \end{aligned} \tag{2.8}$$

The other probable reaction which takes place in reforming includes:

$$C + H_2O \rightarrow CO + H_2 \quad \Delta H^0 = +131.3 \text{ kJ/mol} \tag{2.9}$$

The thermodynamic studies of SMR reaction have revealed that the methane decomposition reaction (2.3) resulting in coke deposition occurs at relatively lower temperature around 500 °C. However, the SMR (2.2) and carbon gasification (2.9) reactions occur at higher temperatures (> 700 °C). Thus, the reactor design as well as materials to be used in reactor, pipelines, etc., are the important factors while designing a reactor.

1.1.1.6 Catalyst and Support

A catalyst is a substance which increases the rate of the reaction and is not consumed or remains unchanged in the reaction. There are three important terms associated with the catalyst—activity, conversion, and selectivity. Activity of a catalyst is its ability to increase the reaction rate, and it is influenced by temperature, pressure, concentration, available residence time, and several other factors. Catalyst should introduce new pathways to accomplish the reaction with reduced activation energy, at a faster pace, to get the maximum conversion of reactants toward the desired products. Selectivity refers to the ability of the catalyst to direct the reaction in such a manner that the reactants interact and convert to give the desired product, which is hydrogen in this case. There can be several other reactions that can take place other than the reforming reactions, these can result in formation of other products, and high selectivity ensures that these products should not be favored in the presence of catalyst.

Requirements from a Catalyst

In any catalytic process, the reactant(s) gets adsorbed on the active sites of catalyst surface and various bonds are formed or broken, with several intermediates formed and finally leading to the products formation and catalyst returning back to its initial state. In order to have fast reaction rates, the active sites and thus surface area need to be maximized. The basic steps involved in a catalytic reaction are: (1) bulk diffusion of reactants to reach the surface of the catalyst, i.e., diffusion from the porous media to active sites of the catalyst, (2) chemisorption on the active sites or nearby sites, (3)

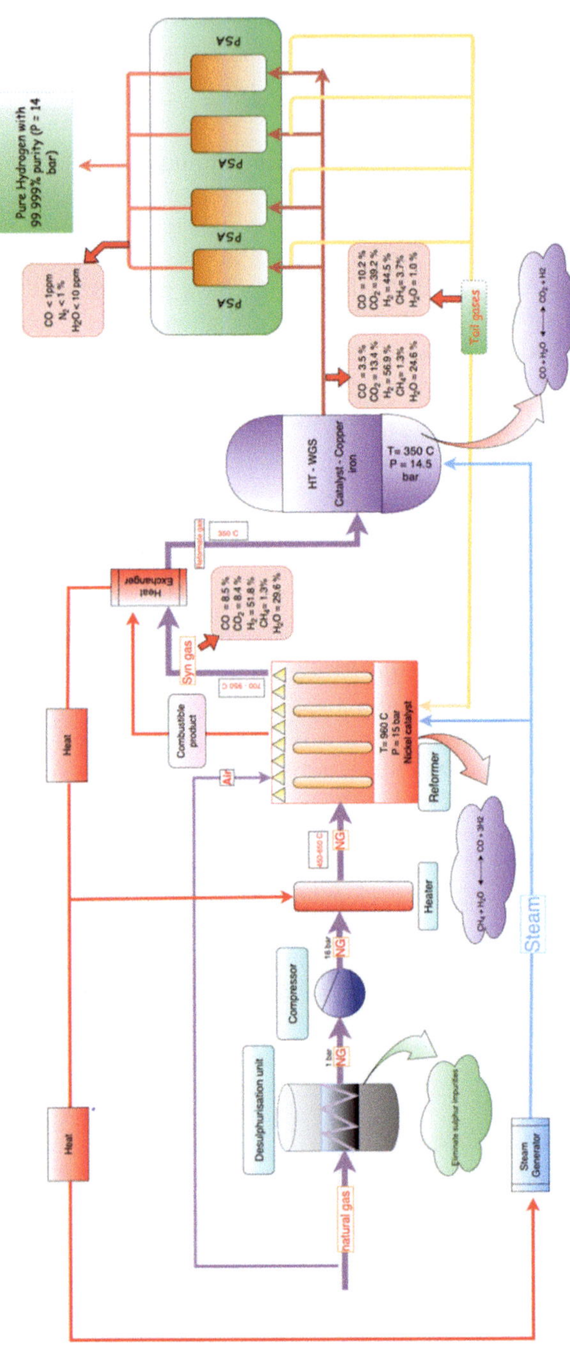

Fig. 2.5 Complete process flow diagram for steam methane reforming with the process parameters and all inputs

on the active sites formation of the product, (4) desorption of the formed product from these sites, (5) diffusion of these products through the porous media to the surface of the catalyst structures, (6) bulk diffusion of products to the main stream. The steps 1, 5, and 6 depend on the physical properties and geometry of the catalyst and support. The steps 2, 3, and 4 are the chemically activated processes involving intermediates being formed in the process and finally those resulting into the formation of products and thus depends on the chemical properties or nature of interaction between reactants, products, and active sites. Any of these steps can be the rate-limiting step in the catalytic reaction.

The industrial scale SMR is performed over Ni-based catalyst; Ni is widely used because of several advantages like low cost and its good activity (which depends on the external surface area). Typically 16–20% NiO supported on Al_2O_3 or calcium aluminate is taken in the form of pellets with sizes in the range of 3/8″–5/8″; on in situ reduction it yields the Ni catalyst with the required active sites. The major requirements from a good catalyst are its low cost, robustness toward extreme operating conditions (inlet temperature of 450–650 °C and outlet temperature of 700–950 °C with the steam pressure of 30 bar), long cycle life, stability under transient conditions, high activity, good heat transfer, low-pressure drop, high thermal stability, mechanical strength, and also acceptability for wide variety of feedstock. The Ni catalyst is usually supported on refractory material like alumina. Usually the lifecycle of such catalysts is reported to be more than 5 years of continuous operation. Besides several advantages, Ni-based catalysts suffer from issues like deactivation due to coke formation and sintering. In order to overcome these disadvantages, global research is focused on various aspects like different possible support materials, inclusion of a promoter and the synthesis route of catalyst. The other catalysts which can be used for the reforming reaction include noble metals like, Rh, Ru, Ir, Pt, Pd, etc. The activity of some of these catalysts being even higher than Ni and coking tendency is lower, especially for noble metals. However, Ni is being cost-effective and more abundant, as such it is the most widely used catalyst in commercial hydrogen production plants.

The major factors involved in catalyst design include the shape, size, and surface area, pressure drop involved, and the heat transfer from the tube wall to the gas. The shape of the catalyst is equally important and should have a high surface-to-volume ratio, and the reason being the access to the active surfaces like the inner surface can reduce with the diffusion-related limitations at high-temperature conditions. The pressure drop inside the catalyst bed depends strongly on the void fraction of the packed bed and decreases with increasing particle size; thus, the requirement is high external diameter and high void fraction. The shapes used include rings, cylinders with several holes, etc. The optimum shape of the catalyst is multichannel wheels, which offers better heat and mass transfer characteristics and minimum pressure drop under the industrial operating conditions. Other shapes considered in literature include gear wheels, spoked wheels, rings with holes, etc. Size of the catalyst also matters as the small size will lead to better heat transfer characteristics but at the same time higher-pressure drop, so an optimum catalyst is essential for better performance.

There are two possible diffusions: One is across the catalyst bed, and another is inside the catalyst to reach the active sites; to improve these, size can be reduced or loading can be improved. However, decrease in size increases pressure drop and can increase resistance to the flow of reactants. It is observed that larger-sized catalyst are more prone to deactivation and can have reduced activity. Sulfur passivation to reduce the ensemble size is practiced to control the conversion but that can lead to both poisoning as well as reduced activity of the catalysts. This is especially done in the dry reforming process.

Not only the catalyst, but also the support plays an important role in the catalytic reforming process. The support should provide a high surface-to-volume ratio, ensure good dispersion of the active metal, and help stabilize the catalyst. It must withstand extreme operating conditions, resist sintering, and interact effectively with the catalyst to minimise deactivation and sintering. Various support materials have been researched including MgO, $MgAl_2O_3$, ZrO_2, TiO_2, and different perovskites. These support materials have different advantages like Al_2O_3 has acidic nature which promotes reforming reaction, MgO promotes carbon gasification thereby reducing coke formation, ZrO_2 showed better performance, and CeO_2 reduces coke formation.

A small amount of promoters added to catalyst are known to markedly suppress the coke formation. Promoters are capable of changing the interaction between the active species, can influence the acid/basic surface characteristics, electronic and geometric characteristics, conversion, and selectivity of the catalysts. Some of the alkali (e.g., K), alkaline (e.g., Mg, Ca), transition metal (e.g., Mo, W), and lanthanides (e.g., La, Ce, etc.) have been considered as possible promoters for the catalytic reforming reaction. The carbon formation and catalyst deactivation can be reduced with the use of excess steam, use of promoters, use of pre-reformer in case of higher hydrocarbons, use of noble metal or bimetallic catalysts.

The choice of catalyst, support, and the optimal design can affect the rate of conversion from reactants to products and also the selectivity of the reforming reaction to get the desired product(s). Table 2.2 summarizes the catalysts used for different steps in the SMR process. The catalyst deactivation not only occurs due to poisoning but it also can occur due to sintering of catalyst or support. The high temperatures required for methane conversion can lead to catalyst deactivation by means of sintering. Catalyst breakage can occur during startup, shutdown, and during cooling as thermal stresses (thermal coefficient of expansion and associated effects) and various other forces acting on to the reformer tubes. Pressure induced by coke formation can also contribute to catalyst breakage. Ideally, such breakage should be minimized. Uniform catalyst loading inside the reformer tube is very much essential for long life and stable operation of the reformer. In case of non-uniform loading, the problems like hotspots, creation of uneven pressure, uneven flow of gases inside, and pressure drop can cause damage to the tubes and at the same time reduce the life of the tubes. The catalyst settling can be another possible issue in the reformers.

Since the reforming reaction is highly endothermic, good thermal management is essential to achieve high conversion. The thermal conductivity of the support material can play an important role by providing a better heat transfer, leading to higher conversion, and has least possibility of hotspot formation in the reformer reactor. The major limitations in reformer are heat transfer, temperature, and flue

Table 2.2 Catalysts and adsorbents often used in various processes involved in SMR

Application	Catalyst type
Hydrodesulfurization	Co-Mo/NiMo
Sulfur removal after conversion to H_2S	ZnO
Chloride guard for HCl removal	Activated Al_2O_3
Pre-reforming	Ni/Al_2O_3
Steam reforming (methane, LPG, naphtha)	Ni/Al_2O_3
High-temperature watergas shift	Fe–Cr (Fe_3O_4 promoted with Cr_2O_3)
Low-temperature water–gas shift	Cu–Zn/Al_2O_3 (CuO–ZnO on Al_2O_3)
Methanation	Ni best, others can be Rh, Ru

gas flow. The catalytic activity decides the rate of the reaction. However, the heat and mass transfer are the major rate-limiting steps. The steam methane reformer reactors are better suited to operate under steady state for longer durations. The reactor performance is limited by the heat transfer; as such, these reactors are designed in a manner to work for better heat exchange and thus are large and bulky in nature [2].

In reformer, the heat transfer occurs in stages like (1) directly from burner, (2) by radiation from other tubes as well as reformer walls to the catalysts tubes, (3) by conduction across the tube walls, (4) by convection from tube walls to the catalyst and reactants. Convection from tube walls to the catalyst can be non-uniform, with higher temperature near the walls and less temperature in the core; this will result in lower conversion of reactants near core, resulting in deactivating catalysts and blocking active sites. The reactor tube dimensioning plays an important role in this case, if the radius is reduced the radial heat transfer will improve, but at the same time length has to be increased which results in increased pressure drop. Thus there will be regions near the wall of tubes, where catalyst can get overheated leading to sintering and toward center/core, where insufficient heat flow can cause fouling. Thus, the loss of catalytic activity can result from multiple factors, including deactivation by coke deposition, catalyst breakage, fouling, sintering, and poisoning by impurities like sulfur, etc. Sintering occurs, when the hotspots or certain reactions create unevenly high temperature. Close to the tube walls, the temperature is high, which can result in melting of the catalyst or support, leading to the phase transformation and loss of activity.

1.1.1.7 Efficiency of Steam Reforming

Steam reforming is the most widely used method for hydrogen production, because it is efficient and economical for commercial hydrogen production. The plants are designed to have high energy efficiency. The theoretical energy consumption as reported in literature is roughly around 13 MJ/Nm^3 for a steam reforming plant considered on LHV. It is thus 94% efficient, with a thermal efficiency of

95% including waste heat recovery, wherein the heat carried by both the flue gases and reformate gases is used in a heat exchanger for pre-heating and steam generation. Specific energy consumption for a typical SMR process ranges from 12.3 to 13.8 MJ/ Nm^3 H_2 on the low heating value basis. However, the economics and efficiency are favorable for large-scale steam methane reforming units, but for decentralized natural gas reforming, there are several challenges.

The major challenges in the distributed natural gas reforming include [3]:

1. Reformer Capital Costs—Currently the reformers designed are too cost-intensive. Besides, these are built one at a time, so the units are too large and too expensive and need to be customized as per the urban settings and physical footprint of the site/stations.
2. Operation and Maintenance—O&M costs for distributed hydrogen production via natural gas reforming are too high. Both the frequency and cost of repairs, as well as downtime, need to be minimized. Automation and remote monitoring could further help to cut down the associated O&M costs. Further, minimizing the system components, the amount of material required, and various energy losses involved could further lead to savings.
3. Feedstock Issues—A wide variation in the chemical composition of available feedstocks can be observed depending on the source. For example, in natural gas, the methane content can vary from 80 to 100%. Reformers and catalysts capable of handling variations in feedstock composition and consistently producing hydrogen of acceptable quality are required. Reformers and catalysts capable of handling variations in feedstock composition and consistently producing hydrogen of acceptable quality are required.
4. Control and Safety—The reforming sites need to assure public safety and at the same time take care of environmental issues without any substantial cost increase. Especially the sensors and controls are required to avoid any kind of leakage and accidents and at the same time minimize emissions.

The major needs for distributed natural gas reforming are:

1. Reduce capital costs.
   (a) Improve catalysts (lower catalyst cost, lowering of operating temperature to reduce material cost for reactor and energy requirements for heating, high yields).
   (b) Improve separation and purification technologies (develop improved membranes for separation and cost-effective methods).
   (c) Integrate system components (develop one-step reforming/shift or shift/ hydrogen separation).
   (d) Optimize system turndown capability to manage variable demand (can be improved by onsite hydrogen storage and proper sizing of the reformer, a balance between the two is essential, else cost may escalate).
   (e) Design for manufacture & assembly (DFMA)/high-volume equipment (optimization of subsystem design without change in the processes, reduction in the total amount of material, and replacement by less expensive

materials, integrating more systems together, e.g., compression, storage, and dispensing).

(f) Balance of plant optimization (improved, low-cost sensors, pumps, blowers, and monitoring equipment; by developing common, interchangeable components, automated joining processes, and low-cost stamping and extrusion methods that will permit high-volume, assembly-line production of critical components that are currently machined and welded).

2. Reduce operation and maintenance costs.

   (a) Automate process control (onsite planned maintenance can be avoided by automating and remote monitoring, and also automatic fault detection could reduce the costs and use of low-cost sensors).
   (b) Improve equipment reliability (sustained catalyst activity, reduction in impurities of feedstock, reliability of the components, and reduction in the equipment complexity).
   (c) Minimize material and energy losses (any leakage of hydrogen is to be minimized, which in turn improves both efficiency/yield as well as safety, and any kind of heat loss should be avoided using well-designed insulation and heat integration).
   (d) Capital utilization can be improved by leveling peaks and valleys in hydrogen demand, proper sizing of the reformer, and making use of by-products such as electricity, heat, steam, off-gases.

1.1.1.8 Performance Indices

There are certain parameters which are important in reforming like the ratio of steam to carbon, flow rates, amount of conversion of the reactants, the amount of desired product formed (is defined by the term selectivity, i.e., of a particular desired product over unwanted products), the yield of hydrogen in the outlet stream of product gas, etc.

The flow rate is defined in terms of space velocity (SV), and it is the ratio of volume flow rate $\dot{V}$ and the corresponding reactor volume (V), represented as follows:

$$\text{Space velocity} = \frac{\dot{V}}{V} \tag{2.10}$$

The SV is usually mentioned as gas hourly space velocity (GHSV) for gaseous reactants; the temperature and pressure are mentioned along with GHSV at inlet. For liquid fuels reforming, the term used is liquid hourly space velocity (LHSV), as in the case of methanol reforming.

There are a few more important terms used in reforming, such as the percentage conversion of reactants:

$$\text{Conversion} = \left(\frac{\text{Amount of reactants consumed}}{\text{Reactants supplied}}\right) \times 100\% \tag{2.11}$$

This percentage conversion is not related to the hydrogen product obtained which is the desired product in case of SMR. There may be different products formed and how much is the desired product formed among the other products obtained is defined in terms of selectivity:

$$\text{Selectivity} = \left(\frac{\text{Amount of desired product formed}}{\text{Amount of undesired or other products formed}}\right) \tag{2.12}$$

The selectivity of a product against the other products formed can also be defined as:

$$\text{Selectivity} = \left(\frac{\text{Amount of desired product formed}}{\text{Amount of reactants consumed}}\right) \times \text{Soichiometric ratio of desired product formed to the feedstock used} \tag{2.13}$$

For example, in the case of the steam methane reforming reaction

$$CH_4 + H_2O \rightarrow CO + 3H_2$$

one mole of CH_4 reacts with one mole of H_2O to produce one mole of CO and three moles of H_2, giving a hydrogen-to-methane ratio of 3:1 in the products.

$$\text{Yield} = \frac{\text{Amount of product formed}}{\text{Amount of feedstock consumed}} \tag{2.14}$$

$$\text{For e.g. } H_2 \text{ Yield\%} = \frac{\text{Moles of Hydrogen produced}}{\text{Moles of methane consumed}} \times 100 \tag{2.15}$$

Yield can be obtained as the ratio of the amount of desired product to the amount of the reactant fed.

$$\text{Steam to Carbon Ratio} = \text{Total Steam input/Total carbon in the fuel} \tag{2.16}$$

1.2 Steam Reforming of Higher Hydrocarbons

The major processes and steps involved in the steam reforming of higher hydrocarbons ($> C_1$) remain the same as those for steam methane reforming: feed pretreatment processes involving desulfurization and pre-reforming, catalytic reforming, water–gas shift reaction, and purification of the outlet gas. Since higher hydrocarbons are complex mixtures of various hydrocarbons such as aromatics and olefins, they are more prone toward carbon formation, leading to catalyst deactivation. Hence, pre-reforming becomes essential before the actual reforming step, which can serve multiple purposes like reducing the coke deposition on catalyst; the steam reformer can be operated at lower S/C ratio resulting in better efficiency. Pre-reforming can enable feedstock flexibility by producing lower hydrocarbons that can be reformed in the main reformer, thus reducing size of the main reformer and allowing the use of less expensive catalysts in the main reformer. The pre-reforming is usually carried out at relatively lower temperatures than the main steam reforming reaction, in an adiabatic reformer at 400–550 °C. Catalysts similar to those used in steam reforming can be used in the prerefroming unit. The various reactions that take place in the catalytic reformer include steam reforming, cracking of hydrocarbons, followed by gasification, decomposition to lower hydrocarbons, and steam methane reforming. The reforming of propane and n-butane is of interest as they are the main constituents of LPG which is easily available.

The steam reforming reaction of the ethane, propane, and butane is as follows:

$$C_2H_6 + 2H_2O \rightarrow 2CO + 5H_2 \quad \Delta H^0 = +374.3\ \text{kJ/mol} \tag{2.17}$$

$$C_3H_8 + 3H_2O \rightarrow 3CO + 7H_2 \quad \Delta H^0 = +497.7\ \text{kJ/mol} \tag{2.18}$$

$$C_4H_{10} + 4H_2O \rightarrow 4CO + 9H_2 \quad \Delta H^0 = +651.3\ \text{kJ/mol} \tag{2.19}$$

These reactions are highly endothermic reactions, as evident from the heat of reaction, with endothermicity increasing with increasing carbon content. In addition to the above-mentioned reactions, various other parallel reactions possible include: Cracking

$$C_2H_6 \rightarrow 3H_2 + 2C \quad \Delta H^0 = +84.7\ \text{kJ/mol} \tag{2.20}$$

$$C_3H_8 \rightarrow 4H_2 + 3C \quad \Delta H^0 = +103.8\ \text{kJ/mol} \tag{2.21}$$

$$C_4H_{10} \rightarrow 9H_2 + 4C \quad \Delta H^0 = +126.2\ \text{kJ/mol} \tag{2.22}$$

Carbon gasification

$$C + H_2O \rightarrow CO + H_2 \quad \Delta H^0 = +131.3\ \text{kJ/mol} \tag{2.23}$$

Gasfification

$$C_nH_m + xH_2O \rightarrow yH_2 + (n-1)CO + CH_2 \tag{2.24}$$

Methanation reaction

$$CO + 3H_2 \rightarrow CH_4 + H_2O \quad \Delta H^0 = -206 \text{ kJ/mol} \tag{2.25}$$

Nickel-based catalysts are most widely used, but then coke deposition is the major problem; therefore, several modifications have been introduced for the reforming of higher hydrocarbons, such as the use of promoters and different supports. In the literature, the various promoters studied include K_2O, MgO, CaO, and SrO; reported support materials are MgO, CaO, SiO_2, CeO_2, YSZ, and mixed oxides such as MgO–Al_2O_3, CaO–Al_2O_3, etc. Noble metals like Pt, Pd, Rh, bimetallic, and multimetallic catalysts where noble metals and Ni are used are also demonstrated, on different supports. Noble metals have lower coke deposition, but cost remains the major challenge [4].

1.3 Steam Reforming of the Liquid Hydrocarbons

Steam reforming of various hydrocarbons, such as gasoline, diesel, or jet fuel, can also be used for hydrogen production, especially in refineries, where the heavier products that can't be processed due to stringent fuel norms can, however, be upgraded and used for hydrogen production.

The liquid fuels have quite complicated compositions including various olefins, paraffins, aromatic compounds, and naphthalenes, among others. The simplistic representation of fuels to mention the steam reforming can be done in terms of their main constituents of such fuels like isooctane in gasoline, dodecane in jet fuel, and hexadecane in diesel. The reforming reactions for these representative components can be represented as:

$$C_8H_{18} + 8H_2O \rightarrow 8CO + 17H_2 \quad \Delta H^0 = +1273 \text{ kJ/mol} \tag{2.26}$$

$$C_{12}H_{26} + 12H_2O \rightarrow 12CO + 25H_2 \quad \Delta H^0 = +1866.2 \text{ kJ/mol} \tag{2.27}$$

$$C_{16}H_{34} + 16H_2O \rightarrow 16CO + 33H_2 \quad \Delta H^0 = +2473.9 \text{ kJ/mol.} \tag{2.28}$$

These are highly endothermic reactions that require high temperatures between 700 ºC and 1000 °C, carried out in the presence of nickel or noble metal or metal carbides as catalyst. As mentioned in the previous section, several reactions can occur under the process operating conditions, such as cracking to produce lower hydrocarbons, decomposition resulting into carbon formation, carbon gasification, reforming, water–gas shift, and others. To reduce the amount of coke deposition, a higher steam-to-carbon ratio is being used for the reforming. A pre-reformer is desired to convert higher hydrocarbons to methane, and a desulfurization step to remove sulfur-containing impurities. According to thermodynamics, the carbon deposition is more favored at lower temperatures. Higher hydrocarbons have higher C-content, and a higher S/C ratio can reduce the carbon deposition, but at the same time, it also reduces the efficiency by requiring additional heating. In most of the cases, an S/C ratio between 2 and 3 is ideal for the reforming reaction. Other than Ni, noble metal, metal carbides which are also reported to be used as catalyst. Different supports such as Al_2O_3, $MgO–Al_2O_3$, $CaO–Al_2O_3$, CeO_2, ZrO_2, $CeO_2–ZrO_2$, $CeO_2–Al_2O_3$, perovskites and different promoters to reduce carbon deposition includes K_2O, MgO, CaO, and SrO are being reported in literature. The use of bimetallic or multimetallic materials, as well as variation in catalyst preparation methods, also affects the stability, cyclability, and rate of hydrogen production.

1.4 Steam Reforming of Alcohols

Hydrogen can be produced from alcohols, with methanol and ethanol being of particular interest among these alcohols. There are several reasons for this interest; they are easily availabile and can be produced from renewable sources such as biomass; the reactions involved are simpler than heavier hydrocarbons; they can be transported to the point of use and reformed for distributed onsite hydrogen production; they react at lower temperatures; and fewer processing steps are required, as desulfurization unit and water–gas shift unit are not needed.

Methanol

The steam reforming of methanol can be represented by the reactions:

$$CH_3OH + H_2O \leftrightarrow CO_2 + 3H_2 \quad \Delta H^0 = +49.7 \text{ kJ/mol} \tag{2.29}$$

Methanol decomposition

$$CH_3OH \leftrightarrow CO + 2H_2 \quad \Delta H^0 = +90.7 \text{ kJ/mol} \tag{2.30}$$

WGS

$$CO + H_2O \leftrightarrow CO_2 + H_2 \quad \Delta H^0 = -41.2 \text{ kJ/mol} \tag{2.31}$$

Partial oxidation of methanol

$$CH_3OH + 0.5O_2 \leftrightarrow CO_2 + 2H_2 \quad \Delta H^0 = -192.5 \text{ kJ/mol} \tag{2.32}$$

Methanol reforming can be carried out at lower temperature (240–260 °C) in membrane reactors or in the fixed bed reactors. The reforming catalysts usually used are Cu-based catalysts, for example Cu–ZnO on Al_2O_3 with an S/C ratio between 1 and 2 at operating temperature lying in the range of 200–300 °C. Noble metal catalysts can also be used for reforming reactions; however, cost is the major concern. Various support materials are reported, and promoters could be added to reduce the sintering of the Cu-based catalysts. The reaction is endothermic and yields higher amounts of H_2, around 70–75%. The required heat for this endothermic reaction can be reduced by addition of oxygen. The endothermic steam reforming can be combined with an exothermic oxidation process, and with an appropriate choice of steam and oxygen, making the process either thermoneutral, endothermic, or exothermic. The reaction involved is as follows:

$$CH_3OH + xO_2 + (1 - 2x)H_2O \rightarrow CO_2 + (3 - 2x)H_2 \quad (2.33)$$

Ethanol Reforming

Similar to the methanol reforming, the various reactions for ethanol reforming can be mentioned as follows:

$$C_2H_5OH + 3H_2O \leftrightarrow 2CO_2 + 6H_2 \quad \Delta H^0 = +173.3 \text{ kJ/mol} \quad (2.34)$$

Oxidative steam reforming of ethanol

$$C_2H_5OH + 2.28H_2O + 0.36O_2 \rightarrow 5.28H_2 + 2CO_2 \quad (2.35)$$

Partial oxidation of ethanol

$$C_2H_5OH + 1.5O_2 \leftrightarrow 2CO_2 + 3H_2 \quad \Delta H^0 = -552 \text{ kJ/mol} \quad (2.36)$$

Ethanol decomposition

$$C_2H_5OH \leftrightarrow CO + 2H_2 + CH_4 \quad \Delta H^0 = +49.7 \text{ kJ/mol} \quad (2.37)$$

WGS

$$CO + H_2O \leftrightarrow CO_2 + H_2 \Delta H^0 = -41.2 \text{ kJ/mol} \quad (2.38)$$

The reaction of steam reforming of ethanol is highly endothermic and requires temperatures between 300 and 800 °C. By adding oxygen and controlling the oxygen-to-steam ratio, the process can be made endothermic, exothermic, or thermoneutral. Besides the reactions mentioned above, several undesirable reactions can occur, leading to decrease in the hydrogen yield, such as formation of acetylene or formaldehyde, through various dehydrogenation or decomposition processes. With the use of suitable catalyst, a better selectivity toward producing hydrogen can be achieved. Various catalysts including Ni, Co, Cu, and noble metals have been reported for ethanol reforming. Ni and Rh and few bimetallic catalysts are very promising on CeO_2, ZrO_2, or CeO_2–ZrO_2 support.

1.5 Advanced Methods of Reforming

1.5.1 Sorption-Enhanced Reforming

Conventional steam methane reforming has been used for decades to produce hydrogen and is the most widely used technique for hydrogen production. However, several challenges are associated with the process. The high-temperature requirement due to the highly endothermic reforming reaction results in a large quantity of fuel being required to provide the required heat of reaction. The overall process is complicated, and it involves several steps. There are integration challenges between highly endothermic reforming reaction and exothermic shift reaction. Additionally, catalysts used in three of the processes, i.e., reforming, HTWGS, and LTWGS units, are sensitive to impurities present in the feedstock. Complete recovery of hydrogen during the purification steps is not possible, leading to some loss of hydrogen; however, the off-gas is utilized as fuel in the reformer unit. Furthermore, the reforming reaction is endothermic, and the shift reaction is exothermic; both of these reactions suffer from the limitations of chemical equilibrium, and it is impossible to achieve complete conversion of CH_4 or CO in a single reaction under normal reaction conditions. Thus, to achieve better reactant conversion or product selectivity, and to favor forward reaction equilibrium under the conventional reactor operating conditions, if one of the products either H_2 or CO_2 could be removed, several benefits can be achieved. The benefits can be in terms of lower operating temperature, favorable forward reaction, less energy required, and better energy integration between different units for SMR and WGS reactions. The possible solutions in which either CO_2 or H_2 is removed are sorption-enhanced reforming or the use of hydrogen membrane reactor, respectively. Thus, removing one of the products formed favors forward reaction in the reversible steam methane reforming reaction.

$$CH_4 + H_2O \rightarrow CO + 3H_2 \quad \Delta H^0 = +206 \text{ kJ/mol} \tag{2.39}$$

In sorption-enhanced reforming, the concept is, if product CO_2 can be removed, such that the forward reaction favors and the normal equilibrium limits are displaced, then complete conversion of CH_4 can be closely achieved. In this process, all the three steps, i.e., SMR, WGS, and CO_2 absorption, occur in the same reactor. The CO_2 formed is removed using a solid sorbent which can either react with CO_2 to form another compound or be either physically or chemically adsorbed onto the surface of the sorbent. The general reaction can be written as follows:

$$CO_2(g) + A(s) \rightarrow A.CO_2(s). \tag{2.40}$$

The overall reaction can be written as

$$CH_4 + 2H_2O + A \rightarrow 4H_2 + A.CO_2(s). \tag{2.41}$$

Thus, in this method of hydrogen production, all the three reactions, i.e., (2.2), (2.6), and (2.39) take place in a single reactor such that in the reaction zone H_2 is obtained as product. Depending on the reaction conditions, H_2 product purity can be achieved in the range of 95–98%, while CO and CO_2 content is at ppmv level. Typically, equilibrium hydrogen yield obtained is 3.46 mol/mol of CH_4 as opposed to 4 mol of H_2 /mol of CH_4 in the case of SMR. Out of the three reactions occurring in the same reactor, two, i.e., WGS and CO_2 sorption, are exothermic in nature, and SMR is endothermic, and thus the heat released in exothermic reactions is utilized to provide heat of the reaction for the endothermic reaction, thereby improving the energy efficiency of the entire process. The requirement of supplemental natural gas fuel stream is low in SER; it is 0.37 mol/mol CH_4 as against 0.5 mol/mol CH_4 in the case of conventional SMR process [5]. Several sorbents have been reported in the literature including the calcium-based sorbents, potassium-promoted hydrotalcite (K-HTC) sorbents, and metal oxides of lithium and sodium. If CaO is used as sorbent, the reaction (2.12) is as follows:

$$CO_2(g) + CaO(s) \rightarrow CaCO_3(s) \tag{2.42}$$

Inexpensive sorbents like limestone and dolomite can be used for CO_2 sorption. Limestone shows a higher theoretical sorption capacity of 0.79 g CO_2/gCaO compared to 0.46 g CO_2/gCaO for dolomite. However, as per literature reports, dolomite has higher cycling life. All the three reactions (2.2), (2.6) and (2.39) take place in a fluidized bed reactor (FBR) containing the mixture of reforming catalyst and CO_2 acceptor at a temperature of 600–725 °C. The overall reaction in the SER reactor is close to thermo neutral [6], resulting in an overall energy savings of 20–25% compared to traditional steam methane reforming. In the reaction (2.39), the sorbent is either consumed and forms another product or becomes saturated, and it has to be regenerated back to its initial state. The entire process is dynamic, and the sorbent should be able to maintain its activity for several cycles to make the entire process economically viable. Regeneration of the spent CO_2 acceptor, usually $CaCO_3$, is carried out in an adiabatic fluidized bed reactor at a temperature of 885–975 °C. The pure CO_2 separated can either be sequestered or used for specific applications through CCUS route. The high temperature required during regeneration of the CO_2 acceptor is the major issue with SER. The schematic of sorption-enhanced reforming is shown in Fig. 2.6.

There are several advantages of sorption-enhanced reforming; these include:

(1) Reduction in the number of steps.
(2) No need of shift reactor.
(3) A simpler purification step is achieved since CO_2 is removed, and a very low concentration of CO or CO_2 or CH_4 is left; pure H_2 is obtained in the outlet stream.
(4) Lower-temperature operation.
(5) Cost reduction achieved by replacing high-temperature-resistant material for reformer reactor with less expensive materials of construction.

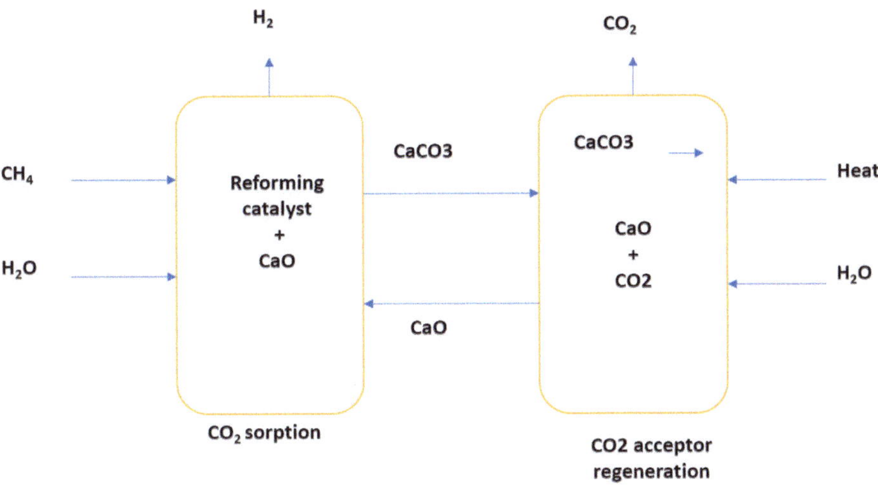

Fig. 2.6 Schematic of sorption-enhanced reforming with Ca-based CO_2 acceptor

(6) Higher methane conversion and better H_2 yield can be achieved at lower temperatures.

As reported [5], the H_2 content in the conventional SMR reaches a maximum of 76% on a dry basis at 900 °C. However, in the presence of a CO_2 acceptor the H_2 content reaches a maximum of about 96% at 650 °C and is $\geq$ 95% for all temperatures till 750 °C. However, above 850 °C, $CaCO_3$ can no longer be formed, and in that case reaction is same for both cases, with or without CO_2 acceptor.

Although there are several benefits of sorption-enhanced reforming as to the conventional SMR, at the same time, there are challenges associated:

1. High regeneration temperature (885-975^{0}C) is required for the regeneration of the CO_2 acceptor.
2. Loss of activity of CO_2 sorbent is observed after repeated cycles due to sintering and structural changes.
3. Separation of the reforming catalyst and CO_2 acceptor.

The sorbent durability is the major bottleneck. The limestone and dolomite contain some quantity of sulfur-containing impurities in the form of sulfates, which under the reducing environment in the reactor can cause sulfur poisoning to the reforming catalyst. There have been several reports in literature, mentioning ways to improve the durability of sorbent, such as use of other options to dolomite example, huntite or use of synthetic calcium-based sorbents, core–shell combination of dolomite core and nickel shell, Li_2ZrO_3, $LiFeO_2$, $LiNiO_2$, Li_2TiO_2, Li_4SiO_4, K_2CO_3 doped Li_4SiO_4, etc. [7]. However, the addition of fresh CO_2 sorbent and removal of spent sorbent is inevitable in sorption-enhanced reforming.

1.5.2 Hydrogen Membrane Reactor

Another option to address the chemical equilibrium limitation and favor the forward reaction in methane reforming is to remove the product H_2. This can be achieved using hydrogen membrane reactor, in which the reaction of hydrogen generation and its separation occurs in the same reactor. The product hydrogen is selectively removed from the reaction side by membrane permeation, thereby shifting an equilibrium-limited reforming reaction toward the product side, and thus achieving higher methane conversion at lower temperatures, according to the Le Chatelier's principle. This "shift effect" allows for two options: for similar conversions as in traditional reactions operating at much milder conditions than SMR or achieving higher conversions under similar operating conditions as in conventional steam reforming reactors (which operate as closed system) [8]. Better methane conversion results in higher hydrogen yield in membrane reactors. In conventional reactors, due to a large difference in operating temperatures between SMR and WGS reactions, they cannot be conducted in the same reactor and thus lead to energy penalty. However, in membrane reactors, due to the shift of the reaction equilibrium toward the product side, higher yield is obtained at milder conditions, and both SMR and WGS reactions can occur in the same reactor simultaneously, reducing the energy penalty. Similar to sorption-enhanced reforming, the membrane reformer performs several processes, including the production and separation of hydrogen in the same reactor. The basic requirements for a membrane to be used in the reactor are high selectivity toward hydrogen as opposed to other products, high chemical and mechanical stability, and high flux. In this method, hydrogen produced in the reforming is selectively removed from the reaction using a hydrogen-permeable membrane, with the driving force being the partial pressure of the species on the membrane side. This driving force is created either with the help of a sweep gas, which can be either an inert gas or nitrogen, or by creating a pressure difference between the two sides of the membrane. A higher pressure in the steam reforming reaction shifts the chemical equilibrium toward reactants side; however, in membrane reactors the increase in reaction pressure enhances the driving force for hydrogen permeation (represented by the hydrogen partial pressure, square root difference between retentate and permeate side). According to Richardson's equation, higher the pressure, higher will be the hydrogen being removed by permeation across the membrane. Thus, of the two competing effects - the thermodynamic effect and the shift effect - the latter dominates, resulting in improved conversion and higher yield. Further, the increase in pressure reduces the CO concentration, since it has a positive effect observed in WGS reaction. Another parameter of interest is the feed flow rate. When this flow rate is lower, it allows for a higher residence time for the interaction between reactants and products, leading to better conversion and better separation.

Thus, the same conversion, as that of traditional reforming can be achieved under a lower exit temperature from the reformer, i.e., less than 700 °C. However, the economy of the process depends on the cost of the membrane versus the savings in terms of fewer processing steps (elimination of PSA and shift reactor). Besides, the hydrogen produced is at a lower pressure so it needs to be compressed later for utilization, which adds up to the cost, or can be used for small-scale applications where low-pressure hydrogen is required. However, the advantage is concentrated CO_2 gas can be obtained which can be beneficial and can be used for CCUS.

The purity of the hydrogen depends on the permeability and selectivity of the membranes used, and with appropriate membrane selection, high-purity hydrogen can be achieved, thus eliminating the need of purification stages. Membranes used for HMR are made from Pd or Pd alloys, ceramic, zeolite, silica, etc., and are several microns thick. Palladium has been widely reported for over a century for its high-temperature hydrogen separation capabilities. However, since palladium is being expensive, alternatives were sought. Those reported include palladium alloys and cermets, i.e., group V elements and their alloys. Palladium is very sensitive toward poisoning, palladium-copper alloys were reported to be tolerant toward impurities especially sulfur, while palladium–silver alloys showed lower phase transition temperature. Alloying with copper or silver resulted in better hydrogen permeabilities while also achieving cost reduction. In the presence of Pd alloy membranes, the reaction temperature is lowered to 500–550 °C, resulting in higher thermal efficiency. This eliminates the need for a separate hydrogen purification unit, thereby reducing the unit's volume by one-third and its area by half compared to the conventional unit. The endurance shown by Pd alloy membranes is high, reaching around 1600 j and 30 start-stops. In literature there are reports targeted toward increasing the life of these membranes [9]. Group V elements had even higher permeabilities than Pd or its alloys; however, embrittlement and poisoning are the major drawbacks [10]. Porous inorganic membranes have advantages over Pd alloy membranes and ion transport membranes, as they can operate at higher pressures, since the flux is directly proportional to pressure for them. In contrast, for Pd, the flux is inversely proportional to the square root of pressure. The permeance for microporous inorganic membranes increases with temperature. Thus, these membranes have the potential to provide high flux at elevated temperatures and pressures. The biggest advantage of these membranes is the low cost. This discussion focuses on the choice of materials for membranes; however, the reactors in which these membranes are used for hydrogen production can have different configurations, such as packed bed, fluidized bed, or micromembrane reactors. Packed bed reactors have a tubular geometry, where the catalytic bed is located between the outer tube and the inner tube, and the membrane is positioned on the inner tube as shown in Fig. 2.7. In fluidized bed MR, the catalytic bed operates in either a turbulent regime or a bubbling regime, and the membrane is located within the reactor, resulting in lesser pressure drop. In micro MRs, with dimensions in the range of 1–1000 μm, better heat and mass transfer characteristics and thus better intensification can be achieved (Fig. 2.7).

Conventional steam methane reformers have thermodynamic constraints, including high exit temperature, which can be addressed with the use of membrane reformers. Important is the availability and good activity of the catalysts at low operating temperatures of around 400 °C. In conventional reformers, the hydrogen exit temperature can be reduced to as low as 700 °C. This results in

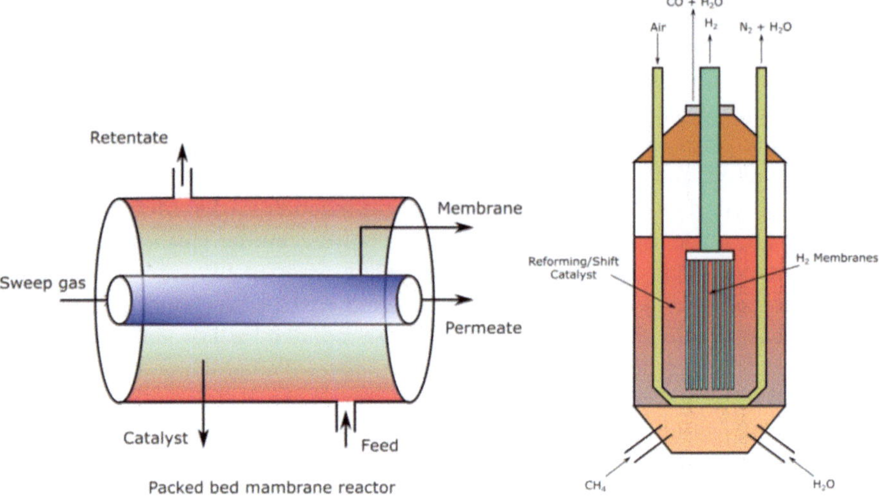

Fig. 2.7 Schematic of hydrogen membrane reactor

a lower outlet gas pressure, which requires additional compression. The resulting low-pressure hydrogen is suitable for low-pressure applications such as fuel cells.

There are several advantages of the process:

1. Hydrogen production is free from the limitations of chemical equilibrium
2. Higher methane conversion achieved.
3. Temperature of operation is lower.
4. Production of pure hydrogen, i.e., separation of other gases and hydrogen is achieved, and thus, an additional purification step is not required
5. The water–gas shift reactor can be eliminated, and thus, poisoning of the shift catalysts is not a concern.
6. Multiple reactions take place in a single reactor, and thus, the process is simplified.
7. Cost, complexity, energy consumption, and footprint requirement of the system are reduced.

The major challenges associated with HMR includes:

1. Membrane pores get blocked with sintered catalyst particles.
2. Mechanical and thermal stability are the biggest concerns for the membranes, due to stringent operating conditions.
3. Dilution results when sweep gas is used.
4. Substantial pressure drop results in HMRs that render method non-economical due to the additional requirement of compression of the product gas, or it has to be used for low-pressure hydrogen applications like fuel cell.
5. Membranes are expensive, and thus, the requirement of research is to explore new inexpensive membranes.
6. Currently, the achievable flux with commercially available membranes is low.

2. Partial Oxidation of Hydrocarbons

As the name suggests, the supply of oxidant is substoichiometric, such that incomplete oxidation of the hydrocarbon occurs. The reason is that if a stoichiometric amount of oxidant for complete oxidation is provided, then the products obtained will be carbon dioxide and water. As opposed to the steam reforming reaction, the hydrogen production from partial oxidation is an exothermic process. However, instead of steam (in case of SMR) either oxygen or air is used as an oxidant. The hydrogen yield in the process as compared to steam reforming is lower; the reason being in steam reforming, some hydrogen content in the product is also obtained from the reactant steam, which is not the case in partial oxidation reaction. Additionally, the CO_2 released is also higher. In the partial oxidation process, high temperatures (even 1300–1400 °C) and high pressures (40–50 bar) can be achieved, which can help address the problem of carbon formation.

Refineries, which were long back considered major producers of hydrogen, have become the major consumers of hydrogen and require hydrogen for various hydrotreatment processes. With increasingly stringent fuel emissions norms, most of the hydrogen produced in refineries is consumed internally and may still have requirement for merchant hydrogen, making hydrogen management a big challenge. Besides hydrogen management, the bottom of the barrel fuel usage is the major concern for refineries; the requirement is upgradation of the heavier fractions or conversion of heavier fractions to valuable products. The partial oxidation method can be used to produce hydrogen from these fractions. The partial oxidation of hydrocarbons is essentially a reaction between the feedstock and a limited amount of oxygen, to produce syngas. The production can be carried out in two ways, either non-catalytically or catalytically. The non-catalytic partial oxidation occurs at higher temperatures, typically 1200–1500 °C, and can be used to produce hydrogen from any possible feedstock, thus providing fuel flexibility. In contrast, the catalytic partial oxidation occurs at a comparatively lower temperature in the presence of a catalyst and usually involves lower hydrocarbons, such as natural gas or naphtha. The POx process has high conversion efficiency and can be used for polygeneration, whereas the syngas obtained can be used for the synthesis of a variety of chemicals.

2.1 Non-catalytic POx

As the name suggests, the partial oxidation of hydrocarbons is carried in the absence of any catalyst; therefore, the reaction takes place at higher temperatures. Since catalyst is not used in non-catalytic POx process, the flexibility of using any feedstock is the biggest advantage. The challenges of catalytic processes such as catalyst deactivation in the presence of higher hydrocarbons, or in case of heavy oils with sulfur or other impurities, coke deposition are absent in this method. With increasingly stringent fuel norms for both refinery products and refinery operations, a growing need for hydrogen could be seen. Large quantities of hydrogen are required for various hydrotreatment processes, such as hydrotreating, hydrodesulfurization, or hydrocracking. Since heavier crudes with high concentrations of sulfur and heavy metals need to be processed, the implementation of partial oxidation of heavy residues can be seen as an integral part of hydrogen production in refineries. Non-catalytic partial oxidation can serve this purpose, thus providing feedstock flexibility used for producing hydrogen. Also, the feedstock with sulfur-containing compounds can be handled in a partial oxidation reactor. Shell and Texaco have commercialized this process. The choice whether to use air or oxygen depends on the use of the hydrogen produced from POx. For example, if it is to be used for ammonia synthesis then air can be used as oxidant (outlet gas will be diluted with nitrogen), while if pure hydrogen is required, then oxygen is used as an oxidant. Besides, the reactivity of the feedstock is another important factor, in deciding the oxidant to be used in PO_X. Certain amount of steam is introduced, which has multiple applications in the process. It acts as a quenching media for soot removal. Additionally, water–gas shift occurs due to the presence of steam inside the reactor. Steam also acts as a moderator to maintain the reactor temperature and avoid hotspots, and ensuring proper mixing of feedstock and oxidant.

The oxygen required for partial oxidation is produced in an air separation unit (ASU), which in fact separates the components into nitrogen and oxygen, through the distillation of air carried out at cryogenic temperatures. ASU produces high-pressure oxygen to be used for the partial oxidation. Low-pressure oxygen can also be used for various sulfur-related processing. The processes involved in a typical ASU are compression, precooling, purification, cold distillation carried out at different pressures, of the ambient air.

In the non-catalytic process, the mixture of oxygen and hydrocarbon is preheated, mixed, and fed to the burner for partial oxidation reaction, resulting in the production of syngas. The reactor temperature reached is very high in the range of 1200–1500 °C with a very low residence time of 0.5–5 s, so as to achieve complete conversion of methane. The hot syngas is passed through a waste heat recovery unit for extracting the waste heat. The particulates or soot are scrubbed using wash quench water to remove any particulate matter from the gas. The slag or soot or ash, depending on the feedstock used in the reactor, is extracted with quenching water used in the reactor. The solid residues extracted from quenching water are fed back to the partial oxidation reactor after treatment, while some wastewater which cannot be reused is discarded, and the remaining treated water is fed back to the reactor. The cooled syngas after waste heat recovery is fed to the water–gas shift reactor, where the carbon monoxide is converted to carbon dioxide and more hydrogen is obtained. The syngas has acid gases (CO_2 and H_2S), and these need to be removed. Removal of acid gases can be done in one step or two steps. In one-step process, it goes through an integrated CO_2 and sulfur removal process, where the syngas is first cooled and then goes through rectisol process with cold methanol as solvent and separates CO_2 and sulfur compounds. The sulfur compounds are treated at the Claus plant to get elemental sulfur. If the sulfur removal is not done prior to WGS step, then the catalyst used should be sulfur tolerant in the process. Else the removal of sulfur is required prior to WGS.

$$C_nH_m + \left(\frac{n}{2}\right)O_2 \rightarrow nCO + \left(\frac{m}{2}\right)H_2 \text{ (exothermic)} \tag{2.43}$$

$$C_nH_mS_x + \left(\frac{n}{2}\right)O_2 \rightarrow nCO + \left(\frac{m}{2} - x\right)H_2 + xH_2S \text{ (exothermic)} \tag{2.44}$$

$$C_nH_m + nH_2O \rightarrow nCO + \left(n + \frac{m}{2}\right)H_2 \text{ (endothermic)} \tag{2.45}$$

$$C_nH_m + nCO_2 \rightarrow 2nCO + \left(\frac{m}{2}\right)H_2 \text{ (endothermic)} \tag{2.46}$$

$$C_nH_m \rightarrow (n - x)C + C_xH_y + \left\{\frac{(m - y)}{2}\right\}H_2 \text{ (endothermic)} \tag{2.47}$$

$$C + \left(\frac{1}{2}\right)O_2 \rightarrow CO \text{ (exothermic)} \tag{2.48}$$

$$C + H_2O \rightarrow CO + H_2 \text{ (endothermic)} \tag{2.49}$$

$$CO + H_2O \rightarrow CO_2 + H_2 \text{ (exothermic)} \tag{2.50}$$

In the two-step process, the sulfur and CO_2 are removed separately. The syngas obtained after POx is cooled down using wash quench water and that also provides the required steam for WGS reaction, hereafter the cooled syngas is directed for scrubbing with cold methanol (– 30 °C), separating CO_2 and H_2S. This sulfur-free gas goes for WGS and thereafter for the CO_2 removal (Fig. 2.8).

There exist two zones in the non-catalytic partial oxidation reactor: first is the flame region, where the major reactions take place between the hydrocarbons and oxygen, and the second is essentially a heat exchanger zone, where the excess heat generated is recovered after the reaction [11]. Large-scale uncatalyzed partial oxidation reactors typically operate in a temperature range of 1150–1500 °C and pressure between 25 and 80 bar. Since there are metallurgical constraints due to the stringent reaction conditions, refractory-lined reactors are used. The oxidant and feedstock are preheated separately and then mixed with steam, which is introduced into the burner where appropriate mixing is ensured, through careful design considerations. Steam acts as a coolant, quenches to control the reaction temperature to desired levels, and is especially used in cases of feedstock involving high-carbon fuels or liquid fuels. Various reaction products are obtained as per the reaction (2.42)–(2.49). A typical gas composition of 46% H_2, 47% CO, and 4% CO_2 on a dry basis and small % of H_2S and N_2 are obtained, if O_2 is used as oxidant from an air separation unit. The soot is formed by decomposition reactions and it is an undesirable product found to be suspended in the product gas. This hot gas is quenched with process water which not only lowers the temperature of gas to the required temperature for

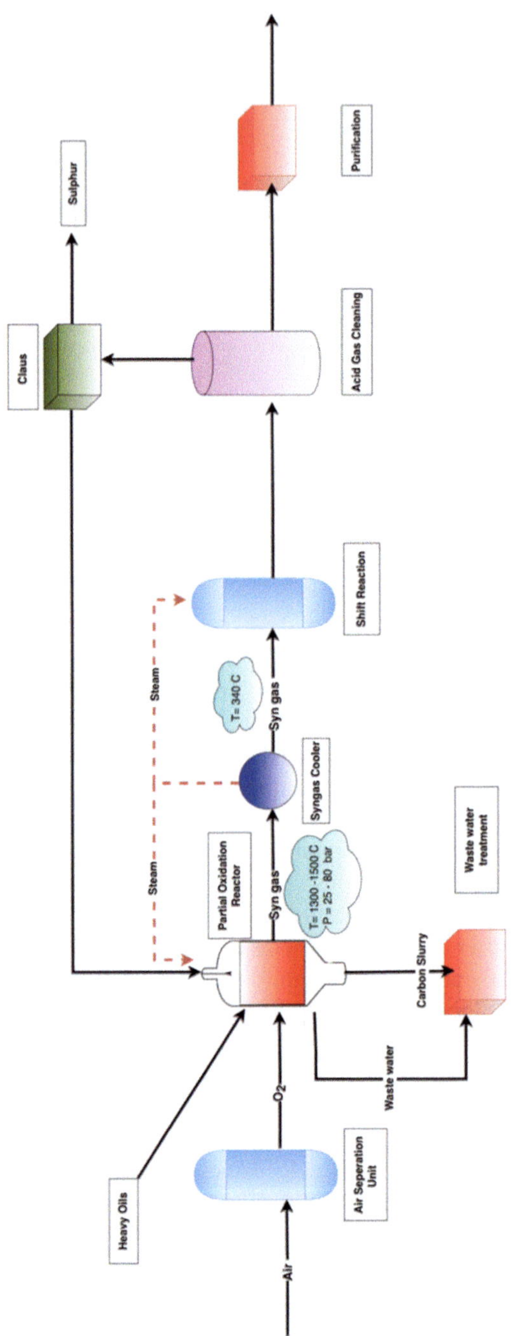

Fig. 2.8 Process flow diagram of partial oxidation process for hydrogen production

water–gas shift, i.e., 300 °C, but also provides the required steam for WGS reaction while removing the soot present in the reactor. This soot is extracted from the wash quench water and after recovery is either fed back along with the oil feedstock or is used as fuel for steam generation. The water from which soot is recovered is again recycled to be used as quench water. The excess steam available during quenching and washing is used in the catalytic conversion of CO to get more hydrogen and CO_2, which occurs in the presence of Co–Mo catalyst, which is a sulfur-tolerant catalyst. The syngas requires cleaning to remove H_2S and CO_2 to obtain pure hydrogen. The solvent-based removal followed by methanator can be used for CO_2 removal. The temperature or the heat generated by the oxidation can be increased or the start-up time can be decreased, by adding more oxygen or increasing the flow rate, but that compromises with increased CO as against the desired product H_2.

Depending on the feedstock used, such as liquid fuels or solid fuels, and also depending on the impurities, present in the syngas due to the feedstock used, the gas cleaning process can be complicated. The major design consideration involved is the burner, is that it ensures proper mixing of fuel and oxidant, reduces soot formation, and avoids any kind of hotspots in the reactor. Now, either pure oxygen, enriched air, or air can be used for the partial oxidation process. If pure oxygen is used, then an air separation unit is required, which makes the POx reactors compact but adds to the capital cost. If, however, air is used as oxidant, then the downstream gas is diluted with nitrogen; thus, gas cleaning is required to get pure hydrogen. However, if the application for the produced hydrogen is the synthesis of ammonia or nitrogen-based chemicals, an appropriate gas composition can be used for the purpose; thereafter, separation is not required to get pure H_2. The two major advantages of non-catalytic partial oxidation are: (1) Any kind of fuel from heavy oils to coal, petcoke, natural gas can be used as feedstock; (2) since the process involves oxidation reaction and the flame temperatures are lower, as such the NO_x and SO_x contents are lower; (3) relatively compact in size compared to reformers. A typical energy efficiency, considering all heat and steam requirements being met internally, ranges from 15.9 to 19.3 MJ/Nm^3.

$$\text{CH}_4 \text{ conversion } (\%) = \left(\frac{\text{CH}_4 \text{ molar flow at inlet} - \text{CH}_4 \text{ molar flow at outlet}}{\text{CH}_4 \text{ molar flow at inlet}}\right) \times 100 \tag{2.51}$$

$$\text{H}_2 \text{ yield } (\%) = \left(\frac{\text{H}_2 \text{ molar flow at outlet}}{2\text{CH}_4 \text{ molar flow at inlet}}\right) \times 100 \tag{2.52}$$

$$\text{COS selectivity } (\%) = \left(\frac{\text{CO molar flow at outlet}}{\text{CH}_4 \text{ molar flow at inlet} - \text{CH}_4 \text{ molar flow at outlet}}\right) \times 100 \tag{2.53}$$

$$\frac{\text{H}_2}{\text{CO}} \text{ molar ratio} = \frac{\text{H}_2 \text{ molar flow at outlet}}{\text{CO molar flow at outlet}} \tag{2.54}$$

Advantages of non-catalytic POx

- Unlike SMR where heat is indirectly supplied to reactor tubes, for pre-heating and steam generation; heat of reaction is directly obtained by partial oxidation within the reactor.
- Heat recovery is poor.
- Heat losses are higher, thermal efficiency is lower.
- Reactions are at higher temperatures; thus kinetics is faster and higher conversions can be achieved.
- More compact, simple, and less expensive reactor, bulky heat exchangers are not required as in case of SMR.
- Efficiency—60–75%.
- Less hydrogen than SMR, but better dynamic response and reduced start-up time than SMR.
- Lower efficiency than SMR.
- Energy Efficiency—15.9–19.3 MJ/Nm$^{3.}$
- Partial oxidation is an important process in refineries, providing fuel flexibility which is the biggest advantage.

The problems associated with the POx are the requirement for an air separation unit, handling oxygen, carbon formation especially during transient conditions, and possible handling issues with the reactor which can lead to explosions.

2.2 Catalytic Partial Oxidation

CPO is the reaction of lower hydrocarbons with an oxidant to produce syngas in the presence of a catalyst. CPO is an exothermic reaction; it is a more energy-efficient process with fast kinetics. The reactor size required is smaller than that for SMR, and the requirement of superheated steam is lower.

Various reactions occurring in CPO are as follows:

$$\text{Partial Oxidation } CH_4 + \frac{1}{2}O_2 \rightarrow CO + 2H_2 \quad \Delta H^0 = -38 \text{ kJ/mol} \tag{2.55}$$

$$\text{Dry reforming: } CH_4 + CO_2 \rightarrow 2H_2 + 2CO \quad \Delta H_{298} = 247 \text{ kJ/mol} \tag{2.56}$$

$$\text{Water gas shift: } CO + H_2O \leftrightarrow H_2 + CO_2 \quad \Delta H_{298} = -41 \text{ kJ/mol} \tag{2.57}$$

$$\text{Boudouard equilibrium: } CO_2 + C \leftrightarrow 2CO \quad \Delta H_{298} = 171 \text{ kJ/mol} \tag{2.58}$$

$$\text{Combustion: } CH_4 + O_2 \rightarrow 2H_2O + CO_2 \quad \Delta H_{298} = -803 \text{ kJ/mol} \tag{2.59}$$

$$\text{Combustion: } 2CO + O_2 \rightarrow 2CO_2 \quad \Delta H_{298} = -294 \text{ kJ/mol} \tag{2.60}$$

$$\text{Combustion: } 2H_2 + O_2 \rightarrow 2H_2O \quad \Delta H_{298} = -286 \text{ kJ/mol} \tag{2.61}$$

Some of these reactions are undesirable as they can lead to coke deposition, reduction in the desired product, and reduction in the purity of the product hydrogen. These competing reactions decide the equilibrium and thus the conversion of reactants, selectivity, and yield of the hydrogen product.

Several studies have been reported in the literature looking at the mechanism of the CPO process. Two mechanisms have been reported, a direct mechanism (direct partial oxidation DPO), wherein methane reacts according to the reaction below on the surface of the catalyst to yield syngas:

$$CH_4 + \frac{1}{2}O_2 \leftrightarrow CO + 2H_2 \tag{2.62}$$

where methane dissociates into carbon and hydrogen and gets adsorbed onto the catalyst surface, the adsorbed carbon then reacts with the surface oxygen of the catalyst to form carbon monoxide. The hydrogen adsorbed on the surface combines to form a hydrogen molecule and desorbs from the catalyst surface.

The second mechanism also known as the indirect mechanism involves combustion and reforming reaction. It is proposed that, firstly the methane and oxygen react to produce CO_2 and H_2O, which further undergo dry reforming and steam reforming to produce syngas [12, 13]as per the reactions:

$$CH_4 + CO_2 \rightleftharpoons 2CO + 2H_2 \tag{2.63}$$

$$CH_4 + H_2O \rightleftharpoons CO + 2H_2. \tag{2.64}$$

Certain experiments reported in the literature [14] indicate that in CPO reactors, there are three zones: the first zone where oxidation reaction occurs, the second zone is the catalytic region, and the third region is the heat exchange region, where excess heat is recovered. The oxygen is consumed in the first few millimeters or 20% of the reactor, undergoing exothermic feedstock oxidation reaction, producing high temperatures of the order of 1000–1250 °C depending upon the ratio of oxygen to the carbon; thereafter the endothermic steam methane reforming occurs reducing the temperature to around 800 °C followed by water–gas shift reaction to further enhance the hydrogen yield. The schematic of partial oxidation reactor is shown in Fig. 2.9.

Catalyst, support, and reactor design play the main role in the CPO. The catalyst used in catalytic partial oxidation is required to enhance the rate of the reaction, allow high gas hourly space velocities, prevent coke formation, prevent formation of

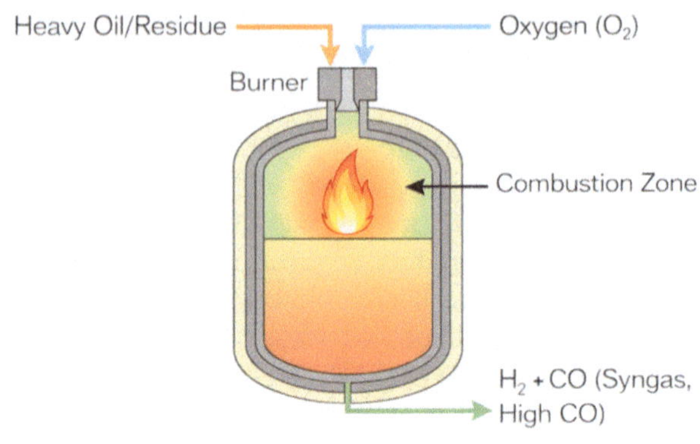

Fig. 2.9 Schematic of the partial oxidation reactor

hotspots and undesired products (like ammonia or HCN) which can lead to the reduction in the desired product hydrogen yield, low-pressure drop across the bed, high conversion of methane, high selectivity toward desired product, syngas. It should be stable under the stringent operating conditions of the CPO ensuring long cycle life without much degradation in the performance and cost-effectiveness.

Nickel is a cost-effective, widely used catalyst, but is prone to deactivation due to coke deposition, reaction of the nickel with the support in presence of acidic active sites, and sintering [15]. The size of the catalyst, the synthesis method, and the nature of support (its surface, uniformity, and acidity or basicity) on which Ni catalyst is dispersed are found to be important. Various oxides have been studied as support material; the most widely studied and used support for Ni catalyst is Al_2O_3, but is prone to coke formation and sintering, leading to the deactivation of catalyst. With alumina support, Ni interacts strongly to form $NiAl_2O_4$, which results in the loss of activity. To avoid this, other supports studied and reported in literature are MgO, La_2O_3, CeO_2, ZrO_2, perovskites (Ca, Sr, Ba titanates, etc.). MgO was found to be effective as with the limited interaction with the Ni catalyst, the stability is better and leading to high activity of the catalyst. The use of La_2O_3 and CeO_2 has shown promising results as the catalyst support interaction favors the carbon removal reaction from the catalyst surface (gasification). However, the combination CeO_2–La_2O_3 together worked more effectively than individually. ZrO_2 was reported to be not very effective in reducing the coke deposition.

Additionally, carbon dissolution in Ni catalyst ultimately leads to detaching the catalyst from the support, which can block the pores and result in a pressure drop. The use of bimetallic catalysts, i.e., Ni with noble metals has shown to reduce carbon deposition by hydrogen spillover from noble metals to non-noble metals leaving the non-noble metal intact. Actually, there are two competing reactions taking place on the catalyst surface, one leading to coke deposition primarily through methane decomposition, and another leading to removal of deposited coke by gasification. The use of bimetallic catalysts enhances the later effect. The acidic nature of the support can also favor methane decomposition, which should be taken into account to reduce catalyst deactivation. Cu being less expensive, but can only be used for lower temperatures, due to thermal stability related issue, this can be addressed with the use of support Al_2O_3, other possible supports are TiO_2, ZrO_2, SiO_2, etc.

Other transition metals like Co and Fe as catalyst have also been studied in the literature. Fe forms multiple oxidation states in oxides which reduces coke formation. Co, however, loses activity at higher temperature and has a much stronger interaction with the support that reduces activity and is expensive.

Various noble metal catalysts like Rh, Ru, Pt, Pd, Ir, Ag, and Re have been reported in literature; these are much more stable than Ni catalyst, are less prone to carbon deposition, and has lower dissolution of carbon into the catalyst. The biggest disadvantage of noble metal catalysts is their cost. Ag is not very good catalyst for the partial oxidation reaction, Ir hs shown sintering issues, Ru could not sustain the reaction conditions, Pd shown to have higher tendency toward coke formation, Re resulted in formation of hotspots, Pt was reported to have lower selectivity giving poor hydrogen yield, and the best-performing noble metal among these, was reported to be Rh with high selectivity toward hydrogen product. Noble metals on various supports like Al_2O_3, MgO, CeO_2, ZrO_2, TiO_2, Ta_2O_5, SiO_2, La_2O_3, Y_2O_3, etc., have been studied. Rh is known to be a better catalyst with high conversion and selectivity.

Bimetallic catalysts, where an inexpensive core such as one made up of Ni, Co, Cu, and a shell of noble metals like Rh, Ru, Pt, can be used as catalyst. Perovskites ABO_3, where A can be La and B can be Fe, Co, Ni, etc., can be used. The major findings from the literature are that the interaction between support and catalyst plays important role toward preventing sintering of the catalyst and maintaining the catalyst in its active oxidation state. Apart from a catalyst, a proper mixing of feedstock and oxidant is essential to reduce the coke formation. The method of synthesis for catalyst also determines its susceptibility toward deactivation and sintering.

The heat and mass transfer across the catalyst bed also plays an important role. Since the reaction is exothermic, limitations in heat transfer across the catalyst bed can lead to the creation of hotspots or overheating problems, which can result in sintering of the catalysts. The difficulty in controlling temperature also affects the product gas composition, as the latter depends on temperature, pressure, and O_2/C ratio. Besides, the transient conditions during reactor start-up and shutdown need to be better controlled to prevent any unwanted situations. This can be achieved by selecting the appropriate O_2/C ratio and following proper protocols while both starting and shutting down the reactor, ensuring that the reactions occur in a controlled manner and do not result in excessive temperatures and pressures leading to any accidents. The process of starting up is such that the catalyst bed is heated to the desired temperature by means of a heated inert gas flow, and later, the O_2/C ratio is fed in a systematic manner, starting from the lower to the finally desired one. Thus, by controlling the temperatures and flow rate of gases, keeping the composition of feed and oxidant below the flammable range, the reactor can be operated within the safe limits. Similarly, safety during the shutdown of the reactor is important. This can be achieved by reducing the oxidant and by either flowing an inert gas or steam to ensure the removal of all oxidants before shutdown.

In the catalytic partial oxidation method

- High T, short residence time, heat and master transfer issues, hotspots are created.
- Gas composition is determined by many factors—including feedstock, thermodynamics, and catalyst.
- CPOx requires a small footprint, modular construction, simple operation, fuel flexibility, and flexibility in production capacity, as well as lower capital cost and energy consumption.

Challenges

- Short lifespan of the CPOx unit, due to carbon formation and catalyst sintering.
- To achieve high purity, undiluted syngas, oxygen is preferable as against air as an oxidant, which requires additional costs for gas separation unit.
- Safety remains a concern as the exothermic nature of the reaction generates extreme hotspots. With elevated temperatures that can ignite a range of CH_4/O_2 mixtures, they should be operated outside the explosive limits.
- There is a lack of catalysts/oxygen carriers suitable for efficient and stable operation while still being economically viable.

Advantages of POx:

1. The reactor is compact and simpler than the steam reforming process.
2. Energy consumption is lower than steam reforming because the reaction is exothermic.

Challenges:

1. Operational conditions are stringent with high temperatures and pressures.
2. Emissions are high.
3. Soot formation takes place.

3. Autothermal Reforming

The autothermal reforming process combines the advantages of both steam reforming and partial oxidation, such that the exothermic partial oxidation reaction provides the energy required for the endothermic steam reforming reaction, making it a thermodynamically neutral process. The basic difference between steam reforming and autothermal reforming is that the former is endothermic, while the latter is thermoneutral. In steam reforming, the heat required for the endothermic reaction is supplied by combustion fuel in burners, outside the catalyst tube bundles, whereas the same heat is supplied in an autothermal reactor via fuel oxidation inside the reactor. The disadvantage of partial oxidation was the lower hydrogen yield; the endothermicity of the steam reforming is mitigated in autothermal reforming. The autothermal reforming has faster response time than SMR as such short start-up time required, better efficiency than partial oxidation, simple design and wide flow ranges possible, hotspots reduced due to steam, lower temperatures than POx results in reduced sintering of catalyst, but have higher risks involved due to exothermic reaction inside reactor as compared to the reforming that is why it is not much being used as compared to SMR. Besides the benefits of ATR mentioned above, there are several other benefits like high purity H_2, fuel flexibility (sulfur tolerance), and coke burnt off during catalyst regeneration [16].

The product gas composition from the ATR reformer is determined by the thermodynamic equilibrium at the exit temperature and pressure, which depend on the composition and flow rates of the feed, as well as the ratios of the oxygen and steam added in the reactor. The general reaction for ATR can be represented as:

$$C_nH_m + n/2O_2 + \text{ nH}_2O \rightleftharpoons nCO_2 + (n + m/2)H_2. \tag{2.65}$$

The autothermal reformer is a refractory brick-lined reactor with three distinct regions or zones. Once the reactor reaches the operating conditions (the temperature of the catalyst bed is increased by exothermic reaction between methane and oxidant), all three, the fuel, oxidant, and steam, are fed together to the reformer. The feed is introduced into the burner along with a substoichiometric amount of oxygen and steam. The burner is the key component of the reformer, which allows for uniform mixing between the feedstock and oxidants, uniform heat throughout, and results in reduced soot formation. The burner has a turbulent diffusion flame, and with the core temperature of the flame being more than 2000 °C, it is desired that the heat backflow from the core region of the flame to the burner or from the combustion gases to the burner should be minimum. A well-designed burner can lead to increased burner life and safer operation [17]. In the combustion region, the feed and oxidants are mixed and combusted. The partial oxidation of methane occurs with a substoichiometric amount of oxygen, where oxygen to CH_4 ratio is around 0.55–0.6. Some uncombusted methane will be left due to the insufficient amount of oxygen and will undergo further reaction in the other zones. The reactions in the combustion zone provide the required heat for the endothermic steam reforming reaction in the catalytic zone, according to the reaction:

$$CH_4 + \frac{1}{2}O_2 \rightarrow CO + 2H_2 \tag{2.66}$$

In general,

$$C_nH_m + n/2O_2nCO + m/2H_2 \tag{2.67}$$

$$H_2 + \frac{1}{2}O_2 \rightarrow H_2O \tag{2.68}$$

$$CO + \frac{1}{2}O_2 \rightarrow CO_2 \tag{2.69}$$

The next zone, the thermal zone, is a part of the combustion chamber where further conversion of the hydrocarbons occurs through homogeneous gas-phase reactions, such as steam methane reforming and water–gas shift reaction. In addition to these major reactions, various pyrolysis reactions involving higher hydrocarbons can also occur.

After the thermal zone is the fixed bed catalytic zone, where the final hydrocarbons conversion occurs through the heterogeneous catalytic reactions, according to the reactions mentioned below:

$$CH_4 + H_2O \leftrightarrow CO + 3H_2 \tag{2.70}$$

$$CO + H_2O \leftrightarrow CO_2 + H_2 \tag{2.71}$$

The front or top of the catalyst bed has a temperature of the combustion zone, i.e., 1200–1400 °C, therefore, stability in the top region of the catalyst and support is essential under these conditions.

The CO content in the ATR product gas is more than in SMR, due to a partial oxidation process. The H_2/CO ratio is comparatively lower than obtained in SMR, thus requiring a higher shift activity in ATR. Besides this, the higher temperatures inside the ATR result in restricted water–gas shift activity (which being an exothermic reaction is favored at low temperatures), thereby resulting in comparatively lower CO_2 at the outlet from the shift reaction. As seen from earlier reactions, some of the hydrogen is being consumed by oxygen in ATR. Although it has a lower steam requirement than SMR, but due to the high shift conversion requirement (high CO content in ATR), steam is still required for shift conversion. Since ATR operates at a higher temperature and the oxidation reaction provides faster kinetics, it can handle multiple fuels and achieve a faster start-up by providing the required heat through an exothermic reaction. The exothermic steps involved in ATR are fast in kinetics, and the response time is also shorter for the start-up of the reactor. The amount of heat available is controlled by the oxidation process. The exothermic reactions occur at the top of the reactor; thus, this assists in fast start-up. Like SMR, the ATR also has heat and mass transfer constraints, and the rate of reaction in the catalytic bed depends on the constraints, including the flow of reactants onto the catalyst surfaces. Autothermal reforming can be carried out for higher hydrocarbons, such as gasoline or alcohols.

By properly controlling the oxidation and reforming reactions in the catalyst bed by adjusting the O_2/CH_4 and H_2O/CH_4 ratio, the heat input requirement can be met internally. This allows the product gas composition to be selected, along with temperature and pressure, thereby determining the conversion and selectivity. Typical H_2O/C and O/C ratios in the ATR are around 1–2 and 0.5–1, respectively.

Well-designed ATR reactor overcomes the heat transfer limitations. The exothermic reaction not only supplies the required heat, but also needs to account for the heat losses. At the same time, if the heat losses from the reactor are more, while an excess heat is generated from the exothermic reaction, it can potentially lead to sintering of the catalyst. The final temperature achieved in ATR is lower than in POX; it is slightly exothermic, as an endothermic SMR reaction taking place inside ATR absorbs the heat produced during an exothermic reaction. This

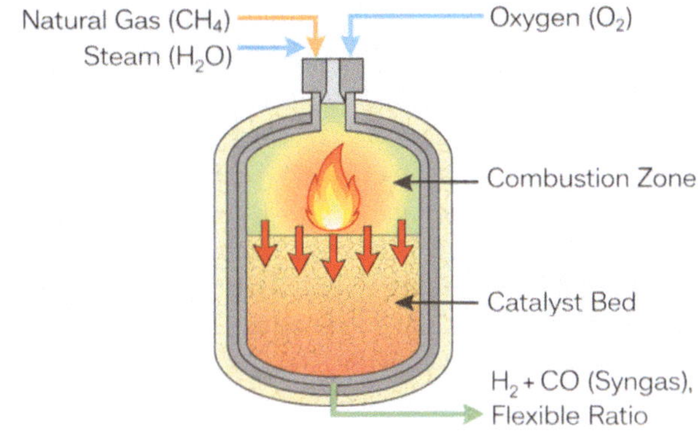

Fig. 2.10 Schematic of an autothermal reactor

has several benefits in terms of reactor design, including reduced thermal management requirements. The size of the reactor is smaller than that of a steam reformer, as the temperature experienced is lower, resulting in lower costs of insulation material, there is a wide choice of materials, and fuel will be required for start-ups. Thus, the reactor is simple, cost-effective, and easier to start up. The schematic of an autothermal reformer is shown in Fig. 2.10 and the process flow diagram is shown through schematic in Fig. 2.11

Control over the extent of oxidation and the reforming reactions is also provided by the catalyst, and an appropriate selection of catalyst is crucial to achieve the desired conversion of reactants and product selectivity. Since the heat generated in the exothermic oxidation reaction is used for the endothermic steam reforming reaction, there is a limit on the maximum rise in reactor temperature. After balancing out the heat requirements as mentioned, the overall reaction is slightly exothermic, which is essential to meet the heat losses in the processes. There are multiple benefits of lower temperature like the complexity of thermal integration reduces, less of fuel consumption to provide the desired heat unlike in the case of indirect heating (SMR), less fuel consumption while starting up phase, wide range of choice of materials possible, reduced manufacturing cost as the metallurgical constraints are relaxed, reduced size of the reactor as well as cost as the requirement of insulation is not very stringent.

The catalyst to be used for autothermal reactor should be thermally and mechanically stable under the stringent operating conditions, should have long life, be less prone to deactivation in the presence of impurities, especially in the combustion zone, reduced coke deposition and that it can carry out both oxidation as well as reforming reactions and should be able to withstand transient conditions like start-up or shutdowns. The catalyst used is the same as that used for SMR and POx reactions. The Ni catalyst (15% NiO) on alumina is the preferred catalyst due to its cost-effectiveness, stability, and good activity. However, several other catalysts, including transition metals, and noble metals, have been reported in the literature, like Cu, Co, Fe Ru, Rh, Ir, Pt, and Pd. Noble metals were found to be more active than Ni, but the cost is major concern. In contrast, Cu as a catalyst and promoter and Co as a catalyst were found to have good catalytic activity (at lower T). Bimetallic catalyst of Ni and with a small amount of noble metals were reported to have better activity, better conversion of CH_4. Perovskite catalysts have also been reported to be ATR catalysts. The support for Ni catalyst other than Alumina is being studied, like MgO, CeO_2, ZrO_2, and mixed like CeO_2–ZrO_2 which shows better thermal stability and surface area. Zirconia provides the required basicity to reduce coke deposition. Ceria provides better stability and reduced coke deposition and sintering; even Al_2O_3–ZrO_2 shows reduced coke formation and better stability. Small amount of promoters like calcium aluminate or potassium oxide added to increase mechanical strength and reduce coke formation, respectively. Besides the requirements of thermal and mechanical stability, selectivity, conversion and activity, the catalysts are required to perform well under transient conditions during start-ups, and also to allow for low-pressure drop and high flow rates. Thus, ATR could be helpful, where faster dynamic response, high efficiency, fuel flexibility, and at the same time compact reactor are required.

The value of GHSV is calculated as the ratio between the total gas volume flow rate (at STP) and the catalyst bed volume.

The percentage conversion of methane is calculated as:

$$\text{methane conversion } (\%) = \frac{C_{CO} + C_{CO_2}}{C_{CO} + C_{CO_2} + C_{CH_4}} \times 100 \tag{2.72}$$

where C is the dry basis concentration of each gas in the exhaust stream

Very concentrated CO_2 can be obtained using ATR, all of that from the single reactor, making it a better choice if integration with CCUS is required. The hydrogen plant energy efficiency is roughly with natural gas 9.2–10.5 MJ/Nm^3 [18].

Chemical kinetics, along with heat and mass transfer, are the major limitations in the reformer. The rate of the reaction varies exponentially with the temperature of the operation. However, higher temperatures can lead to catalyst damage, while lower temperatures will limit the rate of the reaction. Reducing the catalyst size and increasing the loading can increase the reaction rate, but at the same time, it can also increase the pressure drop and reduce the gas flow rate.

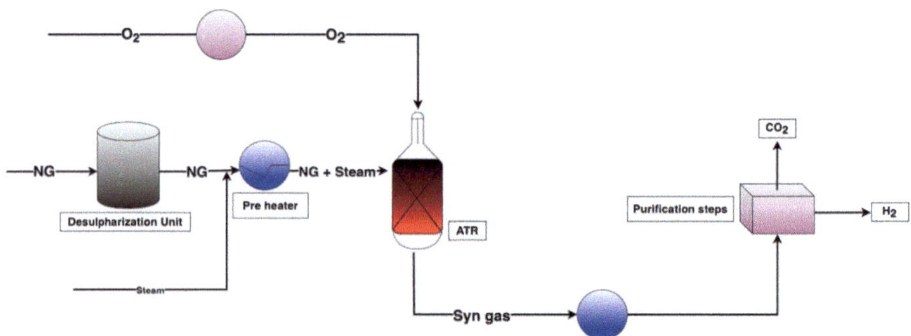

Fig. 2.11 Process flow diagram of autothermal reforming for hydrogen production

The main features of ATR are:

1. The process energy requirement is low.
2. High gas space velocities.
3. The H_2/CO ratio in the outlet stream can be regulated by appropriate selection of $CH_4/O_2/H_2O$.
4. Since CO_2 is produced in a single reactor, much more concentrated CO_2 can be obtained, which is easier for CCUS. The plant cost of ATR with CCUS is lower than that of SMR with CCUS for CO_2 capture rates higher than 90%. As such projects like HyNet and H21 plan to use ATR with CCUS instead of SMR.
5. ATR reactors are more compact than SMR reactors but larger than POx units.
6. ATR uses a higher steam-to-carbon ratio than POx for subsequent SMR and WGS to produce more hydrogen.
7. ATR is most competitive to SMR in producing syngas on an industrial scale, but can be used for applications with a requirement for a lower H_2/CO ratio for the specific applications.

The methods of hydrogen production on an industrial scale, as described in the previous sections, are SMR, POx, and ATR, and a comparison of these methods is provided in Table 2.3.

In addition to the industrial processes for producing hydrogen from hydrocarbons, there are several other methods, that also produce hydrogen, although their applications may differ due to a lower H/C ratio. These methods are described in the later sections.

4. Combined Reforming

The three processes, steam methane reforming, partial oxidation, and autothermal reforming; when any of these processes are combined, it is then called a combined reforming process. In combined reforming, there are two reformers: the first is a primary reformer which is a smaller fire-tubular reactor and the second is a secondary reformer (Fig. 2.12).

In the combined reforming process, the partially reformed gas from the primary reformer is sent to the secondary reformer to achieve the desired gas composition.

The process is similar to what has been described in the earlier sections. The desulfurization unit is used to remove the sulfur-containing impurities. After removal of the sulfur-containing contents, it is preheated and along with steam is passed into the primary reformer, where it undergoes the reforming reaction. The resulting syngas, which is partially being reformed, is then passed through an autothermal reformer, where in the secondary reformer, it undergoes various processes as discussed earlier. The outlet gas from the secondary reformer is to be cooled, and excess heat from the syngas can be utilized for steam generation and pre-heating. After gas cooling, the syngas undergoes the water–gas shift reaction and then undergoes various gas cleaning steps, to finally obtain the desired product. Thus, combined reforming involves two different processes working together to achieve the desired H_2/CO

Table 2.3 Comparison of the SMR, POX, and ATR process for hydrogen production

	SMR	POx	ATR
Thermodynamics	Endothermic	Exothermic	Thermoneutral
Heating required	External heat source	No indirect heat exchanger	No indirect heat exchanger
Thermal efficiency	High efficiency	Lower efficiency as the excess heat cannot be easily recovered, degree of heat integration among various processes is lower than SMR	Higher system efficiency than Pox, the heat generated by the POX reaction is fully utilized to drive the SR reaction. Thus, autothermal reformers typically offer higher system efficiency than POX systems, where excess heat is not easily recovered
H_2/CO ratio	Highest concentration of H_2, i.e., 3–5	2–2.5	2
Utilization of syngas	Preferred when the rich syngas or H_2 is required on large scale for the applications where pure hydrogen is required or high H_2/CO ratio desired	Can be used where lower H_2/CO syngas is required like methanol production of FT processes	A wide variation in the H_2/CO possible by changing the O_2/CH_4 or H_2O/CH_4 ratio; thus these ratios can be adjusted to get the desired syngas for the required application
Reactor size and complexity	Reactor performance limited by heat transfer since reactors designed to promote heat exchange. So large and heavy. Are well suited for long periods of steady-state operation	Small reactor size as no external heating requirement and bulky heat exchangers. However several reactions occur so the system is complicated	Reactor size smaller than SMR but bigger than POX, the autothermal reformer requires no external heat source and no indirect heat exchangers. This makes autothermal reformers simpler and more compact than steam reformers, resulting in lower capital cost
Response time	High endothermicity, since the reactor is externally heated as such start-up and shutdown requires longer time	Short start-up time due fast exothermic reaction. Fast response to load changes	Less NG required to support reaction

(continued)

Table 2.3 (continued)

	SMR	POx	ATR
	Due to high endothermicity, 25% of the total gas consumed is required to support the heating requirement	Can be heated quickly, be increasing the flow rate to increase combustion of fuel, so start-up thus can handle transient loads	Temperature of burner is 2200–2300 °C so high capital and operational cost
	50% of carbon in natural gas converts to CO_2, i.e., carbon efficiency is low	Reactor cost of non-catalytic Pox lower than ATR but carbon efficiency is poor	Carbon efficiency higher than SMR but required expensive gas separation plant to separate C2+ from NG to prevent coking
Transient response	Endothermic process externally heated thus inherently slow process and this slow dynamic response as externally heated thus a long start-up time	Very quick start-up and shutdown possible by initiating the exothermic reaction, but need to follow the procedures so as to avoid any kind of explosions that increases the transient time	The autothermal reforming has faster response time than SMR as such short start-up time required, better efficiency than partial oxidation, simple design and wide flow ranges possible, hotspots reduced due to steam, due to lower temperatures than POX reduced sintering of catalyst
Heat transfer related	If load is decreased, flow rate decreased and reactor heats up leading to sintering and if load increased, reformer cannot supply hydrogen	Hotspots and non-uniform temperatures can lead to possible sintering and need to be operated below the explosive limit	Use of steam controls the temperature, besides the endothermic reforming reaction after the exothermic combustion controls the overall temperature
Overall system	Heavy and large system	Compact but complicated	Simple and compact

(continued)

Table 2.3 (continued)

	SMR	POx	ATR
Catalyst challenges	Carbon deposition due to some reactions but the reaction temperature lower so decomposition reaction less favorable and sintering is lower compared to POX and ATR	Catalyst deactivation problems are not present in a non-catalytic process. However catalyst deactivation and sintering in the catalytic POX process due to both the carbon deposition and high temperatures. Non-uniform heat and mass flow can lead to further sintering, breakage and catalyst consumption is high	Intermediate temperature, both deactivation and sintering could be seen; however due to presence of steam, the deactivation and sintering are lower compared to POX
Operating temperature	Temperature of operation 700–900 °C and 15–30 bar. Lowest temperature requirement	Temperature of operation 1000–1400 °C and 30–80 bar. High temperatures and pressure	Temperature of operation 800–1300^{0}C and 20–80 bar
Thermal management	Thermal management and thermal integration very essential besides steam can corrode and special materials required for reformer tubes to sustain the reactions conditions	Brick lined furnace used due to metallurgical constraints	Less thermal integration and management required compared to POX
Fuel consumption	Fuel consumption, response time and system required for start-up is complicated and higher	Least amount of fuel required to provide the appropriate reaction conditions	Less NG required to support reaction
Efficiency	Efficiency highest	Lowest	Higher than POX but lower than SMR

ratio, which can be used for various applications such as for ammonia synthesis and methanol synthesis. With this method, the steam requirement can be reduced.

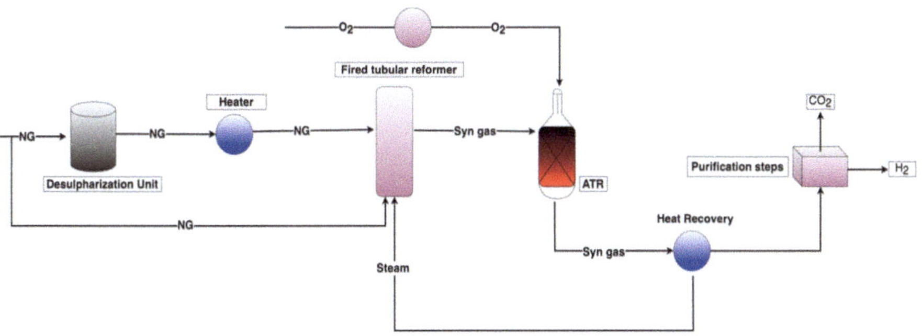

Fig. 2.12 Process flow diagram of combined reforming for hydrogen production

5. Dry Reforming

Another method of reforming is dry reforming, which is also known as carbon dioxide reforming or stoichiometric reforming. The difference between the earlier processes and dry reforming is that here carbon dioxide is used as the oxidant. In the process of dry reforming the feedstock, that is, methane reacts with the oxidant, carbon dioxide, to produce syngas.

The important point here is that both the feedstock and oxidant are greenhouse gases; thus their use means reducing the carbon footprint. The process is used when the natural gas used as feedstock has a higher CO_2 content. In such cases, instead of complex processes involving gas cleaning, it can be directly used for dry reforming process.

$$CH_4 + CO_2 \rightarrow 2CO + 2H_2 \quad \Delta H^0 = +\frac{247\ \text{kJ}}{\text{mol}}. \tag{2.73}$$

However, the reaction occurs between reactants methane, which is a highly reduced form of carbon, and carbon dioxide which is a highly oxidized form of carbon, react together to give the syngas. The H_2/CO ratio as in the product side of the equation (2.73) is ideally 1, but in actual practice it lies between 0.7 and 1.9 and varies based on the oxidant to feedstock ratio, i.e., CO_2/CH_4 ratio. The heat of the reaction suggests that it is endothermic, and the value suggests that it is even more endothermic than the steam methane reforming reaction. The less than 1 value of H_2/CO is not desirable, for hydrogen production.

However, this method of hydrogen production can be used wherein the produced syngas can be directly used for various processes, production of acetic acid, the synthesis of dimethyl ether, or can be converted into olefins using Fischer–Tropsch process to olefin reactions (FTO).

The other possible reactions are:

$$CO_2 + H_2 \leftrightarrow CO + H_2O \quad \Delta H^0 = 41\ \text{kJ/mol} \tag{2.74}$$

$$CO_2 + 4H_2 \leftrightarrow CH_4 + 2H_2O \quad \Delta H^0 = -165 \text{ kJ/mol} \tag{2.75}$$

$$CO + 3H_2 \leftrightarrow CH_4 + H_2O \quad \Delta H^0 = -\frac{206.2 \text{ kJ}}{\text{mol}} \tag{2.76}$$

In these reactions, a certain amount of hydrogen is consumed in the process as shown in reactions (2.73), (2.74), (2.75).

Dry reforming of methane occurs in the temperature range of 900–1200 °C. As the name suggests, dry reforming of methane is a dry process; however, a certain amount of water is obtained on the product side, as seen in Eqs. (2.73), (2.74), (2.75). At the same time, a certain amount of steam is added to the process to reduce the carbon formation.

Since carbon dioxide is used as an oxidant, carbon formation is the major bottleneck. Thus, both the endothermicity of the process and carbon formation are the two major challenges to the industrial use of the process.

This carbon formation can be reduced by using an appropriate catalyst, support, and promoter. It can also be reduced by incorporating steam and by ensuring proper dispersion of the catalyst onto the support, as well as by utilizing promoters. Various catalysts used in the dry reforming process can be either non-noble metal catalysts or noble metals. The nickel or other nickel-based catalysts such as Ni/Al_2O_3, NiMgAl, and NiCe have shown to have a very good methane conversion. The major problem in the nickel-based catalysts is they have a good initial methane conversion, good activity but the coke formation is higher in nickel-based catalyst. Certain promoters, such as potassium, can be added to increase its life and stability. The most widely used catalyst is Ni/Al_2O_3. The other non-noble metal catalysts are Cu, Co, and Fe, which also show good activity and can be used preferably as a bimetallic catalyst together with a noble metal, e.g., $Ni–Cu/Al_2O_3$.

Compared to non-noble metals, noble metals have a higher activity, provide better conversion, have a good thermal stability and mechanical stability and are less prone to coking and sintering. The order of reactivity is Rh > Ru > Ir > Ni and Pt, Pd > Co > Fe > Cu.

Noble metal catalysts exhibit better activity and stability, providing good conversion and selectivity, but the cost is a major problem. To address the cost issues and get the benefits of noble metals' characteristics, a small amount of noble metal like Rh, Ru, Ir, Pt, Pd, Au can be added to the non-noble metal catalysts like Ni, Co, Cu, Fe and can provide together a stability, catalytic activity, higher conversion at a reduced cost.

Several bimetallic catalysts like Ni–Fe and Ni–Co show very good stability. Ni–Co show higher conversion, while Ni–Pt exhibit higher activity and reduced coking. In addition to these catalysts, transition metal-based catalyst and perovskites are also used as catalyst with noble metal doping.

The various supports which can be used for dry reforming of methane include $Al_2O_3, TiO_2, ZrO_2, Nb_2O_5, CeO_2, Ta_2O_5, La_2O_3, ZrO_2$; a combination of the supports $ZrO_2–La_2O_3$, $Mg–Al_2O_3$, SiO_2, yttrium Y_2O_3, zeolites, and SBA-15.

Various combinations of catalyst and support can provide good performance in terms of activity, selectivity, conversion, reduced coking, and sintering.

Various reactions that can result into catalyst deactivation by means of carbon deposition are

1. Boudouard Reaction

$$2CO \leftrightarrow CO_2 + C \quad \Delta H^0 = -172 \text{ kJ/mol} \tag{2.77}$$

   This is an exothermic reaction, which is favorable at lower temperatures.
2. Reduction of CO

$$CO + H_2 \leftrightarrow H_2O + C \quad \Delta H^0 = -131 \text{ kJ/mol} \tag{2.78}$$

   Although it occurs at low temperature, it is not preferred due to its higher Gibbs energy, except in cases with a high partial pressure of H_2.
3. Methane Decomposition

$$CH_4 \leftrightarrow C + 2H_2 \quad \Delta H^0 = +74.6 \text{ kJ/mol} \tag{2.79}$$

   Favored at higher temperatures.
4. Reduction of CO_2

$$CO_2 + 2H_2 \leftrightarrow C + 2H_2O \quad \Delta H = -90 \text{ kJ/mol} \tag{2.80}$$

The reverse of any of these 2.77–2.80 can result in the removal of coke by means of coke gasification reaction. The use of appropriate support, promoter, and catalysts can reduce the carbon formation. The support can provide the desired basicity, which, instead of supporting carbon formation through the decomposition of the various species, can facilitate carbon gasification. Therefore, the appropriate selection of catalyst, support, and promoter can reduce carbon formation. Also the coke formation depends upon the temperature and CO_2/CH_4 ratio.

If CO_2/CH_4 ratio is stoichiometric, i.e., 1, then the carbon formation is lower for temperatures higher than 1100 °C, if ratio is 2, then the temperatures required are above 750 °C for reduced coke formation, and if 3, then temperatures above 700 °C results into lower coke formation. The carbon which is formed can be of different type.

Whisker-type of carbon is formed usually in the catalysts that have aged and undergone several cycles. The whisker carbon forms on to the catalyst surface, diffuses inside the catalyst, reaches the catalyst and support interface, and, over time lifts the catalyst. This can lead to breakage of catalyst and will also block the pores in the catalytic bed after being removed from the support. Whisker form of carbon is the most problematic form of carbon that forms at high temperatures and lower S/C ratios, and in the presence of aromatics. The other form of carbon is pyrolytic carbon, which is usually formed in the case of higher hydrocarbons. Both whisker and pyrolytic carbon are formed at higher temperatures. The third possibility is the gum, which is formed at lower temperatures and has several layers of carbon. These layers are, in fact, graphene-type carbon that is formed on the catalyst surface.

Which of these carbons will be formed depends on the thermodynamics of the process, the catalyst used, and the catalyst-substrate interaction. Besides, the shape and size of catalyst, dispersion, catalyst support interaction, promoters used also determine the coking tendency. With small catalyst size, the surface area is high, dispersion is better, and coke formation will be lower, but this will lead to a higher-pressure drop in the reactor.

Thus, an optimum size is required, which is determined by both the heat transfer and the pressure drop inside the bed.

There are two processes of dry reforming that have reached a commercialization state, namely CALCOR and SPARG processes. In the first process, which is CALCOR process, it is designed to achieve a particular H_2/CO ratio which could be used for various downstream applications. The H_2/CO ratio which is preferred in the process is 0.43 with Ni catalyst being used under the CO_2 partial pressures.

The second process, known as the SPARG process, involves achieving a wide range of H_2/CO ratios through intentional sulfur poisoning, which reduces the ensemble size. However, intentional poisoning can lead to the sulfur poisoning of the downstream process catalyst, and at the same time, reduce the catalyst activity.

6. Bireforming

In bireforming of methane (BRM), the oxidant used is steam and carbon dioxide; these react with feedstock to produce syngas.

$$3CH_4 + CO_2 + 2H_2O \leftrightarrow 4CO + 8H_2 \quad \Delta H^0 = +712 \text{ kJ/mol (BRM)} \tag{2.81}$$

Other possible reactions:

$$CO + H_2O \rightarrow CO_2 + H_2 \; \Delta H^0 = -41 \text{ kJ/mol (WGS)} \tag{2.82}$$

$$CH_4 + CO_2 \rightarrow 2CO + 2H_2 \quad \Delta H^0 = +247 \text{ kJ/mol (DRM)} \tag{2.83}$$

$$CO_2 + H_2 \rightarrow CO + H_2O \quad \Delta H^0 = +41 \text{ kJ/mol (RWGS)} \tag{2.84}$$

$$CH_4 + \frac{1}{2}O_2 \rightarrow CO + 2H_2 \quad \Delta H^0 = -38 \text{ kJ/mol (POX)} \tag{2.85}$$

$$2CO \leftrightarrow C + CO_2 \quad \Delta H^0 = -172 \text{ kJ/mol (Boudouard reaction)} \tag{2.86}$$

$$CH_4 \leftrightarrow C + 2H_2 \quad \Delta H^0 = +74.6 \text{ kJ/mol (Methane Decomposition)} \tag{2.87}$$

The H_2/CO ratio obtained in the process is 2, can be further adjusted by selecting the $H_2O/(H_2O + CO_2)$ ratio, temperature and pressure. The operating temperature in the process ranges from 700 to 950 °C, and the pressure is between 10–30 bar.

The advantage of bireforming is the use of greenhouse gases to produce H_2 or syngas in the desired ratio of H_2/CO as per the required application which can be used directly for the various downstream production processes.

Compared to the dry reforming of methane, here the carbon formation is comparatively less, resulting in reduced catalyst deactivation due to the use of steam as an oxidant in the process.

7. Trireforming

If O_2, H_2O, and CO_2 all three are used together, then the process is called trireforming. Thus, the probable reactions in the trireforming also includes partial oxidation, combustion, steam reforming, or dry reforming given by following reactions.

$$CH_4 + H_2O \rightarrow CO + 3H_2 \quad \Delta H^0 = +206 \text{ kJ/mol} \tag{2.88}$$

$$CH_4 + CO_2 \rightarrow 2CO + 2HO_2 \quad \Delta H^0 = +247 \text{ kJ/mol} \tag{2.89}$$

$$CH_4 + 2O_2 \rightarrow CO_2 + 2H_2O \quad \Delta H^0 = -891 \text{ kJ/mol} \tag{2.90}$$

Other reactions that occur in TRM are as follows:

$$CO + H_2O \rightarrow CO_2 + H_2 \quad \Delta H^0 = -41 \text{ kJ/mol} \tag{2.91}$$

$$CH_4 + \frac{1}{2}O_2 \rightarrow CO + 2H_2 \quad \Delta H^0 = -38 \text{ kJ/mol} \tag{2.92}$$

$$2CO \leftrightarrow C + CO_2 \quad \Delta H^0 = -172 \text{ kJ/mol} \tag{2.93}$$

$$CH_4 \leftrightarrow C + 2H_2 \quad \Delta H^0 = +74.6 \text{ kJ/mol} \tag{2.94}$$

The operating temperature for trireforming is between 700 and 1000 °C, pressure of operation around 20 bar, the H_2/CO can be within 1.2–1.5. The catalysts can be Ni, Pd, Ir, Pt, Rh and support like Al_2O_3, MgO, etc.

8. Hydrogen Production by Methane Decomposition

The process of methane decomposition is a non-oxidative process and the hydrogen obtained is recognized as turquoise hydrogen. In the process of decomposition, the hydrocarbon is subjected to high temperatures, allowing the C–H bonds to break and hydrogen to be produced as a by-product. The amount of energy required for this bond breaking or the decomposition depends upon the type of hydrocarbon used as reactant, whether a saturated hydrocarbon or an unsaturated hydrocarbon. It is highest for saturated hydrocarbons and is less for unsaturated hydrocarbon. In fact, it is exothermic for aromatics like acetylene or for benzene. Considering the methane decomposition reaction:

$$CH_4 \rightarrow C_{(s)} + 2H_2 \quad \Delta H^0 = 75.6 \text{ kJ/mol} \tag{2.95}$$

According to the heat of reaction, the reaction is less endothermic; however, endothermicity is because reactant methane, being a stable, inert, and symmetric compound, lacks polarity in its electronic structure and has no functional groups present.

The product formed in the reaction is solid carbon, a value-added product that can be used for various other applications. Since the product is solid carbon, it is very easy to separate. As the other product is a gas, thus there is no need for bulky separation processes. Besides, since there are no direct emissions in the process of CO, CO_2, etc., there are several benefits of the process. The method is relatively simpler with a lesser number of steps, as the water–gas shift or carbon dioxide removal step is not required in the process. Since the products, i.e., one product is gaseous product, another product is a solid product, it is easier to separate. The efficiency of the process is less, i.e., 58% efficiency. However, the efficiency is similar to that of steam methane reforming integrated with carbon capture, use, and sequestration.

The required temperatures range from 1100 to 1300 °C for the decomposition of methane. The required heat to reach that temperature can be provided by means of either solar heating or through plasma in an electric arc furnace, or using a molten metal in a molten metal bath, or using pulsed microwave power. To reduce the temperature of the reaction, a catalyst is used; however, since the solid carbon obtained as the product, deposit onto the catalyst. Thus, deactivation of the catalyst is the biggest challenge. The methane decomposition reaction is a complex heterogeneous gas–solid phase reaction, and depending on the catalyst being used, different types of carbon products are obtained. Different catalysts show different activity in different temperature ranges as shown in Fig. 2.13. The reaction depends upon the catalyst size, morphology, pore structure, electronic state, dispersion, specific surface area, catalyst preparation method and defect structure, all these impacts on the conversion, stability and products obtained.

The different possible catalysts include:

1. Transition metal-based catalysts.
   (a) Ni, Co, Fe—high activity, moderate operating conditions, and gives valuable carbon products.
   (b) Partially filled 3d orbital promotes methane decomposition.
   (c) Ni is the most active; Co is lower, and Fe has higher resistance to C deposition.

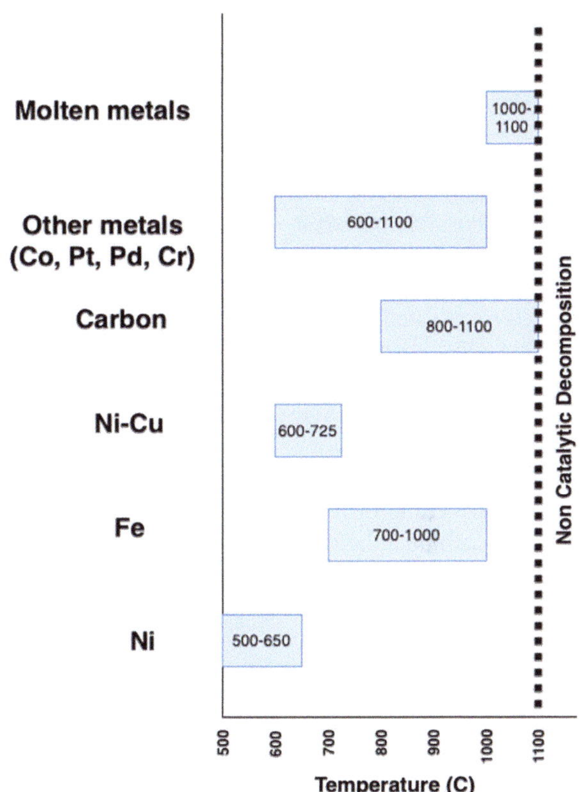

Fig. 2.13 Operating temperature range of different catalysts used for methane decomposition

(d) Ni–Cu/Al_2O_3 better than Ni, with high loading (600–725 °C).
(e) Ni-Cu/MgO stability improved (665–725 °C), forming CNF and achieving a better yield of hydrogen.
(f) Ni-Cu/Al_2O_3 with Fe shows good stability and activity (700–750^{0}C, 70 mol% H2).
(g) 20%Co/Al_2O_3 (600–800 °C), with highest conversion at 800^{0}C.
(h) Pd and Co promoters to Ni delay deactivation and promote decomposition reaction.
(i) Pd–Ni/Al_2O_3 Pd–Co/Al_2O_3 gives higher hydrogen yield.
(j) Ni-Cu/Al_2O_3 higher conversion, Pd–Ni, Pd–Cu on Al_2O_3 higher hydrogen yield.
(k) The order of activity is Co, Ru, Ni, Rh > Pt, Re, Ir > Pd, Cu, W, Fe, Mo.

2. Carbon-based catalysts

(a) Metal catalysts show higher activity and thus lower operational temperatures, but deactivation occurs due to encapsulation.
(b) Carbon catalyst, although lower in activity, has high cycle life.
(c) With carbon catalysts, the operational $T = 800$–1000 °C is high.

(d) Advantages—Low cost and economic feasibility, improved resistance to catalyst deactivation, not susceptible to sulfur poisoning and toxic metal contamination, excellent stability even at high temperatures, no requirement of burn-off of carbon for regeneration, the carbon formed may itself exhibit catalytic activity, and encapsulated carbon encapsulated retains and may enhance catalytic activity. Fuel flexibility.
(e) Possible **catalyst**—Disordered (carbon black, activated C, amorphous C, char, etc.), intermediate order (pyrolytic C, glassy C, fullerene soot, fullerene, CNT, CMK, etc.), highly ordered (diamond, graphite).
(f) Activated C has better activity, while the C black has better stability.
(g) With disordered catalysts used in methane decomposition, the impacting parameters are—specific surface area, defect concentration, dislocations, vacancies, lower coordination sites, pore volume, size, and pore size distribution.
(h) Source of activated carbon can be—coconut, palm, olive, almond, etc.
(i) Blocking of the pore mouth is the major problem in the case of activated carbon, although it gives higher conversion.
(j) Carbon black is less prone to the pore-blocking issue.
(k) Deactivation slows down at higher temperature as there is better diffusion of the reactant.

The type of carbon product obtained from methane decomposition depends on the reaction conditions and the catalyst used. Non-catalytic processes mostly result in amorphous carbon or carbon black. With Ni as a catalyst carbon fibers are obtained. With Fe, the carbon nanotubes are obtained. When Co or Ni is added to Fe, then multi-walled carbon nanotubes are obtained. For, Fe catalysts above 900 °C, either amorphous carbon, carbon nanofibers, or carbon flakes are obtained. With carbon-based catalysts like activated carbon, the product can be carbon black, carbon nanofibers, or layered carbon, and which one these depends upon the source from which the activated carbon is obtained, i.e., the raw material used to produce activated carbon.

Carbon black used as a catalyst produces amorphous carbon as the product. Thus, depending upon the catalyst, the type of carbon product obtained may vary. In the methane decomposition process, the catalyst deactivation is inevitable. In SMR, the catalyst life is typically 5 years of continuous operation. In methane pyrolysis, the life of the catalyst is very short, less than 10 hours, as carbon is the major product obtained and it deposits on the catalyst surface.

The activity of a catalyst that is lost can be regained or regenerated through the various processes.

$$C_{(s)} + H_2O_{(g)} \rightarrow CO_{(g)} + H_{2(g)} \quad \Delta H^0 = +131.39 \text{ kJ/mol} \tag{2.96}$$

$$C_{(s)} + O_{2(s)} \rightarrow CO_{2(g)} \quad \Delta H^0 = -393.78 \text{ kJ/mol} \tag{2.97}$$

$$2C_{(s)} + O_{2(s)} \rightarrow 2CO_{(g)} \quad \Delta H^0 = -221.2 \text{ kJ/mol} \tag{2.98}$$

$$C_{(s)} + CO_{2(s)} \rightarrow 2CO_{(g)} \quad \Delta H^0 = +\frac{172.58\ \text{kJ}}{\text{mol}}. \tag{2.99}$$

The deposited carbon can undergo gasification, partial oxidation, or complete oxidation. In presence of CO_2, it can convert into CO. Thus in process (2.95) the catalyst can get regenerated, and the surface carbon converts into CO or CO_2 with steam resulting in an additional product which is hydrogen.

Thus, the most commonly used method for regeneration is gasification and oxidation. However, the oxidation is an exothermic reaction, resulting in excessive heat, which can lead to sintering of the catalyst and at times complete powder formation. Regeneration with steam is preferred, as it yields additional hydrogen obtained in the product, and maintains the surface morphology of the catalyst.

Parameters affecting the methane decomposition process—The methane decomposition reaction is an endothermic process and is preferred at a higher temperature. However, at higher temperatures, deactivation will be higher, and sintering occurs. The reaction is preferred at a lower pressure. Gas hourly space velocities, if higher, will reduce the residence time, thus decreasing the hydrogen produced. Consequently, the catalyst deactivation increases as there is insufficient time for the carbon product to diffuse inside the catalyst.

Various types of reactors such as fluidized bed, packed bed, spouted bed, monolithic, molten metal, and salt bath are used for the CDM process. Primarily, a fluidized bed is used because it allows for the easy flow of gases, provides uniform flow and uniform temperature. Additionally, reactants and the catalyst can be easily supplied and removed from the bed. A packed bed has its own disadvantage, as the blocking of the pores is more severe in the case of a packed bed reactor.

Methane Decomposition with Molten Metals

Metals or salts in molten state can be used to decompose methane, producing hydrogen and carbon products. The products are easier to separate, due to density differences. Molten metal is in a liquid state, carbon is produced in a solid state, and hydrogen is in the gaseous state. Since the molten metals are in a liquid state, better heat transfer can be achieved. The different metals that can be used include Fe, Al, Sb, Pd, etc. These systems operate at a higher temperature $T > 1000$ °C and there are corrosion-related challenges in the process. Another option is the use of low-melting-point materials like Ni-Bi and Cu-Bi (most active), which have a good activity while also being operational at a lower temperature compared to molten metals.

The metal salts are less expensive, and the problems associated with metal loss are reduced, when metal salts like $MnCl_2$ or KCl are used. The heat required to take it to higher temperatures can be provided by means of solar heating.

9. Alternate Energy Sources for Reforming

The steam methane reforming is a highly endothermic reaction; the heat required for the reaction is supplied by the combustion of the methane, which also serves as fuel in the various burners. The natural gas combustion results in emissions, which are usually released into the environment during unabated hydrogen production. The other possibility is that the required heat of the reaction can be provided by a renewable source, allowing us to reduce the associated emissions in the process. The energy required for reforming can be provided by means of any alternate source, such as nuclear energy (high-temperature gas-cooled reactor), or solar energy (solar thermal, either direct or indirect concentration), or electrical energy (either thermal or non-thermal plasma).

Solved Problem 1. Trireforming of Methane

In a trireforming process, a feed mixture contains 1 mol of methane (CH_4), 0.5 mol of steam (H_2O), 0.3 mol of carbon dioxide (CO_2), and 0.2 mol of oxygen (O_2). The reactions occurring are:

1. **Steam Reforming**: $CH_4 + H_2O \rightarrow CO + 3H_2$
2. **Dry Reforming**: $CH_4 + CO_2 \rightarrow 2CO + 2H_2$
3. **Partial Oxidation**: $CH_4 + 0.5O_2 \rightarrow CO + 2H_2$

Assume the reactions proceed to completion based on the limiting reactant for each reaction, and the reactions occur sequentially in the order listed above (steam reforming first, then dry reforming, then partial oxidation). Calculate:

(a) The moles of CH_4, H_2O, CO_2, O_2, CO, and H_2 remaining or produced after all reactions are complete.
(b) The H_2/CO molar ratio in the syngas produced.

Solution

Step 1: Initialize the Molar Quantities

Initial moles:

- $CH_4 = 1.0$ mol
- $H_2O = 0.5$ mol
- $CO_2 = 0.3$ mol
- $O_2 = 0.2$ mol
- $CO = 0$ mol (product)
- $H_2 = 0$ mol (product)

The reactions are assumed to occur sequentially, and each reaction proceeds to completion based on the limiting reactant.

Step 2: Steam Reforming ($CH_4 + H_2O \rightarrow CO + 3H_2$)

Stoichiometry: 1 mol CH_4 reacts with 1 mol H_4O to produce 1 mol CO and 3 mol H_2.

Available Reactants

- $CH_4 = 1.0$ mol
- $H_2O = 0.5$ mol

Limiting Reactant

- For 1 mol CH_4, 1 mol H_2O is needed.
- H_2O is limiting since only 0.5 mol is available.
- Therefore, 0.5 mol H_2O reacts with 0.5 mol CH_4.

Moles Consumed

- CH_4: 0.5 mol
- H_2O: 0.5 mol

Moles Produced

- CO: 0.5 mol
- H_2: $3 \times 0.5 = 1.5$ mol

Moles Remaining After Steam Reforming

- CH_4: $1.0 - 0.5 = 0.5$ mol
- H_2O: $0.5 - 0.5 = 0.0$ mol
- CO_2: 0.3 mol (unchanged)
- O_2: 0.2 mol (unchanged)
- CO: $0 + 0.5 = 0.5$ mol
- H_2: $0 + 1.5 = 1.5$ mol

Step 3: Dry Reforming ($CH_4 + CO_2 \rightarrow 2CO + 2H_2$)

Stoichiometry: 1 mol CH_4 reacts with 1 mol CO_2 to produce 2 mol CO and 2 mol H_2.

Available Reactants

- $CH_4 = 0.5$ mol
- $CO_2 = 0.3$ mol.

Limiting Reactant

- For 1 mol CH_4, 1 mol CO_2 is needed.
- CO_2 is limiting since only 0.3 mol is available (less than 0.5 mol required for 0.5 mol CH_4).
- Therefore, 0.3 mol CO_2 reacts with 0.3 mol CH_4.

Moles Consumed

- CH_4: 0.3 mol
- CO_2: 0.3 mol

Moles Produced

- CO: $2 \times 0.3 = 0.6$ mol
- H_2: $2 \times 0.3 = 0.6$ mol

Moles Remaining After Dry Reforming

- CH_4: 0.5 − 0.3 = 0.2 mol
- H_2O: 0.0 mol (unchanged)
- CO_2: 0.3 − 0.3 = 0.0 mol
- O_2: 0.2 mol (unchanged)
- CO: 0.5 + 0.6 = 1.1 mol
- H_2: 1.5 + 0.6 = 2.1 mol

Step 4: Partial Oxidation ($CH_4 + 0.5O_2 \rightarrow CO + 2H_2$)

Stoichiometry: 1 mol CH_4 reacts with 0.5 mol O_2 to produce 1 mol CO and 2 mol H_2.

Available Reactants

- CH_4 = 0.2 mol
- O_2 = 0.2 mol

Limiting Reactant

- For 1 mol CH_4, 0.5 mol O_2 is needed.
- For 0.2 mol CH_4, O_2 needed = 0.5 × 0.2 = 0.1 mol O_2.
- O_2 available = 0.2 mol, so CH_4 is limiting.
- Therefore, 0.2 mol CH_4 reacts with 0.1 mol O_2.

Moles Consumed

- CH_4: 0.2 mol
- O_2: 0.1 mol

Moles Produced

- CO: 1 × 0.2 = 0.2 mol
- H_2: 2 × 0.2 = 0.4 mol

Moles Remaining After Partial Oxidation

- CH_4: 0.2 − 0.2 = 0.0 mol
- H_2O: 0.0 mol (unchanged)
- CO_2: 0.0 mol (unchanged)
- O_2: 0.2 − 0.1 = 0.1 mol
- CO: 1.1 + 0.2 = 1.3 mol
- H_2: 2.1 + 0.4 = 2.5 mol

Step 5: Final Molar Quantities (Part A)

After all reactions are complete, the moles of each species are:

- CH_4: 0.0 mol
- H_2O: 0.0 mol
- CO_2: 0.0 mol
- O_2: 0.1 mol
- CO: 1.3 mol

- H_2: 2.5 mol

Answer for Part A

- Moles remaining or produced $CH_4 = 0.0$ mol, $H_2O = 0.0$ mol, $CO_2 = 0.0$ mol, $O_2 = 0.1$ mol, $CO = 1.3$ mol, $H_2 = 2.5$ mol.

Step 6: Calculate H$_2$/CO Molar Ratio (Part B)

The syngas consists of H_2 and CO. The molar ratio is:

$$\frac{H_2}{CO} = \frac{\text{Moles of } H_2}{\text{Moles of CO}} = \frac{2.5}{1.3} = 1.92$$

Answer for Part B

- H_2/CO molar ratio = 1.92 (rounded to two decimal places).

Solved Problem 2. Partial Oxidation of Methane

In an industrial reactor, methane (CH_4) is partially oxidized with oxygen (O_2) to produce synthesis gas ($CO + H_2$) according to the following reaction

$$CH^4 + {}^1\!/_2O^2 \rightarrow CO + 2H^2 \quad \Delta H = -36\frac{\text{kJ}}{\text{mol}}$$

A feed stream of 500 mol/h methane and 300 mol/h of oxygen is supplied at 25 °C and 1 atm. Assume:

Only the above reaction occurs (no full oxidation or reforming).

The reactor operates at complete conversion of methane.

The specific heat capacities Cp of all gases = 29 J/mol K (assumed constant for simplicity).

Reaction temperature at outlet is 900 °C.

Tasks:

1. Determine the limiting reactant.
2. Calculate the extent of reaction (in mol/h).
3. Find the amount of each product formed (in mol/h).
4. Estimate the temperature rise (ΔT) of the system due to the exothermic reaction.
5. Estimate the outlet gas temperature using energy balance.

Solution

Step 1: Limiting Reactant

Given:

- Methane: 500 mol/h
- Oxygen: 300 mol/h

Reaction:

$$CH_4 + {}^1\!/_2O_2 \rightarrow CO + 2H_2$$

Mole ratio required: 1 mol CH_4: 0.5 mol O_2
Required O_2 for 500 mol $CH_4 = 0.5 \times 500 = 250$ mol.
Available $O_2 = 300$ mol → excess.
Limiting Reactant: CH_4

Step 2: Extent of Reaction (ξ)

The extent of reaction (ξ) represents the moles of reaction that occur per unit time, based on the limiting reactant.

- Reaction stoichiometry: $CH_4 + {}^1\!/_2O_2 \rightarrow CO + 2H_2$
- Limiting reactant: CH_4 fully consumed (500 mol/h).
- Extent of reaction:

For 1 mol CH_4 consumed, the extent of reaction $\xi = 1$ mol/h.
For 500 mol/h CH_4 consumed: $\xi = 500$ mol/h.
Since CH_4 is limiting and fully converted:
$\xi = 500$ mol/h.

Step 3: Product Moles

From balanced equation:

$$CH_4 + {}^1\!/_2O_2 \rightarrow CO + 2H_2$$

Using the extent of reaction and stoichiometry, calculate the moles of reactants consumed and products formed.

Stoichiometric coefficients (with signs for consumption/production):

- CH_4: − 1 (consumed)
- O_2: − 0.5 (consumed)
- CO: + 1 (produced)
- H_2: + 2 (produced)

Moles consumed/produced = ξ × stoichiometric coefficient.

- CH_4 consumed: $500 \times 1 = 500$ mol/h
- O_2 consumed: $500 \times 0.5 = 250$ mol/h
- CO produced: $500 \times 1 = 500$ mol/h
- H_2 produced: $500 \times 2 = 1000$ mol/h

Moles Remaining

- CH_4: $500 - 500 = 0$ mol/h
- O_2: $300 - 250 = 50$ mol/h
- CO: $0 + 500 = 500$ mol/h
- H_2: $0 + 1000 = 1000$ mol/h
- CO_2: 0 mol/h (no side reactions)
- H_2O: 0 mol/h (no side reactions).

Step 4: Temperature Rise (ΔT)

To estimate the temperature rise due to the exothermic reaction, assume an adiabatic reactor (no heat loss) and that all heat released by the reaction raises the temperature of the product gases (and unreacted O_2). The heat of reaction is used to calculate ΔT.

Enthalpy released by reaction:

$$\Delta Hr \times n = -36\ \frac{\text{kJ}}{\text{mol}}$$

or 500 mol/h of CH_4:

$$Q_{\text{released}} = 500 \times 36 = 18000\ \frac{\text{kJ}}{\text{h}}$$

Heat absorbed by outlet gases

Total molar flowrate of outlet gas = 500 (CO) + 1000 (H_2) + 50 (O_2) = 1550 mol

$$Q_{\text{released}} = n \times C_p \times \Delta T = 1550 \times 29 \times \Delta T\ \ \text{J/h} = 1.8 \times 10^7\ \text{J/h}$$

Solving for ΔT: $\Delta T = 1.8 \times 10^7/(1550 \times 29) \approx 400.2$ K.
Temperature Rise = 400.2 K.

Step 5: Outlet Temperature

Initial Temp = 25 °C = 298 K.
Outlet Temp = 298 + 400.2 = 698.2 K ≈ 425.2 °C.
Outlet Temperature = 425.2 °C.
Given in problem: reactor operates at 900 °C.

Discrepancy

- The calculated adiabatic outlet temperature (425.4 °C) is lower than 900 °C, suggesting that the reactor is not truly adiabatic or that external heating is applied to maintain 900 °C.

  To reconcile:

- Calculate additional heat required to reach 900 °C from 425.4 °C:

$$\Delta T_{\text{additional}} = 900 - 425.4\ °\text{C}.$$

$$Q_{\text{additional}} = n_{\text{total}} \cdot C_p \cdot \Delta T_{\text{additional}} = 1550 * 29 * 474.6 = 21{,}363{,}210\ \text{J/h}$$

- This indicates that approximately 21.36 MJ/h of additional heat is needed to reach 900 °C, possibly supplied externally or due to inaccuracies in the simplified C_p assumption.

Answer for Part E

- The adiabatic outlet temperature is approximately 425.4 °C.
- To achieve the specified 900 °C, additional heat input of ~ 21.36 MJ/h is required, or the system is non-adiabatic.

Final Answers

1. Limiting Reactant: CH_4
2. Extent of Reaction (ξ): 500 mol/h
3. Products:
   (a) CO: 500 mol/h
   (b) H_2: 1000 mol/h
   (c) O_2 (unreacted): 50 mol/h
4. Temperature Rise (ΔT): ~ 400.2 K
5. Outlet Temperature: ~ 425.2 °C

Solved Problem 3. Steam Reforming or Steam Methane Reforming

Steam reforming or steam methane reforming is a chemical process to produce syngas. The main purpose of this technology is for hydrogen production. The chemical equation representing the steam reforming of methane (SMR) processes is mentioned below:

$$CH_4(g) + 2H_2O(g) \rightleftharpoons CO_2(g) + 4H_2(g)$$

Use the thermochemical equations mentioned below to solve for the overall enthalpy change (ΔH_0) of the given above-mentioned target reaction:

1. $H_2(g) + \frac{1}{2}O_2(g) \rightarrow H_2O(g) \quad \Delta H_0 = -244$ kJ/mol
2. $2H_2(g) + C(s) \rightarrow CH_4(g) \quad \Delta H_0 = -75$ kJ/mol
3. $C(s) + O_2(g) \rightarrow CO_2(g) \quad \Delta H_0 = -395$ kJ/mol

Solution

To solve for the enthalpy, change of the steam reforming of methane (SMR), we need to use the given reactions and their enthalpy changes to derive the target reaction:

Target Reaction:

$$CH_4(g) + 2H_2O(g) \rightleftharpoons CO_2(g) + 4H_2(g)$$

Given reactions and their enthalpy changes are.

1. $H_2(g) + \frac{1}{2}O_2(g) \rightarrow H_2O(g) \quad \Delta H_0 = -244$ kJ/mol
2. $2H_2(g) + C(s) \rightarrow CH_4(g) \quad \Delta H_0 = -75$ kJ/mol
3. $C(s) + O_2(g) \rightarrow CO_2(g) \quad \Delta H_0 = -395$ kJ/mol

Concept: We need to manipulate these reactions to match the target reaction, using Hess's Law, which states that the total enthalpy change of a reaction is the same regardless of the path taken. It needs to express the target reaction as a combination of the given reactions, ensuring the species match (including their states and coefficients) and canceling out intermediates.

1. Manipulate the Given Reactions

Reaction 2 (to get CH_4 on the reactant side):

The target reaction has $CH_4(g)$ as a reactant, but in reaction 2, $CH_4(g)$ is a product. Therefore, we reverse reaction 2:

$$CH_4(g) \rightarrow 2H_2(g) + C(s) \quad \Delta H_0 = +75 \text{ kJ/mol}$$

When reversing a reaction, the sign of changes ΔH_0.

Reaction 3 (to get CO_2 on the product side):

Reaction 3 already has CO_2 as a product, which matches the target so no need to do anything:

$$C(s) + O_2(g) \rightleftharpoons CO_2(g) \quad \Delta H_0 = -395 \text{ kJ/mol}$$

Reaction 1 (to handle H_2O):

The target reaction has 2 mol of H_2O (g) as a reactant. Reaction 1 has H_2O (g) as a product, so we reverse it to get H_2O on the reactant side:

$$H_2O(g) \rightarrow H_2(g) + \frac{1}{2}O_2(g) \quad \Delta H_0 = +244 \text{ kJ/mol}$$

Since we need 2 mol of H_2O, we multiply this reaction by 2:

$$2H_2O(g) \rightarrow 2H_2(g) + O_2(g) \quad \Delta H_0 = +488 \text{ kJ/mol}$$

2. *Combine the Manipulated Reactions*

Now, add the modified reactions together

$$CH_4(g) \rightarrow 2H_2(g) + C(s) \; \Delta H_0 = +75 \text{ kJ/mol}$$
$$C(s) + O_2(g) \rightarrow CO_2(g) \quad \Delta H_0 = -395 \text{ kJ/mol}$$
$$2H_2O(g) \rightarrow 2H_2(g) + O_2(g) \quad \Delta H_0 = +488 \text{ kJ/mol}$$

Adding them, this simplifies to:

$$CH_4(g) + 2H_2O(g) \rightleftharpoons CO_2(g) + 4H_2(g)$$

Calculate the total enthalpy change; add the ΔH_0 values of the manipulated reactions:

$$\Delta H_0 = +75\ \frac{\text{kJ}}{\text{mol}} + \left(-395\ \frac{\text{kJ}}{\text{mol}}\right) + 488\ \frac{\text{kJ}}{\text{mol}}$$

$$\Delta H_0 = 168\ \frac{\text{kJ}}{\text{mol}}$$

The enthalpy change for the steam reforming of methane is + **168 kJ/mol.**

References

1. A. Luengnaruemitchai, S. Osuwan, E. Gulari, Comparative studies of low-temperature water–gas shift reaction over Pt/CeO_2, Au/CeO_2, and Au/Fe_2O_3 catalysts. Catal. Commun. **4**, 215–221 (2003)
2. S. Ahmed, M. Krumpelt, Hydrogen from hydrocarbon fuels for fuel cells. Int. J. Hydrogen Energy **26**, 291–301 (2001)
3. Hydrogen Production Roadmap-Technology pathways to the future—January 2009 by Freedom Car and Fuel partnership. Last accessed on 2 Feb 2025
4. P.K. Cheekatamarla, C.M. Finnerty, Reforming catalysts for hydrogen generation in fuel cell applications. J. Power Sources **160**, 490(2016)
5. B. Balasubramanian, A. Ortiz Lopez, S. Kaytakoglu, D.P. Harrison, Hydrogen from methane in a single step process. Chem. Eng. Sci. **54**, 3543–3552(1999)
6. D.P. Harrison, Sorption enhanced hydrogen production: a review. Ind. Eng. Chem. Res. **47**, 6486–6501 (2008)
7. M. Kato, K. Essaki, S. Yoshikawa, K. Nakagawa, H. Uemoto, Novel CO_2 absorbents using Lithium containing oxides, in *Proceedings of the 6th Conference on Greenhouse Control Technologies*, Kyoto, Japan, October 2002
8. T.Y. Amiri, K. Ghasemzageh, A. Iulianelli, Membrane reactors for sustainable hydrogen production through steam reforming of hydrocarbons: a review. Chem. Eng. Process.-Process Intensif. **157**, 108148 (2020)
9. L. Barelli, G. Bidini, F. Gallorini, S. Servili, Hydorgen production through sorption-enhanced steam methane reforming and membrane technology: a review. Energy **33**, 554–570 (2008)
10. A. Wunsch, P. Kant, M. Mohr, K. Haas-Santo, P. Pfeifer, R. Dittmeyer, Recent developments in compact membrane reactors with hydrogen separation. Membranes **8**, 107 (2018)
11. S.A. Ghoneim, R.A. El-Salamony, S.A. El-Temtamy, Review on innovative catalytic reforming of natural gas to syngas. World J. Eng. Technol. **4**, 116–139 (2016)
12. D. Dissanayake, M.P. Rosynek, K.C.C. Kharas, J.H. Lunsford, Partial oxidation of methane to carbon-monoxide and hydrogen over a Ni/Al_2O_3 catalyst. J. Catal. **132**, 117–127 (1991)
13. D.A. Hickman, L.D. Schmidt, The role of boundary-layer mass-transfer in partial oxidation selectivity. J. Catal. **136**, 300–308 (1992)
14. R. Horn, K.A. Williams, N.J. Degenstein, L.D. Schmidt, Syngas by catalytic partial oxidation of methane on rhodium: mechanism conclusions from spatially resolved measurements and numerical simulations. J. Catal. **92**, 242 (2006)
15. C.A. Galvan, M. Mayra, L.R. Matas, J.L. Eslava, R.M. Navarro, M. Ahmadi, B.R. Cuenya, J.L. G. Fierro, Partial oxidation of methane to syngas over nickel-based catalysts: influence of support type, addition of rhodium, and preparation method. Front. Chem. **7**, 104 (2019)
16. G. Nahar, V. Dupont, Recent advances in hydrogen production via autothermal reforming process (ATR): a review of patents and research articles. Recent Patents Chem. Eng. **6**, 8–42 (2013)
17. I. Dbykjaer, Tubular reforming and autothermal reforming of natural gas—an overview of available processes. Fuel Process. Technol. **42**, 85–107 (1995)
18. IEAGHG, Technical review (Part1): current state of art technologies for hydrogen production, [September 2016]. Last accessed on 2 Feb 2025

Chapter 3
Hydrogen Production from Biomass

Biomass is a solid fuel and thus has its own challenges and is less convenient than the gaseous or liquid fuels used for producing hydrogen. Biomass is a form of solar energy, wherein plants absorb around 2–6% of the solar insolation and convert it to chemical energy through photosynthesis. In this process, hydrogen in water and carbon dioxide from the atmosphere are converted into higher hydrocarbons or sugars. Biomass is a renewable biological resource that is available and can have various origins, including forestry, agriculture, industry, fisheries, municipal wastes, etc. However, the use of biomass has several advantages, such as being a low cost resource with low sulfur content, and the process of producing hydrogen remains carbon neutral. In fact, biomass is a complex biogenic solid made up of organic and inorganic components and can be categorized into natural, i.e., as grown or generated by animal and human digestion, or it could be technogenic, i.e., obtained after processing the natural biomass mentioned earlier. As per the 2009/28/EC European Directive, the biomass is defined as "the biodegradable fraction of products, waste and residues of biological origin which could be from agriculture, forestry or related industries, industrial, and municipal wastes. This could include agricultural or animal substances, fisheries, and aquatic biomaterials." Classification can be made based on its origin, which can be wood or wood-based biomass, agricultural biomass, aquatic biomass, human and animal waste, industrial waste-based biomass, or a combination of these. Depending upon its source, biomass can be either wet biomass, like that obtained from manures or molasses, or it can be dry biomass like that obtained from agricultural wastes, paddy, wood, etc.

Biomass contains organic compounds with major elements carbon, hydrogen, and oxygen, with a mass fraction of hydrogen being 6 wt% (672 L of hydrogen can be obtained per kg of biomass). The various types of biomass are typically composed of cellulose, hemicellulose, lignin, and proteins. Their relative fractions differ based on the origin of biomass. Among the constituents of biomass, cellulose and hemicellulose consist of long polymeric chain molecules that are bonded by saccharides, which in fact make up the fibers of the biomass. While lignin, which

P. Sharma, *A Textbook on Hydrogen Energy*,
https://doi.org/10.1007/978-981-95-4601-5_3

consists of phenolic polymers, acts as a glue to the fibers and provides strength to the biomass, it also provides stiffness to the structural proteins. However, proteins are mainly present in herbaceous biomass [1].

Processing of biomass can provide gas, liquid, or solid biofuels. Biological processes, such as fermentation, can be used to produce alcohols or, through digestion route, methane from wet biomass. However, to produce energy or heat, as well as valuable liquid or gaseous fuels from biomass, the moisture content in biomass should be less than 50%.

The simplest route is direct combustion, which involves the burning of biomass and has long been used for cooking in rural areas, providing heat alone. There are two more processes which can yield valuable gaseous, liquid, and solid fuels: These are pyrolysis and gasification. The pyrolysis process can yield oils, charcoal, tars, and gases, while the gasification process can give gases that can be used as fuel. The process of gasification can be used to produce heat, fuel, and power.

The technology of gasification dates back to 1798 and was widely used in the mid-nineteenth century. It was further commercialized in the 1940s, when, during World War II, around a million of gasifiers were used in Europe to power vehicles, trains, lighting, and cooking. However, after the war, with the availability of less expensive fuel options such as gasoline, the gasification technology experienced a setback. Interest was again found at the time of increasing oil prices, particularly for countries where the availability of fossil fuels was a concern, given the huge import cost that affected their economies. The technology has again revived due to the widespread interest received toward adoption for hydrogen production and especially in countries like India, where there is ample biomass available, and at the same time, local air pollution is observed at certain time frames resulting from the combustion of biomass. The primary advantages of using biomass for hydrogen production are its low cost, local availability, lower ash and sulfur content, and its overall carbon neutral nature as a feedstock. The process can be applied to both isolated or distributed power generation, as well as hydrogen production.

Biomass is, in fact, another form of solar energy, which is stored via photosynthesis, while coal is derived from biomass. Biomass is a renewable resource that includes a wide range of feedstocks and grows naturally in the environment. As such, there are many commonalities between biomass and coal, as well as the processes used for hydrogen production from both. Biomass is a general term used to describe a wide variety of biological materials that can be in different physical forms or have varying origins. The moisture content and other properties may vary significantly depending on their origin. Based on the characteristics and properties of biomass, the gasifier design and operating conditions must be tailored to achieve better performance, efficiency, and yield of the desired products. Typically, biomass contains 55–65 g of hydrogen per kg of the dry biomass, along with carbon and oxygen in the chemical bonds. However, the wet biomass contains moisture that is not chemically bonded. Depending on its source or origin, biomass may be derived from any types of waste or, alternatively, from energy crops cultivated and harvested specifically for use as feedstock in the gasification process. Examples of energy crops include sugarcane grown in Brazil and sweet corn in the USA, used for the production of bioethanol. At the same time, the waste biomass could include forest waste in the form of logs, leaves, barks, chips, etc., as well as municipal waste, agricultural waste, and agro-industrial waste. Forest waste, also known as tree waste, has been used for decades in rural areas for cooking and keeping warm during winters. The agro-industrial waste could be from the paper mills, molasses from the sugarcane industry, from refineries, or the food industry such as biscuit, ketchup, chips industries. For example, in the sugar industry, when sugarcane juice is converted to sugar, the by-product obtained is molasses, which is used for bioethanol production. An example of agricultural waste is the residue left over after extracting a useful grain or agricultural product such as rice husk, wheat paddy, or corn stock. In India, the burning of paddy straw in fields leads to significant air pollution and is a major concern during the winter months, as it contributes to smog and particulate matter, resulting in poor air quality. Similarly, municipal solid waste, including biodegradable materials such as paper, inert materials like fabric, and non-degradable materials like plastic (although not considered biomass), can be used to produce hydrogen. (Fig. 3.2)

Energy generation from biomass or the production of valuable chemicals and fuels from biomass can be regarded as carbon neutral if the feedstock used is waste. The reason is that, if forest waste or agricultural waste is left to decompose in fields, it will be consumed by microorganisms, which digest it to meet their energy requirements. In this process, they will again produce carbon dioxide, which will be released into the environment. While replacing natural degradation, an energy harvesting process can produce fuel or chemicals and release carbon dioxide in environment, it is similar to the process carried out during natural decomposition. Additionally, the carbon dioxide released into the environment when producing energy from biomass can be absorbed by the plants, allowing the cycle to continue and making the process carbon neutral. Thus, natural biomass and the biofuels derived from it do not contribute to emissions and are considered CO_2-neutral. This waste biomass is abundantly available globally and can be harnessed to produce energy; in current case, the interest is in producing hydrogen. Converting waste biomass into energy is, in fact, a way to add value and manage waste while minimizing its contribution to climate change.

However, deforestation to grow energy crops is not considered as carbon neutral, as the carbon dioxide absorbing potential of the forest would be higher than that of the grown energy crops.

Thus, the most significant advantages of using biomass for energy is that the natural biomass is renewable, CO_2 neutral and can produce fuel and chemicals; with less of ash, carbon, sulfur, nitrogen and trace element; while high on hydrogen, calcium, magnesium, oxygen, and phosphorus; wide availability and cheap resource; good reactivity toward conversion; with other applications like fertilizers, sorbents, building materials, etc. Valorization of waste biomass can not only provide sustainable energy production but also helps toward waste management and reduction of greenhouse gase (GHG) emissions [2]. However, there are disadvantages associated with it, such as a high moisture content and the presence of some trace elements, associated processing costs, variable composition based on the source, competition with food, and regional availability.

The biggest challenge in producing fuel or chemicals from biomass is the significant variability between the different biomass varieties and within a single variety, which arises due to differences in growth and harvesting conditions. In contrast, the conversion process requires, a uniform feedstock. The variability of products obtained from converting biomass to fuels or chemicals depends on the quality of biomass, which in turn depends on the variability of the biomass composition, production, harvesting, collection, and storage conditions.

Generating energy from biomass can be done through various routes and using different energy conversion processes as depicted in Fig. 3.1. Biomass conversion to energy is not a new technology; the combustion of biomass, conversion to bioethanol, and biogas are well-known and have been in use for centuries. Among the different energy conversion processes for harvesting energy from biomass are chemical, thermochemical, and biological routes. The example of a biological route is the conversion of waste biomass, such as molasses from the sugar industry, to bioethanol. Bioethanol is currently blended with gasoline in India to provide transportation fuel for vehicles. The simplest of the three biomass conversion processes is combustion (thermochemical process), which is the direct burning of biomass (in its dry form) to produce heat and light and has been used since the beginning of civilization. The two routes for producing bioethanol through biological and thermochemical processes have also been known for a very long time.

However, the focus of this chapter is on producing hydrogen from biomass, which is essentially achieved through the third route, i.e., thermochemical processes including pyrolysis and gasification. Through these two methods, we can obtain valuable products in addition to hydrogen, including char, bio-oil, and syngas. The syngas produced can be cleaned to extract hydrogen. The last process is the chemical conversion route, which is a relatively recent method for extracting fuel from biomass. In this process, oils extracted from seeds (jatropha or karanja) and waste cooking oils are converted into biodiesel through transesterification reaction, in which triglycerides and fatty acids are transformed into fatty acid alkyl esters (FAMEs). The conversion of fats, oils, and fatty acids in biomass is a catalytic process.

In this chapter, the two processes considered for hydrogen production are gasification and pyrolysis.

Biomass Composition

The composition of biomass is highly variable, with significant differences in terms of the moisture content, ash, and inorganic matter based on the origin of the biomass. This variability is much lower when ultimate and proximate analyses are carried out on a dry basis or a dry ash-free basis. The order of elemental abundance in biomass generally follows the trend: C > O > H > N > Ca > K > Si > Mg > Al > S > Fe > P > Cl > NA > Mn > Ti [3]. The composition of biomass differs significantly from that of coal. Biomass is rich in elements such as Mn, K, P, Cl, Ca, Mg, Na, and O, and

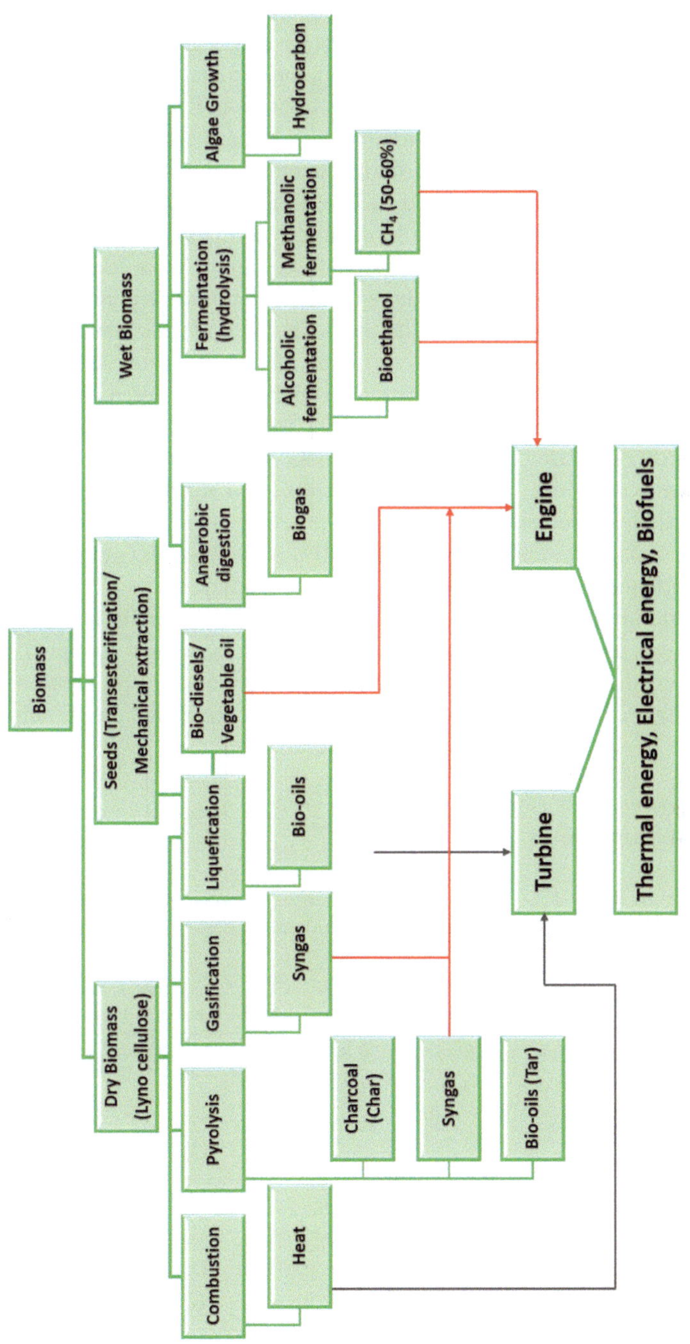

Fig. 3.1 Different processes, products, and applications of biomass-based energy conversion

has moisture and volatile matter contents as well as have higher hydrogen content, but lower ash, carbon, Fe, Al, Si, S, N, and Ti contents compared to coal.

Proximate and ultimate analyses are important for characterizing biomass in terms of its physical, chemical, and fuel properties. The proximate analysis is about determining the moisture (M) content, fixed carbon content (FC), ash (A), and volatile matter (VM) in the biomass. At the same time, ultimate analysis involves determining the chemical composition and higher heating value of the fuel, in which the C, H, O, S, N, and ash contents are reported on a dry basis in weight percent (wt%) as mentioned in Table 3.1. In the chemical composition, five components are considered, which are cellulose, hemicellulose, lignin, volatiles, and ash. The major component in biomass is cellulose, which is composed of polysaccharides of glucose monomers held together with linkages. The second most abundant component is hemicellulose, a heteropolymer consisting of carbohydrates such as glucose, xylose, and mannose. Cellulose is resistant to hydrolysis, whereas hemicellulose can readily undergo hydrolysis. The third major component of biomass is lignin, which is primarily composed of an array of aromatic alcohols and is interwoven with hemicellulose and cellulose, providing the strength and rigidity of lignocellulosic parts in a tree, such as the bark. These three major components, including cellulose, hemicellulose, and lignin, make up to the 80–90% of the biomass constituents. The remaining components in the biomass are the volatiles and ash.

Thus, not all varieties of biomass are suitable for all conversion processes. For example, in biogas production, animal excreta can serve as a major feedstock, where microbes convert the waste into biogas. Similarly, in bioethanol production, waste from sugar industries, such as molasses, can be used as feedstock. In the same way,

Table 3.1 Proximate and ultimate analysis of different biomass

Feedstock composition	Biomass type							
	Wood or woody biomass			Herbaceous biomass and agricultural biomass			Waste	
Proximate analysis	Wood	Sawdust	Forest residue	Switch grass	Corn straw	Coconut shells	Paper waste	Sewage sludge
VM	77.5	55.1	34.5	70.8	67.7	70.5	76.8	45
M	7.8	34.9	56.8	11.9	7.4	4.4	8.8	6.4
FC	12.2	9.3	7.3	12.8	17.8	22	6.8	5.3
A	4	0.7	1.4	4.5	7.1	3.1	7.6	43.3
Ultimate analysis								
C	49.6	49.8	52.7	49.7	48.7	51.1	52.3	50.9
O	44.1	43.7	41.1	43.3	44.1	43.1	40.2	33.4
H	6.1	6	5.4	6.1	6.4	5.6	7.2	7.3
N	0.1	0.5	0.7	0.7	0.7	0.1	0.2	6.1
S	0.006	0.02	0.1	0.11	0.08	0.1	0.08	2.33

other biomass-based processes for producing fuels or chemicals are selective with respect to the feedstock used. In the case of wood or woody biomass, the major constituents are cellulose, hemicellulose, and lignin, which make up to 87–95% of the material, with the remaining portion comprising of other components including, fats and proteins. However, in seeds, fruits, and peanuts, fats are the major component of their composition. Typically, in a plant, the composition is as follows: cellulose in the range of 38–50%, hemicellulose around 23–32%, lignin is 15–25%, and others 5–13%. Cellulose, which is generally composed of glucose, a straight chain oxygenated hydrocarbon, can be used for biogas generation, while simple sugars can be used for bioethanol production. Since cellulose is not digestible by humans or animals, it is excreted as waste in feces, which can be acted upon by different microbes in a biogas plant to produce biogas. Hemicellulose primarily consists of pentose sugars, while lignin is basically a phenylpropyl-based component. Lignin provides the strength to the plant and is a constituent of the bark, which is not digestible by microbes except for fungi. Details of the different important properties, their formula, and some of the key aspects required for understanding the biomass-based hydrogen production are given in Table 3.2

Thus, in the biological route for biomass to fuel, cellulose (a complex sugar structure) or simple sugar-based feedstock is used to produce biogas or bioethanol. In the chemical route, through transesterification or acid esterification process, the fats,

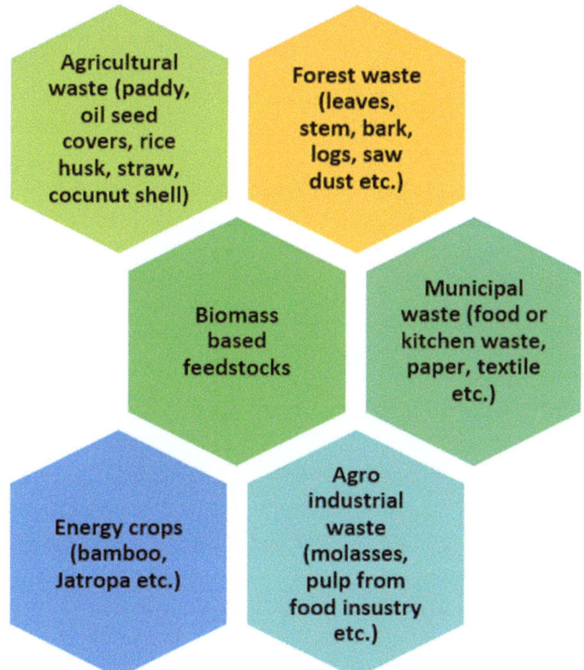

Fig. 3.2 Classification of biomass-based feedstocks

oils, or fatty acids are converted into biodiesel. In thermochemical processes, any variety of biomass can be used; however, the requirement is to use dry biomass.

Understanding various biomass-based energy conversion processes requires knowledge of the different physico-chemical properties of biomass. The differences among various biomass materials are primarily due to variations in moisture content, ash content, inorganic content, and chemical composition. The possible elements present in biomass include carbon, oxygen, hydrogen, nitrogen, calcium, potassium, silicon, magnesium, aluminum, sulfur, iron, chlorine, sodium, and manganese.

Biomass Energy Conversion Processes

Hydrogen can be produced from biomass using a biological route or a thermochemical route. However, the biological route scalability is the major bottleneck, and lab-scale demonstrations have been reported. Thus, the primary route for hydrogen production from biomass is through the thermochemical conversion of biomass. The thermochemical conversion of biomass involves processes in which lignocellulosic biomass is converted into biofuels. These processes include pyrolysis, gasification, and liquefaction. Liquefaction primarily results in liquid fuels; however, when the requirement is to produce hydrogen, the best-suited route is gasification, as the primary product with high yield is syngas. Through gasification, the maximum yield of hydrogen is obtained and it has higher efficiency. Syngas produced can be used for various applications, including power generation, heating, and the production of synthetic chemicals, such as methanol, ammonia, and dimethyl ether. Syngas, after purification and separation, can yield hydrogen, which can be used for fuel cells and other applications.

Combustion is the complete oxidation of biomass to produce CO_2, H_2O, and heat/light, and remaining inorganic residues, i.e., ash (which is the non-combustible part). Dried biomass, which could be forest or agricultural waste, can be used as feedstock for combustion.

The term pyrolysis stands for “pyro” meaning heat and “lysis” is breakdown, i.e., breaking down into components when heat is supplied. In the process of pyrolysis of biomass, when heat is supplied to biomass, it results in the breakdown of the constituents of the biomass (typically hemicellulose at $T > 200$ °C, cellulose $T > 250$ °C and lignin at $T > 350$ °C) to give volatiles and charcoal. The volatile consist of hydrocarbons, C2 +, C3 +, CH_4, CO, CO_2 and H_2, a small amount of inorganic residues depending on the feedstock used, which exist in vapor form at higher temperature. The higher hydrocarbons condense when brought back to ambient temperature, while the gases and lower hydrocarbons remain in vapor form. The solid component obtained through pyrolysis is char, which typically contains ash and other inorganic components. Typically, volatiles make up to 70–80% of the product, and charcoal accounts for 20–30% of the product of biomass pyrolysis. Vapor can be condensed to produce bio-oil, and also yields combustible gases (H_2, CO, CH_4, C2-C3 compounds) and char as products of pyrolysis. More than 200 types

Table 3.2 Properties, their formula, and some of the key aspects required for understanding the biomass-based hydrogen production

S. no	Property		Formula	Remarks
1	Density	(a) True density (ρ_t) $= \frac{\text{total mass}}{\text{solid volume excluding pore volume}}$ (b) Apparent density (ρ_a) $= \frac{\text{Total mass}}{\text{Apparent volume(solid+internal pores}-\text{interstial volume)}}$ (c) Bulk density (ρ_b) $= \frac{\text{Total mass}}{\text{Bulk volume occupied by particle}}$	$\rho_a = \rho_t(1 - \epsilon_{p)}$ $\rho_b = \rho_a(1 - \epsilon_{b)}$ $\epsilon_p = \text{Porosity} = \frac{\text{Pore volume}}{\text{Total volume}}$ $\epsilon_b = \frac{\text{Intersrial volum}e}{\text{Total packed volume}}$	ρ_b using ASTME -873–06 method
2	Thermal conductivity	$Q = \text{KA}\frac{\Delta T}{\Delta x}$		
3	Heating value	LHV = HHV-hg $\left(\frac{9H}{100} - \frac{M}{100}\right)$ H = % hydrogen M = % moisture Hg = Latent heat of steam $\text{HHV}_{\text{ar}} = \frac{Q}{M_f}\frac{\text{Kj}}{\text{Kg}}$ $\text{HHV}_d = \frac{Q}{M_f - M_w}\frac{\text{Kj}}{\text{Kg}}$ $\text{HHV}_{\text{daf}} = \frac{Q}{M_f - M_w - M_{\text{ash}}}\frac{\text{Kj}}{\text{Kg}}$	$\text{HHV} = \frac{\text{HHV}_{\text{ar}}}{1-M}$ $\text{HHV} = \frac{\text{HHV}_{ar}}{1-M-\text{ash}}$ $M = \frac{M_w}{M_f}$ Ash(wet basis) = $(1 - M)\text{Ash}_d$ HHV = $337 * C + 1442\left(H - \frac{O}{8}\right) + 935$ When the oxygen content < 10% using Dulong's formula	Depends upon the composition, moisture content, density, porosity, and temperature • QkJofheat • $M_f$kg of • fuel • $M_f$kg of • Moisture • $M_{\text{Ash}}kgof$ • ash
4	Ultimate analysis	C + H + N + S + O + ash = 100% on dry basis C + H + N + S + O = 100% on dry& ash freebasis		ASTME -D3176- method

(continued)

Table 3.2 (continued)

S. no	Property		Formula	Remarks
5	Proximate analysis	Fixed carbon + volatile matter + ash + moisture On dry and ash-free basis = FC + VM Dry basis = FC + VM + ash as received = FC + VM + ash + moisture		• ASTME -D3176- method • NRFL process • Klass method 1998
6	Atomic ratio	$\frac{H}{C}\text{ratio} = \frac{\frac{\%w_H}{MW_H}}{\frac{\%w_C}{MW_c}}$; ratio $= \frac{\frac{\%w_O}{MW_O}}{\frac{\%w_C}{MW_c}}$ $\%w_{H,c}$ is percentage wt of hydrogen or carbon MW is molecular weight of wet element		

of hydrocarbons are obtained, which are collectively referred to as liquid bio-oil. The simplest equation for the pyrolysis of cellulose can be written as:

$$\underset{}{3C_6H_{10}O_6} \rightarrow \underset{\text{Bio}-\text{oil}}{C_6H_8O+} \underset{\text{combustible gases}}{(2CO + 2CO_2 + H_2 + CH_4)+} \underset{\text{water of pyrolysis}}{8H_2O+} \underset{\text{biochar}}{7C} \tag{3.1}$$

In this process, the amount of hydrogen obtained is very small(roughly 15–20 g of hydrogen per kg of biomass), and the rest of the hydrogen is in compound form as water or in bio-oil. Additionally, depending on the temperature and rate of heating, the product phase composition will vary.

For slow pyrolysis carried out at low temperatures with a very long residence time, the process yields roughly 35% of gaseous, 30% of liquid, and 35% char as products. In case of fast pyrolysis carried out at moderate temperatures (500–700 °C) and with shorter residence times, the product gas composition is 13% gaseous, 75% liquid, and 12% char. For high temperatures (>800 °C) and long residence times, the major constituents are gaseous products (85%), liquid (5%) and char (10%). Thus, the processes being considered for hydrogen production should have either high temperature and long residence times (producing more gaseous products) or fast pyrolysis at moderate temperatures and short residence times (producing liquid products that can be converted to gaseous products). Actually, a long residence time allows several primary, secondary, and tertiary reactions to occur, ultimately breaking the long-chain molecules into shorter, lighter hydrocarbons and producing more gaseous products. However, the short residence time will allow only primary reactions to occur, leading to the breakdown of primary constituents into higher hydrocarbons and aromatics, which then condense at room temperature, resulting in more liquid products. For low temperatures and long residence times, the temperature is insufficient to allow lignin to convert into liquid/gas products, and it is obtained as a solid product, char. At high temperatures and with long residence times, all biomass, including lignin, is converted into gaseous products. Thus, depending on the products requirements, the appropriate conditions can be selected.

When we get liquid-like pyrolysis oil or bio-oil (under moderate temperature and short residence time), it can undergo catalytic steam reforming, producing hydrogen, followed by water gas reaction, similar to what was mentioned in the previous chapters.

Biomass gasification is a thermochemical process in which organic materials are converted into a high-value gas, known as syngas and solid phase, referred to as char. Both of these products have their own energy and economic value. The composition of the product gas or species obtained depends on various factors, including the nature of feedstock and its characteristics, the gasifying agent used (e.g., oxygen, air, steam, or carbon dioxide), operational parameters such as temperature and pressure, the catalysts used, and the gasification technology. During the gasification process, thermal degradation of lignocellulosic biomass occurs to produce bio-oil at the intermediate stage and syngas in the final stage [4]. The process can be catalyzed with the use of catalysts or other additive chemicals to produce the desired biofuels. Thus, gasification, which occurs at high temperatures and with a long residence time, produces hydrogen-rich gas and is being used for hydrogen production. Gasification can be referred to as partial oxidation or indirect combustion, where solid fuel or biomass can be converted into syngas in the presence of a limited amount of oxidant, i.e., less than that required for stoichiometric combustion or oxidation. A small portion of the fuel is combusted to supply the required heat for gasifying the remaining amount of the solid fuel or biomass. The air-to-fuel ratio in case when air is used as gasifying agent can lie between 1.5:1 and 1.8:1. The product gas obtained may be contaminated with organic components, particulate matter, tar, alkali metal, chlorides or sulfides and nitrogen (when air is used in the process) and thus gas cleanup is required to get syngas or finally produce hydrogen through a sequence of processes. The lower heating value of the syngas obtained from the gasifier can range from 4 to 13 MJ/Nm^3, depending on the operational conditions of the gasifier, the particular gasification technology used, and primary feedstock employed. In addition to the gases or vapors, a solid residue, i.e., char is produced, which is a mixture of unconverted organic fractions, and primarily consists of carbon and ash. The amount of organic fractions obtained depends on the gasification technology used and the operational conditions, whereas the ash content depends on the type of feedstock used. The lower heating value of the resulting char ranges from 25 to 30 MJ/kg.

The major reactions occurring in biomass gasification are endothermic, and the required energy for carrying out these reactions comes from the partial oxidation of a portion of the biomass through either an allothermal or an autothermal process. In the allothermal process, the required energy for endothermic reactions is supplied externally, whereas in autothermal processes, the required energy is obtained internally through partial combustion of the biomass.

Gasification consists of a complex sequence of physicochemical reactions. The complete series of processes can be categorized into three main stages: primary, secondary, and tertiary reaction stages. While the main steps involved in the gasification process are as follows:

(1) Drying and pretreatment (endothermic),
(2) pyrolysis (endothermic),
(3) oxidation (exothermic),
(4) reduction (endothermic).

It is essential that every step should be carried out under optimum conditions to achieve the best possible and desired product yield and quality. In a typical gasifier, all the four above-mentioned processes occur in a single unit.

(1) Drying—Since the product gas quality heavily depends upon the characteristics of the feedstock used, pretreatment is essential to homogenize the feedstock in terms of size, composition, and characteristics. In this step, drying involves evaporation of moisture present in the feedstock and reducing the moisture content. The amount of heat required for drying is proportional to the moisture content present in the feedstock. Since the drying process requires energy, it

makes the process energy-intensive, thereby reducing process efficiency. As we supply heat to the biomass at temperatures below 200 °C (to evaporate free moisture and the most tightly bound moisture before pyrolysis), the moisture present in the biomass is removed. Typically, the allowed moisture content in biomass for the gasification process is 25–30%; beyond this, the process is neither feasible nor economical. Therefore, green biomass with a higher moisture content should be dried prior to gasification. The required heat for drying is obtained from the other stages of the process.

(2) Pyrolysis—On heating the biomass to further higher temperatures, it undergoes pyrolysis, i.e., thermochemical decomposition of the carbonaceous materials involving breaking of chemical bonds, resulting in the formation of lower molecular weight species or volatiles. In the pyrolysis step, volatiles (the gaseous fraction), condensate (the liquid phase), and the solid fraction are obtained. The exact proportion or composition of the three different fractions or phases obtained depends on the type of biomass feedstock and the operating conditions. In a typical fluidized bed gasifier, the solid fraction is around 5–10 wt%, while in a fixed bed gasifier, it lies in the range of 20–25 wt%. This solid fraction has a high carbon content, and the inert materials present in the biomass feedstock are in the form of ash. The solid fraction obtained is called char. The liquid fraction is tar, and its amount varies between different types of gasifiers from 1 wt%, 1–5 wt%, to 10–20 wt% for the downdraft, bubbling bed, and updraft gasifiers, respectively. This liquid fraction consists of complex organic substances that condense at ambient conditions.

The gaseous fraction, which typically accounts for 70–90 wt%, is a mixture of gases that are non-condensable under ambient temperature conditions. This gaseous fraction is referred to as the pyrolysis gas and consists of hydrogen, carbon monoxide, carbon dioxide, methane, and lighter hydrocarbons, such as C2-C3 hydrocarbons, as well as small amount of acid gases or inert gases. The pyrolysis process occurs in a temperature range of 250–700 °C, and the reactions involved are endothermic in nature. The heat required for these reactions is obtained from the oxidation stage of the gasification process. The process can be written in a simple equation as follows:

$$\text{Biomass} \rightarrow \text{Volatiles} + \text{Char}$$

Or

$$C_nH_mO_p \rightarrow H_2 + CO + CO_2 + CH_4 + H_2O + \text{Tar} + \text{Char}. \quad \text{(Endothermic)} \tag{3.2}$$

Usually, the major component of biomass is cellulose (around 50% by weight), and for cellulose, the pyrolysis occurs between 600^0 and 700 °C. The pyrolysis process is a complex phenomenon and involves several reactions in series. Besides chemical reactions, heat transfer and product diffusion through the pores of biomass to the bulk gas phase are also involved. At low temperatures, the reaction kinetics is the rate-limiting step, while at higher temperatures, product diffusion and heat transfer could be the rate-limiting steps. During the pyrolysis process, both drying and loss of molecular weight occur initially. Loss of moisture content occurs until 200 °C, while close to 300 °C, the reduction of molecular weight of biomass due to the decomposition of amorphous cellulose to form carbonyl or carboxyl group radicals starts, and during this process, carbon dioxide and carbon monoxide are formed. When the temperature exceeds 300 °C, the crystalline cellulose produced decomposes, yielding char, tar, and gaseous products. At the same time, the hemicellulose decomposes into a soluble polymer, resulting in tar, char, and volatile gases. Lignin decomposes in the range 300–500 °C, yielding methanol, acetic acid, acetone, and water. The reactions occurring up to 300 °C are exothermic, and those beyond 300 °C are endothermic. Thus, for char production, the required temperature of 300 °C can be achieved without external heating, whereas external heating or oxidation of partial biomass is required to increase the yield of the required fuel components. The pyrolysis products can be a solid fraction, and their amount depends on the gasifier used, like 5–10 wt% in fluidized gasifier to 20–25 wt% in fixed bed gasifiers. This solid fraction, called char, has high carbon content and a high heating value and also yields a certain quantity of inert ash. The amount of liquid fraction resulting is again dependent on the gasifier type and referred to as tar. Amount of tar produced in downdraft gasifier is < 1 wt%, while in bubbling bed gasifier, it is in the range of 1–5 wt%, and for updraft gasifiers, it lies in the range of 10–20 wt%. Tar is primarily composed of organic substances that condense at low temperature. The gaseous product obtained in the pyrolysis process is also called pyrolysis gas and consists of a mixture of gases that do not condense at ambient temperature. Its amount in the process lies in the range of 70 – 90 wt%. The major constituents of pyrolysis gas include carbon monoxide, carbon dioxide, hydrogen, methane, C2-C3 compounds, and small quantities of acid and inert gases.

(3) Oxidation process—The required thermal energy to conduct various endothermic reactions in the gasifier is obtained by oxidation of a part of the biomass. This oxidation of some part of biomass is carried out under a limited amount of oxygen or less oxygen than the stoichiometric amount, leading not only to the oxidation of a part of biomass but also the production of sufficient heat to carry out the endothermic reactions in the gasification process. The volatiles of the pyrolysis step undergo oxidation to produce CO_2 and H_2O. Thus, the major products of the oxidation stage are CO, CO_2, and H_2O, along with heat, with maximum temperature of 1100–1500 °C being produced, which is required for all the endothermic processes. The oxidation step determines the quality and type of final products of gasification. The yield of product gas is determined by the gasifying agent used and operating conditions, i.e., temperature and pressure. With the use of steam as a gasifying agent the process of gas reforming is supported. When oxygen is used as a gasifying agent, the resulting product is highly suitable for use in power generation. However, the use of air as the gasifying agent results in larger quantities of nitrogen oxides

being present in the product gas, thus decreasing, the heating value of the obtained gas. Typical hydrogen yield in air gasification is 35–40 g H_2/kg of biomass, and the produced gas is highly diluted with nitrogen. However, when the oxidant is oxygen, the problem of nitrogen dilution is not there. When steam is used as a gasifying agent, biomass is converted to combustible gases, like H_2, CO, CO_2, CH_4 and lower hydrocarbons. The use of steam as gasifying agent can produce hydrogen-rich syngas; the yield of hydrogen depends on the operating conditions of the gasifier and steam to biomass ratio. The major reactions occurring in the biomass steam gasification involves:

$$\text{carbon gasification---C} + H_2O \rightarrow CO + H_2 \tag{3.3}$$

$$\text{water gas reaction} \quad CO + H_2O \rightarrow CO_2 + H_2 \tag{3.4}$$

$$\text{steam reforming of methane} \quad CH_4 + H_2O \rightarrow 3H_2 + CO \tag{3.5}$$

$$\text{hydrocarbon reforming} \quad C_xH_y + xH_2O \rightarrow xCO + \left(x + \frac{y}{2}\right)H_2 \tag{3.6}$$

The hydrogen content in the syngas can be further enhanced by using catalysts that reduce the reaction temperature, increase rate of gasification, and enhance hydrogen yield. The different catalysts used can be either alkaline earth metal catalysts (NaOH, KOH, K_2CO_3, Na_2CO_3, $Ca(OH)_2$, $KHCO_3$, Na_3PO_4 and MgO, etc.), metal or metal oxides (Ni, Cu, Fe, Pd, Pt, CoNi-Fe, CeO_2-FeO_3, $NiAlO_2$, etc.), natural mineral-based catalysts (dolomite, olivine, etc.) and hybrid catalysts (ZrO_2-Ni-M, where M = Ca, Mg, Na or K oxides). Still higher yields can be obtained using oxy-steam gasification, with a typical yield exceeding 100 g H_2/kg of biomass. In the case of oxy-steam gasification, hydrogen is derived not only from biomass, but also from steam. The possible reactions include steam reforming of hydrocarbons in biomass, water gas reaction, and the water gas shift reaction.

(4) Reduction step—The products obtained in the earlier oxidation and pyrolysis steps undergo reduction. Here, char plays a dominant role, acting as a reducing agent that results in the formation of syngas. So, the various possible reactions in the reduction zone include:

$$C + CO_2 \leftrightarrow 2CO \quad \Delta H = 172\ \text{kJ/mol} \text{Boudouard reaction} \tag{3.7}$$

$$C + H_2O \leftrightarrow CO + H_2 \quad \Delta H = 131\ \text{kJ/mol} \quad \text{Reforming of the char} \tag{3.8}$$

$$CO + H_2O \leftrightarrow CO_2 + H_2 \quad \Delta H = -41\ \text{kJ/mol} \quad \text{water gas shift reaction} \tag{3.9}$$

$$C + 2H_2 \leftrightarrow CH_4 \quad \Delta H = -75\,\text{kJ/mol} \quad \text{Methanation reaction} \tag{3.10}$$

The first two reactions are endothermic, while the later two reactions are exothermic; overall, the reduction step is endothermic. The required energy for the reduction reactions comes from the prior partial oxidation step. The above reactions are temperature-controlled, and chemical equilibrium favors product formation at higher temperatures for the first two reactions and at lower temperatures for last two reactions. Thus, the temperature of the reduction reaction determines the syngas composition, its heating value, and amount of solid residue obtained. At higher temperatures, the reduction of solid products, i.e., char, is favored and thus the formation of tar is reduced. The typical temperatures for the gasification process range from 800 to 1100 °C, while the steps that utilize oxygen in the gasification occur within a temperature range of 500 to1600 °C.

The solid and the gas/vapor end products of biomass gasification differ. The solid phase, i.e., ash, consists of inert materials present in the feedstock and the residue unreacted char (<1 wt%). The gas phase or vapor phase can be partially condensed under the ambient condition, which is the case for higher hydrocarbons, and partly remains as a gaseous phase or syngas even under these conditions. Thus, the gaseous products obtained are the gases which actually do not condense at ambient conditions, like H_2, CO, CO_2, light hydrocarbons, CH_4, C2-C3 compounds, inert gas N_2 if air is used in the oxidation stage, and in that case some amount of NH_3 and minor amounts of acid gases like H_2S and HCl. The LHV of syngas obtained lies in the range 4-15 MJ/m^3 and its amount may vary in the range 1–3 Nm^3/kg of the dry biomass; values vary based on the feedstock used, gasification technology, and the operating conditions of the gasifier. The condensable phase, commonly referred to as tar, is a complex mixture of higher hydrocarbons whose composition depends on the biomass feedstock used, the gasification technology employed, and the operating conditions of the gasifier. Tar consists of very high molecular weight species, including oxygenated condensable and aromatic compounds. It is formed through a series of complex chemical reactions that depend on the reaction conditions. The formation of tar begins during the pyrolysis stage, where cellulose and hemicellulose-containing oxygenated species undergo reactions to produce the primary tars, which are oxygenated organic compounds like alcohols, ketones, aldehydes, and acids. While the pyrolysis of lignin results in aromatic compounds, primarily phenols such as cresol, xylenol. As the temperature rises beyond 500 °C during the oxidation stage, with oxygen, air, or steam used as oxidant, the primary tar undergoes a series of stages including dehydration, decarboxylation, and decarbonylation, etc., to form secondary tars which are alkyl mono and di aromatics like furan, dioxin, thiophene, and pyridine. When temperature rises further beyond 800 °C, the decomposition and recombination of secondary tar result in the formation of tertiary tars, which are primarily aromatic and polynuclear aromatics hydrocarbons (PAHs), such as benzene, naphthalene, pyrene, and benzopyrene [1] (Fig. 3.3).

The tar produced in the gasifier must be removed or converted into useful compounds; various methods are employed, including end-of-pipe tar syngas cleanup, thermal cracking, and catalytic cracking of tar.

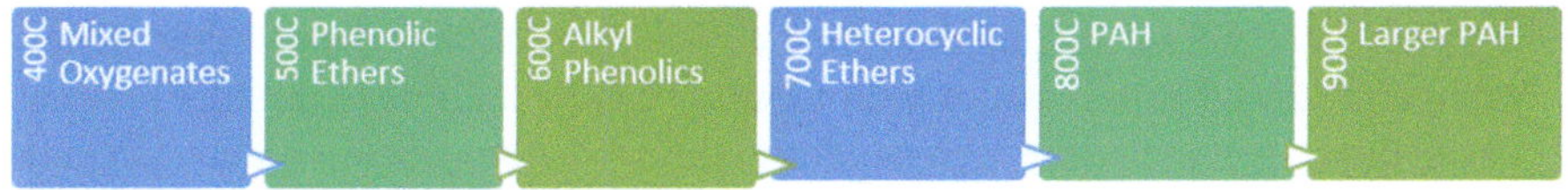

Fig. 3.3 Tar formation at different temperatures

In summary, the processes occurring in a gasifier can be represented as:

Drying: Heat (T < 200 °C) + Biomass → removing H_2O.

Pyrolysis: Heat (T > 200 °C, in the absence of oxidant) + Biomass → Volatiles (70–80%) + Char (20–30%).

Oxidation/Combustion: Volatiles + Oxidant (sub-stoichiometric amount of oxidant) → H_2O + CO_2 (T rises to > 1100 °C)

The gas phase reactions (volatiles and oxidant) occur much faster than gas–solid reactions(oxidant and char)

Reduction: $CO_2 + H_2O$ + Char + heat → H_2 + CO (T = 800–950 °C)

WGS: $CO + H_2O \rightarrow CO_2 + H_2$

When solid fuels contain organic sulfur compounds, then these are reduced to H_2S in the gasifier. Finally, unwanted components such as ash, tar, residual char, and H_2S are separated from the syngas before further downstream processes, including water gas reaction, carbon dioxide removal, and methanation.

Pretreatment of biomass—Since the product gas quality heavily depends upon the characteristics of the feedstock used, pretreatment is essential to homogenize the feedstock in terms of size, composition and characteristics.

1. Drying of the biomass—Moisture present in the biomass can affect the quality of the product gas obtained from the gasification process. As such, drying of biomass is required to ensure the moisture content in the feedstock is below 25–30 wt%. A higher moisture content in feedstock will reduce the quality of syngas, decrease the throughput of the gasifier, increase the thermal budget of the gasifier, and increase tar formation during gasification. The moisture present in biomass is either the unbounded moisture present in feedstock or the chemically bound water in the chemical formula of the biomass. For drying, waste heat or solar energy can be used to reduce the thermal energy required for the process.
2. Pelletization of biomass—Biomass is a lean fuel in terms of energy and exhibits a wide diversity based on its sources. Not only the chemical composition, but also the shape and size, and its properties vary based on the origin. As such, the cost of transporting biomass per unit of its energy content is high. The requirement is to improve its energy density and homogenization in terms of size, and to some extent, composition. Thus, to improve the density, biomass is compressed into pellets or briquettes. The pelletization process increases bulk density by mechanically compressing the material into pellets. At times, steam conditioning is carried out, as the biomass contains a high amount of fibers. This conditioning

softens those fibers, making densification easier, and it also helps in the pelletization process. Various issues with biomass can be addressed through pretreatment processes, including high moisture content, heterogeneous nature, irregular shape, lower bulk density, and high storage and transport costs. However, pelletization can be made more effective by either torrefaction before pelletization or the use of bioadditives. Typically, biomass is torrefied, cooled, ground to the required size, and then densified.

3. Torrefaction—Torrefaction is a thermochemical process carried out in the absence of oxygen and at a temperature in the range of 200–300 °C. In this process, the biomass loses some of the low-temperature volatile and low-energy-dense compounds, along with chemically bound moisture, through chemical reactions, thereby increasing its energy content. This process increases the energy density, for example, in the case of wood, from 10.5 to 20.7 MJ/kg on a mass basis and from 14 to 18.5 GJ/m^3 on a volume basis. Through torrefaction and pelletization, the handling and transportation of biomass becomes easier and also competitive with coal.
4. Use of bioadditives for pelletization—Adding different bioadditives helps to bind firmly and increase the mechanical stability and calorific value of the biomass during densification process. Various natural bioadditives can be used during pelletization including lignosulfate, starch, sawdust, glycol, castor bean cake, biopolymers, and gasification residue.
5. Hydrothermal upgrading—In the hydrothermal upgrading process, the biomass is decomposed to produce biocrude oil in the presence of water. This pretreatment process consists of two distinct steps. Firstly, biomass is treated with water at a temperature range of 200–250 °C and pressure of 30 bar. In the second step, the pressure of water is increased to 120–180 bar, and its temperature ranges from 300 to 350 °C. The bio-oil produced has a lower calorific value of 30–36 MJ/kg and can be used for the synthesis of fuels, chemicals, or as a secondary fuel in the coal-based power plants.

Gasification Technologies

Biomass gasification is carried out on an industrial scale in reactors, where pre-treated biomass is used in the process and is called gasifiers. There are different gasifiers used for the process and they differ from each other based on:

1. Contact between feedstock and oxidant,
2. rate of heat transfer,
3. residence time of feedstock.

Different technological variations exist primarily based on the mode of contact between the biomass feedstock and the gasifying agent used; these options can be either co-current, counter-current, cross-flow types. The heat required to carry out the process could be supplied externally or obtained from the partial oxidation of a portion of the biomass within the gasifier. The residence time can vary from minutes to hours, depending upon whether it is a fixed bed or fluidized bed gasifier.

Different Types of Gasifiers

Different types of gasifiers are used for gasification; they can be categorized, based on the method of feedstock feeding, the direction of biomass and oxidant flow, and the method of heat generation for endothermic reactions. The design of a gasifier can also vary based on the type of feedstock, its moisture content, the ash content, and the required end-use application

The major types of gasifiers used for the biomass gasification process are:

(1) Fixed bed,
(2) fluidized bed,
(3) entrained bed,
(4) rotary kiln type,
(5) plasma gasifier,
(6) supercritical water gasification.

1. Fixed bed gasifier—In a fixed bed gasifier, the gasifying agent flows through the solid fuel particles bed either in the upward or downward direction. The gasifiers are made of concrete or steel and operate at a slow gas speed, with a long residence time and higher carbon conversion.

The two major technological variations in fixed-bed gasifiers are updraft and downdraft gasifiers; schematics are shown in Fig. 3.4. In the updraft gasifier (also known as countercurrent gasifier), the solid biomass moves from top to bottom or downward with respect to the gasifying agent, while the syngas produced moves in the upward direction, i.e., in a counter-current manner. Biomass enters through a lock hopper from the top in the rising stream of hot gases. The first section is the drying section where moisture is removed. Thereafter is the pyrolysis section, where the hot gases rise and pyrolyze the biomass to produce tar, charcoal, and gases. The next zone is the reduction zone, where the char produced is reduced in the presence of rising CO_2 and H_2O to form syngas. In the final section, the incoming hot air burns the char to produce CO_2 and the required heat for the endothermic reactions. This is the combustion or oxidation zone, which provides the heat required for the endothermic reduction, pyrolysis, and drying zones that are present above it. Depending on the pyrolysis conditions, a wide range of vapor compositions is produced in the gasifier. Since the gasifying agent and feed move in a counter-current direction, the gas, pyrolysis products, and steam are released together.. Evaporation of steam in the combustion zone can not only result in better-quality syngas but also prevent the gasifier from overheating. Higher thermal efficiency is achieved through improved heat exchange between the hot gases as they flow through the bed and exit at a lower temperature, even providing the required heat for drying of feedstock and the production of steam. The major advantages of fixed-bed gasifiers are that these are simple and robust, with higher thermal efficiency, lower pressure drop, limited slag production, the ability to use different-sized materials, good contact between feed and oxidant, the capability to handle high-moisture-content feed, and the potential for scaling up. However, the major disadvantages include higher tar content in the product gas (often more than 20%), low syngas yield, the need to improve tar cracking, the requirement of uniform feed properties, higher sensitivity to feed characteristics, the need for grates, the requirement for gas cleaning, and the possible need for external heating to provide the required energy.

In the downdraft gasifier (co-current gasifier), both the solid biomass and the gasifying agent move from top to bottom, i.e., downward. Like updraft, the first stage is the drying zone, followed by the pyrolysis zone. Oxidant contacts the pyrolyzing biomass in these gasifiers before it comes into contact with the char, and as such, the heat from the oxidation of volatiles maintains the pyrolysis process. Since the decomposition products obtained after pyrolysis and drying must flow through the oxidation zone, the volatile materials undergo thermal cracking, resulting in improved fuel quality in the downdraft gasifier and reduced tar production. The limited amount of air or oxidant is consumed very fast. In these gasifiers, air must interact with biomass before interacting with char, thereby accelerating the flame and maintaining the pyrolysis process. At the end of pyrolysis, gaseous products consisting of CO, H_2, CO_2 and H_2O are formed, along with tar and char. Since the flame is in the limited air supply, it is also called "flaming pyrolysis," and it results in end products that are mostly in the gaseous phase, and the process consumes 99% of tar, which is the biggest advantage of such gasifiers. For this, the pyrolysis products interact with each other as well as with the gasifying medium to form the final gasification products. After the oxidation zone, which takes place at 800–1000 °C and involves partial combustion, cracking, and other reactions in the presence of an oxidant, the gas proceeds to the reduction zone. The different products formed in the oxidation zone undergo reduction in the reduction zone at ~ 800 °C to form syngas and ash. Thus, the product of gasification typically consists of: CO (19±3%), H_2(18±2%), CH_4(3±1%), CO_2(10±2%), and a balance nitrogen with air as the oxidant. The higher heating value of the gas lies in the range of 4-7 MJ/m^3. The major advantages of the downdraft gasifier include a simple and robust design that can be easily scaled up, higher conversion of feedstock, lower tar production, and a longer residence time of the solid. The major disadvantages of the downdraft gasifier include the requirement for uniform-sized pellets not exceeding 100mm in size, a low-moisture-content feed, lower heat transfer, difficulty in the start-up, issues with temperature control, scaling up could be an issue, and deactivation of catalysts can occur.

2. Fluidized bed gasifier—In a fluidized bed gasifier, the gasification agent is sent from bottom up through a distribution grid at a velocity of 1–3 m/s. Under these conditions, the solid bed behaves like a fluid. The flow of gas ensures uniform conditions within the bed, both in terms of heat and mass between the solid and gas mediums. The region of the distribution grid determines the heat and mass transfer between the gasifying agent and the solid bed. There are two main types of fluidized beds: bubbling type and recirculating type.

The bubbling bed gasifier bed consists of an inert granular material, such as sand, maintained in fluidized condition, and gasifying agent (1–3 m/s) flows upward across the bed through a distribution grid. Due to the continuous movement of the gasifying agent, fluidization conditions ensure uniform heat and mass transfer; however, the efficiency of heat and mass transfer also depends on the grid.

In the bubbling type, the upper part of the gasifier usually is gas only. In contrast, the recirculating type has two regions. One is similar to the bubbling bed, where the combustion reaction to produce the heat for gasification occurs, and other is where high-speed gas (5–10m/s) drags solids. In this region, the pyrolysis and gasification reactions take place. In a recirculation type of fluidized bed, the cyclone at the outlet separates solids, which are then fed back into the gasifier. This improves the efficiency of carbon conversion as compared to a bubbling bed gasifier. These gasifiers can operate at higher gas velocities, under pressurized conditions, and with higher energy throughput per unit cross-sectional area. The major challenge with circulating bed gasifiers is the formation of high levels of tar and dust (Fig. 3.5).

The fluidized bed gasifier is the most commonly used technology, as it provides uniformity, good heat, and mass transfer, ensuring a uniform flow and temperature across the bed. The principle of operation is based on fluidization principle, where both the fluid and the inert solid particles of the bed exhibit fluid-like properties. The bed is typically composed of silica, with other options sand, olivine, glass beads, and dolomite. The gasifying agent used can be air, steam, or a combination of both. The primary concept in such gasifiers is to achieve improved gasification through

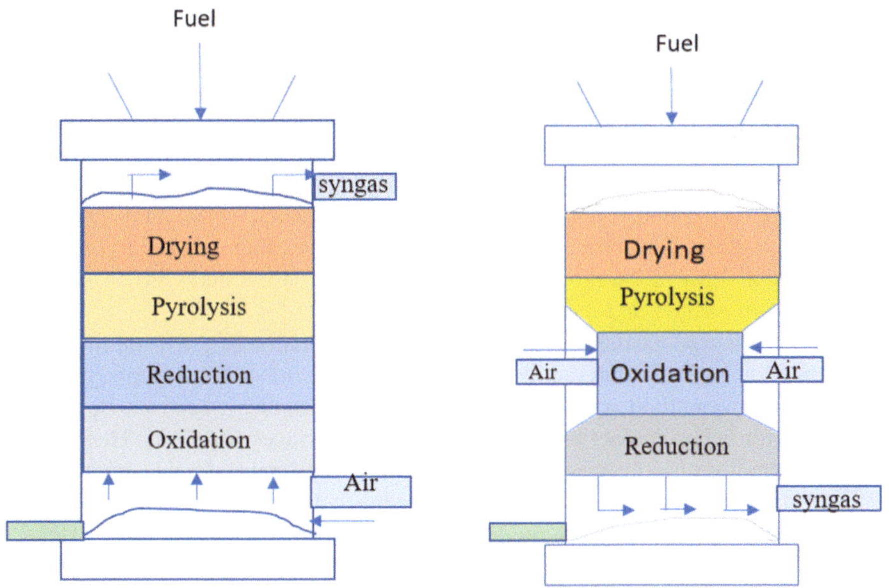

Fig. 3.4 Schematic of updraft and downdraft biomass gasifiers

enhanced heat transfer, allowing the bed to operate under nearly isothermal conditions. The choice of bed material determines the maximum operating temperature of operation which typically ranges from 800 to 900 °C. This temperature is low, and as such, a catalyst is required to achieve better performance. Limited residence time also drives the reaction away from chemical equilibrium.

Thus, the major advantages of a bubbling fluidized bed are improved feed and oxidant contact, enhanced temperature control, higher conversion to obtain the desired products, fuel flexibility, reduced tar formed, easier start-up and shutdown, improved control, and easier to scale up. At the same time, the disadvantages include the requirement of pretreatment and the loss of carbon in the form of ash. The advantages of the circulating fluidized bed gasifier are higher conversion and lower tar production, less residence time, fuel flexibility, and easier scale-up. While the disadvantages include the loss of carbon in the form of ash, feed pretreatment is required to ensure the particle size is smaller than 100 mm. The technology is not simple, and controlling parameters is challenging, resulting in reduced interaction between the solid and gas phases and higher costs.

3. Entrained bed gasifier—In an entrained bed gasifier, both the fuel and gasifying agent are fed co-currently (Fig. 3.6). The feed is in the form of fine particles in a size range of 0.1–1 mm. Such gasifiers operate at high temperatures, ranging from 1300 to 1500 °C, and pressures of 25–30 bar. The feed can be in the form of dry feedstock, which is pneumatically injected into the gasifier. If it is in slurry form, it is atomized and fed in a pulverized form. Entrained flow gasifiers can have ash leaving in the form of liquid slag, which is called slagging type, or slag is not produced in non-slagging type, where an ash content of $\leq 1\%$ is allowed. The major advantages of an entrained flow gasifier include high feed conversion to desired products, uniform temperature throughout the gasifier, fuel flexibility, good control over the gasifier's parameters, very low tar production, and easier scale up. At the same time, the major challenges of this technology include feedstock preparation and reduction of particle size to a range of 0.1–1 mm, reduction of moisture content to < 15 wt%, large quantity of oxidant required, heat recovery essential to improve efficiency as large quantity of sensible heat in the product gas stream, high CAPEX and OPEX, and short lifespan of gasifier and its components.
4. Rotary kiln-type gasifier—In such gasifiers, the rotation of the drum mixes and allows for gas–solid interaction for exposure of solid surfaces to the gasifying agent. In this method, a slightly tilted cylindrical chamber is used whose rotation allows for better contact between gas and solid surfaces (Fig. 3.7). However, in this technology, the heat and mass transfer are not optimal and can be improved by using certain barriers inside the drum to enhance contact and transfer. Typically, the feed is introduced from the top, and the gasifying agent is added from the bottom, ensuring a counter-current flow. The major advantages of rotary kiln-type gasifier include higher conversions, flexibility in feedstock, minimal feedstock pretreatment requirements, shorter residence times, lower tar production, and possibility of scale up. At the same time, the major disadvantages include the

requirement, that the size of feed particles should be less than 100 mm, complicated technology, difficulty in controlling the operational parameters, higher OPEX and CAPEX, higher ash content, and the need for specialized materials.

5. Plasma gasifier—The arc produced between the two electrodes ionizes the gas producing high-temperature species (Fig. 3.8). The gases used can be Ar, N_2, H_2, or a mixture of gases, and either DC or AC arc plasma torch generators can be employed. In a typical plasma gasification process, the plasma can be used either to provide the required heat for gasification or for tar cracking after the standard gasification. In the presence of an oxidizing agent, the temperature rises for the required gasification process; in its absence, it is pyrolyzed at a higher temperature. Feedstock enters and first passes through a low-temperature zone, where drying occurs, followed by the high-temperature zone created by the arc, where the rest of reactions can occur. This high-temperature region creates the low-temperature zone above, which removes excess moisture content in the biomass. The major factors that impacts the quality of the product gas include particle size, the type of feedstock used, and the input power to the plasma torch. The primary application of plasma gasification is in the treatment of hazardous waste, which cannot be upgraded by other techniques. It allows for feedstock flexibility, produces inert and non-leachable waste products, has a short residence time, and can be scaled up. The advantage of the plasma gasifier is high CO and H_2 content, low tar concentration, and high heating value of the syngas. However, the limitations are in terms of energy efficiency, which is lower than that of other techniques, and CAPEX and OPEX are higher. Additionally, there are problems with material consumption, such as electrodes and refractory materials, and molten materials can solidify in ducts. The batch process also presents challenges, as well as issues with start-up and shutdown, which involve heat shocks.
6. Supercritical water gasification—The supercritical water gasification process can be used for biomass that is high in moisture content. Thus, this method can be applied to wet biomass and as such pre-drying before gasification is not required (Fig. 3.9). The moisture or water present in the biomass provides the medium for the supercritical gas to reach the critical temperature under the operating conditions. The major benefit of this process is its flexibility in utilizing various feedstocks, including sewage sludge, food waste, agricultural waste, lignocellulosic biomass, and industrial waste. This method can also be applied to types of coal, asphalt, petcoke, oils, and organic solvents. The product gas obtained from supercritical water gasification includes CO, H_2, CO_2 and CH_4. The tar and coke formation can be controlled in this particular method

Thus to summarize, restrictions in terms of feedstock moisture content, composition, and particle size are more pronounced in entrained bed gasifier; however, feedstocks without any pretreatment can be used in plasma gasifiers and rotary kiln-type gasifiers. Similarly, fixed-bed and fluidized-bed gasifier can allow a wider range of particle sizes and moisture content than the entrained-bed gasifier. For both fixed and fluidized-bed gasifiers, the feedstock with a low ash melting temperature should be used. The best quality syngas is obtained using a plasma gasifier, while an entrained

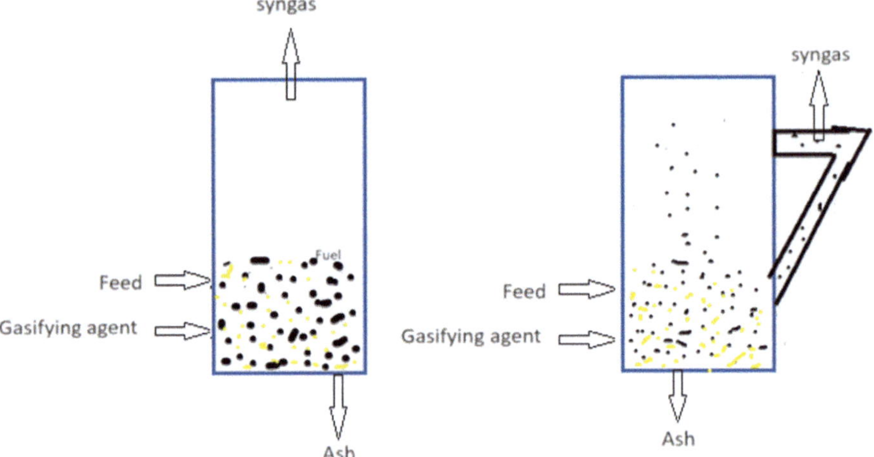

Fig. 3.5 Schematic of fluidized bed gasifiers

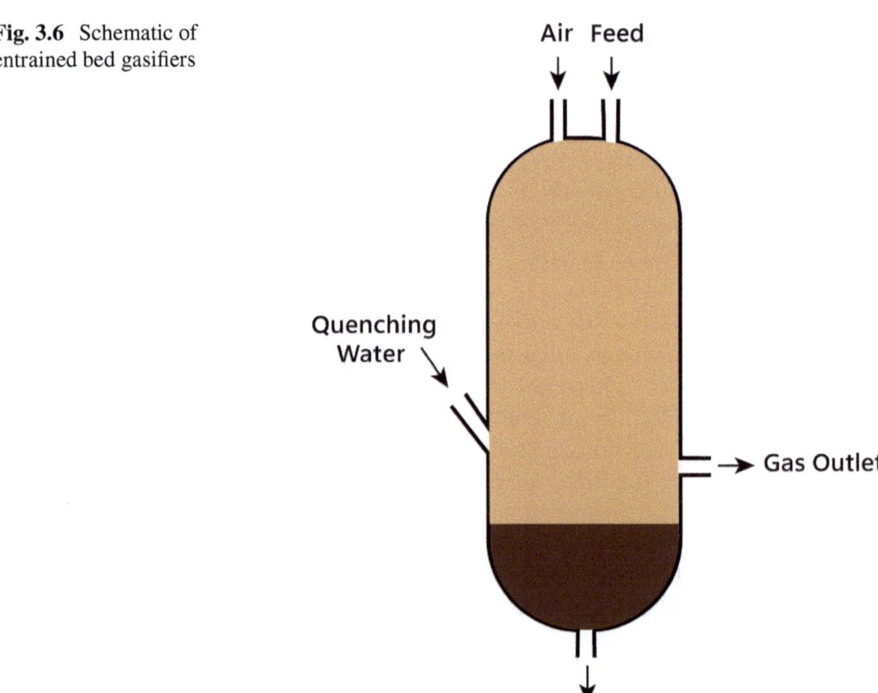

Fig. 3.6 Schematic of entrained bed gasifiers

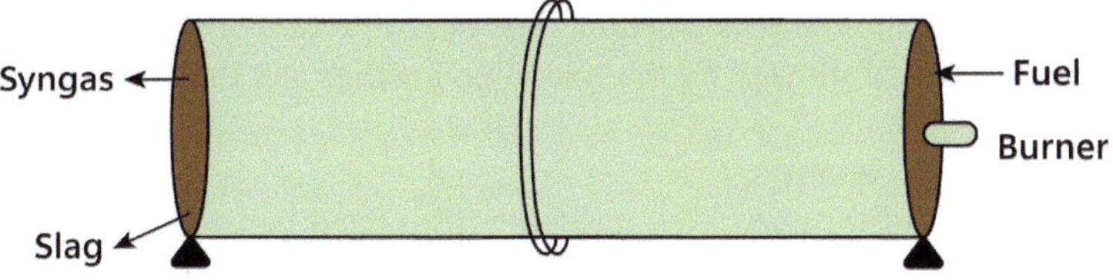

Fig. 3.7 Schematic of rotary kiln-type gasifier

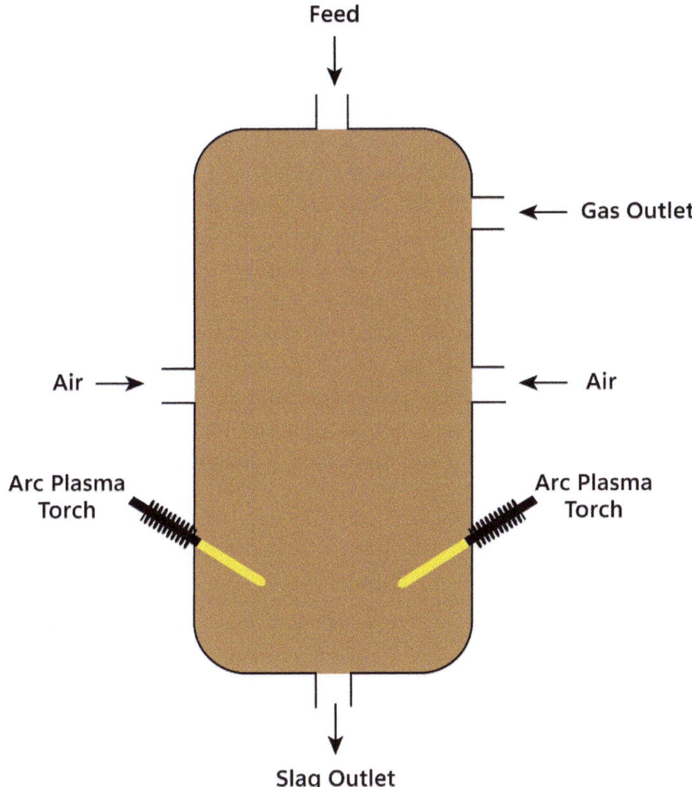

Fig. 3.8 Schematic of plasma gasifier

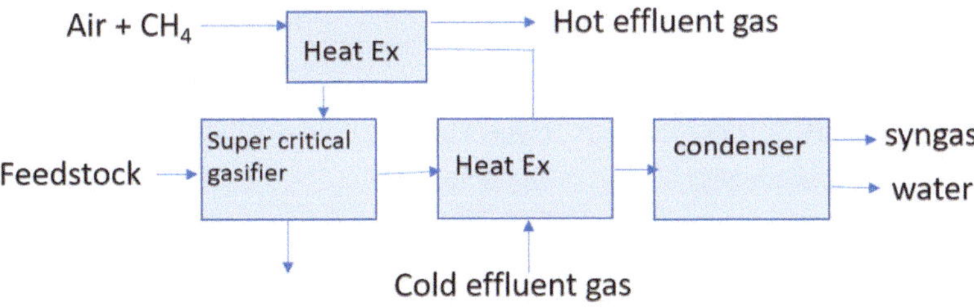

Fig. 3.9 Schematic of supercritical water gasifier

bed gasifier provides a lower tar content. The next in terms of syngas quality is the fluidized bed gasifier, followed by the fixed bed gasifier and the rotary kiln gasifier, in decreasing order.

In terms of syngas quality:

Plasma gasifier > Entrained flow gasifier > Fluidized bed > Fixed bed > Rotary kiln gasifier.

In terms of easier scale-up, entrained bed gasifiers have huge potential for commercialization and do the fluidized bed gasifiers. In contrast, fixed bed and rotary kiln gasifiers are the least recognized due to poor syngas quality, while plasma gasifiers are energy and cost-intensive.

In terms of efficiency, entrained bed, fluidized bed, and rotary kiln gasification technologies demonstrate high efficiency, while the fixed bed gasifier has moderate efficiency, and the plasma gasifier shows low efficiency. In terms of capital and maintenance costs, the rotary kiln has low CAPEX and high OPEX, while the fixed bed has limited CAPEX and OPEX; the entrained bed has moderate costs, and the fluidized bed and plasma gasifier have high CPAEX and OPEX. Some of the typical reactions that occur in the gasification can be written as follows (Table 3.3):

Most of the reactions mentioned above either yield hydrogen or methane; the produced methane can further undergo either steam or dry gasification to yield syngas. To further upgrade the syngas, various downstream processes can be employed. These downstream processes include syngas cooling using heat exchangers or quenching systems, syngas cleaning (acid gas removal and particulate matter removal, condensing the moisture), gas conditioning (adjustment of H/

Table 3.3 Summarizing the typical reactions that occur during the gasification

S. no	Reactions	Description	Heat of reaction (kJ/mol)
1	$C + CO_2 \rightleftharpoons 2CO$	Boudouard reaction	172
2	$C + H_2O \rightleftharpoons CO + H_2$	Water gas reaction	131
3	$C + 2H_2 \rightleftharpoons CH_4$	Hydrogasification	−74.8
4	$C + 1/2O_2 \rightarrow CO$	Oxidation	−111
5	$C + O_2 \rightarrow CO_2$	Oxidation	−394
6	$CO + 1/2O_2 \rightarrow CO_2$	Oxidation	−284
7	$CO + H_2O \rightleftharpoons CO_2 + H_2$	Water gas shift reaction	−41.2
8	$2CO + 2H_2 \rightleftharpoons CO_2 + CH_4$	Methanation reaction	−247
9	$CO + 3H_2 \rightleftharpoons H_2O + CH_4$	Methanation reaction	−206
10	$CO_2 + 4H_2 \rightleftharpoons 2H_2 + CH_4$	Methanation reaction	−165
11	$CH_4 + 2O_2 \rightleftharpoons CO_2 + 2H_2O$	Oxidation	−803
12	$CH_4 + CO_2 \rightleftharpoons 2CO + 2H_2$	Dry reforming	247
13	$CH_4 + H_2O \rightleftharpoons CO + 3H_2$	Steam reforming	206
14	$CH_4 + 1/2O_2 \rightleftharpoons CO + 2H_2$	Partial oxidation	−35.7
15	$H_2 + 1/2O_2 \rightarrow H_2O$	Oxidation	−242

C ratio as per the application and to get the desired syngas composition) through water gas shift reaction and CO_2 capture, product gas purification (separation of hydrogen using PSA or membrane-based method).

Effect of various parameters - The major parameters of interest that determine the yield and composition of the product gas includes:

1. Type of feedstock used, moisture content, and its properties
2. operating conditions of the gasifier, including temperature, pressure, and residence time
3. particle size and density of biomass
4. air equivalence ratio
5. catalyst used
6. which gasification technology or gasifier is being used

Equivalence ratio—The gasification process occurs in a limited amount of oxidant or air/oxygen-deficit environment; as such, the equivalence ratio for air/oxygen and its flow rate plays a key role. The equivalence ratio is defined as the actual air/oxygen provided in the process to the stoichiometric amount of air/oxygen required for the gasification process. It is usually represented by the symbol λ and is represented as

$$\lambda = \frac{m_{\text{ox, act}}}{m_{\text{ox, stoic}}} \tag{3.11}$$

A value of λ less than 0.2 leads to incomplete gasification and incomplete char-to-gas conversion, thereby resulting in more char formation and a gas of lower calorific value. However, a value of λ being higher than 0.3 can lead to increased combustion, thereby reducing overall efficiency, increasing emissions, and lowering the calorific value of the product gas. Thus, the optimum value of λ lies in the range of 0.2–0.3.

Various biomass-based thermochemical processes including combustion, pyrolysis, and gasification can produce primary products such as gaseous fuel, liquid, char, and heat. These can be processed to produce the secondary products such as ammonia, methanol, gasoline, diesel, electricity, which can then be used as substitutes for the other fossil fuel-based energy products. A value of $\lambda = 1$ means that an adequate amount of oxygen is available for complete combustion. Pyrolysis occurs in the absence of oxygen, whereas gasification requires less than stoichiometric oxygen, with an optimum value of around 0.2–0.3, as mentioned above.

Effect of operating temperature—The gasification process aims to have a product gas with desired characteristics, achieve higher carbon conversion, and low tar formation. In achieving these, the operational temperature of the gasifier plays a crucial role in impacting the chemical reactions that occur within it. It is reported that a temperature above 800 °C is desirable, and the yield of various oxygenated species, including phenol, cresol and benzofuran, is lower. Also, the formation of one-ring or two-ring aromatics is restricted (benzene, naphthalene, and phenanthrene). When the temperature exceeds 850 °C, the destruction of these aromatics occurs; however, the formation of three- or four-ring aromatics increases. When the operating temperature is increased from 700 to 900 °C, the tar yield decreases by 40%. At temperatures close to 900 °C, the formation of naphthalene is observed to increase, with its content rising by 40%. When the temperature is increased from 700 to 850 °C in a bubbling fluidized bed gasifier, the amount of hydrogen increases from 5 to 10 vol%, while the amount of carbon monoxide increases from 12 to 18 vol %. In contrast, the carbon dioxide content decreases from 16 to 13 vol%; however, the ammonia and acetylene contents remain unaffected with this temperature rise.

Effect of operating pressure—It is reported that with the increase of operating pressure, the tar content reduces but the PAH fraction increases. The operating pressure affects wood char gasification; when the pressure of CO_2 increases from 1 to 15 bar at 750 °C, a slight reduction in char gasification is observed. While the increase of H_2O pressure from 1 to 15 bar at 850 °C results in increased char gasification. For the whole tree chips-based biomass, the operating pressure of 21.4 bar results in the elimination of the formation of phenols.

Effect of gasifying agent—The various gasifying agents being studied in the gasification process include, oxygen, air, steam, steam and oxygen, and carbon dioxide. The choice of gasifying agent decides on the product gas composition and its heating value. It is reported in the literature that the use of air as a gasifying medium results in a reduced heating value of the syngas, which is found to be in the range of 4–7 MJ/Nm^3 and is diluted with inert nitrogen. When oxygen is used as a gasifying agent, a high-heating value syngas is obtained, with the values as high as 28 MJ/Nm^3. The use of steam and oxygen as a gasifying agent yields an intermediate heating value syngas, with a value lying in the ranging from 10 to 18 MJ/Nm^3, characterized by higher quantities of hydrogen in the syngas. When steam is used as a gasifying agent, then the syngas is rich in hydrogen (50% H_2 in syngas). An increase in the steam-to-biomass ratio from 0.5 to 2.5 increases, this hydrogen content up to 60%, while a decrease of carbon monoxide from 35 to 10% and an increase in carbon dioxide from 10 to 30% are observed. A higher steam-to-biomass ratio results in lower tar formation, as several additional reactions, such as gas reforming, water gas shift, and cracking take place. When carbon dioxide is used as a gasifying agent, dry reforming occurs, leading to reduced tar formation. The use of both steam and carbon dioxide as gasifying agents has been reported to result in maximum carbonization, leading to the highest activity char and high ash content [4].

Effect of residence time—Residence time has a significant impact on both tar yield and syngas quality. It is reported that higher residence time at the pyrolysis stage results in further degradation of secondary phase species, leading to the formation of additional gaseous species, as well as water, formic acid, and acetic acid.

Effect of catalyst—Usually, on an industrial scale, the production of hydrogen from solid fuels is non-catalytic, since these feedstocks contain various impurities (sulfur and heavy metals) which could poison the catalyst. However, studies have been conducted using appropriate doping or catalysts to increase the char gasification, obtain a higher yield of syngas with a higher calorific value, enhance tar reforming, thereby reducing tar formation, and achieve an increased hydrogen content. Usually, the bed material in a gasifier is sand due to higher temperature resistance. The catalysts studied include nickel on alumina, limestone, and orthosilicates of Fe and Mg, etc. However, Ni-based catalysts are preferred due to lower cost.

Products of Biomass Gasification

The end products of biomass gasification include solid and gaseous phases. The solid fraction consists of unreacted char and ash (inert material left out of the biomass feedstock). The char content is $< 1\,\mathrm{wt\%}$, since the major aim of gasification is conversion of carbon into gaseous species. The gas or vapor fraction consists of a condensable and an incondensable phase. The incondensable gas fraction at ambient temperature includes CO, CO_2, H_2, CH_4, and C2-C3 compounds. If air is used as a gasifying agent, then nitrogen is also present in syngas. Some of the other minor constituents of product gas can be ammonia, acid gases such as hydrogen sulfide and hydrochloric acid. The condensable fraction is tar, which consists of organic compounds excluding C1-C6, like bituminous oil, etc. Tar is highly undesirable, and its content depends on the reaction conditions, forming as a result of the pyrolysis step, which can further undergo recombination and decomposition reactions. However, the removal of tar can be carried out through primary methods, which involve controlling the reaction conditions to enable such tar cracking using high temperatures, a catalyst, or using non-thermal plasma. Secondary methods to eliminate tar downstream of the gasifier include thermal or catalytic cracking after gasification, or through the use of mechanical separation techniques using cyclones, filters, scrubbers. In addition to tar, a small fraction of alkali or alkaline earth metals can be present in the product from the gasifier. Mostly, alkali metals are present in herbaceous biomass, while alkaline metals in woody biomass, or they can be introduced through the use of catalysts. The presence of elements like Na and K can result in corrosion or fouling and can be removed through condensation and adsorption techniques. When air is used as gasifying medium, the syngas produced can be contaminated with nitrogen and its compounds, such as NH_3 and HCN. The nitrogen-derived compounds can crack in the presence of a catalyst used in a gasifier, particularly at high temperatures within the gasifier. The resulting nitrogen can be removed using a cold gas cleaning technique, such as the absorption of ammonia in water. Sulfur could be found in the form of hydrogen sulfide, carbonyl sulfide, and carbon disulfide in the syngas. Usually, the sulfur content is very low in biomass as compared to coal. These sulfur-containing species can be removed through either physical or chemical adsorption through sorbents, chemical or physical absorption, chemical redox, or biological techniques. Some traces of hydrochloric acid may also be present in the outlet stream from gasifier, which can lead to corrosion, clogging, and fouling in downstream utilities. Syngas can be cleaned for halides through sorbent-based adsorption or cold gas cleaning, followed by wet scrubbing and absorption.

Other Routes for Hydrogen Production from Biomass

The other method for producing hydrogen from biomass involves the fermentation route. In this route, various types of bacterial enzymes are used to convert biomass into hydrogen. This can be carried out in the dark or in the presence of light and is accordingly referred to as either dark fermentation or photo-fermentation process. In dark fermentation, anaerobic bacteria are used to produce hydrogen under dark conditions, and it is usually cost-effective. It produces hydrogen from carbohydrate-rich biomass under conditions where the temperature ranges from 30–80 °C. The resulting hydrogen contains impurities such as methane or H_2S and needs to be purified for further use. In photo-fermentation, it has similarities to photosynthesis; however, it is an inefficient process. Typically, most of these bacterial-based bioconversion processes have poor efficiency and have been demonstrated on a laboratory scale. Scalability and poor efficiency are the biggest hurdles in the commercialization of such processes.

Solved Problem 1

A power plant burns 100 metric tons of wood chips per day. The wood chips have a heating value of 18 MJ/kg, and the power plant operates at an efficiency of 30%. Calculate the daily electricity generated in kilowatt-hours (kWh).

Solution:

Step 1: Convert the mass of wood chips to kilograms.
100 metric tons = 100 × 1000 = 100,000 kg/day
Step 2: Calculate the total energy content of the wood chips.
Heating value = 18 MJ/kg
Total Energy Input = 100,000 kg/day × 18 MJ/kg = 1,800,000 MJ/day
Step 3: Calculate the useful energy output using efficiency.
Efficiency = 30% = 0.30
Useful Energy Output = 1,800,000 MJ/day × 0.30 = 540,000 MJ/day
Step 4: Convert MJ to kWh (1 MJ = 0.27778 kWh).
Electricity Generated = 540,000 MJ/day × 0.27778 kWh/MJ ≈ 150,001.2 kWh/day

Final Answer:

The power plant generates approximately 150,001.2 kWh of electricity per day.

Solved Problem 2

A biomass gasification plant processes two metric tons of dry biomass per hour. The biomass has an average empirical formula of $CH_{1.4}O_{0.6}$ and a heating value of 17 MJ/kg. The gasifier produces syngas with a composition of 40% H_2, 25% CO, 20% CO_2, and 15% CH_4 by volume. The gasifier's cold gas efficiency (CGE) is 75%, and the hydrogen is separated from the syngas stream with 90% recovery.

Assume standard temperature and pressure (STP) conditions.

Calculate the amount of hydrogen (in kg) produced per hour.

(Use: Molar volume at STP = 22.4 L/mol, molar mass of H_2 = 2 g/mol, average heating value of syngas = 10 MJ/Nm^3).

Solution:

Step 1: Calculate total energy input from biomass per hour

Mass of biomass = 2 tons = 2000 kg

Energy input = $2000 \times 17 = 34{,}000\,\text{MJ/h}$.

Step 2: Calculate the energy content of syngas (cold gas output)

Cold Gas Energy Output = $34{,}000 \times 0.75 = 25{,}500\text{MJ/h}$.

Step 3: Syngas has average heating value = 10 MJ/Nm^3 (typical for biomass gasification)

Volume of syngas produced = $2550\,\text{Nm}^3/\text{h}$.

Step 4: Calculate the volume of H_2 in syngas

H_2 volume = $40\%\text{ of }2550 = 0.40 \times 2550 = 1020\,\text{Nm}^3/\text{h}$.

Step 5: Calculate moles and mass of H_2

- 1 Nm^3 = 1000 L
- Moles of $H_2 = 1020 \times \frac{1000}{22.4} = 45,536\,\text{mol}$

Mass of $H_2 = 45{,}536 \times 2 = 91,072\,\text{g} = 91.07\,\text{kg}$.

Step 6: Apply H_2 recovery efficiency

Recovered $H_2 = 91.07 \times 0.90 = 81.96\,\text{kg/h}$.

Solved Problem 3. Given ***Ultimate analysis result*** of the biomass sample, calculate the empirical elemental formulae of given Biomass.

C–45 %

H – 6%

O – 49%

All values in weight percentage.

Solutions

To determine the **empirical elemental formula** of the given biomass from its **ultimate analysis**, we follow these steps:

Solution

Given:

- C = 45%
- H = 6%
- O = 49%

We will assume **100 g** of biomass to make the math easier.

Step 1: Convert weight percentages to moles

We divide each mass by its atomic weight:

Element	Mass (%)	Atomic mass (g/mol)	Moles = mass/atomic mass
C	45 g	12.01	45 / 12.01 = 3.747 mol
H	6 g	1.008	6 / 1.008 = 5.952 mol
O	49 g	16.00	49 / 16.00 = 3.063 mol

Step 2: Divide each by the smallest number of moles

We divide all mole values by the **smallest mole value**, which is **3.063 (for O)**:

Element	Moles	Divided by 3.063	Result
C	3.747	3.747 / 3.063	1.22
H	5.952	5.952 / 3.063	1.94
O	3.063	3.063 / 3.063	1.00

Step 3: Multiply to get whole numbers

We need to eliminate decimals. Try multiplying all by **5** (since 1.22 and 1.94 are near 6.1 and 9.7 when multiplied by 5):

Element	Approx. ratio × 5	Rounded
C	1.22 × 5 = 6.10	**6**
H	1.94 × 5 = 9.70	**10**
O	1.00 × 5 = 5.00	**5**

Step 4: Write the empirical formula

Thus, the **empirical formula** is: $C_6H_{10}O_5$

Solved Problem 4. A biomass gasifier is designed for the conversion of 100 kg/h of dry biomass. It is working with air as a gasifying medium and generates 250 m^3 of producer gas (PG) per hour.

Ultimate analysis of biomass:

C	45 g
H	6 g
O	49 g

The gas analysis of the PG gives following gas composition:

$$H_2 - 20\%$$

$$CO{-}20\%$$

$$CO_2{-}12\%$$

$$CH_4{-}2\%$$

$$N_2 - \text{Rest } (46\%)$$

Calorific value of different gases

$$H_2{-}10.8\ \text{MJ/m}^3$$

$$CO{-}12.6\ \text{MJ/m}^3$$

$$CH_4{-}35.8\ \text{MJ/m}^3$$

i. How many kgs of PG will be produced per hour?
ii. How many grams of H_2 will be produced per kg of biomass?
iii. What is the H_2 conversion efficiency?
iv. What is the energy conversion efficiency of the gasifier?

Assume that the calorific value of biomass = 17 MJ/kg

Solution

Given:

- Biomass feed rate = 100 kg/h
- Producer gas (PG) volume = 250 m^3/h
- PG composition:
  - $H_2 = 20\%$
  - $CO = 20\%$
  - $CO_2 = 12\%$
  - $CH_4 = 2\%$
  - $N_2 = 46\%$

Step 1: Molar Masses of Gases (g/mol)

Gas	Molecular weight (g/mol)
H_2	2.016
CO	28.01
CO_2	44.01

Gas	Molecular weight (g/mol)
CH_4	16.04
N_2	28.01

Step 2:

(i) Mass of Producer Gas generated per hour

We assume PG behaves like an ideal gas.
Use **average molar mass** of PG from the composition:

$$M_{\text{avg}} = \sum(\text{yi} \cdot \text{Mi})$$

where:

- y_i: mole fraction
- M_i: molar mass of each gas

$$\begin{aligned} M_{\text{avg}} &= (0.20 * 2.016) + (0.20 * 28.01) \\ &+ (0.12 * 44.01) + (0.02 * 16.04) + (0.46 * 28.01) \\ &= 0.4032 + 5.602 + 5.2812 + 0.3208 + 12.8846 = 24.49\,\text{g/mol} \end{aligned}$$

Now convert **250 m³/h to mass** using molar volume at NTP (1 mol = 22.4 L = 0.0224 m^3):

- Total moles of gas:

$$n = 250/0.0224 = 11160.71\,\text{mol/h}$$

- Total mass of PG:

$$M = n * M_{\text{avg}} = 11160.71 * 24.49 \approx 273325.54\,g/h = 273.3\,\text{kg/h}$$

Alternative way

$$\text{density, d} = M_{\text{avg}}\ /\text{Volume of gas}$$

$$d = 24.49\,\text{g/mol}/0.0224\,\text{m}^3/\text{mol} = 1093.3\text{g/m}^3 = 1.093\,\text{kg/m}^3$$

Mass of PG generated is

$$\text{Mass of PG} = 250\,\text{m}^3/\text{h} \times 1.03\text{kg/m}^3 = 273.3\,\text{kg/h}$$

Answer (i): 273.5 kg/h of PG is produced

Step 3:

(ii) Grams of H_2 per kg of biomass

Mass fraction of $H_2(x)$ = (mole fraction of H_2 * molar mass of H_2)/mass of producer gas

$$\text{Mass fraction of } H_2(x) = \frac{0.2 * 2.016}{24.49} = 0.01646 \text{ or } 1.646\%$$

H_2 generation = 273.5 * 0.01646 = 4.5 kg/h.
Answer (ii): 4.5 g of H_2 per kg of producer gas
H_2 generation per kg of biomass = 4.5/100 = 0.045 kg H_2/kg biomass

Step 4:

(iii) H_2 Conversion efficiency

- Biomass contains 6% H by mass (from ultimate analysis)
- H in 1 kg biomass = 6% = 60 g

Only 45 g of H_2/kg biomass is recovered.
So, H_2 conversion efficiency = H in H_2 produced/H in biomass = 45/60 = 75%
Answer (iii): 75% H_2 Conversion efficiency
Step 5: iv. Energy conversion efficiency of gasifier

$$\text{Energy conversion efficiency} = \frac{\text{Energy in PG}}{\text{Energy in biomass}}$$

$$\eta = \frac{\text{Calorific value of PG}\left(\frac{\text{MJ}}{\text{m}^3}\right) * \text{PG}\left(\frac{\text{m}^3}{\text{Kg of biomass}}\right)}{\text{Calorific value of biomass}\left(\frac{\text{MJ}}{\text{Kg}}\right)} * 100$$

$$\text{Calorific value of PG}\left(\frac{\text{MJ}}{\text{m}^3}\right) = \sum \text{cv}_i * y_i$$

$$\text{Calorific value of PG}\left(\frac{\text{MJ}}{\text{m}^3}\right) = \text{cv}_{H_2} * y_{H_2} + \text{cv}_{CH_4} * y_{CH_4} + \text{cv}_{CO} * y_{CO}$$

$$\text{Calorific value of PG}\left(\frac{\text{MJ}}{\text{m}^3}\right) = 10.8 * 0.2 + 35.8 * 0.02 + 12.6 * 0.2 = 5.396\,\text{MJ/m}^3$$

$$\eta = \frac{5.396\left(\frac{\text{MJ}}{\text{m}^3}\right) * 2.5\left(\frac{\text{m}^3}{\text{Kg of Biomass}}\right)}{17\frac{\text{MJ}}{\text{kg}}} * 100 = 79.35\%$$

Answer (iv): 79.35% energy conversion efficiency

References

1. A. Molino, S. Chianese, D. Musmarra, Biomass gasification technology: The state of the art review. J. Energy Chem. **25**, 10–25 (2016)
2. L. Cao, I. K. M. Yu, X. Xiong, D. C. W. Tsang, S. Zhang, J. H. Clark, C. Hu, Y. H. Ng, J. Shang, Y. S. Ok, Biorenewable hydrogen production through biomass gasification: A review and future prospects, Environ. Res. **186,** 109547 (2020)
3. S.V. Vassilev, D. Baxter, L.K. Andersen, C.G. Vassileva, An overview of the chemical composition of biomass. Fuel **89**, 913–933 (2010)
4. M. Yan, N. Afxentiou, P. A. Fokaides, The state of art overview of the biomass gasification technology, Curr. Sustain. Renew. Energy Rep. **8**, 282–295 (2021)

Chapter 4
Hydrogen Production from Coal

Coal, like biomass, is a solid fuel and thus presents its own challenges, being less convenient than the gaseous or liquid fuels used for producing hydrogen. Coal is the most abundant fossil fuel currently and the biggest contributor of greenhouse gas (GHG) emissions. Combustion in coal-based power plants is the primary method of utilizing coal and a means of producing electricity. On combustion of coal not only CO_2 but also NO_x and SO_x are released, and as such, the separation and sequestration of CO_2 are capital- and energy-intensive. Besides, since the conversion of thermal energy from coal combustion to power generation using the Rankine cycle is less efficient, with an efficiency less than 40%, this route of coal utilization is not very effective and efficient, while also being very polluting. However, coal gasification technology is comparatively more efficient and cleaner than coal combustion and can result in a variety of products, including fuels, chemicals, as well as electricity and energy output. Since the emphasis is on clean hydrogen production routes, hydrogen production units utilizing coal gasification should integrate carbon capture and sequestration or storage to benefit from producing hydrogen in a cleaner manner. In biomass-based hydrogen production, other than gasification, there are other routes like anaerobic fermentation or supercritical water gasification (higher moisture content in biomass); however, the primary method in the case of hydrogen production from coal is coal gasification. The general formula for solid fuels can be represented as $C_aH_bO_c$, where, in the case of biomass $a < b$. while in the case of coal, $a > b$.

Based on the basic characterization of solid fuels, those derived from fossils are called fossil fuels, whereas those derived from organic matter of biological origin are referred to as biomass (originated from plants, animals, microorganisms, etc.). The fossil fuels are formed when the organic matter buried under the earth's surface is subjected to high temperature and pressure over millions of years.

So, the solid fuels can be biomass, which is described in more detail in the earlier chapter, as well as peat and fossil fuels. From biomass, typical hydrogen production lies in the range of 55–65 g of hydrogen per kg of biomass, while for coal,

P. Sharma, *A Textbook on Hydrogen Energy*,
https://doi.org/10.1007/978-981-95-4601-5_4

it is approximately 20 g of hydrogen per kg of coal. Among fossil fuels, examples include coal, oil sands, and shale oil. Peat is formed in waterbodies over a period of time, during which dead plants undergo decomposition and carbonization in the absence of oxygen. The coal is formed from peat beds, which have undergone sedimentation of minerals and organic matter, resulting in coalification under high temperatures and pressures at depth within the earth's surface over millions of years. There are several variants of fossil fuels, including coke, and oil shale or oil sand. Coke consists of mainly of carbon (90–95%), with traces of ash and sulfur, formed from the carbonization of coal. That is, when coal is heated in absence of an oxidant at high temperature (~ 1000 °C), it causes the removal of the volatiles, leaving behind coke. Shale oil is a fine sedimentary rock that contains a large amount of organic matter called kerogen, formed over millions of years by submerged water-based species, including algae, planktons, etc. Kerogen contains hydrocarbons that are rich in carbon and low in hydrogen, with oxygen, nitrogen, and sulfur also present.

Classes of Coal

Coal can be classified into different types based on its rank, which is determined by the degree of progressive change or metamorphism undergone, as per American Society for Testing and Materials (ASTM, 2011). The different types of coal based on increasing rank and increasing order of alteration are:

1. Lignite,
2. sub-bituminous,
3. bituminous,
4. anthracite.

Thus, the highest-ranked coal is anthracite, and the lowest-ranked is lignite. Based on the rank or alteration, various properties of coal vary like the HHV, volatile matter, moisture content, ash content, and fixed carbon content (Fig. 4.1).

In gasification, a wide range of feedstocks can be utilized to produce fuels, chemicals, and energy. Gasification is a thermochemical process (conversion of solid fuel into liquid or gaseous products upon heating) by which any carbonaceous feedstock can be converted to a gaseous product that has appreciable calorific value and can be used as a fuel. Steps involved in gasification include pyrolysis, partial oxidation, and hydrogenation. Pyrolysis occurs at temperatures around 600 °C, followed by partial oxidation and hydrogenation occur at much higher temperatures (700–1700 °C). Initially, when a solid fuel is subjected to heat, it loses the moisture content through drying. Heating further to higher temperatures can lead to the release of volatile matter present in the solid fuel, leaving behind solid residues like char (mostly carbonaceous) and ash (mostly consisting of inorganic mineral substances, which are non-combustible matter). This process is known as pyrolysis or devolatilization. Further heating in the presence of sub-stoichiometric oxygen leads to partial oxidation of volatiles and char; however, most of the oxygen present is consumed by

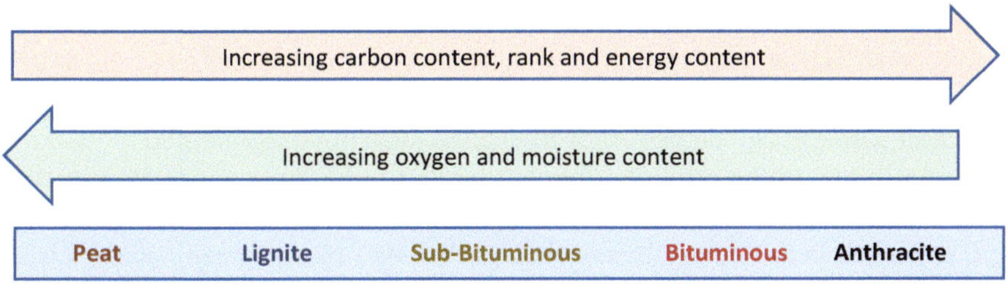

Increasing carbon content, rank and energy content →

← Increasing oxygen and moisture content

	Peat	Lignite	Sub-Bituminous	Bituminous	Anthracite
Burial depth	<0.2km	0.2-1.5km	1.5-2.5km	2.5-6km	>6km
World Reserves		17%	30%	52%	1%
Bulk Density (kg/m^3)		641-865	640-900	673-913	800-929
Moisture (%w/w)		20-60	16-36	2-16	1-16
Fixed Carbon (%w/w)		28-34	42-49	45-47	80-86
Ash (%w/w)		4-5	5-9	3-12	9-20
C (%w/w)		70	75	85	94
H (%w/w)		5	5	5	3
O (%w/w)		25	20	10	3
S (%w/w)		0.4	0.4-0.5	0.4-0.7	0.6-0.8
HHV (MJ/kg)		14.6-19.3	19.3-26.7	26.7-32.5	32.5-36.0
Formation (years)	~10 – 15k	~ 6 crore		10-30 crore	35 crore

Fig. 4.1 Classification of coal on the basis of various parameters

volatiles, with less consumed by char, producing carbon dioxide and water vapor. Later, under reduction conditions, these undergo reduction to produce syngas and small amounts of methane. The amount of hydrogen present in syngas or producer gas can be improved with the use of steam.

Coal gasification can be carried out using a gasifying agent such as air, water, or oxygen, resulting in a synthetic gas composed of CO, H_2, CH_4, etc.

As the coal feedstock is solid, pulverized coal is used in coal gasification systems. Since solid–gas reaction rates can be slow, the particle size required is small to achieve a higher surface area and improve the reaction rate. Thus micron-size coal particles are used. The exothermic oxidation process of coal provides the necessary heat for the subsequent processes. The Boudouard process, which is endothermic, and the water gas shift reaction, being mildly exothermic, enhance the amount of hydrogen in the product gas. The product gas is then cleaned, and particulate matter, dust, or ash is removed. Sulfur needs to be removed so that it does not poison the catalyst used in stage two, which is the water gas shift reaction, thereby enhancing the hydrogen content in outlet product stream. In coal, the sulfur content is typically much higher than in biomass, which presents a challenge and makes it essential to have desulfurization prior to the water gas shift reaction. Later CO_2 and H_2O need to be removed; H_2O can be removed simply by condensation through temperature reduction, while CO_2 can be removed using a solvent-based method.

As described in the previous chapter, similar to biomass gasification, coal gasification can be carried out using various gasification technologies, including fixed bed, entrained bed, floating bed, fluidized bed, and plasma gasification. The fixed, fluidized, and floating bed designs allow for gasification at temperatures < 1200 °C, while the entrained bed and plasma gasification are carried out in the range of 1200–1700 °C. However, the conversion and utilization of coal have significant environmental effects due to its high carbon content, resulting in GHG emissions, predominantly CO_2, and contributing to global warming. Especially in coal-based power plants, where emissions include CO_2, N_2, and SO_2, carbon capture not only adds cost but also leads to energy losses due to differences in combustion temperature and initial temperature of the Rankine cycle, resulting in power plant efficiency lower than 40%. However, when compared with coal combustion, the coal gasification process is much cleaner and cost-effective for producing various chemicals and energy, utilizing coal as the feedstock.

Coal gasification can be used for high-ash and high-moisture content coal; the process is efficient, and the syngas produced has a higher calorific value. Coal gasification is the primary method for hydrogen production in regions where coal is abundant, while natural gas and oil are scarce. However, this method is not economical and is much more polluting than the steam reforming of methane. Syngas produced via coal gasification can be used for combined cycle (IGCC) to improve the overall theoretical power generation efficiency to more than 45%.

Origin of Coal

The origin of coal is basically from biomass. The major components of the biomass include hemicellulose, cellulose, and lignin, as mentioned in the previous chapter. Among these, lignin provides the structural strength to trees and is the primary contributor to the majority of the coal we have today.

Looking back into history, the carboniferous period is credited with producing the bulk of the coal that we have now. At the beginning of the era of evolution, most life originated in the sea, while later it evolved on land. When present in sea, these plants were held upright by the buoyancy force of water and did not fall, remaining floating. When the plants grew on land earlier, they were in the form of shrubs or grasses and did not have lignin. However, later, as the forest cover received ample sunlight, the plants evolved. Over the course of millions of years, the compound lignin evolved to give strength and structural stability to trees, protecting them from heavy winds, animals, and other impacts. Thus, lignin is present in all big plants and trees, which are not being digested by microorganisms. As a result, wood has longer life and does not decompose like leaves, branches, or fruit, etc. Again, with evolution, fungus was found which could digest and decompose lignin, which is found in wood. Thus, these are some of the evolution processes that took millions of years and are known as the carboniferous period. During this period, various microbes and insects were capable of decomposing most of the components, except for lignin. Since lignin was not decomposed or degraded by the microorganisms, the forests accumulated with tree logs and trunks. With earthquake or floods, this lignin component of hard trees and trunks went below the earth's surface. Over thousands of years, being exposed to high pressures and temperatures, it converted into coal. In this manner, depending upon temperature and pressure, and depth to which it was exposed below the earth's surface, coal of different ranks was obtained as mentioned in Fig. 4.1. The process of coal formation is very slow and has taken place over millions of years. The earliest stage, which is approximately 10–15 thousand years old, is peat, which still retains 60% of its organic matter that has not completely decomposed. Peat is, in fact, the plant matter that was buried in the marsh and not completely decomposed, because the rate of accumulation was greater than the rate of decomposition. This peat undergoes coalification process when subjected to high pressure and temperatures during burial. Depending on the burial depth and period, different types of coal are obtained. The longest buried period and depth are for anthracite, which, due to its exposure to the highest temperature and pressures, is the highest rank coal, containing 80–95% of carbon. The moisture, oxygen, and hydrogen content decreases as the burial depth and duration increase from peat to anthracite. Since coal has less hydrogen than biomass, the hydrogen yield from coal is less compared to that from biomass. As the rank of coal increases, the hydrogen content present also decreases. To enhance the hydrogen yield, steam gasification or two-stage gasification with water gas shift reactions can be employed. The process of coal gasification dates back to the nineteenth century, when coal gas was used for street lighting.

Considering the chemical composition of coal, it consists of aromatic clusters with single or multifused aromatic rings, which are connected by aliphatic bridges and have side chains that involve aliphatic groups. These side-chain molecules may or may not be part of the bridges. All this makes the coal structure very complicated, with a very high molecular weight and differing properties depending on the structure.

Coal Gasification for Hydrogen Production

Gasification, in general, refers to the incomplete combustion of solid feedstock, with the aim of converting the non-ash fraction of the feedstock to a higher calorific value gas that can be used for various applications. The ash fraction has a highly variable composition or mixture of minerals that melts over a wide range of temperatures. In a fluidized bed gasifier that operates at 1000 °C, ash slagging or sticky ash formation occurs as the operating temperature is close to the ash softening temperature. While in entrained bed or moving bed gasifiers, as they operate at higher temperatures, ash melts to form a fluid slag. The difference between combustion and gasification lies in the oxygen concentration, which is sub-stoichiometric in gasification. The gases from gasification can be used to produce energy or other products, whereas the gaseous products of combustion cannot be further combusted. Coal gasification is a complex process involving multiple reactions, including pyrolysis, gasification, reforming, cracking, combustion, etc., all occurring within a single reactor. It is challenging to manipulate any specific reaction to achieve higher conversion, lower emissions, better syngas quality, or other desired characteristics.

Steps Involved in the Gasification

The first step in the gasifier is the removal of moisture content, as coal, being porous, can contain sufficient quantities of moisture that need to be evaporated, consuming a significant amount of the combustion heat ($\Delta H_{\text{vap}} = 43.99$ kJ/gmol). The vaporization of water becomes critical when low-grade, high-moisture content coal is used or when coal slurry is used as a feedstock.

Pyrolysis of coal occurs to produce coke and volatiles. When temperatures are higher than 320 °C, various bonds, including C–C and C–O, C–S, C–N, etc., break, yielding unstable compounds that further pyrolyze to form stable compounds. Upon cooling to room temperature, coal pyrolysis yields gaseous products, a liquid fraction (tar), and a solid fraction (soot). The non-volatile fraction can either further undergo cracking to produce lower molecular weight fragments or form char and smaller fragments. In gasifiers where the feedstock and oxidant inlets are located close together, the volatiles undergo combustion or partial combustion to produce products such as CO_2, CO, H_2O, H_2, H_2S and SO_2. For example, in fluidized bed gasifiers, which operate at 1000 °C, very little tar is produced, and more gaseous products are formed; a similar case is observed for entrained flow gasifiers (which operate at 1400–1500 °C). However, in a moving bed gasifier (operating at 700–800 °C), the flow directions are counter-current and the entire oxygen is consumed before it reaches pyrolysis zone, resulting in large quantities of tar formation.

Various possible reactions that can take place in a gasifier are as follows:

Oxidation reactions

$$2C + O_2 \rightarrow 2CO \quad \Delta H^0 = -221.3\ \frac{\text{kJ}}{\text{gmol}} \tag{4.1}$$

$$C + O_2 \rightarrow CO_2 \quad \Delta H^0 = -393.9\ \frac{\text{kJ}}{\text{gmol}} \tag{4.2}$$

$$2CO + O_2 \rightarrow 2CO_2 \quad \Delta H^0 = -566.6\ \frac{\text{kJ}}{\text{gmol}} \tag{4.3}$$

$$2H_2 + O_2 \rightarrow 2H_2O \quad \Delta H^0 = -484.2\ \frac{\text{kJ}}{\text{gmol}} \tag{4.4}$$

$$2CH_4 + O_2 \rightarrow 2CO + 4H_2 \quad \Delta H^0 = -71.4\ \frac{kJ}{gmol} \tag{4.5}$$

Char combustion reactions

$$C + H_2O \rightarrow CO + H_2 \quad \Delta H^0 = +131.5\ \frac{kJ}{gmol} \tag{4.6}$$

$$C + CO_2 \rightarrow 2CO \quad \Delta H^0 = +172.7\ \frac{kJ}{gmol} \tag{4.7}$$

$$C + 2H_2 \rightarrow CH_4 \quad \Delta H^0 = -74.9\ \frac{kJ}{gmol} \tag{4.8}$$

Water gas shift reaction

$$CO + H_2O \leftrightarrow CO_2 + H_2 \quad \Delta H^0 = -41.2\ \text{kJ/mol} \tag{4.9}$$

Steam methane reforming

$$CH_4 + H_2O \leftrightarrow CO + 3H_2 \quad \Delta H^0 = +206.2\ \frac{kJ}{gmol} \tag{4.10}$$

Some of the reactions are endothermic, while others are exothermic; the exothermic reactions provide the required energy to drive endothermic ones. Combustion reactions, being exothermic, provide the required heat of reaction for the gasification reactions; however, the combustion products themselves do not retain any calorific value. In contrast, gasification reactions are endothermic, and the product gases possess a higher calorific value or exist in a higher energy state, than the reactants, with the difference in energy being supplied by the combustion reaction to carry out the non-spontaneous endothermic reactions. Essentially, the feed to a gasifier consists of coal, oxygen, and water; their ratios can be controlled to regulate the various reactions, operating conditions, and thus the product gas composition. However, in addition to the (O_2 + H_2O)/coal ratio, the type of gasifier, feed, and oxidant flow directions, as well as the operating temperatures, and type of oxidant used, are important parameters that determine the syngas composition. If air is used instead of oxygen, the syngas becomes diluted with nitrogen and argon, and additionally, energy is consumed in heating gases to the reaction temperature. Therefore, air separation units are helpful in applications where higher hydrogen or syngas purity is required.

Gasifier Technologies

Various gasification technologies are being studied and demonstrated, which can be categorized into conventional technologies and new technologies (Fig. 4.2). Conventional gasification technologies include fixed bed (updraft or downdraft), entrained flow bed, and fluidized bed gasification, which are technologically quite mature and have reached to large-scale industrial use. Advancements required in the coal gasification include the ability to process different coal types, cost reduction (CAPEX), achieving higher efficiency, and carbon conversion, and improved environmental performance, scale up to larger capacities, effective regulation of pollutants, and energy- and cost-effective separation of NO_x, SO_x, CO_x, as well as utilization of the solid residues generated during the process to enhance overall cost-effectiveness.

The new technologies include plasma gasifier, supercritical gasifier, chemical looping, and decoupling gasification. These recent technologies have advantage of better energy conversion efficiency, producing hydrogen-rich syngas and lower CO_2 emissions. Another technology is underground coal gasification, which enables more efficient conversion and utilization of coal.

The gasifying agent can be O_2, air, CO_2 or steam. Although steam has been identified as producing a hydrogen-rich gas, the use of CO_2 as the gasifying agent offers a dual advantage in terms of resource utilization and emissions reduction. Therefore, the use of CO_2 as gasifying agent has garnered significant interest and is recognized as a method for enabling the clean and efficient utilization of carbon [1]. The syngas produced from coal gasification contains H_2, CO_2, CH_4, H_2S and other gases. This syngas is passed through a heat exchanger to recover the high-grade heat from the product gas and undergoes gas cleaning steps to obtain pure hydrogen for various applications. However, the hydrogen-rich syngas can be used for other applications without separation and purification, such as the production of other fuels and chemicals, including methanol, dimethyl ether, and synthetic fuels using the Fischer–Tropsch process. Syngas produced from coal gasification can also be utilized in combined cycle technology (IGCC), leading to more efficient electricity production.

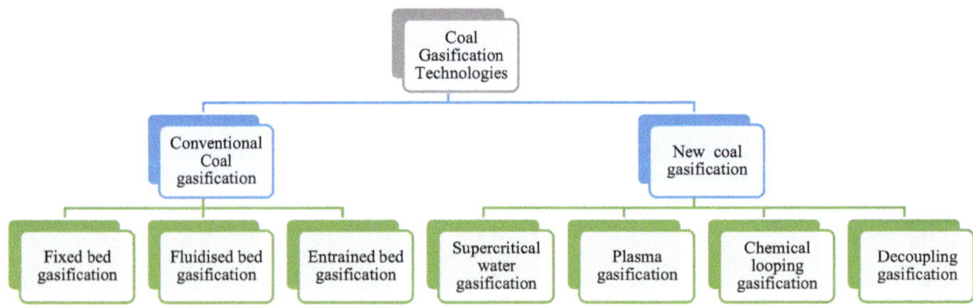

Fig. 4.2 Classification of coal gasification technologies

Conventional Coal Gasification

In the conventional coal gasification technologies, different coal types can be used as feedstock, and there are primarily three types of gasifiers: fixed bed, fluidized bed, and entrained bed gasifiers. These technologies are quite mature, and numerous studies have been conducted on the effects of temperature, pressure, coal type, gasifying agent, gasifier technology, and catalysts on their impact on the final product, i.e., syngas composition.

1. Fixed Bed Gasifier

The fixed bed gasifiers can be of two types depending upon how the feed and oxidant is loaded and the flows.

In an updraft-type fixed bed gasifier, fuel is loaded from the top while the gasifying agent is introduced from the bottom in a counter-current flow manner. The produced syngas is collected from the upper part of the gasifier while the ash is removed from the bottom. The gasification process, as mentioned in the previous chapter, consists of four steps: drying, pyrolysis, reduction, and oxidation. The updraft gasifier is suitable for producing a high yield of tar, where tar is the major product required for specific applications.

In a downdraft gasifier, coal is fed from the top, while the gasifying agent is passed through channels in the middle concurrently with the feed The produced syngas is collected from the bottom of the gasifier. The advantage of this type of gasifier lies in its low tar yield, improved carbon conversion, and relatively clean product gas. The lower tar yield is due to the presence of high-temperature zone in the lower part of the gasifier, which leads to thermal cracking of larger tar molecules.

Examples of fixed bed gasifiers include the Lurgi gasifier and UGI gasifier. The basic requirement of a fixed bed gasifier is a uniform particle size of around 5 cm. These can operate at high temperatures, around 1000–1100 °C. The major disadvantage of these gasifiers is the low coal feeding rates and the restriction on feedstock particle size. The Lurgi gasifier was modified by Sasol and is known as the Lurgi–Sasol gasifier, which can be used with various types of coal. Coal is initially fed into the bunker, which is at atmospheric pressure. The valve at the bottom of bunker allows coal to enter the coal lock. Thereafter, the valve is closed, the region is pressurized, and finally, the valve opens to discharge coal into the gasifier. Coal and oxidant (oxygen and steam) flow in a counter-current direction. Char is gasified at temperatures ranging from 615–760 °C. The hot syngas rises due to a density difference and is cooled by the cold coal above the gasification zone and by quenching water, exiting at the temperatures of 370–590 °C. As the hot syngas rises, it pyrolyzes the coal, producing tar and char. Coal is dried in the upper region of the gasifier. The Lurgi–Sasol gasifier operates at a relatively lower temperature than other gasifiers and therefore produces an appreciable amount of tar. However, the high-temperature gasifiers produce a relatively lower tar content (Fig. 4.3).

2. Fluidized Bed Gasifier

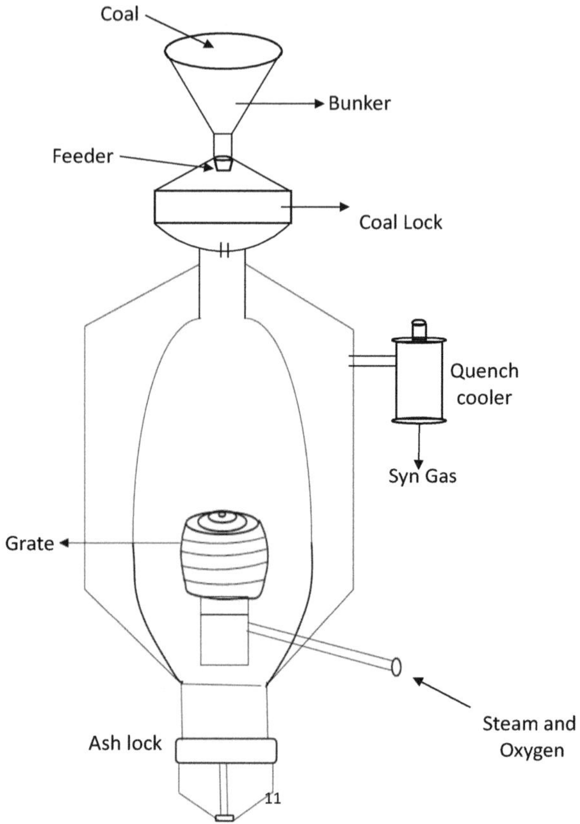

Fig. 4.3 Schematic of a fixed bed coal gasifier

Fluidized bed gasifiers are the most widely used gasifiers, with several design variations that classify them into various types, including bubbling bed, spouted bed, dual bed, and circulating bed gasifiers. Among these, the most commercially developed types of gasifiers are the bubbling and circulating types.

In a fluidized bed gasifier, the fuel is typically loaded from the side, the gasifying agent from the lower part, and the syngas produced is collected from the top. The obtained ash or slag is collected from the bottom of the gasifier. In this type of gasifier, the feed is introduced only when the desired temperature is reached. The coal particles are then mixed with the bed material very quickly, allowing them to reach the bed temperature in a relatively short time span. The coal particles achieve stable fluidization when the weight of the coal particles equals the thrust force of the gas. The operating temperature of a fluidized bed gasifier is usually lower than that of a fixed bed gasifier, as these reactors operate at around 800–1000 °C. However, the primary requirement with such gasifiers is a uniform particle size of around 0.6 mm to ensure uniform temperature throughout the bed via rapid heat and mass transfer across the bed. Some of the representative examples of this technology include Kellogg Rust Westinghouse (KRW), U-gas, Ende, and high-temperature Winkler (HTW).

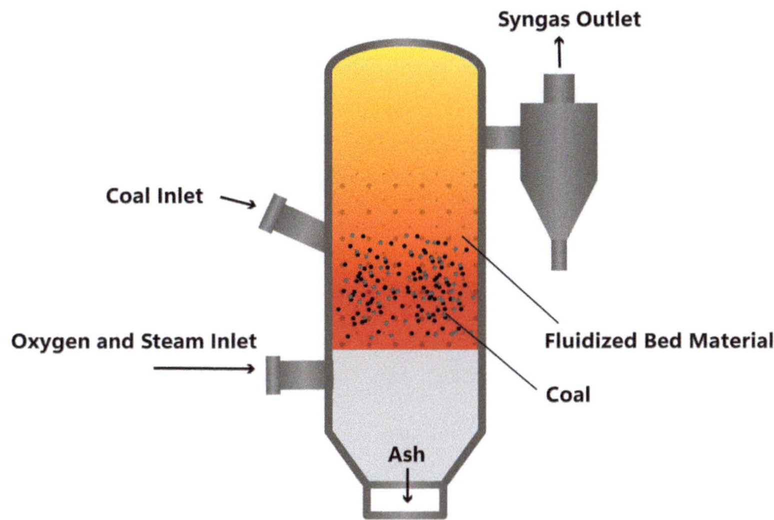

Fig. 4.4 Schematic of a fluidized bed coal gasifier

The Winkler gasifier is the oldest type of bubbling fluidized bed gasifier. In this gasifier, the bed operates at near atmospheric pressure; therefore, a simple technique is used in which coal is first fed into the bunker and then injected into the bed from the feeder. However, the gasification agent, which is steam and air or oxygen, is fed below the grate, fluidizing and making coal particles less dense and smaller. In this gasifier, the major challenge is that a significant quantity of carbon remains unreacted and is discharged with the ash. The gasifier initially operates at around 1000 °C, which is below the ash softening temperature. When the temperature is further increased to 1200 °C due to the reaction of oxygen with the solid feed, the ash partially melts, thereby maximizing carbon conversion. A hydrocyclone, combined with cooling at the top of the gasifier, facilitates the separation of ash for further carbon conversion. However, these gasifiers suffer from lower carbon conversion and are therefore not in much use. This design was further modified in the high-temperature Winkler gasifier, where pressurization of the feed is achieved through a different coal feeding mechanism compared to the conventional Winkler gasifier. In the high-temperature Winkler gasifier, coal is loaded at an atmospheric pressure and then transported through lock hoppers to the gasifier. In this process, there are two lock hoppers. One is depressurized for take charge of coal, and then it closes for pressurization at an operational pressure of around 10 bar. It is then charged to the main gasifier through a filter and a charge bin. Filters separate gases and send them back to the compressor for repressurization.

U-gas gasifier differs from Winkler gasifier in the nature of the grate where instead of rotating grate it's a conical one.

3. Entrained Bed Gasifier

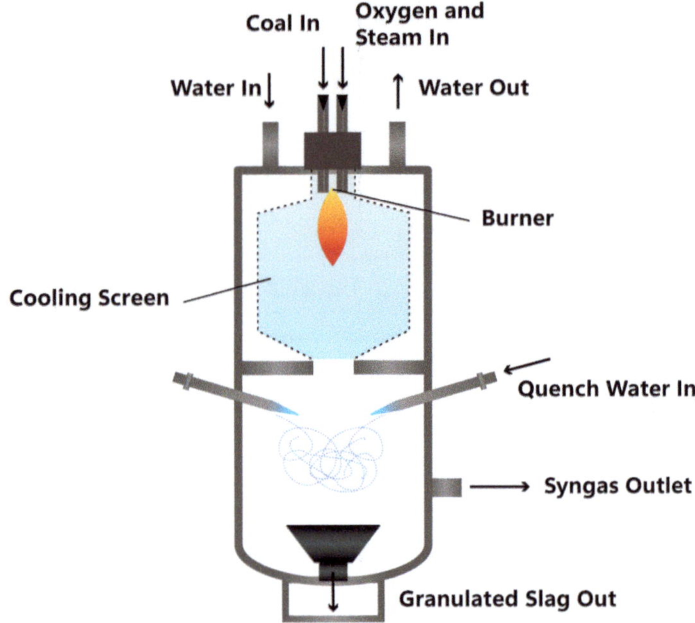

Fig. 4.5 Schematic of an entrained bed coal gasifier

In an entrained bed gasifier, the feed and the gasifying agent are supplied from the top, where coal is fed using a specialized nozzle. The syngas produced leaves the gasifier from the middle, and ash is removed from the bottom of the gasifier. High conversion, along with low residence time, can be achieved in such gasifiers due to the high temperatures. The operating temperature of the gasifier ranges from 1200–1700 °C, resulting in higher carbon conversion. The major restriction is on the particle size, which must be around 0.15 mm. Typical examples of entrained flow gasifiers include those developed by Siemens, Shell, and Texaco. The most significant advantage of such gasifiers is that hydrogen-rich syngas is produced, with very little solid residue generated due to improved conversion at high temperatures. Texaco designed the gasifier, which received coal slurry and oxygen as inputs at the top and produces gas, slag, and some solid products collected from bottom. It employs different heat recovery methods, either through water quenching followed by heat exchange to obtain cooled gas and slag, or through a radiant cooling mechanism using various steel tubes carrying steam.

The Shell gasifier uses dry, ground coal injected through the sides near bottom of the gasifier, with oxygen fed below these points and slag removed from the bottom. The temperature is controlled by introducing quench gas to maintain it below the tolerance level of the gasifier.

Studies on conventional gasification technologies have reported that, depending on the temperature, the product gas composition can exhibit wide variability:

Table 4.1 Comparison of different conventional gasification technologies, temperature dependence on the product gas composition

S. No.	Gasification technology	Temperature (°C)	H_2 (%)	CO (%)	CO_2 (%)
1	Fixed bed gasifier	500–1000	2.97–67	2–23.9	4.4–33.7
2	Fluidized bed gasifier	700–1400	4.5–59	2–43	4.2–45.1
3	Entrained bed gasifier	827–2249	1–32.5	5–65	0.8–80

However, considering the coal type and more specific temperature ranges, the product gas compositions are summarized in Table 4.1.

The major challenges associated with conventional coal technologies relate to efficiency, emissions, scaling up of gasifiers, and management of the solid residues obtained. Consequently, there is significant interest in newer technologies to improve efficiency, achieve higher conversion to hydrogen-rich syngas, enable cleaner coal conversion, scale gasifiers to larger capacities, and eliminate pollutants, including CO_x, NO_x, and SO_x, while utilizing the solid residues generated during gasification. Collectively, these objectives aim to achieve cleaner coal conversion, higher efficiency, upscaling gasifier plants to larger capacities, broader acceptance toward a diverse coal types, and reduced CAPEX and OPEX.

Modern Coal Gasification Technologies

1. Supercritical Water Gasification

In supercritical water gasifiers, supercritical water is used as the gasifying agent, providing a rapid and uniform reaction medium for improved and cleaner carbon conversion, without the NO_x and SO_x emissions. This is achieved due to low density, low dynamic viscosity, and higher diffusion of supercritical water, along with weaker hydrogen bonding [1]. Supercritical water gasification (SCW) is performed at high pressures, ranging from 24–29.7 MPa, and at lower temperatures, 500–850 °C. Thus, the advantages of this technology are that gasification is carried out at a much lower temperature than other gasification technologies. Various contaminants present in coal, including sulfur, nitrogen, phosphorous, and alkali metals are converted to inorganic salts, which can be easily removed, resulting in almost negligible SO_x and NO_x content. Additionally, the use of supercritical water yields higher carbon conversion and lower solid residues, including tar and coke. The product gas is rich in hydrogen, as hydrogen is derived not only from coal but also from water, and CO_2 emissions are reduced in SCWG. Different parameters that influence the degree of SCWG and the composition of the product gas are temperature, type of coal, coal-to-steam ratio, size of coal particles, and residence time. Among these, temperature is the primary parameter that influences the most; an increase in temperature results in increased hydrogen yield and provides better coal conversion. The use of catalysts like K_2CO_3 has been shown to result in faster kinetics by creating mesoporous structure with a higher specific surface area and pore volume. Additionally, it has been demonstrated that the use of an alkali salt catalyst can lower the required gasification temperature. The addition of alkaline catalysts, such as K_2CO_3, KOH, NaOH, Na_2CO_3, etc., results in an increased hydrogen yield. Furthermore, the increase in temperature, the increase in catalyst loading, the decrease in the feedstock concentration, and the optimum flow rate of water all increase the yield of hydrogen in the product gas. The impact of an increase in coal concentration in the slurry is a decrease in the hydrogen yield in the syngas composition. Designing the SCWG requires optimizing operational parameters, and for this, a better understanding of the reaction mechanisms and kinetics is important. The advantages of SCWG over the other conventional methods of coal gasification include a higher hydrogen yield (often exceeding 50% in concentration), lower CO_2 release (typically in the range of 22–50%), and lower NO_x and SO_x emissions, as well as higher efficiency. At the same time, SCWG has many disadvantages. Since it occurs at low temperatures (500–850 °C), complete coal conversion may not occur, leading to reactor blockage. Designing a catalyst that can operate at the mentioned temperature is important. Corrosion of the reactor needs to be addressed through proper design, as well as the selection of a reactor material that can resist harsh conditions and provide longer equipment life. The SCWG mechanisms are still not well understood; thus, a proper design of the reactor with higher efficiency, better conversion, high yield, and longer life will require a better understanding of reaction mechanisms and kinetics, optimization of process parameters, and catalyst development. Currently, the demonstrations are on pilot scales; however, to scale up, several issues need to be resolved, including better design, optimizations, and system integration. At the same time, studies on the economic and environmental feasibility of this approach are needed.

2. **Plasma Gasification**

Plasma gasification is a relatively new technology that is efficient and produces fewer residues, providing a higher yield of hydrogen. In the plasma gasification technique, a plasma torch is used to decompose the raw material at very high temperatures, ensuring higher conversion efficiency and releasing products, such as syngas and slag. The plasma gasification process offers several advantages over conventional

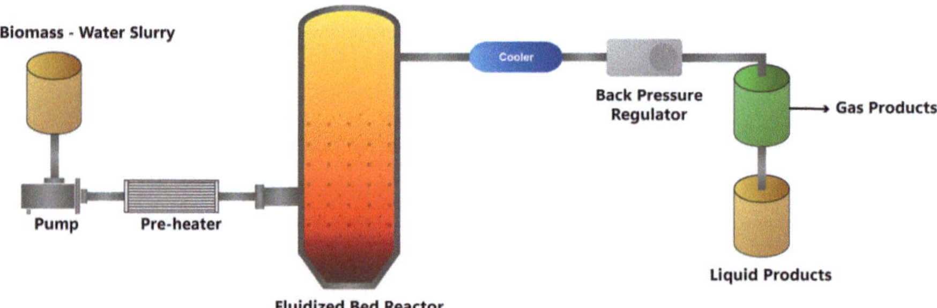

Fig. 4.6 Schematic of a Supercritical gasifier

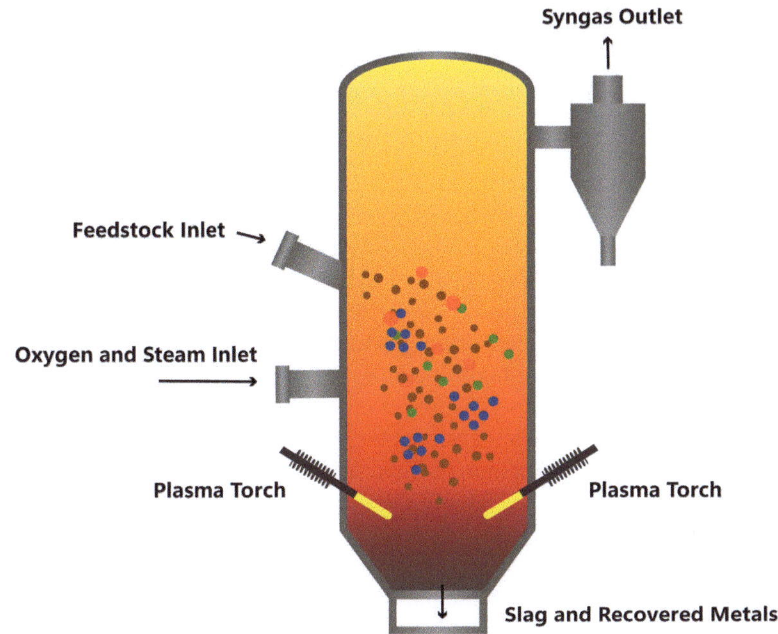

Fig. 4.7 Schematic of a plasma gasifier

technologies. It is more efficient and environmental friendly, has a better carbon conversion rate with reduced tar production, and enables the recovery of waste metals. The toxic residues of gasification can be converted into useful materials, such as glassy materials or slag, which can be utilized as construction materials. Additionally, plasma gasification yields a higher hydrogen content (0.1–0.17 kg_{H_2}/kg_{coal}) in the product gas, resulting in an overall cleaner process.

During plasma gasification, the required high temperature for gasification is achieved using a DC plasma torch. The inner wall of the gasifier is ceramic-coated to resist the thermal degradation at such high temperatures. Depending on the gasifying agent used—air, oxygen, carbon dioxide, or steam—the process is called as air plasma gasification, oxygen plasma gasification, carbon dioxide plasma gasification or steam plasma gasification. When air is used as gasifying agent, it is supplied under pressure. In contrast, when steam is used as a gasifying agent, it is fed at temperatures above 250 °C. Plasma consists of ions, radicals, and energetic electrons. The feedstock, i.e., coal is introduced into the gasifier from the top at a location close to the plasma torch. In gasification, primarily syngas and a small quantity of slag are obtained. The produced syngas is passed through a heat exchanger, where heat is recovered from the hot syngas and can be used for steam generation. After heat recovery, the syngas is cleaned to obtain pure hydrogen using gas separation and purification methods such as pressure swing adsorption, membrane separation, or cryogenic separation. In the plasma gasification process, the typical equivalence ratio is maintained below 0.32, operating temperatures exceed 1400 °C, and catalysts such as Na_2CO_3, Li_2CO_3 and K_2CO_3 may be used; however, the overall gasification efficiency is less than 57%. Literature reports show a considerably higher hydrogen gas concentration in syngas when steam is used as the gasifying agent (up to 60%), compared with 20% when air is used. With thermal steam plasma, the hydrogen content was reportedly even higher than 80%. Simulations have shown that the quality of syngas, reaction rates, and gasification temperature depend upon the plasma used. Various studies also report the influence of coal type, coal feeding rate and temperature on the carbon conversion, hydrogen yield, and efficiency. However, most studies on plasma gasification have been conducted at the laboratory scale, and large-scale implementations are not yet available. Due to its advantages—such as complete coal conversion at high operating temperature (1200–2500 °C), higher syngas yields (30–96%), increased hydrogen content in syngas (9–57%), and lower of CO_2 emissions (< 18%)—plasma gasification is an efficient, cleaner, and promising technology for hydrogen production. In addition to the mentioned advantages, plasma gasification offers several other benefits, including reduced tar formation and the conversion of organic waste into value-added products, etc. However, despite these advantages, challenges are associated with syngas cooling, including high energy demand and a shorter lifetime for the plasma electrodes. The syngas produced is at very high temperatures, typically on the order of 1400 °C or higher; consequently, significant technical and materials challenges exist in designing heat exchangers to extract the high-grade heat. In the case of thermal plasma systems, electrode erosion is substantial, and the associated energy requirements are high. Thus, there is a trade-off in terms of coal conversion, hydrogen yield versus energy demand, running costs, investment, as well as the cost associated with post-processing the product gas to obtain hydrogen.

3. **Chemical Looping Gasification**

The process of chemical looping gasification (CLG) has its origin in chemical looping combustion. In this process, coal is converted into hydrogen-rich syngas through partial oxidation using an oxygen carrier, resulting in a higher calorific value, high-purity hydrogen, and lower tar content. It is a type of metal oxide-mediated hydrogen generation mechanism where a reduction–oxidation reaction occurs, in which the metal oxide in its higher valence state is reduced to its lower valence state, releasing syngas. In the regeneration process, the metal oxide regains its higher oxidation state via air oxidation. The process involves two reactors: a fuel reactor, where the reduction reaction of metal oxide occurs, and an air reactor, where it is regenerated back to its initial state. In the fuel reactor, coal reacts with metal oxide in a higher valence state, producing syngas and metal oxide in a lower oxidation state, as well as other hydrocarbons via reaction:

$$\text{Coal} + x\text{M}_p\text{O}_q \rightarrow x\text{M}_p\text{O}_{q-1} + \text{CO} + \text{CO}_2 + \text{H}_2 + \text{C}_m\text{H}_n. \tag{4.11}$$

In the air reactor, the reduced form of the metal oxide is oxidized by air to its higher valence state. This reaction is exothermic, providing the necessary heat for the first reaction.

$$M_pO_{q-1} + \frac{1}{2}O_2 \rightarrow M_pO_q \tag{4.12}$$

The advantages of chemical looping gasification include a higher hydrogen yield, the elimination of separation step as the two reactions occur in separate reactors, lower tar generation, improved coal conversion, and reduced emissions due to the reducing environment in the fuel reactor, resulting in lower of NO_x and SO_x formation Numerous studies have been conducted and published in the literature on syngas production using chemical looping gasification with various catalysts, under different operating conditions, and employing different oxygen carriers, with a focus on the reaction mechanism of the process. The major role in CLG is played by the oxygen carrier, which must have high activity and a long lifespan. Among the various carriers, Fe_2O_3 has been widely studied. Wherein, the transition between the Wüstite and haematite states results in the production of syngas and regeneration of the metal oxide. Other mixed oxides were studied, including $CaFe_2O_5$, $CeFeO_3$, $COFe_2O_4$, $NiFe_2O_4$, etc. The use of CaO has been found to enhance syngas production when combined with aforementioned oxides. Bimetallic oxygen carriers, including Fe–X (X = Cu, Co, Ni, Ba, Ca, etc.), were studied, and it was found that Fe–Ni oxides on Al_2O_3 exhibit excellent thermal stability, selectivity toward syngas generation, and improved coal conversion. Various processes are used for the synthesis of oxygen carriers, including sol–gel, wet impregnation, etc., which are employed in different reactors, such as fixed bed or fluidized reactors, and exhibit distinct characteristics in CLG. The use of CuO with haematite increases the gasification rate and improves the syngas generation. Various modeling tools have been utilized to optimize the process parameters, including gasification temperature, C/O ratio, flow rates, etc., thereby enhancing gasification efficiency, optimizing reactor geometry, reducing emissions, and increasing syngas production.

The fuel and air reactors typically operate at temperatures between 700 and 1100 °C, and the syngas yield ranges from 57–89%, with hydrogen contents of 16–48%, and CO_2 contents 9–48%, depending on various operating factors. The CLG technique offers coal flexibility, allowing it to operate with various types of coal and produce hydrogen-rich syngas. However, most studies are currently limited small-scale demonstrations and pilots systems, and several challenges still remain to be addressed. The major challenges include: (1) appropriate selection of oxygen carriers with high stability, long life, high activity, and selectivity, which are resistant to poisoning under reaction conditions involving high-ash content, solid and toxic residues, and harsh operating conditions; (2) optimized design of the gasifier, although various gasifiers studied including fixed bed, fluidized bed, moving bed, etc., but fluidized bed has been reported to be the best performing but still have issues with gas crossovers, circulation of oxygen carriers between the fuel and air reactors, long-term stability, and performance controls, etc., and (3) heat transfer and efficient heat utilization between the two reactors, i.e., heat transfer from the oxidation reaction to the regeneration unit.

4. **Decoupling Gasification**

Decoupling gasification is a relatively new technique that involves the separate handling of different types of coal. The process involves the pyrolysis of coal to produce high-value volatiles and char at temperatures ranging from 550–800 °C. The volatiles are cooled and separated, resulting in a high hydrogen content pyrolysis gas. The coal tar obtained can be further upgraded to fuel oils through hydrogenation, or different high-value chemicals can be extracted from it, including naphthalene, anthracene, phenols, etc. Coal char can also be gasified to obtain syngas. This technique has several advantages, including near complete conversion of coal, high efficiency, improved economy, and the production of high-quality hydrogen-rich syngas, allowing a variety of coal types to be used. A variety of products can be obtained through pyrolysis gas, char gasification, and tar hydrogenation, including hydrogen, fuel oils, phenol, methanol, etc. Different simulation and experimental studies have been conducted to understand the thermodynamics and economics of the process. Two-step and three-step methods have been studied and reported in the literature. In the two-step process, coal is first pyrolyzed, and the tar and pyrolysis gas obtained is separated in the next step. The tar obtained is employed to produce fuel oil by hydrogenation; the char obtained is gasified to produce hydrogen-rich syngas. The pyrolysis gas and syngas produced can be used for various applications. In the three-step process, coal is pyrolyzed to produce hydrogen-rich pyrolysis gas at temperatures ranging from 800–1100 °C. The coal char is gasified with CO_2 (steam, oxygen, or air can also be used as a gasifying agent) to produce CO, which then undergoes the water gas shift reaction to produce CO_2 and H_2. The heat required for pyrolysis and gasification is provided by coal combustion. Since, in this process, CO and H_2 are produced separately, instead of syngas, a decoupling of C and H takes place, and thus it is named as decoupling gasification. If steam is used as the gasifying agent, a hydrogen-rich gas is produced directly. In contrast, when CO_2 is used as the gasifying agent, a high CO content gas is produced via the Boudouard reaction, which is subsequently converted into a hydrogen-rich gas through water gas shift reaction. However, the complex coal pyrolysis reaction is not well understood and requires further study using simulation tools. During coal char gasification, since char is less reactive than coal, there are challenges that require special techniques, such as catalytic or microwave gasification. A lot of optimization is required for operational conditions and parameters to enhance the char gasification process. Besides, in coal pyrolysis, the design of efficient reactors, the removal of solid and liquid products, including oil and dust, and the reduction of pollutants, etc., need to be addressed. Above all, the integration of pyrolysis and gasification to achieve improved polygeneration with higher energy efficiency and better economic viability needs to be studied [1].

5. **Underground Coal Gasification**

Another economical and simple method of gasification is underground coal gasification, in which coal is gasified in situ by providing reactant gases directly at its geological location, rather than being mined, processed, transported to the surface,

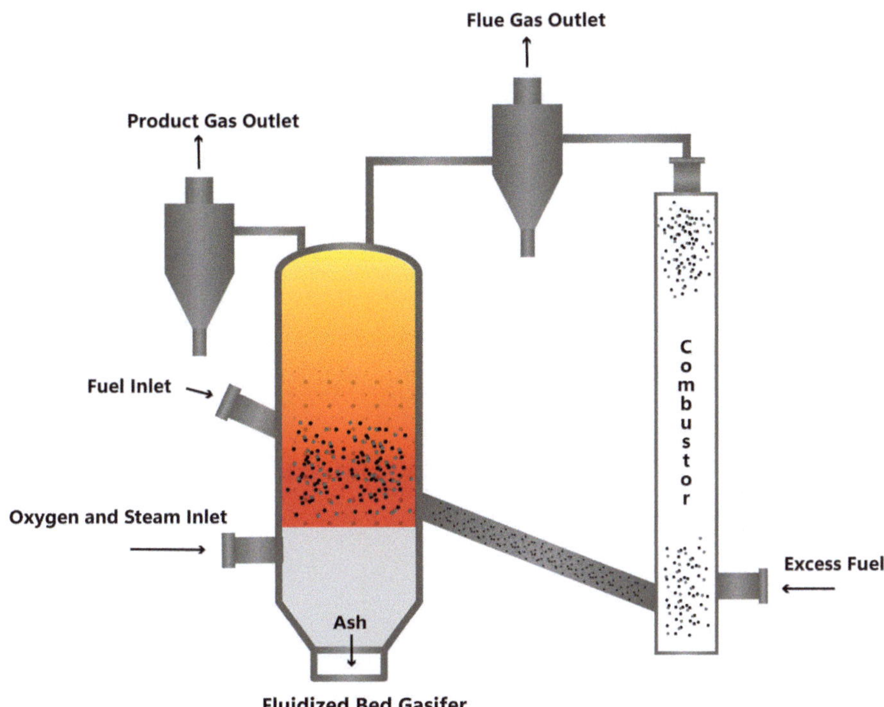

Fig. 4.8 Schematic of a decoupling gasifier

and then gasified. Deep coal seams that are not economically viable for mining can be utilized for underground coal gasification. In this method, the ground itself acts as a support for the reactor. Inside the earth, a reactor is created by excavating the coal seam and supplying the gasifying agent to produce the product gas; alternatively, the mine itself acts as the reactor. The idea is that once we excavate coal, cavities are created underneath the ground, and these cavities can act as a reactor with an ample amount of coal and a high surface area available for gasification. Whatever surface which is exposed will be able to participate in the gasification process and act as a reaction site. An injection well is used to supply high-temperature steam as well as a limited amount of oxygen and to collect product syngas. The syngas can be cleaned with sulfur removal followed by a water gas shift reaction, thereby enhancing its hydrogen content. The construction cost of a normal coal gasifier is much higher than the cost of coal; as such, this technique has the advantage that the cost involved in drilling air injection and syngas removal wells is substantially less than installing a gasifier of such capacity.

In case of underground coal gasification, there are vast reaction sites underground, so that the reaction must be carried out in a highly controlled manner. The technology is used in Russia and China. India has vast coal reserves, but the quality is not suitable for use in the steel industry due its high-ash content. However, such coals can be

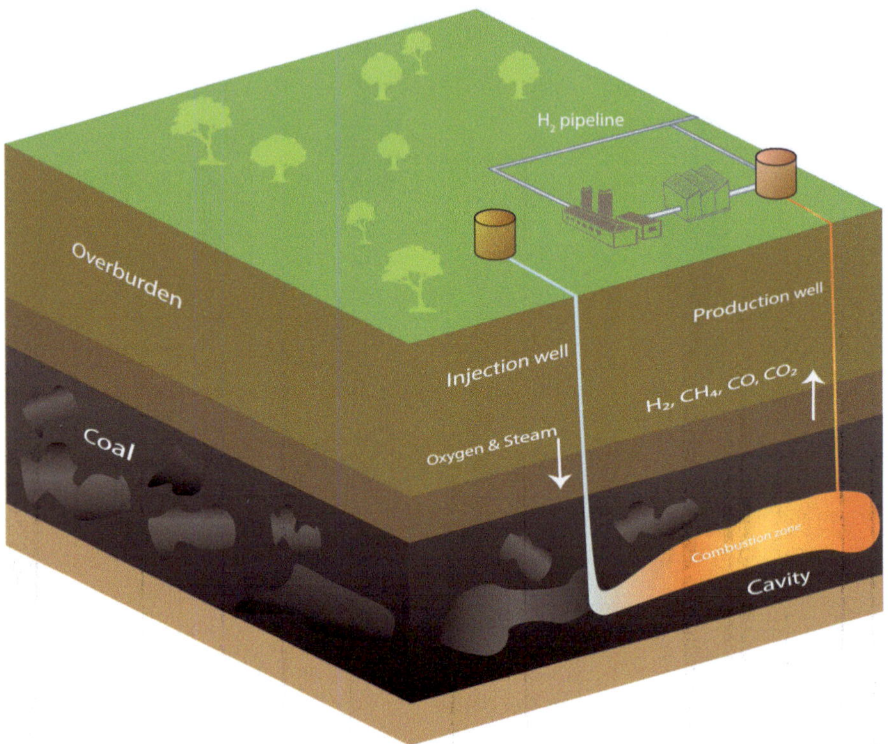

Fig. 4.9 Diagram depicting underground coal gasification technology

utilized in gasification processes, and underground coal gasification offers a route to produce low-cost syngas, which can be obtained, allowing the plant to operate for an extended period. The syngas produced can be converted into methanol, or liquid fuels, or purified to obtain hydrogen.

Numerous modeling studies, coupled with experimentally obtained measurements of parameters such as temperature, pressure, gas flow rate, and gas composition at injection and recovery wells, have been conducted to understand the dynamics of the constantly changing underground gasification cavity. These studies aim to improve the design and operation of underground coal gasification.

After Treatment Processes

The inorganic component of the gasifier feedstock ultimately exits as ash or molten slag, which must be removed from the gasifier; however, when gasification is carried out at low temperatures, tar formation occurs.

Since coal has a higher sulfur content than other hydrocarbon feedstocks, it is present in the form of organic compounds and inorganic pyrites. During coal gasification, sulfur in the syngas is converted to H_2S under the reducing environment, which must be removed to prevent catalyst poisoning in downstream applications and to avoid SO_x emissions. H_2S and CO_2 are referred to as acid gases and are typically removed from syngas using a solvent-based removal method. If small quantities of COS are present, they are hydrolyzed to convert to H_2S. This reaction occurs as a side reaction in the sour gas shift reactor, and a separate reactor is not required when sulfur-tolerant catalysts are used. The syngas is subsequently cooled using quench water. This water, after becoming contaminated with various impurities, turns black and is referred to as black water, which can be either recycled for quenching or used in the preparation of coal slurry. However, if solid particles are not present in syngas, the resulting wastewater is referred to as called sour water, as it contains dissolved acid gases, such as H_2S and NH_3. Sour water can be treated in a sour water stripper to obtain treated water. The Claus process can be used to recover elemental sulfur, or the difference in affinity of H_2S and NH_3 can be used to separate the two (Table 4.2).

The conventional method for removing acid gases (H_2S and CO_2) is a solvent-based removal process, in which an absorption tower is used for separation. The syngas is fed from the bottom, and the solvent is fed from top. After separation, the purified syngas exits from the top, and the solvent containing dissolved acid gases leaves from the bottom. The solvent regeneration is achieved by depressurizing and heating to remove the absorbed gases and recover the solvent. Depending on the nature of the solvents, the process may involve either physical absorption or a chemical reaction-based removal. The difference between these two mechanisms lies in the affinity of the acid gases toward the solvent. When the affinity is weak, and stronger bonds are not formed, absorption is slower, but solvent regeneration requires less energy. In contrast, chemical solvents exhibit a stronger affinity and bonding with acid gases, resulting in faster absorption; however, higher energy is required for regeneration. The well-known and the oldest acid gas removal method, called the Rectisol process, is based on a physical solvent and has been widely used in the Lurgi gasifiers, where, in addition to acid gases, small amounts of lower hydrocarbons are present due to lower operating temperatures. The solvent used in the Rectisol process is methanol, which is operated at – 25 °C. The separation of syngas components is based on their relative solubility in methanol. Another well-known physical solvent used to separate acid gases is dimethyl ether, derived from polyethylene glycol oligomers, and the process is called the Selexol process. The method is based on relative solubility, with H_2S having a higher affinity for the solvent than CO_2; therefore, H_2S is absorbed first, and the gas stream is subsequently sent to the lean Selexol to remove CO_2. In addition to these physical solvents, there are chemical solvents that can remove acid gases. The well-known solvents in this category are organic amines, such as monoethanol amine, which operate via acid base reactions.

Solved Problem 1

Table 4.2 Comparison of all hydrogen production technologies

Technologies	Advantages	Disadvantages	Efficiency
SMR	Most viable, mature and widely used method, currently lowest cost, infrastructure exists, can operate continuously year around, thus better for providing baseload with high load factors (around 95%), no oxygen requirement, lower operating temperature than ATR and POX, high H_2/CO ratio	External heating required, catalyst deactivation problem, high CAPEX and OPEX, high maintenance costs, requirement of low-cost and efficient purification technology, crucial is design of the plant and feedstock pretreatment optimization, reliability, cannot be turned on and off readily takes a long time, CO_2 emissions, so SMR + CCUS	74–85%
POX	Proven technology, reduced requirement of pretreatment, fuel flexibility, lower content of unreacted hydrocarbons, compact design, easy to start and stop, can upgrade dirtier feedstocks	Low H_2/CO ratio, high operating temperatures, complex handling process due to hydrogen oxidants and very high temperatures, requirement of ASU, in catalytic partial oxidation problem of catalyst deactivation	60–75%
ATR	Lower operating temperature than POX and lower content of unreacted hydrocarbons, more compact than SMR, lower oxygen requirement than POX and lower steam requirement than SMR	Limited commercial units and experience, requirement of oxygen and steam	60–75%
Coal gasification	Syngas produced at a lower cost, abundant and cheap feedstock available	Reactor costs high, lower system efficiency, wide variety of feedstock and impurities present, tolerance toward impurities, requirement of low-cost and efficient purification system, hydrogen purity obtained, fluctuating yield of hydrogen due to presence of impurities in feedstock, tar formation	35–55%
Methane decomposition or pyrolysis	No emissions, fewer steps involved, value-added exotic carbon products, easier separation and purification	Deactivation of catalysts, and use of carbon products as the present demand is not high	
Biomass pyrolysis	Syngas produced at lower cost, abundant and cheap feedstock available, carbon neutral	Tar formation, availability of feedstock of all types, variation in hydrogen produced due to impurities present in the feedstock	35–50%

(continued)

Table 4.2 (continued)

Technologies	Advantages	Disadvantages	Efficiency
Biomass gasification	Syngas is produced at a lower cost, abundant and cheap feedstock available, carbon neutral	Reactor costs high, lower system efficiency, wide variety of feedstock and impurities present, tolerance toward impurities, requirement of low-cost and efficient purification system, hydrogen purity obtained, fluctuating yield of hydrogen due to presence of impurities in feedstock, tar formation, continuous availability of feedstock	30–40%
Dark fermentation	Simple, no requirement of sunlight, CO_2 neutral, can be used for waste recycling	Larger reactor required, need for elimination of fatty acids, low hydrogen yield and rate of production, and getting high capacity	60–80%
Photo fermentation	CO_2 neutral, can be used for waste recycling, can use a variety of wastes	Large reactor required, poor efficiency and low rate of hydrogen production, requirement of sunlight, lower yield	0.1–10%
Thermochemical cycles	Waste heat/nuclear/solar thermal to provide the required energy, scalable, can allow large-scale hydrogen production, robust, low-cost materials requirements, reliable, clean, and sustainable method, and chemicals are recycled	Corrosion and toxicity of materials, requirement of high temperatures, requirement to design a low-cost reactors and to design low-cost, high-volume, flexible system with efficient heat transfer	Nuclear: 45% Solar 17–20%
Electrolysis	Mature technology, no emissions if renewable power is used, infrastructure exists, high-purity hydrogen and oxygen products	Scaling up, high capital cost, reducing cost, use of less expensive materials, integration, storage, and transportation problems, cost-effective membranes and catalysts development	60–80%
Photo electrolysis	Clean and sustainable method, abundant feedstock, oxygen and hydrogen products, high purity	Corrosion problems, expensive and toxic materials, high capital costs, low efficiency, lack of materials for efficient conversion, losses in the process, and better control over the process	0.06%

(continued)

Table 4.2 (continued)

Technologies	Advantages	Disadvantages	Efficiency
Photolysis	No emissions, abundant feedstock produces both O_2 and H_2, high purity	Low efficiency, lack of materials for efficient splitting	6.00%
Biophotolysis	Consumption of CO_2, mild operating conditions, pure oxygen, and hydrogen produced	Lower yield of hydrogen, larger area reactor required, cost is high, low rate of hydrogen production	10–11%

Calculate the empirical formula of the compounds (coal, and printed circuit board (PCB)) having the following percentage composition as shown in table:

Proximate analysis, ultimate analysis, and calorific value of the fuels.

Sample	Proximate analysis (%)				Ultimate analysis (%)				
	M	VM	FC	Ash	C	H	N	O	S
PCB	0.1	28.1	2.5	69.3	18.2	1.6	0.5	10.2	0.2
Coal	11.2	34.7	52.0	2.1	78.3	5.5	2.4	12.1	1.7

Where *M* Moisture content, *VM* Volatile matter, *FC* Fixed carbon, *C* Carbon, *H* Hydrogen, *N* Nitrogen, *O* Oxygen, *S* Sulfur

Solution

Step 1: Understanding the Data

The ultimate analysis provides the percentage composition of elements (C, H, N, O, and S) in the samples. The molecular formula requires the molar mass, which is not directly given, so we will calculate the empirical formula first and assume the molecular formula is a multiple of the empirical formula.

Step 2: Calculate Moles of Each Element

Assume 100 g of sample for simplicity, so percentages are equivalent to grams. For each sample, convert the percentage of each element to moles by dividing by the atomic mass of the element:

- C: 12 g/mol.
- H: 1 g/mol.
- N: 14 g/mol.
- O: 16 g/mol.
- S: 32 g/mol.

Elemental composition	For coal (mol.)	For PCB (mol.)
C	78.3 g/12 = 6.525	18.2 g/12 = 1.517
H	5.5 g/1 = 5.5	1.6 g/1 = 1.6
N	2.4 g/14 = 0.171	0.5 g/14 = 0.036
O	12.1 g/16 = 0.756	10.2 g/16 = 0.638
S	1.7 g/32 = 0.053	0.2 g/32 = 0.006

Step 3: Determine the Simplest Mole Ratio (Empirical Formula)

Divide each mole value by the smallest number of moles to get the simplest whole-number ratio.

For Coal

Smallest moles = 0.053 (S)

- C: $6.525/0.053 \approx 123.1$
- H: $5.5/0.053 \approx 103.8$
- N: $0.171/0.053 \approx 3.2$
- O: $0.756/0.053 \approx 14.3$
- S: $0.053/0.053 = 1$.

Based on the mole ratio, the empirical formula is $C_{123}H_{104}N_3O_{14}S_1$ (for whole numbers based on original moles scaled).

For PCB

Smallest moles = 0.006 (S)

- C: $1.517/0.006 \approx 252.8$
- H: $1.6/0.006 \approx 266.7$
- N: $0.036/0.006 = 6$
- O: $0.638/0.006 \approx 106.3$
- S: $0.006/0.006 = 1$.

Adjust by dividing by 0.006:

Based on the mole ratio, the empirical formula is $C_{253}H_{267}N_6O_{106}S_1$ (approximate whole numbers).

Step 4: Empirical Formula.

The empirical formula of a coal and PCB is

- **Coal**: $C_{123}H_{104}N_3O_{14}S_1$
- **PCB**: $C_{253}H_{267}N_6O_{106}S_1$.

Reference of Proximate and Ultimate Data

Rai, C. and Bhui, B., 2022. Techno-economic analysis of e-waste based chemical looping reformer as hydrogen generator with co-generation of metals, electricity and syngas. International Journal of Hydrogen Energy, 47(21), pp. 11177–11189.

Solved Problem 2

A sample of coal has the following mass percentages by weight:

- Carbon (C) = 75%
- Hydrogen (H) = 5%
- Oxygen (O) = 10%
- Sulfur (S) = 1%
- Nitrogen (N) = 2%
- Ash = 7%.

1. Find the empirical formula of the coal (excluding ash).
2. The coal is gasified with steam and oxygen to produce the following dry product gas composition by volume:

   $CO = 40\%$
   $H_2 = 35\%$
   $CO_2 = 15\%$
   $CH_4 = 5\%$
   $N_2 = 5\%$.

Assuming 1 kg of coal is used, calculate the total volume of dry product gas at STP (0 °C, 1 atm).

(Molar volume at STP = 22.4 L/mol).

Solution

Step 1: Exclude ash and normalize the composition

Ash is non-combustible. So we will only consider the combustible portion:

$$\text{Total} = 100\% - 7\% = 93\%.$$

Adjusted mass percentages:

C: (75/93) × 100 = 80.65%
H: (5/93) × 100 = 5.38%
O: (10/93) × 100 = 10.75%
S: (1/93) × 100 = 1.08%
N: (2/93) × 100 = 2.15%.

Step 2: Convert to moles (per 100 g of sample)

C: 80.65/12 = 6.72 mol
H: 5.38/1 = 5.38 mol
O: 10.75/16 = 0.67 mol
S: 1.08/32 = 0.034 mol
N: 2.15/14 = 0.154 mol.

Step 3: Divide all by smallest number of moles

Smallest is 0.034 (S), so we divide each by 0.034:

C: 6.72/0.034 ≈ 198
H: 5.38/0.034 ≈ 158
O: 0.67/0.034 ≈ 20
S: 0.034/0.034 = 1
N: 0.154/0.034 ≈ 5.

Step 4: Final empirical formula

Empirical Formula: $C_{198}H_{158}O_{20}N_{55}S$

Step 5: Calculate molar mass of coal (approx.)

Molar mass = (198 × 12) + (158 × 1) + (20 × 16) + (5 × 14) + (1 × 32) = 2956 g/mol.

Step 6: Gas yield from gasification

Given 1 kg (1000 g) of coal, the organic mass is:

$$= 0.93 \times 1000 = 930 \text{ g}.$$

Ash (70 g) does not contribute to the gas. Calculate moles of each element in 930 g:

- C: 0.75 × 930 = 697.5 g, $\frac{697.5}{12} = 58.12$ mol
- H: 0.05 × 930 = 46.5 g, $\frac{46.5}{1} = 46.5$ mol
- O: 0.10 × 930 = 93 g, $\frac{93}{16} = 5.81$ mol
- S: 0.01 × 930 = 9.3 g, $\frac{9.3}{32} = 0.29$ mol
- N: 0.02 × 930 = 18.6 g, $\frac{18.6}{14} = 1.33$ mol.

The dry product gas composition by volume (mole fractions) is:

- CO: 40% (0.40)
- H_2: 35% (0.35)
- CO_2: 15% (0.15)
- CH_4: 5% (0.05)
- N_2: 5% (0.05).

Let n_{total} be the total moles of dry product gas. Moles of each gas:

- $n_{\text{CO}} = 0.40n_{\text{total}}$
- $n_{\text{H}_2} = 0.35n_{\text{total}}$
- $n_{\text{CO}_2} = 0.15n_{\text{total}}$
- $n_{\text{CH}_4} = 0.05n_{\text{total}}$
- $n_{\text{N}_2} = 0.05n_{\text{total}}$.

Elemental Balances

Carbon (in CO, CO_2, CH_4):

$$
\begin{aligned}
&n_{\text{CO}} + n_{\text{CO}_2} + n_{\text{CH}_4} = 58.125\\
&0.40n_{\text{total}} + 0.15n_{\text{total}} + 0.05n_{\text{total}} = 58.125\\
&0.60n_{\text{total}} = 58.125\\
&n_{\text{total}} = 58.125/0.60 = 96.875\\
&\boldsymbol{n_{\text{total}}} = \mathbf{96.875\ mol}.
\end{aligned}
$$

Elemental balance can be performed for any element, such as C, H, or N.

Volume at STP

$$V = n_{\text{total}} \times 22.4 = 96.875 \times 22.4 = 2170\,\text{L}.$$

Step 7: Calculate the volume of each gas

- CO: 40% of 2170 L = 868 L
- H_2: 35% of 2170 L = 759.5 L
- CO_2: 15% of 2170 L = 325.5 L
- CH_4: 5% of 2170 L = 108.5 L
- N_2: 5% of 2170 L = 108.5 L.

Reference

1. F. Dai, S. Zhang, Y. Luo, K. Wang, Y. Liu, X. Ji, Recent progress on hydrogen-rich syngas production from coal gasification. Processes **11**, 1765 (2023)

Chapter 5
Hydrogen Separation and Purification

Depending on the feedstock and the process used for hydrogen production, the composition of the product gas at the outlet of the hydrogen production system will be different (Table 5.1). However, the most widely used method of hydrogen production is steam reforming, and the reformate gas from reformer contains CO, CO_2 and unreacted CH_4 along with the main product hydrogen (Table 5.1). To obtain pure hydrogen, it is essential these gases near to be separated from the desired product hydrogen, and high-purity hydrogen obtained for various applications. This chapter discusses various methods for hydrogen purification.

Following four methods are employed for hydrogen separation from other gases present in the product stream obtained from a hydrogen production plant:

1. Adsorption, in which all the gaseous species except for hydrogen are adsorbed, thus producing high-purity hydrogen,
2. Absorption, where the major constituent CO_2 is selectively absorbed, producing pure CO_2 and separating it from hydrogen,
3. Membrane-based separation, which enables bulk separation of hydrogen from the gas mixture,
4. Cryogenic separation, which allows the separation and recovery of the individual gas species.

Since the 1980s, pressure swing adsorption (PSA) is the most widely used and employed technique to obtain high-purity hydrogen by separating it from other gases in the product stream. The different methods used for hydrogen separation and purification are shown in Fig. 5.1.

P. Sharma, *A Textbook on Hydrogen Energy*,
https://doi.org/10.1007/978-981-95-4601-5_5

Table 5.1 Typical product gas composition of the outlet gas from different hydrogen production processes

Sr. No.	Method	H_2 (%)	CO (%)	CO_2 (%)	CH_4 (%)	Other (%)
1.	Natural gas reforming	70–75	10–15	10–15	1–3	N_2
2.	Methanol reforming	75–80	0.5–2	20–25	–	–
3.	Coal gasification	25–35	35–45	15–25	0.1–0.3	N_2, S, H_2O
4.	Biomass gasification	25–35	30–40	10–15	10–20	N_2, S, O_2
5.	Coke oven gas	45–60	5–10	2–5	25–30	N_2, S, O_2

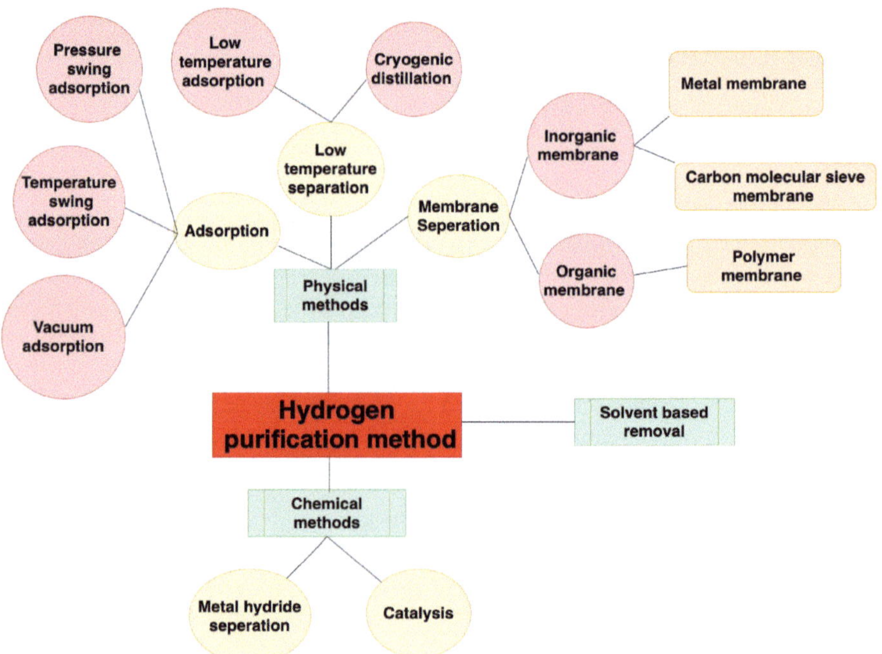

Fig. 5.1 Classification of hydrogen purification methods

Pressure Swing Adsorption

Pressure swing adsorption is the most widely used technology for hydrogen separation and purification. Modern PSA plants are capable of producing ultrapure hydrogen with purity up to 99.999% and with a recovery level of approximately 90%, making it the most efficient purification technique. The process exhibits high selectivity of sorbents toward impurities and can be used for large-scale purification of hydrogen, with a plant capacity of 265,000 Nm^3/h has been reported.

PSA cycle was invented by Charles Skarstrom in 1960 and originally consisted of four steps in the cycle. The first commercial-scale PSA plant integrated with a steam reforming unit was installed in 1966, featuring four beds and a drum for tail gas.

Later, multibed PSA plants were commissioned in the 1970s, with a 10-bed system installed in Germany in 1977.

With advancements in technology, it is now possible to process a wide variety and mixtures of gases. Steam reformation can have two technological variations, where after reforming, the options are: (1) HTWGS and PSA (2) HTWGS, LTWGS followed by CO_2 absorption and methanation. Out of which, option 1 is prominently used. PSA is not only used for hydrogen purification, but also for other industrial processes, such as separation of branched and linear paraffins, air fractionation, and air drying. It is also reported for its use for biogas upgrade and carbon capture.

The working principle behind pressure swing adsorption is that the variation in pressure selectively adsorbs and desorbs all other impurities in hydrogen product stream, so that only hydrogen remains. Thus, in the process, all the other constituents of the gas stream other than hydrogen are adsorbed on a microporous or mesoporous solid adsorbent such that the gas left out in pores or voids is the less adsorbing or non-adsorbing hydrogen, which can be collected. Once the sorbent becomes saturated with the other impurities (all the other gases other than hydrogen), the pressure is varied so as to remove these impurities and regenerate back the sorbent. As the name suggests, the change in pressure, i.e., pressure swing, is used for adsorption and desorption of the adsorbed species in the bed. On desorption of the bed, the gases released are those which adsorb more strongly on the sorbents (Fig. 5.2).

The adsorption process is a spontaneous process where a molecule from a gas stream experiences attraction upon coming close to the surface of the sorbent. Since the adsorption is a surface phenomenon, materials with a high surface area per unit mass are required for effective gas sorption. Since the different components

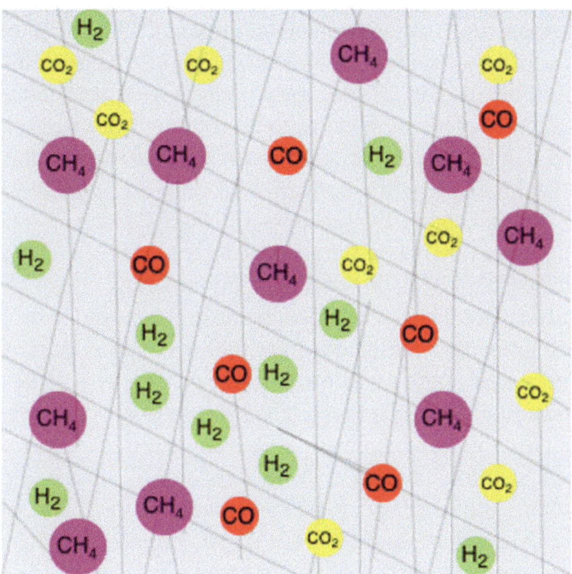

Fig. 5.2 Representation of the gas distribution in a typical adsorption process where hydrogen is non-adsorbing while rest being adsorbing gases

of the gas stream exhibit different interactions, attractions, and bonding strengths, the constituents present in the gas can be separated. When the sorbent is exposed to the reformate gas or product gas, adsorption of components occurs until the equilibrium is achieved. This equilibrium, at a particular temperature, pressure, and composition, determines the thermodynamic limit of the loading during adsorption on the sorbent. This adsorption equilibrium information is essential to design the process and system. The PSA process operates at ambient temperature and can achieve hydrogen recovery up to 90%. Varying plant sizes exist from as small as 30 Nm^3/h to 400,000Nm^3/h with the separation columns ranging from 1 to even 16 columns being used in parallel. The feed gas input can be obtained from SMR, POx, methane decomposition, gasification, petrochemical or refinery off-gas or coke oven gas, with hydrogen content varying from 30 to 99 vol%. Integration of PSA with SMR is cost-effective, by reducing the number of steps, processes, and units required, as well as improves the energy efficiency of the off-gas or tail gas from PSA. The tail gas obtained from PSA can meet around 60–90% energy demand of burners in SMR, providing the required heat for the reaction. This tail gas typically has a lower heating value (LHV) in the range of 4000–8000 kJ/Nm^3. Furthermore, PSA systems exhibit low operation and maintenance costs and possess long operational life, exceeding 30 years, as the sorbents do not suffer from significant deactivation or poisoning under normal operating conditions.

Fundamentals of the PSA

The feedstock and process employed for hydrogen production determine the syngas composition, which typically contains CO, CO_2, unreacted CH_4, and minor quantities of N_2, Ar, etc. This gas mixture must be treated in PSA system to obtain high-purity hydrogen. In addition to syngas with high concentration of hydrogen, refinery off-gases containing as low as 30 vol% hydrogen can also be effectively purified using PSA, thus a wide variation in the composition of the syngas still be purified using PSA to obtain pure hydrogen. PSA units are available over a wide range of capacities, from small-scale to large industrial plants. In the PSA process, hydrogen is obtained at high pressure, while the adsorbed impurities are released near-ambient pressure. The basic principle is that certain solid materials attract some molecules on their surface selectively. Thus, when a gas mixture is exposed to such solid materials, their surfaces attract certain molecules over the rest. The requirement for such adsorption is the presence of high surface area as the process is a physical adsorption, which occurs on the surface of the sorbent. The process of adsorption is spontaneous with adsorption energies ranging from 40–50 kJ/mol and it is an exothermic process. Consequently, adsorption is favored at lower temperatures and higher pressures, while the desorption requires energy input. The amount of species adsorbed at a constant temperature and varying pressure defines its adsorption capacity and can be represented by the adsorption isotherms obtained under equilibrium. Depending upon the binding of the adsorbate on the adsorbent, the isotherm has different shapes. Ideally the isotherm should be linear for better operation. The reason being, in case of a nonlinear isotherm, a portion of the adsorption occurs at low pressure, making desorption difficult and often requires vacuum, which in turn increases energy consumption and operational costs. When the adsorption bed is exposed to feed gas, it selectively adsorbs the impurities, while hydrogen being weakly adsorbed is left behind unabsorbed, as the product gas. As the bed reaches equilibrium, it gets saturated and reaches to its maximum adsorption capacity for the impurities. Thereafter, it needs to be regenerated back to its initial state. If regeneration is achieved by varying pressure, the process is termed as pressure swing adsorption (PSA), whereas if regeneration is by variation of temperature then termed as temperature swing adsorption (TSA). PSA response time is very fast compared to the TSA, while TSA cycles can extend to several hours due to the large amount of thermal energy that must be supplied and subsequently removed during each cycle. (Fig. 5.3).

During a PSA cycle, the feed gas enters into a fixed packed-bed adsorption column, where impurities are adsorbed until the bed reaches equilibrium, i.e., saturation. Hydrogen exhibits negligible affinity toward the adsorbents used; it passes through the bed and is recovered as the product gas at high pressure. In contrast, the adsorbed impurity species are ultimately released at pressures close to or slightly higher than ambient. Once the bed gets saturated with impurities, the pressure is reduced to desorption pressure, releasing the adsorbed impurities, and the bed gets recycled to its initial state, ready for next cycle of adsorption. Although PSA is a batch process, it can be made continuous (cyclic) by the use of more number of beds, as the requirement of refineries or other plants is a continuous supply of hydrogen. In steam methane reforming integrated with PSA, the tail gas from PSA is used for providing

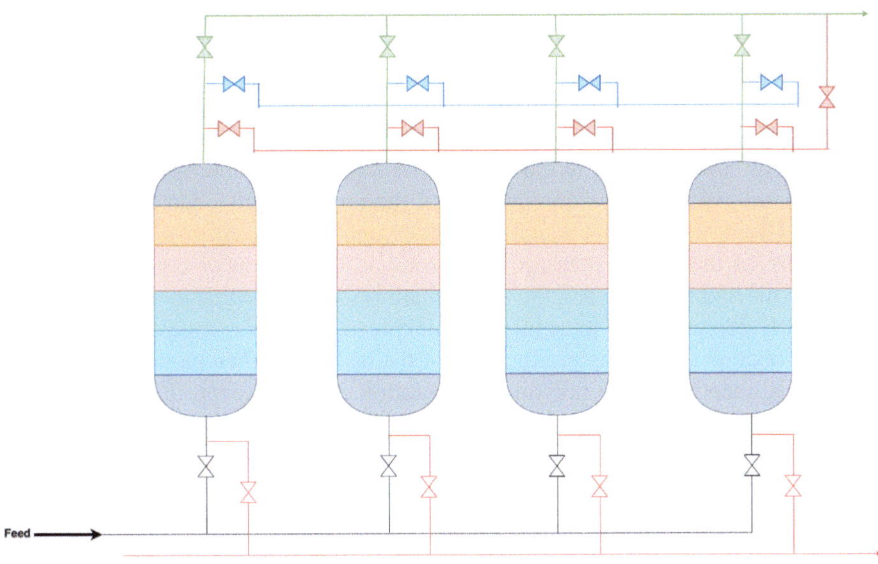

Fig. 5.3 Schematic of a 4-bedded pressure swing adsorption process

fuel for burners. Combustion of this tail gas provides the required heat of reaction for the reforming in SMR. To ensure a continuous supply of tail gas, the PSA has tail gas drum along with adsorption columns. The tail gas obtained is comparatively at low pressure, and hydrogen is obtained at higher pressures. PSA cyclic adsorption–desorption process is carried out in fixed beds under dynamic conditions, where the composition, temperature, and pressure vary inside the bed at different times as well as at different positions inside the bed. The steps involved in a simple four-step cycle of pressure swing adsorption unit are as follows:

1. **Adsorption**—The feed gas enters the adsorption bed at high pressure, where the contaminants get adsorbed in the different sorbent layers, while the hydrogen remains unadsorbed and is obtained at high pressure. This is the highest pressure step in the process.
2. **Counter-current blowdown**—As the bed gets saturated, the feed is stopped and the pressure in the column is reduced in steps, to remove the contaminants, and finally the lowest pressure of the cycle is achieved; this is called depressurization step.
3. **Counter-current purge**—Once the contaminants are desorbed by reducing the pressure in the earlier step of depressurization, the bed is prepared for the next cycle, in this purging step. The bed is purged with hydrogen in the light recycle step, thereby reducing the contaminants, decreasing their partial pressure, and further desorbing the remaining impurities from the bed at a low pressure in the system.
4. **Pressurization**—To prepare the bed for the next cycle, the pressure in the bed is increased using the recycled purified gas in a counter-current direction to that of the feed.

An additional pressure equalization step is added to reduce the hydrogen loss in the second step or blowdown step, where there is a substantial amount of hydrogen present. To enhance the hydrogen recovery, two beds are placed in contact at different pressures. Gas flows from the bed at higher pressure to the bed at lower pressure in a co-current direction relative to the feed, in contrast to the counter-current flow used during blowdown. The flow continues until the pressure gets equalized in the two columns. During the pressure equalization step, since the process involves a reduction of pressure of the higher pressure bed, to avoid desorption of contaminants and polluting the hydrogen stream, the feed stream is stopped. In this method, the bed is co-currently depressurized to an intermediate pressure. In this process, the hydrogen stream is obtained, from the product end of the bed and is then passed into another bed, that is undergoing the pressure equalization step. However, this increases the number of columns required in the PSA unit. Another method to improve hydrogen recovery is "providing purge," which occurs in counter-current direction, and flow is from the product to feed direction. In this step, the bed is depressurized to an intermediate pressure level, and hydrogen is passed from the product side to another bed undergoing the purge step. These two steps need to occur prior to blowdown step (Fig. 5.4).

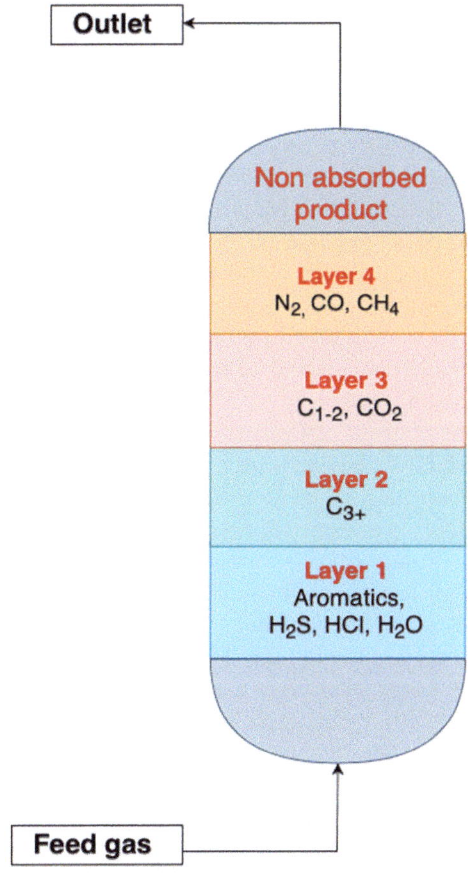

Fig. 5.4 Schematic representing the different adsorption layers and the adsorbing species in these layers

Depending upon the impurities present, or the constituents of the gas to be separated, in order to obtain pure hydrogen which in turn depends upon the feedstock, hydrogen production method, and end-use application purity, the appropriate sorbents are selected. Different materials have different affinities toward various molecules in the gas; as such, the choice of sorbents is based on the constituents to be removed, and the order of their placement must be decided, while designing the PSA unit. The most widely used materials in PSA are alumina, silica gel, molecular sieves, zeolites, and activated carbon. Usually, different materials exhibit affinity toward different impurities, and a specific sequence of removal is followed. Typically the first impurity to be removed is water or moisture followed by CO_2, then CH_4, then CO and, finally the remaining components, including nitrogen or argon. Studies reported in the literature indicate that the order of adsorption, from the most adsorbed to least adsorbed, is: $H_2O > CO_2 > CH_4 > CO > N_2 > H_2$ for activated carbon, and $CO_2 > CO > CH_4 > N_2 > H_2$ for zeolite. Kinetic studies carried out on these materials showed that the dominant mechanism was identified to be micropore diffusion. The diffusivity of the species into the micropores of activated carbon follows the order from fastest or slowest as $H_2 > N_2 > CO > CO_2 > CH_4$, while in zeolite it follows as $H_2 > N_2 > CH_4 > CO > CO_2$ [1]. Since no single material can be selective toward all possible contaminants, several layers of the adsorbents are used; besides, it can improve the purity level to even ppm or ppb. In order to increase the adsorption capacity of the PSA bed, several layers of adsorbents are used, which can adsorb different components.

Usually, the first layer from the bottom, i.e., from the feed side, is alumina or silica gel to adsorb water, as the feed stream to the PSA contains a substantial water content. As such, the bottom-most layer consists of alumina or silica gel. Alumina can adsorb around 30 mol/kg, as compared to silica gel which can adsorb around 15 mol/kg. Although activated carbon can also adsorb water, its capacity is lower than that of alumina (around 20 mol/kg), and it also has the disadvantage of lower volumetric loading; hence, it is less preferred for water adsorption. Silica gel is also used for the removal of C_{3+} hydrocarbons, as these get strongly adsorbed in the next layer which is activated carbon. However, activated carbon shows affinity toward several impurities, as mentioned above. Activated carbon, being less expensive and quite stable in presence of a wide range of impurities, along with having low adsorption of hydrogen, offers several advantages for its use as a sorbent in a PSA. Thus, next step after removal of water from the feed stream is the removal of CO_2, unreacted CH_4, H_2S, and any remaining C1–C2 hydrocarbons over activated carbon, with the order of removal being first H_2S, then hydrocarbons, followed by CO_2 and finally CH_4. This step takes place at high pressure, i.e., at the feed gas pressure, and thus enables high adsorption of CO_2 on activated carbon. In addition, the isotherms of activated carbon are comparatively less nonlinear, allowing desorption or regeneration of bed at close to atmospheric pressure, compared to zeolites. Moreover, activated carbon is less expensive than zeolites; however, its CO adsorption capacity is lower at the partial pressures of CO in the feed gas. As such, a certain amount of CO is adsorbed by activated carbon, and the residual amount is removed by a final layer of zeolite, where higher volumetric loading can be achieved, compared to activated carbon. This layer of zeolite must be kept free of CO_2 and H_2O, and it is essential that these gases are adsorbed prior to entering the zeolite layer, as their adsorption on zeolite is very strong and would require either high temperature or vacuum for desorption. Thus, the zeolite is the last layer from the feed side and lies toward the product side. The zeolite layer can reduce the CO content to below 10 ppm level. The length ratio of the activated carbon and zeolite layers is selected such that CO_2 or H_2O do not reach the zeolite layer [2].

The final layer of zeolites can be used to selectively remove the remaining contaminants from hydrogen, such as nitrogen, argon, CO, etc., but not water, as zeolites form strong bonds with water and thus cannot be desorbed easily. The requirements for a sorbent are that it should be stable under the presence of contaminants, have a long lifetime, cost-effective, possess high loading capacity, and be capable of desorption under optimum conditions without requirement of vacuum or high temperature. In addition, since the PSA column heights are several meters, the sorbents should have high mechanical stability. The cost and size of the PSA plant are largely determined by the productivity which is defined as the ratio of the rate of hydrogen produced to the amount of adsorbent used.

Several modified sorbents, such as the modified activated carbon or activated carbon impregnated with saline solution, as well as zeolites 5A and 13X modified with Na^+ ion, have been reported to exhibit high selectivity and higher loading of impurities to get high purity and yield of hydrogen. Metal organic frameworks have also been studied like MgMOF-74, UiO-66, and combinations of zeolites with activated carbon or zeolites with MOFs have shown promising results. Sorbent hollow fibers have been reported to provide a hydrogen purity of 99.2% and recovery of 88.1% [3, 4].

The performance of a PSA plant can be improved through proper design, optimization, and use of materials with high loading capacity. In order to achieve higher purity and higher recovery of hydrogen, multiple adsorption, pressure equalization, and provision of purge beds can be implemented; however, these measures increase the complexity and cost of the system.

Membrane Based Separation Method

Membrane-based separation is a simple and flexible method; it is compact, has low energy consumption, and is environmental-friendly [3]. Usually membrane-based separation is used for niche applications where the off-gas or feed stream is rich in hydrogen, or it can also be used for adjusting the syngas ratio for specific applications. Membrane-based separation offers low energy consumption, ease of operation, cost-effectiveness; thus lower investment is required, and it can be operated continuously. However, this method is economical, primarily for lower-capacity requirements, and the hydrogen produced using this method is obtained at a lower pressure; as such, compression after separation becomes essential (Fig. 5.5).

History

The history of membrane-based gas separation dates back to 1829, when Thomas Graham first demonstrated gas transport through polymeric membranes; in 1855, Fick gave the quantification of such transport through boundary layers. In 1866, Thomas Graham came up with the concepts related to gas permeation. Contribution from Fick, Graham, and Martin Knudsen has been significant toward understanding the governing parameters and their relationship toward performances of membranes evolved. The first commercially available membrane ways back to the 1930s, wherein the contribution of Karplus has been significant. Thereafter, several polymeric membranes found applications for drinking water filtration. The 1960s marked the beginning of the modern membrane era, which was further accelerated after the 1980s when synthetic membranes were first adopted for commercial large-scale gas separation [5].

Fig. 5.5 Schematic showing the process of separation using membrane

Membranes can be classified based on the membrane materials, including metallic membranes (pure metals or alloys), inorganic membranes (oxides, glasses, ceramics, and zeolites), composite or hybrid membranes, carbon-based porous membranes, and organic polymer-based membranes as mentioned in Fig. 5.6. The key performance indicators for membrane used for hydrogen separation include: thermal, mechanical, chemical stability; selectivity and permeability; membrane lifetime; ease of synthesis; transport mechanism; and the possible maximum hydrogen flux. [6].

The major requirements from a membrane-based separation of hydrogen are: (1) low cost of material, synthesis, and hydrogen separation, (2) durability and long life, (3) high hydrogen flux, (4) less requirement of auxiliary power units.

The most important parameters in membrane-based gas separation are selectivity and permeability of these membranes. An effective membrane should selectively allow specific gas molecules to permeate from a mixture of gases, thereby separating them from the remaining constituents. The three important terms used in gas separation are selectivity, permeability, and flux. The selectivity means the separation capability of a certain gas with respect to other gases of a membrane. Permeability is a measure of the amount of gas permeating through a membrane. Flux (J) is defined as the amount of gas permeating per unit time per unit area through a membrane.

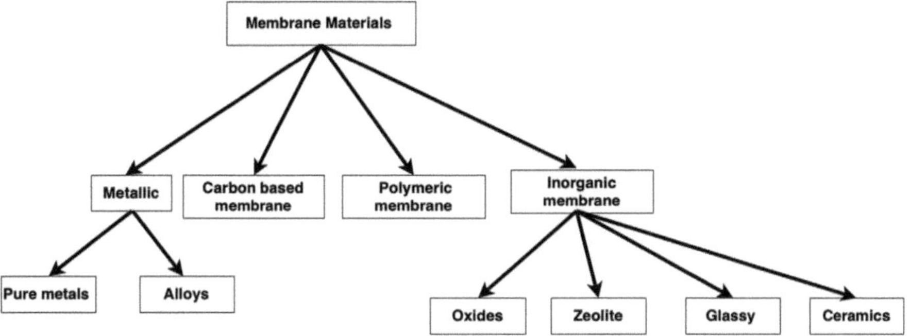

Fig. 5.6 Classification of different membranes used for hydrogen purification based on their material of construction

According to Fick's first law:

$$J = -D\nabla C,$$

where D is the diffusion coefficient and C is the equilibrium concentration.

Metallic Membranes

Historically, Pd-based membranes have been widely used to separate hydrogen due to their catalytic activity toward hydrogen dissociation. Hydrogen separation using dense sheets or films of metallic materials is based on dissociative chemisorption of hydrogen molecules on the membrane surface, where hydrogen is split into protons and electrons. The electrons are conducted through the metallic lattice, followed by adsorption, while the diffusion of H^+ ions occurs through the membrane. On the permeate side, protons recombine with electrons to form molecular hydrogen, resulting in effective separation of hydrogen from the remaining gas species. The remaining constituents or impurities due to their larger molecular size and inability to dissociate are impermeable through the dense membrane. These membranes have high thermal stabilities and can be operated at high temperatures and are mechanically stable when operated in the presence of hydrogen. Other requirements for membranes are that these should have high selectivity toward hydrogen, thus can obtain pure hydrogen at the outlet, high permeability, diffusivity, and solubility for hydrogen and high catalytic activity to dissociate hydrogen.

Pt and Pd-based membranes are expensive, the less expensive membrane materials include: Ta, Nb, V, Ni, Ti, Zr, Hf, etc. Pd membranes face several issues like, non-reversible poisoning with impurities, embrittlement when exposed to hydrogen. Alloying Pd with metals such as Cu or Au can mitigate embrittlement and improve stability; however, the high material cost remains a significant challenge. As a result, an alternative, the metallic membranes based on non-noble metals are actively studied. The mechanism of hydrogen permeation is via solution-diffusion method which involves steps like: (1) flow of feed stream to reach membrane surface, (2) hydrogen getting chemisorbed on to the surface, (3) dissociation of hydrogen into H^+ and electrons, (4) adsorption of these ions on the surface, (5) diffusion of H^+ ions and electrons through the membrane, (6) desorption of H^+ ions from surface into product (7) reassociation of these ions with electrons to form hydrogen molecules, (8) finally diffusion of hydrogen molecules form surface into the product stream.

To enhance membrane performance, in terms of permeability, selectivity, mechanical strength, and resistance to embrittlement, various metallic alloys have been developed, including V-Al, V-Al-Ni, N-B-Ti-Ni, V-Ti, V-Co, Ni-Nb-Zr, etc.

Polymeric Membranes

Two types of polymers—glassy or rubbery—can be used for synthesizing polymeric membranes. Glassy polymers generally exhibit high selectivity toward hydrogen but results in lower flux, whereas rubbery polymers provide higher flux at the expense of selectivity, indicating a trade-off. Usually, in polymeric membranes, the membranes with high selectivity have lower permeability and the ones with high permeability have lower selectivity. However, among the glassy and rubbery polymers, the glassy polymers are preferred due to high selectivity, and they can separate even when small difference in size of the constituents, making them more size and shape-selective compared to rubbery polymers. The mechanism of separation in polymeric membranes can include the following: (1) Solution diffusion: Gas molecules are absorbed on membrane surface, dissolve and transport across membrane due to driving force which is chemical potential. (2) Surface diffusion—Gas molecules adsorb on surface and undergo movement, subjected to the driving force which is concentration gradient. (3) Knudson diffusion: Hydrogen molecules undergo more frequent collisions with pore walls than the rest of the gases in the mixture. (4) Capillary condensation: Diffusing gas molecules condense on pore walls resulting in capillary forces (5) Molecular sieving: Separation occurs based on differences in molecular size and shape relative to the pore dimensions.

Porous membranes have mechanism based on points 2–4, while in non-porous based on point 1. (Fig. 5.7).

Polymeric membranes used for hydrogen separation include polysulfone and polyimide. Examples include P84 copolymide (BTDA-based) membranes, which exhibit high selectivity, along with high thermal, mechanical, and chemical stability. Matrimid 5218 showed high hydrogen permeability and higher selectivity for hydrogen in a hydrogen methane mixture, while PBI membranes showed high selectivity toward hydrogen in a hydrogen and carbon dioxide mixture. Other polymers investigated for hydrogen separation include polyesters, ethers, urethanes, etc. It is relatively easier to separate CH_4 and H_2, but difficult to separate CO_2 and H_2 due to their very similar permeability coefficient. Both hydrogen and carbon dioxide are often referred as fast components; the reason being hydrogen has high diffusivity

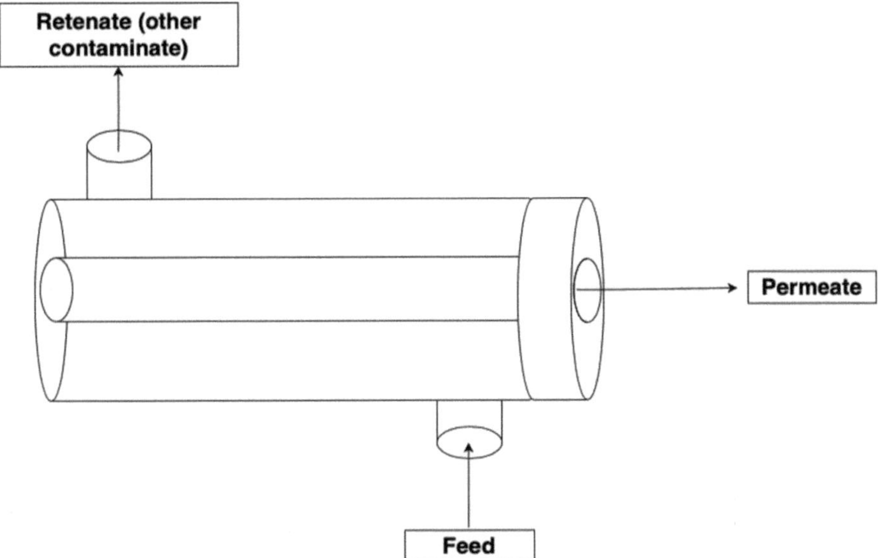

Fig. 5.7 Schematic of the hydrogen membrane-based purification reactor

while carbon dioxide has high solubility. The next important and difficult separation is of hydrogen and carbon monoxide. These membranes can be symmetric, i.e., made up of single material or asymmetric—a thick porous layer and a selective layer coating. The upper layer provides high gas flux while the inner porous layer provides mechanical stability. The membranes of choice for industrial applications include polysulfones, polyimides, cellulose acetate, etc.

Inorganic Membranes

Silica membranes are inorganic membranes with a microporous network structure. They are cost-effective, easy to fabricate, and scalable. These membranes are asymmetric membranes with three layers: a support, an intermediate layer, and the top membrane layer and the separation occurs by hopping diffusion. The intermediate layer is composed of zirconia, alumina, or silica, and support is composed of alumina. These designs can have additional layer which is a coating of metal on silica layer.

Zeolites

The mechanism of separation in zeolites is by adsorption and diffusion; however, the presence of intercrystalline pores is a significant drawback, as these pores have larger pore sizes than the intrinsic zeolite pores. The various zeolites studied for separation include MFI, ZSM-5, MOR, and FAU, among others. The process of separation involves adsorption on the surface, diffusion within the zeolite, transport to the opposite surface, then to the other surface, and finally desorption from the surface. Thus, molecules with a smaller size permeate, while large molecules do not.

Carbon-Based Membranes

The mechanism for carbon-based membranes is different; in this, the impurities permeate, rather than, hydrogen. While hydrogen is collected as retentate. The commonly considered carbon-based materials include: carbon nanotubes, carbon membranes, and molecular sieves. These membranes are less expensive than polymeric membranes and selectively block to enhance hydrogen separation. Carbon membranes are lower in strength and are brittle. Molecular sieves, with their optimum pore dimensions, allow high selectivity toward separation. The separation occurs by the solution-diffusion method. The major challenges with carbon membranes are their cost, stability, and selectivity.

Mixed Matrix Membranes

Mixed matrix membranes are hybrid membranes in which metal-polymer, ceramic-polymer, and metal-ceramic mixed systems are used to incorporate the advantages of each component. Typical examples include embedding nanostructured porous materials in a polymer matrix, to enhance permeability and selectivity. These can exhibit high stability, better transport, and low cost. Modified zeolites, porous titanosilicates, mesoporous silica, carbon molecular sieves (CMS), and carbon nanotubes (CNTs) are other possible options.

CO_2 Absorption Systems

Although most gas cleaning plants are now PSA-based, when carbon capture and sequestration is the objective, PSA does not allow this, as the carbon species separated are burned in the off-gas, resulting in emissions again. As such, when carbon capture is desired, CO_2 absorption selectively is a better option for hydrogen purification. However, since the purity of hydrogen obtained is lower in this method and if the application's requirement is higher purity, then a smaller PSA unit can be integrated after CO_2 absorption system to upgrade the purity of hydrogen. In solvent-based absorption, the major component CO_2 present along with hydrogen in the product stream from SMR, can be removed using a liquid sorbent, that is highly selective toward absorbing CO_2 compared to other gases in the mixture. The major requirements for such sorbents are the stability of sorbent, as it should not get poisoned by other species present in the gas mixture. It should have low vapor pressure to avoid solvent losses, should be cost-effective, non-corrosive, and easy to regenerate. The two well-known methods for CO_2 absorption include removal by chemical absorption using amine, carbonate, and alkali metal hydroxides or physical absorption using methanol (Rectisol process) and polyethylene glycol (Selexol process). The CO_2 is absorbed from the mixture at lower temperatures and higher pressures, while it is regenerated at higher temperatures and lower pressures. Although chemical absorption is highly efficient, it requires more energy for regeneration and upon repeated cycling undergoes degradation. Physical sorbents although regenerate at lower energy conditions. But they degrade with cycling. The most commonly used sorbents include monoethanolamine, diethanolamine, triethanolamine, diisopropanolamine, methyl diethanolamine, and diglycoamine. Absorption occurs at 26–60 °C and pressure of around 60 bar, while regeneration occurs at approximately 115 °C and pressure of $\leq$ 1 bar. Amine-based removal is more expensive than PSA and involves operational issues, corrosion problems, and sorbent losses. It can also be separated using sorption enhanced reforming method using CaO, Na_2CO_3 or Li_4SiO_4,, etc.

Cryogenic Separation

This method is based on low-temperature separation of the individual components from the gaseous mixture. Difference in the boiling points of the individual components is used to separate the condensable components of the feed gas stream. This method is useful to obtain the individual gas components, when required for specific applications. Since the separation occurs at lower temperatures, such a method is not economical and is challenging to operate.

To summarize, chemical solvent absorption-based processes can be used for large-scale hydrogen purification, but the purity level is typically limited to 95–97%, and a significant amount of steam is required for the solvent regeneration. The cryogenic separation method is a highly energy-intensive technology since it requires extremely low operating temperatures. Additionally, the hydrogen obtained is at low pressure, necessitating further energy input for compression. The hydrogen purity achievable through cryogenic separation generally lies in the range of 90–99%. Membrane-based technology is preferred for small-scale or medium-scale purification, although the maximum achievable purity can be as high as 90–98%, with hydrogen recovery depending on membrane material, typically in the range of 85–95%. Hybrid systems with PSA and membrane-based separation have also been explored and can increase the hydrogen recovery by 2–6%. However, this improvement comes at the expense of increased capital expenditure (CAPEX), operating expenditure (OPEX), and additional compression requirements.

Solved Problem 1. A simplified pressure swing adsorption (PSA) system is used to separate CO_2 from a combustion exhaust gas stream, where the purge step is neglected. The exhaust gas consists of 13 mol% CO_2 and 87 mol% N_2 at a temperature of 45 °C. The PSA unit achieves a CO_2 capture efficiency of 92% and produces a CO_2 product with 96% purity. The feed gas is compressed adiabatically from 0.1 MPa to 1.2 MPa, and the compressor operates at an efficiency of 86%. The heat capacity ratio (k) for the gas mixture is 1.33.

Determine the following:

(a) The theoretical minimum work required per mole of CO_2 captured (in kJ/mol),
(b) The actual work needed for compression per mole of CO_2 captured (in kJ/mol), based on the given conditions,
(c) The Second Law efficiency of the process, considering only the compression step,

Solution

Given:

- Feed gas: 13 mol% CO_2 and 87 mol% N_2
- Temperature: 45 °C = 318.15 K.
- Initial pressure, $P_1 = 0.1$ MPa.
- Final pressure, $P_2 = 1.2$ MPa.
- CO_2 capture efficiency: 92%
- CO_2 purity: 96%
- Compressor efficiency $\eta = 0.86$.
- Heat capacity ratio (k) = 1.33.

(a) **Theoretical minimum work** ($W_{\min}$).

The theoretical minimum work of separation is calculated based on the change in Gibbs free energy required to separate CO_2 from the gas mixture. For ideal gas mixtures, the theoretical minimum work per mole of CO_2 captured is given by:

$$W_{\min} = RT\left[\text{In}\left(\frac{yCO_2, feed}{yCO_2, product}\right) + \left(\frac{1 - yCO_2, feed}{yCO_2, feed}\right)\text{In}\left(\frac{1 - yCO_2, feed}{1 - yCO_2, product}\right)\right]$$

To solve this problem, we will go step by step for each part:

$T = 318.15$ K
R = 8.314 J/mol. K

$$y_{CO_{2,feed}} = 0.13$$

$$y_{CO_{2,product}} = 0.96$$

Step 1: Plug in the values

$$W_{\min} = 8.314 \times 318.15\left[\text{In}\left(\frac{0.13}{0.96}\right) + \left(\frac{1-0.13}{0.13}\right)\text{In}\left(\frac{1-0.13}{1-0.96}\right)\right]$$

$$W_{min} = 8.314 \times 318.15 \times [-2.0 + 6.692 \times 3.079]$$

$$W_{min} = 8.314 \times 318.15 \times [-2.0 + 20.61] = 8.314 \times 318.15 \times 18.61$$

$$W_{min} = 49225.29 \text{ J/mol of } CO_2$$

(b) **Real work of compression (adiabatic process)**

Equation for adiabatic compression

$$W_{real} = \left(\frac{k}{k-1}\right)RT_1\left[\left(\frac{P_2}{P_1}\right)^{\frac{k-1}{k}} - 1\right]\frac{1}{\eta}$$

where

- $k = 1.33$
- $R = 8.314$ J/mol·K
- $T_1 = 318.15$ K
- $P_1 = 0.1$ MPa and $P_2 = 1.2$ MPa
- $\eta = 0.86$.

Step 1: Calculate pressure ratio

$$r = P_2/P_1 = 1.2/0.1 = 12$$

Step 2: Plug into formula

$$W = \left(\frac{1.33}{1.33-1}\right) \times 8.314 \times 318.15 \times \left[(12)^{0.33/1.33} - 1\right] \times \frac{1}{0.86}$$

$$W_{real} = 10568.033 \text{ J/mol of feed gas}$$

Now, to find **per mole of CO_2captured**, recall:

- 13% CO_2 in feed
- 92% capture = 0.92 × 0.13 = **0.1196 mol CO_2captured/mol feed**

$$W_{real} = \frac{10568.033}{0.1196} = 88361.48\,\text{J/mol of CO}_2$$

(c) **Second Law Efficiency**

$$\eta_2 = \left(\frac{W_{\min}}{W_{real}}\right) \times 100 = \frac{49225.59}{88361.48} \times 100 = 55.71\%$$

Solved Problem 2

A simplified pressure swing adsorption (PSA) system is used to separate H_2 from a steam methane reforming product stream, where the purge step is neglected. The exhaust gas consists of 20 mol% H_2 and 80 mol% CO_2 at a temperature of 45 °C. The PSA unit achieves a H_2 capture efficiency of 95% and produces a H_2 product with 99% purity. The feed gas is compressed adiabatically from 0.1 MPa to 1.8 MPa, and the compressor operates at an efficiency of 88%. The heat capacity ratio (k) for the gas mixture is 1.33.

Determine the following:

(a) The theoretical minimum work required per mole of H_2 captured (in kJ/mol),
(b) The actual work needed for compression per mole of H_2 captured (in kJ/mol), based on the given conditions,
(c) The Second Law efficiency of the process, considering only the compression step.

Solution

Given:

- Feed gas: 20 mol% H_2 and 80 mol% CO_2
- Temperature: 45 °C = 318.15 K
- Initial pressure, P_1 = 0.1 MPa
- Final pressure, P_2 = 1.8 MPa
- H_2 capture efficiency: 95%
- H_2 purity: 99%
- Compressor efficiency $\eta = 0.88$
- Heat capacity ratio (k) = 1.33.

(a) **Theoretical minimum work** (W_{min})**per Mole of** H_2**Captured**

The theoretical minimum work of separation is calculated based on the change in Gibbs free energy required to separate H_2 from the gas mixture. For ideal gas mixtures, the theoretical minimum work per mole of H_2 captured is given by:

The minimum work is calculated using the change in Gibbs free energy

$$W_{min} = RT\left[\ln\left(\frac{y_{H2,feed}}{y_{H2,product}}\right) + \left(\frac{(1 - y_{H2,feed})}{y_{H2,feed}}\right)\ln\left(\frac{1 - y_{H2,feed}}{1 - y_{H2,product}}\right)\right] \tag{5.1}$$

Given:

R = 8.314 J/mol·K.

T = 318.15 K

$$y_{H_{2,feed}} = 0.20$$

$$y_{H_{2,product}} = 0.99$$

Step 1: Plug in the values in Eq. (5.1), we get.
$W_{\min} = 8.314 \times 318.15 \times \left[\ln\left(\frac{0.2}{0.99}\right) + \left(\frac{1-0.2}{0.2}\right)\ln\left(\frac{1-0.2}{1-0.99}\right)\right]$

$$W_{\min} = 8.314 \times 318.15 \times (-1.59 + 4 \times 4.382)$$

$$W_{\min} = 42.16\,\mathrm{kJ/mol\,H_2}$$

(b) **Real work of compression (adiabatic process)**

Equation for adiabatic compression

$$W_{real} = \left[\frac{k}{k-1}\right]RT_1\left[\left(\frac{P_2}{P_1}\right)^{\frac{k-1}{k}} - 1\right]\frac{1}{\eta} \tag{5.2}$$

where

- k = 1.33
- R = 8.314 J/mol K
- T_1 = 318.15 K
- P_1 = 0.1 MPa
- P_2 = 1.8 MPa
- η = 0.88.

Step 1: Calculate pressure ratio

$$r = \frac{P_2}{P_1} = \frac{1.8}{0.1} = 18$$

Step 2: Plug into formula

$$W_{real} = \left[\frac{1.33}{1.33-1}\right] \times 8.314 \times 318.15 \times \left[(18)^{\frac{1.33-1}{1.33}} - 1\right] \times \frac{1}{0.88}$$

$$W_{real} = 12703.096 \text{ J/mol of feed gas}$$

Now, to find **per mole of H_2captured**, recall:

- 20% CO_2 in feed

- 95% capture = 0.95 × 0.20 = **0.19 mol H_2captured/mol feed**

$$W_{real} = \frac{12703.096}{0.19} = 66.86\,\text{kJ/mol of } H_2$$

(c) **Second Law Efficiency**

$$\eta_2 = \left(\frac{W_{min}}{W_{real}}\right) \times 100 = \frac{42.16}{66.86} \times 100 = 63.06\%$$

Solved Problem 3

A gas mixture containing 80% hydrogen (H_2) and 20% carbon dioxide (CO_2) by volume is passed through a PSA unit to purify hydrogen. The PSA system operates at a feed pressure of 5 bar and atmospheric pressure (1 bar) during desorption.

Given:

Adsorbent selectively adsorbs CO_2.
Feed flow rate = 10 Nm^3/h.
CO_2 adsorption capacity of adsorbent = 0.1 mol/kg.
PSA cycle time = 1 h.
Molar volume at STP = 22.4 L/mol.

Calculate:

1. The mass of adsorbent required to capture all the CO_2 in 1 h.
2. The volume of purified H_2 obtained in 1 h at STP.

Step-by-Step Solution:
Step 1: Calculate the moles of CO_2in the feed gas per hour

Feed gas flow rate = 10 Nm^3/h = 10,000 L/h
Composition of CO_2 = 20%
CO_2 volume = 0.2 × 10,000 = 2,000 L/h
Moles of CO_2 = Volume / Molar volume = 2000 / 22.4 ≈ 89.29 mol/h

Step 2: Use adsorption capacity to find the adsorbent mass

Adsorption capacity = 0.1 mol CO_2 per kg of adsorbent
Adsorbent required = 89.29 / 0.1 = 892.9 kg.

Step 3: Calculate purified H_2output

H_2 in feed = 80% of 10,000 L = 8000 L = 8 Nm^3
Since CO_2 is fully adsorbed, all hydrogen passes through.
Purified H_2 = 8000 L/h.

Final Answers:

1. Mass of adsorbent required = 892.9 kg
2. Purified hydrogen obtained per hour = 8000 L (or 8 Nm^3).

References

1. F.V.S. Lopes, C.A. Grande, A.M. Ribeiro, J.M. Loureiro, O. Evaggelos, V. Nikolakis, A.E. Rodrigues, Adsorption of H_2, CO_2, CH_4, CO, N_2 and H_2O in activated carbon and zeolite for hydrogen production. Separation Sci. Technol. **44**, 1045–1073 (2009)
2. M. Luberti, H. Ahn, Review of polybed pressure swing adsorption for hydrogen purification. IJHE **47**(20), 10911–10933 (2022)
3. Z. Du, C. Liu, J. Zhai, X. Guo, Y. Xiong, W. Su, Guangli, A review of hydrogen purification technologies for fuel cell vehicles. Catalysts **11**, 393 (2021)
4. R.P. Lively, N. Bessho, D.A. Bhandari, Y. Kawajiri, W.J. Koros, Thermally moderated hollow fiber sorbent modules in rapidly cycled pressure swing adsorption mode for hydrogen purification. Int. J. Hydrogen Energy **37**, 15227–15240 (2012)
5. P. Pandey, R.S. Chauhan, Membranes for gas separation. Prog. Polym. Sci. **26**, 853–893 (2001)
6. N.W. Ockwig, T.M. Nenoff, Membranes for hydrogen separation. Chem. Rev. **107**, 4078–4110 (2007)

Chapter 6
Thermochemical Cycles

Currently, most of the hydrogen produced globally is through fossil fuels, resulting in significant environmental emissions. However, to produce hydrogen in a clean and sustainable manner, the production facilities must either be integrated with the CCUS technologies in short term or rely on water electrolysis powered by renewable energy sources in the long term. A middle path in the intermediate time frame could involve meeting the energy required for the endothermic processes for producing hydrogen from fossil fuels using renewable energy sources, instead of burning the fossil fuels to provide the required heat of the reaction.

Hydrogen can be produced without emissions from water splitting, and this is the route for green hydrogen production. However, since water is in a lower energy state than hydrocarbons, thus the production of hydrogen from water requires much more energy as compared to the energy required for the production of hydrocarbons. When hydrogen is produced from fossil fuels, the resulting emissions at the point of production make the method environmentally polluting unless carbon capture and sequestration become essential at the point of production for sustainable hydrogen production. Although most of the hydrogen produced today still comes from fossil fuels, but in the long-term reduction of emissions and decarbonization necessitate the substantial increase in the share of green hydrogen. In the near term Steam Methane Reforming (SMR) is expected to remain dominant, but in medium term hydrogen production from biomass or electrolysis is likely to rise. In the long term, hydrogen production solely from electrolysis, thermochemical and photoelectrochemical routes will become more prominent. Green hydrogen can be produced from either biomass or water using renewable electricity or heat from nuclear and solar energy.

Hydrogen can be produced from water splitting through a wide range of methods. Some historical methods include steam splitting with the use of metallic iron. This method was later replaced by the conventional chemical water splitting involving the chemical redox reaction, in which water reacts with carbon or carbon monoxide during coal gasification or water gas shift reactions respectively. However, the process

P. Sharma, *A Textbook on Hydrogen Energy*,
https://doi.org/10.1007/978-981-95-4601-5_6

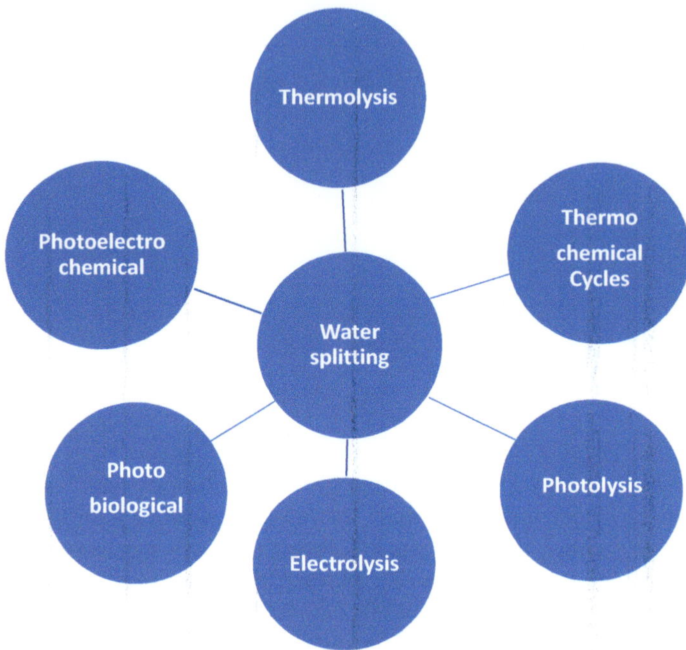

Fig. 6.1 Different methods for hydrogen production from water splitting

results in emissions. These reactions have either a negative or slightly positive enthalpy of reaction, also valence state of oxygen remains unchanged during the chemical splitting of water.

The other possible methods for water splitting can be categorized based on the source of the required dissociation energy, the process is called thermal (involves thermolysis or thermochemical cycles), mechanical (sonochemical methods using ultrasound), biological (biochemical methods), photonic (photoelectrochemical or photocatalytic), and electrical (electrolysis using energy form solar or wind) [1] (Fig. 6.1).

The Gibbs free energy for a single-step thermochemical water-splitting reaction is zero at about 4300 K under a pressure of 1 bar. Thermal decomposition of water can be carried out in a single step without the requirement of any additional redox system and can occur completely at 4500 K, to a reasonable level at 2500 K under 1 bar pressure, while the amount of dissociation reaches about 64% at 3000 K [2]. However, several challenges are associated with this reaction, including metallurgical constraints, separation of products at such high temperatures, recombination of the products, and intermediates [3] and provision of the required heat.

Since the temperature required for direct water splitting is extremely high, the process becomes economically unfavorable and technically challenging. The theoretical work required in a process involving multiple chemical reactions is equal to the sum of the Gibbs free energy changes, and this concept can be used to reduce the required temperature through multiple reactions, some operating at higher temperatures and others at relatively lower temperatures such that the overall ΔG becomes zero [4]. Accordingly, an alternative approach to reduce the temperature of the thermolysis is the coupling of different chemical reactions which occurs at different temperature levels, thus allowing the water splitting to occur at significantly lower temperatures than the thermolytic water splitting. This alternative method is called the thermochemical water splitting and involves thermochemical cycles where the water is split into hydrogen and oxygen at temperatures that can be managed from the materials perspective. In thermochemical water splitting, multiple step reaction occur; in fact, a series of reactions take place in which certain chemicals undergo a series of processes and finally get recycled back to their initial state, however in the whole process water being fed in the reaction undergoes splitting to form hydrogen and oxygen.

Thermochemical Cycles

Thermochemical cycles involve water splitting through a series of chemical processes where the input requirements are water and heat, and the output obtained is hydrogen and oxygen; the process involves use of several chemicals, during which intermediates are formed that are ultimately recycled back to the initial chemicals at the end of the cycle. These cycles may involve two or more reaction steps, and as the number of steps increases, the maximum temperature required for water splitting within the cycle reduces; however, as the number of steps increases, the process engineering gets more complicated. One key advantage of thermochemical cycles is that since hydrogen and oxygen are produced in separate steps, thereby eliminating the challenge of the recombination that is encountered in direct thermolysis. The required temperatures during various steps in the cycles can be obtained from solar-based technologies; however, due to the relatively low efficiencies of solar thermal-based technologies, there is a need to develop materials and catalysts to improve the overall efficiency. The required high temperature for thermochemical water splitting can be obtained using central tower, parabolic dish, or double concentration solar thermal systems. Instead of having electrolysis driven with electricity from any solar thermal power plant, the direct thermochemical water splitting using solar thermal technology is potentially more efficient; the advantage of eliminating one more step of conversion of thermal to electricity will reduce efficiency losses.Moreover, thermochemical process can be more cost effective than the electrolysis alone or other water-splitting methods. This is also an effective way to convert the concentrated solar energy into chemical energy of hydrogen, which can subsequently be used for a wide variety of applications. The major disadvantages of the thermochemical cycles include slow reaction rates, the requirement for large-sized reactors, the use of hazardous or corrosive chemicals, and a lower level of technological maturity compared to electrolysis.

Other splitting technologies include electrolysis powered by wind energy or solar energy, photolytic water splitting, photoelectrochemical water splitting; these technologies are expected to play an important role in future of green hydrogen production [5].

History of Thermochemical Water-Splitting Cycles

The research on thermochemical cycles for water splitting started as early as 1960s, where Funk and Reinstrom showed that this method could be more efficient than the electrochemical route for hydrogen producion [1]. The efficiency measure was defined as the ratio of the higher heating value of hydrogen produced to the heat required for it's generation. A very strong interest in thermochemical cycles emerged between 1974 and 1986 resulting in around 800 publications during this period. Most of these studies were dedicated toward various aspects of thermochemical cycles, including theoretical and practical aspects, experimental results, pilot scale plants, and techno-economic analysis. The important parameters considered while studying the different thermochemical cycles included the thermal efficiency of the cycle, chemical reactions, intermediates and side reactions, toxicity and corrosion effects of chemicals, materials compatibility at operating temperatures, maximum temperature requirement, and associated heat transfer challenges, as well as availability and costs of the chemicals used in the processes.

In 1969, during an international round-table conference on Direct Production of Hydrogen with Nuclear Heat, held at Ispra, Italy, the 24 Mark cycles were considered for further studies during the period 1970–83. A program was approved by European Communities for the period from 1973 to 1983, providing a significant boost to thermochemical water-splitting research. This program focused on thermodynamic studies to find the suitable reactions and assess their theoretical feasibility, construction of demonstration systems, and economic and comparative analysis. All these studies provided guidelines for selection of cycles and development of process flow sheets for the possible cycles. In Phase I of the research program, the focus was toward the Marchetti and DeBeni's Mark 1 cycle (in the year 1972), which was based on Hg, Ca, and Br. However, the scalability was a challenge due to inclusion of mercury. In Phase II of the program, the research was focused on Fe-Cl-based family of cycles, such as the Mark 15 cycle but the biggest challenge was economic viability of these cycles. In Phase III of the program, S-based cycles were considered, where the major reaction is sulfuric acid decomposition step followed by second step which was carried out differently by Westinghouse, General Atomics and Ispra. Considerable work on sulfuric acid decomposition was carried out in this Phase III.

Significant research contributions were carried out by the Joint Research Centre of European Union in Ispra, Italy; by Westinghouse and General Atomics in United States; and by the Japan Atomic Energy Research Institute (JAERI). At General Atomics, extensive research conducted between the middle of 1970s and 1980s led to the developement of the Mark 16 cycle also referred as the General Atomics process. Demonstration studies are carried out along with the cost and efficiency calculations. The efficiency of the process was reported to be 47%. The work on reduction in cost of the major cost-intensive step involving separation of HI from water was carried out at JAERI, Japan and Aachen, Germany. In addition to these cycles, a gas-solid reactions-based cycle, UT-3 process, was developed at University of Tokyo, Japan, in 1976. A bench-scale pilot plant named MASCOT was demonstrated for this cycle.

The UT-3 cycle, which used Ca, Br, and Fe as the main components, achieved an efficiency of 49%. A detailed process thermal efficiency, cost analysis, and process feasibility were conducted based on this pilot plant.

Around 200 thermochemical cycles were identified by General Atomics; however, very few were taken up and out of those the Sulfur-Iodine (I-S) cycle is the most studied one. Several Mark cycles were investigated, and research projects were taken up in China, Canada, Japan, US, and the EU. There was reduced interest for a certain duration from mid-1980s to 2000, due to widespread adoption of steam methane reforming drop in oil prices and since most of the research on TWSCs (Thermochemical Water-Splitting Cycle) was toward integrating with nuclear reactors, the accidents with the nuclear reactor facilities resulted in setbacks. Nevertheless, Japan continued research efforts during this period, particularly on the UT-3 cycle and the S–I cycle developed by General Atomics.

The interest in TWSCs again revived in 2003, with the launch of DOE's solar thermochemical hydrogen production research program (STCH), further many projects in China, EU, and Canada were started. The important developments like UT-3 named after University of Tokyo and Sulfur-Iodine (S-I) cycle by General Atomics, were the major achievements and also led to gain the global interest in thermochemical cycles.

Since the early 2000s, extensive research work has been carried out on Cu-Cl and hybrid Cu-Cl cycles in Canada and US; on UTC (Uranium three-step cycle) in China and US; metal oxides cycles in Germany (German Aerospace Research Centre DLR), France, Switzerland (Paul Scherrer Institute) and South Korea; S-I cycle in China, Japan, US, and EU; while on UT-3 in Japan.

Atomic Energy of Canada Limited (AECL) along with University of Ontario Institute of Technology and Argonne National Laboratory initiated joint research program in 2007 on Cu-Cl cycle and its integration with nuclear reactors. Besides, Ontario Institute of Technology also worked toward other cycles including the Mg-Cl, with the aim at improving the efficiency of the cycle.

Overall, more than 3000 thermochemical cycles have been identified and evaluated theoretically; however, only a small fraction could reach to experimental validation and demonstration stages. The selection of the cycles for advancement depends on the factors such as chemical viability, safety, materials availability, thermodynamics, number of reaction steps, and the simplicity of the cycle, material abundance, efficiency, and cost. The later surge of interest has been toward cycles where the required heat input can be utilized from solar thermal with point focusing concentration solar technologies. Several demonstrations have employed central receiver systems with heliostat fields concentrating solar radiation onto receivers located on top of the towers to achieve the high temperatures required for thermochemical cycles. Among the most studied cycles include, metal oxides-based cycles and sulfur-iodine cycle. When a heat source other than thermal energy is used to drive a cycle, it is classified as hybrid thermochemical cycle.

Thermochemical Cycles for Water-Splitting

In thermochemical water-splitting, the water dissociates during a series of chemical reactions involving formation of several intermediate compounds, such that all materials used in the process return to their initial state, leaving only hydrogen and oxygen as the final products, which is one of the highly endothermic step. Thus, the energy required for thermochemical cycles is supplied in the form of thermal energy. In contrast, hybrid cycles may utilize thermal energy in combination with other forms of energy like electrical or photonic. The required heat for the cycles can be obtained from either the waste heat from a nuclear reactor, from solar thermal or partly as electrical, or photonic energy. The theoretical efficiencies of these cycles are high, i.e., more than 50% but actual efficiencies are relatively low due to losses and irreversibility of the materials involved.

The splitting of water in the thermochemical cycles is carried out in two or more steps, since for the one-step decomposition the heat requirement is extremely high, thus unfavorable thermodynamics of the process, and separation of gases at such high temperatures to avoid the formation of an explosive mixture are the biggest challenges. At temperature exceeding 2000 K, not only materials and metallurgical challenge, but providing the required heat of reaction and having efficient heat exchange mechanism along with economic feasibility also become a major concern. Therefore, considering technical viability if upper limit is considered to be 1000 K and lower limit 300 K, then the maximum achievable thermal efficiency, considering Carnot cycle, is approximately 88%.

These thermochemical cycles can have high conversion efficiencies and thus can be useful for large-scale production of hydrogen. However, it is to be noted that in order to reduce the temperature required for the maximum temperature step of the process compared to the water thermolysis, the number of steps required are two or more than two, but with the increase in the number of steps the cycle gets complicated in terms of process engineering, leading to the reduction in efficiency.

Metal Oxide-Mediated Water Splitting

A study carried out by Abraham and Schreiner [6], based on entropy change calculations, has shown that for any thermochemical cycle operating below 1000 K, the required number of steps is at least three. However, for temperatures beyond this, the constraints for compound formation via thermochemical route are relaxed, and in such cases the two-step cycles can be carried out.

Interest toward metal oxide-based thermochemical cycles arises from their simplicity and smaller number of steps involved. Using different metal oxides, the splitting of water can be achieved in two steps; however, the required temperature is very high, typically in the range of 1700–3000 K, which makes these cycles less attractive. In two-step thermochemical cycles, the fundamental reactions are reduction–oxidation reactions involving metal oxides and water to produce oxygen and hydrogen, respectively. In the first step, the metal oxide in its higher valence state is reduced to a metal oxide in a lower valence state, liberating oxygen; in some cases, the metal oxide may even be reduced to the metallic state (zero-valence). This step is usually the rate-limiting step. In the second step, the metal or metal oxide in its lower oxidation state is oxidized back to the parent metal oxide in the higher valence state using water and producing hydrogen. The first step is a reduction, regeneration, or activation step which is endothermic, and the required energy can be provided by either solar thermal or nuclear sources, whereas, the second step is a hydrolysis step or a water-splitting step which is exothermic in nature (Fig. 6.2).

Since hydrogen and oxygen are produced in two separate steps, as such no separation of the products, i.e., hydrogen and oxygen are required. Several metal oxide redox pairs have been considered for the process including Fe_3O_4/Fe, ZnO/Zn, SnO_2/SnO, CeO_2/Ce_2O_3, Mn_2O_3/MnO, CdO/Cd, GeO_2/GeO, Co_3O_4/CoO, etc. [1].

Using Fe_3O_4/FeO pair of metal oxides, the reaction that takes place is:

$$Fe_3O_4 \rightarrow 3FeO + \frac{1}{2}O_2 \quad T > 1873\,\text{K Endothermic reaction} \tag{6.1}$$

$$3FeO + H_2O \rightarrow Fe_3O_4 + H_2 \quad \text{Exothermic reaction.} \tag{6.2}$$

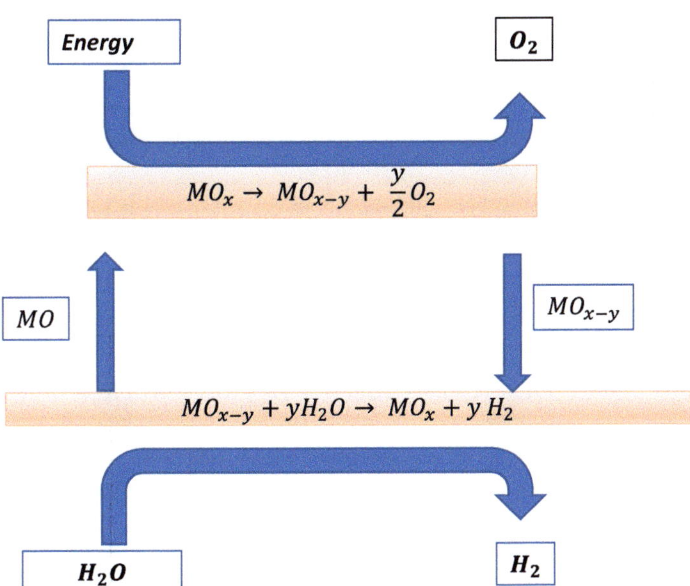

Fig. 6.2 Metal oxide-mediated two-step thermochemical cycle

The theoretical efficiency of the water-splitting reactions lies between 50–62%, while the actual efficiency using the concentrated solar energy (at 2300 K) is limited between 20 and 25% only, primarily due to thermal losses, and additional challenges such as oxide volatilization and material degradation. Since the temperatures required for iron oxide-based two-step cycles are extremely high, so in order to reduce the temperature involved, the iron in the metal oxide can be substituted by other metals (for e.g., Ni, Zn, Co, Mn, etc.) forming mixed metal oxides, thus forming the reduced and oxidized phases $(Fe_{1\text{-}x}A_x)_{1\text{-}y}O$ and $(Fe_{1\text{-}x}A_x)_3O_4$ respectively.

Zn/ZnO-based redox pair is another attractive option which was reported to have a maximum solar thermal energy efficiency of 40%. The process involves two steps, dissociation of ZnO to Zn, producing oxygen which takes place at 2273 K and hydrolysis, where Zn returns to its initial state and ZnO releases hydrogen at a lower temperature of 723 K.

$$ZnO \rightarrow Zn + \frac{1}{2}O_2 \tag{6.3}$$

$$Zn + H_2O \rightarrow ZnO + H_2 \tag{6.4}$$

Among the mentioned redox pairs, the ZnO/Zn is mostly studied, the energy requirement is fulfilled from solar energy. The cycle with ZnO/Zn has been tested as early as 1990 but several challenges were identified including sluggish kinetics, separation of zinc, and oxygen leading to recombination, reverse reactions, economics (hydrogen cost from the cycle \$5/kgH_2) etc., which makes it difficult to develop on industrial scale. Another process where carbon was added to the cycle is being developed for reduction of ZnO using a 300kW$_{th}$ reactor. Multiple demonstrations integrate the ZnO-based cycles with solar thermal in order to provide the required high-temperature heat for systems form 10 kW to 90 MW plant size with solar to fuel efficiencies lying in the range of 36–40%.

The another metal oxide pair studied is CeO_2/Ce_2O_3 and is the most promising one for the two-step oxidation reduction-based thermochemical cycle.

$$2CeO_2 \rightarrow Ce_2O_3 + \frac{1}{2}O_2 \quad 2273\,K \tag{6.5}$$

$$Ce_2O_3 + H_2O \rightarrow 2CeO_2 + H_2 \quad 673 - 873\,K \tag{6.6}$$

Among the various redox pairs, the ones which undergo stoichiometric phase change (ZnO, Co_2O_3) shows high specific energy storage compared to those undergoing partial reduction which have faster kinetics and lower temperature operation but lower specific energy storage [7].

Another class of compounds used for two-step TWSC includes multicomponent oxide with multiple valence states, such as perovskites or spinels with the general formula ABO_3 or AB_2O_4 respectively. These multivalent metal oxides can form a wide range of non-stoichiometric compounds. The reduction step in these materials is partial reduction (e.g., perovskites and CeO_2, etc.) where the reduction occurs with the formation of oxygen vacancies and not leading to any change in crystal structure, the thermodynamics of the process is governed by the equilibrium between oxygen gas and oxygen vacancies. These materials can allow movement of oxygen, i.e., reduction and oxidation, without impacting the crystal structure. Such materials can utilize or store lesser specific energy for hydrogen production; however, they can operate at relatively lower temperature than ZnO or Co_3O_4 but still can have faster kinetics. With the stoichiometric counterpart the temperature of the reduction step is higher and possesses slow kinetics, the process is called phase change reduction involving complete phase change along with release of oxygen (e.g., of such materials ZnO, Fe_3O_4, Co_3O_4, etc.).

Some cases of binary oxides like Cu–Fe, Cu–Mn, Cu–Co or Co–Fe, or Fe–Mn can exhibit higher cycle life, better kinetics, and lower reaction temperatures.

In summary, in the two-step cycles, the major issue is the high temperature requirement, although these are simple and have fewer steps in the cycle. Another challenge with the high temperature reduction step is evaporation of volatiles and reduction in the concentration causing changes in the composition, thereby resulting in to reduced activity. However, when relatively low-temperature reduction step is considered, complete reduction doesn't occur, which results in lesser hydrogen production and reduced efficiency. In order to reduce the temperature of the step having highest temperature, cycles with three or four steps were developed.

Three-Step Thermochemical Cycles

In order to avoid the high-temperature step of the two-step process, additional step is introduced where the highest temperature step of the two-step process can be replaced by two steps to reduce the required temperature. The entire process steps involved can be represented as

$$H_2O + X \rightarrow \frac{1}{2}O_2 + XH_2 \tag{6.7}$$

$$XH_2 + Y \rightarrow \frac{1}{2}H_2 + XY \tag{6.8}$$

$$XY \rightarrow X + Y. \tag{6.9}$$

Or,

$$H_2O + X \rightarrow H_2 + XO \tag{6.10}$$

$$XO + Y \rightarrow \frac{1}{2}O_2 + XY \tag{6.11}$$

$$XY \rightarrow X + Y. \tag{6.12}$$

Sulfur-Iodine Cycle—The most widely studied and known cycle involving three steps is the sulfur-iodine cycle (S-I), which is also known as General Atomics Cycle (GA cycle).

The required high temperature can be supplied by concentrated solar or nuclear reactors. The steps involved in S-I cycle are as follows:

1. Sulfuric acid decomposition which occurs at high temperature 800 °C in vapor phase, producing oxygen, water, and SO_2, this is an endothermic reaction

$$H_2SO_4 \rightarrow H_2O + SO_2 + \frac{1}{2}O_2(1120 - 1200\,K). \tag{6.13}$$

Here, H_2SO_4 acts as oxygen carrier.

2. SO_2 formed in the earlier process along with iodine and water produces again the sulfuric acid and HI, the reaction is exothermic and occurs at less than 120 °C

$$2H_2O + SO_2 + I_2 \rightarrow H_2SO_4 + 2HI \quad (300 - 400\,K). \tag{6.14}$$

This reaction is called as Bunsen reaction. It is exothermic in nature and constitutes a crucial step in the cycle. However, corrosion issues and the separation of the two acids are the major challenges. Their separation by distillation significantly reduces the overall cycle efficiency, as it requires substantial input of energy. Therefore, there is a need to reduce the energy required for the separation of two acids, as well as to minimize the consumption of both iodine and water. An excess of water is required in the process, while an excess of iodine is needed during the separation step, because significant amount of iodine is lost in the process.

3. The HI produced in step 2 undergoes decomposition which is an endothermic reaction, producing hydrogen and iodine.

$$2HI \rightarrow I_2 + H_2 \quad (400 - 1000\,K) \tag{6.15}$$

The important part of the cycle is optimization of the step 2 in this reaction, HI acts as hydrogen carrier. However, the major challenges associated with the cycle are: the requirement of high temperature (> 800 °C), corrosive and highly reactive nature of the reactants, separation of HI from sulfuric acid, requirement of an excess amount of water and iodine, and loss of iodine. The energy required for the reaction can be supplied from nuclear reactors. Several studies have been carried out to address the mentioned challenges, one such study proposes avoiding the sulfuric acid decomposition step, with the required SO_2 supplied from refinery flue gases, thereby simultaneously achieving desulfurization. In this modified process, both hydrogen and sulfuric acid are obtained as products, which have an added market value and improve the overall economics of the process. This modification is referred as an open loop S–I cycle [8, 9].

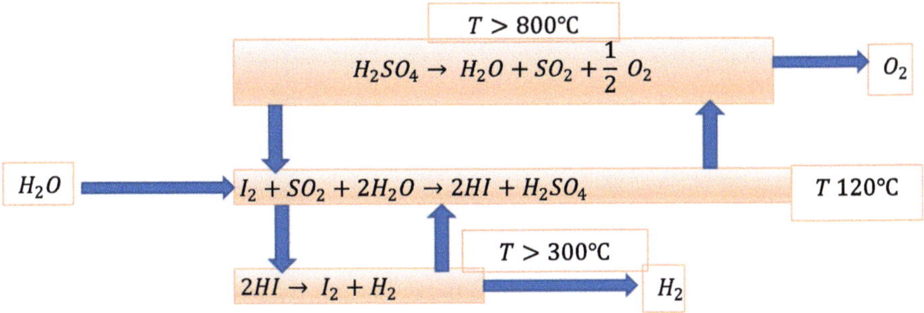

Fig. 6.3 Three-step iodine–sulfur thermochemical cycle

S–I cycle is the most widely studied three-step thermochemical cycle with required energy being supplied from solar energy [10] (Fig. 6.3).

The other variations to S–I cycles are hybrid cycles like Mark 11 or Mark 13V2, etc.

In hybrid Ispra Mark 11 or Westinghouse cycle, which is a hybrid S-I cycle, the energy required for the sulfuric acid formation is provided by electrical energy. The steps in the cycle involved are as follows:

$$H_2SO_4 \rightarrow H_2O + SO_3 \quad T > 723\,K \tag{6.16}$$

$$SO_3 \rightarrow SO_2 + \frac{1}{2}O_2 \quad T > 1123\,K - 1423\,K \tag{6.17}$$

$$2H_2O + SO_2 \rightarrow H_2SO_4 + H_2 \quad T \sim 353\,K \text{ electrochemical step.} \tag{6.18}$$

The first two reactions in fact constitute a single step making it hybrid two-step cycle. The last step is an electrochemical step which requires ~ 0.17 V, is much less than that required for electrolysis of water (1.23 V) to produce hydrogen. Theoretical efficiency of the process is 40%. The project employing scaling up of this process is "Solar to Hybrid Hydrogen Cycle" also called as SOL2HY2, with the estimated cost of hydrogen produced being 3–3.5 euro/kg.

Another version of the S-I hybrid cycle is Mark 13 V2 is again a hybrid cycle with steps as follows:

$$H_2SO_4 \rightarrow H_2O + SO_2 + \frac{1}{2}O_2 \quad T = 1112\,K - 1393\,K \tag{6.19}$$

$$SO_2 + Br_2 + 2H_2O \rightarrow H_2SO_4 + 2HBr \quad T = 373 - 413\,K \tag{6.20}$$

$$2HBr \rightarrow Br_2 + H_2 \quad \text{electrochemical step.} \tag{6.21}$$

The theoretical efficiency of this cycle is 37%.

Thermochemical Cycles with Four or More Steps

The concept underlying multistep thermochemical cycles is that increasing the number of steps reduces the maximum temperature required for the cycle. If this maximum temperature could be further lowered to a level where waste heat or any process heat can be utilized, the efficiency and economics of the process can be markedly improved. This forms the basis for four-step or higher-order thermochemical cycles. Initial research in thermochemical water splitting focused on supplying the required high-temperature heat from nuclear reactors; however, as such reactors were not readily available for widespread experimentation or practical utilization, the required heat was provided from solar thermal. Further reduction in the required heat or temperature for water splitting enabled the use of industrial process heat, waste heat, or other low-temperature energy source to drive the thermochemical water-splitting reactions. In this context, the first cycle studied was Fe–Cl cycle which belongs to the chlorine family of thermochemical cycles. In general, the four steps involved in any four-step thermochemical cycle are as follows:

1. Hydrolysis reaction

$$H_2O + XY \rightarrow XH_2 + YO \tag{6.22}$$

2. Hydrogen release reaction

$$XH_2 \rightarrow H_2 + X \tag{6.23}$$

3. Oxygen release reaction

$$YO \rightarrow Y + \frac{1}{2}O_2 \tag{6.24}$$

4. Regeneration or recycle reaction

$$X + Y \rightarrow XY. \tag{6.25}$$

Argonne National Laboratory studied the various chlorine family cycles and identified five cycles on the basis of parameters like thermal efficiency, thermodynamic feasibility, the chemical used, viability, and cost competitiveness. These were Cu-Cl, Fe-Cl, Ce-Cl, V-Cl, and Mg-Cl cycles. Among all these cycles the Cu-Cl and V-Cl cycles showed higher efficiencies, i.e., 43% and 46% respectively, while the Fe-Cl cycle was considered to be studied further in order to improve its efficiency.

Fe–Cl cycle is among the 24 cycles identified by European Community Joint Research Centre and named as Mark 15 [11]. This cycle was first studied in 1976 by

Funk and is especially attractive due to the high abundance of iron compared to other elements used in four-step cycles, as well as the well-established chemistry of iron oxides.

$$3FeCl_2 + 4H_2O \rightarrow Fe_3O_4 + 6HCl + H_2 \quad T = 650\,°C \tag{6.26}$$

$$Fe_3O_4 + 8HCl \rightarrow FeCl_2 + 2FeCl_3 + 4H_2O \quad T = 125\,°C \tag{6.27}$$

$$2FeCl_3 \rightarrow 2FeCl_2 + Cl_2 \quad T = 425\,°C \tag{6.28}$$

$$Cl_2 + H_2O \rightarrow 2HCl + \frac{1}{2}O_2 \quad T = 650\,°C \tag{6.29}$$

Reactions (6.26) and (6.29) are endothermic, while (6.27) and (6.28) are exothermic in nature. The major challenges of this cycle are: (a) the poor reactivity of the step involving thermal decomposition of $FeCl_3$ (which actually exists as Fe_2Cl_6) (b) lowering the temperature of reverse Deacon reaction, i.e., reaction (6.29) and the hydrolysis (6.26) of $FeCl_2$) the separation of the HCl and solid products. Due to these challenges the research was more focused toward other cycles like Cu–Cl or UT3 cycles (Fig. 6.4).

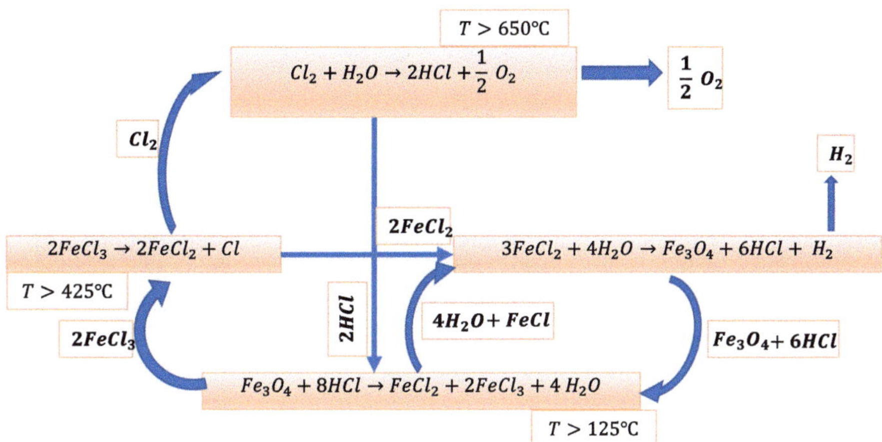

Fig. 6.4 Four-step iron-chlorine thermochemical cycle

Cu–Cl Cycle

The well-known cycles based on the four steps are Cu–Cl and V–Cl, and have high efficiencies. The other four-step cycles include Ce–Cl, Mg–Cl cycles. Some intermediate steps are involved in these cycles to lower down the temperature and at the same time to improve the efficiencies. The Cu–Cl cycle has the lower maximum operating temperature (<550 °C), thus the required energy can be obtained from industrial waste heat recovery units. Besides that, it has the advantage of high efficiency and high yield of hydrogen produced. This cycle was studied by Atomic Energy of Canada Limited and Ontario Institute of Technology Canada.

Hybrid Thermochemical Cycles

In the hybrid thermochemical cycles, the required temperatures are lower, such that waste heat or other lower temperature energy sources can be utilized for the processes. Consequently, hybrid cycles generally exhibit higher energy efficiencies, typically ranging from 48–50%.

In these cycles, both heat and work is utilized, with electricity contributing toward one of the steps; therefore, they are often referred as thermochemical-electrochemical combined cycles.

Hybrid Cu–Cl Cycle

Three Cu–Cl cycles have been proposed, namely the three-step, four-step, and five-step cycles. Among these, the four-step cycle offers the advantage of reduced complexity while simultaneously requiring the lower operating temperatures; therefore, these were considered for further studies. Cu–Cl cycle involves four steps and is being very widely studied. Among the four steps involved; three are, thermal, i.e., the hydrogen, oxygen, and hydrochloric acid generation; and one step is electrochemical which is the $CuCl_2$ formation step.

$$2\mathrm{Cu} + 2\mathrm{HCl} \rightarrow 2\mathrm{CuCl} + \mathrm{H}_2 \quad 700 - 750\,\mathrm{K} \tag{6.30}$$

$$4\mathrm{CuCl} \rightarrow 2\mathrm{CuCl}_2 + 2\mathrm{Cu} \quad \text{electrochemical } 300 - 350\,\mathrm{K} \tag{6.31}$$

$$2CuCl_2 + H_2O \rightarrow CuO.CuCl_2 + 2HCl620 - 673K \tag{6.32}$$

$$\mathrm{CuO} \cdot \mathrm{CuCl}_2 \rightarrow 2\mathrm{CuCl} + \frac{1}{2}\mathrm{O}_2 \quad 720 - 800\,\mathrm{K} \tag{6.33}$$

The electrochemical step requires a low voltage of 0.4–0.6 V, which is very less compared to 1.23 V required for electrolysis of water. The heat required for maximum temperature step can come from concentrated solar thermal. The theoretical efficiency of the cycle is 41–43%. Both Ontario Institute in Canada and Argonne National Lab have worked extensively toward developing and demonstrating the cycle.

Westinghouse Cycle

The Westinghouse cycle also known as two-step hybrid sulfur cycle or ISPRA Mark 11 cycle, is a hybrid thermochemical electrochemical cycle and is well-known with only two steps. The first being thermal decomposition of sulfuric acid and second being regeneration of sulfuric acid by oxidation of SO_2 using water with electrical input, resulting in the formation of sulfuric acid and evolution of hydrogen in the process [12]. The process steps are:

$$H_2SO_4 \rightarrow H_2O + SO_3 \quad > 723\,\mathrm{K} \tag{6.34}$$

$$SO_3 \rightarrow SO_2 + \frac{1}{2}O_2 > 1123\,\mathrm{K} \tag{6.35}$$

$$2H_2O + SO_2 \rightarrow H_2SO_4 + H_2 \quad 353\,\mathrm{K}\ \text{electrochemical step.} \tag{6.36}$$

The third reaction (6.36) mentioned above is electrochemical oxidation of SO_2, with a voltage requirement of 0.17 V. The second reaction (6.35) which is essentially a sub step, can be carried out at 1123 K in the presence of an appropriate catalyst (generally iron oxide), while without catalysts the required temperature is 1423 K. The process step of decomposition which requires more energy can be located close to the heat source or reactor while the electrolysis step, which produces hydrogen, can be carried out at a safer distance from the plant [13].

The theoretical overall thermal efficiency of the process is 40%. However, experimental testing using solar furnace yielded an efficiency of about 28%. Further improvements, such as optimizing the flow rates and use of multichamber reactors can offer higher efficiency and scaling-up to projects of the order of a few hundred kW_{th}. The major challenges with this cycle are the highly corrosive environment and the high temperature requirement for the SO_3 reduction. This cycle has been studied at Savannah River National Laboratory (SRNL), the Commonwealth Scientific and Industrial Research Organization (CSIRO), and DLR Germany (Fig. 6.5).

The Mark 13 V2 cycle developed at IRC IPSRA, Italy, combines sulfuric acid and HBr; however, the separation of oxygen and SO_2 as separate gas stream is the major challenge along with cost, efficiency, and corrosion being the other associated problems [14]. The cycle involves the following reactions:

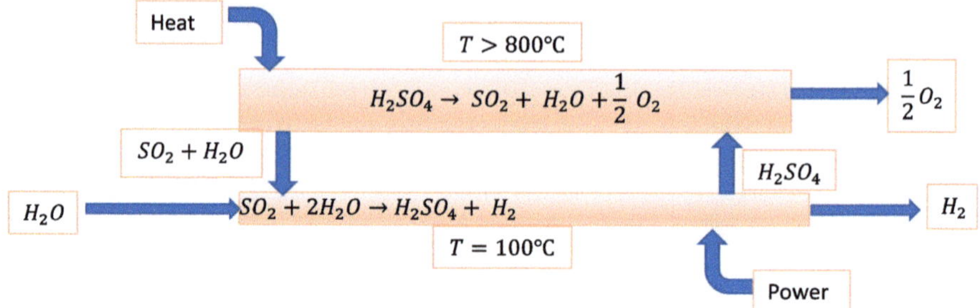

Fig. 6.5 Two-step hybrid sulfur cycle called Westinghouse cycle

$$SO_2 + Br_2 + 2H_2O \rightarrow 2HBr + H_2SO_4 \quad 373 - 413\,K \tag{6.37}$$

$$2HBr \rightarrow H_2 + Br_2 \quad \text{electrochemical step} \tag{6.38}$$

$$H_2SO_4 \rightarrow H_2O + SO_2 + \frac{1}{2}O_2 \quad 1123 - 1393\,K. \tag{6.39}$$

Mark 13A developed by JRC IPSRA, Italy, was used for desulfurization in refineries and had an open loop, where sulfuric acid was not recycled and the energy coming with flue gases was used to produce hydrogen.

Hybrid Sulpfur-Ammonia Cycle (HySA Cycle)

Some of the other hybrid cycles have the thermal and other energy input like photonic. The HySA cycle developed used ammonia as a reagent and is a photoelectrochemical cycle. The processes in the cycle are as follows:

1. Chemical absorption (25 °C)

$$SO_2 + 2NH_3 + H_2O \rightarrow (NH_4)_2SO_3 \tag{6.40}$$

2. Electrolytic oxidation (80–150 °C)

$$(NH_4)_2SO_3 + H_2O \rightarrow (NH_4)_2SO_4 + H_2 \tag{6.41}$$

3. Solar thermal conversion (400 °C)

$$(NH_4)_2SO_4 + K_2SO_4 \rightarrow K_2S_2O_7 + 2NH_3 + H_2O \tag{6.42}$$

4. Solar thermal conversion (550 °C)

$$K_2S_2O_7 \rightarrow K_2SO_4 + SO_3 \tag{6.43}$$

5. Solar thermal conversion (850 °C)

$$SO_3 \rightarrow SO_2 + \frac{1}{2}O_2. \tag{6.44}$$

The oxygen production step is the high-temperature step, the cycle has not yet been demonstrated completely and is being studied at Florida Solar Energy Centre.

The biggest advantage of this cycle is that the UV and visible portion of the light is utilized for hydrogen evolution step, while the infrared is utilized for oxygen evolution step.

UT-3 (Ca/Fe/Br Cycle)

The UT-3 cycle is named after the University of Tokyo, where the cycle was proposed, demonstrated and a lot of research has been carried out. This cycle involves four steps which are gas–solid reactions, out of these two are Ca-based and rest two are Fe-based [15].

$$CaBr_2 + H_2O \rightarrow CaO + 2HBr \quad 950 - 1000\,K \tag{6.45}$$

$$CaO + Br_2 \rightarrow CaBr_2 + \frac{1}{2}O_2 \quad 750 - 800\,K \tag{6.46}$$

$$Fe_3O_4 + 8HBr \rightarrow 3FeBr_2 + 4H_2O + Br_2 \quad 500 - 550\,K \tag{6.47}$$

$$3FeBr_2 + 4H_2O \rightarrow Fe_3O_4 + 6HBr + H_2 \quad 950 - 1000\,K \tag{6.48}$$

The theoretical thermal efficiency of the cycle is approximately 45%; however, the reported efficiency is lower, around 13%. Separation in this cycle is relatively easier because it involves solid-gas phase separation. Consequently, current research is focused toward finding stable materials and developing membrane-based separation techniques. Altough this cycle is less mature than S-I cycle, however the advantage against S-I cycle is easy separation of solid and gas as compared to S-I, where liquid reagents need to be separated.

The interest toward these thermochemical cycles emerged in early 1970s due to oil price shocks. However, this interest declined with the reduction in oil prices and the availability of efficient and low-cost steam reforming technologies for hydrogen production. In recent years, interest toward thermochemical water-splitting cycles has revived due to the growing interest toward the green hydrogen production and the potential for better efficiencies when these cycles are integrated with solar or nuclear plants. However some of the key points to be mentioned includes:

- These cycles are yet not commercialized and are still under development stage even after more than half a decade of studies.
- Thermochemical cycles have the potential to produce hydrogen on a large scale in a sustainable manner.
- The selection of a suitable cycle depends on thermodynamic feasibility, cost, efficiency, and sustainability considerations.
- Among chlorine-based cycles, V–Cl exhibited high efficiency, while the Cu–Cl and Mg–Cl cycles offer lower cost and better integration at lower temperatures; Fe–Cl shows the lowest exergy efficiency [1].
- I–S can be integrated with nuclear (high T required) but corrosion remains the major challenge.
- Metal oxide-based cycles are well-suited for integration with solar thermal systems.
- ZnO/Zn cycle results in higher cost of hydrogen production due to expensive reactants used.
- Sulfur-based cycles exhibit relatively lower cost of hydrogen production, i.e., \$2/ kg of hydrogen produced.
- Severe corrosion issues and high-temperature SO_3 decomposition step reduce the attractiveness of sulfur-based cycles.
- Cu–Cl cycle emerged as a promising candidate for large-scale hydrogen production, due to its lower temperature reactions, lower cost of hydrogen production, and easy integration with the different heat sources.

Solved Problem 1. For the thermochemical water-splitting reaction:

$$H_2O(l) \rightarrow H_2(g) + \frac{1}{2}O_2(g).$$

The standard enthalpy and entropy changes at 298 K are given as:

$\Delta H^\circ = 241.8$ kJ/mol.
$\Delta S^\circ = 188.8$ J/mol K.

Calculate the equilibrium constant K_2 at 1500 K given that $K_1 = 10^{-42}$ at 298 K.
Solution.
Using the Van't Hoff equation given as follows:

$$\ln\left(\frac{K_2}{K_1}\right) = -\frac{\Delta H^0}{R}\left(\frac{1}{T_2} - \frac{1}{T_1}\right).$$

Rearranging the Van't Hoff equation

$$K_2 = K_1 e^{-\frac{\Delta H^0}{R}\left(\frac{1}{T_2} - \frac{1}{T_1}\right)}.$$

Substituting the values,

$$\Delta H^0 = 241{,}800\,\text{J/mol K}.$$

R = 8.314 J/ mol K.
$K_1 = 10^{-42}$, $T_1 = 298$ K, $T_2 = 1500$ K

$$\ln\left(\frac{K_2}{K_1}\right) = 78.2$$

$$K_2 = 10^{-42} * e^{78.2} = 9.1 * 10^{-9}$$

References

1. F. Safari, I. Dincer, A review and comparative evaluation of thermochemical water splitting cycles for hydrogen production. Energy Convers. Manage. **205**, 112182 (2020)
2. I. Dincer, C. Acar, Review and evaluation of hydrogen production methods for better sustainability. IJHE **40**, 11094–11111 (2015)
3. T. Kodama, N. Gokon, Thermochemical cycles for high temperature solar hydrogen production. Chm. Rev. **107**, 4048–4077 (2007)
4. J.E. Funk, Thermochemical hydrogen production: past and present. IJHE **26**, 185–190 (2001)
5. R. Pinsky, P. Sabharwall, J. Hartvigsen, J. O'Brien, Comparative review of hydrogen production technologies for nuclear hybrid energy systems. Progr. Nuclear Energy **123**, 103317 (2020)
6. B.M. Abraham, Thermochemical water splitting cycles: impact of thermal burdens and kinetics. Int. J. Hydrogen Energy **5**(5), 507–513 (1980)
7. B. Bulfin, J. Vieten, C. Agrafiotis, M. Roeb, C. Sattler, Applications and limitations of two step metal oxide thermochemical redox cycles; a review. J. Mater. Chem. A **5**, 18951 (2017)
8. USDOE, *Hydrogen Energy Strategy Plan* (2011). Last accessed 02–07–2025
9. R. Liberator, M. Lanchi, G. Caputo, C. Felici, A. Giaconia, S. Sau, P. Tarquini, Hydrogen production by flue gas through sulphur iodine thermochemical process: economic and energy evaluation. IJHE **37**, 8939–8953 (2019)
10. P. Charvin, S. Abanades, F. Lemort, G. Flamant, Hydrogen production by three step solar thermochemical cycles using hydroxides and metal oxide systems. Energy Fuels **21**, 2919–2928 (2007)
11. F. Safari, I. Dincer, A study on the FeeCl thermochemical water splitting cycle for hydrogen production. IJHE **45**, 18867–18875 (2020)
12. E. Bilgen, Solar hydrogen production by hybrid thermochemical processes. Solar Energy **41**, 199–206 (1988)
13. R. Pinsky, P. Sabharwall, J. Hartvigsen, J. O'Brien, Comparative review of hydrogen production technologies for nuclear hybrid energy systems. Progr. Nuclear Energy **123**, 103317 (2020)
14. G.E. Beghi, A decade of research on thermochemical hydrogen at the joint research centre at ISPRA. IJHE **11**, 761–771 (1986)
15. S. Yalcin, A review of nuclear hydrogen production. IJHE **14**, 551–561 (1989)

Chapter 7
Hydrogen Production from Electrolysis of Water

Introduction

Electrolysis is an electrochemical process, that involves splitting water into its constituents, hydrogen, and oxygen, by means of providing electrical energy. Thus, the device, "electrolyzer", converts electrical energy into chemical energy, producing hydrogen. The process is considered a clean process, and if the electricity used for producing hydrogen comes from a renewable source, then it can be the cleanest route, with the hydrogen produced called green hydrogen. The renewable source used can be either solar photovoltaic, wind, hydroelectric, or geothermal energy. The electrolysis process is a non-spontaneous process with a positive Gibbs free energy over the practical range of temperatures and pressures. It gets negative above 2000 K only, which is a very harsh condition to operate with.

The history of water electrolysis [1] ways back into 1789, when J.R. Deiman and A.P. van Troostwijk for the first time observed the evolution of gases, on placing gold wires in a tube containing water, and electricity was passed through the wires. Thereafter, in the year 1800 Alessandro Volta invented the voltaic pile, which was used just a week later by Nicholson and Carlisle to split water. With Faraday's laws of electrolysis, the relationship between electrical energy consumption and the amount of gases produced was identified. Zanobe Gramme was the one to develop the Gramme machine in the year 1869, to produce hydrogen economically. Dmitry Lachinov developed the industrial hydrogen production technology in the year 1888 using electrolysis and which resulted in more than 400 industrial electrolyzers operating by the year 1902. Electrolysis technology has seen the maximum development in the period between 1920 and 1970. In the year 1939, the first high-capacity water electrolysis plant was established and became operational (10,000 Nm^3/h). The polymer electrolyte-based electrolyzer was first developed by General Electric in the year 1966, the first solid oxide electrolyzer was developed in the year 1978, and the first advanced alkaline electrolyzer was also developed in the same year. There were several large installations like 100 MW in Canada in 1920, with a 162 MW

P. Sharma, *A Textbook on Hydrogen Energy*,
https://doi.org/10.1007/978-981-95-4601-5_7

capacity electrolyzer to generate 32,400 m^3/h by Aswan, and Brown Boveri developed a modular electrolyzer to produce 4300 m^3/h of hydrogen. Thereafter, when the hydrocarbon-based hydrogen production from steam methane reforming or coal gasification started, and these routes were found to be more economical, leading to reduced interest in electrolytic hydrogen production [2]. However, a surge of interest has been seen over the last decade with the objective to abate greenhouse gas emissions, address climate change issues, meet the requirement of net zero by various countries globally, and due to the availability of cheaper renewable power and large-scale renewables deployment.

Electrolysis is used for various applications like metal plating and materials finishing; energy conversion devices; production, extraction, and purification of metals; synthesis of various chemicals; water purification, etc. The two major industries that use electrolysis are the chlor-alkali industry, used for chlorine and caustic soda production, and the aluminum industry, used for electrowinning. Chlor-alkali industry is a major industry where the electrolysis of aqueous brine solution results in the production of by-product hydrogen. In India approximately, 0.2 MTPA of hydrogen is obtained from this industry.

Basics of Water Electrolysis

The basic unit of an electrolyzer is an electrochemical cell in which the ions move across the liquid or solid phase electrolyte which should have high ionic conductivity, and the electrons through the solid phase between the conducting materials called electrodes, which are connected electronically (Fig. 7.1).

A basic electrolytic cell consists of three major components; two electrodes, an electrolyte, and a separator. The requirement for electrodes is that they need to have good electronic conductivity, corrosion resistance, excellent catalytic properties, and structural integrity. The requirements for the electrolyte are that, it should have good ionic conductivity, should not erode or consume the electrodes, and should not undergo any change during the process of electrolysis. Separator or diaphragm used should be such that it separate the two electrodes, preventing shorting, should be able to provide high ionic conductivity, should not allow the two product gases, hydrogen and oxygen, to mix, and should have good chemical and mechanical stability in the operating medium.

The mechanism, in simple terms, can be expressed as on passage of DC current through an electrolyzer, the electrons flow through the negative terminal and reach at cathode, these electrons are taken up by the hydrogen ions to produce hydrogen gas.

For acid electrolyte

Cathode:

$$2H^+ + 2e^- \rightarrow H_2 \tag{7.1}$$

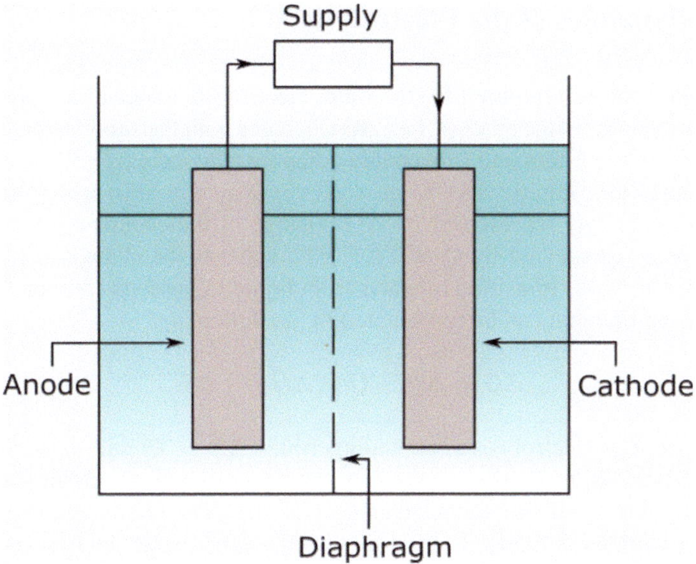

Fig. 7.1 Schematic of a basic electrochemical cell

Anode:

$$H_2O \rightarrow \frac{1}{2}O_2 + 2H^+ + 2e^- \tag{7.2}$$

For alkaline electrolyte:

Cathode:

$$2H_2O + 2e^- \rightarrow H_2 + 2OH^- \tag{7.3}$$

Anode:

$$2OH^- \rightarrow \frac{1}{2}O_2 + H_2O + 2e^- \tag{7.4}$$

The water is split on the anode to produce positive ions and electrons, liberating oxygen at the anode. The electrons flow through the external circuit, reaching the cathode to liberate hydrogen, while the hydrogen ions permeate through the separator and reach the cathode. The gas receivers collect the gases formed at the anode and cathode, respectively. Thus, the reduction half reaction occurs at the cathode, liberating hydrogen, and the oxidation half reaction at the anode, liberating oxygen.

The overall cell reaction can be written as:

$$H_2O \rightarrow H_2 + \frac{1}{2}O_2. \tag{7.5}$$

Thermodynamics of the Electrolysis

The amount of energy required for the water electrolysis to occur is represented by the enthalpy change in the process, i.e., ΔH (includes both thermodynamic reversible energy for water splitting and irreversible component associated with latent heat of phase change from liquid water to gaseous components), this energy required is partly electric energy represented by ΔG (change of the chemical potentials) and partly thermal represented by Q or $T\Delta S$, where ΔS is the change in the entropy associated with phase transition of water from liquid to gaseous components. Thus, the energy requirement can be represented by the following:

$$\Delta G = \Delta H - Q = \Delta H - T\Delta S. \tag{7.6}$$

The process of electrolysis is an endothermic ($\Delta H > 0$) and non-spontaneous ($\Delta G > 0$) process, while the reverse of the process, which occurs in fuel cells, is exothermic ($\Delta H < 0$) and spontaneous ($\Delta G < 0$).. The Gibbs free energy under standard conditions is referred to as the lower heating value, and the enthalpy at standard conditions as the higher heating value.

Thus, galvanic cells like batteries or fuel cells have chemical reactions, such that their Gibbs free energy is negative and they produce power, i.e., electricity generation. In the case of electrolyzers, since the Gibbs free energy is positive, the process is thermodynamically unfavorable, and it requires energy to carry out the process, and that required energy is fed using electrical energy to such devices to carry out the non-spontaneous reaction. The amount of energy fed or the potential difference that is applied will determine the charge transfer or ionic flux. The standard enthalpy of formation is 286.03 kJ/mol, and the ideal gas entropy is 0.163 kJ/mol K for gaseous water, as such, the Gibbs free energy of water at 298 K is given by:

$$\Delta G^0 = \Delta H^0 - T\Delta S^0 = 286.03 - (298 * 0.163) = 237.46\,\text{kJ/mol}. \tag{7.7}$$

The minimum required cell voltage for the electrolysis to take place is denoted by V_{rev} and can be represented by the following equation:

$$V_{rev} = \frac{\Delta G}{nF} = 1.23\,\text{V} \tag{7.8}$$

where n is the number of electrons transferred, and for electrolysis, two electrons are required to split water (n= 2), and F is Faraday's constant, which is the charge on one mole of electrons, has a value of 96,485 C/mol. This is the required voltage for electrolysis, when all the components reacting are in a gaseous state, which means the water consumed in the process is in a vapor state. However, in the conventional electrolyzer, where the process occurs at atmospheric pressure and 80 °C, an additional energy input is required to raise the temperature of water and thus there is a need to provide the required thermal energy to convert water into vapor. If the required thermal energy is also supplied by electrical energy input which is usually the case in most of the electrolyzers, then it is called thermo-neutral voltage

and represented by V_{th} and given by expression:

$$V_{th} = \frac{\Delta H}{nF}. \tag{7.9}$$

Thus, ΔG is the reversible thermodynamic work and the corresponding V_{rev} is the difference between the equilibrium potentials of the two electrodes, while ΔH includes the reversible thermodynamic work for splitting, and the energy associated with the phase transition from liquid to gases, while the corresponding voltage is called V_{th}.

Under standard temperature and pressure conditions the value of $\Delta G^0 = 237.21$ kJ/mol, $\Delta S^0 = 0.163$ kJ/mol K and $\Delta H^0 = 285.84$ kJ/mol. Substituting these values in the above equations gives the values of the reversible and thermo-neutral voltage and are obtained as $V_{rev}^0 = 1.229\,\text{V}$ and $V_{tn}^0 = 1.481\,\text{V}$, at standard conditions.

Thus, the electrolysis of water to produce hydrogen with only electrical energy input becomes feasible at 1.48 V, but in actual practice, it will take place at an unacceptably slow rate. In order to accelerate the reaction certain amount of additional energy is required, which can be provided by supplying a certain extra voltage, called overvoltage. This overvoltage results from the high activation energy required for splitting water, along with the poor conductivity of water. The lower conductivity of water makes it essential that acids, bases, or salts should be used, while the requirement of high activation energy is addressed with the use of appropriate electrocatalysts. Thus, the main objective in these devices is to minimize the overvoltage, which is important toward increasing the efficiency of the process.

In real cases, due to the presence of various irreversibilities involved in the process, the actual required voltage is higher than V_{th} represented by the above equation 7.9. These irreversibilities arise because of the lower temperature and pressure of the supplied water, and the various thermal losses involved in the process.

The Nernst equation, derived from Gibbs free energy, can be represented by the following:

$$dG = \sum_I \mu_I dn_I = \sum_I (\mu_I^0 + RTlna_I)dn_I \tag{7.10}$$

where a is the activity, m is the chemical potential, and n is the number of moles corresponding to the species I and μ_I^0 is the standard chemical potential under standard conditions. For a general reaction

$$aA + bB \leftrightarrow cC + dD. \tag{7.11}$$

The Gibbs free energy change on molar basis can be represented by

$$\Delta g = \left[\left(c\mu_C^0 + d\mu_D^0\right) - \left(a\mu_A^0 + b\mu_B^0\right)\right] + RTln\frac{a_C^c a_D^d}{a_B^b a_A^a} \quad (7.12)$$

$$\Delta g = \Delta g^0 + RTln\frac{a_C^c a_D^d}{a_B^b a_A^a}. \quad (7.13)$$

Using the expression

$$\Delta g = -zFV_N, \quad (7.14)$$

the Nernst voltage is given by

$$V_N = V^0 - \frac{RT}{zF}ln\frac{a_C^c a_D^d}{a_B^b a_A^a}. \quad (7.15)$$

For electrolysis process, this can be rewritten as

$$V_N = V^0 - \frac{RT}{zF}ln\frac{a_{H_2}^1 a_{O_2}^{\frac{1}{2}}}{a_{H_2O}^1}. \quad (7.16)$$

In terms of partial pressure, this Nernst equation can be written as

$$V_N = V^0 - \frac{RT}{zF}ln\frac{p_{H_2}^1 p_{O_2}^{\frac{1}{2}}}{p_0^{3/2} a_{H_2O}^1} \quad (7.17)$$

where $a_{H2} = p_{H2}/p_0$ and $a_{O2} = p_{O2}/p_0$ and $a_{H2O} = 1$ and at standard pressure $p_0 =$ 1 bar, gives

$$V_N = V^0 - \frac{RT}{zF}ln\frac{p_{H_2}^1 p_{O_2}^{\frac{1}{2}}}{1} = V_{rev} - \frac{RT}{zF}ln\frac{p_{H_2}^1 p_{O_2}^{\frac{1}{2}}}{1} \quad (7.18)$$

p_{H2} is the partial pressure of hydrogen on the cathode side, and p_{O2} the partial pressure of oxygen on the anode side.

For a cell voltage less than V_{rev}, the electrolysis reaction will not occur. For cell voltage ($V_N \leq V_{cell} < V_{tn}$) less than V_N, the reaction is endothermic and requires heat, while for cell voltage $V_{cell} > V_N$ it is exothermic, releasing heat.

Components of an Electrochemical Cell

The different components of an electrochemical cell include, the two electrodes, electrolyte, and separator. These components need to have certain requirements so that the performance of the cell is high.

The requirements for the different components are:

Electrodes:

- Should have good electrical conductivity
- Good catalytic properties
- Resistance toward corrosion
- Structural integrity

Electrolyte

- Stable and should not change on reacting with electrodes
- Should not erode the electrodes

Diaphragm or separator

- To avoid recombination of H_2 and O_2
- Prevent shorting of electrodes
- Should have good ionic conductivity
- Physically and chemically stable.

In the electrochemical cell, the following reaction occurs:

$$H_2O \rightarrow H_2 + \frac{1}{2}O_2.$$

The two half-cell reactions can be written as follows:

Reduction Half Reaction

- On the cathode, electrons polarize negatively
- Hydrogen is generated at cathode

Oxidation Half Reaction

- At anode, electrons polarize it positively
- Oxygen is generated at anode.

Depending on the type of electrolyte, the reactions on the anode and cathode sides are as follows:

For acid electrolyte:

- Cathode: $2H^+ + 2e^- \rightarrow H_2$
- Anode: $H_2O \rightarrow \frac{1}{2}O_2 + 2H^+ + 2e^-$.

For alkaline electrolyte:

- Cathode: $H_2O + 2e^- \rightarrow H_2 + 2OH^-$

- Anode: $2OH^- \rightarrow \frac{1}{2}O_2 + H_2O + 2e^-$.

Barriers to the Cell Reactions and Impact of Components of an Electrolytic Cell

There are several barriers to the reactions that occur in the electrolytic cell. These barriers arise due to boundary layers being formed at the electrode surface, which depend on the electrode phase and its characteristics, the characteristics of the electrolyte involved, and the different components of the cell.

Between the electrode and electrolyte is the interface, and differences occur due to the different surfaces, i.e., one is a solid surface (electrode), and another may be a liquid or a solid surface (electrolyte). The differences at this interface, includes:

(1) The electrolyte velocity changes, forming a hydrodynamic boundary layer.
(2) The concentration of the active and electro-active species varies, forming a concentration boundary layer.
(3) The electric potential varies from the surface of the electrode to a certain distance from the electrode, forming a capacitance boundary layer.

Thus, each of these differences gives rise to a boundary layer at the interface (Fig. 7.2).

Figure 7.2 shows a negative electrode on whose surface are positive ions forming a compact layer. Let the potential at the electrode surface be ϕ_e. The remaining region, which is away from the electrode, will have a diffuse layer, which has ions distributed than those in the electrolyte closer to the electrode, i.e, at the interface. Thus, due to this spatial variation, there is a decrease in the potential, as move away from the surface. The potential at the electrode surface is ϕ_e, and in the solution is ϕ_s. In the compact layer, potential varies linearly, and thereafter in the diffuse layer it varies exponentially. Now, because of the presence of this boundary layer, it appears as if there is capacitance operating; one which is because of the compact layer and another because of the diffuse layer. Thus, a capacitance boundary layer is formed, which is due to the electric potential changes at the interface of electrode and electrolyte, and the charge separation is the driving force for the existence of the difference in the electric potential.

The second boundary layer is because of the change in the velocity of electrolyte in the vicinity of the electrode surface gives rise to a hydrodynamic boundary layer. The driving force for the hydrodynamic boundary layer is the difference arising from the convection in the process.

The third boundary layer occurs due to the change in the concentration of electroactive species called the concentration boundary layer. The driving force is diffusion. Based on these boundary layers, there will be different resistances offered to the flow of charges. Thus, the kind of separator used and the different electrical resistances involved in the process will determine the behavior of current voltage characteristics, efficiency, and the amount of hydrogen being produced from a particular

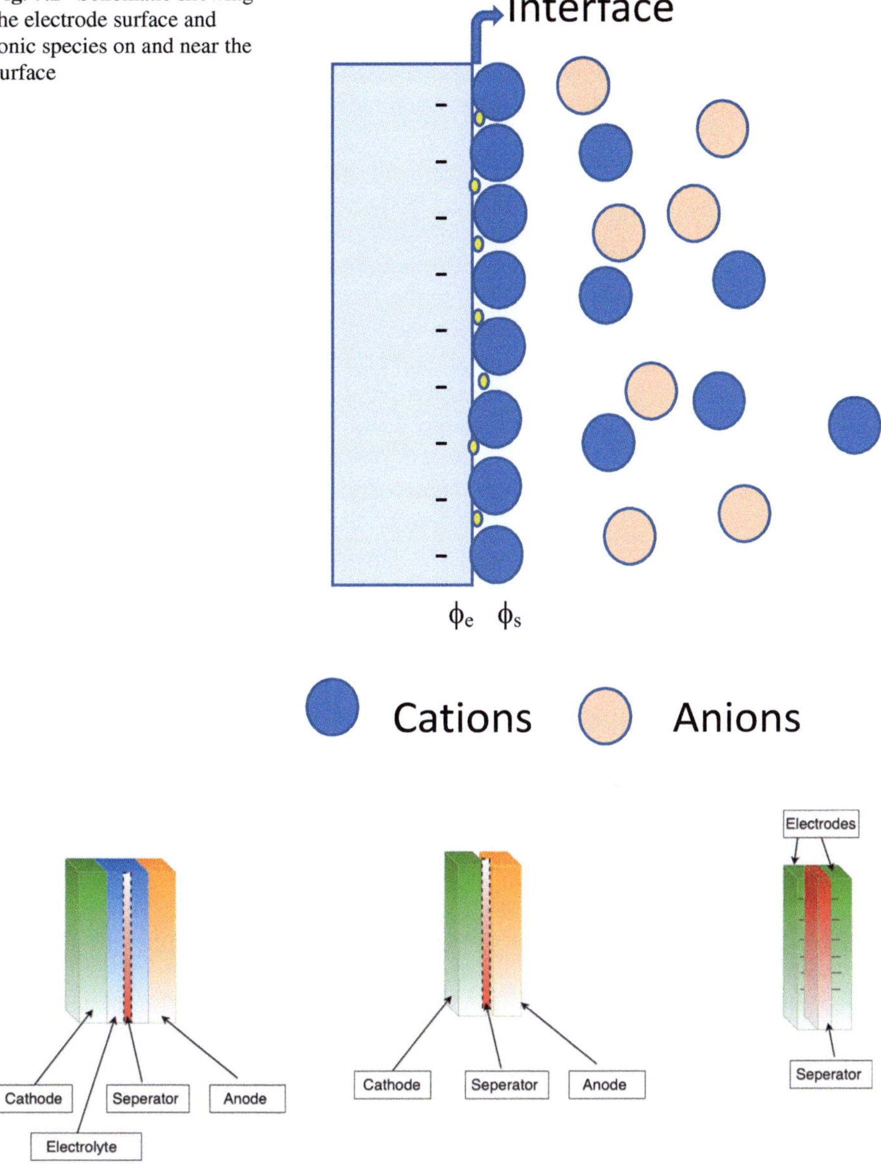

Fig. 7.2 Schematic showing the electrode surface and ionic species on and near the surface

Fig. 7.3 Different configurations of an electrolytic cell

electrolytic cell. From the different phases in the cell, there are certain characteristics they should have, which are detailed in the following section.

Requirements from the Electrode Phase

The requirement for the electrode phase is that it should be mechanically stable (it should not undergo erosion) when acted upon by different reactants, products, and in the presence of different ions.

Electrode should be physically and chemically stable, should be easily integrable in an electrolytic cell, i.e., should be easily disconnected when required for inspection and maintenance. The shape and surface of the electrode phase play an important role, as on the surface various electrochemical reactions occur. The essential requirement from the electrode is that it should allow the reactants to reach the interface and react. After the reactions, the products formed should be easily removed from the surface, in the process. The electrodes should promote the desired reaction leading to the formation of the gaseous product and avoid unwanted chemical reactions from taking place. The desired reactions should occur with a high rate, which defines the efficiency of the process, and the current density involved in the process. The electrode should have good electrical conductivity, low cost, and a long life.

There are two types of processes involved on the electrode surface, where it acts as an electron source or sink, and reaction kinetics are independent of electrode material. Second, the electrode surface acts as a catalyst, the type of surface as well as the rate of reaction depends on interactions between the electrode and electrolyte species. A rough surface can provide a high and active electrode area to increase the electrocatalytic effect. In water electrolyzers, an electrocatalytic electrode surface is required to promote hydrogen and oxygen evolution at a high rate and have a lower overvoltage.

Electrolyte Phase

The electrolyte phase has three components: (1) a solvent which should remain stable under the reaction conditions, (2) an electrolyte, and (3) the electroactive species, the reactants. The requirements from the solvent are that it should be cost-effective, stable, easy to handle, and easy to dispose of and recycling should be safe.

Water is the most promising solvent due to its low cost, stability, and formation of an oligomer. Besides, due to the high dipole moment, the interaction of the electroactive species occurs. The splitting up of water or the ionization of water is very small into H^+ or OH^{-I}, i.e., the self-ionization of water is relatively very small, thus it remains stable under the operating conditions.

In general, the requirements from an electrolyte are a high ionic conductivity in the solution phase, leading to low resistance, and a high ionic strength, which results in equal and constant activity coefficients for the reactants and products.

Electrolytic conductivity is given by

$$k = Dc \tag{7.19}$$

where D is the molar electrolytic conductivity and c is the electrolyte concentration.

$$k = Cs \tag{7.20}$$

where C is the conductance and s is the cell constant.

Typically, electrolytic conductivity lies in the range of 5–25 S/m when the typical concentration is of the order of 1 mol/dm^3.

Separator

The separator plays an important role, i.e., it keeps the two electrodes physically separated to avoid shorting. The gaseous products formed in the electrochemical reactions are separately produced and the seperator prevents their intermixing. Independent cathode, catholyte; anode, anolyte can be used in the presence of a separator. It also prevents the parasitic redox shuttle, i.e,. when the species undergo reduction at one electrode, and then undergo re-oxidation at another electrode. The separator has to be selective toward one ion, i.e., allow only that particular ion to transport through it.

There are different types of separators which are available. Porous separators, also called spacers, provide support to fragile electrodes, can provide support to membranes, and promote turbulence at the electrode surface. However, they cannot prevent the movement of solvent, nor prevent mixing of anolyte and catholyte. Usually, the pore size in the separators lies in the range of 0.5–12 mm.

Another class of separators is micro porous seperators, where the pore size is in micron dimensions. Microporous separators or diaphragms have a pore size lying in the range of 0.1–50 microns range. They allow the transport of solvent, solute, and ions through it by means of hydraulic permeability. Convection and diffusion type barriers exist with such separators.

Using a separator adds to the cost, complexity of the system, and also increases the resistance to the flow of ions.

Another class of separator is of the ion exchange type, which selectively allows a particular type of ion to flow through them. They divide the electrochemical cell into two compartments, thus, separate analyte or separate catholyte can be used. They have pore size of 10^{-8}–10^{-9} m. The membranes should have high chemical stability under the reaction conditions when subjected to different reactants, products, and ions. They should be mechanically stable and provide both strength and flexibility. The membranes should have high ionic conductivity, but a lower electronic conductivity, and should be able to operate under variable current densities (at high current density when it is operating under full operation or it should also be able to operate under lower current densities). Membranes should be homogeneous in structure with no local variations, i.e., should have the same uniform current densities throughout. They should be easily available, convenient to handle, economical and have a long life. They act as a barrier to convection and diffusion.

Different Configurations of an Electrolytic Cell

An electrolyzer is a device that has hundreds of electrochemical cells connected. As mentioned in the previous sections, the different components in an electrochemical cell and their requirements are discussed. These components can be connected to make a single unit of an electrolytic cell in different ways. Between the two electrodes and an electrolyte, there are two interfaces where the two half-cell reactions take place. During operation, when an external power supply is connected, the adjustment of the fermi levels of the electrodes takes place for the two redox couples (protons/ H_2 at cathode and water/O_2 at anode) to react. Electrons are charge carriers which flow through the external circuit and ions flow through the electrolyte internally. To avoid direct contact of electrodes and the product gases formed, a separator that is permeable to ions but not electrons, is placed between the electrodes. There are three concepts or arrangements of these components as shown in Fig. 7.3.

The first one is called a gap cell, which is the simple and most conventional concept where two electrodes (a less expensive electron-conducting substrate surface covered with a relatively expensive electrocatalyst) are placed in an electrolyte. The separator, generally a diaphragm impregnated with electrolyte, is integrated between the two electrodes. The distance between the diaphragm and electrodes is sufficient enough so as to allow the product gases to evolve. The major drawback of this configuration is when the gases or bubble concentration increases and are not removed at the pace at which they are being formed, then a thin layer of resistive gas film is formed over

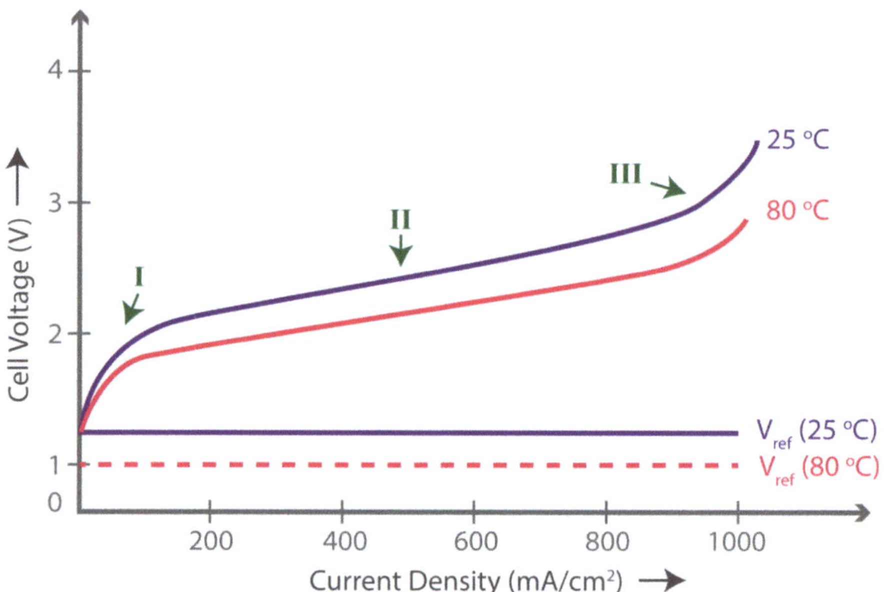

Fig. 7.4 Typical current voltage characteristics for an electrolyzer

the electrodes, and thus limiting the current density that can be achieved with this particular configuration. This arrangement cannot be operated for current densities > few hundred mA/cm^2.

The second concept is called zero gap cell, in this configuration porous electrodes and separator have zero gap or electrodes are pressed against separator reducing ohmic losses; the gaseous products formed are released at the backside from the pores, thus higher current densities can be achieved with lower voltage losses. The liquid electrolyte fills the interpolar region and gives a homogeneous current density. This concept is used in alkaline water electrolyzers.

The third concept is the one used in the PEM electrolysis cell, in which the separator is a proton conducting membrane that not only acts as a separator but also conducts ions in the process. The membrane has porous catalytic layers on both surfaces and then pressed with the current collectors on both sides, with the electrolyte being confined in the membrane. Such an arrangement can have high current densities.

Current–Voltage Characteristics and Overvoltage of an Electrolytic Cell

As mentioned in the earlier section, an additional potential or overvoltage is required, and usually the requirement of a device is to reduce the overvoltage and increase the efficiency of the process. The reversible cell voltage is 1.23 V, and the thermo-neutral cell voltage is 1.48V; however, due to irreversibilities and resistances, the actual cell voltages of 1.8–2.05 are generally required to reach significant levels of hydrogen production rates, leading to energy conversion efficiencies less than one. This overvoltage increases the cost of electrolysis; besides, in PEM, the use of expensive catalysts (due to the need for corrosion resistance under acidic conditions) and membranes, together with the need of deionized water, increases the cost significantly. Under practical operating conditions at any given T and P, there are several irreversibilites involved during the charge transfer process at each of the interfaces and transport of charge carriers inside the cell, which gives rise to overvoltage.

Voltage drops occur due to (1) internal resistances inside the electrodes; (2) across the electro-active layers which do not always have good electrical conductivity; (3) at interfaces where charge transfer occurs; (4) in solution across the diffusion layers near electrode surface where gaseous species are formed or mass transfer limitations occurs; (5) during transport of ions in bulk electrolyte; (6) within the cell separator.

Thus, the electrolysis cell voltage is given by:

$$U_{cell} = V_{rev} + V_{ohm} + V_{act} + V_{con}. \tag{7.21}$$

The V_{ohm} represents the overvoltage due to ohmic losses in the cell arising from the resistance of various cell elements like electrodes, interconnects, current collectors, etc., and is proportional to the electric current flowing in the cell. V_{act} is the activation overvoltage and arises due to the electrode kinetics. Activation overvoltage is the energy barrier that the charge has to overcome to move from the chemical species, i.e., reactants, onto the electrodes or vice versa, and it depends on the electrocatalytic properties of the electrodes. This crossing of the energy barrier requires energy, which is added overvoltage called activation overvoltage, and has a logarithmic nature. The anode side activation overvoltage is much higher than the cathodic half reaction, and thus anodic half reaction is much more demanding. The V_{conc}, which is the concentration overvoltage, arises due to mass transport processes like convection and diffusion; the increase in product concentration and decrease in reactant concentration at the electrode electrolyte interface result in concentration overvoltage.

The I–V plot for an electrolysis cell shows three regions, as shown in Fig. 7.4. The first region shows a logarithmic rise of voltage with current, this region is dominated by activation losses (Region I). Thereafter, a linear rise in voltage is observed with current, this region is dominated by the ohmic losses arising because of the various cell components (Region II). The third region (Region III), is the mass transport region.

Ideally, the minimum required cell voltage for electrolysis is 1.23, which is the reversible cell voltage. Above this, the current starts to flow but there is a need for additional electric power to overcome the internal resistances, and that is dissipated as heat. At low current densities, the ohmic losses are less, but charge transfer overvoltage is maximum, which leads to a logarithmic nature of the polarization curve. In electrolysis, the anodic overvoltage is significantly higher than the cathodic, the reason being anodic oxidation reduction reaction is very demanding, and its kinetics are much slower than the hydrogen oxidation reaction, or hydrogen evolution reaction. Thereafter, as the current density increases, the charge transfer overvoltage decreases or resistance to charge transfer reduces, and the shape of the polarization curve becomes linear. The shape of the curve is determined by the ohmic resistance of the cell (both electronic and ionic resistance), which is a key kinetic parameter and is especially important in large area industrial electrolyzers where parasitic ohmic resistances are high and result in a lot of energy consumption of the cell mainly in PEM, where no liquid electrolyte is involved. Also, ohmic losses can be due to inadequate electrical contacts between various cell components, including current collectors, bipolar plates, etc.

Various thermodynamic potentials calculated are considered at equilibrium conditions; however, when higher current densities (to decrease capital expenditure, CAPEX) are desired, the required electrical work provided to the cell also increases, and thus the contribution of the energy dissipation via heat, increases the inefficiency of the cell as well as operating costs. The three major phenomena that are sources for energy degradation are electronic resistances, charge transfer processes, and the resistance to the ions in the electrolyte.

The cell voltage is significantly higher than the theoretical reversible cell voltage due to the irreversibilities. The Tafel equation represents the relation between the overvoltage and current density, and represented by

$$\eta = A + B\log I \tag{7.22}$$

where I is the current density, η is the overvoltage, and A and B depend upon the anode and cathode material.

Hydrogen Generation from an Electrolyzer

Using Faraday's laws, the rate of hydrogen production from an electrolyzer is given by:

$$f_{H_2} = \eta_F \frac{N_{cell} I_{cell}}{zF} \frac{22.41}{1000} 3600 \tag{7.23}$$

where

η_F is Faraday efficiency, also called current efficiency, is defined as:

$$\eta_F = \frac{\textit{electric charge required for production of a certain amount of hydrogen under ideal case}}{\textit{real electric charge consumed in producing the same amount of hydrogen by electrolyzer}}. \tag{7.24}$$

Its value is < 1 due to parasitic currents which flow but don't contribute toward hydrogen production, either by not participating in the reactions or participating in unwanted reactions; or it produces hydrogen that is lost before its collection. Under rated power operation conditions, its value is usually more than 0.95.

The specific energy consumed by the electrolyzer is given by [3, 4]:

$$C_E = \frac{\int_0^{\Delta t} N_{cell} . I_{cell} . V_{cell} dt}{\int_0^{\Delta t} f H_2 dt} \tag{7.25}$$

where

C_E is the specific energy consumption expressed in the units of kWh/Nm^3 of H_2 produced
N_{cell} is the number of cells in the electrolyzer module
I_{cell} is the cell current in A.

In an actual process, there are several auxiliary components involved in the hydrogen production via electrolysis, which consume energy like pumps, sensors, valves, cooling systems, purification systems, and the losses in the power etc., thus increasing the energy requirement. At the same time, there are various processes that lead to losses of hydrogen in the process; and these could be losses from valves, manifolds, sensors, during purification, etc. When both the additional energy requirement and losses are considered in the expression, then we get the specific energy consumption in the global electrolysis systems for hydrogen production.

Finally, the efficiency of the electrolyzer is the amount of energy contained in the hydrogen produced and the energy required for electrolysis to produce that hydrogen. It can be represented as

$$n_E = \frac{HHV\ of\ hydrogen\ produced}{C_E} * 100. \tag{7.26}$$

The energy efficiency of water electrolysis is the ratio of the minimum amount of energy required to split one mole of water under equilibrium conditions to the real amount of energy required under constant T,P operating conditions:

$$\varepsilon_{\Delta G} = \frac{E(T,P)}{U_{cell}(T,P)} \tag{7.27}$$

or

$$\varepsilon_{\Delta H} = \frac{V(T,P)}{U_{cell}(T,P)} \tag{7.28}$$

$$\varepsilon_{cell} = \frac{energy\ required\ under\ reversible\ conditions}{energy\ required\ under\ irreversible\ conditions} = \frac{W_{rev}}{W_{irr}}. \tag{7.29}$$

At equilibrium (I $=$ 0) $W_{irr} = W_{rev}$ and $\varepsilon_{cell} = 1$; however, under the non-equilibrium conditions (I is not equal to 0) $W_{irr} > W_{rev}$ and $0 \leq \varepsilon_{cell} \leq 1$. The reason for higher irreversible work is that when the charge carrier flows across the cell, there is dissipation of energy, and the fraction of kinetic energy converts into heat due to friction. In low-temperature water electrolysis, since liquid water is electrolyzed, so HHV of hydrogen is to be considered, while in high temperature electrolysis, water vapor is used so LHV needs to be considered.

All the parasitic losses (those in the electrolysis cell or stack, like energy and current losses, heat losses, energy required for auxiliary components, pumps, heat exchangers, etc.) are being considered in the denominator. The energy efficiency term considers all these values. The term in the numerator of energy efficiency is the energy content of the gases produced (H_2 and O_2). Thus

$$\varepsilon_{cell} = \frac{energy\ content\ of\ products}{total\ energy\ requirements} = \frac{W_{rev}}{W_{irr}}. \tag{7.30}$$

To improve the energy efficiency, it is essential to decrease the irreversible losses, i.e., the electrochemical overvoltages and the overall cell resistance. To reduce electrochemical overvoltages, the selection of appropriate electrocatalytic materials especially, on the anode side for OER, is required. Anode side OER is much more demanding than the HER cathode side.

Faradaic Efficiency

Faradaic efficiency is the measure of current flowing through the cell. It is different at both the interfaces and can be given as:

$$(\varepsilon_F)^{anodic} = F\frac{\left(\frac{dn_{O_2}}{dt}\right)}{I} * 100 \tag{7.31}$$

$$(\varepsilon_F)^{cathodic} = F\frac{\left(\frac{dn_{H_2}}{dt}\right)}{I} * 100. \tag{7.31}$$

The gaseous products released at the electrodes should not be allowed to recombine as it is a loss of the electrical work supplied to the cell and it goes in the form of heat due to the recombination; this needs to be avoided with the use of separators and ideally there should not be any crossover but in reality it occurs at high current density and high operating pressures. The solubility of gases is not completely zero, and this nonzero permeability of the separator results in the crossover, both in alkaline and PEM; however, it is found to be less with ceramic separators. So the hydrogen formed on the cathode side reaches the anodic side and either reacts with oxygen chemically or re-oxidizes, and similarly for oxygen, this reduces the Faradaic efficiency.

Kinetics of Water Splitting

In the absence of mass transport limitations, the kinetics of charge transfer at the interface in an electrolysis cell is given by the Butler-Volmer relation:

$$I = I_0\left\{exp\left[\frac{\beta nF}{RT}\eta\right] - exp\left[-\frac{(1-\beta)nF}{RT}\eta\right]\right\} \tag{7.32}$$

where, I_0 is exchange current density in A/cm^2 of half-cell reaction, and is the main kinetic parameter which determines the efficiency of charge transfer processes and is an intensive property. The higher this current density more reversible is reaction and the more efficient is the process. The best electrocatalysts should give a high exchange current density.

β is the asymmetry factor,
η is the charge transfer overvoltage.

Arrangement of Electrochemical Cells in an Electrolyzer

In an electrolyzer, the basic unit is an electrolytic cell; several hundred of these units can be connected to manufacture an electrolyzer. These can be connected in series, parallel, or hybrid. In the parallel configuration, each electrode has only one polarity and is thus called a monopolar or unipolar arrangement. While when connected in series, then each electrode has two polarities and this arrangement is called a bipolar arrangement. Each bipolar electrode has two electrodes, an anode and a cathode, and is physically connected by means of a current collector. Current collectors are usually made up of a plate or a wire mesh. The commonly used materials are graphite and stainless steel. The cells can also be arranged in a mixed, i.e., both series and parallel arrangement, but such an arrangement is not preferred in the stack. A bipolar arrangement is called filter pressed, while unipolar as tank module. Unipolar arrangement requires more space, low voltage and high current, and involves more losses due to additional connections. Bipolar configuration is preferred, although more complex than a series, but has advantages of being compact, low current, and high voltage, and, can achieve higher current densities. Cooling of the stacks can be either done with air in case of small systems or water in larger capacity electrolyzers (Fig. 7.5).

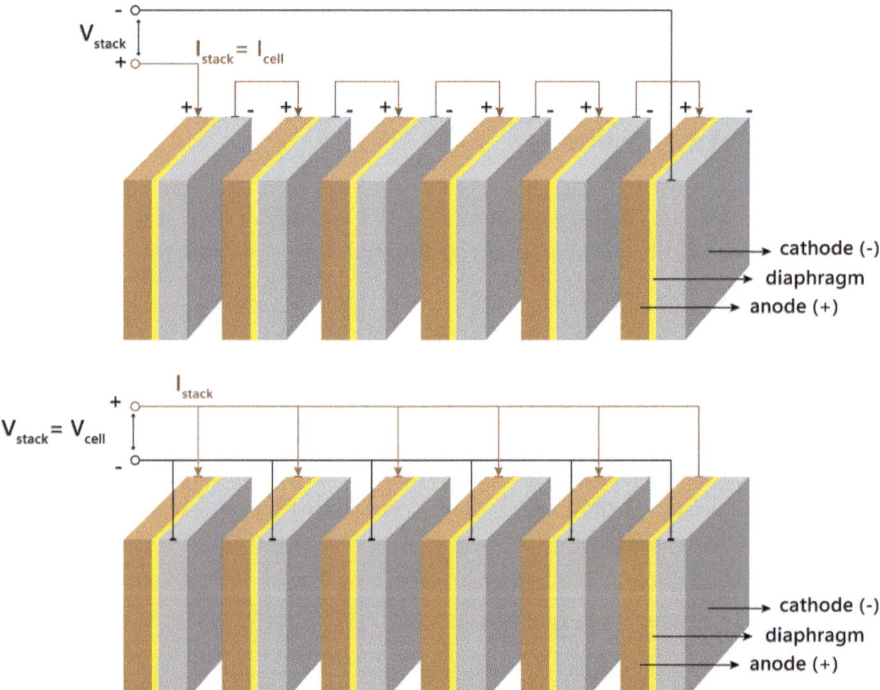

Fig. 7.5 Unipolar and bipolar configurations representing the arrangements of electrolytic cells in an electrolyzer

Different Electrolyzer Technologies

The three main electrolyzer technologies are alkaline electrolysis, proton exchange membrane electrolysis, and solid oxide electrolysis, which have been in use, and one new technology of anion exchange membrane electrolyzer, which is still on the path to commercialization.

Alkaline water electrolysis was the first to be on the market because the alkaline electrolytes are less corrosive than acidic electrolytes. It was in the 1960s when the concept of proton exchange membrane electrolysis emerged, using which high current densities can be achieved. This technology found its role where compactness is essential (e.g., submarines, or spacecrafts) and higher current densities, as well as higher pressure, are a requirement. However, the interest in PEM has seen an unprecedented increase recently.

Alkaline Water Electrolysis (AWE)

Alkaline electrolysis is a mature and commercial technology which has been used since 1920s for hydrogen production in the chemical and fertilizer industries. Plants of large capacity, even up to 165 MW electrical, were built at places where the hydroelectric power could be used for electrolysis, such installations were in India, Canada, Egypt, Norway, etc. However, these were decommissioned when SMR tookover in the 1970s.

The alkaline water electrolysis technology is reliable, safe, and comparatively requires low capital cost due to the use of inexpensive materials and is known to have lifetime of 30 years. AWE has been used extensively on a commercial level globally. Over a period of time, advancements have improved the efficiency of alkaline electrolyzers, thereby reducing the operational costs, i.e., electricity consumption, and the operating current densities have increased which resulted in a decrease in the CAPEX. The major advancements seen have been in terms of using zero gap cells, advanced materials for diaphragms, and better electrocatalytic materials to reduce overvoltages.

Operating Principle of Alkaline Electrolysis.

The liquid electrolyte used is an aqueous solution of 25–30% KOH (use of concentrated solution) to maximize current density.

Cathode:

$$2H_2O + 2e^- \rightarrow H_2 + 2OH^-. \tag{7.33}$$

Anode:

$$2OH^- \rightarrow \frac{1}{2}O_2 + H_2O + 2e^-. \tag{7.34}$$

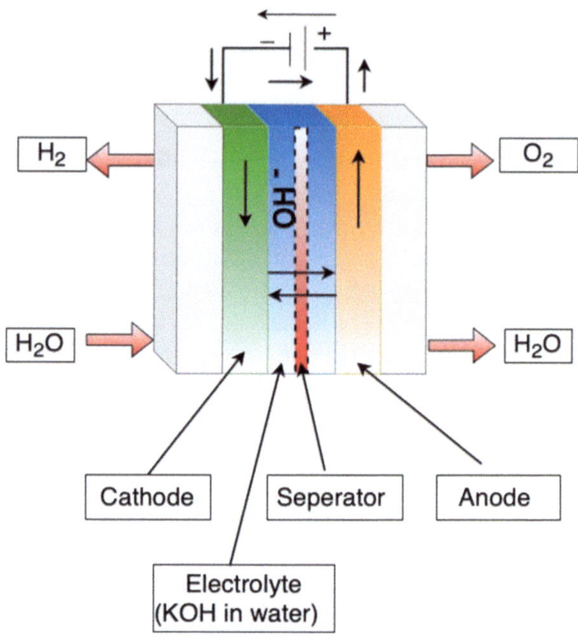

Fig. 7.6 Schematic of an alkaline water electrolyzer

Hydrogen gas is evolved on the cathode side, where water is reduced, resulting in hydroxyl ions and hydrogen. The hydroxyl anions move across the diaphragm to the anode side under the external field applied. On the anode side, the hydroxyl ions recombine to produce oxygen gas, liberating electrons which move through the external circuit, closing the electric circuit. Thus, hydrogen is evolved on the cathode side and oxygen on the anode side. The electrode-electrolyte interface plays a key role in the evolution of the gaseous products from the electrodes (concentration overvoltage), thus maximizing the interfacial area is important. Thus, the electrodes are made porous, having several perforations. Besides, a convective circulation is established in the electrolyte due to temperature gradients and the liquid conveyed with the evolution of gases. This has multiple benefits, it results in homogenization of the concentration of species inside the pores of electrodes, facilitates gas evolution, and improves heat transfer rate (Fig. 7.6).

Electrolyte is usually a 25–30% aqueous solution of potassium hydroxide (KOH), which is pumped to the cell, and gaseous products bubbling out are collected. However, the natural circulation obtained due to temperature gradients and resulting in a buoyancy effect is used only for low current densities. Purity of hydrogen lies in the range of 99.5–99.9% and of oxygen in 99–99.8% range. Anode is made up of nickel or nickel-based catalysts or other combinations (Co, Mn, NiMoFe, etc.) or even noble metals (Pt, Pd, IrO_2, RuO_2), while the cathode is made of nickel or nickel-based catalyst. The advantage of non-noble metals is cost benefit, and these are also stable under an alkaline environment. Diaphragm is robust and ZrO_2-based, Ni-coated stainless steel for electrodes. Porous diaphragm is used to enable the electrochemical reactions but this even allows the intermixing of gaseous products formed, which gets dissolved in the electrolyte, and this limits its use to the lower power operating applications, and restrictions are for high pressures. To reduce this, thicker membranes are used (0.252 mm), but that increases the resistance and thus results in reduced efficiency [1]. In the current advanced electrolyzers, the zero gap concept is being used, with a thin diaphragm and better electrocatalysts, which can result in higher current densities in the alkaline electrolyzers. The response to the change in power fluctuations is less fast than PEM due to ion transport in the liquid electrolyte, which shows higher inertia for this transport (OH^-).

The lifetime of alkaline water electrolyzers is found to be higher, and the annual maintenance costs are lower compared to a PEMEL system.

Polymer Electrolyte Membrane Electrolyzer (PEM)

The PEM electrolyzer was first developed in the 1960s by General Electric; however, ion exchange membranes were used by the end of the 1950s. This technology is known as polymer electrolyte membrane or proton exchange membrane, or even solid polymer electrolyte electrolysis.

In PEMFC, the anodic compartment is fed with liquid deionized water. The water molecules are oxidized at the catalytic anode (usually made of unsupported Ir-based oxide particles) according to the half-cell reaction (Fig. 7.7):

$$H_2O \rightarrow \frac{1}{2}O_2 + 2H^+ + 2e^- \tag{7.35}$$

$$2H^+ + 2e^- \rightarrow H_2. \tag{7.36}$$

Oxygen evolves in the gaseous form from the anode side, electrons circulate in the external circuit, and hydrated protons migrate across the polymeric membrane to the cathode which is made up of unsupported Pt particles or carbon-supported Pt nanoparticles., where they are reduced by the electrons from the external DC supply, producing hydrogen on the cathode side, resulting in an overall reaction of:

$$H_2O \rightarrow H_2 + \frac{1}{2}O_2.$$

The most commonly used membranes are Perfluorosulfonic acid polymer membranes such as Nafion®, Fumapem®, Flemion®, and Aciplex®. These membranes have several advantages like high strength, high efficiency, and high oxidative stability. Dimensionally, they are thermally stable, have good durability, and high proton conductivity. However, the most widely used membranes currently are Nafion® membranes (Nafion® 115, 117, and 212) due to their additional advantages of being able to operate at higher current densities (2 A/cm^2), high durability,

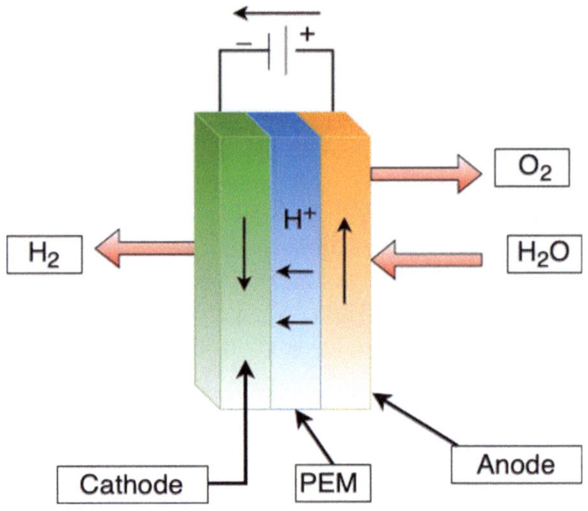

Fig. 7.7 Schematic of a polymer electrolyte membrane electrolyzer

high proton conductivity, and good mechanical stability. The most popular membrane is Nafion, perfluorosulfonic acid (PFSA) ionomer, electrocatalysts for HER/HOR are Pt nanoparticles dispersed on large surface area carbon cloth (e.g., Vulcan XC72R), while for OER/ORR are Ir-based oxides dispersed on a corrosion proof support (e.g., Ti mesh or foam). The anode, cathode, and membrane make the membrane electrode assembly. The membrane is thin (0.2 mm) PFSA-based, and electrodes are designed with having with advanced architecture to achieve lower resistance and higher efficiency. The advantage of PEM is that the perfluorosulfonic acid (PFSA) membrane is chemically and mechanically robust, which allows for high pressure differentials. Thus, the PEM electrolysis cells can operate at up to 70 bar or higher on the hydrogen side with the oxygen side at atmospheric pressure. The acidic environment provided by the PFSA membrane, high voltages, and oxygen evolution in the anode create a harsh oxidative environment, demanding the use of materials that can withstand these conditions. Titanium-based materials, noble metal catalysts, and protective coatings are required for long-term stability and provide the required electronic conductivity and higher efficiency. The use of expensive materials for the reasons mentioned and to avoid the material corrosion when subjected to an acidic environment that makes the PEM stacks more expensive than alkaline electrolyzers. But the PEM electrolyzers are compact in design, offer high current density, and higher hydrogen pressures due to compact nature of electrolysis module, and structural properties of MEA (membrane electrode assemblies consists of membrane, ionomer solution, and anode, cathode electrocatalysts) which allow high-pressure difference between the two electrode compartments along with higher purity hydrogen and can be obtained due to very limited crossover as the membrane being least permeable for crossovers. The major components PEM water electrolysis cell are membrane electrode assemblies (MEAs), current collectors (gas diffusion layers), and separator plates. The heart of the electrolysis cell is MEA, which separates the cell into two half cells (anode and cathode). Currently, electrode areas have been 2000 cm^2 or higher, but still lower as per the requirements of future concepts of large MW or GW stack units using single stack concepts. The important characteristics of PEM as against alkaline electrolyzers are that they can operate under variable power feeding conditions, this is because of the quick response to any power fluctuations, which is because of the faster response of proton transport across the membrane. This makes them ideal for integration with renewables based power. PEMEL systems are preferred when dynamic operation requirements are important due to the short start-up time, and a broad load flexibility range is obtained [3, 5].

Solid Oxide Electrolyzers (SOEC)

This is an advanced technology of water electrolysis where steam electrolysis occurs at high temperatures (700–850 °C). Thus, SOEC has higher efficiency and lower electrical energy requirement than PEM and AWE. At the same time, due to high temperature operation, a favorable kinetics for electrochemical reactions is an advantage. Besides, the electrical energy demand is reduced, and thermal energy required can be provided by waste heat and thus SOEC has a higher overall efficiency (Fig. 7.8).

The operating principle is:

$$H_2O(g) + 2e^- \rightarrow H_2(g) + O^{2-} \tag{7.37}$$

$$O^{2-} \rightarrow \frac{1}{2}O_2(g) + 2e^-. \tag{7.38}$$

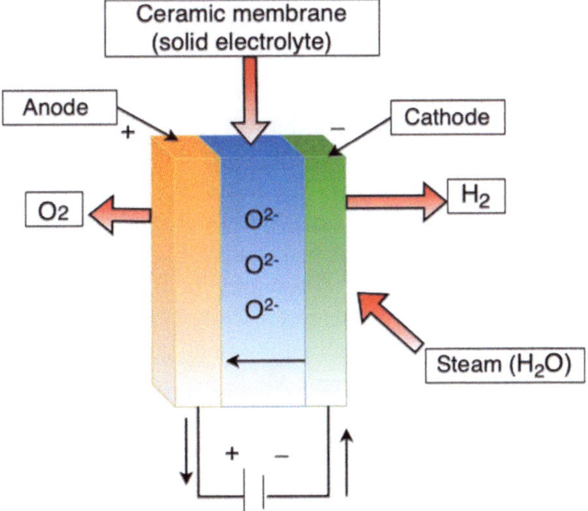

Fig. 7.8 Schematic of a solid oxide electrolyzer

Steam is fed on the cathode side, where it gets reduced to produce hydrogen. The oxide ions produced move through the solid electrolyte to the anode side, where they recombine to form electrons, and oxygen is evolved. The electrons flow through the external circuit. Since the electrodes in this case come in contact with the vapor or gas phase, maximizing the interfacial area between electrodes and the chemical species is important. Thus, the electrodes in SOE need to be porous. The electrolyte is yttria stabilized zirconia (YSZ), which has high-oxide ionic conductivity at high operating temperatures. The cathode used is a cermet, generally Ni and YSZ, anode is a composite of YSZ and perovskites like $LaMnO_3$, $LaFeO_3$ or $LaCoO_3$, partially substituted with Strontium. However, the disadvantages associated with SOEC are related to the high-temperature operation, leading to the stack degradation, including challenges related to sealing at higher differential pressure; electrode contamination by silica used as sealants; and other additional contaminant sources from piping, interconnects, and sealing. This technology is still at a developmental stage where there are projects at the kW-scale, although some current demonstration projects have already reached 1 MW.

Anion Membrane Electrolysis (AEM)

Recently, the anion exchange membrane technology received interest where the hydroxyl conducting polymers are used, these are impregnated with molar KOH, and could be operated in cells fed with only deionized water. The first class of polymer of interest is with poly-aryl-ether-ketone backbones and cyclic quaternized ammonium cationic moieties, which provide good mechanical properties (~70 mΩ/cm2 for a 60 micron thick membrane) and limited hydrogen crossover. Performance of these outperforms nafion-115, Pt group metals (PGM) can be used as electrocatalysts, but the advantage is that due to mild pH even non-noble metals, i.e., cheaper transition metals or metal alloys or oxides can be used [3]. Electrocatalysts for alkaline electrolysis includes Ni, NiFe, $NiMn_2O_4$, $NiCO_2O_4$ spinel oxides and Ni phosphide for OER; and Ni, NiMo, Mn-doped Ni_2P and Ni/CeO_2 on CNT for HER.

Anion membrane electrolysis is the latest technology, and very few companies are commercializing it. The advantage of the technology is that it combines the positives of both PEM and AWE, i.e., the advantage of alkaline electrolyzers having a less harsh environment and that of PEM having high efficiency and simplicity at the same time, use of non-noble metal catalysts, Ti-free components can operate under differential pressure conditions due to the use of membrane. But the major disadvantage of the technology is related to the chemical and mechanical stability problems of the membrane, which can lead to unstable operation and reduced lifetime. The performance reported is low due to the low conductivity of AEM and slow catalyst kinetics. Use of supporting electrolyte with KOH or $NaHCO_3$ has been tried to improve membrane conductivity, but that has adverse effect on durability. Since the transport of OH^- ion is three times slower than H^+ ion, the requirement is of thinner membranes (Fig. 7.9).

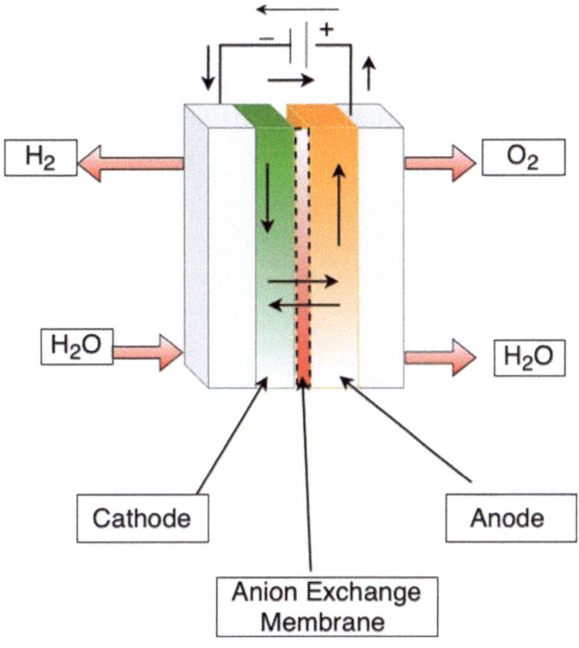

Fig. 7.9 Schematic of an anion exchange membrane electrolyzer

$$4OH^- \rightarrow O_2 + 2H_2O + 4e^- \quad (7.39)$$

$$4H_2O + 4e^- \rightarrow 2H_2 + 4OH^- \quad (7.40)$$

An anion exchange membrane is used as electrolyte and separator, ionic conduction is by means of transport of hydroxyl ions. Anion exchange membrane (such as A-201 from Tukuyama) is pressed against anode and cathode containing the electrocatalysts. Typical electrocatalysts used at the anode for OER are metal transition oxides (Co_3O_4, $CuCo_3O_4$, etc.). Rare earth metal oxides, e.g., Ni/CeO_2-La_2O_3 deposited on microporous carbon paper are used at the cathode for HER. Water is fed at the anode side, where it is reduced to produce hydroxyl ions by the addition of electrons from an external DC power supply. Hydroxyl ions migrate through the membrane to the cathode, while electrons go to the cathode via the external circuit. In the anode chamber, hydroxyl ions are oxidized into water and oxygen by losing electrons. Oxygen is released from the anode.

Typical operational cell voltage for alkaline and PEM electrolysis are 1.85–2.05 and 1.75 resp. at 70–90 °C. For anion membrane electrolysis cell, it is 1.9 V for a current density of 400 mA/cm^2 at 40 °C, and typical current densities range in 200–500 mA/cm^2.

AEM technology has several advantages compared to PEM and AWE:

1. Only transition metal catalysts are used instead of platinum group metals (PGM).

2. Distilled water or a low concentration of alkaline solution (1%K_2CO_3 solution) can be used as electrolyte instead of concentrated KOH. The use of a low concentrated alkaline solution eliminates the precipitation of K_2CO_3 (a side reaction resulting from dissolution of CO_2 of air into KOH).
3. The quaternary ammonium ion exchange group containing membrane, which is used in AEM electrolysis, is less expensive than the Nafion membrane.
4. Interaction between CO_2 and AEM is low due to the absence of metal ions in AEM.
5. Absence of corrosive liquid electrolyte has advantages of absence of leakage, volumetric stability, ease of handling, reduction in size, and weight of electrolyzer.

A comparison of the various electrolyzer technologies is presented in Table 7.1.

Other Water-Splitting Processes for Hydrogen Production—Photoelectrochemical Water Splitting (PEC)

The other method of hydrogen production from water splitting is the photoelectrochemical process. The process is inspired by photosynthesis in plants, in which plants use solar energy and water to convert it into chemical energy using a biochemical process. In the process of photosynthesis, the water is split into hydrogen-related species and oxygen using oxidation reduction reaction which occurs in the separated photosystems. In the year 1971, Honda and Fujishima came up with the demonstration of water splitting using TiO_2 and the reaction can be represented by:

$$2H^+ + 2e^- \rightarrow H_2 \tag{7.41}$$

$$2H_2O \rightarrow O_2 + 4H^+ + 4e^-, \tag{7.42}$$

which requires a voltage of 1.23 eV, is a non-spontaneous process and requires energy to proceed. As in the case of electrolysis, in PEC also the oxygen evolution step is more demanding than the hydrogen evolution reaction; it is kinetically and thermodynamically more challenging (Fig. 7.10).

A photoelectrochemical cell consists of photoelectrodes (photoanode and photocathode), which are selected based on their bandgap, to absorb sunlight. Other than electrode, there is an electrolyte and charge conducting surface. When the cell is exposed to sunlight, the electrons and hole pairs are generated at the photoelectrodes. The electrons and holes are transferred to the conduction and valence bands, respectively. The electrons generated move to the cathode side to reduce H^+ ions, producing hydrogen. Holes move to the anode side to oxidize water evolve oxygen. Although a voltage of 1.23 V is required but in actual practice, the reaction at this

Table 7.1 Comparison of various electrolyzer technologies

Electrolyzer technology	Current status	Efficiency (%)	Anode	Cathode	Electrolyte	Seperator	Electrocatalysts	Voltage (V)	Current (mA/ cm^2)	Lifetime of the stack (h)	Major companies
Alkaline water electrolyzer	Most mature	50–78	Ni-Mo	Ni–Co oxides	KOH	Ceramic oxides, e.g., asbestos, ZrO_2, PPS mesh, etc.	Co-based	1.4–3	0.2–0.8	More than 60 k	Accelera by Cummins, Hydrogen Pro, John Cokerill
Polymer electrolyte membrane electrolyzer	Commercialized	56–80	IrO_2	Pt	Solid electrolyte is acidic in nature	Perflouro sulfonic acid membrane (Nafion)	Pt, Ir, Ru-based	1.4–2.5	02-Jan	50–80 k	Horizon, Ballard, Bloom
Solid oxide electrolyzer	R&D and demonstration level	74–81			Yttria stabilized Zirconia (YSZ)		Pervoskites, Ni/ YSZ	1–1.5	0.3–1	Still under R&D but around 20 k	Elcogen
Anion exchange membrane electrolyzer	R&D with few demonstrations and some companies coming for commercialization	57–60	Ni–Fe, Ni-Co–Fe	Pt–Ni, Pt/C, Ni–Mo,	$NaHCO_3$, polymer with KOH	Ammonia polysulfone	IrO_2, Ni–Fe alloys, $Cu_{0.7}CO2.3O_4$, $Pb_2Ru_2O6.5$, Pt-black, Ni–Mo, Ni/ CeO_2–La2O_3/ C, Ni, CuCoOx, graphene	1.4–2	0.2–2	Still under R&D but in the range of 20–60 k	EvolOH, Enapter

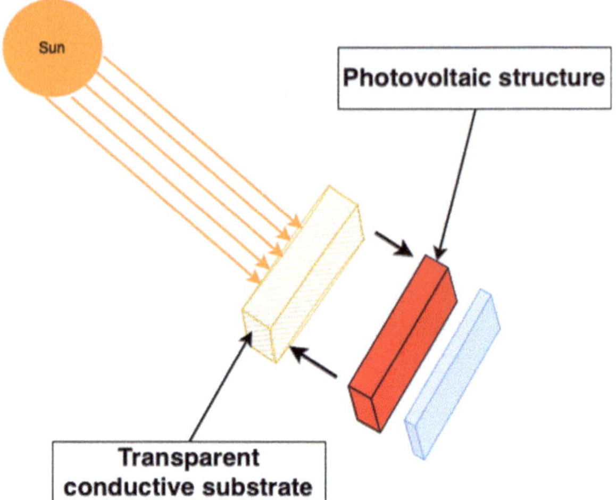

Fig. 7.10 Very simple schematic to show photoelectrochemical water splitting

voltage occurs at a very slow pace, and catalysts are required to improve the reaction kinetics, which are present on the electrodes surface.

Under the standard conditions, the reaction of water splitting required 1.23 eV as the Gibbs free energy change is ($\Delta G^0 = 237.1$ kJ/mol). Thus, in PEC cell, the total bandgap of the semiconductors used should be higher than this voltage, but in real conditions, due to various irreversibilities and losses or overvoltages present at electrode or in the solution to the flow of ions, the actual required potential difference is higher than 1.6 eV.

Similar to the electrolysis, the electrodes should have good electrical conductivity, stability, a higher surface area for the photoelectrochemical reactions to occur, good crystallinity, etc.

The PEC reaction occurs at the electrode electrolyte interface. The semiconductors are known to absorb the solar energy, but the range of the solar spectrum that the semiconductors absorb is very restrictive, thus limiting the use of the entire solar spectrum. The important requirement is an appropriate selection of semiconductor materials that could enable the photoelectrochemical water splitting, utilizing the broad solar spectrum; thus, an appropriate bandgap of the semiconductor is required, which should be consistent with the redox potential of the water.

The lower edge of the conduction band should lie above the potential for hydrogen evolution H^+/H_2 (i.e., > -4.6 eV relative to vacuum) to enable the reduction reaction between electron and proton to release hydrogen. While the topmost level of valence band should be below the redox potential for oxygen evolution O_2/H_2O (≤ 5.9 eV relative to vacuum), so that the hole transfer could occur and oxidation of water can result into oxygen evolution. Most of the semiconductors do not have a suitable bandgap, conduction, and valence band locations, and are thus not suitable for water splitting when illuminated. Among the several semiconductors, TiO_2 is most studied material and has several advantages to be used for PEC water splitting, such as a bandgap being adequate, required energy band positions being favorable, stability

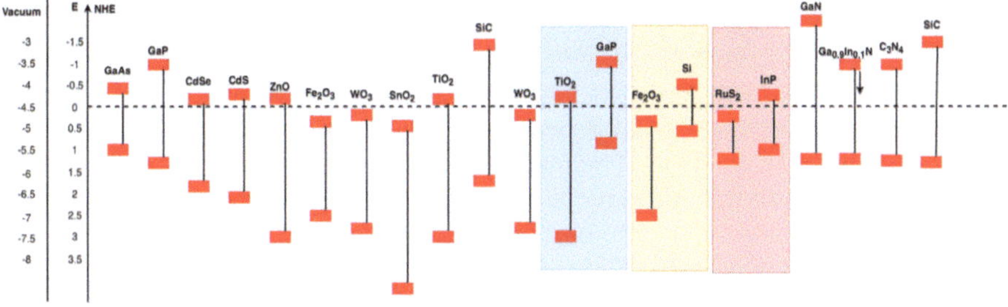

Fig. 7.11 Different possible materials for PEC

against light-induced corrosion, non-toxic and reasonable cost, making it quite suitable for use in photoelectrochemical water splitting. But the problem or disadvantage with TiO_2 is that being a wide bandgap semiconductor (higher bandgap improves stability of electrodes in electrolyte, less corrosive) with bandgap lying between 3.0 and 3.2 eV has a restricted absorption of the sunlight, which is limited to the ultraviolet region of the solar spectrum. The intensity in this UV wavelength region is low since only 4% of the entire solar spectrum comprises UV radiation, along with a high recombination rate of electrons and holes when considering water splitting. Other than TiO_2, as can be seen in the Fig. 7.11 other semiconductors like GaN, SiC, C_3N_4, $Ga_xIn_{1-x}N$ can be used for water splitting due to the correct positions of CB and VB in these materials. Further, the TiO_2 can either be modified such that the positions of CB and VB can be changed, or it needs to use some low bandgap materials which could provide better performance for water splitting. Thus, the selection of an appropriate photosensitive material to absorb sunlight and favor photoelectrochemical reactions is the major requirement. Making hybrid materials that have advantages of different materials to enhance the efficiency of the PEC process can be done by designing and engineering suitable material, avoiding their disadvantages. Generation of H_2 via the splitting of water is of great importance in the current renewable energy scenario.

Principle Behind PEC Water Splitting

Water splitting is a non-spontaneous process, with positive Gibbs free energy and requires energy. The required energy could be provided by the sunlight or photons. With the use of semiconductor, that has a conduction band more negative than water reduction potential, and a valence band more positive than water oxidation potential, water splitting can be achieved. In the presence of such a semiconductor, when light falls on this semiconductor surface with energy greater than its bandgap, the electron present in the valence band gets sufficient energy to be excited to the conduction band, leading to holes in the valence band. These photon-generated charge carriers, electrons and holes need to be separated before they recombine, else they will recombine, releasing the energy in the form of photons. Thus, efficient and effective separation of charge carriers is essential in the process. This charge separation is essential for water splitting and is being enhanced by photocatalysts. These charges, being separated, reach the surface of the semiconductors to participate in the oxidation and reduction reactions at the two interfaces of the electrodes and electrolyte. The electrocatalysts used enhance the hydrogen production by reducing the activation energy, providing the required active sites for the reaction. Besides the thickness of the photocatalyst also plays an important role; if the diffusion length of charge carriers is less than the thickness of the photocatalyst, the recombination of the charge carriers can occur even before they reach the surface of the semiconductor. Thus, the PEC process uses photocatalysts to separate and catalyze water splitting. When the photon energy, which is higher than the Gibbs free energy, is provided or the photon energy is higher than the bandgap of the semiconductor, electrons in the conduction band and holes in the valence band are created. These holes reach the electrode surface at the electrode electrolyte interface, oxidize water molecule, resulting in the generation of O_2 and H^+ ions. The electrons pass through the external circuit and reach the counter electrode, where at the interface the H^+ ions are reduced, producing hydrogen.

Thus, there are steps involved in the photoelectrochemical water splitting and these are:

(1) Photocatalysts absorbs the sunlight or photons that have energy higher than the bandgap of the semiconductor.
(2) Electrons and hole pairs are created.
(3) The charge carriers generated need to be separated, and they migrate to the nearest recombination surface. There are two processes that can happen: either they can recombine and get lost or get separated in the material.
(4) The separated charge carriers reach the surface and undergo the electrochemical reactions. Reduction–oxidation reactions occur with involvement of these charge carriers at the respective electrode surface, resulting in the production of hydrogen and oxygen gases.

Thus, the surface morphology, structure, and electronic properties of the photocatalysts used and the proper choice of photocatalyst play a dominating role in these reactions and especially the step numbers (1) and (2). Thus, the major requirements for the PEC reaction are: (1) providing the energy for the reaction, i.e., photon energy greater than the bandgap of the semiconductor, i.e., bandgap >1.23 eV; (2) High surface area to facilitate the reaction at the interface, as the higher the surface area more will be the interaction of the reacting species; (3) Although the requirement of a higher surface area is essential for the reaction but at the same time, when the surface area is higher, then higher will be the surface recombination on the defect centers in the photocatalyst. Thus, higher the surface area faster the reaction, but more of recombination. The less the surface area slower the reaction but less will be the recombination of the charge carriers; (4) Use of noble metals as cocatalyst, like Pt or Ru can be dispersed on the surface to enhance the active sites, reduce the activation energy, and enhance the charge transport of the reaction.

Hole scavengers can be used so that the recombination can be reduced and electrons can generate hydrogen. The most widely studied hole scavenger is EDTA (ethylene diamine tetra acetic acid), others can be alcohols like methanol or ethanol or acetic acid, thus leading to an increase in the hydrogen production.

Ideally, the required photon energy for water splitting should be greater than 1.23 eV, bandgap of the semiconductor material should be greater than this. Under ideal conditions, like at AM 1.5 and when no losses occur, the conversion efficiency from the solar energy to hydrogen production theoretically is 47.4%. However, in real conditions, there will be several losses at the interfaces, charge transport losses, and recombinations leading to loss in the form of heat. As in electrolysis, an additional voltage is required to compensate this loss which is around 0.8 eV, thus a minimum of 2.0eV is required, or the bandgap of the semiconductor should be higher than 2 eV, in wavelength terms this corresponds to around 650 nm and to reach substantial current density the required bandgap is still higher, thus lying in range of 2.0–2.4 eV. The photoanode is an important part of the photoelectrochemical cell, because it is more demanding than even the photocathode. The material used for the photoanode should absorb light, be stable, should not corrode, have bandgap higher than 2 eV to take care of overvoltages, conduction and valence bands should be suitably placed, less loss involved in transport of charges within the semiconductor or at the interface, and it should have low cost and longer life.

The different materials used as photoanode include TiO_2, WO_3, ZnO, Cu_2O, Fe_2O_3, GaAs, CdS, GaP, etc., however oxides show better performance toward oxygen evolution reaction. Among the various studied materials, TiO_2 has stability, an appropriate bandgap, band positions, and non-toxicity, but can absorb in the UV region only. Thus, to improve the absorption and extend the absorption in the visible range of the solar spectrum, modifications by adding suitable materials, doping, heat treatment, and nanosizing are being used.

Reference:

Q1. Calculate the hydrogen production rate from an electrolyzer in $\mathbf{Nm^3/hr}$ for an $\mathbf{N_{cell} = 350}$, $\mathbf{I_{cell} = 2A}$ and $\eta_F = \mathbf{0.95V}$

Given:

- $N_{cell} = 350$
- $I_{cell} = 2\,A$
- $\eta = 0.95\,V$
- $z = 2$
- 1 mol of $H_2 = 22.4\,L = 0.0224\,Nm^3$ (at STP)
- Faraday constant $F = 96{,}500\,C/mol$.

Formula:

Hydrogen production per second can be calculated from Faraday's Law using the formula:

$$f_{H_2} = \eta_F \times \frac{I \cdot N_{cell}}{2F}(mol/s).$$

To convert to Nm^3/hr:

$$\dot{V}_{H_2} = \frac{N_{cell} \cdot I_{cell} \cdot 22.4 \cdot 3600}{2 \cdot 96500 \cdot 1000} = \frac{0.95 \cdot 350 \cdot 2 \cdot 22.4 \cdot 3600}{2 \cdot 96500 \cdot 1000} = 0.284\,\text{Nm}^3/\text{hr}$$

Q2. Calculate the ohmic overvoltage of an electrolyzer stack with the following characteristics

- Current density: $j = 300\ \text{mA/cm}^2$.
- Membrane thickness: $t = 0.005$ inch
- Conductivity of membrane: $\sigma = 20\,\text{mS/cm}$.
- Electronic resistance: $R_{\text{electrode}} = 0.2\,\Omega$.
- Electrode area $= 1\ \text{cm}^2$.

Given:

- Current density: $j = 300\,\text{mA/cm}^2$.
- Membrane thickness: $t = 0.005$ inch $= 0.0127$ cm
- Conductivity of membrane: $\sigma = 20\,\text{mS/cm}$.
- Electronic resistance: $R = 0.2\Omega$.
- Electrode area $= 1\ \text{cm}^2$.

To find ohmic overvoltage:

Ohmic Overvoltage = (Ionic Resistance + Electronic Resistance) × Current Density × Electrode Area.

$$\text{Ionic Resistance} = \frac{1}{\text{Specific Conductance}} = \frac{\text{membrane thickness}}{\text{conductivity of membrane x area of membrane}} = \frac{0.0127}{0.02}.$$

Ionic Resistance $= 0.635\Omega$,

$\therefore$ Ohmic Resistance= Ionic Resistance + Electronic resistance $= (0.635 + 0.2)\Omega = 0.835\Omega$.

Ohmic Overvoltage $= (0.835) \times 300 \times 1 = 250.5\text{mV} = 0.25$ V..

Q3. If power consumed by an electrolyzer operating under thermo-neutral voltage condition with an efficiency of 85% consumes 3 W power, find:

- **Voltage, current, and heat rejection in kJ/day**
- **Hydrogen production in kg/s**

(Given: $N_0 = 6 \times 10^{26}$kmol^{-1}, $V_0 = 1.48$V,,charge of an electron $q = 1.6 \times 10^{-19}C$,,molecular mass of hydrogen $= 2$g/mol.)

Given:

- Power = 3 W
- $\eta = 0.85$

- V_0 = thermo-neutral voltage, that is, no heat rejected or accepted by cell

1. Voltage $= \frac{1.48}{0.85} = 1.74$ V
2. Current $= \frac{\text{Power}}{\text{Voltage}} = \frac{3}{1.74} = 1.724$ A
3. Heat Rejected = Overvoltage × Current = (Voltage-Thermoneutral Voltage) × Current $= V - V_0 \times$ Current $= (1.74 - 1.48) \times 1.724 = 0.44824$ W
4. Heat Rejected (per day) $= 0.448424 \times 60 \times 60 \times 24 = 38.73\,\text{kJ/day}$
5. Hydrogen Production in kg/s:
   Current $= n_e \times N_0 \times q \times H_2$ produced

$$= \frac{\text{Current}}{n_e \times N_0 \times q} = \frac{1.724}{2 \times (6 \times 10^{26}) \times (1.6 \times 10^{-19})} = 1.8 \times 10^{-8}\,\text{kg - H}_2\text{/s..}$$

Q.4. The V − I relation in an electrolyzer is given by the relation:

$$V = V_0 + IR_{\text{int}}$$

where

- ***V* is terminal voltage,**
- **V_0 is the voltage that is required for electrolyzer to start conducting,**
- ***I* is the current passing through it,**
- **R_{int} is the internal resistance.**

Given the voltage overvoltage is 0.1 V. If a current of 50 A can cause a voltage drop of 1.48 V, and assuming reversible cell voltage is 1.23 V:

(a) **What is the internal resistance of the electrolyzer?**
(b) **If 125 electrolytic cells are connected in series, what is the required input energy in kWh/ day?**
©) **Also find the cost of the production per kg of H_2 (in kgH_2 /day), if fixed costs are Rs. 75,000/ day and variable costs are at the rate of Rs. 15/kWh.**

Solution:

(a) Calculation of Internal Resistance

Given:

- Reversible cell voltage: = 1.23 V
- Overvoltage: = 0.1 V
- Current: = 50 A
- $V = 1.48$ V
- Time: 24 hrs = 86,400 s

To find: Internal resistance of the electrolyzer R_{int}

$$V = V_0 + IR_{\text{int}}$$
$$V_0 = V_{\text{rev}} + \text{Over voltage} = 1.23 + 0.1 = 1.33\,\text{V}$$
$$V = V_0 + IR_{\text{int}}$$

$$1.48 = 1.33 + 50 \cdot R_{\text{int}}$$

$$R_{\text{int}} = \frac{1.48 - 1.33}{50} = \frac{0.15}{50} = 0.003\,\Omega = 3\,\text{m}\Omega.$$

(b) Input energy in kWh/day.

Formula:

$$\text{Input energy} = V \times I \times \text{no. of cells} \times \text{time}\,.$$

Substituting:

$$\text{Input energy} = 1.48 \times 50 \times 125 \times 86{,}400 = 799.2\,\text{MJ}$$
$$\text{Input energy} = \tfrac{799.2}{3.6} = 222\,\text{kWh/ day}\ (\because\ 1\,\text{kWh} = 3.6\,\text{MJ}).$$

©) Cost of Hydrogen Production

Hydrogen produced per day:

$$\begin{aligned} H_2\ \text{produced/day} &= \frac{I \times t \times N_{\text{cells}}}{2 \times q \times N_0} \\ &= \frac{50 \times 86{,}400 \times 125}{2 \times 6 \times 10^{26} \times 1.6 \times 10^{-19}} \\ &= 2.8125\,\text{kmoles/ day} \\ &= 2.8125 \times 2 = 5.625\,\text{kg} - H_2/\ \text{day}\,. \end{aligned}$$

Cost of $\mathbf{H_2}$ produced:

$$\text{Cost} = \tfrac{\text{Fixed cost} + \text{Variable cost}}{H_2\ \text{produced / day}}$$
$$\text{Variable cost} = 222\,\text{kWh/day} \times 15\ \text{Rs./kWh} = \text{Rs. } 3330\text{/day}$$
$$\text{Cost of } H_2 \text{ production} = \tfrac{75{,}000+3330}{5.625}$$
$$\text{Cost of } H_2 \text{ production} = \text{Rs. } 13{,}925.33\text{/kg} - H_2.$$

References

1. D.M.F. Santos, A.C. Cesar, Hydrogen production by alkaline water electrolysis. Sequeira Quim. Nova. **36**, 1176–1193 (2013)
2. J. Brauns, T. Turek, Alkaline water electrolysis powered by renewable energy: a review. Processes **8**, 248 (2020)
3. C. Lamy, P. Millet, A critical review on the definitions used to calculate the energy efficiency coefficients of water electrolysis cells working under near ambient temperature conditions. J. Power Sources **447**, 227350 (2020)

4. A. Ursua, L.M. Gandia, P. Sanchis, Hydrogen production form water electrolysis: current status and future trends, in *Proceedings of the IEEE*, vol. 100, p. 410–426 (2012)
5. Md. Mamoon Rashid, M.K Al Mesfer, H. Naseem, M. Danish, Hydrogen production by water electrolysis: A review of alkaline water electrolysis, PEM water electrolysis and high temperature water electrolysis. Int. J. Eng. Adv. Technol. (IJEAT) **4**(3) (2015). ISSN: 2249 – 8958

Chapter 8
Introduction to Hydrogen Storage

Whether hydrogen is produced in a centralized or a decentralized plant, the demand does not always occur at the point of production. Moreover, the demand and supply are often not in synchronization, making storage an essential component of the hydrogen value chain. Except for few applications in which hydrogen production and utilization are fully co-located and operate at a consistent load, hydrogen storage is required in nearly all other scenarios. Hence, hydrogen storage is required at various places, such as hydrogen production sites, refueling stations, onboard vehicles, and points of end use in industrial and stationary applications.

Hydrogen storage is the major bottleneck in the hydrogen energy value chain as it has the lowest gas density of 0.089 kg/m^3 under ambient temperature and pressure conditions. Hydrogen also has high diffusivity which makes it difficult to store. Components used in hydrogen storage, such as valves, fittings, and seals are more prone to leakage compared to other gases. Comparison of hydrogen with other conventional fuels is mentioned in Table 8.1; hydrogen has an advantage in terms of gravimetric energy density but suffers from significant disadvantage in volumetric energy density,posing several technical and economic challenges.

From Table 8.1, it can be understood how the low volumetric density of hydrogen means a significantly high volume requirement to store hydrogen at NTP. Thus, in order to reduce the volume required to store hydrogen, the possible methods are either to compress and store at higher pressure, or liquify and store as liquid hydrogen, or to store in solid-state form. To store 1 kg of hydrogen, 11 m^3 of space is required at STP, which is enormous. However, to store an equivalent amount of energy using gasoline the volume required is 3200 times lower than that required by hydrogen, i.e., 3.4 L or 0.00343 m^3. Thus, conformable storage of hydrogen is the biggest challenge when it comes to hydrogen energy value chain.

For a light duty vehicle, roughly 1 kg of hydrogen is required to drive around 100 km which may vary depending on the type of vehicle. So, to have a range of 500 km, 5 kg of hydrogen is required to be stored in a hydrogen container (also known as a vessel or tank). Now, if equivalent amount of energy, stored in 5 kg of

P. Sharma, *A Textbook on Hydrogen Energy*,
https://doi.org/10.1007/978-981-95-4601-5_8

Table 8.1 Comparison of the fuel properties of hydrogen with the conventional fuels

On mass basis			On volume basis	
Fuel	On LHV basis (MJ/kg)	On HHV basis (MJ/kg)	Fuel	On volume basis
Hydrogen	120	142	Hydrogen at 1 bar, 15 °C	0.084 kg/m^3, 0.01 MJ/L, 3 Wh/L
Methane	50	55.6	Hydrogen, 200 bar	15 kg/m^3, 1.8 MJ/L
			Hydrogen 300 bar	20.77 kg/m^3, 0.69kWh/L
			Hydrogen 350 bar	23.69 kg/m^3
LPG	46	50	Hydrogen, 700 bar	40.2 kg/m^3, 4.82 MJ/L, 1.3 kWh/L
Gasoline	44	47	Methane, 200 bar	8.38 MJ/L
Diesel		45	Methane,700 bar	15.8 MJ/L
Hydrogen	33.3 kWh/kg		LPG at 5 bar	26 MJ/L
Gasoline	12 kWh/kg		Gasoline	32 MJ/L, 8.8kWh/L
			Diesel	36 MJ/L, 9.8 kWh/L
			Liquid hydrogen, 1 bar	8.5 MJ/L, 70.9 kg/m^3, 2.36 kWh/L
			Solid-state hydrogen	> 150 kg/m^3

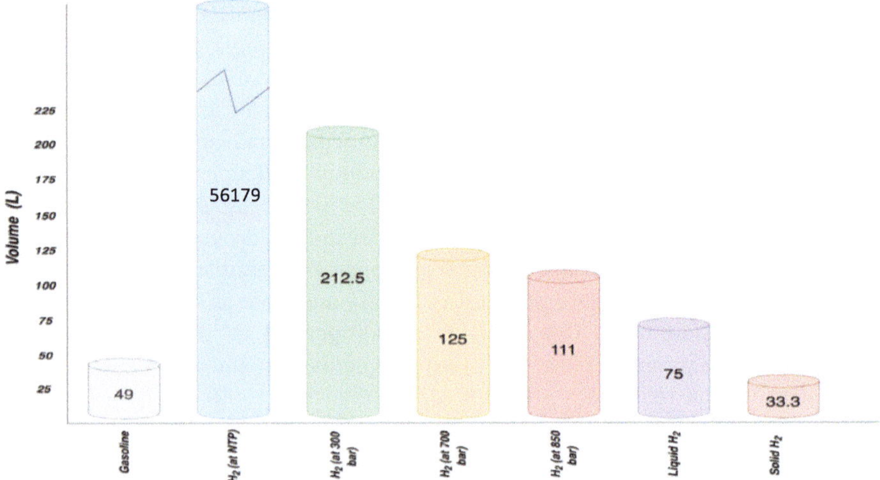

Fig. 8.1 Comparison of volume required to store 5 kg of hydrogen using different hydrogen storage technologies and as against energy equivalent amount of gasoline

hydrogen is to be supplied by gasoline in the same light duty vehicle, the volume of the tank would be 49 L.

The same energy, if to be stored in the form of hydrogen under normal temperature and pressure (NTP) conditions, would require 56179 L of volume, which is very high compared to that of the gasoline tank, making hydrogen unfeasible for various applications. Hence, to store hydrogen conformably, is to force hydrogen molecule in a given volume, by compressing hydrogen to high pressure, or to decrease the temperature below its critical point to liquefy. Alternatively, the repulsive forces between the hydrogen molecules can be reduced by interaction with other elements in the periodic table or their alloys. If hydrogen is compressed, to say, 300 bar, then the volume of the tank would reduce to 212 L. If it is further compressed to 700 bar and 850 bar, the tank volume would reduce to 125 L and 111 L, respectivley. The pressures involved are very high, necessitating ways to further improve the volumetric density, by liquefying hydrogen to store in cryogenic form. When hydrogen is in the liquefied form in a cryogenic tank, the volume required to store 5 kg of hydrogen is 75 L. Further higher densities can be achieved when stored in solid-state form, and the volume required will be 33.3 L. Figure 8.1 shows a relative comparison of tank sizes/volume to store 5 kg of hydrogen in different forms as well as the gasoline tank size needed to provide a range of roughly 500 km in a light-duty vehicle.

The storage requirement differs for various applications. For instance, in stationary applications, the weight and volume considerations may not be as important as those in mobile or vehicular applications. Important parameters to consider for stationary applications are the cost, refilling time, duration of storage without loss (dormancy), cyclability, fast response, life and material of the storage tank. For mobile applications, the important factors include: conformability, weight and volume, safety, life, refilling time, response time, transient operation, and cost, with dormancy being relatively less important.

Important Properties of Hydrogen for Storage

Before understanding the various methods of hydrogen storage, it is crucial to consider the key properties of hydrogen. Considering the phase diagram of hydrogen, as shown in Fig. 8.2, the triple point exists at 13.96 K and 7.36 kPa. The equilibrium lines separate the different phases: solid and liquid, solid and gas, and gas and liquid. The critical point is at 33.15 K and 1.3 MPa which occurs at a low temperature due to the strong repulsive interaction between hydrogen molecules. Beyond this point, liquid and gas phases are indistinguishable, and further increasing the pressure will not yield any liquid hydrogen. A small region, which starts at the triple point and ends at the critical point and marked in Fig. 8.2, is the region where liquid hydrogen exists at a temperature of 20 K and with a density of 70.8 kg/m^3. Thus, reducing the temperature to 20 K at ambient pressure gives liquid hydrogen, and further reduction to at 14.2 K at ambient pressure is where solid phase exists with a density of 70.6 kg/m^3. When the temperature is between 20 K and 14.2 K at atmospheric pressure, a state called slush hydrogen is obtained. In this state, solid and liquid phases coexist at a vapor pressure lower than the atmospheric pressure. Thus, it is important to note that while storing this slush hydrogen, any air inflow into the gas needs to be avoided. As can be seen from Fig. 8.2, beyond the critical point lies the supercritical hydrogen, which exhibits characteristics of both liquid and gas, i.e., it can flow like a liquid and is compressible like a gas. Beyond the temperature of 20 K lies solid region and the solid-phase hydrogen exists even at very high pressures. At pressures up to10^{11} Pa and temperatures around 200 K, solid hydrogen has been identified, its found to exist in four different phases: hexagonal close packing (HCP), face-centered cubic (FCC), orthorhombic, and monoclinic. For further higher pressures, i.e., beyond 8×10^{11} Pa and within the temperature range of 10^2–10^3 K, hydrogen transitions to metallic hydrogen and to liquid metal at even high pressures. The presence of this liquid metallic phase was predicted way back in 1935, but it was experimentally observed in 1996. This phase is found on the planets Saturn and Jupiter. Hydrogen exists in the plasma phase at further higher temperatures of 10^4–10^5 K, which is found in stars and in sun.

The state of hydrogen, like any pure substance, can be described using two intensive state variables. The rest of the variables can be found using the equation of state, which relates temperature, pressure, and volume.

The equation of state which relates *P*, *T*, and *V*

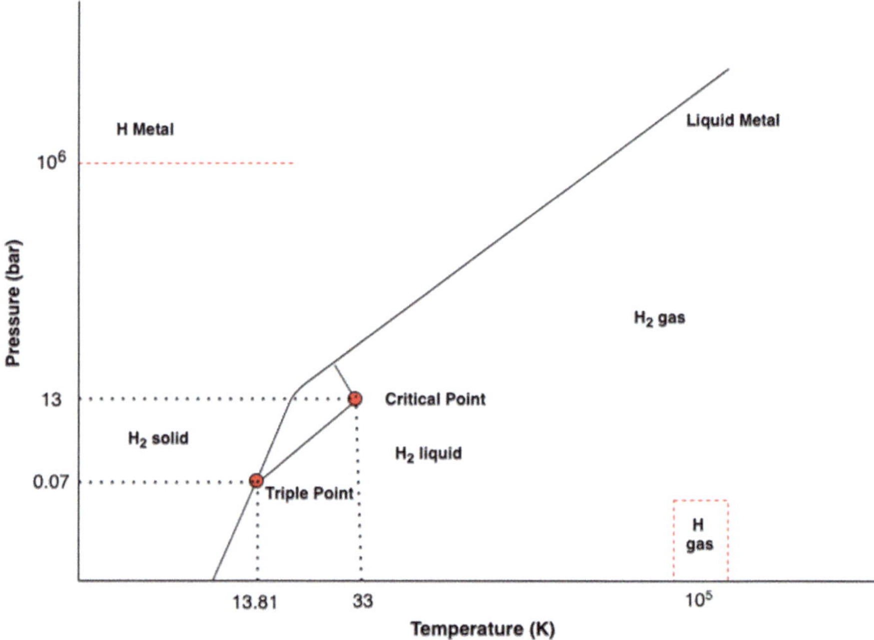

Fig. 8.2 Phase diagram of hydrogen

$$f(T, P, V) = 0. \tag{8.1}$$

Calorific equation of state

$$f(u, T, v) = 0,\ f(h, T, P) = 0 \tag{8.2}$$

and

$$f(T, s, v) = 0 \text{ or } f(s, T, P) = 0. \tag{8.3}$$

These relationships have been experimentally determined, and their tables and graphs are available for reference in the literature.

At ambient temperature and pressure, hydrogen satisfies the ideal gas equation:

$$PV = nRT. \tag{8.4}$$

R is the gas constant = 8.314 kJ mol^{-1} K^{-1}. This relation is valid for low pressure and high temperature conditions, i.e., temperatures beyond the critical point. The ideal gas assumes that the forces between the molecules are considered to be negligible except during collisions, and the molecules are considered as solid, hard, infinitesimally small, and incompressible spheres. However, at high pressures and low temperatures deviations from ideal gas behavior are observed and that can be accounted by other models such as Van der Waals, which considers the intermolecular forces in the pressure term and the volume deviation by including the molecular volume. The relation is

$$\left(P + \frac{a}{V^2}\right)(V - b) = RT. \tag{8.5}$$

For hydrogen, $a = 0.025$ m^6 Pa mol^{-2} is the constant related to dipole interaction, also known as the repulsion constant and $b = 2.66 \times 10\text{–}5$ $\text{m}^3\text{mol}^{-1}$ corresponds to the volume occupied by the hydrogen molecules.

Another way to consider the deviation from ideal gas behavior is by using the compressibility factor, a dimensionless quantity given as

$$Z = \frac{n_{\text{ideal}}}{n_{\text{real}}} = \frac{m_{\text{ideal}}}{m_{\text{real}}}. \tag{8.6}$$

The deviation of the value of Z from 1 is a measure of the deviation from ideal gas behavior. Its value has been determined experimentally for different gases and is available as graphs in the literature. By replacing, pressure by reduced pressure, P/P_{cr} and temperature by reduced temperature, T/T_{cr}, a generalized compressibility factor can be obtained for all gases. As seen from Fig. 8.3, the value of Z is 1.2 at ambient temperature and 300 bar pressure, which can be observed even at lower temperatures. Therefore, it is important to account for the deviations at higher pressures and consider real gas behavior. If the ideal gas equation were used (ambient T,

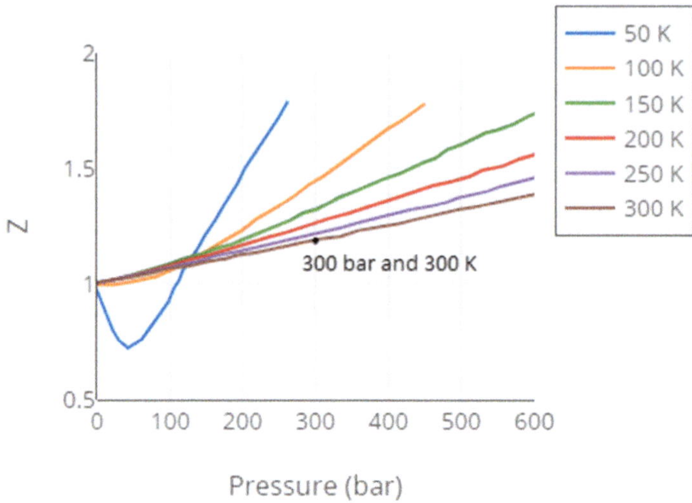

Fig. 8.3 Variation of compressibility factor with pressure at different temperatures

300 bar) to calculate the mass contained in a tank, it would result in an overestimation of approximately 20%.

Hydrogen is a diatomic molecule over a wide range of temperature and pressure. It consists of a single bond by the overlap of 1 s atomic orbitals of the two hydrogen atoms, each having one electron. This results in a lower-energy occupied bonding molecular orbital and a higher-energy unoccupied anti-bonding molecular orbital. Thus, the bond formed in molecular hydrogen is strong and stable with a very high dissociation energy of 435 kJ/mol. The spins of the two electrons occupying the bonding molecular orbital are antiparallel, however, the nuclear spins can be different. The property that is important to consider for liquid-state hydrogen storage is the spin of the elementary particle about its own axis. The higher-energy ortho state of hydrogen (o-H_2) is characterized by parallel nuclear spin, while the para state of hydrogen (p-H_2) has antiparallel nuclear spins and is the lower-energy state. Thus, p-H_2 is energetically more stable than the o-H_2. The two forms of hydrogen have slightly different properties. Normal hydrogen, that is, hydrogen at NTP, consists of 75% ortho and 25% para, also called as equilibrium hydrogen. This relative composition of ortho to para depends on the temperature. When cooled down to 20 K to convert it into liquid hydrogen, the equilibrium composition is 99.821% para and 0.179% ortho hydrogen, and this conversion is a very slow process (takes days to convert). Besides, since the (p-H_2) is the lower energy state, a conversion from room temperature normal hydrogen to equilibrium composition at 20 K releases the extra energy; thus, the process is exothermic (Fig. 8.4).

Another important property to be considered when understanding hydrogen storage is its Joule–Thomson coefficient. Hydrogen has a negative Joule–Thomson coefficient above 193 K, which is its inversion temperature. As a result, at temperatures above 193K, when hydrogen expands its temperature

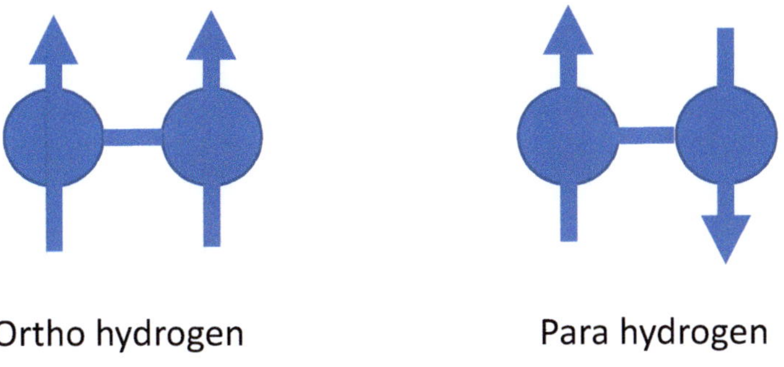

Fig. 8.4 Two different spin forms of hydrogen

increases rather than showing a cooling effect. This inversion temperature has two effects. Firstly, during liquefaction, excess heat released due to the ortho-to-para conversion must be removed. Secondly, if there is a sudden depressurization of the system, it can lead to ignition.

The heat released due to the ortho-to-para conversion at 20 K is 670 kJ/kg, which is higher than the latent heat of vaporization which is 446 kJ/kg at 20 K. Consequently, additional cooling is required to remove that heat generated during expansion, as well as the heat released during ortho-para hydrogen conversion for liquefaction. A positive inversion temperature makes it essential to pre-cool hydrogen during liquefaction to temperatures below 193 K to obtain a cooling effect during expansion through a Joule–Thomson valve or a porous plug.

Different Ways of Hydrogen Storage

Hydrogen storage methods can be broadly categorized into physical storage and chemical storage methods. Among the physical storage methods, the most widely known and used method is the compressed state hydrogen storage at high pressures. Hydrogen can also be stored physically in liquid-state hydrogen by cooling it to 20 K and storing it in cryogenic containers. In both of these methods, hydrogen molecule do not chemically react with the storage medium. Lastly, among the physical methods is the adsorption of hydrogen molecule on the surface of the high-surface-area materials. This method of storage takes place at temperatures $T < 100$ K, where a weak Van der Waals bond is formed between hydrogen and the host material. In contrast, chemical hydrogen storage, involves chemical interaction between hydrogen and the storage medium through various mechanisms. The mechanisms can be: (a) absorption of hydrogen at the interstitial sites in a metal or a metal alloy forming metal hydrides; (b) chemical bonding of hydrogen in ionic or covalent compounds, forming complex or chemical hydrides; (c) hydrolysis of reactive metals leading to oxidation; (d) storage in high energy density compounds known as liquid organic hydrogen carriers (LOHCs) (Fig. 8.5).

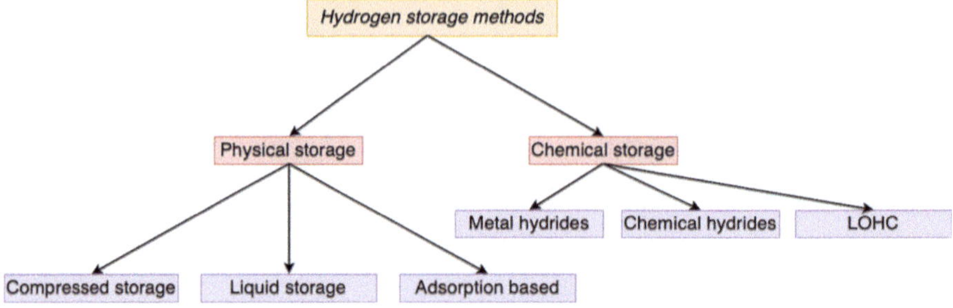

Fig. 8.5 Classification of different hydrogen storage methods

More details of these methods of storage are described in the later chapters; a brief comparison of operating conditions and density of storage in these methods of hydrogen storage is presented in Table 8.2.

If we look at the history of gas storage, it dates back to the 1880s when it was stored in cylinders made of wrought iron at 120 bar, weighing over 500 kg. Later, the pressure vessels were made of stainless steel, this was a practice that continued till 1960, when the high-pressure tanks could store up to 150 bar. After 1960, the storage pressure increased to 200 bar. Thereafter, the use of high-pressure tanks expanded to applications outside military or space in 1970s. The containers used in 1970s were mostly composite tanks with metallic or polymeric liner. In the 1980s, such pressure tanks with pressures in the range of 150–300 bar began for uses such as diving paintball, or storing gases. Several advancements in terms of reduction in weight and increases in storage pressure have led to the development of present-day tanks.

Table 8.2 Comparison of different methods of hydrogen storage

Method of storage	Mass density (ρ_m in wt%)	Volume density (ρ_v in kg H_2 m^{-3})	T (K)	P (bar)
Compressed gaseous hydrogen storage	13	< 40	RT	700
Liquid hydrogen storage	Depends on size	70.8	20 K	1
Storage on high surface area materials via adsorption	≥ 2	20	> 77	100
Storage via absorption in intermetallic alloys	≤ 2	100–150	*RT*	1
Complex hydrides	< 18	150	> 100	1
Hydrolysis of complex hydrides or metals	< 40	> 150	*RT*	1

Targets for Hydrogen Storage

Hydrogen can be used for power generation either by combustion in IC engine (ICE), producing mechanical power, or electrochemically by using a fuel cell, producing electrical power. In both these cases, hydrogen reacts with oxygen to produce water. In case of fuel cells, water is the only product, but for ICE, combustion at very high temperatures produces NO_X along with water. Hence, ICE must run with lean mixtures to prevent NO_X emissions. Hydrogen can have a major impact in transport sector because hydrogen fuel cells have higher energy density, thus extending the vehicle's driving range. However, the use of hydrogen for onboard vehicular applications is more challenging as compared to stationary applications due to the low density of hydrogen, making hydrogen storage a major issue in transitioning to hydrogen economy. The US DOE has set tough targets for hydrogen storage for the development of hydrogen energy-based vehicles. These targets reflect the importance of both gravimetric and volumetric capacity of the storage medium, as automotive OEMs or consumers will not accept storage systems that are excessively heavy (limiting the driving range of the vehicle) or too voluminous (limiting the luggage space). The summary of these DOE targets is included in Table 8.3.

Solved Problem 1. (a) Calculate the mass of hydrogen stored (in kg) in a 120 L vessel at 700 bar and 298 K, considering hydrogen as an ideal gas and real gas (equation of state), respectively. (Compressibility factor, $Z = 1.452$, Molecular mass of $H_2 =$ 2.016 g/mol, Universal gas constant = 8.314 J/mol K).

Soln.

Table 8.3 DOE targets for hydrogen storage

Storage parameter	Units	2025	Ultimate
System gravimetric capacity	kWh/kg (kgH_2/kg system)	1.8 (0.055)	2.2 (0.065)
System volumetric capacity	kWh/L (kg H_2/L system)	1.3 (0.040)	1.7 (0.050)
Storage system cost (and fuel cost)	$/kWh net ($/kg H_2)	9 (300)	8 (266)
Durability/operability			
Operating ambient temperature	°C	– 40/60 (sun)	– 40/60 (sun)
Min/Max delivery temperature	°C	– 0.4706	– 0.4706
Cycle life (1/4th tank to full)	Cycles	1500	1500
Min. delivery pressure from tank	Bar	5	5
Max delivery pressure from tank	Bar	12	12
Charging/discharging rates System fill time (for 5 kg)	Min	5-Mar	5-Mar
Fuel purity	% H_2	Meet or exceed SAE J2719	

Given:

Pressure, $P = 700$ bar $= 700 \times 10^5$ Pa.
Volume, $V = 120$ L $= 120 \times 10^{-3}$ m^3
Temperature, $T = 298$ K
Molecular mass of $H_2 = 2.016$ g/mol
Compressibility factor, $Z = 1.452$
Universal gas constant $= 8.314$ J/mol K

To find:
Mass of hydrogen stored (in kg) following ideal & real gas equation of state.

(a) **Ideal gas**:

$$PV = mR_sT$$

and specific gas constant of H_2,

$$\begin{aligned} R_s &= \frac{R}{M} \\ &= \frac{8.314}{2.016 \times 10^{-3}} \\ &= 4124 \text{ J/kg K} \end{aligned}$$

$$\begin{aligned} m_{\mathrm{H_2,ideal}} &= \frac{PV}{R_sT} \\ &= \frac{700 \times 10^5 \times 120 \times 10^{-3}}{4124 \times 298} \\ M_{\mathrm{H_2,ideal}} &= 6.835 \text{ kg H}_2. \end{aligned}$$

(b) **Real gas**:

$$PV = m_{H_2,real}ZRT$$

$$\begin{aligned} m_{\mathrm{H_2,real}} &= \frac{PV}{ZR_sT} \\ &= \frac{700 \times 10^5 \times 120 \times 10^{-3}}{1.452 \times 4124 \times 298} \\ m_{\mathrm{H_2,real}} &= 4.707 \text{ kg H}_2 \end{aligned}$$

$$\begin{aligned} \therefore \mathrm{M_{H_2ideal}} &= 6.835 \text{ kg H}_2 \\ m_{\mathrm{H_2real}} &= 4.707 \text{ kg H}_2 \end{aligned}$$

Solved Problem 2. Calculate volume required to store 5-kg H_2 at 350 bar, 298 K considering hydrogen as an ideal gas and a real gas ($Z = 1.06$). Also, calculate the percentage deviation from ideal gas volume? (Molecular mass of $H_2 = 2.016$ g/mol).

Soln:

Given:

Mass of $H_2 = 5$ kg H_2
Pressure, $P = 350$ bar $= 350 \times 10^5$ Pa
Temperature, $T = 298$ K
Compressibility factor, $Z = 1.06$

(a) Specific

$$PV = mZR_sT$$

$$\begin{aligned} \text{Specific gas constant, } R_s &= \frac{R}{M} \\ &= \frac{8.314}{2.016 \times 10^{-3}} \\ &= 4124 \text{ J/kg K} \end{aligned}$$

$$\begin{aligned} PV &= mZR_sT \\ \therefore V_{\text{real}} &= \frac{mZR_sT}{P} \\ &= \frac{5 \times 1.06 \times 4124 \times 298}{350 \times 10^5} \\ V_{\text{real}} &= 0.186 \text{ m}^3. \end{aligned}$$

(b) Percentage deviation from ideal gas

$$\begin{aligned} \therefore V_{\text{ideal}} &= \frac{mR_sT}{P} \\ &= \frac{5 \times 4124 \times 298}{350 \times 10^5} \\ V_{\text{ideal}} &= 0.176 \text{ m}^3 \end{aligned}$$

$$\begin{aligned} \therefore \% \text{ deviation} &= \frac{V_{\text{real}} - V_{\text{ideal}}}{V_{\text{ideal}}} \times 100 \\ &= \frac{0.186 - 0.176}{0.176} \times 100 \\ &= \frac{0.01}{0.176} \times 100 \end{aligned}$$

$$\therefore \text{ \% deviation} = 5.681\ \% \approx 6\ \%\ .$$

Solved Problem 3. Calculate the volume occupied by 1 mol of hydrogen gas at 298 K and 50 bar using Van der Waals equation.

$$\left(P + a\frac{n^2}{V^2}\right)(V - nb) = nRT,$$

where Van der Waals constant for H_2:

$a = 0.025$ Pa m^6/mol^2
$b = 2.66 \times 10^{-5}$ m^3/mol

Soln:

Mole of H_2 gas, $n = 1$ mol
Pressure, $P = 50$ bar $= 50 \times 10^5$ Pa
Temperature $= 298$ K
$a = 0.025$ Pa m^6/mol^2
$b = 2.66 \times 10^{-5}$ m^3/mol

Set

$$\left(P + a\frac{n^2}{V^2}\right)(V - nb) = nRT$$

Substituting the known value in above equation

$$\left(50 \times 10^5 + \frac{0.025}{V^2}\right)\left(V - 2.66 \times 10^{-5}\right) = 8.314 \times 298$$
$$\left(50 \times 10^5 + \frac{0.025}{V^2}\right)\left(V - 2.66 \times 10^{-5}\right) = 2477.572.$$

The above nonlinear equation can be solved iteratively or by using numerical method.

However, considering approximate solution using ideal gas volume as a starting guess.

$$V_{\text{ideal}} = \frac{nRT}{P} = \frac{1 \times 8.314 \times 298}{50 \times 10^5}$$
$$V_{\text{ideal}} = 4.955 \times 10^{-4}\ \text{m}^3.$$

Substituting value of V_{ideal} as V in Eq. (8.1)

$$\left(50 \times 10^5 + \frac{0.025}{\left(4.955 \times 10^{-4}\right)^2}\right)\left(4.955 \times 10^{-4} - 2.66 \times 10^{-5}\right) = 2477.572.$$

Solving LHS approximately

$$\frac{50 \times 10^5 \times \left(4.955 \times 10^{-4}\right)^2 + 0.025}{\left(4.955 \times 10^{-4}\right)^2} \times \left(4.955 \times 10^{-4} - 2.66 \times 10^{-5}\right) = 2477.572.$$

(I) Taking a guess $V = 5 \times 10^{-4}\ \text{m}^3$

$$\frac{50 \times 10^5 \times \left(5 \times 10^{-4}\right)^2 + 0.025}{\left(5 \times 10^{-4}\right)^2} \times \left(5 \times 10^{-4} - 2.66 \times 10^{-5}\right) = 2414.34.$$

(ii) Take another guess $V = 5.05 \times 10^{-4}\ \text{m}^3$

$$\frac{50 \times 10^5 \times \left(5.05 \times 10^{-4}\right)^2 + 0.025}{\left(5.05 \times 10^{-4}\right)^2} \times \left(5.05 \times 10^{-4} - 2.66 \times 10^{-5}\right) = 2438.89.$$

(iii) Take another guess $V = 5.1 \times 10^{-4}\ \text{m}^3$

$$\frac{50 \times 10^5 \times \left(5.1 \times 10^{-4}\right)^2 + 0.025}{\left(5.1 \times 10^{-4}\right)^2} \times \left(5.1 \times 10^{-4} - 2.66 \times 10^{-5}\right) = 2463.46.$$

Therefore,
LHS is approximately equal or close to RHS value so,

$$\%\ \text{deviation} = \frac{2477.572 - 2463.46}{2477.572} \times 100 = 0.57\%.$$

Therefore, volume occupied by 1 mol of the gas is approximately

$$5.1 \times 10^{-4}\ \text{m}^3 \text{ or } 0.51\ \text{L}.$$

Chapter 9
Underground Hydrogen Storage

Hydrogen is being recognized as a potential energy carrier that could decarbonize several sectors, including a few hard-to-abate ones. It can provide clean energy solutions and reduce greenhouse gas emissions in various sectors inlcuding transportation, heating, chemical industries, refineries, the power sector and many others. As global concerns over climate change become increasingly understood, and the various implications are seen worldwide such as flash floods, drought, and acid rains, there is an increase in renewable capacity worldwide to combat the ill effects of fossil fuels. However, the intermittency of renewables, whether solar or wind, makes it essential to have energy storage associated with the renewable power plants. The requirement for storage varies from daily to seasonal, depending on changes in solar or wind energy availability patterns. In large electricity networks, where the contribution or energy output of renewables is low, the deficit or variation can be easily balanced by the grid. However, for systems with higher output capacity, reliable storage will always be required. Although options such as compressed air, flywheels or batteries, may be considered for small-scale and short-term applications. In such cases, hydrogen can play a crucial role to support renewable energy systems by addressing supply-demand gap to balance the intermittency and enhancing grid stability. The conversion of surplus renewable electricity in the form of hydrogen for terawatt-hour (TWh) scale in economically feasible and favorable geographical locations presents a promising pathway for large-scale energy balancing. During periods of excess renewables-based generation and to provide daily grid balancing, hydrogen can be used as efficient storage medium to alleviate the challenges associated with renewables. The added advantage of converting to hydrogen and transporting it in the form of gas is that the energy transport losses are significantly lower (0.1%) compared to power network losses (8%) [1]. The daily demand can be met by options for hydrogen storage, including pipelines and tanks which can provide MWh-scale and storage for hours and days. However, when seasonal storage of relatively large quantities of hydrogen on scales of GWh or TWh is required to satisfy the storage needs for weeks or seasons, then underground storage of hydrogen is the

P. Sharma, *A Textbook on Hydrogen Energy*,
https://doi.org/10.1007/978-981-95-4601-5_9

most promising option as large volumes are available for storing at higher pressures resulting in higher energy densities.

Underground gas storage refers to the storage of gas in a natural environment below the ground at depth of several hundred meters. This method is economically feasible in places with available geological structures, large renewable installations, a need for storage, proximity to end-use industries, and other commercial utilities. Storage of gas underground in such structures is safe, economical, and less susceptible to fire or attack, also it requires less space than storing an equivalent amount of gas above the surface.

The unique properties of hydrogen such as low molecular weight, low viscosity, high diffusivity, and high penetrability makes it difficult to store. It's low density and high mobility complicate conventional storage methods; however, the same characteristics are advantageous for underground hydrogen storage, as hydrogen's high mobility results in lower residual amount of hydrogen remaining in a porous media after its withdrawal cycles. This method can provide a large quantity of storage over longer durations through underground storage in various geological formations such as salt caverns (excavated or solution mined), including salt, coal, igneous, or metamorphic rocks, depleted oil and gas reservoirs, and saline aquifers.

These geological structures have different chemical and physical characteristics defined by the working gas capacity, the requirements in terms of cushion gas, site preparation, monitoring of the structures, cost of maintenance, release rate or production rate and the chances of leaks. These are the essential factors to understand a particular structure and site quantity and duration of storage. The higher penetrability of hydrogen can be a problem when it comes to salt cavern-based storage compared to depleted oil and gas reservoirs and aquifers due to the lower sealing capability of the rocks in caverns. The low solubility of hydrogen in water (0.00079 mol/mol at 25 °C) is an advantage when considering underground hydrogen storage.

Geological structures have been widely used for storing natural gas (by various petroleum companies for hundreds of years), carbon dioxide (for carbon capture and storage), compressed air, or fluids like thermal water or oil. However, there are limited examples of storing hydrogen in these structures. Although hydrogen has different characteristics from these gases, the experience gained from their storage provides valuable insights for the development of underground hydrogen storage in such structures. However, differences in the properties of hydrogen compared with natural gas or carbon dioxide affect operational procedures, site preparation, economics, suitability of sites, and thus a detailed analysis is required when it comes to hydrogen storage.

There are primarily three options of such geological structures for storing hydrogen: salt caverns, depleted oil and gas reservoirs, and saline aquifers. Additional potential options under considered include abandoned salt mines, limestone mines as well as rock caverns [2].

The primary consideration in identifying a site for underground hydrogen storage will be safety, efficiency, and losses due to gas leakage, which in turn depend on the tightness and quality of the caprock. Therefore, selecting an appropriate geological structure requires a thorough assessment of site-specific geological conditions, including reservoir surface area, depth, thickness, tightness, permeability, porosity, pressure, geomechanical properties, and rock characteristics. In summary the various technical, economic, safety, and environmental criteria need to be studied before selecting a particular site.

History of Underground Storage [3]

A year-wise progress in the field of subsurface storage is listed below:

- 1915 Ontario gas field was used for natural gas storage
- 1976 Walters identified that there are no insurmountable or environmental problems with the underground hydrogen storage (UHS)
- 1979 Carden & Patterson found that injected hydrogen losses were only 1% per cycle, while the use of cushion gas reduced first-cycle losses to approximately 0.4% was, improving the storage efficiency
- 1985 Lindblom proposed a network of man-made tunnels for hydrogen storage-based on the economics considerations
- 1986 Taylor et al., studied the feasibility and economic viability of UHS in salt caverns, rock caverns, and depleted natural gas
- 2004 Schaber et al., compared the UHS with high-pressure storage of hydrogen above the ground, and studies revealed that the UHS could be preferred due to comparatively larger storage capacity
- 2018 Tarkowski and Czapowski introduced the potential salt domes for UHS in Poland
- 2018 Heinemann et al., studied the UHS in Midland Valley site, UK, and their study was based on the storage capacity and geological uncertainties
- 2019 Lemieux et al. identified the favorable sites for seasonal storage of hydrogen in Ontario, Canada
- 2020 Lankof and Tarkowski provided a series of maps of salt caverns with the potential of UHS, including maps of storage capacity, maps of energy, and maps of calorific value of hydrogen
- 2020 Narayanamoorthy et al. proposed a normal wiggly dual hesitant fuzzy set (NWDHFS) for UHS site selection based on the five criteria, including the technique of operation, investment cost, social, economic, and risk factors
- 2020 Shi et al. investigated the potential of storing a hydrogen and natural gas blend into a depleted gas/oil reservoir, which has already been used for natural gas storage through a series of experiments.

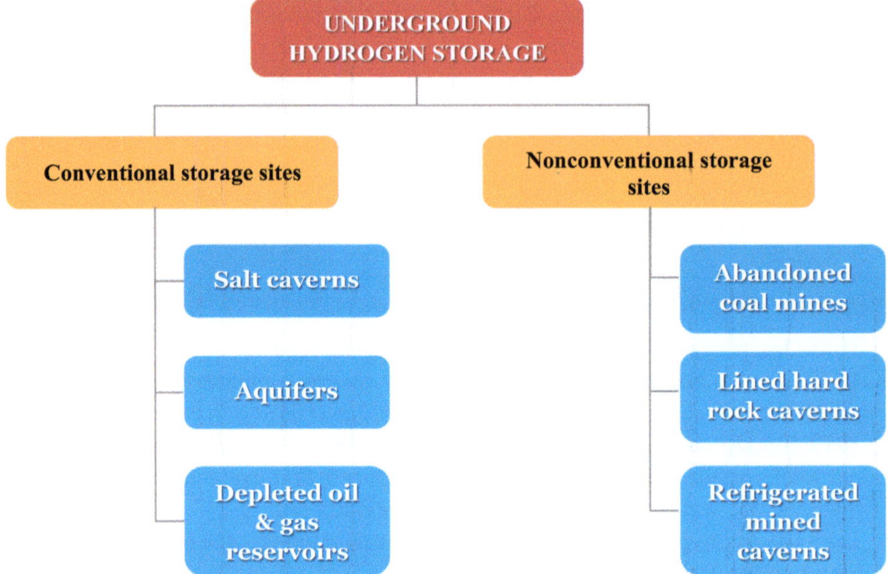

Fig. 9.1 Classification of different underground storage technologies

Different Types of Underground Gas Storage

Underground hydrogen storage is not a recent concept; these technologies have been in use since 1916, with the Ontario gas field being first site. Most of the current gas storage capacity is in the form of natural gas, which amounted to 429 billion m^3 in 2022. However, global conflicts and rising uncertainties have since driven a significant expansion in the storage capacity. Different types of underground hydrogen storage methods are listed in Fig. 9.1 and shown via schematic in (Fig. 9.2).

Hydrogen Storage in Salt Caverns

Compared with depleted oil and gas reservoirs and water aquifers, salt caverns are better suited for hydrogen storage due to their very low permeability, this is because of the chemical inertness of salt toward hydrogen. These have been used for storing hydrogen since the 1970s in UK and 1980s in USA. Salt cavern-based storage is most promising underground hydrogen storage method, considering the experience, technology maturity, possibility of large-scale storage, flexibility and faster cycling capability. The walls of salt cavern are impermeable to hydrogen and the plasticity of salt prevents the formation and propagation of cracks leading to reservoir tightness. This results in favorable rheological and mechanical properties which provide effective stress distribution, high cycling performance, and high storage efficiency [4]. The

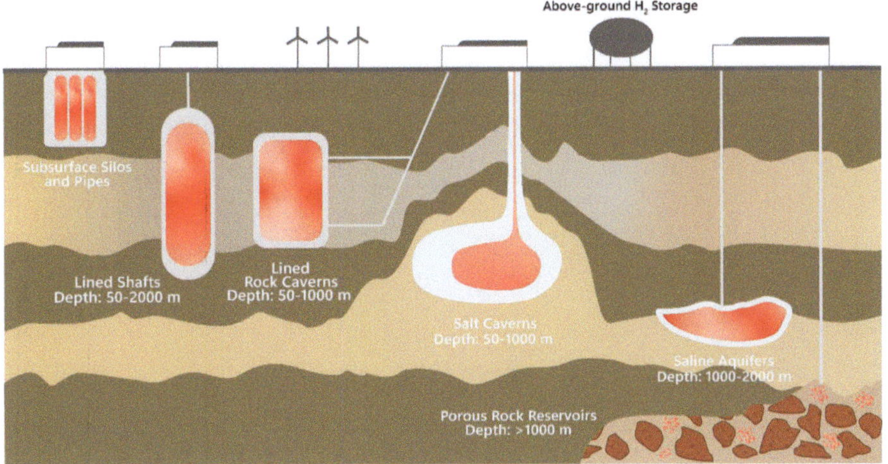

Fig. 9.2 Schematic to show the different types of underground storage options

high salinity of salt caverns suppresses microbiological activity, which is crucial because such activity can consume hydrogen and enable formation of products such as CH_4, causing contamination of the stored gas.

Subsurface salt formations are formed by the evaporation of seas, leading to crystallization at places where the input of water to the site is less than the evaporation rate, resulting in salt precipitation under arid climatic conditions. These deposits may vary largely in salt composition, thickness, purity, and depth. This evaporation enhances the sealing behavior, making salt caverns ideal for short- and medium-term storage (Fig. 9.3).

Creation of Salt Caverns

Salt caverns for hydrogen storage can be created by pumping water in the regions with large salt deposits or salt domes to dissolve the salt and removing the resulting brine solution. This process is called solution mining. Salt domes, being more rigid and homogenous facilitate the development of a robust storage structure. In contrast, bedded salt deposits are less firm, thinner, more heterogenous, have less stability, and typically found at shallower depth as compared to the salt domes. For long-term storage, the serviceability and stability are critical and depend on the site-specific characteristics, requiring a detailed understanding of these sites prior to usage. Salt caverns typically have a depth of approximately 2000 m, with cavern heights ranging from 50--500 m, diameters of around 50–100 m, and volumes of about 1,000,000 m^3. Probable operating pressures range from 30 to 80% of lithostatic pressure. Once the required cavern volumes are reached during solution mining, hydrogen gas is injected to displace the brine solution, which can be

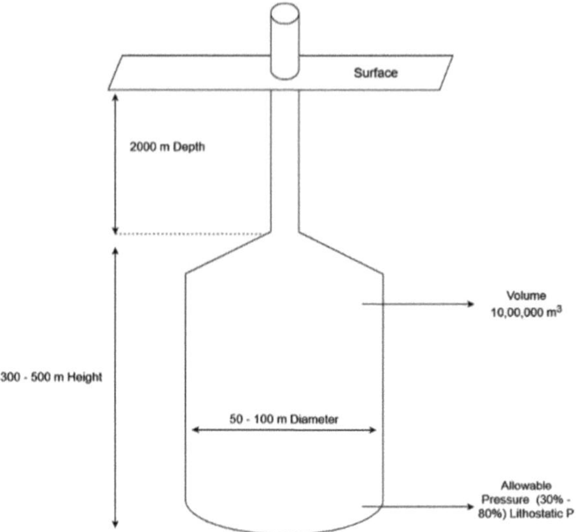

Fig. 9.3 Schematic of a salt cavern-based hydrogen storage

reused for salt production. Salt caverns construction is relatively simpler, faster, and less expensive than other mentioned geological structures used for hydrogen storage. These structures were widely used for storing oil, natural gas, and compressed air. For hydrogen storage, there are three commercial salt cavern sites worldwide: one at the UK (Teeside) and three in the USA (Clemens, Moss Bluff and Spindletop). Another facility exists at Kiel, Germany, used for storing town gas. Renewed interest in underground hydrogen storage has emerged due to the anticipated increase in demand of hydrogen across various sectors and the necessity for large-scale storage storing to bridge supply-demand gap. As a result, many new sites are being developed or repurposed to store hydrogen. Examples of new projects include the jointly developed SSE Thermal and Equinor and Aldbrough Hydrogen Pathfinder project in the UK; Advance Clean Energy Storage (ACES) project in Utah, USA; HyPOS inBad Lauchstadt, Germany; Gasunie and Veendam in Netherlands; and HyPSTER in Etrez, France. Other European and international collaborative research projects on large-scale hydrogen storage include HyGeo, Nouvelle-Aquitane, Roads2HyCOM, HyStorPor, HyUnder, H2 STORE, ANGUS+, SUN STORAGE, Hychico, and ADMIRE [4]. Several studies have also investigated the potential of hydrogen storage in salt caverns [5–12].

The most critical aspect—whether the salt caverns are to be abandoned or actively used for gas storage—is their stability. It is necessary to understand the heterogeneity, geometrical complexities, and stress conditions when selecting a site and later while it is in service. It is equally important to understand salt bed thickness, depth, composition, and presence of non-salt rocks, which can lead to loss of hydrogen at interfaces. The state of stress plays a key role when durnig gas storage, continuous mechanical stress can lead to creep deformation. Creep deformation in rock salt cavern can occur initially as a primary or a transient creep, followed by secondary or steady-state creep, and eventually tertiary deformation. In literature, various scientific explanations and experiments attributing creep deformation to crystal structure changes and grain boundary effects within salt caverns are reported.

There is a possibility of the presence of various impurities, including dolomite, gypsum, calcite, pyrite, quartz, etc. These impurities lead to chemical reactions such as the conversion of $CaSO_4$ to H_2S. However, the self-healing and low permeability of salt caverns makes the storage structure much more advantageous than other geological structures. Salt caverns can be of various depths, usually about 1000–1500 m. Additionally, salt domes and thick horizontal-bedded rock salt deposits are most suited for gas storage. Another major issue could be the loss of hydrogen, which can be due to leakage, dissolution, adsorption, and chemical reactions. However, it is reported that the loss is usually less than 3.2% of the hydrogen storage capacity [13]. The hydrogen loss due to chemical reactions can occur through reaction with the rock salt minerals like gypsum, carbonate, carnalite, sulfate, oxides, and also through microbial activities. The environment in hydrogen storage caverns is highly reducing, converting the abovementioned salts: carbonatesto CO_2; CH_4S^{6+} to S^{2-}; Fe^+ to Fe^{2+}. All these factors impact the purity of hydrogen stored and corrode the pipelines used at the site and reduce their life. The corrosion effect can be reduced by using coatings, inhibitors, and providing the required insulations. Additionally, the various microorganisms present can convert H_2 to CO_2 or CH_4 (methanogenic bacteria) or H_2S (sulfate-reducing bacteria), which necessitates the need for detailed studies on interaction of hydrogen with rocks and the installations to avoid any future hazards.

In underground gas storage there are two types of gases: working gas, which is injected and withdrawn during operation, and cushion gas, which remains as a permanently till the facility's lifetime. The role of cushion gas is two-fold: to provide the minimum pressure to prevent any water inflow into the storage space and to provide the optimum set of conditions for injection. Here, cushion gas serves as buffer in the form of unextracted volume of gas used that maintains the required pressure and supports effective site management [14]. The choice of cushion gas is based on its wettability and cost. For example, nitrogen can be cost-effective while also having high wettability, leading to easier separation during injection and withdrawal cycles in a sandstone or basaltic-type of structures. Other selection criteria include the difference in densities of the working and cushion gas. The amount of cushion gas directly affects storage efficiency and depends on the site-specific parameters of the geological site like the depth, shape, and permeability. However, it is reported in the literature that approximately 22–33% of volumetric capacity is sufficient to serve the requirements of max of 78% working gas. In some literature, even the use of hydrogen as a cushion gas has been suggested, aiming to save the cost of separation and purification. However, a detailed study is required to examine both technical and economic challenges for such usage. The storage process involves first injecting a cushion gas, such as, nitrogen or methane, before hydrogen injection. This serves as a permanent inventory in the storage unit and does not participate in the process. Multiple cycles of compression and expansion are performed during the injection and withdrawal processes to achieve the desired pressure and deliverability rate [3]. The primary requirement of the cushion gas is to achieve pre-pressurization of the geological formation, allowing for the desired pressure and flow rate during the hydrogen withdrawal. While the presence of cushion gas can reduce hydrogen losses, separation of hydrogen from the cushion gas becomes necessary when high-purity hydrogen is required for end use. Usually one well per salt cavern is used for both exploration and operation of the storage facility. The requirement for cushion gas depends on the depth, capacity, and pressure of the storage; typically, cushion gas constitutes about 22–33% of the cavern volume, with the remaining being the working gas. Since hydrogen is the working gas, the various components or pipes to be used for casing should be resistant to hydrogen embrittlement and cement slurry is used for ensuring gas tightness.

There are several demonstrations especially in the US and the UK, on geological storage in salt caverns. However, key challenges for salt-based hydrogen storage include integration with large-scale production plants for fast cyclic injection and withdrawal, achieving high-purity hydrogen, limited availability of suitable geological formations, access to cost-effective storage sites, water requirements for solution mining, safety, brine disposal, and the need for robust regulatory frameworks. Additional constraints arise from cavern location, particularly the distance from hydrogen production facilities and pipeline infrastructure.

For the safe use of salt caverns for gas storage, from injection to abandonment, it is essential to consider the structural stability of the structure. Various stress concentrations may be created due to inhomogeneities, geometries, and nonlinear creep deformations. Creep deformation can arise due to continuous stress concentrations resulting from long-term storage and may lead to permanent deformations with time due to mechanical stress. Depending on storage duration and cavern geometry, creep may occur in primary (transient), secondary (steady-state), or tertiary (accelerated) stages. These deformations must be evaluated in advance through laboratory experiments, microstructural analysis, and numerical simulation. Such assessments, combined with considerations of cavern shape, depth, salt bed thickness, size and creep properties, are critical for reliable cavern design.

Depleted Oil and Gas Reservoirs

The conventional hydrocarbon reservoirs, which were used for gas storage, when become unfeasible or uneconomical after the gas extraction, can be repurposed for hydrogen storage. These geological structures can, however, be used for hydrogen storage. The depleted reservoirs, which previously trapped hydrocarbon from migrating due to the underlying rock formations, are typically porous and permeable, located thousands of feet beneath the surface and often underlined by aquifers. Such structures are simple and easy to operate and maintain, as they have been benefited from years of operational history. The sites' geological structure, integrity, composition, and infrastructure are well-understood. Hence, they act as gas storage container with the desired permeability to allow the flow of gas being stored, while also serving to trap the stored gas, preventing its leakage or migration to other regions.

The biggest advantage of such reservoirs is the availability of residual hydrocarbon gas, which can act as a cushion gas and can contribute to achieving the desired pressure during operation. The requirements for cushion gas in case of depleted reservoirs are roughly 50–60%, which is higher than in salt caverns (22–33%) and lower than in aquifers (approximately 80%). However, a major challenge in using the residual hydrocarbon gas as cushion gas is a contamination and affects the purity of the injected hydrogen, necessitating post-withdrawal purification when high-purity hydrogen is required. Nevertheless, the depleted reservoirs remain well-suited for long-term and seasonal hydrogen storage.

When residual hydrocarbon gas is used as cushion gas, it is often economical to stop further hydrocarbon extraction from a site when the requirements for volumetric capacity of hydrogen storage becomes viable, economically and will save time. Since the gas pressure stored is higher than the initial reservoir pressure, such structures can accommodate large amounts of gas. However, any water that invaded in the sites after the exploration was stopped, must be removed, and commissioning a fully function underground hydrogen storage typically takes up to five years. If, however, there is difference, if an underground depleted oil storage facility is to be used for hydrogen storage, a thorough understanding of the possible reactions between oil and stored hydrogen is essential to prevent unexpected loss of hydrogen, purity degradation hydrogen gas stored, and integrity of the structure. In particular, chemical reactions between oil and hydrogen leading to methane formation pose a significant challenge and must be carefully assessed.

Aquifer

Aquifers are a subsurface layer of porous and permeable rock structures, found in the sedimentary basins globally. They are typically located at depths of several hundred feet, making the chances of gas leakage to surface very low. These are either filled with fresh or saline water and have storage mechanism similar to the ones discussed above. Aquifers are more commonly available subsurface bodies than salt caverns and depleted oil and gas reservoirs, offering an advantage for UHS. Their use requires host rock that allows hydrogen injection while preventing gas migration. This migration has to be prevented with an impermeable seal to avoid the buoyancy-driven vertical leakage and traps to limit lateral migration of the stored hydrogen.

Since the aquifers already contain water, the difference between the densities of liquid and gas displaces water in the aquifers, when injected with pressurized hydrogen This displacement of water occurs either vertically or laterally, creating the space required for gas storage within the geological structure. However, the presence of the liquid-gas interface leads to issue of water escaping out along with stored hydrogen during the withdrawal cycle. Another disadvantage of storing hydrogen in aquifers is the storage volume; the cushion gas requirement in aquifers is the highest compared to salt caverns and depleted gas reservoirs, resulting in a lower volume of stored hydrogen as compared to cushion gas. The cushion gas requirement is reported to be as high as 80% of the total structure volume (compared to 50–60% in depleted gas reservoirs and 22–33% in salt caverns-based storage), resulting in 20% by volume of working gas storage. The biggest hurdle in using aquifers is the lack of site and structure specific knowledge, leading to potential hydrogen losses by migration, biological activities, and reaction of hydrogen with minerals present in rock formations. Thus, a detailed sites study, involving exploratory drilling, and careful assessment of subsurface are crucial. Aquifers-based storage is the most recent method being explored for storing hydrogen and structures which could store 100% hydrogen is still not developed. The various aquifer gas storage projects are dedicated to storing town gas, some examples of such projects include, Beynes in France, Lobodice in the Czech Republic and Ketzin in Germany.

Non-conventional Methods of Underground Hydrogen Storage

In regions where the conventional underground geological structures mentioned above are unavailable; such as the Pacific Northwest, Northeast, and the Southeast Atlantic; non-conventional options such as abandoned coal mines, hard rock caverns or refrigerated mining caverns can be used.

Abandoned Coal Mines

Coal mines where further mining is uneconomical, and the operations are ceased, can be used for hydrogen gas storage. However, the residual coal will adsorb the gas injected and thus needs to be considered when injecting the gas for storage. The existing knowledge about the site details, structure, and available volume for gas storage, plays a crucial role in determining the suitability of operating a UHS site. Notable examples of such sites include abandoned coal mines of Peronnes and Anderlus in Belgium and Leyden in United States, all which have since been decommissioned. These sites have a long successful history of storing natural gas for peak demand, avoiding peak gas pricing. The site located at Belgium, i.e. Anderlus (capacity of natural gas storage was 180 Mm^3) was used for natural gas seasonal storage between 1857 and 1969, while Peronnes (capacity of natural gas storage was 120 Mm^3) was operational between 1860 and 1969. The site at Colorado, i.e., the Leyden mine, was built to supply gas during peak hours and meet daily fluctuations in the natural gas supply to Denver. It is essential that such sites should have an impermeable core to stop gas leakage and with secure site from all sides to avoid migration of stored gas. If naturally such protective impermeable layer is not available then concrete or gravel is primarily used to protect gas leakage by sealing the sites.

Lined Hard Rock Caverns

As mentioned in previous sections, sedimentary deposits like salt caverns can be used for underground hydrogen storage, or similar sedimentary structures can be excavated to create gas storage facilities. Beyond, sedimentary deposits, the metamorphic and igneous geological structures can also be utilized by constructing storage caverns through techniques such as blasting or shaft sinking. These engineered structures can be sealed using protective casing or linings to prevent gas leakage and providing containment. The permeability associated with an enclosed casing or lining can allow for higher-pressure gas storage. It is mentioned in literature that concrete can be used to reduce strain and stress points between the rock of the uneven structures and caverns. Also, polymer or steel liners could be used, but cost will be the determining factor. The primary purpose of having such lining is to ensure gas tightness and prevent any possible chemical reaction that could lead to losses, as well as avoid deformations to the host rock surfaces. Besides, groundwater ingression must be controlled which can be done by installing proper drainage system around periphery of cavern to prevent excessive hydrostatic pressure that could cause damage to liner during depressurization. Hard rock caverns are structurally stable and need of cushion gas is minimal. Although they have been widely used for natural gas storage, their application for hydrogen storage has not yet been realized; nevertheless, they show strong potential. However, the need for lining or casing, as well as the construction of site and the installation of suitable infrastructure, is capital-intensive.

Refrigerated Mined Caverns

In this method of underground storage, the gas to be stored at the geological site is first refrigerated, and then the cooled gas is stored. This chilled gas or cooling reduces the gas volume, thereby reducing the gas storage space requirement and increasing the amount of gas storage. These mined caverns could be preferred for igneous or metamorphic rock sites, to avoid lining and save on the cost of liner. The requirement of such sites is to be impermeable, without joints, faults, and stress regions. Refrigerated mined caverns require lesser quantity of cushion gas and lie at the depths of abount 2500–3000 m. They have enhanced storage potential, improved delivery rates, good cycling capability, and sufficient capacity to meet the seasonal demands.

The choice of UHS also depends on factors such as, whether ultrapure hydrogen or has impurities, and whether it will be used for fuel cell applications, gas turbines, chemical industries, or blended into gas pipelines. For example, when ultrapure hydrogen of fuel cell grade is to be stored, salt caverns are preferred over other geological structures because of the least contamination. Applications where such high-purity hydrogen is not required such as gas turbines, chemical industries, or pipeline, blending any of the above-mentioned method geological structures can be used.

Possible Chemical Reaction-Based Loss of Hydrogen in Underground Storage

Hydrogen stored in geological structures may be consumed or converted into other gaseous hydrocarbons due to the presence of pre-existing microorganisms, residual gases, present prior to injection, minerals, or ions in subsurface region. Therefore, it is very important to study the geological site and understand the possibilities of potential reactions that could cause not only loss of hydrogen but also, contamination, or safety risks. Abiotic (inorganic) reactions between hydrogen and brine, minerals, or dissolved ions consume hydrogen and can alter the structure's composition, stability, permeability, and pores structure. Similarly, the presence of microorganisms can consume hydrogen resulting in change in permeability and composition. Methanogenic bacteria can consume hydrogen to produce methane, and acetogenic bacteria convert CO_2 and H_2 to acetic acid. Similarly, archaea microorganisms consume hydrogen to produce H_2S, while the iron reducing bacteria consumes hydrogen during the reduction of iron oxides, leading to hydrogen loss. All these processes not only lead to the consumption or loss of hydrogen, but can also alter the reservoir characteristics by changing mineral characteristics through dissolution or precipitation resulting in changes in porosity or permeability of the formations and thus leading to safety issues.

Table 9.1 details of some of the global underground hydrogen storage facilities.

The underground hydrogen storage has a long, successful history, with the Teeside UK site being operational since 1972, storing 95% H_2 at a depth of 400 m. The Conoco Philllips site at Texas US stores 95% H_2 at depth of 1000 m. More such sites are enlisted in Table 9.1.

In conclusion, selection of a suitable geological site depends on multiple factors: economics feasibility, site availability, and proximity to production or utilization locations. Key technical considerations include site stability and safety; geological structure (depth, reservoir volume, capacity, porosity, permeability, and integrity) lithography and geothermal gradient existing knowledge of geology of the site, and environmental and social constraints.

Table 9.1 Current global status of underground hydrogen storage

Field/project name	Storage type	H_2 (%)	Working condition	Volume (m^3)	Status
Teesside (UK)(1972)	Bedded salt	95	45 bar	210,000	Operating
Clemens (USA)	Salt dome	95	70–137 bar	580,000	Operating
Moss Bluff (USA)	Salt dome	95	55–152 bar	566,000	Operating
Spindletop (USA)	Salt dome	95	68–202 bar	906,000	Operating
Kiel (Germany)	Salt cavern	60	80–100 bar	32,000	Closed
Ketzin (Germany)	Aquifer	62	Not reported	Not reported	Operating with natural gas
Beynes (France)	Aquifer	50	Not reported	3.3×10^8	Operating with natural gas
Lobodice (Czech Republic)	Aquifer	50	90 bar/34 °C	Not reported	Operating
Diadema (Argentina)	Depleted gas reservoir	10	10 bar/50 °C	Not reported	Not reported
Underground Sun storage (Austria)	Depleted gas reservoir	10	78 bar/40 °C	Not reported	Operating

Important Parameters Impacting Site Selection

The major parameters that influence the underground hydrogen storage at a geological site are parameters associated with solid, i.e., reservoir characteristics and solid–fluid interactions. The reservoir-related characteristics include permeability and porosity; fluid-related parameters are density, viscosity, solubility, diffusivity, and interface characteristics, while solid–fluid interaction parameters encompass relative permeability, wettability, capillary pressure due to, absorption-desorption behavior, interfacial characteristics, and mobility. (Table 9.2).

Besides the mentioned parameters the viability of underground storage is decided based on safety, large-scale, and long-term storage capability and favorable economics. Economics of such storage is governed by cost of transportation, hydrogen injection/withdrawal, site monitoring, and infrastructural costs. Cost considerations can be further expanded into laying the foundation, electricity requirements, compressor operations, handling cost including injection and withdrawal, construction and infrastructure costs, excavation and reinforcement expenses, equipment replacement due to embrittlement or corrosion, workforce and labor costs, and cushion gas costs. The use of geological sites for hydrogen storage, in general, are dictated by legal and societal aspects, often requiring government support through favorable policies and financial incentives. For widespread adoption of the

Table 9.2 Description of the important parameters impacting underground hydrogen storage

S. No.	Parameter	Definition	Relation	
1	Permeability	Ease with which a single fluid can flow through reservoir rock	$k_a = \frac{Q\mu\Delta L}{A\Delta P}$ where k_a is absolute permeability, Q is volumetric flow rate, ΔL is distance between injection and withdrawal well, A is area of cross section, ΔP is pressure difference	Depends on pore size distribution, given by Darcy's law
2	Porosity	Void space present in the rock or geological formation	$\varphi a = \frac{\text{Total pore volume}}{\text{Bulk volume}}$ $\varphi\text{eff} = \frac{\text{Interconnected pore vol}}{\text{Bulk volume}}$, where φ_a is the absolute porosity and φ_{eff} is the effective porosity	Connected pores contribute to hydrogen flow
3	Hydrogen storage capacity	Storage capacity of the geological structure	$C = \text{LwtS}\ \varphi_{\text{eff}}$ where L is length, w is width, t is thickness, φ_{eff} is effective porosity, and S is saturation defined as ratio of pore volume filled with hydrogen to total pore volume	C depends on effective porosity, effective stress, and lithography
4	Density of fluid		Fluid density changes with pressure and temperature	Difference between hydrogen and liquids such as brine results in gravity segregation
5	Fluid viscosity	Measure of capacity of fluid flow, depends on temperature and pressure	Fluid viscosity and relative permeability provides mobility ratio	Important in understanding interfacial stability among fluids and gas leakage
6	Solubility	Solubility of hydrogen in fluids in the UHS	Solubility changes with change in pressure, temperature and the salinity of the reservoir	Increases with increase in pressure and temperature, with solubility being an order of magnitude more in non-aqueous fluids than that in aqueous fluids

(continued)

Table 9.2 (continued)

S. No.	Parameter	Definition	Relation	
7	Diffusivity	Quantity of hydrogen that diffuses through a rock surface	Depends on temperature, pressure, and the type of fluid	Diffusivity is higher in aqueous solution than in hydrocarbons so more important in aquifers
8	Wettability	Capability to adhere to the rock surface, i.e., hydrophobicity or hydrophilicity considered in presence of another immiscible or partially miscible fluid	Wettability is considered of that of gas, water or oil on rock surface determined by adhesive or cohesive forces between different fluid phases, such as gas, water, or oil on the rock surface	Differences when considering water, oil, and gas and their wettability on rock. Depends on pore structure, rock composition, and chemistry. It can be measured using contact angle
9	Fluid–fluid or solid–fluid interface tension	Tension between immiscible fluids is determined by phase boundary which depends on intermolecular interaction. Solid–fluid tension determines hydrogen diffusion	Depends on interface between rock and hydrogen, rock and brine, and brine and hydrogen in salt caverns	Depends on pressure, temperature, and acidity
10	Sorption	Adhesion or removal of atoms, molecules, or ion on the surface	Depends on pressure and temperature	Micropore-based phenomena with meso/ macro pores providing transport pathways

UHS methods, it is essential to consider allgeological, societal, financial, safety, regulatory, technological, and economical factors.

References

1. N.S. Muhammed, B. Haq, D. Al Shehri, A. Al-Ahmed, M.M. Rahman, E. Zaman, A review on underground hydrogen storage: insight into geological sites, influencing factors and future outlook. Energy Rep. **8**, 461–499 (2022)
2. R. Tarkowski, L. Lankof, K. Lubon, J. Michalski, Hydrogen storage capacity of salt caverns and deep aquifers versus demand for hydrogen storage: a case study of Poland. Appl. Energy **355**, 122268 (2024)
3. D. Zivar, S. Kumar, J. Foroozesh, Underground hydrogen storage: a comprehensive review. IJHE **46**, 23436–23462 (2021)
4. J. Miocic, N. Heinemann, K. Edlmann, J. Scafidi, F. Molaei, J. Alcalde, Underground hydrogen storage: a review. Geol. Soc. London, Special Publications **528**, 73–86 (2023)

5. A. Ozarslan, Large-scale hydrogen energy storage in salt caverns. Int. J. Hydrogen Energy **37**, 14265–14277 (2012)
6. M. Bai, K. Song, Y. Sun, M. He, Y. Li, J. Sun, An overview of hydrogen underground storage technology and prospects in China. J. Petroleum Sci. Eng. **124**(2014), 132–136 (2024)
7. I. Iordache, D. Schitea, A.V. Gheorghe, M. Iordache, Hydrogen underground storage in Romania, potential directions of development, stakeholders and general aspects. Int. J. Hydrogen Energy **39**, 11071–11081 (2014)
8. J. Michalski, U. Bunger, F Crotogino, S. Donadei, G. Schneider, T. Pregger, K. Cao, D. Heide, Hydrogen generation by electrolysis and storage in salt caverns: potentials, economics and systems aspects with regard to the German energy transition. Int. J. Hydrogen Energy **42**, 13427–13443 (2017)
9. R. Tarkowski, G. Czapowski, Salt domes in Poland—potential sites for hydrogen storage in caverns. Int. J. Hydrogen Energy **43**, 21414–21427 (2018)
10. D.G. Caglayan, N. Weber, H.U. Heinrichs, J. Linßen, M. Robinius, P.A. Kukla, D. Stolten, Technical potential of salt caverns for hydrogen storage in Europe. Int. J. Hydrogen Energy **45**, 6793–6805 (2020)
11. A. Lemieux, A. Shkarupin, K. Sharp, Geologic feasibility of underground hydrogen storage in Canada. Int. J. Hydrogen Energy **45**, 32243–32259 (2020)
12. W. Liu, Z. Zhang, J. Chen, D. Jiang, F. Wu, J. Fan, Y. Li, Feasibility evaluation of large-scale underground hydrogen storage in bedded salt rocks of China: a case study in Jiangsu province. Energy **198**, 117348 (2020)
13. S. Zhu, X. Shi, C. yang, Y. Li, H. Li, K. Yang. X. Wei, W. Bai, X. Liu, Hydrogen loss of salt cavern hydrogen storage. Renew. Energy **218**, 119267 (2023)
14. R. Tarkowski, Underground hydrogen storage: characteristics and Prospects. Renew. Sustain. Energy Rev. **105**, 86–94 (2019)

Chapter 10
Compressed State Hydrogen Storage

In the previous chapter, underground hydrogen storage in geological structures was described. Underground hydrogen storage is useful in providing large-scale and long-term storage solutions, that can enhance grid stability and assists in seasonal balancing. With storage capacities exceeding 5 GWh and discharge times of several days or more, it can effectively support grid stability. However, when the requirement is for small-scale storage, daily or short-term hydrogen storage, storing it in either compressed or liquid form would be a better choice for both stationary or portable applications, including vehicular applications. Thus, the choice of storage technology is decided based on the requirements and boundary conditions of the end-use applications. Prior to delving into depth of compression or liquefaction, few fundamentals related to the properties of hydrogen need to be reiterated, i.e., the thermodynamics of hydrogen compression and expansion process need to be considered.

Joule–Thomson Expansion—Adiabatic throttling of a fluid at constant enthalpy through an insulated porous plug is called Joule–Thomson expansion. Depending upon the state of the fluid, the temperature on expansion can either increase, decrease, or stay constant; this is called Joule–Thomson effect. The Joule–Thomson coefficient is given by:

$$\mu_{JT} = \left(\frac{\partial T}{\partial P}\right)_h \tag{10.1}$$

Corresponding change in temperature is given by

$$\Delta T = T_2(h, P_2) - T_1(h, P_1) = \int_{P_1}^{P_2} \mu(T, P)\mathrm{d}P \tag{10.2}$$

For an ideal gas, no temperature change is observed in isenthalpic expansion process, i.e.,

P. Sharma, *A Textbook on Hydrogen Energy*,
https://doi.org/10.1007/978-981-95-4601-5_10

$$\mu_{JT} = \left(\frac{\partial T}{\partial P}\right)_h = 0 \tag{10.3}$$

For real gases, the temperature of the gas changes on isenthalpic contraction or expansion and the Joule Thomson coefficient is non-zero. In real gases, the molecules of the gas interact by means of attractive or repulsive forces. In case of expansion, when the pressure reduces or volume increases, and if there is an attractive interaction between the molecules, the potential energy also increases. Thus, in order to keep the internal energy constant, the kinetic energy decreases, thereby the temperature of the gas decreases, and a cooling effect is observed during expansion. In this case, the Joule–Thomson coefficient (μ_{JT}) is positive.

However, if the interaction between the gas molecules is repulsive, the potential energy will decrease with increase in volume during expansion, and thus to keep the internal energy constant, the kinetic energy will increase, thereby the temperature will increase on expansion. In this case, the Joule–Thomson coefficient (μ_{JT}) will be negative.

For hydrogen, the inversion temperature is around 200 K. During hydrogen liquefaction, when it is passed through Joule–Thomson valve or a throttle valve, then the temperature of the gas needs to be less than 200 K; else on expansion, heating effect will be produced. At room temperature, μ_{JT} for hydrogen is negative with a value of – 0.30 K/MPa.

Equation of State

In thermodynamics, there are different independent state variables, which are related and can be expressed in terms of equation of state. The equation of state is used to define the state of a system, this equation relates the various thermodynamic state variables, and it provides fundamental basis for describing the nature of gas and tells about how it behaves.

1. Ideal Gas

$$PV = mR_iT = nRT, \tag{10.4}$$

where m is mass, n is molar mass, R_i is specific gas constant, and R is the universal gas constant.

Isobaric specific heat is given as $c_p^0 = f(T)$..

Isochoric specific heat is given as $c_v^0 = c_p^0 - R_i$..

Isentropic exponent for ideal gas is $k^0 = \frac{c_p^0}{c_v^0}$..

For an ideal gas, the isentropic exponent is the ratio of isobaric specific heat to the isochoric specific heat.

Enthalpy and internal energy of an ideal gas at a temperature T can be obtained using,

$$h^0(T) = \int_{T_0}^{T_1} c_p^0(T)\mathrm{d}T + h_0, \quad \text{with } h_{0=0}. \tag{10.5}$$

Here, h_0 is the initial enthalpy, since the initial enthalpy is zero, the last term in above equation (10.5) becomes zero.

Considering ideal gas model, several phenomena can be explained with accuracy.

$$u^0(T) = \int_{T_0}^{T_1} c_v^0(T)\mathrm{d}T + u_0, \quad \text{with } u_0 = -R_i T_0. \tag{10.6}$$

2. Real Gas

An ideal gas is a theoretical concept where the molecules are considered to be hard, solid spheres with negligible volume and no interaction between the molecules. For some cases, the ideal gas behavior can be a reasonable assumption, and the corresponding equation of state mentioned above (10.4) can be used to predict the behavior of gas molecules to a reasonable degree of accuracy. However, for high pressure and low temperatures, the ideal gas assumption is not very accurate and can lead to errors, and thus it is not valid. The deviations from ideal gas behavior increase as the critical point in the phase diagram is approached. The density of hydrogen under NTP is 0.08 kg/m^3; thus, to store 1 kg of hydrogen under NTP would require approximately 11m^3 of volume, and to reduce this space, the hydrogen gas molecules need to be packed together compactly by applying pressure. Thus, as the pressure increases, the density also increases, and this density can be found as a function of pressure using the thermodynamics equations. At elevated pressures above 100 atm, the behavior of hydrogen deviates from ideal gas behavior significantly, because at increased pressure, the polar nature of hydrogen results in attractive forces between the gas molecules. Thus, for describing the behavior of hydrogen in compressed state, a different equation of state than ideal gas equation is required. There are many such equations for real gases, such as the Van der Waals gas, the virial equation, and the compressibility factor inclusion etc.

In order to consider this deviation from the ideal gas behavior, there are various models. One of them is considering a dimensionless (Z) quantity, which is given as a parameter called the compressibility factor.

$$PV = mZR_iT = nZRT. \tag{10.7}$$

Several models are used to predict the real gas behavior. It is observed that near the critical point, the rate of change of compressibility factor is higher and it is also a function of temperature and pressure. For hydrogen gas, the compressibility factor is found to increase linearly with pressure and temperature. The dependence on temperature becomes more pronounced at higher pressures (Fig. 10.2). This is very important when we consider compression of hydrogen as this deviation can

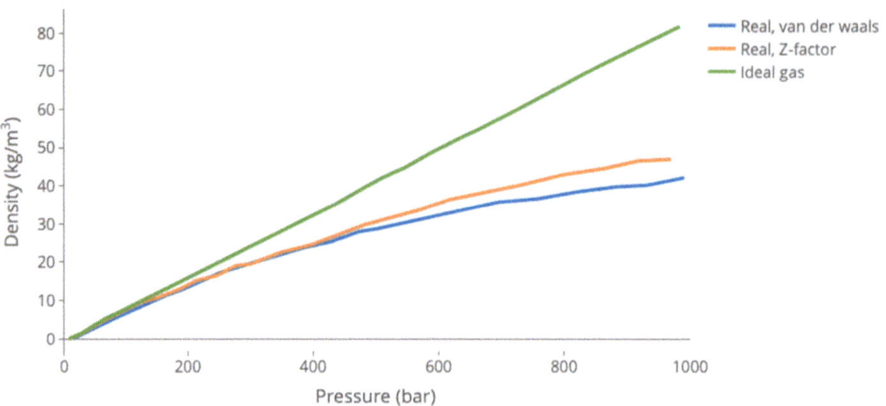

Fig. 10.1 Variation of the density of hydrogen with pressure for different gas models

lead to error, e.g., consider a storage tank of volume 1 m^3 at 20 °C and 250 bar, if ideal gas model is applied then mass of hydrogen stored is 21 kg, and while applying compressibility factor, the mass of storage comes out to be 18 kg which is the actual mass. Thus, applying the ideal gas assumption would have led to an overestimation of mass by 14%. To predict the behavior of hydrogen at different pressures, different equations of state must be used. Figure 10.1 shows the behavior of hydrogen when different gas models are used to predict the density variation with pressure. Till a pressure of about 150 bar, the three models are almost identical. Above 150 bar, the ideal gas model overestimates the density of hydrogen; however, the two real gas models gives identical results until 400 bar. Above 400 bar, even the two real gas models gives different values. It is challenging to decide which model is better, as the values of the variables in the different equation of state are fitted in such a way that the prediction lies close to the experimentally observed values. To avoid this ambiguity, a hydrogen-dedicated equation of state or calorific equation is required.

Isentropic Exponent

An isentropic process is defined as a process where the entropy of the system remains constant. An isentropic coefficient, along with different state variables, can be used to describe the state of the gas or change in its state. The relationship between state variables and this exponent is required to describe the compression and expansion of the gas. Different real gas equations of state can be used to determine the required compression and expansion work in a compressed state hydrogen storage. Considering the different independent state variables, we can have different equations of state. By incorporating the isentropic exponent, these equations can be used to describe various thermodynamic processes.

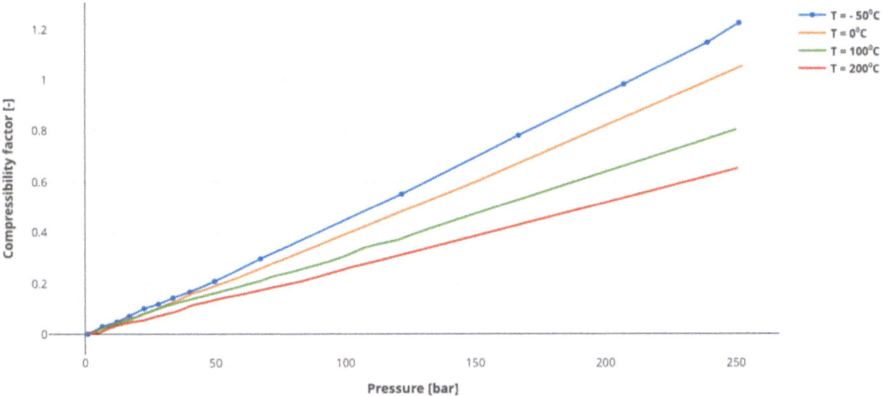

Fig. 10.2 Variation of compressibility factor of hydrogen with pressure for different temperatures

The relation for an ideal gas is given by:

$$PV^k = \text{constant} \tag{10.8}$$

$$T^k P^{1-k} = \text{constant} \tag{10.9}$$

$$TV^{k-1} = \text{constant} \tag{10.10}$$

$$k = \frac{c_p^0}{c_v^0}. \tag{10.11}$$

k depends on temperature or remains constant.

For real gas, the relations between state variables and the different isentropic exponents can be described by equations of state. These equations for real gas can be given by:

$$PV^{k_1} = \text{constant} \tag{10.12}$$

$$T^{k_2} P^{1-k_2} = \text{constant} \tag{10.13}$$

$$TV^{k_3-1} = \text{constant}. \tag{10.14}$$

The relation between different state variables considers three different isentropic exponents. All three exponents, k_1, k_2, and k_3, behave differently with varying temperature and pressure, depending on the type of gas. Only one of these three equations is required to describe the state of the real gas. Considering the equation,

$$PV^{k_1} = \text{constant}. \tag{10.15}$$

The temperature–pressure relation for an isentropic change can be written as:

$$\frac{T_2}{T_1} = \left(\frac{Z_2}{Z_1}\right)\left(\frac{P_2}{P_1}\right)^{\frac{k_1-1}{k_1}}. \tag{10.16}$$

The assumption in the above equation is that while the state of system changes from 1 to 2, the isentropic constant remains the same. However, under actual conditions, the exponent is a function of temperature and pressure. The isentropic exponent depends upon the nature of the gas, as well as the temperature and pressure.

Considering that the exponent varies linearly, the average of the two values corresponding to the states 1 and 2 can be considered. However, in cases where this assumption of linear variation doesn't hold, the use of such relation will lead to erroneous results and a deviation from the actual values. However, the exponent is observed to vary linearly with pressure, while the slope of the curve relating the exponent to temperature decreases as the temperature increases. Figure 10.3 shows the variation for the isentropic coefficient k_1, used for the equation of state relating pressure and volume. The k_1 exponent varies with pressure at different temperatures, and the variation is linear. The isentropic exponent could have a different value at different states. The consideration of linear variation simplifies the calculation, and the arithmetic mean can be used as the value of the exponent between two different states to determine the exponent at a particular state.

As the temperature increases, the slope also decreases for the isentropic exponent as seen in Fig. 10.3.

With the understanding of above-mentioned thermodynamic fundamentals required to calculate the work required for compression, the next section of the chapter is dedicated to calculating compression or expansion work.

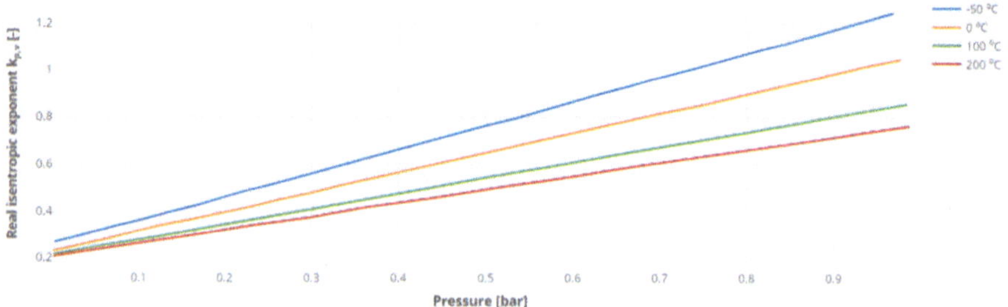

Fig. 10.3 Variation of the isentropic coefficient for hydrogen with pressure for different temperatures

Work of Compression for Compressing Hydrogen

1. **Direct Method**

To calculate the work of compression required to compress hydrogen, the exact equation of state or calorific equation is required. However, in the absence of this information, various models can be used, as described in previous sections. To simplify calculations, compression could be considered an isentropic process, i.e., the process of compression is reversible, and no heat exchange occurs between compressor and outside environment (which is an approximation). In that case, the work of compression can be calculated using the following equation:

$$W = \int_{V_1}^{V_2} p\mathrm{d}V, \tag{10.17}$$

where, V is the volume of the gas, and P is the pressure at which the gas is compressed. Now, different gas models or equation of states as mentioned in previous section can be used to calculate the work of compression. If we assume the gas to be ideal in that case, the compression work to increase the pressure of one mole of hydrogen from initial pressure P_1 to final pressure P_2 will be given as:

$$W_{\text{isentropic}} = \frac{\gamma}{\gamma - 1} RT_1 \left[\left(\frac{p_2}{p_1} \right)^{\frac{\gamma-1}{\gamma}} - 1 \right], \tag{10.18}$$

where, T_1 is the temperature of hydrogen at P_1, and γ is the specific heat ratio of the gas (C_p/C_v) which has a value of 1.4 for hydrogen. The basic assumption in considering the process to be isentropic was reversibility of the process. But, actual processes, even if considered adiabatic, are not reversible; in that case, the deviation from this assumption leads to the introduction of a factor called isentropic compressor efficiency represented by $\eta_{\text{isentropic}}$ (value that lies between 75 and 85%). Including this in the work of compression modifies the equation as follows:

$$W_{\text{act}} = \frac{W_{\text{isentropic}}}{\eta_{\text{isentropic}}}. \tag{10.19}$$

Besides, the isentropic efficiency of various components and accessories involved in the compression process needs to be included in the above equation. The assumption in calculation of compression work was the ideal gas equation of state, which holds up to 150 bar without significant errors. Beyond this pressure, deviations are huge, so a real gas equation of state needs to be considered. Further, the process of compression of hydrogen can be approximated to be isothermal to make the calculations simpler. The work of compression for one mole of gas from initial pressure P_1 to final pressure P_2, considering isothermal process, will be given as:

(a) For ideal gas:

$$W_{\text{isothermal}} = RT \ln \frac{V_2}{V_1} \tag{10.20}$$

(b) For Van der Waals gas:

$$W_{\text{isothermal, Van}} = RT \ln \frac{V_2 - b}{V_1 - b} + \frac{\alpha}{V_2} - \frac{\alpha}{V_1} \tag{10.21}$$

(c) For compressible gas:

$$W_{\text{isothermal,}Z} = ZRT \ln \frac{V_2}{V_1}. \tag{10.22}$$

Depending on whether the gas is considered ideal or real, and whether the process is isentropic or isothermal, different sets of results are obtained for work of compression, as shown in Fig. 10.4. However, the general feature of the curve is that irrespective of the gas model used and the thermodynamic process assumed, the variation of work of compression with pressure is parabolic in nature. This observation is very important when calculating the energy required in compression, and it shows that the initial pressure of the gas in the compression process is the key factor. Thus, the energy required to compress the gas from ambient to 350 bar is higher than that required to raise the pressure from 350 to 700 bar. Therefore, the work required to compress hydrogen by a given pressure increment depends on the initial (suction) pressure; the higher the suction pressure, the lower the energy demand for compression. This parabolic relationship highlights the importance of production methods that deliver hydrogen at a higher pressure, as a significant amount of energy for compression can be saved if the electrolyzer or reformer delivers hydrogen at a higher pressure.

According to the thermodynamics, least amount of energy is required for isothermal processes, or the compression work is lowest. In contrast, the energy required to compress hydrogen will be highest for an isentropic compression process. Thus, isothermal and isentropic processes represent the lowest and highest amount of work required for compression. However, in actual case, the process of compression is neither isothermal nor isentropic; rather, it lies between the two. It can't be considered isothermal because, during compression, the temperature of the gas increases (actual process can be mapped closer to isothermal by providing intermediate cooling; however, keeping it completely isothermal is not possible in realistic conditions). Similarly, the actual process can't be isentropic because the process can't be completely reversible; there will be irreversible losses involved during compression (due to energy losses due to frictional, mechanical movements, and viscous forces). The process can't be adiabatic either, since heat will flow from the compressor during the compression process to the surroundings, resulting in temperature reduction. Thus, the actual compression work lies between the two extremes of isothermal

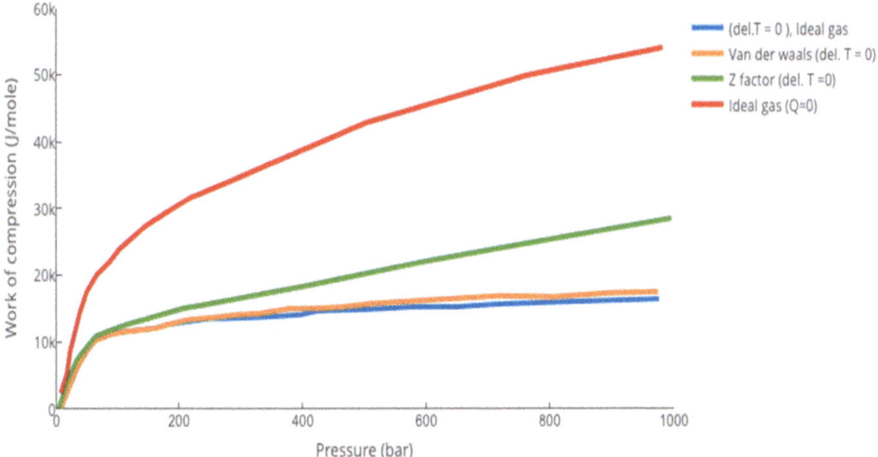

Fig. 10.4 Required work of compression for compressing to different pressures on the basis of the gas model used

and isentropic processes, which can be closely approximated to a polytropic process, for which:

$$PV_n = \text{constant} \tag{10.23}$$

(when $n > \gamma$, heat is supplied to the gas during compression, and for $n < \gamma$, heat is rejected by the gas during compression),

where n is the polytropic index that depends upon the compression process itself and the nature of the gas.

Work of compression for a polytropic process when a mole of hydrogen gas is compressed from an initial pressure P_1 to a final delivery pressure P_2 is given as:

$$W_{\text{polytropic}} = \frac{n}{n-1} RT_1 \left[\left(\frac{p_2}{p_1} \right)^{\frac{n-1}{n}} - 1 \right] \tag{10.24}$$

This expression is for an ideal polytropic process, but in the actual case, the work of compression is larger than given by above equation; the irreversibilities and losses are accounted for by the polytropic efficiency of the process (η_p) which is known. The polytropic efficiency, specific heat ratio, and polytropic index are related by:

$$\frac{n}{n-1} = \frac{\gamma}{\gamma - 1} \eta_P \tag{10.25}$$

The value of η_p is specified by manufacturers on the compressors.

During the compression process, the temperature of the gas increases from T_1 to T_2 which can be given by:

$$\frac{T_2}{T_1} = \left(\frac{P_2}{P_1}\right)^{\frac{n-1}{n}} \tag{10.26}$$

According to this relation, the final temperature changes with the pressure ratio. If cooling is included between the compression processes to mimic closer to isothermal process, the required work of compression will decrease. Ideally, the heat should be removed at the same rate as it is produced, but implementing such a heat exchanger is practically impossible and would prove to be very costly, with issues related to energy efficiency. However, multistage compression with intercooling between the stages can be more effective. During the intercooling between the stages, the temperature of the gas can be made the same as the initial temperature prior to compression, and this can be done at each stage of the compression. Adding more stages can help achieve near-isothermal conditions for compression, thereby reducing the required work of compression; however, this may not be practically viable due to increased hardware costs. Thus, a trade-off exists between energy savings and added cost due to each stage.

If we consider two-stage compression, then the isentropic work for such compressor is given by:

$$W_{\text{isentropic}} = \frac{\gamma RT_1}{\gamma - 1}\left[\left(\frac{p_2}{p_1}\right)^{\frac{\gamma-1}{\gamma}} - 1\right] + \frac{\gamma RT_1}{\gamma - 1}\left[\left(\frac{p_3}{p_2}\right)^{\frac{\gamma-1}{\gamma}} - 1\right] \tag{10.27}$$

where P_1 and P_3 are the suction and the discharge pressures, and P_2 is the optimal intermediate pressure, its value ($P_2 = \sqrt{P_3 P_1}$) should be such that the pressure ratio is the same in each compression stage; which is a requirement to achieve the conditions of minimum work.

Figure 10.5 compares the work of compression required for a single-stage and multistage compression of hydrogen, where with the addition of stages, the work of compression reduces; however, beyond a certain number of added stage, the required compression work is least affected. It is also evident that the impact of initial pressure is high; the higher the suction pressure, less is the required work of compression.

In this section, the direct method for calculating the work of compression from the equation of state is described. This method is very simple for practical cases.

However, there is another method that can be used for calculating the work of compression. The method that uses the calorific state equation for calculating work of compression is called indirect method. This method is more accurate and involves calculating calorific values, but is slightly difficult in practical cases compared to using the direct method.

2. **Indirect Method**

The work of compression for an isentropic process is given by:

$$W_{\text{isentropic}} = (h_2 - h_1) - q_{12}. \tag{10.28}$$

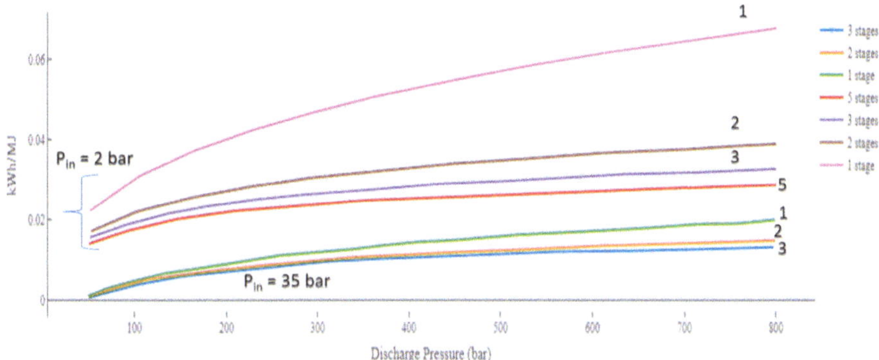

Fig. 10.5 Required work of compression for compressing to different pressures for multistage compression

For an isentropic process, there is no heat flow, thus the above equation can be written as:

$$W_{\text{isoentropic}} = (h_2 - h_1) - T(s_1 - s_2). \tag{10.29}$$

As calculated in the previous section, the work of compression for an isothermal process is given by:

$$W_{\text{isothermal}} = R_i T \int_{p_1}^{p_2} \frac{Z}{p} \mathrm{d}p = R_i T Z_{av} \ln\left(\frac{p_2}{p_1}\right), \tag{10.30}$$

where Z is the compressibility factor, which is known to vary between different states corresponding to the different pressures P_1 and P_2, for simplification, an average value corresponding to these states can be considered as that for the corresponding state

$$Z_{av} = \frac{Z(p_1, T) + Z(p_2, T)}{2}. \tag{10.31}$$

In certain cases, the Z, which is a function of pressure, can be considered as constant to simplify, and a suitable mean value is used.

Isentropic work of compression using indirect method is given by:

$$W_{\text{isentropic}} = R_i T_1 Z_1 p_1^{\frac{1-k_{p,v}}{k_{p,v}}} \int_{p_1}^{p_2} p^{\frac{-1}{k_{p,v}}} \mathrm{d}p = \frac{k_{av\,p,v}}{1 - k_{av\,p,v}} R_i T_1 Z_1 \left[\left(\frac{p_2}{p_1}\right)^{\frac{k_{av\,p,v}-1}{k_{av\,p,v}}} - 1\right] \tag{10.32}$$

$$k_{av\,p,v} = \frac{\left(k_{p,v}(T_1, p_1) + k_{p,v}(T_2, p_2)\right)}{2}. \tag{10.33}$$

Similarly, the isentropic index can be taken to be the mean value between the two states and considered to be constant at that value for simplicity. Considering an isothermal process for compression, heat generated during the process of compression needs to be removed to maintain constant temperature T.

$$W_{\text{isothermal}} = (h_2 - h_1) - T(s_2 - s_1) \tag{10.34}$$

$$= \left[h(T, p_2) - h(T, p_1)\right] - T\left[s(T, p_2) - s(T, p_1)\right]. \tag{10.35}$$

During isentropic compression, no heat is being removed; thus,

$$W_{\text{isentropic}} = (h_2 - h_1) = h(s_1, p_2) - h(s_1, p_1). \tag{10.36}$$

For a compression ratio of 5, isentropic compression work required is 2545 kJ/kg of hydrogen, which is 2.12% and 1.69% of the LHV of hydrogen for isentropic and isothermal compression, respectively.

If we compare the direct method (using calorific equations of state) and indirect method (using a state equation calculated), there is very little deviation. The deviation increases with increasing compression ratio. Similarly, the deviations between ideal and real gas behaviors are more pronounced for higher compression ratios. The ideal gas law underpredicts the compression work.

Isenthalpic and Isentropic Expansion

For liquefaction of hydrogen, the cooling effect during expansion plays the critical role. On the similar basis as compression, the expansion processes are considered. Two different expansion processes that can be considered include: isenthalpic (Joule–Thomson expansion, in a throttle valve) or isentropic expansion (in a turbine). Similar to the previous section, both direct and indirect methods can be considered, and as mentioned above, they can be used for expansion.

Indirect Method—Calculation of Joule–Thomson coefficient (function of T and P) is considered, and using this, the final temperature can be calculated in steps of pressure variation:

$$T_2 = T_1 + \left[\frac{\mu(T_1, p_1) + \mu(T_2, p_2)}{2}\right](p_2 - p_1). \tag{10.37}$$

Direct Method—Since the enthalpy in the process remains constant for isenthalpic process, the temperature can be calculated for the constant h and for a final pressure P_2:

$$h(T_1, p_1) = h(T_2, p_2) \tag{10.38}$$

$$T_2 = f(h, p_2). \tag{10.39}$$

For isenthalpic process, rather than pressure ratio, the pressure difference is important. Deviations using ideal gas behavior are huge; however, when real gas behavior is considered, the deviations between the direct and indirect methods are very less. For an isentropic expansion process, the pressure ratio is the decisive factor.

Modeling the Storage Process

Considering the law of mass conservation:

$$\frac{\mathrm{d}m_s}{\mathrm{d}t} = V_s\frac{\mathrm{d}\rho_s}{\mathrm{d}t} = x\frac{\mathrm{d}m_{\mathrm{in}}}{\mathrm{d}t} - y\frac{\mathrm{d}m_{\mathrm{out}}}{\mathrm{d}t}, \tag{10.40}$$

where dm/dt are the mass flow rate, V_s is the geometric storage volume, $x = 1$, $y =$ 0 for filling, and $x = 0$, $y = 1$ for withdrawal process. $\left(\frac{\mathrm{d}m_{\mathrm{in}}}{\mathrm{d}t}\right)$ is the rate of change of mass and $\frac{\mathrm{d}\rho_s}{\mathrm{d}t}$ is the rate of change of density inside the storage.

Energy balance for the storage system, neglecting the kinetic and potential energy changes and onsidering it as an open system is given as:

$$\frac{\mathrm{d}U_s}{\mathrm{d}t} = x\left(h_{\mathrm{in}}\frac{\mathrm{d}m_{\mathrm{in}}}{\mathrm{d}t}\right) - x\left(h_{\mathrm{out}}\frac{\mathrm{d}m_{\mathrm{out}}}{\mathrm{d}t}\right) + z\left(\frac{\mathrm{d}Q}{\mathrm{d}t}\right). \tag{10.41}$$

$\frac{\mathrm{d}U_s}{\mathrm{d}t}$ is the rate of change of internal energy inside storage system, and $\frac{\mathrm{d}Q}{\mathrm{d}t}$ is the rate of heat exchanged with the surroundings, while h_{in}, h_{out} are the specific enthalpies entering in and leaving the system, respectively, $z = 1$ for diathermal process and $z = 0$ for adiabatic process. Also

$$U_s = m_s u_s \tag{10.42}$$

$$\frac{\mathrm{d}U_s}{\mathrm{d}t} = \left(u_s\frac{\mathrm{d}m_s}{\mathrm{d}t}\right) + \left(m_s\frac{\mathrm{d}u_s}{\mathrm{d}t}\right). \tag{10.43}$$

To solve this one time rate of change should be known, which then becomes the controlling equation of the system. There are two ways to solve this: (a) Mass or volume flow entering or leaving through a storage system is given as a function of time, and (b) change of one of the thermodynamic variables inside the storage vessel is known, e.g., the pressure increase in the vessel is measured. Considering the case

(a) which is more general and assuming a constant mass flow rate:

$$m_s(t) = V_s\rho_s(t) = m_s(t=0) + x\dot{m}_{\text{in}}t - y\dot{m}_{\text{out}}t \tag{10.44}$$

where $m_s(t = 0)$ is the initial fill or withdraw, and the $\dot{m}_{\text{in}}(t), \dot{m}_{\text{out}}(t)$ are the constant mass flow rates entering and leaving the system, respectively. Combining the above equations:

$$\frac{du_s}{\mathrm{d}t} = \frac{\left[h_{\text{in}}x\dot{m}_{\text{in}} - h_{\text{out}}y\dot{m}_{\text{out}}\right] - u_s(x\dot{m}_{\text{in}} - y\dot{m}_{\text{out}}) + z\dot{Q}(t)}{m_s(t=0) + x\dot{m}_{\text{in}}t - y\dot{m}_{\text{out}}t}. \tag{10.45}$$

For constant enthalpy during filling and no heat exchange, this equation can be simplified to:

$$\frac{\mathrm{d}u_s}{\mathrm{d}t} = \frac{[h_{\text{in}}\dot{m}_{\text{in}}] - u_s(x\dot{m}_{\text{in}})}{m_s(t=0) + \dot{m}_{\text{in}}}. \tag{10.46}$$

Integrating with time, we get:

$$u_s = \frac{[h_{\text{in}}\dot{m}_{\text{in}}]t + m_s(t=0)u_s(t=0)}{m_s(t=0) + \dot{m}_{\text{in}}t}, \tag{10.47}$$

where $u_s(t = 0)$ is the initial specific internal energy of hydrogen in the storage system.

Heat Transfer Equations

Heat exchange with surroundings is considered using steady-state overall heat transfer coefficient. The rate of change of heat transfer across the walls of the storage system can be given by:

$$\dot{Q} = U_{\text{heat tr}}A_{\text{heat tr}}(T_\infty - T_s), \tag{10.48}$$

where $U_{\text{heat tr}}$ is the overall heat transfer coefficient, $A_{\text{heat tr}}$ is the heat transfer area of the storage system, and T_∞ and T_s are the temperatures of the surroundings and inside the storage system, respectively. Since the temperature inside the storage system changes, the heat flow is unsteady. Generally, storage systems are cylindrical, and the overall thermal resistance can be calculated using:

$$R_{\text{heat tr}} = \frac{1}{U_{\text{heat tr}}A_{\text{heat tr}}} = \frac{1}{\pi L}\left[\frac{1}{\alpha_i d_i} + \frac{ln\left(\frac{d_O}{d_i}\right)}{2\lambda} + \frac{1}{\alpha_O d_O}\right], \tag{10.49}$$

where L is the cylinder height or length, λ is the thermal conductivity, d_i and d_0 are the inner and outer diameters, respectively, and α_i and α_0 are the heat transfer coefficients of the gas inside and outside the storage system, respectively. More details will be discussed in the metal hydride-based storage systems' chapter.

In summary, the required work of compression differs depending upon the type of compression used and presence of cooling. Isothermal compression is the ideal process, being the most energy-efficient process with an energy requirement of 1.5 and 2.0 kWh/kg for compressing hydrogen to 253 and 1013 bar, respectively. Considering the various losses in actual system, the energy consumption can be 2.5 kWh/kg and 4.0 kWh/kg which is 7.5% and 12% of the LHV of hydrogen. In literature reports, the energy requirements of 5% and 20% for compressing hydrogen to 350 and 700 bar has been reported.

Hydrogen compressors are widely used in the chemical and oil industries. The compressors widely used due to the specific characteristics of hydrogen are the reciprocating displacement compressors such as the piston, diaphragm, and ionic compressors. Among these, the reciprocating-type compressors are used for high volume, and the piston or diaphragm type is used for small-volume applications. Other compressors considered are non-mechanical types, including the high-pressure electrolyzers, metal hydrides, and electrochemical compressors. More description about the different types of compressors is included in the next chapter.

Chapter 11
Hydrogen Compressors

Hydrogen compression and compressors are an integral, indispensable, and inevitable part of hydrogen energy value chain. When hydrogen is being produced in a centralized industrial scale hydrogen plant, then compressors are required prior to the distribution network wherein the distribution could be either by means of pipelines or tube trailers. However, if hydrogen is produced in a decentralized plant, then it is required downstream, after the electrolyzer or reformer and prior to the storage.

Mechanical compressors are the most widely used compressors in the hydrogen energy value chain and are in fact the most mature technology. They can also provide very high hydrogen flow rates of the order of thousands of Nm^3/h. The major challenges associated with hydrogen compression are because of the unique properties of hydrogen, i.e., it is the lightest element with lowest density and a lot of energy is required to compress it to a higher pressure. At the same time, when metals are exposed to hydrogen over a period of time at a certain set of temperature and pressure conditions, the phenomenon of hydrogen embrittlement is observed due to which various components, accessories, and infrastructure used can lose mechanical strength and lead to failures.

Hydrogen, being the smallest molecule, has very high diffusivity as compared to other gases; thus, it is much more prone to leaks. Hydrogen can leak from the seals, end joints, connects, welds, etc., which is a major concern when considering the mechanical compressors. This loss of hydrogen through components, seals, joints, etc., can not only result in a reduction in pressure, but can also lead to a reduction in the efficiency of the process and at the same time, can lead to safety concerns due to the hydrogen properties like its low ignition energy and wide flammability limit. Thus, leakage is highly undesirable while the compression of hydrogen is being considered, and is an important consideration in the case of the different hydrogen compressors.

The choice of compressor technology requires considering the various parameters or process variables like the desired capacity or amount of hydrogen to be compressed, flow rate, the suction pressure at which hydrogen is available from the

P. Sharma, *A Textbook on Hydrogen Energy*,
https://doi.org/10.1007/978-981-95-4601-5_11

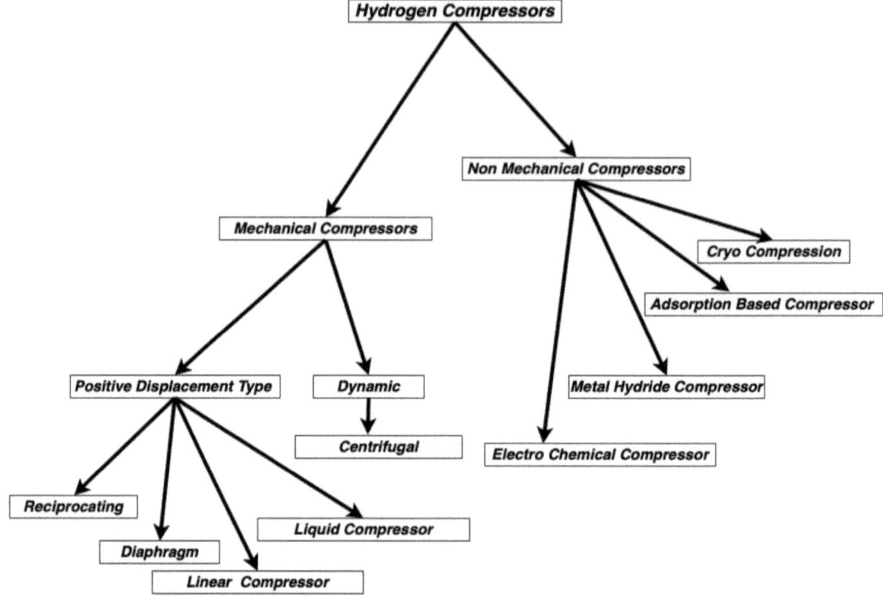

Fig. 11.1 Different types of mechanical and non-mechanical hydrogen compressors

production plant, the discharge pressure, and the requirements on the demand side for that particular end-use application.

As per the thermodynamics of the compression process as mentioned in the previous chapter, the compressor technology that requires the minimum amount of work is considered to be the most preferred one.

We can broadly classify the different compressor technologies into mechanical and non-mechanical types of compressors (Fig. 11.1).

Mechanical compressors can be further subdivided into positive displacement-type compressors and dynamic compressors. The class of dynamic compressors includes the centrifugal type of compressor, while the positive displacement-type compressor includes reciprocating type, diaphragm type, linear compressor, and the liquid compressors.

The non-mechanical compressors can be subdivided into electrochemical compressors, metal hydride-based compressors, adsorption-based compressors, and cryo-compressors.

Mechanical Compressors

The mechanical compressors are the class of compressors that are most widely used, and it's a mature technology. The principle of compression in these compressors is to convert the mechanical energy into gas energy. Among the various mechanical compressors, the positive displacement topology is one of the most widely used and has the simplest working principle. The mechanism behind positive displacement compression is that a fixed amount of gas is pushed or compressed, such that the volume associated with or occupied is decreased by using a piston or any other similar arrangement. During the process of compression, due to the pushing more amount of gas and reduced volume available, the impact or the collision of gas molecules onto the walls of the compressor chamber increases, leading to an increased pressure. The detailed mechanisms of compression for this category of compressors are included in the following section.

Reciprocating Compressors

The simplest of the mechanical or positive displacement type of compressor is a single-stage reciprocating compressor. Such a reciprocating machine consists of a piston, cylinder, an inlet or suction valve, and a discharge or delivery valve; these are the automatic valves. The piston is connected by means of a rod to a crankshaft. The compression is achieved by the pulsating action of the piston. The linear motion of the piston is converted to the rotating motion due to the rod connected to a crankshaft; thus, the device is known as a reciprocating type of machine (Fig. 11.2).

In a single-stage compressor, when the piston moves from the lower position, called the bottom dead center, toward the higher position, i.e., top dead center, compression is achieved.

Working Principle of the Single-Stage Reciprocating Compressor

During the suction stroke, the inlet valve opens, and the piston moves from the top dead center toward the bottom dead center, taking in the gas charge. This process is known as the suction stroke. During the compression, the piston moves from the bottom dead center toward the top dead center and thereby reduces the confined volume in which the gas is enclosed. When it reaches the top dead center, the delivery valve or the discharge valve opens and that pressurized gas is discharged through the outlet valve, known as the discharge or delivery stroke. So, this completes one cycle of the reciprocating compressor, and similarly, several cycles are carried out so as to compress, to get the desired pressure for a particular amount of hydrogen.

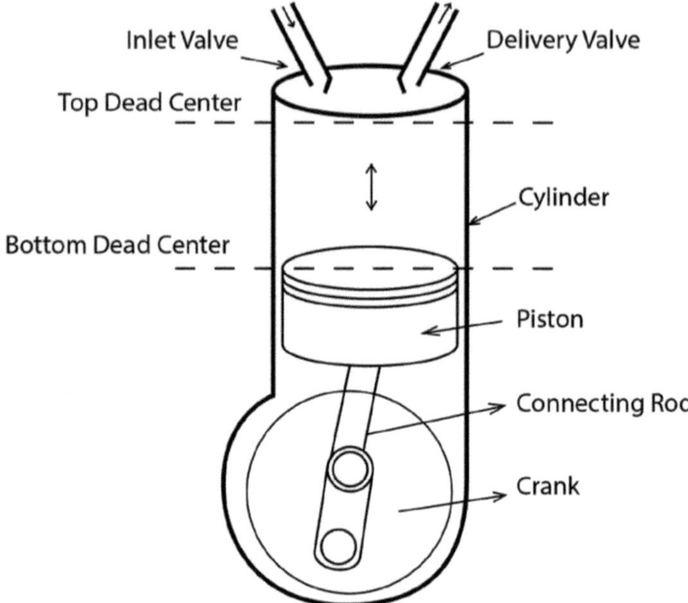

Fig. 11.2 Schematic of a typical reciprocating type of compressor

The major limitation in reciprocating piston cylinder-type compressors is the maximum achievable compression ratio using a single stage. With a single-stage compressor, the compression ratio of 3:1 or 4:1 can be achieved, while with a well-designed compressor, a compression ratio of 6:1 can be achieved. This maximum achievable compression ratio is limited by the cylinder dimension or the speed of compression; however, another important factor that controls the compression ratio in such a compressor is related to the maximum allowable temperature of the gas during the compression process. This maximum allowable temperature sets a limit on the pressure ratio of approximately 2.5–3.0 per cylinder.

During the compression process, the gas temperature increases and the maximum allowable temperature of the gas (usually around 250 °C) is limited in the compression, which in turn puts a limit on the maximum compression pressure or compression ratio achievable. Another important and limiting parameter is the achievable gas flow rate in the compression process using a single-stage reciprocating compressor. The achievable flow rate is limited by the cylinder dimension and speed of compression (i.e., number of cycles per unit time). In order to overcome this limitation and achieve a higher flow rate, if the cylinder dimension is increased, in that case the system will become heavier and bulkier, leading to increased inertia forces. At the same time, for a higher cylinder dimension, if the speed of compression is increased, then that will lead to more vibration and noise, and result in increased mechanical stresses. This makes it essential to limit the speed of compression, in turn limiting the flow rate in case of a higher cylinder dimension. In short, a better speed of compression can be achieved for smaller cylinders, but then there is a limit on the maximum achievable flow rate.

Hydrogen is a peculiar molecule with characteristics leading to embrittlement when in contact with metals in general under a given set of conditions, making it essential to select special materials for which the embrittlement will be as low as possible. Thus, both the selection of appropriate material and the design of such type of compressors are very important.

The different materials that can be used include either special steel, which could be alloy steel, cast iron, or cast, or forged steel. The reciprocating type of compressors has moving parts, leading to a lot of wear and subsequently requires substantial maintenance. In order to avoid the usual wear in such compressors, an inner coating is applied. In fact, a liner coating onto the cylinder walls in the inner surface of the cylinder not only helps in reducing wear but also assists in the maintenance, i.e., by providing the option of replacing this layer in case it gets worn out, reduces the overall cost of maintenance. With the presence of such a coating, major replacement of the moving parts is not required. At the same time, with the incorporation of such a coating, the diameter of the cylinder can be changed as desired and can provide a lot of versatility and flexibility in the operating conditions. Hence, having a coating on the inner surface of the cylinder has several advantages.

As mentioned, there is a limit to which we can achieve a compression ratio using a single-stage compressor. However, if higher compression ratios are desired, then this can be achieved by adding more stages to the reciprocating compressor. Besides, adding a number of stages not only leads to achieving higher compression ratios but also reduces the work of compression as per the thermodynamics of the process. However, there lies a trade-off between the amount of energy saved by adding additional stages and against the added CAPEX and OPEX with every added stage, and there lies an optimum depending upon which technology for compression is being used.

Figure 11.3 is a schematic showing a two-stage compression process for a reciprocating type of compressor with Figure (a) showing the first stage of compression, and (b) showing the second stage of compression. The working principle of such a two-stage compressor can be explained as follows: During the start of the process, a gas, which is hydrogen in this case, at low pressure is taken into the first stage of the compressor. The first-stage cylinder has a higher diameter compared to the second-stage cylinder. During the suction stroke, when low-pressure hydrogen is taken inside the first-stage cylinder, the piston moves toward the right-hand side in the first-stage compression, and this movement is driven by a hydraulic fluid. Once the piston reaches its maximum displacement position, the compression stroke starts with the movement of the piston toward the left-hand side, leading to a decrease in volume and thus an increase in pressure. Once it reaches the top maximum displacement position on to the left-hand side or achieves the highest pressure for the first stage, the delivery valve opens to discharge the gas from the first stage to the second stage of compression. While the gas charge is being transferred from the first stage to the second stage of compression, an intercooler stage can take away the heat of compression and thus make the process near isothermal and, as per thermodynamics, can reduce the work of compression, since the minimum work is required for isothermal compression. Near isothermal conditions can be achieved by adding the number of stages to the compression process, thus having a multistage compression with intercooling in between. In addition to the intercooling between stages, the compressor can be designed to have a cooling arrangement in the walls of the compressor, to remove the heat of compression, thereby with both these arrangements will make process nearly isothermal and reduce the required work of compression or energy to compress the gas.

Once the gas is compressed to the desired pressure in stage 1, it is discharged from this stage for further compression to higher pressure into stage 2. During entry into stage 2, i.e., during the suction stroke, in this stage, the piston moves toward the left-hand side to take the gas charge. During the compression stroke in stage 2, the piston

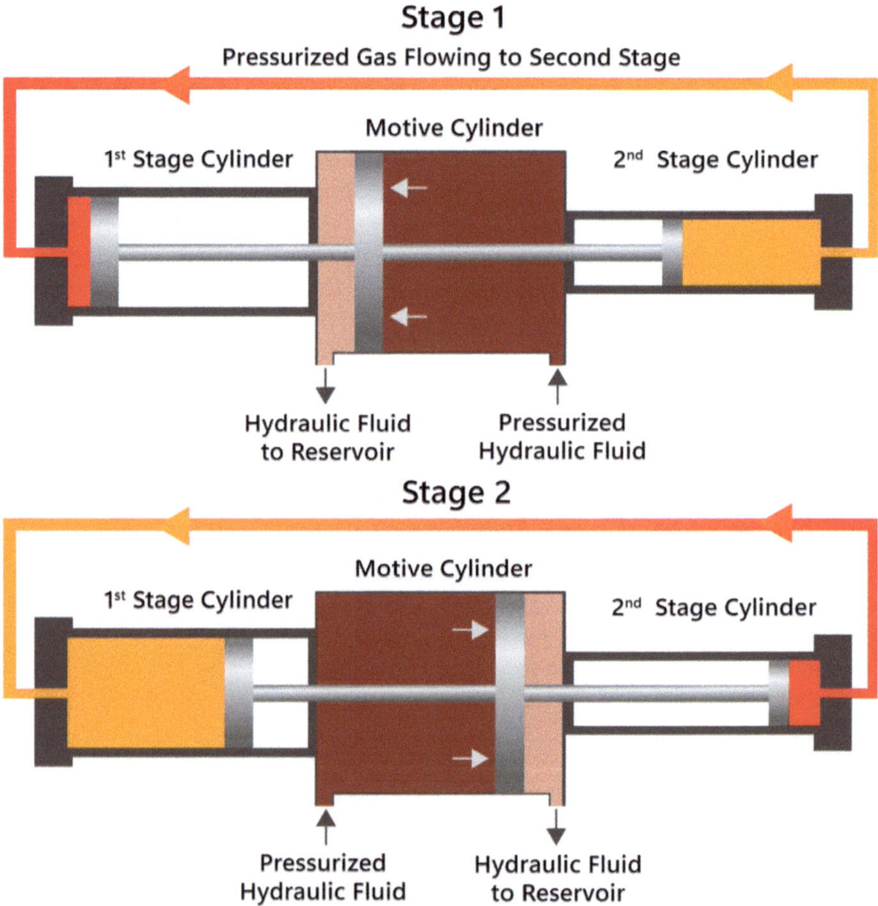

Fig. 11.3 Schematic of a two-stage reciprocating type compressor [4]

moves to the right-hand side to decrease the volume and increase the gas pressure. Once the piston reaches the maximum displacement position, the gas achieves the highest pressure of that stage and is discharged from stage 2. The movement of the piston is provided by a hydraulic fluid circuit.

The biggest disadvantage of such reciprocating type of compressors is the presence of several moving parts due to the very basic nature of compression, leading to wear associated with pulsating motion as well as friction losses. Both wear and friction can be reduced in these types of compressors with the use of lubricating oils. The use of lubricating oils not only reduces wear and friction losses but also improves the reliability, life, and durability of such compressors, and at the same time minimizes the chances of leakage of gas through these compressors. Hydrogen has a tendency to leak from airtight seals and joints, etc., and this leakage can lead to safety issues, loss of pressure, and efficiency. With the use of lubricating oils, this can be reduced. The general observation is that the life of lubricating oil-based compressors is higher than that of non-lubricating type of compressors.

With all these associated advantages of lubricating type compressors, the basic disadvantage of such compressors is that the presence of lubricating oil can cause contamination of the compressed hydrogen, as the oil can enter the compression chamber. The hydrogen then requires to be purified in order to get ultra-pure hydrogen. Thus, the non-lubricating type of compressors will have friction losses and wear, while the lubricating type of compressors may have gas contamination-related issues. Another option is to introduce a non-lubricating layer of thermoplastic material, or to fit rider bands made of such material onto the piston or piston rings in the compression chamber. The thermoplastic materials used should have high thermal and chemical stability. The introduction of such thermoplastic rings or rider bands not only reduces the wear and improves the life of the compressor, but at the same time, the presence of these will also prevent the leakage of hydrogen. The objective of introducing these thermoplastic layers is to prevent oil-based contamination as seen in lubricating compressors, but even with these thermoplastic materials used in non-lubricating compressors, abrasive particles of these materials, due to pulsating actions, can enter the hydrogen gas stream and can cause some amount of contamination. If the life of all these three types of reciprocating compressors is compared, then the longest life is for the lubricating type of compressors.

With the use of completely oil-free compressors, higher pressures can be achieved, but then there are chances of early failure of ceiling rings. The reason for such failures is because there will be a non-uniform distribution of pressure inside. However, to address this pressure non-uniformity, there could be a compartment distance piece that can be used to take care of the gas venting, finally leading to the reduced embrittlement. For improving the performance, the important points could be decreasing the piston speed, selecting appropriate materials for the rider band or piston rings, and adopting a proper design. With all these, the maximum achievable pressure can be increased to 1000 bar and can get flow rates of 300 Nm^3/h.

Oil-free compressors are used where a moderate flow rate and high pressure are desired, although they exhibit higher power consumption.

Table 11.1 Commercially available reciprocating compressors and their parameters

Compressor technology	Company	Flow rate (Nm^3/h)	Suction pressure (bar)	Delivery pressure (bar)
Reciprocating compressors	Hydropac	4820	350	859
	Hitachi infrastructure system	300	6	1000

In summary, with the use of the reciprocating type of compressor, a wide range of capacities, flow rates, and delivery pressures can be achieved. The major requirement is in terms of an appropriate selection of material as well as the design of the compressor. By reducing the speed of compression, a higher discharge pressure can be achieved, and as per literature reports, as high as 100 MPa pressure can be achieved with flow rates of thousands of Nm^3/h. Usually, this class of compressors is preferred for high pressures and moderate flow rates. Some of the important parameters for such compressors are as follows [1]:

$$\text{Average Piston Speed(m/s)} = 2 \times \text{Piston Stroke} \times \text{Rotation Speed}(1/\text{s}).$$

The upper limit of average piston speed for the different types of reciprocating compressors is as follows:

Lubricated Cylinders: 4–4.5 m/s (up to 5 m/s for high-speed types)
Oil-Free Cylinders: 3.5 m/s (common for most hydrogen services).

The piston stroke is given as follows:

Large Machines: 300–333 min^{-1}
Medium Machines: 450–375 min^{-1}
Small Machines: 500–600 min^{-1}.

There have also been reports where very high flow rates could be achieved, around 10,000 Nm^3/h with an outlet pressure of 25 MPa. However, research in the case of reciprocating compressors is required in the area of identifying better materials for rings and seals; coatings to the inner wall of compressor chambers, piston, and rod; use of materials and design which could reduce wear and friction; and continuous monitoring of such compressors for an early detection of faults in such compressors (Tables 11.1 and 11.2).

Diaphragm Compressors

Disadvantages of piston cylinder type of compressors, as mentioned in the previous section, are the presence of rotating parts, which leads to increased wear and friction, heat generation, losses, and contamination from lube oils or abrasive particles due

Table 11.2 Advantages and disadvantages of reciprocating type of compressors and their application areas

Compressor technology	Advantages	Disadvantages	Applications areas
Reciprocating compressors	• Wide range of capacities • Wide flow rates • Wide range and higher delivery pressure can be achieved • Flexibility in size and capacity • Mature technology • Widespread usage • Good performance	• Moving parts leading to complexity, noise, wear of components, and friction losses • Higher maintenance requirements, difficulty in maintenance, and higher cost of maintenance • For lubricating oil-type compressors, contamination issues • For non-lubricating with inner wall coating contamination from abrasive particles • Complexity in manufacturing • Thermal management issues related to heat transfer during compression • Difficult to achieve efficient cooling, due to high speed of compression • Due to moving parts, more heat is being produced due to friction, and that needs to be removed • Reduced system life • Pressure fluctuation arising in the compression chamber, which can be highly detrimental because that can induce different vibrations and noise into the system	• Hydrogen refueling station • Compressed gaseous hydrogen tanks-based storage • Onboard filling of tanks on vehicles • Hydrogen plants • Catalytic reformers • Pipeline-based hydrogen transport

to contact of gas with either lube oil or coating particles. However, another class of compressors is diaphragm compressors, which takes into consideration the challenges with the piston cylinder type of compressors. Although this class of compressors is still piston driven, they have no moving parts in the compression chamber, and the oil and gas circuits are separated.

Figure 11.4 shows a schematic of a diaphragm compressor. The diaphragm separates the gas to be compressed and the hydraulic fluid. The motion of the diaphragm for the suction and compression stroke is provided by the hydraulic pressure of the fluid in the hydraulic chamber, and this pressure is created by the movement of the piston. But since the gas and hydraulic fluid chambers are separated, there are no issues of contamination in case of such compressors. The working principle of a diaphragm compressor can be briefly explained as follows:

The movement of the piston creates pressure by reducing volume in the gas cavity, increasing the gas pressure, and the movement of the piston is driven by the hydraulic fluid circuitry, wherein the piston on the hydraulic fluid side moves to transfer movement to the hydraulic fluid, thereby onto the diaphragm. During the suction stroke, the

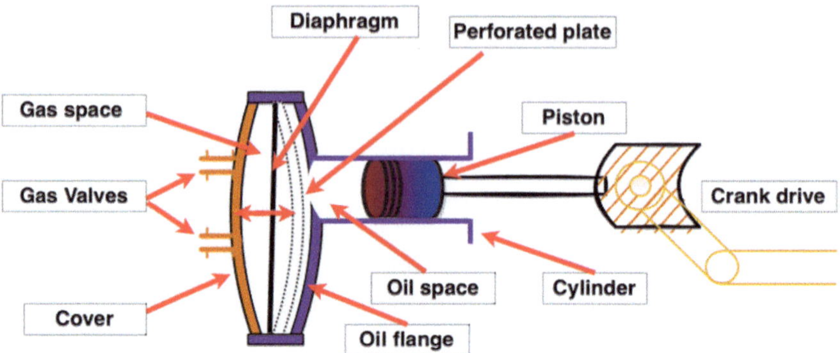

Fig. 11.4 Schematic of a typical diaphragm compressor

piston moves on to the right-hand side, deflecting the diaphragm to the right to take the gas charge through the inlet valve. Oil space drives the movement of the diaphragm in turn to the gas space. During the compression stroke, the piston moves to the left-hand side, pressurizing the hydraulic fluid and deflecting the diaphragm to the left and thereby compressing the gas in the gas chamber. Once the maximum pressure is reached, the delivery valve opens, delivering the compressed hydrogen at the desired pressure. The principle of operation is simple. However, the problem arises during compression: when the diaphragm reaches its maximum deflection position toward the gas chamber, the pressure on the hydraulic side is higher, even though the piston has not yet reached its top dead center. This pressure has to be restricted; the pressure difference between the gas space and the hydraulic space needs to be limited. Thus, with diaphragm compressors, there is always a hydraulic pressure limiter that controls this pressure. At the same time, to increase the life of this diaphragm, there is a perforated plate which acts as an oil distributor and provides a uniform pressure across the diaphragm. Both the oil pressure control and the presence of a perforated plate are very much required to not only improve the efficiency of the process, but at the same time to increase the life of the compressor.

In brief, the movement of the piston creates pressure in the hydraulic chamber, which is then transferred by the movement to the diaphragm, and from the diaphragm, that movement is transferred to the gas; and the gas gets compressed. The gas compression is achieved in the region, which is separated from the hydraulic chamber or from the piston. The major component of such a compressor is the diaphragm. The construction of the diaphragm is such that it is a thin metallic membrane consisting of three plates. These three plates are such that: one plate, which is on the hydrogen side, is known as the process plate, another plate, which is on the hydraulic oil side, is known as the hydraulic plate. The middle or the intermediate plate provides strength to the diaphragm and, at the same time, prevents hydrogen leakage.

The major benefits of the diaphragm compressors include their high throughput, less cooling requirement, less power requirement, they are very effective for hydrogen compression, also specially fit for handling chemically pure gases, as the oil and gas sections are separated by a diaphragm. Better cooling can be achieved in the case of a diaphragm compressor than the reciprocating compressors, because of multiple reasons. The diaphragm surface area is higher, the gas chamber surface area is higher, and cooling which can be provided in the cover of the diaphragm, makes heat management better in such compressors. Not only these, but oil, which is supplied in the hydraulic circuit, can also be cooled to achieve near isothermal conditions, thereby improving the efficiency and reducing the work and energy required for compression. Higher compression ratio can be achieved with such compressors, with less number of stages. The most important component of such a compressor is the diaphragm so the appropriate selection of materials and design of the diaphragm are crucial. The diaphragm should be highly durable and be made up of corrosion-resistant and embrittlement-resistant materials. Diaphragm plates are usually made up of stainless steel, chrome nickel stainless steel, copper beryllium alloys, or duplex steel. In such compressors, there are no seals or joints prone to leakage, no moving parts in the compression chamber, and the diaphragm is hermetically sealed; therefore, the possibility of hydrogen or gas leak is significantly lower than in the reciprocating type of compressor.

Another variation to the piston-driven hydraulic fluid via diaphragm movement could be a variant where the movement of the diaphragm is supported by an electrostatic field supplied by DC voltage. Such a compressor is called an electrostatic diaphragm compressor. The principle of such compressors is almost the same, with the major difference being that DC voltage is applied between the diaphragm and compression chamber (Fig. 11.5).

This DC voltage applied causes movement of the diaphragm by creating an electric field in the compressor chamber, which deflects the diaphragm. During the suction stroke, the diaphragm moves away from the surface of the compression chamber, and

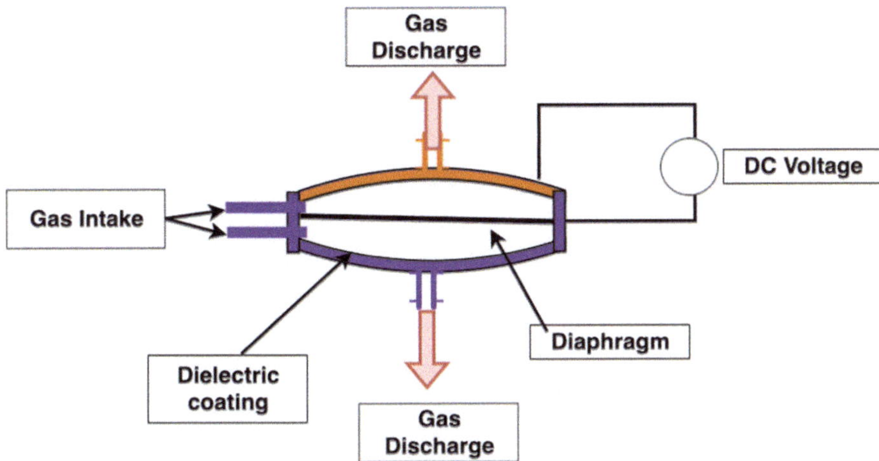

Fig. 11.5 Schematic of an electrostatic diaphragm compressor (inspired from [5])

Table 11.3 Commercially available diaphragm compressors and their parameters

Compressor technology	Company	Flow rate (Nm^3/h)	Suction pressure (bar)	Delivery pressure (bar)
Diaphragm compressors	PDC machines	50–280	350	51.7
	Hofer Hochdrucktechnik	300	5–35	3000

the gas charge is taken in. During the compression stroke, the diaphragm will move toward the surface of the compression chamber, compressing the gas, and at the end of this stroke, the gas will be discharged at a higher pressure. Thereafter, the polarity of the voltage changes to have the compression and expansion in the two gas cavities. With the reversal of the polarity in the two gas cavities, the complete cycle can be achieved. A possible issue in such compressors could be that when the diaphragm moves toward the surface of the compressor chamber, it can come in contact with the compressor chamber and can cause electrical sparking. In order to avoid this, a dielectric material coating is applied on the inner surface of the compressor chamber.

In summary, the diaphragm type of compressors are better suited for small capacities. A higher compression ratio can be achieved with a reduced number of stages. However, the biggest challenge is the durability of the diaphragm. Both the material selected for diaphragm plates and the design are very important to achieve longer life and durability, as well as corrosion resistance of such compressors. Since in the process of compression the diaphragm undergoes several mechanical stresses which could lead to diaphragm failure, the design and appropriate selection of material used for the diaphragm are required to avoid diaphragm failure.

Other challenges with such compressors include the flow rate limitation and the complexity of the system. The flow rate in such compressors is limited by the size of the diaphragm (Tables 11.3 and 11.4).

Linear Hydrogen Compressor

Linear compressors are primarily used for cryogenic applications, these are driven by stirling cycle coolers that use hydrogen or helium, and they are also used for domestic refrigeration.

They have a very small number of moving parts as compared to the compressors, discussed in the previous sections. In these compressors, the crankshaft and the connecting rod are absent (Fig. 11.6).

Since the rotating components are not present, the design of linear compressors is relatively simpler and has cost savings as compared to the other reciprocating type of compressors. The construction of such compressors is such that they have an inlet valve, an outlet or the discharge valve, a cylinder and a piston, a spring, a linear motor with a magnet or a coil connected to a DC voltage. The linear motor is responsible for providing movement to the piston, and it is connected to the piston by means of

Table 11.4 Advantages and disadvantages of diaphragm-type compressors and their application areas

Compressor technology	Advantages	Disadvantages	Applications areas
Diaphragm compressors	• Contamination-free gas compression so ideal for handling pure gases • Lower cooling demand • Lower power consumption • High throughput • Higher pressure achievable • Higher efficiency possible with less number of stages • Compact design	• Complexity of design • Failure of diaphragm due to mechanical stresses • Limitation on flow rate due to the size of the diaphragm	• Hydrogen refueling stations • Filling of onboard compressed hydrogen storage tanks in vehicles

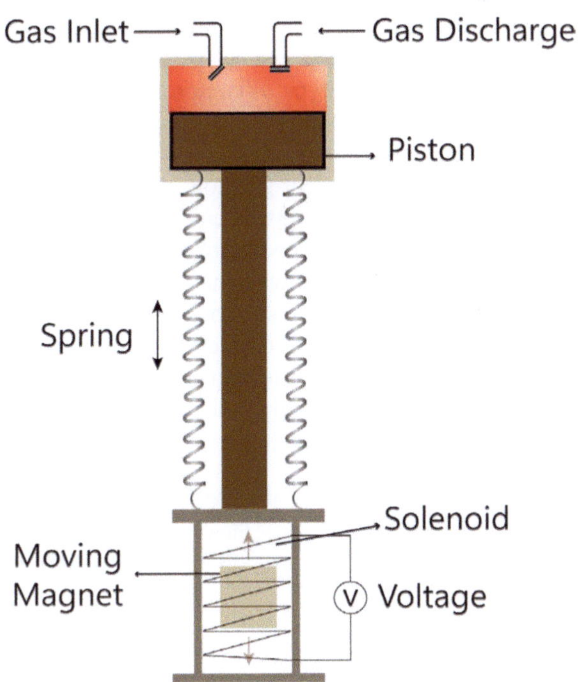

Fig. 11.6 Schematic of a linear compressor (inspired from [5])

a resonating spring system. These linear motors are of magnetic type and can be of moving coil and moving magnet type. The moving coil type of linear motors have very high efficiency, long life with lower noise and vibrations. A series of hollow coils are kept in a strong magnetic field provided by a permanent magnet, and these coils move in the axial direction. The requirement of strong magnetic fields is essential for their operation, or in turn, large permanent magnets are required, making them restricted to be used for limited applications. In order to address this disadvantage of moving coil type linear motors, the other option is the moving magnet type of linear motors. A moving magnet controls the movement of the piston in the axial direction, and with the reversal of the polarity of the applied voltage, the direction of the piston can be changed during the suction or discharge stroke. In moving magnet type of linear motors, the requirement is for a small magnet, and the heat losses are lower and have higher reliability.

The linear compressors operate under optimum conditions, determining the mechanical resonance frequency, and any change from this frequency needs to be controlled by modifying the frequency control systems. Thus, such compressors can achieve higher efficiency, and a lot of interest has been recently seen in such class of compressors. The absence of a crank rod system and the presence of a gas bearing system between the piston and cylinder, acts as a pad to avoid friction and losses. As against the lubricating type of compressors, the outlet pressurized gas is free from contaminants in such compressors. The variation in the pressure on the compressor chamber side and constant pressure on the motor side can lead to piston drift: Besides, the temperature, voltage, or gas flow can change the conditions of operation for the piston. In such a case, the piston could even get deflected by the gas properties, like if there is a change in temperature, the gas flow, or the supply voltage, there will be a deflection associated with the piston. To control the piston movement and to achieve stable operation, piston position control devices are required for accurate operation, which involves complex and cost-intensive designs. At the same time, it has to operate under the resonant condition. The reported efficiencies for linear compressors are usually higher than 73% (Tables 11.5 and 11.16).

Table 11.5 Commercially available linear compressors and their parameters

Compressor technology	Company	Flow rate (Nm^3/h)	Suction pressure (bar)	Delivery pressure (bar)
Linear compressors	At laboratory scale demonstration	> 112	20	850–950

Table 11.6 Advantages and disadvantages of linear compressors and its application areas

Compressor technology	Advantages	Disadvantages	Applications areas
Linear compressors	• High efficiency • Less number of moving parts • Compact and reliable • Long life • Less noise and vibration	• Oscillating pressure due to the difference in pressures on the compressor chamber and motor side • Requirement of piston position control devices makes it complex in design • Requirement to operate at the resonance frequency making the system complex and to avoid deviations of frequency from resonance conditions is required	• Still at research scale • Aerospace-related applications • For cooling electronic devices • Domestic refrigeration • Cryogenic applications

Liquid Compressors

Liquid compressors are another type of positive displacement type of compressors and are suitable for hydrogen gas compression. Since these compressors don't have moving parts or moving seals, and the compression is achieved using the liquids under near isothermal conditions, thus, a lower amount of compression work is required, resulting in an efficient and cost-effective operation. In the process of compression, both the gas and liquid undergo compression together, and the heat produced during the compression is taken away by the liquid, which has higher heat capacity and density as well as by the walls of the compression chamber. This eliminates the need for external heat exchangers, which reduces the overall cost of the system, while improving the efficiency of the system, with maximum efficiency of more than 83% (Fig. 11.7).

Like in the other positive displacement type of compressors, the gas that is in the compression chamber above the liquid surface is compressed by the movement of the liquid, which acts as a piston in this case. Movement of the liquid column inside the chamber via the pump changes the volume of the gas above the liquid head and compresses the gas. Considering the construction of these compressors, there are two chambers that are connected by means of a pump which controls the flow of liquid between the two chambers. There are both liquid and gas in these compression chambers. During the suction stroke, when the gas is taken inside the chamber at low pressure, the pump drives the fluid from one chamber to the other chamber, while the reverse happens in the compression stroke. During the compression stroke, the liquid level rises, reducing the volume available to the gas in the chamber, and thus compression is achieved, and gas is released at a higher pressure. A series of compression and expansion can be achieved by the movement of the liquid column between the two chambers via a pump. Liquid compressors are mostly used for compressed air storage when integrated with the renewable power plants and are capable of delivering compressed air at pressures of 200–300 bar.

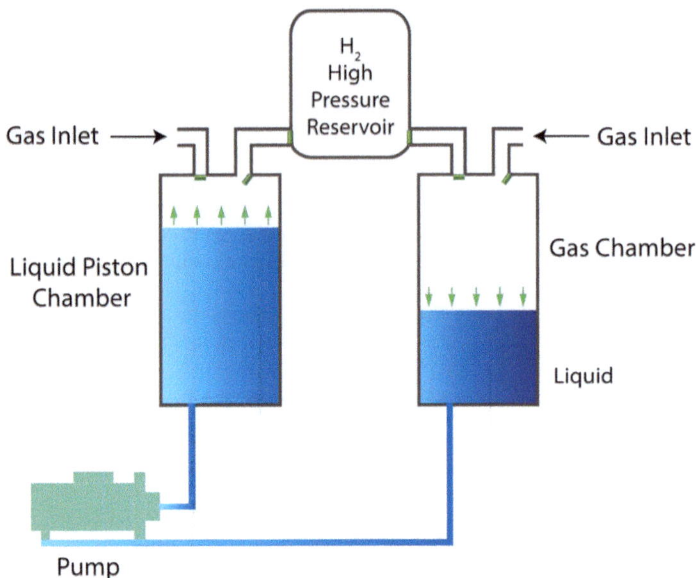

Fig. 11.7 Schematic of a liquid piston compressor (inspired from [5])

As mentioned, the heat of compression is taken up by the liquid that has good heat transfer characteristics, which can be further enhanced by using porous media. This porous medium, due to the presence of interconnected pores, provides a high surface area for the flow of liquid, enhancing the heat dissipation and heat transfer across the column. Introduction of porous media can have multiple advantages like improving efficiency, enhancing heat transfer, making the process nearly isothermal, reducing the dead volume, and also reducing the chances of fluid leakage. Further enhanced heat dissipation can be achieved employing an internal liquid spray cooling technique, which will provide a higher surface area for better heat transfer and then can finally condense and return to the liquid.

Instead of the back-and-forth linear motion, liquid rotary type of compressors can also be used to compress a gas, wherein the impeller on a stator with the blades is present. The liquid, usually water, is fed to the cylindrical housing having the rotor. As the rotor spins, water is forced outward by centrifugal force, creating a void in the center that draws in air. The resulting motion of the liquid compresses the gas introduced at the rotor's center. Such compressors are suitable only for gases having high liquid content, and they have poor efficiency, as such are not very suitable for compressing hydrogen gas. The major problem in using a centrifugal compressor is that the molecular weight of hydrogen is very low; thus, in order to compress hydrogen to an appreciable pressure, a high tip speed will be required. High tip speed will result in vibrations and challenges associated with the rotor dynamics, there could also be impeller failure issues. Contamination issues can also arise from the seals in such compressors. So, the volumetric type of compressors is preferred compared to the centrifugal type of compressors for hydrogen compression among the mechanical type of compressors [5].

The third type of liquid compressors, known as ionic liquid compressors, are specifically suitable for compressing hydrogen. The construction of these compressors is the same as mentioned above; only that the liquid used in the chambers is ionic liquid. Ionic liquids are nearly incompressible, low melting point salts that are in the form of liquid state at room temperature. The gas is compressed by the up and down movement of the ionic liquid column, and the hydrogen gas solubility is poor in these liquids. There is no requirement for seals or bearings in such compressors. The advantage of using ionic liquids is that they have very good thermal properties, good chemical stability, high ionic conductivity, moderate viscosity, and very good lubricating characteristics to reduce friction losses and leakages, when used for high-pressure compression. These ionic liquids are safe to operate with (are fire retardant) almost negligible volatility with low vapor pressure, and at the same time have almost negligible solubility for hydrogen. The biggest advantage of such compressors is that they are highly volumetrically efficient and can provide a high compression ratio for hydrogen.

Efficiency of ionic compressors is high, i.e., more than 65% and even reported to be as high as 83%, this higher overall efficiency as against the other reciprocating type compressors is due to better heat removal, better lubricating and coolant properties, less of losses due to lack of moving parts in compression chamber and lower friction. The energy consumption is low, and the life is long for such compressors, the pressure achievable is higher, and throughput is also high compared to reciprocating compressors. However, the biggest challenge is corrosion issues associated with ionic liquids, which could reduce the life and strength of materials used and thereby increase maintenance costs, reduce efficiency and contamination of the gas with the impurities from the corrosion products. Thus, the use of appropriate materials is essential with ionic liquids, such as AISI 316L, which is corrosion resistant and resistant to hydrogen embrittlement, is well suited for hydrogen compression in such compressors. Some other challenges with such compressors include liquid being carried out along with the gas from the compression chamber, gas becoming trapped due to the cavitation phenomenon, gas being driven into the liquid in the low-pressure areas of the chamber which can lead to loss of hydrogen (Tables 11.7 and 11.8).

Table 11.7 Commercially available liquid compressors and their parameters

Compressor technology	Company	Flow rate (Nm^3/h)	Suction pressure (bar)	Delivery pressure (bar)
Liquid compressors	At laboratory scale demonstration	> 112	20	850–950
Ionic liquid compressors	Linde	375–750	5–200	1000

Table 11.8 Advantages and disadvantages of liquid compressors and their application areas

Compressor technology	Advantages	Disadvantages	Applications areas
Linear compressors	• High efficiency, more than 80% • Less number of moving parts, 8 parts as against around 500 in reciprocating type compressors • Compact and reliable • Long life, in fact around 10 times more than reciprocating • Less of noise, vibration, emissions and mechanical losses as less of moving parts • Low energy consumption, approx. 20% lower than reciprocating • Due to insolubility of gas in liquid, around 1000 bar pressure is also achievable; thus high compression ratio can be achieved • No need of seals or bearings • Heat generated is removed by the liquid and coolant flow in the body of the chamber, near isothermal operation thus increasing efficiency, reducing compression work and better energy efficiency and high volumetric efficiency • Good lubricating properties of ionic liquids • Pure hydrogen obtained with no contamination	• Corrosion issues reducing the strength of the materials used and life and increases maintenance cost, which may also lead to contamination of the pressurized gas by corrosion products • During delivery of the compressed gas, liquid may also comes out along with the gas thus making use of liquid traps essential • Some amount of gas may get trapped in the liquid due to cavitation effect	• Hydrogen refueling stations

In summary, ionic liquid compressors have very high efficiency, even higher than 70%. They can compress gas to high pressures, ranging from 5 bar suction to 1000 bar at the delivery. The number of steps involved are less and energy efficient, and the energy required for compression is 2.7 kW/kg, opposed to 4–5 kW/kg for many of the reciprocating compressors. Ionic liquids have a low vapor pressure of the liquids, have low solubility of hydrogen, and have good tribological properties, making them very favorable for a hydrogen compressor.

Major challenges associated with the mechanical compressors are the moving parts, leading to noise and vibration, higher maintenance requirements, and shorter life; contamination of the gas because of the lubricating oil use or seal materials orfrom the abrasive particles or corrosion products; and the design of these compressors is complex; there are challenges associated with the reliability of these compressors.

Due to moving parts, a higher cost associated with maintenance, lower reliability, chances of leakage, and contamination, the need for regular maintenance and monitoring of these devices to avoid failure is required. At the same time, a lot of energy is consumed in compression, typically 5–15% of the energy content of hydrogen, accounts for the compression energy requirement; this varies depending upon which compression technology is being used. The size of mechanical compressors is large, requiring larger footprint area. Higher power requirement for driving the compression process, leads to a higher cost of the compressed hydrogen and thereby, higher cost of the hydrogen delivered in the hydrogen energy value chain. Typically, the CAPEX of a hydrogen compressor at a hydrogen refueling station accounts for 58% percent of the capital expenditure in the hydrogen refueling station, 65% of the operating cost is for maintenance, 28% of the required energy consumption goes for compression, and 18% of the operating and maintenance cost is due to the mechanical compressors in a hydrogen refueling station along with long maintenance hours and frequency of incidents being 18% due to compressors. In fact, hydrogen compression is a major contributor to the associated cost not only in hydrogen refueling stations(HRS) but also in the complete hydrogen energy value chain. Compressors in HRS can contribute to significant downtime and the time required for maintenance can be long.

Thus, with all these issues and challenges being mentioned, the possible alternatives are non-mechanical compressors.

Non-mechanical Hydrogen Compressors

In the earlier section, various disadvantages of mechanical compressors were mentioned, and the alternative to mechanical compressors is non-mechanical compressors. Non-mechanical compressors don't have moving parts: The operation is quiet with no vibrations, and operational and maintenance costs are lower. The design of these compressors is simple; they are more compact. Compressors which are thermally driven, the required energy input for compression can be provided from the waste industrial heat. Non-mechanical compressors are based on various physical phenomena, like evaporation of liquid. In the case of cryogenic compressors, absorption and desorption of hydrogen, in hydride-forming alloys for metal hydride compressors, adsorption and desorption processes on the surface of high surface area materials in adsorption-based compressors, and motion of charges and ions subjected to electrical energy in the electrochemical compressors. However, the non-mechanical compressors are still not very mature, and much more research and improvements are required to make them cost competitive with mechanical counterparts, and increase the discharge pressure, productivity, and capacity.

Cryogenic Compressors

Cryogenic compression is a combination of hydrogen liquefaction and compression, and has advantages and disadvantages of both the methods of hydrogen storage. The benefit of compressing at low temperature instead of compression at the ambient temperature is the major advantage, which makes the systems more compact, increases volumetric density, and makes higher pressure achievable. In this method of compression, liquefied hydrogen is pressurized and stored in a cryo-compressed tank; these tanks are capable of holding higher pressures, and are super insulated to hold cryogenic liquid hydrogen. Liquid-state hydrogen storage will be elaborated further in a later chapter. Such cryo-compressed tanks have dual characteristics of holding fluid at high pressures, so the inner vessel is made up of carbon fiber composite wrapping with a metal liner, and there is multilayer insulation in the vacuum space in between the inner and outer walls of the tank. Cryogenic pumps are used to compress liquid hydrogen to high pressures, around 850 bar and a hydrogen density of 80 kg/m^3. These tanks have the advantage of lower hydrogen loss due to boil-off, and can store 2–3 times more hydrogen compared to storage at ambient temperature in a compressed gas tank. For example, in order to store 4.1 kg of hydrogen, 100 L of volume is required at 750 bar for ambient temperature compressed gas, while the same amount and same volume can be stored at much reduced pressure of 150 bar when cooled down to liquid nitrogen temperature. The overall system consists of a liquefier to liquefy hydrogen, a cryo-compressed tank to store this hydrogen, a cryogenic pump increasing the pressure of liquid hydrogen to the desired pressure, and an evaporator or vaporizer to convert the cryogenic liquid to the high-pressure hydrogen gas required at the desired pressure. The system is complex and sophisticated; the cost of liquefaction and maintaining all the lines and fittings at that temperature being high, continuous monitoring is required for thermal insulation, and long-term vacuum stability must be achieved. All these make the overall system complicated: The biggest advantage of such systems is high volumetric efficiency, leading to a compact system with a lower footprint area (Tables 11.9, 11.10 and Fig. 11.8).

Table 11.9 Commercially available cryo-compressors and their parameters

Compressor technology	Company	Flow rate (Nm^3/h)	Suction pressure (bar)	Delivery pressure (bar)
Cryo-compressed compressors	Linde	> 1000	–	350–900

Table 11.10 Advantages and disadvantages of cryogenic compressors and their application areas

Compressor technology	Advantages	Disadvantages	Applications areas
Cryo-compressed compressors	• Higher compression ratios • High volumetric density of hydrogen • High volumetric efficiency • High gravimetric efficiency	• Requirement of low temperature • Cost and energy intensive liquefaction • Short life in terms of thermal insulation and vacuum stability • Sophisticated design and complex systems • Performance depends upon the choice of material and geometry of cryo-compressed tank • Cost of the cryo-compressed tanks	• Automotive applications, demonstrated in BMW, tested in Toyota Prius • Hydrogen refueling stations

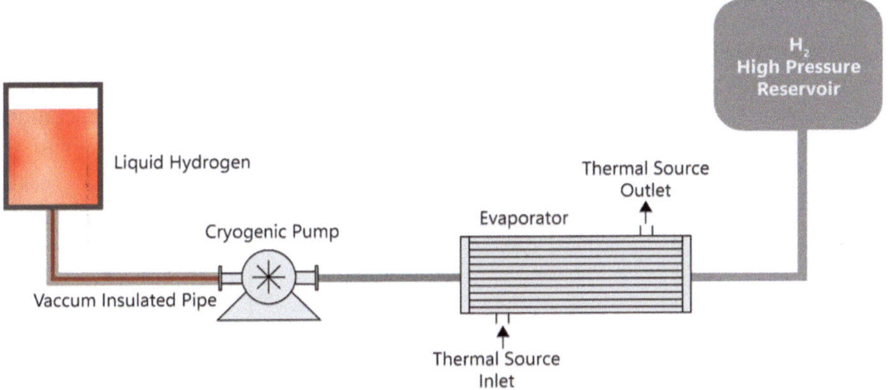

Fig. 11.8 Schematic of a cryogenic compressor (inspired from [5])

Metal Hydrides Based Non-mechanical Compressors

Metal hydride-based compressors are very simple in design and operation, with the least number of moving parts; less noise, vibration, and wear; safe and reliable; and with the use of waste heat for desorption, they have decreased operational costs and improved efficiency.

The first such compressors were reported in the year 1970 via a patent by Wiswall and Reilly [2].

The metals and alloys absorb hydrogen, forming metal hydrides; the formation reaction is exothermic, while the reverse process, i.e., desorption of hydrogen from metal hydrides, is an endothermic process. The charging or absorption and discharging or desorption of hydrogen in such alloys can be represented as follows:

$$M(s) + \frac{x}{2}H_2(g) \rightleftarrows MH_x(s) + Q. \tag{11.1}$$

Forward reaction or formation of metal hydrides takes place at lower temperatures and low hydrogen pressures with the release of heat Q, which is equal to the hydrogenation enthalpy (– ΔH). The reverse process, which involves the release of hydrogen from metal hydrides, is carried out at a higher temperature as it is an endothermic reaction, results in hydrogen being released at higher pressure, and requires the same amount of heat as ΔH. With an increase in temperature, the hydrogen equilibrium pressure increases exponentially; this is the principle of operation of such metal hydride-based compressors.

The basic principle of operation of a metal hydride-based compressor operating in a cycle consists of four process steps and is demonstrated in Fig. 11.9.and low pressure

1. Hydrogen is supplied to metal hydride 1 through the central artery at low temperature T_L and low pressure P_L, as depicted by the red curve in the PCI, which is the absorption curve for metal hydride 1. The absorption by the metal hydride takes place until the equilibrium pressure corresponding to that hydride is equal to the supply pressure. This process, where the metal hydride gets saturated and reaches its capacity, is represented by Point X on the curve. The hydrogen content in the alloy at this point is c_X.

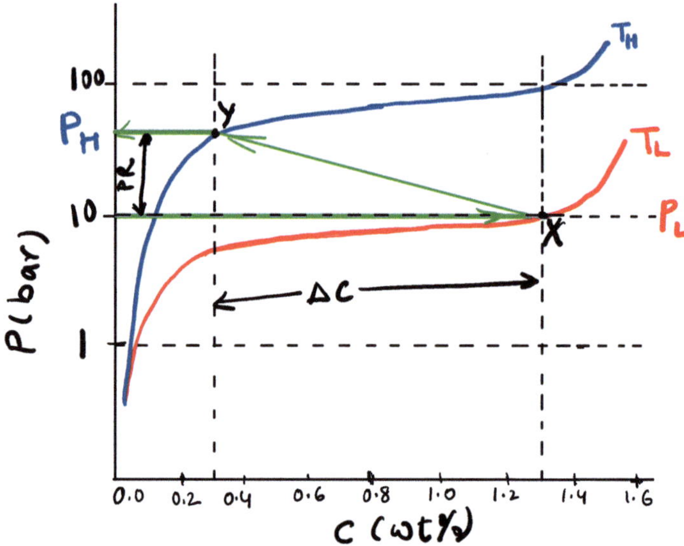

Fig. 11.9 P–C–T diagram for a typical compression process using metal hydrides

2. During the absorption process, heat is released, which raises the temperature; this corresponds to the sensible heating.
3. During desorption, the required heat is supplied, and the heating occurs from low temperature T_L (Point X) to high temperature T_H (Point Y); thereby the equilibrium pressure rises from P_L to P_H (along the path XY). This is the compression process (as the same volume and desorption occur, increasing the pressure), and the hydrogen content at this point is represented by c_Y. In this step, the desired pressure and higher P_H are reached.
4. Since desorption is an endothermic process, thus hydrogen desorption leads to sensible cooling from T_H to T_L.

These four steps make the complete compression cycle, and this has two important characteristics in terms of the compression ratio, which is given by pressure ratio (PR), and another one is the reversible hydrogen sorption capacity of the metal hydride, which is related to the amount being compressed or the productivity of the compressor and given by Δc. These parameters are defined as: PR $= P_{\mathrm{H}}/P_{\mathrm{L}}$ and reversible capacity of material $\Delta c = c_X - c_Y$. This sequential heating and cooling, controlling the absorption and desorption process, is entirely heat transfer driven and requires a very efficient thermal management system. The impact of various design parameters and material characteristics on heat transfer will be discussed in the chapter on solid-state hydrogen storage.

Since the entire compression process involves two discrete operations of absorption and desorption to make the entire compression process continuous, two metal hydride canisters are used such that absorption carried out in one canister, simultaneously, desorption takes place in the other. This is done by periodically heating and cooling the two canisters, which are connected by various valves to maintain the flows and perform the relevant operations.

The same hydrogen compression cycle can be shown on a T–S plot as shown in Fig. 11.10, and the processes can be described as follows:

The ideal compression cycle on a T–S plot with isobars corresponding to P_{L} and P_H, and temperatures T_L and T_H marked on the plot is shown in figure 11.10. During the isothermal absorption process at temperature T_L and pressure P_L, the hydrogen uptake on the plot is represented along the path from point 1 to point 2. The second step involves a polytropic process that is the heating process, wherein the system moves from 2 to 3, and thus the heating is carried out to raise the temperature from T_L to T_H along the process from 2 to 3. The third step involves an isobaric and isothermal desorption along the path from point 3–4, which occurs at a temperature of T_H and pressure of P_H. The final step is a polytropic cooling process, which is represented by the path along the points 4–5 and then to 1, respectively. This last step involves polytropic cooling from T_H–T_L and then isobaric cooling along P_H, which is represented by the path 4–5–1.

However, as seen in the PCI curve in Fig. 11.9, the achieved pressure ratio or compression ratio through these steps is limited, and the maximum pressure difference achieved is P_H–P_L subjected to $T_H - T_L$. Using a single-stage compressor, the

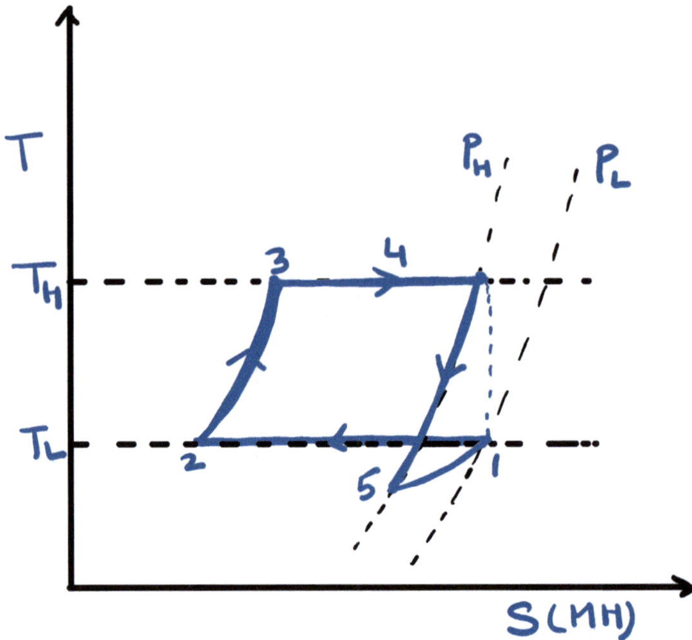

Fig. 11.10 Hydrogen compression cycle on a T–S plot

final pressures attained are not very high. Thus, in order to achieve even higher pressures or compression ratios, we have to couple a greater number of stages. The operation of such multistage compression couples the desorption of the previous stage to the absorption of the next stage, and achieves the ultimate pressure on desorption from this stage, which is the maximum achievable pressure using this compression system. The successive absorption and desorption cycle takes place but in different reaction beds. These are matched such that the low-pressure hydride desorbs at a temperature T_H that is synchronized with the absorption of the next stage. The second stage will be superimposed and synchronized in such a manner that the desorption temperature of the first stage matches the absorption temperature of the second stage, and in this way, the two stages are coupled together. In multistage compression, it is important to select appropriate metal hydrides with desired thermodynamic properties. The ΔH, ΔS values, the required heat capacity, and the kinetic properties should ensure that the compression process has a better productivity and faster kinetics.

To explain the operation of a two-stage compressor with two different metal hydrides having different equilibrium pressures, the schematic in Fig. 11.11 depicts the processes involved.

For a two-stage metal hydrides-based compression, two metal hydrides are required, represented by MH_1 and MH_2 for explaining the working principle. The red and blue curves in Fig. 11.11 show absorption and desorption, respectively, for

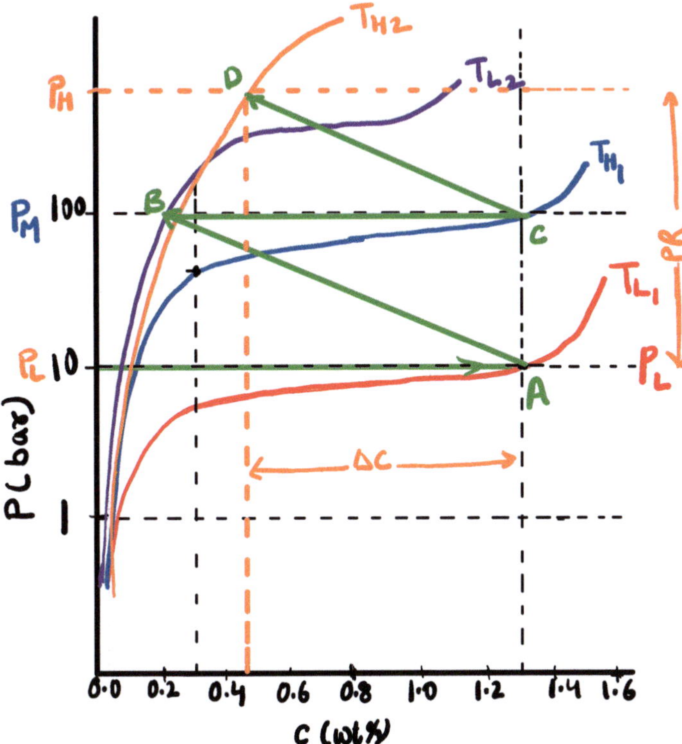

Fig. 11.11 Multistage compression cycle for a metal hydride-based compressor

MH_1; while purple and orange are the absorption and desorption curves, respectively, for MH_2. The steps involved are as follows in two-stage metal hydride-based compressors:

1. Hydrogen is absorbed by MH_1 at low temperature T_{L1} and low pressure P_L and reaches saturation at point A.
2. Desorption from stage 1 is achieved by heating MH_1 to a higher temperature T_{H1}, such that compression in stage one is achieved to an intermediate pressure P_M, and this occurs along the path AB.
3. The hydrogen desorbed from stage 1 from MH_2 at temperature T_{H1} is absorbed by the second metal hydride MH_2 at the same temperature and depicted by path BC.
4. Desorption from MH_2 via heating it to a temperature T_{H2} results in an increase in the pressure from P_M to P_H, and this process occurs via path CD.

Thus, through the two-stage compression process, the final pressure achieved is higher, and the compression ratio is $PR = P_H/P_L$, while the compressor productivity or the reversible hydrogen capacity achieved in two stages is given by $\Delta c = c_A - c_D$. As in the case of single-stage compression, for continuous operations, two canisters of MH_1 and two canisters of MH_2 are required, such that heating in one and cooling in another result in desorption in the first and absorption in the second. This continuously compresses hydrogen in the process. The most important part of metal hydride-based compression is the appropriate choice of metal hydrides (MH_1 and MH_2) such that the thermodynamics of the absorption and desorption processes match and produce the desirable final pressure from the multistage compression. Thus, coupling of the desorption cycle of the first with the absorption cycle of the second alloy plays a critical role in the design. Further, as per the requirement, a large number of stages can be added to achieve further higher pressures. The different types of hydride-forming materials used for compression applications can be broadly classified into four categories, with details listed in Table 11.11.

In selecting the appropriate materials for compression applications, the pressure–composition–temperature diagrams and Van't Hoff plots determined from PCT play the key role. These curves provide information about the absorption and desorption temperature and pressure, achievable compression ratio, and process productivity in terms of reversible hydrogen storage capacity of the materials used. These curves can be plotted from the experimental results, or using models which can predict these characteristics to a high degree of accuracy. These models are developed to predict enthalpy and entropy of formation, hysteresis, plateau length, plateau slope, and any deviations. Other than these characteristics, the cyclic stability, which is the number of cycles of absorption and desorption that the material can undergo without much performance degradation, is another important characteristic that plays a key role when it comes to the life of the compressor. Usually AB_5 type of alloys have a higher

Table 11.11 Typical metal hydride characteristics used for hydrogen compressor applications

S. No.	Hydride-forming materials	Crystal structure	Hydrogen occupancy	Examples
1	AB_5 type of intermetallics	$CaCu_5$ type structure hexagonal	In tetrahedral interstitials in the lattice and there is no change in symmetry	$La_{0.8}Ce_{0.2}$ Ni_5 (A = RE and B = Ni, Co, Sn, Al, etc.)
2	AB_2 type of intermetallics	Hexogonal C14 Laves phase	Tetrahedral sites without change in symmetry	$Ti_{0.65}Zr_{0.35}Cr_{0.84}Mn_{0.28}Fe_{0.65}Ni_{0.20}$ (A = Ti and/or Zr, B = TM like Cr, Mn, Fe, V, etc.)
3	AB type of intermetallics	BCC (CsCl type)	Forms an interstitial hydride with orthorhombic deformation in the lattice	FeTi
4	BCC solid solution alloys	FCC dihydrides		Ti–V–Cr systems, Ti–V–Mn alloys

cycle life compared to AB_2 type of alloys. AB_5 alloys show better performance at moderate temperatures, but more than 100 bar is difficult to achieve in a single stage. With the use of Ti-based AB_2 alloys in multistage compression, even 700 bar could be achieved. Ti–V-based alloys with Nb and Mn addition can deliver up to 650 bar pressure under moderate temperature. AB type of materials has issues with activation, although they have a low cost. However, the performance degradation occurs due to the disproportionation, impurities present in the hydrogen and some of the components in the alloy form strong hydrides, while some form weak hydrides. Disproportionation can lead to lattice strain and the creation of defects such as hydrogen trapping sites, formation of non-hydride forming intermetallics, and amorphization, all of which impact the reversible hydrogen storage capacity, reduce the plateau pressure, increase hysteresis, and consequently decrease compression productivity over time. In addition, the effects associated with the metal atoms being subjected to high temperature, the phase change that occurs between unhydrided and hydrided phases in every cycle, alongwith the associated volume changes (contraction and expansion), can lead to a loss of capacity or productivity over long term operation. However, most of these issues can be reversed, and capacity can be recovered for most of the alloys by heating the material at high temperature, either in vacuum or under hydrogen at low pressure.

Besides materials, the design of such compressors and their operation are equally important, which determines the efficiency, productivity, cost, pressure ratio, and safety. It is observed that there is a trade-off between productivity and efficiency, and optimization is required to get the values for both. The efficiency can be improved with better heat recovery, but in turn that reduces the efficiency due to longer duration required for the compression cycle, which will reduce the productivity. The filling density of metal hydrides in the canisters should be high to improve the dynamic performance, the kinetics of absorption and desorption, and productivity, but puts stress on the metal hydride canister during cycling and reduces gas permeability. Further increasing the number of stages adds to the cost; besides, there lies a trade-off between the highest compression pressure and the overall efficiency achieved. Metal hydride-based hydrogen compressors (where heat energy is directly converted to gas energy) have higher efficiency than mechanical compressors.

Advantages of such compressors are reliability, scalability, compactness, safety, high purity of hydrogen, high discharge pressures, simple design and operation, modular design, and utilization of low-grade heat for the process. Since waste heat available from industrial processes can be used for desorption, and these are thermally driven devices, the OPEX is lower than the mechanical compressors, where electricity is required for operation. However, the technology is at prototype demonstration level. Some of the laboratories in South Africa, USA, Europe, Australia, Russia, have demonstrated such prototypes(Table 11.12).

Usually, the low-pressure alloys can compress hydrogen to 150 bar and high-pressure alloys to more than 800 bars. The selection of suitable metal hydrides and the design of the metal hydride tank are the most important. The alloys used for low-pressure compression to around 150 bar are AB_5 type, while the AB_2-type alloys can compress to very high pressures; even 1000 bar is achievable. In case when

Table 11.12 Advantages and disadvantages of metal hydride-based hydrogen compressors

Compressor technology	Advantages	Disadvantages	Applications areas
Metal hydrides-based hydrogen compressors	These are thermal driven devices and waste heat or low-grade heat can be used for compression Less maintenance and lower power consumption Simple design Modular design Reliable Scalable High pressure achievable High volumetric efficiency Compact Safe Higher efficiency than mechanical compressors	Challenges associated with materials characteristics in terms of hysteresis, disproportionation, loss of capacity, and degradation of the alloys With the increase in the number of stages, the cost increases	Compression for laboratory use Compression of hydrogen isotopes for nuclear-related applications In situ synchrotron experiments Hydrogen refueling station Hydrogen liquefaction process In integrated hydrogen systems with the entire hydrogen value chain

the delivery pressure P_H is around 100–150 bar, the compression ratios achieved can be 10–50, but when the P_H increases, the compression ratio decreases and is limited to 5–10. Compression efficiency depends on plateau width, plateau slope, and the hysteresis. The requirements of appropriate metal hydrides are tuneable PCT characteristics to get flexibility and range so that the desired compression ratio could be achieved when the temperature is changed from T_L to T_H, high reversible storage capacity to achieve higher productivity, fast kinetics to reduce the cycle duration, low plateau slope to have consistent and uniform characteristics of the material, low hysteresis, high cyclic stability for long life of the material without appreciable degradation, tolerance toward impurities to resist loss of capacity in presence of any contaminants, low cost, and ability to scale up synthesis.

In general, the tank is cylindrical in shape with a tubular configuration having a central porous filter or artery to distribute hydrogen inside the reactor and stop the metal powder from coming out of the tank into the valves or tubings, etc.

Usually, these metal hydrides have poor thermal conductivity: As such, the requirement is to improve their thermal conductivity or have a better heat exchange through heat transfer surfaces, so that the reaction does not stall.

Metal hydride compressors have less than 25% efficiency at 423 K. This low efficiency is due to several losses involved during heat transfer, and the requirement of heating and cooling further lowers the overall efficiency. The metal hydride-based non-mechanical compressors are mostly at the developmental or demonstration stage and have not been widely commercialized. HySA systems in South Africa have developed several metal hydride-based compressors, and those have been demonstrated to be integrated for hydrogen refueling stations to meet the hydrogen demand for fuel cell-powered forklifts.

Electrochemical Compressors

Electrochemical compressors do not have any moving parts, and the driving force for compression is the electrical energy. The device configuration is as shown in Fig. 11.12 with the structure and components same as that of a polymer electrolyte membrane fuel cell.

It has membrane electrode assembly, where the two electrodes and membranes are integrated with electrocatalyst to form an assembly MEA (with Pt nanoparticles dispersed on the solid electrolyte, typically 0.1–0.2 mg/cm^2), and other components are the gas diffusion layers, bipolar plates, current collectors (rigid porous Ti on the anode side and porous carbon paper on cathode side), and the end plates. The major difference between an electrochemical hydrogen compressor and a polymer electrolyte membrane fuel cell is thathydrogen is not passed on to the cathode side. As a result, the reaction of hydrogen with oxygen to produce water, which occurs in a fuel cell, does not take place in the electrochemical compressor. The principle of operation of electrochemical compression is that a low-pressure hydrogen gas is allowed to pass on to the anode side, which undergoes a hydrogen oxidation reaction on to the anode side, producing protons and electrons. The electrons produced migrate through the external circuit while the protons permeate through the proton-conducting membrane

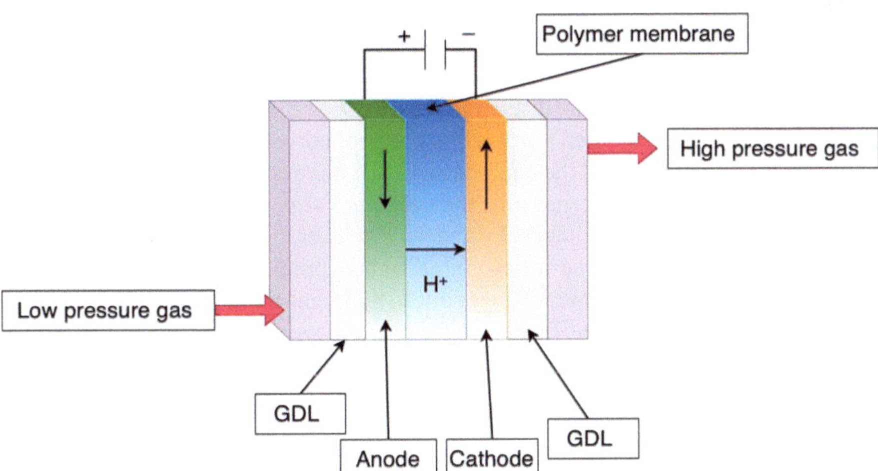

Fig. 11.12 Schematic of an electrochemical device used for a hydrogen compressor

onto the cathode side. Onto the cathode side, the hydrogen evolution reaction occurs.

$$H_2 \rightarrow 2H^+ + 2e^-(\text{on Anode}) \tag{11.2}$$

$$2H^+ + 2e^- \rightarrow H_2(\text{on Cathode}). \tag{11.3}$$

So, protons and electrons recombine to form hydrogen. Now, a continuous pumping of low-pressure hydrogen gas in a closed system will increase the pressure on the cathode side, and when the desired delivery pressure is attained, the high-pressure hydrogen is obtained as discharge out of the compressor. This process will continue until the driving force, which is the electrical energy or current, is being fed into the system. The higher the current being applied, the higher will be the rate of protons being transferred from the anode side onto the cathode side. The rate at which the pumped hydrogen on the anode side is converted to protons can be represented by Faraday's law, with n being the inlet gas flow in moles of hydrogen

$$\frac{dn}{dt} = \frac{I}{2F}. \tag{11.4}$$

The performance of such compressors also depends upon the voltage being supplied. That is given by the Nernst voltage.

$$E_{\text{Nernst}} = E_0 + \frac{RT}{2\text{F}}\left(\frac{P_H}{P_L}\right). \tag{11.5}$$

The E_{nernst} is the Nernst voltage, E_0 is the standard cell voltage under the standard condition, R is the gas constant, T is the temperature, the number of electrons being transferred is two, F is the Faraday's constant, and P_H and P_L are the high pressure and the low pressure in the system, i.e., the discharge pressure and the inlet pressure, respectively.

According to Eq. (11.5), the discharge pressure or the maximum achievable pressure depends upon the applied voltage. As the applied voltage increases, the rate at which protons recombine on the cathode side also increases; thus, a higher pressure is achieved. Nernst voltage represents the minimum voltage required to achieve the desired pressure output or compression ratio. However, due to various losses in the compressor, the actual required voltage is higher than predicted from the Nernst equation. The anode and cathode overvoltages due to hydrogen oxidation and reduction are negligible, and the major contribution to the overvoltage is due to ohmic losses. The ohmic losses are primarily due to the membrane and its proton conductivity, to have a high ionic conductivity of the membrane is desirable. Usually, the membrane used is the perflourosulfonic acid-based membrane, i.e., Nafion 117, which is operated at near-ambient temperature and requires hydration when used for electrochemical compression. The reason for the optimum hydration requirement is that, unlike fuel cells, water is not obtained in the electrochemical reaction and needs to be supplied in the required amount. Nafion is expensive, and other options to this are being studied by various researchers, like sulfonated poly ether ether ketone (SPEEK) or cross-linked polystyrene sulfonic acid (CrPSSA) or sulfonated fullerenol, but the performance is not as good as that of Nafion. The discharge pressures in case of an electrochemical cell can be 25–30 MPa at a current density between 0.1 and 1 A/ cm^2, and the life of such a compressor can be higher than 20,000 h. Compression using a single electrochemical cell can be carried out from ambient to 50 bar, with a current density of 0.2 A/cm^2, cell voltage 140 mV, and an energy consumption of 0.3 kWh/Nm^3. Higher values of current density are favorable for reducing CAPEX but have higher OPEX as more energy consumption and vice versa for lower current densities.

However, obtaining high delivery pressures have both practical as well as economic challenges. Major issues are: (1) higher pressure difference between the electrodes results in permeation of hydrogen from cathode to anode, resulting in reduced compressor efficiency, (2) high ohmic losses reduce efficiency, and (3) higher thickness of membrane is required for reducing permeation, which in turn increases the ohmic losses. Thus, the maximum achievable pressure is limited by a single electrochemical cell, and the value demonstrated is 168 bar. The efficiency for higher pressures as well as for higher flow rates reduces for a single electrochemical cell. In order to further increase the delivery pressure, many such cells need to be stacked; the demonstrated pressure with multicell arrangement is even up to 1000 bar. Very high efficiency of around 90% can be achieved for low voltage and current density up to 1A/cm^2 for a low flow rate of hydrogen. Since the operation of electrochemical compressors occurs under near isothermal conditions, the efficiency for final delivery pressure of 10 bar can be as high as 95%; however, it can drop to around 60% when the final delivery pressure is 100 bar.

These compressors require less energy of compression for low-to-moderate delivery pressure, but are not very efficient and economically advantageous for higher pressures. The major advantages of this class of non-mechanical compressors are: simple operation, noise free, no vibrations; process being near isothermal so the energy consumed is lower for compression; quite robust and low maintenance required; lower CAPEX; can be used for even purification of hydrogen as it has capability of delivering pure hydrogen even if the input gas is a mixture of gas (membrane used is selective toward proton permeation).

However, the major disadvantages include: the efficiency decrease with increase in delivery pressure, and the permeation or back diffusion of hydrogen, which occurs with increased pressur. In order to get higher pressure, there are both practical as well as economic challenges, and there is a requirement of sealing so as to avoid the hydrogen leakage.

If we consider the current status of these compressors, then there are several projects which have demonstrated the use of electrochemical compressors, like the PHAEDRUS project wherein the output pressure achieved was 100 MPa at a flow rate of 0.93 Nm^3/h. Another project that was DON QUICHOTE project that offered an output pressure of 40 MPa and a flow rate of 28 Nm^3/h (Table 11.13).

Table 11.13 Advantages and disadvantages of electrochemical compressors and their application areas

Compressor technology	Advantages	Disadvantages	Applications areas
Electrochemical compressor	• No moving parts, noise, or vibration • High compression efficiency • Low OPEX • High-purity hydrogen obtained	• Low efficiency at high pressures • Sealing is required, difficult to get perfect sealing • Hydrogen permeation or back diffusion • Multicell manufacturing difficulty	• Hydrogen purification • Hydrogen refueling station • With fuel cells and electrolyzers

Adsorption-Based Compressors

Adsorption-based non-mechanical compressor works on the principle of compression via adsorption–desorption and is thermally driven by means of temperature swing. Since the bonds formed during the adsorption process are weak Van der Waals bonds, thus low temperature is required for storage, and adsorption being an exothermic process, is favored at lower temperatures. At room temperature, the bond energy can be of the same order of magnitude as the thermal energy of hydrogen molecules and is therefore sufficient to desorb the hydrogen molecules. As such, in order to increase the adsorption capacity, a lower temperature is required. Usually, adsorption is carried out at liquid nitrogen temperature of 77 K, which is easy to attain and economically feasible. Thus, low temperature and high pressure are ideal for adsorption. Hydrogen is adsorbed on the surface of high surface area materials, and when the temperature is increased, desorption of hydrogen gives rise to the release of higher pressure hydrogen. There are several materials which can be used for sorption of hydrogen, like carbon-based materials, metal organic framework (MOF), zeolites, etc. NU-100 which is an MOF was observed to adsorb 9.9 wt% of hydrogen at 77 K. The corresponding surface area was measured using BET and was recorded to be 6143 m^2/g with a pore volume of 2.82 cm^3/g [3]. A lot of work has been reported in various carbon-based structures; these materials have the advantage of low cost, high surface area and pore volume, along with good stability. For activated carbon, a hydrogen storage capacity of 6.4 wt% has been reported at 77 K and 40 bar, while a capacity of 1.6 wt% is reported at room temperature and 700 bar. Since the adsorption is exothermic and desorption is endothermic, thermal management plays a key role in the operation of such compressors. The adsorption has to be carried out at low temperature, the removal of heat generated during the exothermic adsorption reaction is essential. The energy carried by hydrogen molecules entering the system also needs to be dissipated, and thus precooled hydrogen needs to be introduced to the tank storing sorbent material. Besides, the poor thermal conductivity of sorbents is the major challenge toward thermal management. An appropriate design of a tank for storing these sorbents is required with proper heat augmentation methods. Studies have been carried out on activated carbon as a sorbent, including numerical studies, and it has been reported that pressures as high as 700 bar can be achieved from an initial pressure of 40 bar when the temperature is increased from 77 to 298 K in such compressors. Upon addition of expanded natural graphite (ENG), thermal conductivity was found to increase from 0.25 to 2.0 W/mK. In a well-designed sorbent storage tank, which is a closed system when a low-pressure gas is introduced, the tank is filled with high surface area sorbent material having high porosity, hydrogen gets adsorbed onto the surface of these materials. Once it reaches saturation, then the desorption process can be carried out to get the high-pressure hydrogen. For desorption, heat energy input has to be provided, which increases the temperature of the system, which increases the temperature of the adsorption bed, thereby desorbing hydrogen; both the adsorption and desorption occur at a specific set of temperature and pressure conditions. Since hydrogen, when it is in the adsorbed state, is in a very dense state, during desorption the desorbed hydrogen enters the bulk gas phase which results in an increase in the pressure. The maximum achievable pressure from such devices depends upon the mass of absorbent used, the final temperature or the higher temperature to which heat is supplied, and the volume which is available for the bulk gas. The important requirement in case of adsorption-based compressors is high surface area materials and the porosity of the sorbent material.

Adsorption–desorption process is completely reversible process with the interaction energy involved being 0.01–0.1 eV. The adsorption process takes place on the surface of the high surface area materials, and it forms a mono layer on the absorbent, and requires low temperature.

These are thermally driven devices like the metal hydrides-based compressors, and the waste heat available, like the industrial waste heat, can be utilized for desorption. The advantage of such compressors are that they do not have any moving parts, so there is no vibration or noise, sealings are not required, they have a very high volumetric energy density, the volumetric efficiency is high, the cost of material is low, the overall cost is less, and thus, both CAPEX and OPEX costs are lower.

The higher pressure can be achieved using these adsorption-based devices with less number of stages. But the major disadvantages with these devices is the requirement of an efficient thermal management system; thus both the heat removal and heat supply mechanisms are required to have a uniform bed temperature. Poor thermal conductivity is further detrimental to thermal management. The requirement of low-temperature operation adds to the cost and reduces efficiency.

These adsorption based compressors are an emerging technology with few laboratory-scale demonstrations available in literature, wherein, with the input pressure of 0.1–0.8 MPa, an output pressure of 10–35 MPa with flow rates between 3 and 560 Nm^3/h is reported.

Comparison Between the Various Compressors

Table 11.14 includes the comparison of all the compressor technologies discussed in the previous sections.

Comparing the various compressor technologies, it is evident that, for mechanical compressor, the CAPEX and OPEX are both high, they have many moving parts, higher maintenance cost, and lower efficiency. Ionic liquid compressors have high efficiency and higher compression ratios (even 100 MPa achievable in five steps), low specific energy, i.e., 2.7 kWh/kg. The positive displacement type of reciprocating compressors are widely used for their reliability and capability to achieve high pressures. Turbine-driven reciprocating compressors are being considered for large-scale hydrogen transport and are used in the oil and gas industry, can provide high throughput and high pressure. The motor-driven compressors give high energy efficiency with low operational cost and are known to be suitable for small-scale hydrogen compression [1].

Metal hydride compressors have both CAPEX and OPEX lower than the mechanical compressors, which require less power for compression of around 5 kW/kg, as against 20 kWh/kg for the reciprocating type. Besides, they have the advantage of weight and volume, around 100 kg and 400 L required for a metal hydride compressor as against the requirement of 6000 L and 3600 kg for a reciprocating compressor, thus requiring a larger footprint area. However, the specific energy consumption for metal hydride compressors is high as the thermal conductivity of metal hydride is poor. The high heat of absorption, needs to be removed alongwith the high heat of desorption that needs to be supplied, making efficient thermal management essential. The metal hydride compressors due to absorption and desorption processes being thermally driven are slow in operation and have a limit on the rate of release.

For the electrochemical compressors, the CAPEX is low,t around USD 170 per unit of hydrogen compression (kg/day), as against the mechanical compressor with 2300 USD per unit of hydrogen compressed. The OPEX for the electrochemical hydrogen compressors is low, and its value is around 1 USD/kg as compared to 1.75–2.3 USD/kg for reciprocating type of compressors [1].

Adsorption-based compression is a relatively new technology, it is predicted to have lower CAPEX and high OPEX due to the requirement of low temperature.

Table 11.14 Comparison of the different hydrogen compressor technologies

Type of compressor	Flow rate (Nm^3/h)	Efficiency (%)	Cost	Energy consumption (kWh/kg)	Advantages	Disadvantages
Piston compressor	~ 10,000	~ 45	170,000 USD	< 5	• Mature • Adaptable to high flow rate • High discharge pressure • Contamination by lubricating oil	• Moving parts • Complexity • Heat transfer management difficult • Vibration and noise • Embrittlement • Not good for high pressure
Diaphragm compressor	< 1000	~ 45	~ 2300 USD/kg/day	< 5	• Low power requirement • Low cooling • Fit for handling pure gas	• Diaphragm failure • Complex design • Restricted throughput
Ionic liquid compressor	< 1000	> 70	–	2.7	• High efficiency • High compression ratio • Low energy consumption • Reduced wear and friction • Long life • Low noise, quite operation • Nearly isothermal • Less number of moving parts • No gas contamination	• Liquid can enter gas chamber • Cavitation of gas in liquid • Corrosion

(continued)

Table 11.14 (continued)

Type of compressor	Flow rate (Nm^3/h)	Efficiency (%)	Cost	Energy consumption (kWh/kg)	Advantages	Disadvantages
MH-based compressor	< 10	< 10	150,000 USD OPEX—1000 USD	10	• Thermally driven so waste heat can be used • Absence of moving parts • No noise • Compact and simple • High-purity hydrogen • Safe	• Appropriate choice of material • Good thermal management • Low efficiency • High weight • Low compression ratio
Electrochemical compressor	< 10	~ 60	170 USD/kg/day	–	• Low-cost operation • No moving parts • High-purity hydrogen • High compression efficiency	• Manufacturing difficulties • Problem in sealing (getting good seals) • High cell resistance • Low compression ratio • Low discharge pressure, better to operate at
Adsorption-based compressor	–	–	–	–	• No sealing required • No moving parts • No vibration and noise • Low-cost material	• Low thermal conductivity of materials • Problem in thermal management • Low -temperature operation • High OPEX

References

1. Hydrogen compression and long-distance transportation: Emerging technologies and applications in the oil and gas industry—a technical review, Yumihito Otsubo. Energy Conversion Manag. **X25**(2025), 100836 (2025)
2. R.H. Wiswall, J.J. Reilly, Method of storing hydrogen. Patent US3516263 (1970)
3. M.P. Suh, H.J. Park, T.K. Prasad, D.W. Lim, Hydrogen storage in metal organic frameworks. Chem. Rev. **112**(2), 782–835 (2012)
4. E. Tzimas, C. Filiou, S.D. Peteves, J.-B. Veyret, Hydrogen storage: state of art and future perspective. Printed in the Netherlands, Institute for Energy—JRC IE, PR & Communication
5. G. Sdanghi, G. Maranzana, A. Celzard, V. Fierro, Review of the current technologies and performances of hydrogen compression for stationary and automotive applications. Renew. Sustain. Energy Rev. **102**, 150–170 (2019)

Chapter 12
Compressed Hydrogen Storage Tanks

In the previous chapters, the thermodynamics of the compression process and different types of compressors have been discussed. However, since the density of hydrogen is very low, thus compression to very high pressures is required to achieve high volumetric energy density. After the compression, the compressed hydrogen must be stored in the highly pressurized tanks. This chapter examines the different types of compressed hydrogen tanks available commercially, their design, and materials used for construction of these tanks. Toward the end of the chapter, the safety parameters and challenges with compressed storage methods are discussed.

History of Pressurized Gas Storage Tanks

Low-pressure gas storage systems operating below 20 bar, with smaller capacity storage and made up of copper were available as early as the middle of the nineteenth century. However, the high-pressure vessels for various applications emerged later. Considering the history of these compressed vessels, the oldest pressure vessels date back to the late-nineteenth century, somewhere in 1870–1880, when the typical steel type I tanks were used for storing carbon dioxide. These tanks were used in the beverages' industry and, in fact, started the beginning of industrial gas business.

Thereafter, in 1880, hydrogen began to be stored at 120 bar pressure in wrought iron tanks, primarily for military applications. In 1885, the use of seamless pressure vessels, manufactured from plates and tubes, came into existence (British patent by Lane & Taunton). The specialty of the manufacturing process was drawing the neck of the vessel, during the fabrication. However, it took a long time to increase the storage pressure to 150 bar. Up to the 1960s the working pressure was 150 bar, and thereafter, it gradually increased to 200 and then to 300 bar with advancement in storage tanks.

P. Sharma, *A Textbook on Hydrogen Energy*,
https://doi.org/10.1007/978-981-95-4601-5_12

In the year 1960, the composite high-pressure vessels came into existence, and these were used for military and space applications; however, there were several challenges at that time because of the high cost of the vessels, and the absence of associated regulatory framework. These tanks had a metallic or polymer liner and wrapping of glass fiber composite. At that time, the applications and the manufacturing of these tanks was limited. As the new applications were identified, in sectors such as by firefighters and for leisure games like paintball and diving, the demand for such vessels began to increase. The numbers increased from a few ten units to hundreds of units in the 1960s, the manufacturing started to grow, and the widespread demand from the non-military application areas increased. During this period, operating pressure range was between 250 and 300 bar. It was only in the twenty-first century that the composite overwrapped pressure vessels were developed, and the handling pressures were increased to the range of 350–700 bar for gas storage. Thus, over a period of time, the application areas have grown, the weight of the tanks reduced, the life of tanks increased, better materials have been used to provide high-pressure storage, and most importantly, the standards and regulations have been established to govern its design, manufacturing, and use.

Compressed Hydrogen Storage Tanks

The compressed gaseous hydrogen storage tanks are being used for a wide range of industrial and commercial applications, for automotive onboard hydrogen storage in different FCEVs, and for space explorations. The requirements for different applications can be different, ranging from bottled hydrogen cylinders to very large-scale storage tanks, depending upon the capacity requirements.

Since the pressures involved are very high, the design of these pressure vessels, their manufacturing, and their use are all very important; all of these along with maintenance are governed by various standards, and the guidelines and protocols to be followed.

Considering the diverse applications where the compressed hydrogen storage tanks will be utilized, the capacity requirements and the constraints in terms of weight and volume are different, especially when the storage is for stationary or vehicular applications. The requirements of onboard hydrogen storage in a vehicle are very stringent, and the weight of the storage system along with the space occupied on vehicle becomes the critical factor.

The density of hydrogen under NTP is 0.08 kg/m^3; therefore to store an appreciable amount of hydrogen in a given volume, compression becomes essential. This requires forcefully packing more number of molecules of hydrogen into a given volume. When the compressed hydrogen storage vessel pressure is 350 bar, the density of stored hydrogen is 23 kg/m^3, while for 700 bar pressure, the density is 38 kg/m^3. However, with the compressed hydrogen tanks, there are several important requirements to ensure the safety of tanks being integrated with different applications. Since the tanks hold extremely high pressures, the strength and durability of the tanks are important, and the material with which these tanks are made plays a key role. The tank design and routine monitoring are equally important. Since hydrogen has smallest molecule size and shows properties like high diffusivity, materials exposed are susceptible to hydrogen embrittlement, this necessitates the use of special materials and vessel design.

Commercially available compressed hydrogen storage tanks can be categorized into four categories: type I, type II, type III, and type IV.

Design Considerations

The design considerations for high-pressure hydrogen tanks start with deciding on the characteristics or the requirements like desired capacity, service pressure, factor of safety, weight, volume, and intended applications whether stationary or mobile. Based on these parameters, the cylinder and domes construction is decided, like the cylinder radius and dome profile. Since the shape of the tank dome affects mechanical strength, quasi-elliptical domes are commonly considered in tank design. The dome depth (which may be hemispherical or elliptical) is particularly important and typically lies between 0.2 and 0.775 of the dome diameter, as it influences both mechanical strength and volume efficiency, providing a stronger structure and higher internal volume.

As mentioned earlier, the design of the compressed hydrogen tank and selection of appropriate materials are critically important. While designing the tanks, the important considerations are:

1. Operating and test pressure, i.e., factor of safety.
2. Internal stresses, due to the pressure of the gas.
3. Possible external stress which could be due to the environmental conditions or possible accidental scenarios: for example, vibrations, impacts, aggressive media, temperature to which it is exposed, weight of peripherals including valves, connectors, etc.
4. Long-term aging effects, which can lead to failures like fatigue, plastic deformation, buckling, creeping, for metals; and delamination, fiber cracking and rupture, other aging effects, etc. for composites

Construction of Compressed Hydrogen Tanks

The type I tanks are completely made up of metal. These tanks are heavy but exhibit high tensile strength. These tanks can store hydrogen at a pressure ranging from 200 to 300 bar, although the maximum sustainable pressure can go as high as 500 bar. In type I tank design, as the pressure increases, in order to meet the requirement of increasing the mechanical strength, the thickness of the wall needs to be increased which in turn increases the weight of the tanks. Type I tanks are available in different sizes, from 2.5 to 50 m^3. Such tanks are very good in terms of strength, but the major challenge is the tank weight. The cost of such tanks is the lowest compared to the other types, and they are used for industrial and commercial purposes.

The second category is type II tank, these are made up of a thick inner metal liner, which provides gas tightness to eliminate any gas leakage. In such tanks, the hoop of the tank is wrapped with fiber resin composite. The fiber resin composite provides resistance to metal liner failure, fatigue, and the residual compressive stresses; at the same time it imparts strength to the tank, enabling it to withstand external environmental and stress conditions. The pictorial view of type II, type III, and type IV compressed hydrogen tanks are shown in Figs. 12.1 and 12.2, respectively.

The presence of fiber provides strength to the tank, while the resin present in the composite helps in the binding of the fibers, provides load transfer, and prevents any wear or failure from external impact or aggressive environmental conditions.

The third category is type III tank which has a thin inner metallic liner and is completely wrapped with fiber resin composite. Unlike type II tanks, that have a thick metal liner, and are only wrapped around the hoop direction, type III tanks have a thin metal liner and is fully wrapped with the fiber-reinforced composite. The liner provides gas tightness, while the wrapping with the fiber resin composite provides strength and the required stiffness. When the category changes from type I to type IV, the weight of the tank reduces, but the cost increases. In fact, the weight of the type III tank is half that of the type I tank, but the cost is twice that of the type II tank, which is the major drawback.

Fig. 12.1 Metal liner for type II and type III compressed hydrogen tanks (image credit, courtesy Swagelok.com)

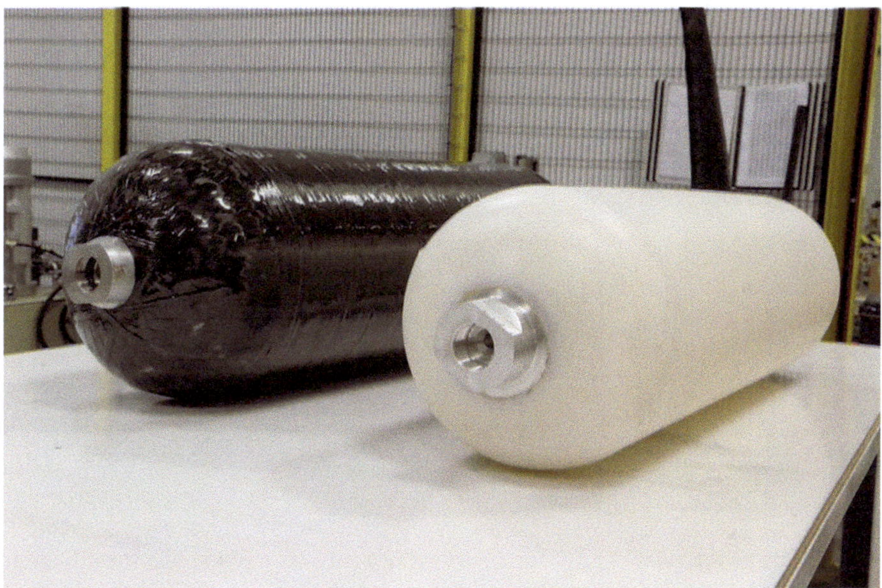

Fig. 12.2 Plastic liner for type IV tanks (courtesy of image: AMRC)

The next category of compressed hydrogen tanks is the type IV tank. Type IV tanks are made up of polymer liner and are completely wrapped with fiber resin composite materials. Thus, the major difference compared to type I, type II, or type III tank is that the liner is made up of polymer instead of metal. The complete wrapping with fiber-reinforced composites provides both mechanical strength and high load-bearing strength. The port is used for filling and emptying; therefore the, interface between the liner and the boss must be leak-free, which is major challenge in the design and manufacture of such tanks.

Another type, which is still in the developmental stage, is the type V linerless, all-composite tank, which will have the lowest weight, and lower permeation than type IV, while, having complex manufacturing; especially are being designed for use in reusable launch vehicles, space applications, and cryogenic applications.

Usually, the compressed hydrogen tanks are cylindrical in shape and consist of a cylindrical section and two domes, with an opening in the form of the filling port. There can be one or two openings, depending on the application. Selection of appropriate tank, out of the type I to IV compressed hydrogen tanks, depends upon the weight and volume requirement for the particular application. In case of stationary application, the requirements are not stringent, but for vehicular or portable application, the weight and volume of the storage tanks are the determining factors. Thus, for vehicular or portable application, the type III and type IV tanks are the preferred choice because the weight and volume requirements are lower for these tanks. Besides the requirements of weight and volume, other important considerations include safety, cost, and end-use applications, which are the determining factors

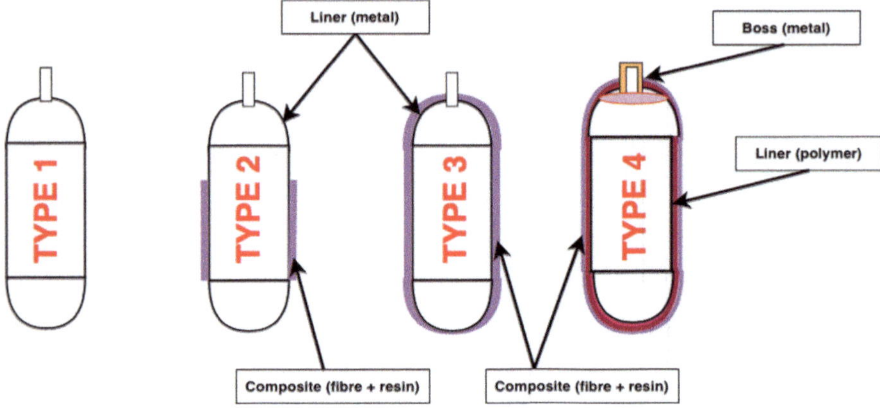

Fig. 12.3 Schematic of type I, II, III, IV compressed hydrogen tanks (reproduced from [1] with permission)

when selecting the type of tank for a particular utility. The schematic flow diagram of type I, II, III, IV compressed hydrogen tanks is shown in Fig. 12.3.

Important Parameters or Requirements in Designing the Tanks

The compressed hydrogen tanks are designed to withstand a certain pressure, which is called service pressure. For example, the type III tanks are designed to store at 350 bar, and type IV at 700 bar. This pressure is operational pressure or the pressure to hold till the entire service life. However, the test pressure is typically 1.5 times the nominal working pressure, i.e., lies in the range of 525–1050 bar for service pressure of 350 or 700 bar, respectively. Therefore, the tanks should be designed to meet the desired test pressure with the factor of safety of 1.5 or higher. During the service life, the tanks may experience external stresses, possible impacts with other structures, vibrations during their operation, exposure to aggressive media, high or low external temperature, and additional loads due to connectors and accessories. Thus, the tank should be capable of bearing those loads and, at the same time, should have a long service life, i.e., they should be able to sustain a large number of filling and emptying cycles, ideally lasting at least as long as the life of the vehicle or utility. Throughout the entire service life, the tanks should be safely operable and should not fail.

The possible failure modes in the case of metal liners or all-metal tanks include fatigue, buckling, creeping, or the plastic deformation. Similarly, in case of type II, III, and IV tanks, issues such as cracks, delamination, or fiber ruptures should not occur in the composite materials used, and these materials should have long life, high strength, stability, and durability. All these aspects must be considered during the tank design. Other possible failure modes in the composite tanks include the liner blistering or liner collapse, which can occur when hydrogen gets trapped between the liner and composite, due to pressure difference during the emptying or filling, as both the service pressure and the internal tank pressure are dynamic and may vary.

Since the compressed hydrogen tanks are required to sustain very high hydrogen storage pressures, they must possess high mechanical strength, the selection of materials should ensure resistance to hydrogen embrittlement; which is one of the major challenges in hydrogen storage systems, and associated components through which hydrogen is transported. The tank materials should be compatible with the hydrogen gas.

Considering all the above-mentioned requirements, both the selection of appropriate materials and proper design of tanks are absolutely essential. In case of all-metal tanks (type I), the metals are isotropic throughout; however, for type II, III, and IV tanks, i.e., the composite tanks, the materials are anisotropic, and their mechanical properties are different and vary with the direction. The anisotropy depends on the fiber orientation and winding pattern (hoop or polar wrapping), as well as the anisotropy between the metal liner and the composite layers. The failure modes can differ, the aging behavior varies, and all these must be considered while designing the tanks. In metallic vessels, i.e., the type I tank, the geometry consists of a cylindrical section and two domes, the domes are oversized and designed to reduce stress concentrations.

In case of type II tanks, which have a thick inner metal liner and are only hoop-wrapped to withstand higher pressures, it is essential that initial simplified design calculations are carried out to estimate the lay-up design at the preliminary stage, and followed by a detailed finite element analysis to obtain the optimized design and the wrapping pattern. Finally, the resulting wrapping design should be coherent with the filament winding machine code. The winding pattern and sectional view of a type IV tank are shown in Fig. 12.4.

Manufacturing of the Compressed Hydrogen Tanks

1. Metallic Liner or Tank or Boss—The simplest, or all-metal tanks, i.e., the type I tanks, can be manufactured from plates, billets, or tubes. Thus, metallic tank or liner manufacturing involves the traditional forming techniques, including the deep drawing or spinning.

    - Manufacturing from Plates—Metallic plates are deep drawn to obtain the desired shape and dimensions, i.e., diameter and thickness, forming the cylindrical section. The neck region in metallic tanks is manufactured from metallic plates using the hot spinning process, the excess metal obtained during spinning to be machined to form filling and emptying ports. In order to get the desired mechanical properties, the drawn structure must be heat treated.

(a)

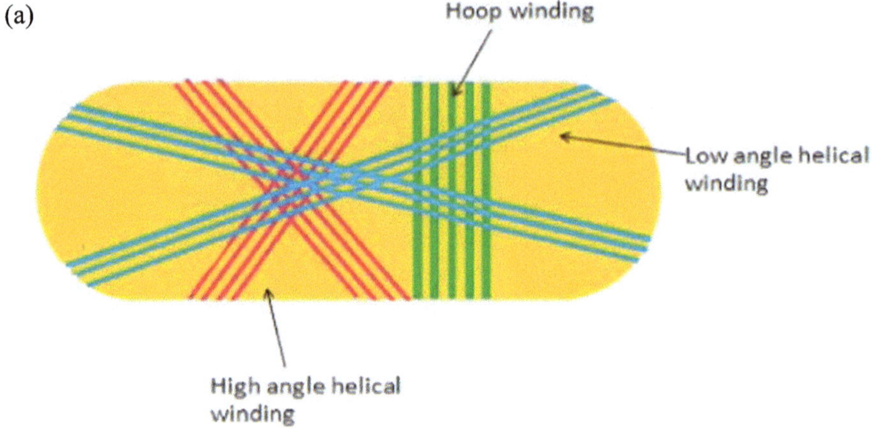

(b)

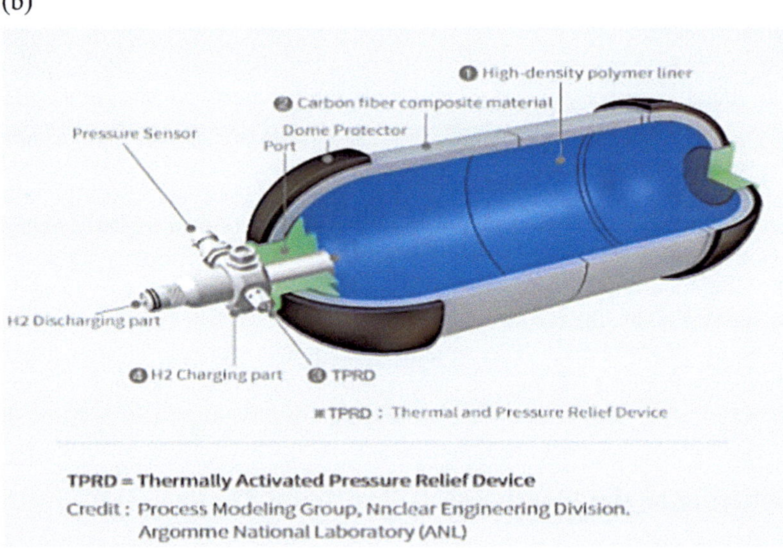

Fig. 12.4 **a** Winding pattern on a type IV tank **b** sectional view (reproduced from [2] with permission)

- Manufacturing from Billets—Initially, the billets are heated to achieve the required conditions for drawing, according to the optimized design shape and dimensions. Thereafter, the process is similar to that used for manufacturing from the plates.
- Manufacturing from the Tubes—The tubes with the required wall thickness are selected, where this tube forms the hoop region. To obtain the domes and filling port, the same process, i.e., hot spinning, is employed to form the port and

domes from the tube. The dimensions must be carefully considered during spinning, and accordingly, tubes of appropriate thickness are selected.

The process of making metallic liners for type II and III tanks is similar as mentioned above. A critical requirement is strict control over each of the manufacturing steps, along with the control of the materials and high precision in following the prescribed procedures.

2. Polymeric liner—In the type IV tanks, the inner liner is made of polymer, which can be either produced from polymers or from the monomers though rotational molding, or blow molding or injection molding processes (welding injected domes to an extruded polymer tube). In the rotomolding process, the polymer or monomer is taken in a mold which is shaped according to the desired liner geometry. The liner is thereafter manufactured by heating and cooling the rotating mold. During heating, the required fusion temperature or polymerization temperature needed to be achieved. The most critical aspect is the integration of metallic boss, which must be leak-tight, with an interface, designed to prevent failure under test or service pressure. The incorporation of boss to the liner either can be included during the rotomolding step itself or during the fiber winding.

In case of extrusion blow molding process used for liner manufacturing, the polymer tubes are produced by blow molding, while the domes with the metallic boss are manufactured separately and later welded together to get the desired final design. The metallic boss can be introduced into the domes either during the forming process or bonded later. In fact the overall design of the tank is influenced by the liner geometry and dome region. In case of compressed hydrogen tanks, the main objective is to minimize the weight and volume of the tank, while maximizing the hydrogen capacity. Although the liner shape affects the tank design, it is also influenced by the fiber placement process, fiber gaps, or overlaps, and reinforcement locations are required to provide adequate strength. Therefore, optimizing the liner shape is the most critical step toward minimizing the weight of the tank and maximizing the internal volume to accommodate higher gas storage.

3. Fiber Wrapping

In types II, III, and IV, the liner which is either metallic or polymeric is hoop wrapped or fully wrapped, depending on tank type on a filament winding machine. Filament winding is a technique used to manufacture hollow or circular tanks, being performed by winding continuous fiber tows on to a rotating mandrel on a specialized winding machine, through a fiber delivery system following a pre-optimized geometrical pattern. The fiber delivery head guides the tow placement and is mounted on a movable carriage. The angle between the fiber direction and the mandrel axis is called winding angle, which determines the strength and stiffness of composite materials. After sufficient winding, the structure is cured with resin to bind the fibers. The shape and size of the tank are determined by the mandrel geometry and thickness of resin composite layers. Figure 12.5 shows the carbon fiber wrapping of the type IV tank.

There are three possible configurations for fiber wrapping, circumferential (hoop), polar or helical. Type II tanks are typically hoop wrapped, with fibers embedded in

Fig. 12.5 Picture of carbon fiber wrapping on a type IV tank (Image Credit: Century Design, Inc.)

resin, either through wet winding, i.e., during winding or by fibers that are preimpregnated with resin. In type III and IV tanks, a combination of helical and polar winding is used to provide the required strength, and in some cases, all the three configurations can also be used. Another advanced method of fiber winding is using the robotic arm, where more precise and controlled winding is possible. Additionally, tape winding is being considered as alternate to fiber winding for various applications.

4. Resin Composite

The resin serves to bind the fibers together, gives strength to make them capable of load-bearing, and protects the fiber from the breakages and damage. The required binding of the fibers with resin is done using the curing process, either by exposure to UV or heating to the required temperature in an oven. Regardless of the curing method, quality control, precision, and continuous monitoring at every step during the manufacturing process are essential.

Fibers may be supplied as preimpregnated towpreg which have high fiber count, improved uniformity, and improved mechanical strength. The towpreg-based dry winding is generally more efficient, provides a cleaner working environment, higher production efficiency, faster processing, and higher deposition rates, while the wet winding, where dry fibers bundle is coated with molten resin during winding, offers lower manufacturing costs but requires extensive cleaning and the use of large quantities of solvents.

After manufacturing, the tanks must undergo several tests, with leak-tightness and burst pressure testing being the most critical. In addition, non-destructive testing and protocols are to be followed to ensure safety and reliability.

Construction Materials for Various Parts of Compressed Hydrogen Tanks

In the previous sections, a list of possible failures and material requirements has been discussed. There are very stringent material requirements for the selection of the liner (metallic or polymeric), fibers, resin used in the tank manufacturing. The compatibility of the materials selected with the hydrogen gas and the impact of the operating conditions, as well as the effect during long service life must be considered while selecting appropriate construction materials. Not only the service and test pressures, but also the exposure to various dynamic operating conditions during filling and emptying cycles, and the quality of the gas are crucial factors in the selection of materials. During long-term use of the tanks, the periodic inspection and monitoring is critically important. Hydrogen embrittlement which can alter the mechanical properties of the metals, leads to cracks, degradation and failures. In steel, this occurs through dissolution and trapping of hydrogen atoms, potentially resulting in burst or catastrophic failure. Therefore all these aspects need to be considered while the selecting of appropriate materials.

For polymer liner, the hydrogen permeation rate is a very important parameter to be considered and is specific to only type IV tanks. This permeation arises due to dissolution and diffusion of hydrogen gas into the polymeric matrix. Since hydrogen is a very small molecule, its permeation and diffusion in polymers is quite common, therefore while designing the liner for type IV tanks, it is essential to restrict permeation rate as low as possible. Another important phenomenon to be avoided in type IV tanks, through proper material selection and design, is liner collapse or cracking due to polymer blistering. This liner deformation is observed in cases where fast emptying of the tanks is carried out, and is attributed to hydrogen diffusion through the liner material followed by accumulation at the liner–composite interface, in voids, or dissolution within the material. This type of deformation depends on the maximum tank pressure, the pressure at the end of emptying, and speed at which the tank is emptied, i.e., the operating conditions. All these processes impact the service life of the tank. A particularly critical region in compressed hydrogen tanks is the polymer liner boss junction, which is subjected to differential temperatures during filling and emptying operations and consequently experiences thermal stresses. Therefore, appropriate material selection and design are required to avoid degradation and any possibility of leakage at this junction.

The composite wrapping provides mechanical strength to the tank against the pressure loads as well as environmental impacts such as mechanical impacts, and accidental impacts, subjected to varying ambient temperature conditions, occasional dropping of the tank, differential pressures during filling and emptying cycles, and damage due to fiber breakage, delamination, or cracking. Reported studies indicate that surface impacts can cause subsurface damage in composites, which may propagate toward the liner and result in liner damage. Hence, periodic assessment and inspection including visual inspection, hydraulic proof test and various non-destructive tests to check the health of the composites are essential. Accordingly, tanks are subjected to a series of tests before first filling, such as burst test, bonfire test, permeation measurements, tensile test, fracture test, disc test, and others, and are also required to undergo regular inspection throughout their service life.

Materials of Construction

The different materials used in the various types of tanks are mentioned below:

1. Metals for Liner, Tank, and Boss—Aluminum alloys for pressure vessels includes AA6061 (Al-Mg-Si) or 7000 series alloys (exposed to water can reduce life), and steel (i.e., inox or Cr–Mo).
2. Polymer for Liner in Type IV Tanks—could be either high-density polyethylene or aliphatic polyamide-based polymers. An important selection criterion is the water content in these polymers which needs to be considered during the drying process, as water has the tendency to desorb from the polymer liner. The polyamide has a higher equilibrium moisture content than polyethylene, hence, an additional drying step is required, when using polyamide. This makes the use of polyethylene preferred.
3. Composite—Glass fibers, aramid, or carbon fibers are used for composite tanks (types II, III, and IV) overwrapping. The difference between these fibers is tensile strength, elongation, and tensile modulus. Considering these properties, the carbon fibers are the preferred choice. Moreover, they have exceptional fatigue performance and resistance to creep and chemical exposure, thereby not only sustaining high storage pressures but also providing long service life without failure. However, they constitute the major cost component of such tanks. Glass fibers are relatively inexpensive but can degrade when exposed to moisture, aggressive chemicals, and sustained mechanical loading. Fibers can be either preimpregnated with resin (these are expensive) to make composite or can be impregnated before winding via wet winding method.
4. Resin—Polyester, epoxy, or phenolic resins can be used. However, among these, the epoxy resin is preferred due to the high chemical stability, good mechanical properties, and its compatibility with fiber winding process.

The advancements in the field of compressed hydrogen storage include research in the field of advanced materials like composites or nanomaterials to achieve enhanced performance, integrating various sensors and IoT-based devices for real-time structural health monitoring, development of efficient and cost-effective manufacturing techniques such as automated fiber placement for composite fabrication, and the establishment of harmonized global standards for the design, qualification, and testing of these tanks.

Testing of Compressed Hydrogen Tanks

A lot of advancements in terms of codes and standard on compressed hydrogen tanks have been developed. Several tests need to be performed on these tanks to test the safe usage of such vessels. These include the following:

- Composite material tests.
- Liner material tests.
- Liner burst test at ambient temperature ($\geq$2.25 times the working pressure, for 700 bar burst pressure test at 1575 bar).
- Hydraulic (proof) test of finished cylinders at ambient temperature.
- Cylinder burst test.
- Resistance to pressure cycles at test pressure and ambient temperature immersion in salt water.
- Exposure to an elevated temperature at test pressure.
- Drop test.
- Flawed cylinder test.
- Extreme temperature cycle test.
- Fire resistance test.
- High-velocity impact (bullet) test.
- Permeability test of cylinders with non-metallic or without liners.
- Test of compatibility of thermoplastic liners with oxidizing gases.
- Torque test.
- Neck strength.
- Cylinder stability.
- Neck ring.

Hydrostatic burst test—The hydrostatic burst test is carried out to determine the burst pressure of the tank which is very important when it comes to the safety and reliability of such tanks and is essential for qualification of tanks to be used for various applications. In this test, the tank is filled with water and is pressurized to 1.5 times of the working pressure to check if there are any leaks at that pressure.

Burst test—In this test, the tank is pressurized usually to more than 2.25 times the working pressure until structure rupture occur.

Flame exposure test—This test is carried out to check for the ability of the tank to withstand fire exposure. In this test, the pressurized tank is exposed an open flame to verify whether the tank safely vents the internal pressure without undergoing explosion. This test is also known as the bonfire test.

Performance durability (hydraulic/gas sequential test) or fatigue test—The hydraulic test is carried out by cyclically pressurizing the tank for around 22,000 cycles or until the leakage develops without catastrophic failure or rupture. Considering a service life of 15 years, the tanks should sustain at least 11,000 pressure cycles without failure.

Permeation test—This test is specifically applicable to type IV tanks, where hydrogen leakage through the liner or shell and the corresponding permeation rate are measured.

Thermal cycle test—Repeated filling and emptying is carried out such that the temperature varies between -40°C to 85°C, and the ability of the tank to bear that thermal cycling is evaluated.

Drop test—In this test, the tank is dropped from a certain height to check its structural strength and mechanical integrity.

Bullet test—The tank is stuck with a sharp object or bullet to check its integrity.

Accelerated stress test—High-pressure tank is subjected to elevated temperature and tested over time to assess the combined effects of these parameters on the long-term aging and durability of the tank.

Additionally standards are being developed and validated (SAE J2579, ISO 15869) for high-pressure hydrogen storage tanks. The international standards followed for testing protocols include UN GTR No.13 (Global Technical Regulation for hydrogen), ISO 19881, ISO 11119, SAE J2579 (North America), EC79/2009 (EU, now harmonized with UN GTR 13), among others.

The tests as mentioned above can be destructive such as the burst test, drop test, and bullet test or non-invasive such as visual inspection and ultrasonic or radiographic inspection.

Q1. Calculate and compare the volume required to store 5 kg-H_2at 1 bar and 700 bar, respectively. Take compressibility factor z = 1.452, molecular weight of H_2 = 2.016 g/mol, and R = 8.314 J/mol K, T = 293 K.

Solution:

Mass of hydrogen to be stored: $m_{H_2} = 5\,\text{kg}\,H_2..$

To find: Volume for hydrogen storage.

At $P = 1$ bar:

$$\begin{aligned} PV &= mR_sT = m\left(\tfrac{R}{M}\right)T \\ V &= \tfrac{5\times8.314\times293}{1\times10^5\times2.016\times10^{-3}} \\ V_{1\,\text{bar}} &= 60.42\,\text{m}^3 = 60416.72\,\text{L}\,. \end{aligned}$$

At $P = 700$ bar: Using compressibility factor z:

$$\begin{aligned} PV &= ZmR_sT \\ PV &= Z\cdot\tfrac{mRT}{M} \\ V &= \tfrac{5\times8.314\times293\times1.452}{2.016\times10^{-3}\times700\times10^5} \\ V_{700\,\text{bar}} &= 0.1253\,\text{m}^3 = 125.32\,\text{L} \end{aligned}$$

∴ Volume required to store 5 kg-H_2 at 1 bar is approximately 482 times higher than the volume required at 700 bar compression.

Q2. Compute the temperature at state 2 for an isentropic compression of hydrogen gas from 5 bar at 298 K to 90 bar.

Solution:

Given:

- $P_1 = 5$ bar
- $P_2 = 90$ bar
- $T_1 = 298$ K.
- Process: Isentropic

To find: T_2, the temperature at state 2 for isentropic compression.

Isentropic relation:

$$\frac{T_2}{T_1} = \left(\frac{P_2}{P_1}\right)^{\frac{\gamma-1}{\gamma}}$$
$$T_2 = T_1\left(\frac{P_2}{P_1}\right)^{\frac{\gamma-1}{\gamma}} = 298\left(\frac{90}{5}\right)^{\frac{1.4-1}{1.4}}$$
$$T_2 = 680.56\,\text{K}$$

Q3. Calculate the intermediate pressure for a two-stage isentropic compression process such that minimum compression work is required to compress gas from 20 to 200 bar.

Solution:

Given:

- $P_1 = 20$ bar
- $P_2 = 200$ bar

Let the intermediate pressure be P_x.

For minimum compression work:

$$P_x = \sqrt{P_1 \cdot P_2} = \sqrt{20 \cdot 200}$$
$$P_x = 63.245\,\text{bar} \quad .$$

Q4. Calculate the isothermal compression work (in kJ/mol) at 303 K for the compressor having suction pressure 35 bar and outlet pressure 150 bar.

Solution:

Given:

- $P_1 = 35$ bar
- $P_2 = 150$ bar
- $T = 303$ K

To find:

Isothermal compression work W (in kJ/mol)

$$W = RT\ln\left(\frac{P_1}{P_2}\right)$$
$$W = 8.314 \times 303 \times \ln\left(\frac{35}{150}\right)$$
$$W = -3.67\,\mathrm{kJ/mol}$$

(Negative sign indicates work is done on the system.)

Q5. Calculate the compression work (in kWh/kg-H_2) for compressing $H_2$from 35 bar, 298 K to 350 bar as follows:

1. **Single-stage isentropic compression**
2. **Two-stage isentropic compression with perfect cooling such that minimum compression work is required**

Also compare both the cases and comment on the observations.

Solution:

Given:

- Suction pressure, $P_1 = 35$ bar
- Delivery pressure, $P_2 = 350$ bar

To find:

(i) Single-stage isentropic compression work (in kWh/kg-H_2)
(ii) Two-stage isentropic compression work with minimum compression work
(i) **Single-stage isentropic compression work**

$$W_{\text{isentropic}} = \frac{\gamma R_s T_1}{\gamma - 1}\left[\left(\frac{P_2}{P_1}\right)^{\frac{\gamma-1}{\gamma}} - 1\right]$$

Here, R_s is the specific gas constant for H_2:

$$R_s = \frac{R}{M} = \frac{8.314}{2.016\times10^{-3}} = 4124\,\mathrm{J/kg\,K}$$
$$W_{\text{isentropic}} = \frac{1.4\times4124\times298}{1.4-1}\left[\left(\frac{350}{35}\right)^{\frac{1.4-1}{1.4}} - 1\right]$$
$$W_{\text{isentropic}} = 1.11\,\mathrm{kWh/kg\text{-}H_2}$$

(ii) Two-stage compression; given, perfect cooling: $T_1 = T_2$

For minimum compression work, intermediate pressure $P_x = \sqrt{P_1 P_2} = \sqrt{35 \times 350}$

$$P_x = 110.68\,\mathrm{bar}$$

$$W_{\text{isentropic,2-stage}} = \frac{\gamma RT_1}{\gamma - 1}\left(\left(\frac{P_x}{P_1}\right)^{\frac{\gamma-1}{\gamma}} - 1\right) + \frac{\gamma RT_1}{\gamma - 1}\left(\left(\frac{P_2}{P_x}\right)^{\frac{\gamma-1}{\gamma}} - 1\right)$$

$$W_{\text{isentropic,2-stage}} = \frac{\gamma RT_1}{\gamma - 1}\left[\left(\left(\frac{P_x}{P_1}\right)^{\frac{\gamma-1}{\gamma}} - 1\right) + \left(\left(\frac{P_2}{P_x}\right)^{\frac{\gamma-1}{\gamma}} - 1\right)\right]$$

$$= \frac{1.4}{1.4 - 1} \times 4124 \times 298\left[\left(\left(\frac{110.68}{35}\right)^{\frac{0.4}{1.4}} - 1\right) + \left(\left(\frac{350}{110.68}\right)^{\frac{1.4-1}{1.4}} - 1\right)\right]$$

$$W_{\text{isentropic,2-stage}} = 3.35\,\text{MJ/kg-H}_2 = 0.93\,\text{kWh/kg-H}_2$$

- Single-stage isentropic compression work: $W_{1-\text{stage}} = 1.11\,\text{kWh/kg-H}_2$
- Two-stage isentropic compression work: $W_{2-\text{ stage}} = 0.93\text{kWh/kg-H}_2$

$$\Rightarrow W_{\text{isentropic,2-stage}} < W_{\text{isentropic,1-stage}}$$

References

1. H. Barthelemy, M. Weber, F. Barbier, Hydrogen storage: Recent improvements and industrial perspectives. IJHE **42**, 7254–7262 (2017)
2. P. Muthukumar, Alok Kumar, Mahvash Afzal, Satyasekhar Bhogilla, Pratibha Sharma, Abhishek Parida, Sayantan Jana, E Anil Kumar, Ranjith Krishna Pai, I.P. Jain, Review on large-scale hydrogen storage systems for better sustainability, International Journal of Hydrogen Energy, **48**(85), 33223–33259 (2023)

Chapter 13
Fundamentals of Liquid-State Hydrogen Storage

Storage of hydrogen in liquid form requires cryogenic temperature. The term *cryogenic* is a Greek word which means ice cold. The importance of storage of hydrogen in liquid form lies in the fact that the density is much higher than gaseous hydrogen. As per reports, gaseous hydrogen requires four times more footprint area than liquid hydrogen, hence liquefaction significantly reduces the required footprint area. Alongwith the lower footprint area, the total weight of storage equipment and peripherals, i.e., supporting structures, required is lower. Thus, when the size and weight of storage systems need to be reduced compared to gaseous storage, the liquid hydrogen storage represents a viable option.

History of Liquefaction

The history of liquefaction dates back to the year 1812, when Michael Faraday carried out small-scale experiments in a test tube with the different gases in order to liquefy them and successfully liquefied ammonia and carbon dioxide. In the year 1840, Gorrie was able to produce ice by means of an air expansion engine.

In 1844, experiments were carried out on large scale for liquefying gases by Thilorier who employed mechanical compressor for this purpose. In 1845 with the use of hydraulic pump, large quantities of liquid and solid carbon dioxide were produced by Faraday. The most significant landmark in the history of liquefaction was the invention of the Joule–Thomson effect in the year 1852. Thereafter, Kirk invented the Stirling cooling system in 1860. The year 1863 marked another milestone, with the introduction of the concept of critical temperature, it was found that gases can't be liquefied without reducing the temperature of the gas below its critical temperature.

In 1877, Pictet and Cailletet individually produced a mist of liquid oxygen based on the knowledge of Joule–Thomson effect and critical temperature. Despite the use of multilevel cascade of pre-cooling, complete liquefaction could not

P. Sharma, *A Textbook on Hydrogen Energy*,
https://doi.org/10.1007/978-981-95-4601-5_13

be achieved. The complete liquefaction was accomplished in the year 1883 by Wroblewski and Olszewski, who successfully liquefied oxygen and nitrogen. In 1884, Wroblewski created a mist of hydrogen for the first time.

Another landmark invention occurred in 1892, when Dewar developed vacuum-insulated vessels for storage of cryogenic liquids, and thereafter the term *cryogenic* was coined in the year 1894. In 1895 Linde patented the air liquefaction using expansion valve and subsequently initiated the industrial-scale production of various gases by establishing the well-known company for liquefaction, Linde. In the same year, recuperative heat exchangers were introduced by Hampson for liquefaction applications, although the concept was developed in 1851 and improved in 1857. Air was liquefied on large scale using a throttle device by Linde, which is considered to be the beginning of a new era of liquefaction technology. In 1898, Sir James Dewar successfully liquefied hydrogen using a very small capacity liquefaction system. Prior to this, Cailletet (1877), Pictet (1878), Wroblewski (1884), and Olszewski (1895) had demonstrated hydrogen mist through separate experiments. In 1902, Claude developed a system for air liquefaction, which used expanders instead of throttling devices, to obtain better efficiency, and was carried out at low pressure. He established a company which is well known as Air Liquide. Cryogenic-based air separation technology was demonstrated by Linde in the year 1902.

In 1908, Onnes demonstrated liquefaction of helium for the first time. From 1910 onwards, large-size liquefiers were constructed. Large capacity liquid hydrogen and oxygen production facilities were developed for various applications after the year 1930.

In summary, during the period 1850–1900, the thermodynamics of liquefaction were established. During the period 1900–1950, various technologies for liquefactions were developed, improved, and led to the large-scale production of these liquefied gases for various industrial applications.

Challenges with Hydrogen Liquefaction

In hydrogen liquefaction, several challenges are involved.

The liquefaction of hydrogen occurs by reducing its temperature to 20 K (− 253 °C). Lowering temperature to the required extremely low temperature is an energy-intensive process.

The second major challenge is the ortho-to-para conversion, which is an exothermic process. Therefore, the heat released must be removed not only to cool hydrogen gas from ambient temperature to 20 K, but also to remove the additional heat released during liquefaction due to the o-p conversion. This makes the liquefaction process both energy intensive, and complex, requiring sophisticated thermal management systems. The process of ortho-to-para conversion is very slow, and in order to accelerate the conversion process, a catalyst is used. If the conversion is not complete, then during storage or at the time of transport significant boil-off losses may occur. Thus, ortho-to-para conversion reactors operating at multiple temperature levels can be incorporated to enhance the conversion rate. Various catalysts used for the catalytic conversion of ortho hydrogen to para hydrogen reported in literature are graphite, charcoal, supported chromium oxide, copper, silver and supported iron oxides, chromia on alumina, iron gel, chromium-doped silica, etc.

The third major challenge is the inversion temperature of hydrogen, which lies below the ambient temperature. As mentioned in previous chapter, in the Joule–Thomson effect, hydrogen exhibits the negative Joule–Thomson coefficient at ambient temperature. In order to cool hydrogen by throttling, its temperature needs to be reduced below its inversion temperature, after which the Joule–Thomson coefficient becomes positive and temperature reduces on throttling. The inversion temperature for hydrogen is around 200 K.

Besides these challenges, the large variation in the specific heat of hydrogen over the cryogenic temperature range poses another issue. The high sonic velocity of hydrogen impacts the design of the turbomachines used for the liquefaction process. Furthermore, cost, safety, and efficiency remain key bottlenecks in hydrogen liquefaction.

Ortho-to-Para Conversion

While measuring the heat capacity of hydrogen, it was observed in the year 1912 that the heating and cooling curves showed a hysteresis at cryogenic temperatures; however, the reason was not clear at that time. In the year 1927, the existence of two different spin isomers was proposed, and this hypothesis was used to explain the hysteresis observed in heat capacity curve of hydrogen. The existence of these spin isomers was proven experimentally for the first time in the year 1929. Hydrogen was found to exist in two different spin isomers, namely, ortho hydrogen and para hydrogen. Ortho hydrogen has parallel nuclear spins, while the para hydrogen has antiparallel spins. The two spin isomers have different sets of optical, magnetic, thermal, and calorimetric properties. The energy level of these two isomeric states is different, with para hydrogen being the lower-energy state, while ortho hydrogen is an excited state with higher rotational energy compared to the para hydrogen. Consequently, during the liquefaction of hydrogen, the high-energy spin state, i.e., ortho hydrogen, is converted to the lower-energy para hydrogen, and the extra amount of energy is released in the process of liquefaction; thus, the process is exothermic. The normal hydrogen at the normal temperature–pressure condition consists of 75% of ortho hydrogen and 25% of para hydrogen. However, during liquefaction at the boiling temperature of 20 K, the equilibrium composition of hydrogen is 99.8% of para hydrogen, and the remaining, i.e., 0.2% is ortho hydrogen. The equilibrium composition shows strong dependence on temperature and negligible dependence on pressure. The complete conversion of 100% ortho hydrogen to para hydrogen releases approximately 703 kJ/kg; however, the normal hydrogen which is 75% ortho and 25% para on conversion to equilibrium hydrogen releases 527 kJ/kg of heat. Both of these exceed the latent heat of vaporization for hydrogen, which is 446 kJ/kg.

Thus, during hydrogen liquefaction by reducing its temperature, the amount of heat released in this ortho-to-para conversion (527 kJ/kg) is sufficient enough to evaporate certain amount of liquid hydrogen (446 kJ/kg). If the stored liquid hydrogen contains ortho hydrogen (i.e., incomplete conversion or if fast cooling has been carried out) and the temperature is reduced to liquefy, then the o-p conversion will occur in liquid phase, which leads to evaporation of hydrogen. The process is known as boil-off, and requirement is to minimize it as much as possible.

The ortho-to-para conversion is an inherently slow process and follows second-order reaction kinetics in the absence of catalyst, requiring several days to week for completion. In order to enhance the rate of this conversion, catalyst is employed. In the presence of catalyst, depending upon the catalyst type, the temperature, and the pressure, this conversion follows the first-order reaction kinetics. However, at liquid hydrogen temperature (at 20 K), the conversion becomes a zeroth-order reaction and takes place very fast. Consequently, once the system is cooled down to liquid hydrogen temperature, the o-p conversion occurs rapidly and is being an exothermic reaction leading to boil-off which is undesirable while long-term storage. Therefore, most of the hydrogen liquefaction cycles incorporate ortho-to-para conversion units prior to liquefaction in order to minimize boil-off losses.

Another important phenomenon in the liquefaction of hydrogen is throttling. The throttling is common to most of the liquefaction cycles and occurs when non-ideality condition is induced. Such non-ideality condition is carried out by increasing the pressure, (i.e. compressing) or by lowering down the temperature (i.e. pre-cooling) or by employing a combination of both compression and pre-cooling. Since the inversion temperature of hydrogen lies below ambient temperature, and in order to get the maximum cooling effect, the gas needs to be cooled down below the maximum inversion temperature, thus making pre-cooling essential. This pre-cooling ensures, that the hydrogen experiences a temperature drop upon throttling.

Hydrogen Liquefaction Process

Prior to 1895, gas liquefaction processes primarily involved compression followed by chilling (temperature reduction) and final expansion through throttling, without the use of pre-cooling or mechanical expanding devices. In these early processes, hydrogen was commonly used as the working medium, and the saturated vapor generated after throttling was not recycled as a coolant. After the throttling process, a certain amount of liquid or mist used to be obtained. After the year 1895, the recuperative cooling was introduced, in which the low-temperature vapor itself acted as a coolant and was circulated to reduce the temperature of the incoming gas, significantly improving efficiency.

The ideal thermodynamics process for liquefaction is a combination of an isothermal compression and an adiabatic expansion. The adiabatic expansion can be carried out either by mechanically extracting the energy from the expanding fluid using an expansion device or expander, in which case the process is isentropic. If

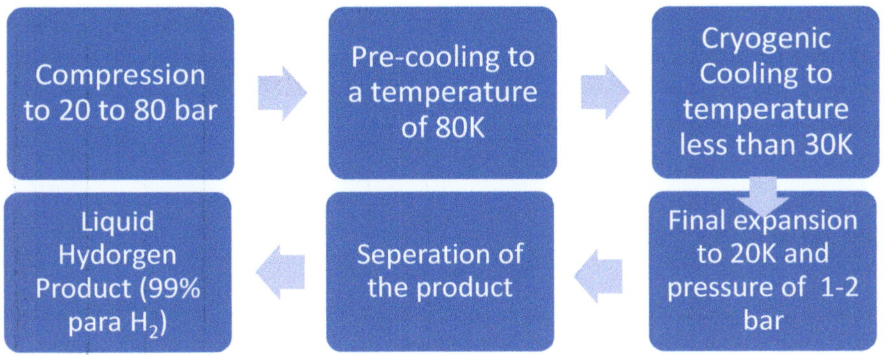

Fig. 13.1 Typical hydrogen liquefaction process

no energy is extracted during expansion, then the process is throttling as illustrated in Fig. 13.1.

The overall process of liquefaction includes compression to a higher pressure, such that the achieved pressure of compressed gas should be higher than the critical pressure. During compression, the gas temperature either increases due to compression or remains close to ambient if an efficient heat removal mechanism is employed. After compression, the next step is to reduce its temperature from ambient or higher than ambient temperature, to around 80 K, which can be achieved using liquid nitrogen cooling and this is called pre-cooling. The subsequent step is the final cooling to around 30 K. This can be achieved using recuperating cooling where the vapor generated after throttling itself, acts as a coolant, via a closed loop refrigeration cycle. This cyclic process involves two steps, an adsorption step involving removal of any impurities, followed by a catalytic conversion step to achieve ortho-to-para conversion. The final step is an expansion, which is carried out through an adiabatic expansion process, where the hydrogen gas at a higher pressure P_H is expanded to lower pressure P_L using the Joule–Thomson valve, i.e., throttling process, or using expansion through a turbine. During this process, the hydrogen partially liquefies. The amount of liquid obtained is collected and the remaining vapor fraction of the gas or the flash gas is allowed to exchange heat in a heat exchanger. This heat exchange cools the incoming hydrogen stream, and the vapor is then recycled back to the main feed stream for recompression. This intermediate process further cool down hydrogen from 80 to 30 K by means of a closed loop refrigeration cycle, and can have other steps like adsorption to remove impurities or a catalytic process to achieve o-p conversion. The liquid hydrogen obtained is stored at 20 K and 1–2 bar pressure. Primarily, these are the three major steps involved in hydrogen liquefaction; however, there are several ways of carrying out these steps using different components depending upon their energy consumption. The required energy can be reduced or optimized, by employing various liquefaction cycles, ranging from simple to complex configurations, including cascaded or hybrid liquefaction cycles. The overall efficiency of the liquefaction process depends on several parameters including the inlet pressure of feed hydrogen, efficiencies of the cooling system, i.e., both pre-cooling system and main cooling system, and efficiencies of the components involved in the liquefaction cycle, i.e., compressors, heat exchangers, expansion system, etc. The various liquefaction cycles are being discussed in the following section.

Liquefaction Cycles

Large-scale liquefaction of hydrogen can be achieved through different liquefaction cycles, the well-known cycles are listed below:

- the simple Claude process,
- Kapitza process, which is a modification of the simple Claude process,
- dual-pressure Claude process,
- pre-cooled Linde–Hampson cycle,
- pre-cooled dual-pressure Linde–Hampson cycle,
- helium pre-cooled Claude cycle,
- pre-cooled mixed-refrigerant cycle,
- cascaded cycle.

These cycles differ depending on the different components used in the cycles. The details of these cycles are elaborated in the following section.

Simple Claude Cycle

Developed in 1902, the process is named after Georges Claude, who used it for the first time to liquefy air using a reciprocating machine and had a combination of expansion engine and Joule–Thomson process. Initial low temperature is produced via isentropic expansion using an expander prior to isenthalpic expansion. This leads to reduction in temperature and makes the liquid nitrogen circuit not mandatory and have less of exergy destruction. However, as reported in the literature, the use of liquid nitrogen cooling can lead to high exergy efficiency other than pre-cooling alone[1]. The liquefaction process based on a simple Claude cycle ideally consists of an isothermal compression followed by an isenthalpic expansion. In the first step, the gaseous hydrogen is compressed near-isothermally in a multistage compressor, with intercooling in between. The compressed hydrogen gas then exchanges heat in a series of heat exchangers, leading to a progressive reduction in temperature. The expanding device installed between heat exchangers can't handle condensed liquid due to corrosion or degradation of the expansion engine; therefore, it is used only to reduce the temperature and liquefy a portion of the gas, which is then mixed with the remaining stream to reduce the overall gas temperature. Finally, when the temperature of the gas drops below its inversion temperature, the compressed gas is expanded using a throttle valve and undergoes Joule–Thompson expansion. Initially,

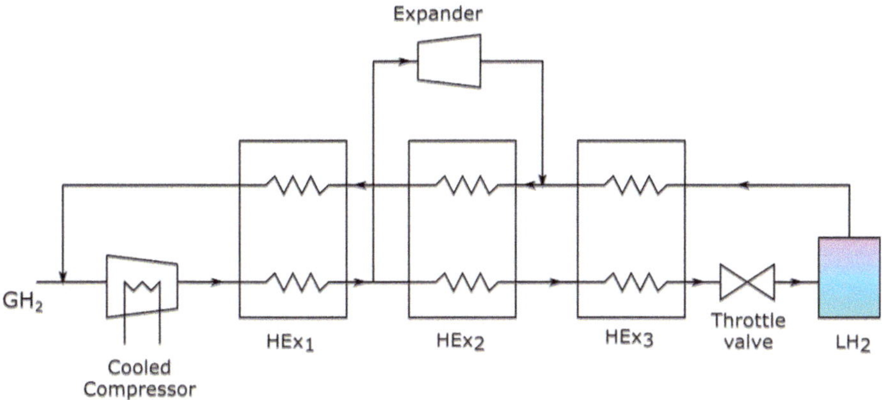

Fig. 13.2 Schematic of simple Claude cycle

this process results in the formation of a small quantity of liquid hydrogen, which is collected. The remaining cold vapor is recirculated to exchange heat with the incoming compressed gas, thereby reducing its temperature through recuperative cooling. After several cycles, the liquefaction of hydrogen is achieved. The schematic of the simple Claude cycle is illustrated in Fig. 13.2.

Thermodynamics of Simple Claude Cycle

Energy and mass balance can be done for the entire cycle. Considering the different points across the cycle, the mass balance of the control volumes can be carried out as shown in Fig. 13.3.

Mass balance for the control volume at steady state is being represented by following set of equations..

After the compressor, $\dot{m}_2 = \dot{m}_{1,}$ let us assume is

$$\dot{m}_2 = \dot{m}_1 = \dot{m}.$$

And the same mass is at point 3, and after the expanding device:

$$\dot{m}_3 = \dot{m}_e.$$

The cold gas which exits at 10, i.e., $\dot{m}_{10}$ can be given by:

$$\dot{m}_{10} = \dot{m}_1 - \dot{m}_f = \dot{m} - \dot{m}_f,$$

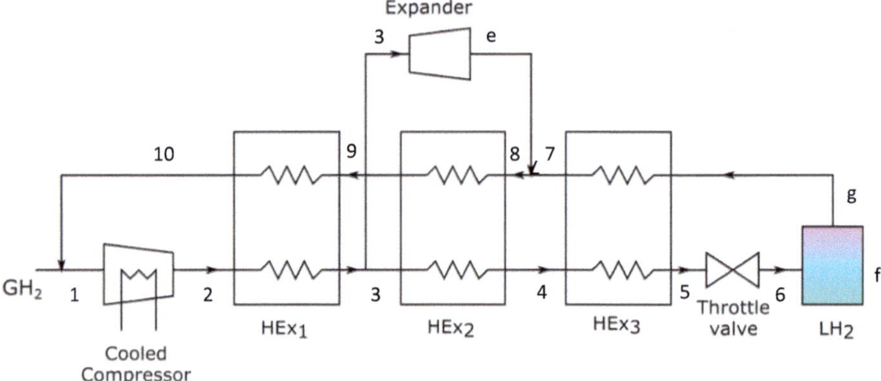

Fig. 13.3 Schematic of simple Claude cycle showing inlets and outlets for carrying out mass and energy balances

where $\dot{m}_1$, which is the mass rate of flow of gas entering and out of which some amount of the gas gets liquefied represented by $\dot{m}_f$. In the above equation $\dot{m}$ is in general represents the mass flow rate of stream "x" and this is expressed in unit kJ/s.

Similarly, the energy balance can be written between the different outlets. Considering the different hardware components such as compressor, heat exchangers, expander, and separator., the energy balance can be written using following equations.

$$\dot{m}_{10}h_{10} + \dot{m}_3 h_3 + \dot{m}_f h_f = \dot{m}h_2 + \dot{m}_e h_e.$$

Since $h_{10} = h_1$, and $\dot{m}_3 = \dot{m}_e$ thus above equation can be rewritten as follows:

$$(\dot{m} - \dot{m}_f)h_1 + \dot{m}_e h_3 + \dot{m}_f h_f = \dot{m}h_2 + \dot{m}_e h_e.$$

The mass fraction at the expander is denoted by "e" which is exiting from the expander and can be written as follows:

$$x = \frac{\dot{m}_e}{\dot{m}}.$$

A certain amount of this mass of gas gets liquefied, and this liquid yield is given by

$$y = \frac{\dot{m}_f}{\dot{m}} = \frac{(h_1 - h_2)}{(h_1 - h_f)} + x\frac{(h_3 - h_e)}{(h_1 - h_f)},$$

where h is specific enthalpy of the stream.

The net work which is required per unit mass of the compressed gas can be calculated, which represents the net work or the net energy consumption in the cycle and is denoted by w and is expressed in kJ/kg. This can be given by the net power consumption in the cycle divided by the mass flow rate

$$w = \frac{\dot{W}}{\dot{m}} = [T_1(s_1 - s_2) - (h_1 - h_2)] - x(h_3 - h_e),$$

where "s" stands for specific entropy of the stream x and is expressed in kJ/kg K.

The specific energy consumption for the process of liquefaction can be calculated and expressed as the energy consumed per unit mass which gets liquefied, i.e., energy required in liquefying 1 kg of hydrogen, and is in unit of kJ/kg or kWh/kg of liquid hydrogen; thus, the specific energy consumption is the net power consumption in a liquefaction cycle divided by mass flow rate of the product, i.e., liquid hydrogen $\dot{m}_y$:

$$\text{SEC} = w_l = \frac{\dot{W}}{\dot{m}_f} = \frac{\dot{W}}{\dot{m}_y}.$$

In this equation, "y" is the fraction which got liquefied.

The coefficient of performance, is represented by the cooling which is provided by the working fluid to the net input power in the cycle:

$$\text{COP} = \frac{\dot{m} \times \left(h_{\text{feed}} - h_f\right)}{\dot{W}},$$

where h_{feed} is the enthalpy of the feed

The exergy analysis of a thermodynamic system provides useful information like the amount of useful energy or loss of work potential in a thermodynamic process, which is important while comparing various thermodynamic systems. The exergy efficiency can be defined as ratio of theoretical minimum specific energy consumption to the actual specific energy consumption and represented as:

$$\eta_{\text{ex}} = \frac{\dot{X}_p - \dot{X}_f}{\dot{W}},$$

where η_{ex} is the exergy efficiency of the liquefaction process, and $\dot{X}_p - \dot{X}_f$ is the difference of exergy flow rate of product liquid hydrogen and gaseous hydrogen feed. Similarly, another important quantity which compares the amount of work done in the liquefaction process under ideal and real conditions is the figure of merit, which is defined as the ratio of ideal work done to the actual work done in the liquefaction process.

In any liquefaction cycle, the yield of liquid hydrogen obtained can be defined as the ratio of liquid hydrogen mass flow rate to the mass flow rate of input hydrogen, i.e.,

$$\text{Yield (in \%)} = \frac{\dot{m}_{\text{liq hydrogen}}}{\dot{m}_{\text{feed hydrogen}}}.$$

Several variations of the simple Claude cycle are described in this section. The variation of simple Claude cycle which is the Kapitza process which uses reciprocating expansion engines, and the process is specially used for large-scale hydrogen production.

In the case of dual-pressure Claude process, instead of compression to a single pressure before the throttling, two compressors are used and compression is carried out at two different pressure levels. The advantage of dual-pressure compression is that the minimum work required for compression is proportional to the pressure ratio, rather than the absolute pressure difference in the cycle. As described in the previous chapter, the compression energy varies with the logarithm of the compression ratio. For example, in a simple Claude cycle with a single primary compressor and a compression ratio of 200:1, the logarithmic term corresponds to a value of 5.3 in the compression energy. However, in the same Claude cycle with dual-pressure levels using two compressors and an effective compression ratio of 200:5, the logarithmic term is reduced to 1.39. This reduction indicates an improvement in performance and a decrease in energy consumption of approximately 26.2% compared to the simple Claude cycle [2]. In an ideal simple Claude cycle, the minimum work of compression decreases because it is proportional to the pressure ratio. Consequently, the expander flow rate, which is also proportional to the minimum work, is reduced. However, the major challenge with the Claude process is related to the use of large number of components, like an expander, and several heat exchangers which increases the capital expenditure (CAPEX) of the system. An alternative liquefaction cycle is the Linde–Hampson cycle, which typically requires lower CAPEX but exhibits higher operating expenditure (OPEX).

Other variations to the simple Claude cycle include liquid nitrogen pre-cooled Claude cycle, and pre-cooled dual-pressure Claude cycle.

In the liquid nitrogen pre-cooled Claude cycle, gaseous hydrogen is first compressed and then pre-cooled using liquid nitrogen before entering the heat exchangers, rest of the process remains unchanged. The exergy efficiency of a pre-cooled Claude cycle is higher, with a higher liquid hydrogen yield, and lower specific energy consumption. Literature reports indicate that for this cycle, the specific energy consumption is 28–39 kWh/kg of liquid hydrogen, and the liquefaction yield increases from 8% in case of simple Claude cycle to 16–20% at an input gaseous hydrogen pressure of 10–30 bar. [3].

Another configuration involves compressing the inlet gaseous hydrogen at two pressure levels, followed by expansion at two corresponding pressure levels. This configuration is referred to as the pre-cooled dual-pressure Claude cycle and includes liquid nitrogen pre-cooling prior to heat exchange and after compression. The primary advantage of this cycle is the lower specific energy consumption, which has been reported to be 10.85 kWh/kg of liquid hydrogen, with an exergy efficiency of 36% compared to 18% for simple Claude cycle and 50–70% in case of pre-cooled Claude cycle. The Claude cycle is employed in most of the large-scale hydrogen liquefaction plants worldwide and has advantage in terms of OPEX.

Linde–Hampson Cycle

Compared to the simple Claude cycle, the pre-cooled Linde–Hampson cycle has a lower CAPEX due to lesser number of components involved, but it has a higher operational cost. The reason for higher OPEX is due to the higher power consumption as compared to the simple Claude cycle. The cycle was invented in 1895 by Carl von Linde and William Hampson to liquefy air and is the simplest and basic liquefaction cycle as shown in Fig. 13.4.

In the Claude cycle, an expanding device is employed; however a liquid nitrogen pre-cooling step is involved in the Linde–Hampson cycle. However, for the liquefaction of hydrogen- which involves a gas with an inversion temperature below ambient - liquid nitrogen pre-cooling is essential. The use of liquid nitrogen loop not only reduces the gas temperature below its inversion temperature, but also improves the yield of liquid hydrogen collected during the complete cycle. The increase in liquid yield is attributed to the reduction in enthalpy of the high-pressure gas at the warm end of the heat exchanger, i.e., of the gas entering into the heat exchanger by cooling. An additional advantage is the reduction in the required high pressure prior to throttling in the liquefaction process.

In the liquid nitrogen pre-cooled Linde–Hampson cycle as shown in figure 13.5, hydrogen is first compressed and the heat of compression is removed. The compressed gas then passes through the heat exchanger, where its temperature drops below the inversion temperature using a liquid nitrogen loop. Since the inversion temperature depends on the pressure, accounting for pressure during pre-cooling is crucial. Following liquid nitrogen cooling, the gas is expanded using a throttling device—a Joule–Thomson valve undergoing an isentropic expansion. During this process, certain amount of liquid hydrogen is obtained which can be collected, while the remaining cold gas is recirculated and exchanges heat with the incoming compressed gas, acting as a coolant and reducing the temperature of the incoming compressed

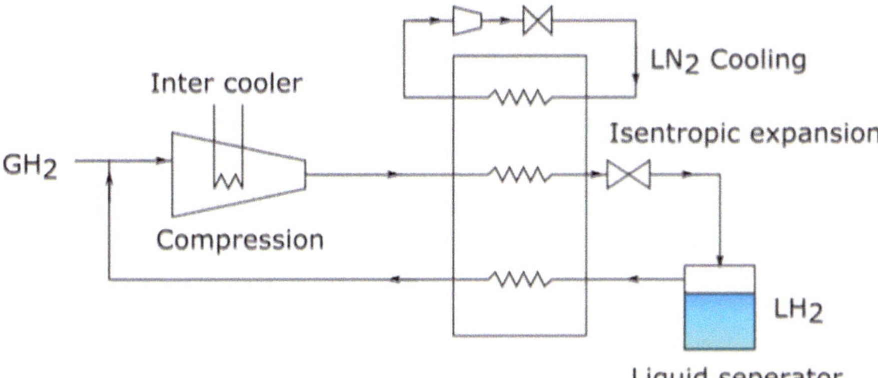

Fig. 13.4 Schematic of a typical Linde–Hampson cycle

feed stream prior to throttling. After heat exchange, the gas is recompressed and undergoes the same cycle.

The pre-cooling step in the Linde–Hampson cycle improves the performance and reduces the required energy input. The specific energy consumption for an ideal Linde–Hampson cycle is 16.24 kWh/kg, while for the Claude cycle it is 22.1 kWh/kg. The performance of Linde–Hampson cycle can be further enhanced by employing two-stage compression instead of single-level compression before throttling, and similarly, two-stage expansion. In this configuration, the first expansion is carried out through a Joule–Thomson valve to an intermediate pressure, producing a certain amount of liquid hydrogen that is collected. The remaining gas then undergoes second-stage compression and is subsequently expanded to atmospheric pressure through another Joule–Thomson valve, yielding additional liquid hydrogen. Thus, the use of a dual-pressure Linde–Hampson cycle results in lower compression work and higher hydrogen yield. When compression is carried out at two different pressure levels, this cycle is called dual-pressure Linde–Hampson cycle. This approach can further reduce the energy requirement in the liquefaction cycle. Pre-cooling before compression leads to the so-called pre-cooled dual-pressure Linde–Hampson cycle, for which the specific energy requirement is 12.12 kWh/kg. However, this ideal specific energy consumption is higher than that of the ideal dual-pressure simple Claude cycle, which is 6.66 kWh/kg. When non-ideal effects such as heat exchanger inefficiencies, ortho-to-para conversion, compressor and expander inefficiencies, liquid nitrogen consumption, and insulation losses are considered, the specific energy consumption of the simple Claude cycle increases to approximately 12.26 kWh/kg of hydrogen.

Studies on liquid nitrogen pre-cooled cycle have shown the specific energy consumption of 72–80 kWh/kg of liquid hydrogen with the exergy efficiency varying between 4.5 and 5% depending upon the feed hydrogen gas pressure [4].

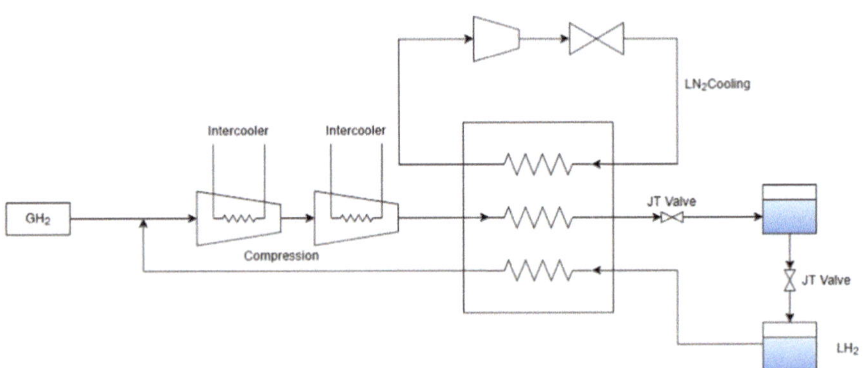

Fig. 13.5 Schematic of a typical pre-cooled Linde–Hampson cycle

Helium Pre-cooled Cycle

The cost breakdown of hydrogen liquefaction indicates that roughly 50% of the total cost is associated with the capital cost, i.e., CAPEX and remaining 50% to the required energy input, operating and maintenance cost. Among the major cost contributors to CAPEX is the compressor and reducing the size of compressor can lead to cost saving. The reduction in outlet pressure, which is related to compressor capacity, can also reduce the capital cost involved in the hydrogen liquefaction process. In order to reduce the required pressure output from the compressor, the required pre-cooling conditions can be altered by reducing the temperature prior to throttling. Consequently, the use of auxiliary refrigeration systems in liquefaction in addition to the liquid nitrogen cooling reduces the compressor output pressure requirement. This additional pre-cooling requirement can be achieved by introducing a helium pre-cooling loop, although the helium refrigerated hydrogen liquefaction cycle is much more complicated than pre-cooled Claude cycle. The role of helium in the cycle is to serve as working fluid, which is not condensed. Thus, in the overall cycle, the helium refrigeration auxiliary system is integrated with the liquid nitrogen pre-cooling to reduce the temperature of the incoming hydrogen prior to throttling. The average specific energy requirement for such a helium pre-cooled cycle is high and has a value of around 39.4 kWh/kg of liquid hydrogen. In this cycle, the helium acts as refrigerant and does not liquefy or condense, but is cooled to the temperature lower than that of hydrogen.

Joule–Brayton Pre-cooled Cycle

In the Joule–Brayton cycle, helium or mixed refrigerant is used in the auxiliary refrigeration system to lower the temperature of the working gas helium below that of hydrogen. In this cycle, the refrigerant is first compressed and then expanded in an isentropic expander to achieve temperatures lower than hydrogen. The final expansion of the working fluid is carried out in an expansion engine to provide the required cooling. The working fluid remains in the gas phase throughout the process and does not undergo liquefaction. The cold refrigerant then exchanges heat with hydrogen to produce liquid hydrogen. A series of heat exchangers may be employed; however, the key requirement is that the heat exchanger operating at the lowest refrigerant temperature exchanges heat with hydrogen at its saturated vapor point. Several configurations of this cycle are possible, and the liquid hydrogen yield can be further enhanced by employing two-stage expansion or by incorporating liquid nitrogen pre-cooling.

Mixed-Refrigerant Pre-cooled Cycles

When a combination of various gases and vapors is used as working fluid in the refrigeration system, the need for a capital-intensive high-pressure compressor can be avoided. In such configurations, a mixed refrigerant is employed to pre-cool hydrogen during the liquefaction process. Using a mixture allows the component with the lowest boiling point to also be liquefied. Consequently, isentropic expanders are used, instead of throttling (isenthalpic) expansion valves to achieve required cooling effect. A mixed-refrigerant process can be integrated with a simple Claude cycle, such that the mixed-refrigerant system serves as the pre-cooling section for the liquefaction cycle. However, the entire cycle involving mixed-refrigerant pre-cooling system is highly complex and presents significant design challenges. Major requirement is a thorough understanding of the thermodynamic properties of the mixed refrigerant, as well as managing the two-phase flows through multiple heat exchangers. The key advantage is that, by employing multiple stages of refrigerant cycles and using a mixture of refrigerant in each stage, the overall system efficiency can be enhanced. Literature reports indicate the efficiencies as high as 55.47%, with specific energy consumption as low as 1.1 kWh/kg of liquid hydrogen. Variations between different mixed-refrigerant pre-cooled cycles arise from the choice and combination of refrigerants, the number of components, and the number of hydrogen Brayton cycles employed. An optimized configuration reported in the literature, with a five-component refrigerant and four-Brayton cycles, achieved an efficiency of 49% and a specific energy consumption of 5.9 kWh/kg of liquid hydrogen.

Cascade Cycles

Cascade cycles are an extension of pre-cooling systems, in which an independent refrigeration cycle pre-cools another cycle, with two or more refrigeration cycles used to pre-cool the hydrogen gas stream. The advantage of such cascade systems is that they more closely approximate ideal liquefaction cycles, thereby achieving a lower-energy requirement for liquefaction. This multilevel cooling approach helps in achieving wider temperature range for cooling, resulting in higher efficiency, improved performance, and increased liquid hydrogen yield.

Collins Cycle

The Collins process is a modification of the Claude process, developed specifically for helium liquefaction and sharing many similarities with the Claude cycle. The primary difference lies in the inclusion of a greater number of expansion devices and

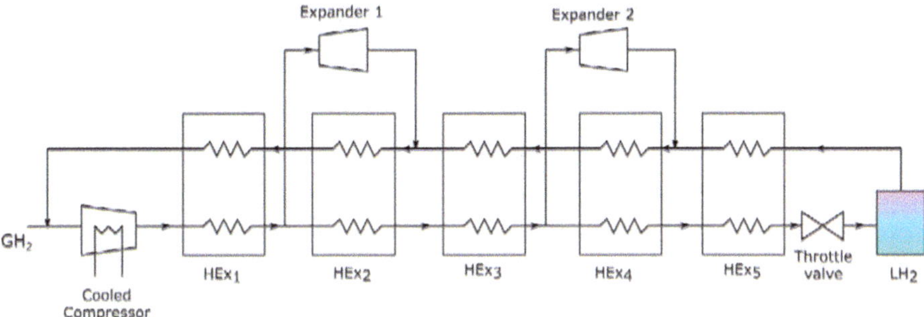

Fig. 13.6 Schematic of Collins process

heat exchangers in the liquefaction cycle, and schematic diagram representing the Collins process is shown in Fig. 13.6.

Gaseous hydrogen is first compressed and the heat released is removed through a series of heat exchangers to reduce its temperature below the inversion temperature before being throttled through a Joule–Thomson valve. During liquefaction, part of the gas undergoes expansion in two different expanding devices operating at different temperature levels, to achieve temperature below the inversion temperature. Once the inversion temperature is reached, the gas is expanded using a Joule–Thomson device, resulting in the formation of a certain amount of liquid hydrogen, obtained in the process is collected. The remaining cold gas is recirculated through the heat exchangers, enhancing cooling and helping in achieving temperature below the inversion temperature.

As the number of components in the cycle increases, both the complexity of the liquefaction process and the overall cost of liquefaction rise.

Magnetic Refrigeration

In magnetic refrigeration-based liquefaction process, a magnetic field is applied to magnetize and demagnetize the magnetic material, resulting in magnetocaloric effect. This effect corresponds to a reversible change in temperature of the magnetic material subjected to change in magnetic field, thus producing low temperatures. Gadolinium is used as working material, and the cycle involves four steps: adiabatic magnetization, isothermal magnetization, adiabatic demagnetization, and isothermal demagnetization. During the adiabatic magnetization step, the magnetic material, inside a thermally insulated system, is subjected to external magnetic field, causing magnetic dipoles to align. This alignment results in a decreased entropy and heat capacity. The process of magnetization is continued under isothermal condition through the process of isothermal magnetization, wherein the heat is removed with the help of another fluid under a constant magnetic field. In this step, isomagnetic enthalpy transfer occurs, such that the alignment of dipoles is not affected, despite

Fig. 13.7 Processes involved in magnetic refrigeration

the removal of heat. The schematic diagram representing the processes involved in magnetic refrigeration is shown in Fig. 13.7.

Next step involves demagnetization under adiabatic condition, wherein the enthalpy remains constant with decreasing magnetic field leading to cooling of the material. The final step is isothermal demagnetization, during which the magnetic field is kept constant, and magnetic material exchanges cooling effect with hydrogen. The principle behind the liquefaction process is that during application of magnetic field, the alignment of the dipoles increases, leading to decrease in the magnetic entropy and thus releases heat. While during demagnetization, the magnetic alignment reduces, leading to increasing in the magnetic entropy and absorption of heat from the surrounding environment or medium. This heat absorption produces a cooling effect in the surrounding environment, which is utilized for hydrogen liquefaction. Since the use of compressor can be completely eliminated, the process is economical and simple. Furthermore, the system using the process is compact, light in weight, easier maintenance and no noise levels.

In summary, a comparison of various liquefaction cycles indicates that the Linde–Hampson cycle has the highest specific energy requirement and lowest efficiency. The Claude cycle performs better than Linde–Hampson cycle in terms of both SEC and efficiency. The more advanced cycles discussed in this section, such as mixed-refrigerant cycles, Collins cycles, and Brayton cycles, outperform the basic Linde–Hampson and Claude cycles in terms of overall performance. Among these, the mixed-refrigerant cycle demonstrates the highest efficiency along with the lowest specific energy consumption. The final cycle mentioned in this section, magnetic material-based refrigeration, although has economic benefit and simple design, but is suitable for small-scale liquefaction.

Global Hydrogen Liquefaction Facilities

Several hydrogen liquefaction plants are currently operational worldwide and are located in various regions, like in India, USA, Japan, Canada, Holland, France, China, and Germany. To cite few representative examples, one of such plants in USA, being installed by Praxair has specific energy consumption between in the ranges of 12.5–15 kWh/kg of liquid hydrogen.

Another plant located in Ingolstadt, Germany, with a capacity of 4.4 tpd, operates on the principle of pre-cooled Claude cycle, have a specific energy consumption of

13–15 kWh/kg of liquid hydrogen produced. This value of specific energy consumption is still very high. The USDOE has set target of achieving a specific energy consumption of 6 kWh/kg of liquid hydrogen for a large-sized plant of capacity 300 tpd. Although the theoretical value of specific energy consumption, based on ideal set of conditions, is about 2.7 kW_{el}/kg of liquid hydrogen at 25 bar, but the actual values under real operating conditions—including all losses and irreversibilities—are significantly higher and currently exceed the USDOE targets. Since the hydrogen liquefaction process is highly cost and energy intensive, approximately 50% of the total cost is associated with capital investment, and the rest is due to operational and maintenance costs. To reduce the specific energy consumption the possible options includes, integrating various processes and improving the cycle performance, adding more and more components to enhance performance. However, adding additional components to improve performance often increases system complexity and overall cost. A lot of literature is available, where different cascaded cycles have been modeled, and the performance and SEC have been computed theoretically.

There are global leading companies such as Praxair (6–35 tpd capacity, many in USA, based on modified pre-cooled Claude cycle), Air Products and Linde (pre-cooled modified Claude cycle, capacity of 4.4 tpd in Germany), having several large-capacity hydrogen liquefaction installations across the globe. Typical average specific energy consumption of the large-scale hydrogen liquefaction plants lies in the range of 10–14 kWh/kg of liquid hydrogen.

In order to reduce the specific energy consumption, there is clear requirement to have more energy-efficient and cost-effective process.

There are different ways to improve the performance, reduce SEC, and improve exergy efficiency and cost-effectiveness. Some of these measures includes: Larger the plant size, lesser is the SEC; reuse or reliquefaction of the boil-off loses occurring during storage and transport; use of any waste cooling effect to provide the cooling for liquefaction cycles; use of more efficient components including compressors, heat exchangers, and insulation systems; use of better refrigerants; and replacement of Joule–Thomson valves with expansion devices, as expanders generally demonstrate superior performance.

Solved Problem.1 Compute the installed liquefier capital cost for liquefied H_2 production having:

- Liquifier Design capacity, **C = 200 MT/day**
- Number of liquefiers, **N = 1**
- Overall Chemical Engineering plant index, **I = 1.16**
  (MT ≡ Metric ton).
  As liquefier design capacity is given in MT/day.

Soln:

Given:

- Liquefier Design capacity, **C = 200 MT/day**
- No. of liquefier, **N = 1**

- Overall chemical engineering plant index, **I = 1.16**

$$\Rightarrow \textbf{Installed liquefier cost} = N \times 10^6 \times 5.6 \times C^{0.8} \times I$$
$$= 1 \times 10^6 \times 5.6 \times 200^{0.8} \times 1.16$$
$$= 450.27 \text{ million } \$$$

Installed liquefier capital cost for producing 200 MT-H$_2$/day = 450.27 million $

Solved Problem.2

Calculate the liquefier design capacity such that:

- Installed liquefier capital cost = **357 million $**
- No. of liquefier, **N = 1**
- Overall chemical engineering plant cost index, **I = 1.16**

Soln:

Given:

- Installed liquefier capital cost = **357 million $**
- No. of liquefier, **N = 1**
- **I = 1.16**

To find:

Liquefier design capacity, **C** in **MT/day**

Soln:

Installed liquefier capital cost $= N \times 10^6 \times 5.6 \times C^{0.8} \times I$

$$357 \times 10^6 = 1 \times 10^6 \times 5.6 \times C^{0.8} \times 1.16$$

$$C^{0.8} = \frac{1 \times 10^6 \times 5.6 \times 1.16}{357 \times 10^6}$$

$$C^{0.8} = 54.957$$

$$0.8 \ln(C) = \ln(54.957)$$

$$\ln(C) = \frac{\ln(54.957)}{0.8}$$

$$\ln(C) = 5.008$$

$$C = e^{5.008}$$

$$C = 149.63\,\text{MT/day}.$$

References

1. M. Aziz, Liquid hydrogen: a review on liquefaction, storage, transportation and safety. Energies **14**, 5917 (2021)
2. M. Aasadnia, M. Mehrpooya, Large-scale liquid hydrogen production methods and approaches: a review. Appl. Energy **212**, 57–83 (2018)
3. T. Zhang, J. Uratani, Y. Huang, L. Xu, S. Griffiths, Y. Ding, Hydrogen liquefaction and storage: recent progress and perspectives. Renew. Sustain. Energy Rev. **176**, 113204 (2023)
4. Ting He, Wensheng Lin, Design and analysis of dual mixed refrigerant processes for high-ethane content natural gas, liquefaction, Chinese Journal of Chemical Engineering, **29**, 354–364, 1004–9541 (2021)

Chapter 14
Liquid-State Hydrogen Storage Tanks

In the previous chapter, different liquefaction cycles were considered. Once the liquid hydrogen is produced through the liquefaction cycles, it need to be stored. The requirement of storage of liquid hydrogen can be for stationary applications, such as at the site of hydrogen production or at facilities where it is utilized for power generation. Storage of liquid hydrogen may be for vehicular applications, such as onboard storage on a fuel cell vehicle. Besides these, the requirement can be for large-scale storage in order to enable transportion of liquid hydrogen between different countries or continents via maritime ships. Transportation of liquid hydrogen on a relatively smaller scale can be carried out through trucks via road, for which appropiate storage containers are required.

There are specific requirements in terms of materials and system design for storing liquid hydrogen due to various challenges associated with this process.

The challenges associated with storing liquid hydrogen include:

Maintaining the 20 K temperature—Since liquid hydrogen needs to be stored at 20 K, which is an extremely low temperature, the production of liquid hydrogen is both energy- and cost-intensive. Besides, the requirement is of cryogenic infrastructure to maintain 20 K temperature, for the storage tanks, pipelines, other peripherals and accessories through processes like chilling. Thus, the chilling process is essential for the entire hydrogen infrastructure involved, and it is essential that all the accessories are able to operate and have long life under such low temperature conditions. The materials used for construction should be capable of withstanding such a low temperature, without undergoing major mechanical degradation in the process. At the same time, icing should not occur on the various storage components, valves, piping, vents, connectors, etc. There should not be any condensation on the outer surface because that would increase the external pressure and mechanical load.

Boil-off—Since the temperature difference between ambient and the stored liquid hydrogen is huge, heat inflow into the system over time is inevitable, especially when long duration storage or long-distance transportation is involved. With the heat

P. Sharma, *A Textbook on Hydrogen Energy*,
https://doi.org/10.1007/978-981-95-4601-5_14

leakage from the surroundings, the liquid hydrogen partially evaporates, and starts to build up pressure in the storage vessel through a phenomenon termed as pressure buildup.

Once this pressure reaches the maximum allowable holding pressure of the tank as per the design specifications, the pressure needs to be reduced inside the tank, or the gaseous hydrogen needs to be released. To release this gaseous hydrogen, a pressure relief valve is essential in liquid hydrogen storage tanks.

Expansion—Considering the expansion ratio, if liquid hydrogen converts into gaseous hydrogen, then the expansion ratio is 848. Thus, if the liquid hydrogen, completely vaporizes, i.e., from liquid H_2 at 1 atmosphere, the pressure that would build up inside a storage vessel will be about 172 MPa, which is substantial. The LH_2 storage vessels are superinsulated and designed to hold and maintain low temperatures such as 20 K, but not very high pressures; the maximum allowable holding pressure is typically around 5–6 bar. The increase in the pressure due to evaporation of LH_2 and conversion to gaseous hydrogen is called the pressure buildup phenomena, and this elevated pressure must be released, leading to the loss of hydrogen. Once the pressure buildup reaches the threshold value, it needs to be released for the safety issues, and the gaseous hydrogen released in to the environment, the process is called as boil-off.

The requirement for liquid hydrogen storage is especially to reduce the boil-off losses, not only to avoid the energy loss and economic penalities, but also to address safety concerns. The possible ways to manage the boil-off loss is to collect and re-liquefy the boil-off gas, if the hydrogen liquefaction plant and the respective hydrogen storage vessel are co-located. Another way is to utilize this boil-off gas released, for certain applications; for example, in ships it could be tapped and used to provide a certain fraction of propelling power requirement, or it can be collected and stored in alternative media such as metal hydrides and utilized later for various applications.

Boil-off Loses

The various reasons that gives rise to boil-off process are as follows:

(a) **Ortho-to-para conversion**—As described in the previous chapter, the normal hydrogen at NTP has 75% of ortho and 25% of para hydrogen, whereas the equilibrium composition at 20 K is approximately 99.8% para hydrogen and with the remaining 0.2% being the ortho hydrogen. The conversion from ortho to para during liquefaction is a slow and exothermic process. However, if the liquefaction to 20 K is carried out rapidly without completing the conversion process, by simply reducing the temperature very quickly, then subsequent conversion from ortho to para will occur over the time, releasing heat and leading to evaporation of hydrogen, which in turn causes pressure buildup, boil-off and loss of hydrogen. In order to avoid this hydrogen loss due to boil-off, it is essential that the o-p conversion is accelerated through the use of catalysts and use of ortho-to-para converters throughout the liquefaction and storage process.

(b) **Heat leakage from the surrounding environment**—Since the liquid-state storage is maintained at 20 K, and the outside ambient temperature is much higher than this, heat inflow or leakage to the storage system is unavoidable. There are several factors on which this heat leakage depends, including the effectiveness of the cryogenic insulation, the shape, and size of the storage vessel, i.e., surface-to-volume ratio. There are various considerations in terms of materials selection and design that can lead to efficient thermal management, improved insulation, and reduced hydrogen loss. The selection of material used for insulation plays a critical role; the choice of material should be such that it exhibits low thermal conductivity. At the same time, all the heat transfer methods which could lead to heat leakage should be addressed. Thus the design of such tanks includes double-walled structures with a vacuum in between the two walls to reduce the conduction and convection modes of heat transfer. The use of multilayer insulation helps to reduce the radiation mode of heat transfer into the liquid hydrogen storage tank. Besides these measures, certain passive measures including the incorporation of cryogenic fluid-based external shields, are also being used to prevent heat inflow and enhance the dormancy period of storage.

(c) **Thermal Stratification**—Due to the heat leakage mechanisms mentioned above, the evaporation of liquid hydrogen can occur, leading to pressure buildup. The non-uniform temperature distribution in the tank arises due to natural convection and heat inflow, leading to a temperature gradient from the tank wall to the depth inside. This can lead to the warmer liquid being closer to tank walls and on the top of the liquid, at surface. This warmer layer, due to its lower density, rises, while the colder liquid sinks to depth, causing further natural convective flow and the creation of layers with different temperatures. The natural convection within the liquid hydrogen storage facility together with buoyancy effects and pressure buildup, as well as density difference between the gaseous and liquid hydrogen, all lead to the formation of a stable vapor—liquid interface. All these processes lead to a different set of thermodynamic properties than liquid hydrogen and contribute toward boil-off and pressure variations. Thus, the need is to have an appropriate design to avoid such a phenomenon.

(d) **Sloshing**—Sloshing of the liquid hydrogen refers to the movement of stored liquid hydrogen inside the tank, which may arise due to motion of the tank (while carrying from one place to another), vibrations or shaking of liquid hydrogen inside the vessel, or acceleration or deceleration of the insulated cryogenic container on a truck. All these movements can lead to the transfer of kinetic energy to thermal energy, resulting in boil-off. Besides the disruption of the liquid vapor phase and natural layering, by increasing the involved surface area at the interface can disrupt the thermal stratification. Such movements can cause dynamic pressure, posing safety concerns, and can impact the structural integrity of the tanks. The low density and high volatility of liquid hydrogen can further increase its susceptibility to the sloshing phenomenon. Thus, it is essential to take care of all these aspects while transporting or storing liquid hydrogen. To avoid the issues associated with the sloshing phenomenon, vibration control methods, such as sloshing dampers need to be incorporated while designing the tanks and storage infrastructure.

(e) **Flashing of liquid hydrogen**—The transfer of liquid hydrogen from one tank to another can lead to phase change and accelerate the evaporation. This transfer of liquid hydrogen from higher to lower pressure, accompanied by a sudden drop in pressure, leads to partial evaporation and is known as the flashing phenomenon, which leads to boil-off losses.

Liquid-State Storage of Hydrogen

Considering all the above issues and constraints, the selection of appropriate materials for storing at liquid hydrogen temperatures, proper design of the storage vessel is essential.

1. **Choice of Material for Liquid Storage**

Materials requirement for liquid-state storage of hydrogen are stringent. A crucial design consideration is that the material should be resistant to hydrogen embrittlement, which affects the material at the atomic level and can lead to macroscopic effects, ultimately resulting in mechanical failure of the system. However, the chances of hydrogen embrittlement are significantly reduced at liquid hydrogen temperature (low temperature) as compared to those in the case of compressed gas storage (high pressure). Besides hydrogen embrittlement, the storage vessel material should be resistant to hydrogen permeation through the tank walls or accessories, which is more prevalent in composite materials. Tank materials should be robust against thermal fluctuations when handling liquid hydrogen, should have high mechanical strength, and exhibit good heat and fire resistance.

The most widely used material for liquid storage vessel construction is stainless steel; the different alloy compositions and grades of stainless steel can be used to ensure that the challenges mentioned above are being addressed. For example, SS 321 with titanium can be used to have good heat and fire resistance, SS316L with molybdenum can be used for marine applications due to good chlorine ion protection. Besides the selection of appropriate grade and composition of stainless steel, the proper choice of material used for the surface coatings, and processes like ion implantation, cathodic protection, and laser shot peening have been reported in literature to address the hydrogen embrittlement-related challenges.

2. **Construction and Design of Liquid Hydrogen Storage Vessel**

The challenges associated with liquid hydrogen storage, arise from the vast temperature difference between ambient and the stored hydrogen, emphasizing the need for efficient thermal management in order to prevent heat leakage into the storage system. There are various factors which contribute toward the heat leakage, like the surface-to-volume ratio of the storage vessel and the use of superinsulation to prevent all three modes of heat transfer. The requirement is to consider these factors when selecting appropriate insulation materials and designing the storage vessels.

Various shapes have different surface-to-volume ratios for a given volume of liquid hydrogen stored. A low surface-to-volume ratio means that a lower surface area is exposed to environmental conditions, resulting in less heat leakage and reduced hydrogen loss due to boil-off. Cylindrical vessels have a higher surface-to-volume ratio compared to a spherical tank. Thus, spherical tanks have lower daily boil-off losses; besides the larger the volume of liquid hydrogen storage, i.e., in the case of higher capacity storage, the boil-off losses are lower. The daily boil-off can be reduced to even less than 0.01% per day for larger installations. Spherical tanks have another advantage compared to cylindrical tanks, there is uniform stress distribution, while in cylindrical tanks there are stress concentrations and higher pressure areas. For cylindrical tanks, to achieve better strength, the required thickness of the tank will be higher. Thus, spherical tanks and larger capacity installations will have lower boil-off losses and a higher dormancy period. However, the construction of spherical tanks is complicated, so cylindrical tanks are used sometimes, which can be kept either vertically or horizontally.

The second important consideration in the liquid hydrogen storage is the tank design and insulation used in order to reduce the heat leakage and boil-off. The design of liquid hydrogen storage vessels consists of double-walled tank with a vacuum and a passive insulation material between the two walls. Both active and passive insulation are required and have been widely studied in the literature. The choice of insulation materials plays a key role, wherein the requirement is to eliminate all three modes of heat inflow from the outside environment to the storage tank. Between the two walls of the tank is multilayer insulation for which various materials that can be used. Multilayer insulation consists of low-emissivity reflectors and low thermal conductivity spacers arranged in layers within a high vacuum region between the two walls of the vessel. The different multilayer insulation materials include hollow glass microspheres, perlite, aerogel, various reflector materials such as aluminized polyester films, and spacer materials like nylon net, spray-on foam insulation, and variable-density multiple-layer insulation. The thermal conductivity of the multilayer insulation is very low in the range of 10^{-5}–10^{-6} W/mK, which can be further reduced by optimizing the thickness of the space along with the number of reflector layers, thereby reducing the overall heat leakage into the storage tank. The other low thermal conductivity materials used include spray on foam insulation having thermal conductivity in the range of 10^{-2}–10^{-3} W/mK or hollow glass microspheres with thermal conductivity of 10^{-3}–10^{-4} W/mK. Besides, the use of passive insulation materials, combined with active insulation technology, such as vapor-cooled shield or liquid nitrogen cooling, is used in the shield or cryocooler, to enhance the thermal insulation performance.

During filling, emptying, and refilling, physical changes such as thermal expansion and contraction occur, due to temperature and pressure changes. Thus, the material used for tank manufacturing, as well as the accessories and peripherals, should also be able to handle such thermal contraction and expansion. The dispenser components used during tank refilling, as well as the connections at the refueling station, fuel lines, return valves, fuel level indicator, fuel limiter, and other similar components, will all be subjected to temperature and pressure changes. During the operation, the required components include pressure regulation valves, heat exchangers, the return valve, and other safety components like the pressure relief valve and the boil-off management system, and all these components should also be capable of operating at liquid hydrogen temperatures. The storage system must have purging or venting arrangement, in order to prohibit any air infiltration into the tank and line. Purging and venting arrangements are important because, during cycles of emptying and filling of the tank, in case of a pressure difference, air can enter and condense inside the tank, and later when it evaporates can cause flammable mixture of air and hydrogen, posing safety issues.

A typical liquid hydrogen storage tank is shown in Fig. 14.1 adapted from Linde. The entire storage system comprises of several components, including a filling port, a liquid extraction port, a gas extraction port, filling lines, and a probe to check the liquid level inside the tank. Other essential components in the liquid hydrogen storage vessel includes safety valve, shut-off valve, cooling water heat exchanger, reversing valve, and electrical heater. An electrical heater is required when gaseous hydrogen is to be supplied in accordance with the demand of the end-use utility.

Considering the global status of such liquid hydrogen storage tanks, different tanks are installed at various places in different countries ranging from very small

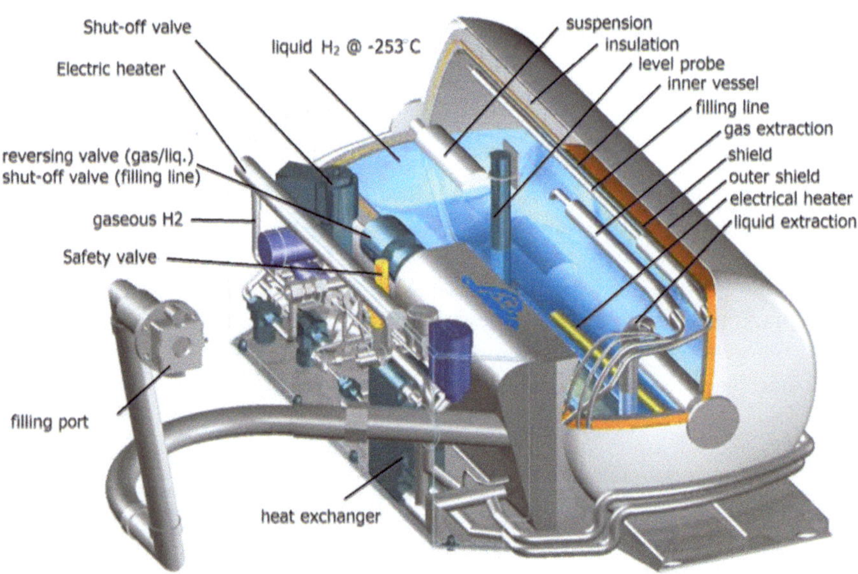

Fig. 14.1 Picture of a liquid hydrogen storage tank (Reproduced with permission from Energy Transition in Aviation: The Role of Cryogenic Fuels, Arvind Gangoli Rao, Feijia Yin and Henri G.C. Werij, Aerospace 2020, 7, 181; https://doi.org/10.3390/aerospace7120181)

size to very large size, and thus exhibiting a wide range of capacity of liquid hydrogen storage. Some of the representative examples are: the largest tank which is located at NASA facility, having a volume of 3800 m^3 and can store 270 tons of liquid hydrogen having an outer radius of 21 m, can hold pressure of 6.2 bar. The tank is of large capacity and spherical in shape, has a very low evaporation rate of 0.06%, and is shown in Fig. 14.2. Another tank at NASA will have a volume of 4732 m^3 and outer diameter of 24 m, and has an evaporation rate of 0.048%.

A liquid hydrogen storage tank at Kawasaki, has a volume of 540 m^3, and can store 38 tons of liquid hydrogen as shown in Fig. 14.3. Similarly, another tank at Kawasaki with a capacity of 11,200 m^3, which can store 710 tons of liquid hydrogen, and has an outer diameter of 30 m, to be installed.

In summary, currently, most of these ground-based liquid hydrogen tanks have insulation, which is perlite, or vacuum insulation along with perlite and without active thermal shielding, leading to relatively higher boil-off losses at a rate of 1–5% per day. However, these boil-off losses can be reduced to 0.01–0.5% per day by using either multilayer insulation or active thermal shielding. In very large-scale facilities, use of the metal hydride for capturing the hydrogen released due to boil-off, or the reliquefication of evaporated and boil-off hydrogen, or its use for various applications can be considered.

Fig. 14.2 Picture of liquid hydrogen storage facility at NASA, US (*Source* https://www.nasa.gov/feature/innovative-liquid-hydrogen-storage-to-support-space-launch-system, last accessed on 07-07-2025)

Fig. 14.3 Picture of liquid hydrogen storage facility at Kawasaki, Japan (Reproduced with permission from—Review on the key technologies and future development of insulation structure for liquid hydrogen storage tanks, Liang Yin, Haonan Yang, Yonglin Ju, International Journal of Hydrogen Energy 57(2024)1302–1315)

Cryo-compressed Storage

A combination of both the compressed as well as liquid state is known as cryo-compressed method of hydrogen storage. This method has the advantages of both compressed and liquid-state storage, and the tanks are designed in a such a way that they can accommodate compressed cryogenic fluids. With cryo-compressed storage, there are several prototypes that have also been demonstrated; BMW demonstrated a prototype in which gravimetric capacity of 5.48% and volumetric energy density of 4 MJ/L were achieved using the cryo-compressed tank with boil-off rate of 4–7 g/h. Another prototype was demonstrated wherein hydrogen was stored at 40–80 K and a pressure of 300 bar.

The cryo-compression method has advantage that the boil-off losses can be curbed. The inner tank is a type III tank and can hold pressure of 300–350 bar, and there is an outer tank, the region in between these two tanks has vacuum and multilayer insulation, so as to prevent the heat losses. These tanks are not designed for very high pressures such as 700 bar (like the type IV compressed hydrogen tank), therefore, the design requirements are comparatively less stringent, resulting in potential savings in material costs.

Since the storage temperature is not at liquid hydrogen conditions of 20 K, but rather the temperatures are above 20 K, the insulation requirements are less stringent, leading to both cost and energy savings. Besides the disadvantages mentioned in the earlier part of this chapter, regarding the liquid-state storage, such as heat inflow leading to evaporation and pressure buildup are partially taken care. With cryo-compressed tanks, the dormancy period can be extended and boil-off losses can be curbed.

Chapter 15
Solid-State Hydrogen Storage—Adsorption-Based Materials

Hydrogen storage can be broadly categorized into physical storage and chemical storage. The physical storage of hydrogen is constrained by the boundaries of a storage medium, such as storage tank or storage vessel. The difference is that the hydrogen in the physical method of storage does not interact with the storage media. The class of physical storage includes the compressed gaseous-state hydrogen storage, liquid-state hydrogen storage, and cryo-compressed method of hydrogen storage. If hydrogen is cooled further below the liquid hydrogen temperature, down to less than 14 K, then solid hydrogen can be obtained. In the physical storage method, hydrogen is stored in the molecular form and it is confined within the boundaries of a storage vessel, and the molecules does not interact with the material of the vessel.

In the another method of hydrogen storage, it is known as the materials-based storage, the hydrogen atoms or molecules interact with the storage media, they react, or they form a bond with the storage material which is being used for storing hydrogen. Based on the type of interaction, there can be different class of materials that can store hydrogen. As shown in Fig. 15.1, chemical storage can be achieved by means of formation of a weak Van der Waals bond with the high surface area materials like carbon nanotubes, carbon nanofibers, metal organic frameworks, zeolites, etc. Hydrogen can also make relatively stronger chemical bonds in another class of materials, like metal hydrides, complex hydrides, chemical hydrides, etc. Different liquid organic hydrogen carriers (including C_6H_6, C_6H_{12}, $C_{10}H_8$, $C_{10}H_{18}$) and reformed fuels (converting into the form of ammonia or different Fischer- Tropsch liquids) can also be used to store hydrogen, the classification is indicated in Fig. 15.1.

Requirements from Materials-Based Hydrogen Storage

Solid-state hydrogen storage materials should have the following major requirements:

P. Sharma, *A Textbook on Hydrogen Energy*,
https://doi.org/10.1007/978-981-95-4601-5_15

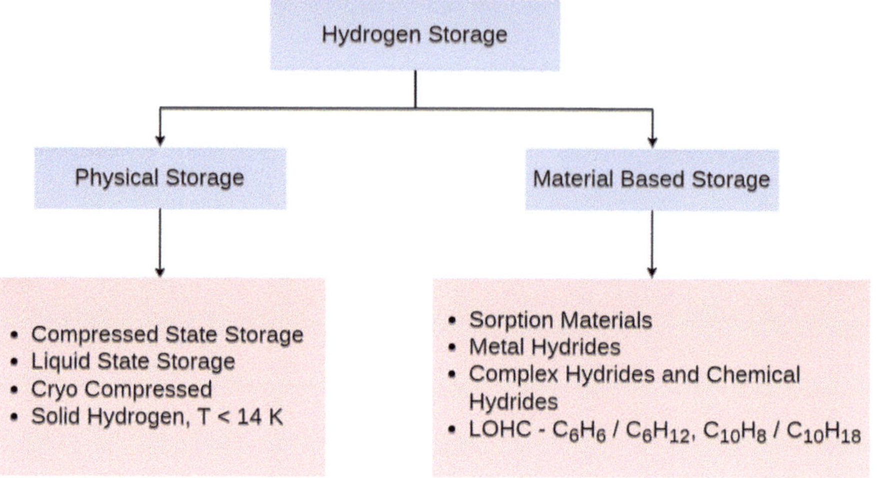

Fig. 15.1 Different methods of hydrogen storage

- High hydrogen capacity both per unit mass and per unit volume (gravimetric hydrogen storage capacity or volumetric storage capacity)
- Store hydrogen under moderate conditions of temperature and pressures
- Moderate heat of formation
- Reversibility
- Fast kinetics
- High stability against oxygen and moisture
- High cycle life
- Low cost
- High Safety.

Different Methods for Storing Hydrogen in Materials

When the atoms or molecules of the adsorbate, attach, adhere, accumulate onto the surface of another material called the adsorbent, the process is known as adsorption. Thus, adsorption is a surface phenomenon, and the higher the surface area, the more sites will be available for adhering and thus more adsorption will take place. However, the adsorption capacity will depend upon parameters like, the temperature and pressure. The process of adsorption is an exothermic process, releasing heat in the process, and is thus favored at lower temperature. The rate of adsorption increases till an equilibrium is attained in the process.

In the absorption process, the molecules or atoms of one substance (absorbate) can get into the bulk of another material(absorbent). As shown in Fig. 15.2b, the hydrogen atoms shown with red color, permeate into the bulk of the material which is shown with blue color. The process of absorption is a bulk phenomenon, and hydrogen in

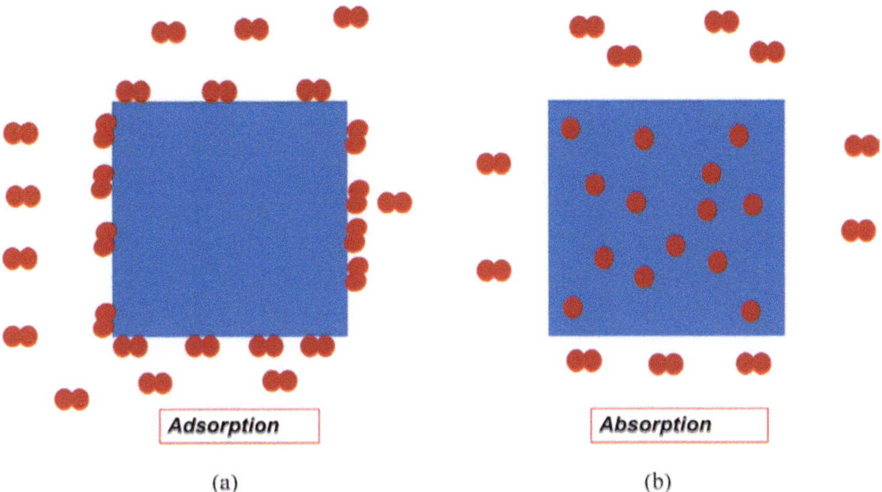

Fig. 15.2 Schematic showing the difference between **(a)** adsorption and **(b)** absorption in materials-based storage

this case is assimilated into the internal volume, or it gets completely diffused into the volume or internal structure, or dissolved into the absorbent material.

Fundamentals of Adsorption-Based Materials

Adsorption Isotherms—The adsorption of different gases can be described by means of adsorption isotherms. The isotherms, represents the amount of adsorbate on the adsorbent as a function of pressure drawn at a constant temperature. The process of adsorption depends upon temperature, pressure, and characteristics of the adsorbate and adsorbent.

The amount of adsorbate uptake needs to be normalized with respect to the mass of adsorbent and can be used to compare the storage capacity of different materials. Different types of adsorption isotherms are known and these can be used to quantify the amount of adsorbate being stored in a particular sorbent material.

Historically, the first adsorption isotherm was given by Herbert Freundlich and is known after his name as Freundlich isotherm. The Freundlich isotherm is a mathematical fit that was provided to the adsorption isotherm using an empirical relationship and is given as follows:

$$\frac{x}{m} = KP^{\frac{1}{n}} \tag{15.1}$$

where x is the mass of the adsorbate, m is the mass of the adsorbent, P is the gas pressure, K and n are the respective constants associated with a pair of adsorbate and adsorbent.

Another set of isotherms was being provided by Irving Langmuir in the year 1916, and he was the first scientist to derive these adsorption isotherms scientifically. He modeled the adsorption of gases onto the solid surface, the model was kinetically based and derived using statistical thermodynamics. There are several assumptions used while deriving the Langmuir isotherm: the adsorption sites are all equivalent and each adsorption site is occupied by a single molecule, there are no phase transitions, the surface is energetically homogeneous, the adsorbents do not interact, and the adsorption takes place in the form of a monolayer. However, there will be deviations in actual practice because of the surface imperfections, the adsorbent may not always be inert, or it may not be possible that a monolayer is always formed. There could be multilayer formation, which is actually being considered in the next set of adsorption isotherms, which is given in the BET theory. At the same time, the conditions when the first molecule gets adsorbed and the last molecule adsorbs will not be similar. Thus, it is not possible that all these assumptions may hold true. The relations for the Langmuir isotherm are:

$$K = \frac{\theta}{(1 - \theta)P} \tag{15.2}$$

$$\theta = \frac{KP}{1 + KP}. \tag{15.3}$$

For very low P, $\theta \sim$ KP

For high P, $\theta \sim 1$,

where P is the pressure, θ is the surface coverage, and K is the equilibrium constant. The surface coverage is the fractional occupancy of the adsorption sites, i.e., the ratio of the volume of gas adsorbed onto the solid to the volume of gas molecules in a monolayer covering the whole surface of the solid and completely occupied by the adsorbate.

The next set of adsorption isotherm was given by Stephen Braunauer, Paul Emmett, and Edward Teller, and is known as the BET theory. The BET theory is in fact, an extension of Langmuir theory, and it is applicable for multilayer adsorption. The basic assumption in the BET theory is that the gas molecules are physically adsorbed in the form of an infinite number of layers. Thus, it is applicable for multilayer adsorption, and these layers are considered as non-interacting and independent; infact the Langmuir model is applicable for each of these independent layers. In order to determine the surface coverage and obtain the isotherms, the carrier gas used is nitrogen. The reason for using nitrogen is that it is an inert, pure gas, and is cost-effective. However, at the ambient set of temperature and pressure conditions, a very small amount of gas gets adsorbed, the reason being weak Van der Waals interaction involved in the adsorption process (the thermal energy at room temperature is sufficient enough to release the adsorbed gas and reduce the capacity). Thus, in order to get a detectable amount of gas being adsorbed, the liquid nitrogen temperatures are being used and at this temperature sufficient adsorption can take place.

The expression that represents the amount of gas adsorbed in this theory is:

$$\frac{1}{x\left[\frac{P}{P_0}-1\right]} = \frac{1}{xc} + \left(\frac{c-1}{xc}\right)\frac{P}{P_0} \tag{15.4}$$

where P/P_0 is the relative pressure, x is the amount getting adsorbed, c is a coefficient which is related to the heat of adsorption, and is known as the BET constant using which the shape of the isotherm can be found. From the BET theory, there are five different types of possible isotherms. Figure 15.3 shows the different BET isotherms.

Type I Isotherm—The trend as evident from the type I isotherm is that, at lower pressure there is a steep rise in the mass of gas adsorbed, and then it flattens to form a plateau. Type 1 is a pseudo Langmuir isotherm and is applicable in general to the microscopic materials; however, it predicts the formation of a monolayer on adsorption.

Type II Isotherm—The type II isotherm shows a sharp, steep rise at very low pressures, then a linear rise, and finally it peaks. The pattern of type 2 isotherm at low pressure predicts the micropores filling on adsorption of the gas, thereafter a

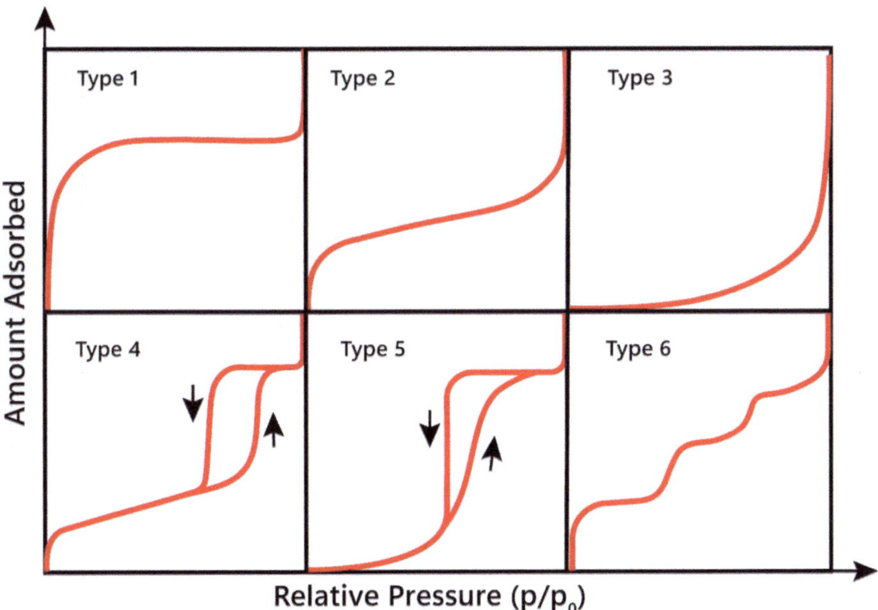

Fig. 15.3 Typical BET isotherms

slight plateau arises because of the monolayer being formed, and at higher pressures the multilayer adsorption occurs, and that multilayer adsorption continues until there is a condensation, which takes place due to capillary forces. This type of isotherm is usually seen in macro or nanoporous materials.

Type 3 Isotherm—The type 3 isotherm shows neither a plateau nor an asymptote. Thus, the type 3 isotherm represents only the multilayer adsorption and indicates that, there is no monolayer formation. From the beginning, the multilayer formation in the adsorption process takes place; it begins before even the monolayer adsorption gets over.

Type 4 Isotherm—The type 4 isotherms represent that initially the capillary condensation occurs, and a monolayer adsorption takes place at the lower pressure, and thereafter this is followed by a multilayer adsorption. Type 4 isotherm is usually seen in the case of mesoporous materials.

Type 5 Isotherm—The type 5 is almost similar to type 4 and shows the presence of capillary condensation. The molecular clustering and later filling up of pores takes place, such isotherms are usually seen in case of materials with complex pore structure, e.g., activated carbon or some of the zeolites.

For materials that store hydrogen onto the surface through weak dispersive forces, (i.e., the Van der Waals forces) and through the adsorption process, i.e., hydrogen is in molecular form (without dissociating into atoms), the above-mentioned isotherms (Type 1--5) can be used to measure the hydrogen uptake. Considering the materials for hydrogen storage i.e., in the sorbent materials, the physisorption is the basic process. In physisorption, hydrogen is stored in the molecular form on the surface of different sorbent materials and these are interacting with the surface of the sorbent by means of weak Van der Waals forces. The weak Van der Waals forces arise in these materials due to the dipoles, which are created because of the random fluctuations in the charge distribution. A hydrogen molecule does not have a permanent dipole moment; the first nonzero moment is the quadrupole moment. However, the hydrogen molecule gets polarized in the presence of external fields, like when it interacts for adsorption on an unsaturated metal center (e.g., MOFs). The adsorption, i.e,. the attraction between the different gas molecules and the surface, or the repulsion between the gas molecules, can be represented by the potential energy curve, which is given by the Lennard Jones potential.

$$L(r) = 4\varepsilon\left[\left(\frac{\sigma}{r}\right)^{12} - \left(\frac{\sigma}{r}\right)^{6}\right] \tag{15.5}$$

where ε is the depth of the potential well, r is the distance between the molecule and the surface atom of the adsorbent, *and* σ is the finite length distance at which $L(r)$ becomes 0. The potential energy curve has its minima at a distance from the surface, which is approximately equal to the sum of the radii of adsorbate and adsorbent. Thus, the stable position for the adsorbate on the adsorbent is one molecular radius from the surface, and at distances higher than that, the potential energy reduces very rapidly. In the case of a micro porous material, the pore size is very small (usually less than 2 nm) and is almost equivalent to that of the adsorbent diameter, due to which the interaction depends upon the pore size and pore dimension. Physisorption, being a surface phenomenon, requires a high specific surface area and is dependent upon the pore geometry, the pore dimension, the pore volume, the porosity of the material, and all these parameters determine the adsorption capacity of the adsorbate onto the adsorbent.

The adsorption process is an exothermic process, because of the low polarizability of the hydrogen molecule. The interaction takes place by means of weak Van der Waals forces, which means that the adsorption energy is very small. The value of adsorption energy typically lies in the range of 1–10 kJ/mol. This value is also referred to as the isosteric enthalpy of adsorption and is different from the adsorption enthalpy. The reason for this difference is, the sites available for adsorption are different and thus the interactions experienced with the surface are different; as such it can't be same from first molecule being adsorbed till the last one getting adsorbed. However, in the class of materials which work on the principle of chemisorption, the bond formed is stronger and the adsorption energy can be even an order of magnitude higher.

The layer of hydrogen adsorbed on the surface of the adsorbent, is in equilibrium with the gas available in the rest of the void volume, i.e., bulk gas. In that case, the Gibbs free energy for both the gas adsorbed and bulk gas, can be considered the same, and the change in enthalpy can be given as follows:

$$H_{\mathrm{ad}} - H_{\mathrm{gas}} = T\left(T_{\mathrm{ad}} - T_{\mathrm{gas}}\right) \tag{15.6}$$

On adsorption, the entropy of the gas reduces, and the enthalpy of adsorption is lower than the enthalpy for the gas phase and thus the adsorption is favored.

Hydrogen Uptake

With all the basic information required being included in the previous section, the amount of hydrogen being adsorbed by the adsorbent can be measured. There are different ways of measuring hydrogen uptake, either gravimetric or volumetric storage capacity could be monitored. The gravimetric capacity is the mass of hydrogen being stored divided by the mass of hydrogen stored plus the material mass (or the mass of the storage media); and the volumetric storage capacity is the mass of hydrogen stored per unit volume of the storage media.

In order to find the hydrogen uptake in the porous materials, the best way could be to find the adsorption isotherms. The amount of hydrogen adsorbed in the porous materials can be obtained by measuring the absolute storage capacity or the excess storage capacity. These terms are used for finding out the amount of hydrogen uptake in a particular class of material.

Excess storage capacity is, in fact, the capacity being measured experimentally and it is defined as the difference of the quantity of gas which is being stored at a particular temperature and pressure in a given volume containing the adsorbent, in the presence and that in the absence of gas-solid interactions, both considered under the similar set of conditions. However, measuring the quantity of gas stored in the absence of gas-solid interaction is very difficult. This quantity can be obtained using a non-adsorbing gas which does not interact with the solid surface, and the volume of this non-adsorbing gas is being measured to find the second term in the excess capacity as mentioned in (15.7).

Another term, absolute adsorption, is difficult to determine and it can be predicted by means of different theoretical calculations. Although this provides a simplistic approach to find the amount of hydrogen adsorbed by a material, and as such a lot of modeling and analytical work has been carried out to estimate its value. Absolute adsorption refers to the amount of hydrogen that is adsorbed on the adsorbent due to adsorption process, and it does not include the hydrogen present in the gas phase (i.e,. the gas filled in the pores or other free space available like the dead space or expansion space in the tank).

$$M_{\text{ad}} = M_i - V_g \rho_g \tag{15.7}$$

M_{ad} amount of gas adsorbed

M_i is the amount of gas introduced into the sample cell.

V_g is the free volume occupied by the gas in the extra space available, as well as including the pore volume of the sample.

ρ_g is the gas density.

The Langmuir adsorption isotherm can be used to plot the hydrogen uptake with pressure. For an equilibrium pressure, the amount of gas adsorbed can be given as follows:

$$m(P) = m_{\text{max}} \left(\frac{KP}{1 + KP} \right) \tag{15.8}$$

where m_{max} is the maximum adsorption capacity, K is the equilibrium constant, and m is the amount of gas adsorbed at the equilibrium pressure P.

It is very difficult to find out the real free volume using the instruments available in laboratories, because what actually is measured experimentally, is the excess adsorption. To find out the free volume, a non-adsorbing gas that does not interact with the adsorbent is taken, and gas expansion is used to find out the skeletal density; generally helium is used for this purpose as an inert gas. The excess adsorption is given by:

$$M_{\text{ex}} = M_i - V_o \rho_g \tag{15.9}$$

where Mi is the hydrogen which is introduced into the sample cell or the measuring cell, V_0 is the helium volume which is being taken up in the absence of gas-solid interaction and the gas density is ρ_g.

The relationship between the experimentally measured excess adsorption and the absolute adsorption is given as follows:

$$M_{\text{ex}} = M_{\text{ad}} - \rho_g (V_o - V_g) = M_{\text{ad}} - \rho_g V_a \quad (15.10)$$

V_o is the He volume in the absence of gas-solid interaction.

V_g is the gas volume in the presence of gas-solid interaction.

V_a is the difference between the real free volume and free volume measured with He gas in the absence of the adsorbed phase, and this is the volume being occupied by the adsorbed phase during adsorption.

The absolute hydrogen uptake at a temperature of 77 K is given by a Langmuir isotherm or Type I type of isotherm, and that indicates the formation of monolayer. Typically, the formation of a monolayer on the adsorbent surface can lead to a maximum of 1.3×10^{-5} mol/m^2 of hydrogen being adsorbed. However, the basic assumption behind the Langmuir model is that all sites are equivalent with the same adsorption potential and also the have same entropy, thus the adsorption on any of the available sites, is independent of how many sites are already occupied, i.e., its independent of the fractional coverage. For hydrogen, multilayers are not formed at 77 K due to the weak interaction between the different monolayers preventing the multilayer adsorption, thus the Langmuir model works well for hydrogen. If the amount of gas adsorbed is plotted as a function of pressure at a particular temperature, then as shown in figure below, a sharp rise is observed and thereafter it becomes constant, obtaining a plateau region, in the case of the absolute adsorption.

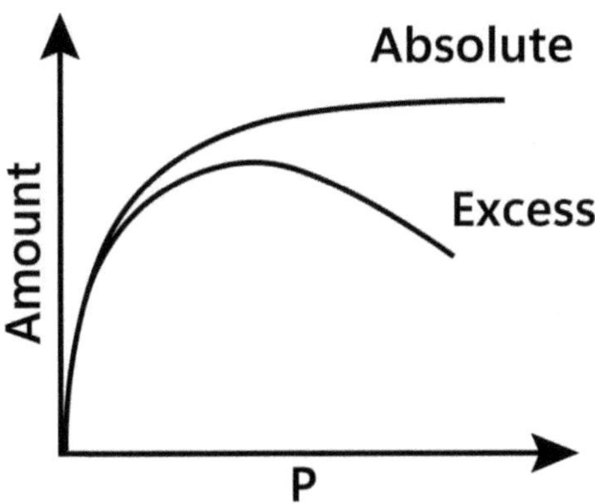

However, the excess adsorption can be measured experimentally, but its plot does not show a plateau. For the initial portion of the plot it remains consistent with the absolute desorption; however, later it peaks, and thereafter it decreases. It can also become zero, or negative. The nature of the curve toward the end, which is a dip, is observed because once the gas is adsorbed in the pores and the pores gets saturated (corresponding to the micropore filling), the gas density still continues to increase and the gas thereafter occupies the free volume or remaining available space in the sample cell. Thus, the decrease in the plot does not corresponds to real physical decrease of the adsorption capacity, rather it is the consequence of the measurement. Since the void volume of the sample is measured using the helium gas expansion in the absence of adsorbed phase, this is the reason of the dip being observed in the adsorbed amount at higher pressure. At moderate pressures, both the absolute and excess adsorption are similar and can be directly represented by the Type I isotherms. However, a difference is observed at a higher pressure, where, for absolute adsorption there is a plateau, while for excess adsorption in the curve after peaking a dip is observed.

The adsorption is carried out at liquid nitrogen temperature, and the isotherms plotted between adsorbed amount and pressure are at 77 K.

At room temperature, the amount of hydrogen uptake is very small in porous materials. The adsorption isotherm in that case, will not show any plateau; a linear rise is observed, which continues till higher pressure, thus a linear isotherm that is a Henry-type isotherm is obtained under such condition. With the increase in temperature, in the case of the physisorption-based material, the adsorption capacity decreases. The reason for the decrease in capacity is that the hydrogen gas molecules are attached on to the surface of sorbent by means of weak Van der Waals forces, at room temperature or higher, the thermal energy is almost equal or higher to the Van der Waals forces. This energy is sufficient enough to release the sorbed hydrogen, thus the capacity or the adsorption potential decreases as the temperature increases. However, to store hydrogen in an appreciable amount, the requirement is to reduce the temperature of the gas, with 77 K being preferably used for storing hydrogen in such adsorption-based materials at ambient pressure (the excess adsorption capacity of many of the materials peaks at around 77 K and 4 MP). The other option is to increase the gas pressure to increase the adsorption potential. The two types of adsorption enthalpies being considered in modeling includes: the one which considers the enthalpy associated with the adsorption of the first molecule on a pristine surface, and that is given by Henry's law; however, this may not correctly predict the value of adsorption enthalpy. If, however, the adsorbate and surface interactions are considered, and also that some sites are occupied, and also how the occupation of certain sites impacts the occupation of the other sites, then the deviation from Henry's law needs to be considered. The value considering all these deviations is defined in terms of the isosteric enthalpy of adsorption, for higher gas loading (which is different from the consideration of the single molecule adsorption in Henry's law case), and its value decreases as the gas loading increases as follows:

$$\Delta H_{is} = R\left(\frac{1}{\left(\frac{1}{T_1} - \frac{1}{T_2}\right)}\right) \ln\left(\frac{P_2}{P_1}\right) \tag{15.11}$$

Another possibility is considering the adsorption with interaction between adsorbate and adsorbent for the case of higher fractional coverage at high pressures, wherein the interaction between near neighbor molecules is considered, model for such a condition was proposed by Fowler and Guggenheim [1].

However, in order to get a realistic and exact picture, it is very much essential to consider the surface inhomogeneities and the difference between the various sites available for adsorption, and these will be different for different materials and thus, it will require an understanding of the surface morphology of the materials considered. Besides, when the hydrogen-solid interaction is also included along with the effect of inhomogenites and site variability, then both the effects will result in a varying enthalpy of adsorption. The variation of enthalpy of adsorption is actually observed in the adsorption-based system, wherein the desorbed amount of hydrogen keeps reducing and changing under the isothermal set of conditions. However, theoretically when the various models are used, they predict that under a given set of conditions, the value of enthalpy of adsorption and desorbed hydrogen remains constant for the adsorption-based materials.

Table 15.1 indicates the comparison of the different materials-based storage with the conventional compressed and liquid-state hydrogen storage. As seen in the table, the solid-state hydrogen storage has a clear advantage of moderate set of operating conditions and high volumetric storage capacity; however, the disadvantage is in terms of gravimetric storage capacity. The compressed state storage definitely has the advantage of high gravimetric capacity of 5.2 wt% and 5.5 wt% at 350 bar and 700 bar, respectively, but the operating pressure are very high and the volumetric density is low. In case of the liquid-state storage of hydrogen, the volumetric density improves but the storage temperature is 20 K; thus, the liquefaction and thereafter maintaining the low temperature along with the associated boil-off losses are the major challenges. Thus, the two conventional methods of hydrogen storage, which are temperature or pressure-controlled storage methods, pose issues related to safety, high cost, and low volumetric capacity. In case of solid-state storage of hydrogen, the storage conditions, i.e., temperature and pressure depend upon the material but there are materials which can store near the room temperature and at slightly higher than ambient pressure. The biggest advantage of solid-state storage of hydrogen is the high volumetric density, the corresponding storage system is in that case compact, smaller in size and requires lower footprint. The solid-state storage of hydrogen is considered as the safest among the different methods of storage. Considering the absorption-based materials, the volumetric capacity can be as high as 150 kg/m^3. Complex hydrides can store at 1 bar or higher pressure and their gravimetric capacity is exceptionally high, volumetric capacity be even higher than 150 kg/m^3, but the temperature of operation is greater than 100 °C. Besides the tuning of composition, structure, porosity in case of materials used for hydrogen storage provides a huge flexibility in terms of operation and utilization for a wide range of applications.

Thus, there are different classes of materials and each one of those has different characteristics, for example, the adsorption-based materials have weak bond formation, thus shows fast kinetics and very good reversibility.

Table 15.1 Comparison of various methods of hydrogen storage

Storage method	Pressure (bar)	ρ_m(wt%)	ρ_v(kg/m^3)	Temperature
Compressed hydrogen storage	350	5.5	23.65	Room temp
Compressed hydrogen storage	700	5.2	40.2	Room temp
Liquid hydrogen storage	1	Size dependent	70.8	20 K
Cryo-compressed	240	Size dependent	87	20 K
Adsorption-based materials	100	2	20	77 K
Absorption-based materials	1	2	150	Room temp
Complex hydrides	1	< 18	150	> 100 °C

The reason for the fast kinetics of such adsorption-based materials is that there is no activation energy involved so the uptake of hydrogen or release of hydrogen is very quick, because the bond formed in these materials is a weak bond and can easily release hydrogen. Also, the reversibility of such materials is very high, because hydrogen does not form a strong bond with the storage media. The reversibility is an important criterion, while selecting a hydrogen storage material. The adsorption-based materials show very good cycle life, the reason being there is no diffusion of hydrogen into the bulk of the material, no permeation of molecules, no formation of strong bonds with the adsorbent, and hydrogen molecule sits on the surface and does not enter into the internal structure of the sorption materials, making the process easier and reversible. Since there are no structural changes taking place (as hydrogen does not enter or leaves lattice during sorption), no electronic changes or charge transfer, no perturbation of the electronic charges occur in these materials, as such the cycle life of such materials is high. Since the heat of adsorption associated with the weak Van der Waals forces involved between the hydrogen gas molecules and the surface is very low, the advantage is observed while designing a hydrogen storage tank and will not involve very complex thermal management.

Adsorption-Based Solid-State Hydrogen Storage Materials

Different adsorption-based materials that can be used for hydrogen storage will be considered in this section.

The design of optimum adsorbent materials is a challenging task. The major challenge with adsorption-based material, as mentioned in the previous section, is that the weak Van der Waals bond between the hydrogen molecule and the surface of the adsorbent, essentially requires low temperature for storing hydrogen, typically 77 K. Sufficient gravimetric capacity is observed at 77 K but as the temperature increases, the capacity reduces and near room temperature becomes negligibly small. Thus, in order to increase the capacity, there are two ways: either to increase the number of adsorption sites per unit weight and per unit volume of the adsorbent, which can be met by selecting materials that have high specific surface area and high pore volume. Another possibility is to increase the heat of the adsorption, which can be done in different ways. An increase in the adsorption energy means that the bond between the hydrogen molecule and the surface of adsorbent gets stronger. The adsorption energy can be increased by either introducing the hydrogen spillover effect, or by through unsaturated adsorption sites, or by synthesizing the adsorption-based materials with narrow pores and small-sized pores, thereby increasing the the adsorption potential.

Considering the first method, i.e., the hydrogen spillover effect, research reports have shown that the effect of the introduction of the transition metals, noble metals like platinum or palladium inside the host matrix however, this has not shown much promising results.

Another option is including unsaturated adsorption sites, by changing the chemical composition of the porous host material by introducing certain other species, and these species can even be metal. Hydrogen, is a molecule with no permanent dipole moment, but in the presence of external field, gets polarized like that happens in the case of MOFs with the unsaturated metal centers, the metal ion can induce a dipole moment in such a way, that the induced dipole moment has a higher interaction energy than the metal ion-quadrupole hydrogen interaction. Thus, the binding affinity of hydrogen increases in the presence of such electrostatic interactions. However, this concept has proved to be only partially helpful. The reason is that the interaction of the hydrogen molecule with the functionalized adsorption sites was not very strong. Thus, adding unsaturated sites also didn't help much; at the same time, the metals introduced are heavy and thereby reduces the gravimetric capacity.

Another possibility is the adsorption of hydrogen into very small or very narrow-sized pores, in this case the hydrogen molecule interacts with multiple pore walls, and that helps onto improving on to the adsorption potential. Thus, various materials with desired pore geometry need to be designed and considered for this type of storage.

Different Materials for Adsorption of Hydrogen

Carbon-Based Materials

Carbon-based materials are well-known to have very good characteristics and are used for different energy storage applications. Carbon exists in various allotropic forms like diamond and graphite both being non-porous. The different allotropic forms of carbon differ in their chemical structures. Other than the non-porous diamond and graphite, there are several other structures that are highly porous with very high specific surface areas, like the different carbon nanostructures. The carbon nanostructures have a structure similar to that of benzene, having hexagons with carbon being sp^2 hybridized. The different carbon nanostructures differ in the way these hexagons are arranged. Thus, the different carbon-based materials can be divided into two classes, one which has a long-range order in terms of arranging these hexagons, the simple examples being carbon nanotubes or carbon nanofibers.

Another class of the carbon materials are those having irregular structure, like the activated carbon, porous carbon, and amorphous carbon. These forms are known to have a higher hydrogen uptake capacity. Activated carbon have very high specific surface area: for example, the regular one can have specific surface area in the range of 800–1700 m^2/g. The highly porous forms of activated carbon like AX-21 can even have specific surface area of around 3000 m^2/g. The different nanostructures of carbon, which are ordered includes, the single-walled carbon nanotubes, which are like the rolled graphene sheets, or the multiwall carbon nanotubes. Other possible carbon structures are onion-like carbon, carbon dots, fullerenes, graphene dots, multi-layered graphitic sheets or graphene, carbon nano ribbons, carbon nano horns, carbon nanofibers, etc. Thus there is a large number of carbon structures known and they find application in a variety of applications. The biggest advantage of carbon-based materials is the high specific surface area, high porosity, tuneable pore structure, low cost, low weight, and good stability (Fig. 15.4).

A lot of interest about two decades back, was due to the exceptionally high hydrogen storage capacity reported for some of the carbon-based materials. However, the non-repetitiveness and errors were observed in the reporting, which were found to be due to the presence of impurities and other forms of measurement errors. However, the carbon-based materials are still known for their higher storage capacity. The activated carbon shows very good hydrogen storage capacity, other than that, the

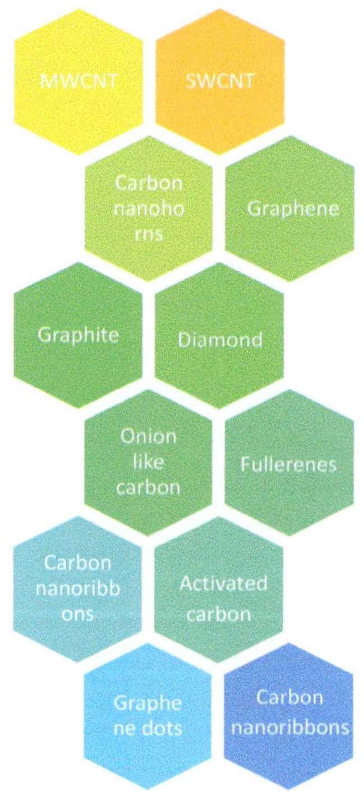

Fig. 15.4 Different forms of carbon which can be used for adsorption of hydrogen

carbon nanotubes, nanofibers, nanohorns, and graphene also show good capacity for hydrogen storage. All these are commercially available and have a very porous microstructure, having high specific surface area (SSA) and pore volume, their microstructures can be microporous, mesoporous, and macroporous. These structures have SSA reported in the range of 10–3000 m^2/g, possess low mass density, thus resulting in high gravimetric capacity. Empirical studies on hydrogen storage in carbon-based materials have been carried out and reported in literature, wherein consideration of charge transfer and orbital hybridization in the DFT studies has given quite reliable results.

Zeolites—An important class of materials under the adsorption-based materials used for hydrogen storage are zeolites. Zeolites are three-dimensional crystalline alumino-silicate materials, that show a microporous structure. These are built up of TO_4 type of tetrahedron structures with corner sharing of oxygen, where T stands for Si^{4+} or Al^{3+}. These materials have an irregular intracrystalline structure with networks of pores and channels and have a high specific surface area. The general chemical formula for these types of zeolites can be written as $M_{x/2}[(AlO_2)_x(SiO_2)_y]\cdot mH_2O$, where M is the non-framework metal cation or exchange metal cation. Zeolites are known for their structure which has cages or channels where cations, such as water or smaller molecules, can get into the structure. Besides, the zeolites are also known for their ion exchange characteristics and good thermal stability, and can be used for hydrogen storage.

In the TO_4 structure, the T–O–T bond is highly flexible, and the flexibility of the bond allows the tetrahedral to get linked in a variety of ways, allowing to form a large number of network topologies. Thus, the increased flexibility of bonds and linkages gives rise to large number of possible zeolite structures. At the same time, the flexibility of the T–O–T bond also makes the framework flexible, such that the framework can adapt to the various changes in temperature, pressure, or chemical surroundings.

However, the electro-neutrality of such structures need to be maintained: for example, if Si^{4+} is being replaced by Al^{3+}, then the electro-neutrality of this structure will be maintained by introducing a non-framework exchanged metal cation. The selection or choice of the non-framework exchange metal cation, depends upon the available or accommodating pore volume and also on the charge on the metal cation. The introduction of the charged metal cation into the framework structure gives rise to an electrostatic force, which increases the polarization effect of the various adsorption sites. Thus, the polarization increases in the channels or pores present in such frameworks, and the developed electrostatic force depends upon the charge on the metal cation. The electrostatic force increases when the charge on the cation increases, and it decreases with the pore volume. The charge balancing cations used are generally alkali or alkaline earth metal ions including Na^+, K^+, Ca^{2+} and Ba^{2+}. The zeolites can be naturally occurring or synthesized in laboratories. However, most of the zeolites are synthetically synthesized. The hydrogen uptake in zeolites primarily depends upon the specific surface area, the type of metal cation being used, and the concentration of the metal cations. Zeolites can be classified on the basis of silica to alumina ratio. Zeolites with low silica to alumina ratio are with Si:Al lying between 1-1.5, intermediate Si:Al ratio is when Si:Al lies between 2-5, and high if Si:Al ratio ranges from 10 to several thousands. Example of zeolite is Zeolite X, wherein the cation can be Na or K. With Na, different types of zeolite having sodalite cages can be obtained. Other important parameters that governs the hydrogen uptake in zeolites includes the micropore geometry, pore volume, and the pore diameter (whether of the order of the kinetic diameter of the hydrogen molecule or higher).

Hollow Glass Microspheres (HGM)—Another class of materials that could be used for hydrogen storage is hollow glass microspheres. This class of materials is similar to the compressed state storage concept, wherein a certain dimension and volume, which is micron-sized, is used like the high-pressure compressed state storage. HGMS are safe to operate, safe to store, and can be carried over longer distances and can be stored for a longer duration.

Hollow glass microspheres (HGMs) are a unique class of adsorption-based materials with interconnected pores in the microsphere walls, which can adsorb, retain, and release hydrogen and other gases. These are synthesized in the form of finely dispersed and free-flowing powder having a wall thickness of ~ 0.5 to 2.0 μm and a diameter of 10–200 μm. Porous-walled HGMs with sphere diameter in the range of 1 to 100 μm can have density in the range of 1.0–2.0 g/cc, and porous-wall structure having pore diameter of 10–1000 Å. HGMs have characteristics like spherical shape, controllable size, low density, relatively high strength in uniform compression, good thermal and dielectric properties and can be used for hydrogen storage up to even 100 MPa.

Hydrogen adsorption takes place due to the diffusion of gas into the HGMs. Diffusivity of hydrogen increases rapidly with the increase in temperature, allowing hydrogen gas to diffuse through the porous walls of the HGMs in a relatively short time. The filled microspheres are then cooled to room temperature, which decreases the diffusivity of hydrogen gas, and finally, the filled hydrogen gas is retained in the HGMs. Hydrogen gas stored within the HGMs can be released by reheating the microspheres.

A lot of parameters used in synthesis affect the morphology and pore structure of HGMS, and a lot of literature reports have mentioned the effect of glass feed particle size as well as feed flow rate, in the formation of HGMs (Fig. 15.5).

Modification of the microsphere wall thickness and porosity could be achieved by incorporating known amounts of blowing agents while preparing the feed glass powder for HGMs. In order to improve the thermal conductivity and thereby increase the hydrogen storage capacity of the product, HGMs can be incorporated with different metal loading in the feed glass powder. The metals used could be cobalt, zinc, magnesium, iron, etc. (Table 15.2).

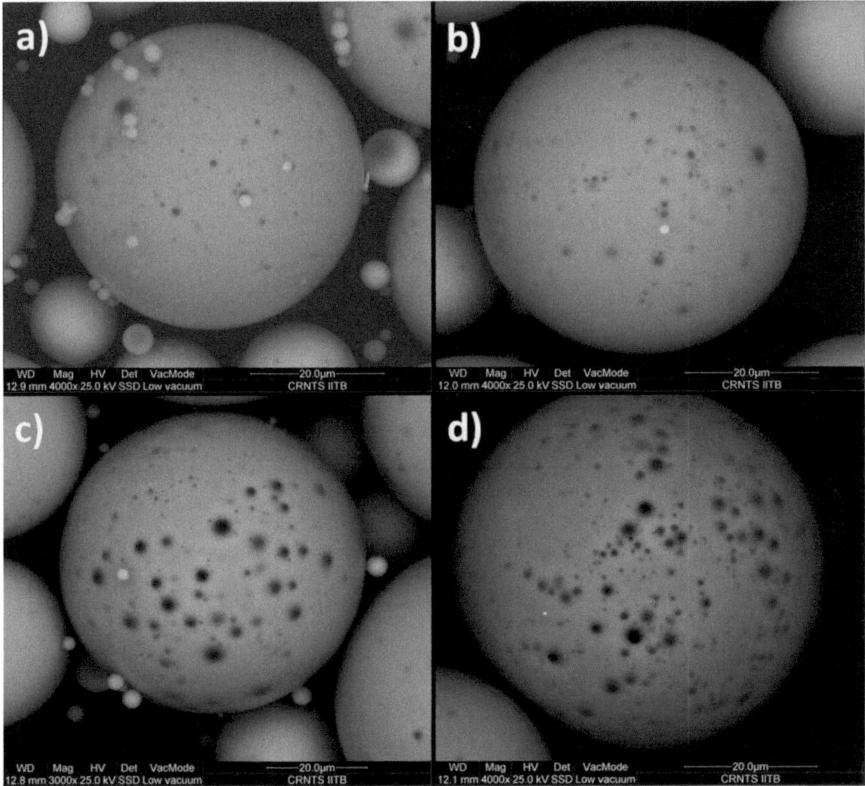

Fig. 15.5 ESEM images of **a** HACo0.2, **b** HACo0.5, **c** HACo1, **d** HACo2 (Reproduced from [2] with permission)

Table 15.2 Comparison of hydrogen storage characteristics of different hollow glass microspheres [2–4]

S. No	Material	Filling parameters		Hydrogen storage capacity (wt%)
		Filling temperature (°C)	Filling pressure (bar)	
1	2wt% Mg loaded HGMs	200	10	2.33
2	2wt% Zn loaded HGMs	200	10	3.26
3	2wt% Co loaded HGMs	200	10	3.31

Materials	Surface area (m^2/g)	H_2 storage	Pressure (bar)	Temperature (K)	References
MOF-210	6240	176 mg/g (Total)	80	77	Science 2010;329:424–428
Ni_2(m-dobdc)	–	11.9 g of H2 per L 23.8 g of H2 per L	100 100	298 223	Chem. Mater. 2018, 30, 22, 8179–8189
MOF-177	4746	7.5 wt%	70	77	J Am Chem Soc 2006; 128:3494–3495
NU-100	6143	164 mg/g (Total)	70	77	Nat Chem 2010;2:944–948
IROMF-20	4024	6.7 wt%	80	77	J Am Chem Soc 2006;128:3494–3495
Li.MIL101(Cr) and Li.Cu_3(BTC)$_2$	1840 and 795	3.39 and 3.50	1	77	Int J Hydrogen Energy 2012;37:946–950
COF-102, COF-103,COF-10	3620, 3530, 1760	72.4, 70.5, 39.2 mg/g	1	77	J. Am. Chem. Soc. 2009, 131, 25, 8875–8883
AC (AX-K5)	3190	7.08 wt%	60	77	J. Am. Chem. Soc. 2009, 131, 20, 7016–7022
SWNT		4.2 wt%	120	298	Science. 1999; 286: 1127- 1129
MWCNT		6.3 wt%	148	298	The Journal of Physical Chemistry B, 2002, 106, 5, 963–966,
CNF	51	6.54 wt%	120	298	Nano Letters, vol. 2, no. 3, pp. 201–205, 2002
NaA, NaX, NaY		1.54, 1.79, 1.81 wt%	15	77	J Alloy Comp 2003;356–357:710–715
ZSM-5		1.97,0.65,0.38 wt%	70	77/195/293	Chin. J. Chem. Phys., Vol. 19, No. 5, 2006, 457–462
MCM-41		2.01, 0.69 wt%	35/10	77/298	Appl Phys A Mater Sci Process 2001;72(5):619–623; Int J Hydrogen Energy 2014;39(16):8749–8753
SiO_2	79	4 (ml(STP)/g)	1	77	Appl. Phys. A 72, 619–623 (2001)
Al_2O_3	233	7 (ml(STP)/g)	1	77	Appl. Phys. A 72, 619–623 (2001)

Table 15.3. Examples of some of the absorption-based materials, their hydrogen storage capacities, and sorption conditions

Metal Organic Frameworks—Another class of adsorption-based material which could be used for hydrogen storage is metal organic frameworks. The hydrogen molecule in MOFs gets polarized due to the coordinatively saturated metal centers. The metal organic framework is a class of compounds that consists of a metal ion or metal oxide cluster or inorganic cluster that has high dimensionality and is coordinated with organic ligand and forms either one, two, or three-dimensional structures. Both the metal and the organic ligand/linker (azoles, di and tricarboxylates), which make up the framework, can be changed, which gives a wide variety of possibilities of MOFs, and these will have different compositions and different structures.

The most studied metal organic frameworks include MOF5, MOF177, ZIF, ZIF100, ZIF21, MIL53, MIL101, etc., and many more. The MOFs have high specific surface areas, which can be as high as 1000–10,000 m^2/g. MOFs have ultrahigh porosity, and the pore size can be tuned depending on the selection, and the internal structure can be modified. They have high thermal and chemical stability and very good adsorption capacity. However, the biggest challenge is to achieve homogeneous material, achieving the same structure in different batches of synthesis is difficult and its also difficult to synthesize on large scale, thus, scalability is a major bottleneck. The hydrogen storage capacity in MOFs depends upon the pore structure and the high dimensionality of the framework, which can be changed by changing the organic linker, metal/ligand ratio, reaction temperature, etc. Thus, MOFs can be designed by changing these parameters, and their structure and properties can be changed. MOFs are characterized by very high porosity, high crystallinity, and high specific surface area.

Example of MOF is MOF-210, has a specific surface area of 6240 m^2/g, can store about 176 mg of hydrogen per gram of the material, at a pressure of 80 bar and 77 K; however, at room temperature the capacities are very low. MOF-177 with a specific surface area of 4746 m^2/g has been shown to store 7.5 wt% of hydrogen at a pressure of 70 bar and 77 K temperature.

NU-100 with 6143 m^2/g of specific surface area can store 64 mg/g of hydrogen at 70 bar pressure and 77 K. IROMF-20 having a surface area of 4024 m^2/g can store 6.7 wt% at 80 bar and 77 K temperature. Recent literature reports have shown that some of the MOFs, like Zr-based and Al-based MOFs, have achieved a high hydrogen storage capacity of more than 9 wt%.

Similarly, there are Covalent Organic Frameworks: COF-102, COF-103, COF-10 with surface areas of 3620, 3530, 1760 m^2/g respectively. They have been reported to store 72.4, or 70.5, 39.3 mg/g of the material respectively at a pressure of 1 bar and 77 K.

Examples of the different classes of adsorption-based materials are included in Table 15.3. Similarly, the different carbon-based nanostructures, for example the activated carbon, e.g., AX-K5 with a surface area of 3190 m^2/g is reported to store 7.08 wt% at 60 bar pressure and 77 K temperature. Single-walled carbon nanotubes can store around 4.2 wt% at a pressure of 120 bar and 298 K, multiwalled carbon nano tubes can store around 6.3 wt% at 148 bar and 298 K. Carbon nanofibers with a very small specific surface area have been reported to store 6.54 wt% at 120 bar and 298 K.

In the case of zeolites like NaA, NaX, or NaY are reported to have a smaller capacity of 1.5 wt%, 1.8 wt%, 1.8 wt% at 50 bar pressure and 77 K, respectively. The reason for the low capacity observed in zeolite is that several of the cages are not accessible to the hydrogen molecule for its uptake.

ZSM-5 has been reported to have a storage capacity of 1.97 wt%, 0.65 wt%, 0.38 wt% at 70 bar pressure, and at different temperatures like 77 K, 195 K, and 293 K, respectively. MCM-41 at 35 bar and 77 K has been reported to have a capacity of 2.01 wt%, while at 10 bar pressure and 298 K the capacity observed is 0.69 wt%.

References

1. R.H. Fowler, E.A. Guggenheim, *Statistical Thermodynamics; A Version of Statistical Mechanics for Students of Physics and Chemistry* (Macmillan, University Press, New York, 1939)
2. S. Dalai, S. Vijayalakshmi, P. Shrivastava, S.P. Sivam, P. Sharma, Effect of Co loading on the hydrogen storage characteristics of hollow glass microspheres (HGMs). Int. J. Hydrogen Energy **39**, 3304–3312 (2014)
3. S. Dalai, V. Savithri, K. Yeon Choo, P. Sharma, Magnesium and iron loaded hollow glass microspheres for hydrogen storage (HGMs). Int. J. Hydrogen Energy **39**, 16451–16458 (2014)
4. S. Dalai, S. Vijayalakshmi, S. Param Sivam, P. Sharma, Preparation and characterization of zinc loaded hollow glass microspheres (HGMs) from amber coloured glass frits. Adv. Mater. Res. 864–867, 668–671 (2014)

Chapter 16
Solid-State Hydrogen Storage—Absorption-Based Materials

The adsorption-based materials used for solid-state hydrogen storage, although they show very good kinetics, good reversibility, and high cyclic life but since the hydrogen surface interactions involve weak Van der Waals forces; thus, the storage is carried out at either 77 K or higher pressure. Thus, the requirement of low temperature or high pressure is a challenge while using such materials. Another class of materials which can be used for solid-state hydrogen storage includes absorption-based materials.

In the absorption-based materials, the hydrogen molecule disintegrates or dissociates into hydrogen atoms and this hydrogen atom undergoes few steps occupying the interstitial sites and form a relatively stronger bond with the material.

As shown in Fig. 16.1a in the case of absorption-based materials, the hydrogen molecule dissociates into hydrogen atom, gets into the bulk of the material and as such hydrogen storage in an absorption-based materials is a bulk phenomenon compared to the adsorption-based phenomena which is a surface phenomenon as shown in Fig. 16.1b. Besides another difference is that the bond formed is stronger compared to that in case of adsorption-based materials.

As mentioned in previous chapters, the solid-state storage method is the most volumetrically efficient method and requires much lesser volume to store the same quantity of energy as compared to the other conventional methods of hydrogen storage. In terms of energy density, gasoline has agravimetric and volumetric energy density of 47.3 MJ/kg and 35.5 MJ/L, repectively, whereas for the solid-state hydrogen storage, these values are 1.944 MJ/kg, and 12.63 MJ/L, respectively, as mentioned in Table 16.1. The gravimetric and volumetric storage capacities are defined as follows:

$$\text{gravimetric storage capacity} = \frac{m_H}{m_{MH} + m_H} \tag{16.1}$$

$$\text{volumetric storage capacity} = \frac{m_H}{V} \tag{16.2}$$

P. Sharma, *A Textbook on Hydrogen Energy*,
https://doi.org/10.1007/978-981-95-4601-5_16

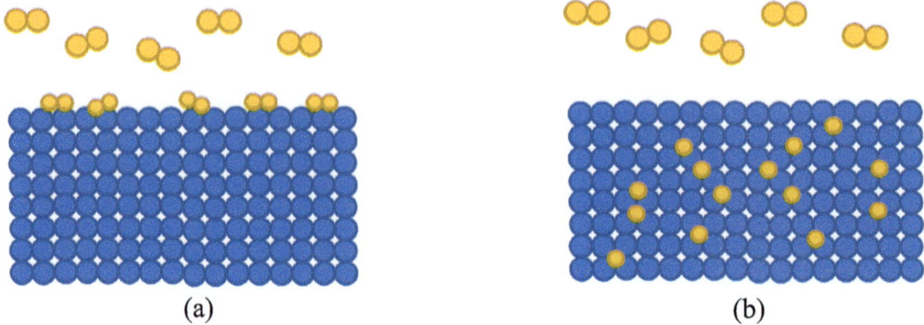

Fig. 16.1 Adsorption and absorption in case of materials for hydrogen storage

Table 16.1 Comparison of the energy density per unit mass and per unit volume for gasoline and different methods of hydrogen storage

Medium	Energy density, MJ/kg	Energy density MJ/L of vol
Solid storage ($LaNi_5H_{7.0}$)	1.944	12.63
Liquid hydrogen	141.9	10.075
Gaseous hydrogen (100 bar)	141.9	1.170
Gasoline	47.3	35.5

Metal Hydrides

Metals and alloys, when exposed to hydrogen, can reversibly absorb hydrogen. Hydrogen molecules on interaction with certain materials dissociate into two hydrogen atoms, diffuses within the metal matrix and form a bond with the metal atoms, forming the hydrides of those metals or alloys. The nature of the bond formed between the hydrogen and metal can be ionic, covalent, or metallic. Hydrogen with its one electron in the 1s orbital is very reactive and can react with almost all elements of the periodic table forming hydrides. Thus, depending upon the nature of the bond formed, these hydrides can be classified into different types. As shown in Fig. 16.2, the elements highlighted in green form ionic hydride, elements highlighted in blue form covalent hydride, those in pink form metallic hydride, in purple form covalent polymeric hydride, while those in yellow form metastable hydride (Fig. 16.3).

Figure 16.3 represents the classification of hydrogen storage materials. The ionic hydrides are formed with the alkali or alkaline earth metals. These are usually formed when the electronegativities of the two atoms are much different. The hydrogen in ionic hydrides usually occurs as a strong negatively charged ion (H^-). Ionic hydrides are compounds in which hydrogen forms an ionic bond with the corresponding element; they exhibit high conductivities, either below or at the melting point, e.g., NaH, CaH_2, $LiAlH_4$, $NaAlH_4$, etc.

The covalent hydrides are formed when the electronegativities of the two elements, i.e., hydrogen and an element of the periodic table are almost similar. In this case,

1	2	3	4	5	6	7	8	9	10	11	12	13	14	15	16	17	18
H 2.20																	He
LiH 0.97	BeH_2 1.47											BH_3 2.01	CH_4 2.50	NH_3 3.07	H_2O 3.50	HF 4.10	Ne
NaH 1.01	MgH_2 1.23											AlH_3 1.47	SiH_4 1.74	PH_3 2.06	H_2S 2.44	HCl 2.83	Ar
KH 0.91	CaH_2 1.04	ScH_2 1.20	TiH_2 1.32	VH VH_2 1.45	Cr 1.56	Mn 1.60	Fe 1.64	Co 1.70	$NiH_{0.5}$ 1.75	CuH 1.75	ZnH_2 1.66	GaH_3 1.82	GeH_4 2.02	AsH_3 2.20	H_2Se 2.48	HBr 2.74	Kr
RbH 0.89	SrH_2 0.99	YH_2 YH_3 1.11	ZrH_2 1.22	NbH NbH_2 1.23	Mo 1.30	Tc 1.36	Ru 1.42	Rh 1.45	$PdH_{0.5}$ 1.35	Ag 1.42	CdH_2 1.46	InH_3 1.49	SnH_4 1.72	SbH_3 1.82	H_2Te 2.01	HI 2.21	Xe
CsH 0.86	BaH_2 0.97	LaH_2 LaH_3 1.08	HfH_2 1.23	$TaH_{0.5}$ 1.33	W 1.40	Re 1.46	Os 1.52	Ir 1.55	Pt 1.44	AuH_3 1.42	HgH_2 1.44	TlH_3 1.44	PbH_4 1.55	BiH_3 1.67	H_2Po 1.76	HAt 1.96	Rn
Fr	Ra	AcH_2 1.00															
			CeH_3 1.06	PrH_2 PrH_3 1.07	NdH_2 NdH_3 1.07	Pm	SmH_2 SmH_3 1.07	EuH_2 1.01	GdH_2 GdH_3 1.10	TbH_2 TbH_3 1.10	DyH_2 DyH_3 1.10	HoH_2 HoH_3 1.10	ErH_2 ErH_3 1.11	TmH_2 TmH_3 1.11	YbH_2 YbH_3 1.06	LuH_2 LuH_3 1.14	
			ThH_2 1.11	PaH_2 1.14	UH_2 1.22	NpH_2 NpH_3 1.22	PuH_2 PuH_3 1.22	AmH_2 AmH_3 1.2	Cm	Bk	Cf	Es	Fm	Md	No	(Lw)	

Ionic hydrides
Covalent hydrides
Covalent polymeric hydrides
Metallic hydrides
Metastable hydrides

Fig. 16.2 Different types of hydrides based on the bonds formed with the different elements of the periodic table

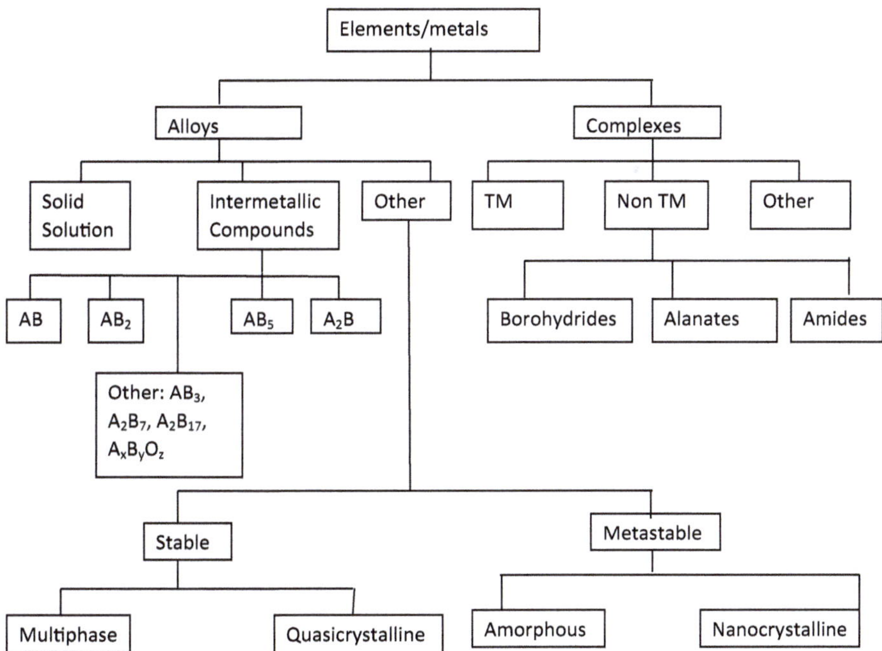

Fig. 16.3 Classification of the absorption-based materials for hydrogen storage (reproduced from [5] with permission)

the sharing of electrons occurs resulting in the formation of hydride. Usually, these hydrides are formed with the non-metals, and the interaction involved is a weak Van der Waals force. Thus, the covalent hydrides are usually not used for hydrogen storage, because these are low melting point solids, low boiling point liquids, or present in gaseous form. These hydrides have toxicity and low thermal stability, thus making them not useful for hydrogen storage. Some of the examples of covalent hydrides are NH_3, H_2S, H_2O, etc.

Metallic hydrides are formed when hydrogen reacts with either transition metals or rare earth metals or lanthanide and actinide series. The bond formed is metallic. This class of alloys exhibits easier uptake/release of hydrogen, good reversible hydrogen storage capacity, and are most widely considered for hydrogen storage. Examples of the metallic hydrides are TiH_2, FeTi, VH_2, etc.

Advantages of Metal Hydrides-Based Hydrogen Storage

The important category of absorption-based materials are the metal hydrides and they are of specific interest due to their good reversible hydrogen storage capacity. These are simple in the sense, that materials or alloys can directly combine chemically with hydrogen at certain pressure and temperature conditions, forming metal hydrides. The uptake and release of hydrogen are a simple process in the case of metal hydrides subjected to temperature and pressure changes and complex systems are not required for analysis.

The volumetric storage capacity of metal hydrides is very high, even higher than that of liquid-state hydrogen storage, approximately by 60%. High volumetric capacity means that more hydrogen can be stored in a given volume as compatred to the compressed or liquid hydrogen storage. Thus the solid-state hydrogen storage systems are very compact and dimensionally smaller in size compared to other conventional hydrogen storage systems.

Hydrogen in the form of metal hydrides can be stored for a longer duration, i.e., larger quantities of hydrogen can be stored without any loss for an indefinite time, catering to durational changes, and there are no dormancy-related challenges as well as boil-off losses as observed in liquid-state hydrogen storage. The metal hydrides used for hydrogen storage should exhibit very good reversibility, cyclic life, and versatility in terms of applications, which is vital toward end use.

The operating conditions for hydrogen storage in case of metal hydrides are moderate, thus neither very high pressure (like 350 to 700 bar in the case of compressed state hydrogen storage) nor very low temperature (like in the case of liquid state hydrogen storage) are required. The operating temperatures may range from near ambient temperature (for vehicular applications) to high temperature (for thermal energy storage applications), and pressure can be close to 1 bar (for near ambient equilibrium pressure) to high pressure (for use in non-mechanical compressors) depending on the selection of different types of metal hydrides with respect to desired output effect or applications. The best part is that the required set of operating conditions (T & P) for any specific applications can be obtained by tuning the composition of the alloys. Thus, the near ambient conditions operation saves in terms of energy and cost. The energy saving, in terms of the compression work required for higher pressures of 350–700 bar or energy for liquefaction to 20 K in conventional storage methods, because metal hydrides operate near ambient conditions. Besides, the cost saving is also because of simple operations, and unlike compressed state and liquid-state storage, they neither require CAPEX-intensive compressors or liquefiers or liquefaction process, nor highly insulated tanks as for liquid hydrogen or carbon fiber-reinforced composite tank as in compressed hydrogen storage. Since the hydrogen charging pressure can range from 10 to 35 bar corresponding to certain class of metal hydrides, thus metal hydrides-based systems can be directly charged with the hydrogen obtained from the electrolyzer output; for example, the pressure output from a PEMFC is around 24 to 30 bar, and up to 35 bar in case of Anion Exchange Membrane Electrolysis and this pressure is sufficient enough to charge metal hydride-based hydrogen storage systems; thus, there is no requirement of highly CAPEX and OPEX-intensive hydrogen refueling stations (HRS) [1–3].

Metal hydrides-based systems are safest among all the methods of hydrogen storage, since the operating temperature and pressure are optimum and close to ambient conditions. Also in case of any leak, the release of hydrogen may stop on its own as the desorption is an endothermic process, absence of a continuous energy supply equivalent to the enthalpy of decomposition either from the thermal mass of the system or surroundings resulting in a cooling effect, and thus the release of hydrogen gets stalled. Thus metal hydrides-based solid-state hydrogen storage systems are very safe to handle and store hydrogen [4].

Mechanism of Hydride Formation

Metals or alloys when exposed to hydrogen, as shown in Fig. 16.4, the molecules of hydrogen in the bulk are far away from the surface as shown in Fig. 16.4a. As the hydrogen molecules approach the surface of metal or alloys, it interacts with the surface atoms as shown in Fig. 16.4b. Initially, the interaction is by means of weak Van der Waals forces; thereafter, the hydrogen molecule dissociates into hydrogen atoms on the surface as shown in Fig. 16.4c. The hydrogen atoms get chemisorbed after dissociation in the third step as shown in Fig. 16.4c. Thereafter, hydrogen enters into the matrix of the material, it occupies the subsurface sites, further it diffuses deep into the bulk and then it occupies the bulk lattice sites as shown in Fig. 16.4d (Figs. 16.4 and 16.5).

$$M + \frac{x}{2}H_2 \leftrightarrow MH_x + Q \tag{16.3}$$

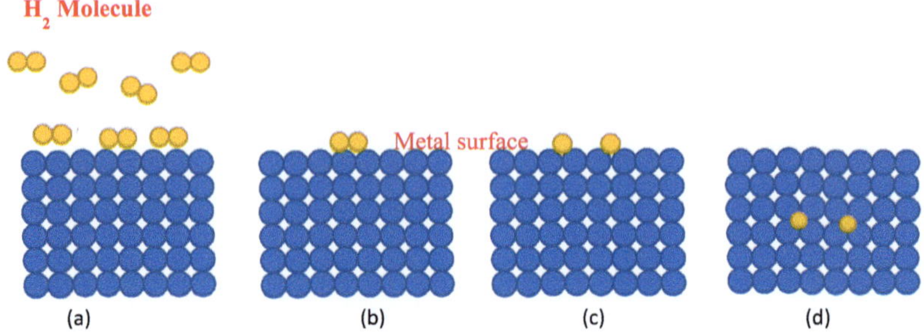

Fig. 16.4 Steps involved in formation of metal hydrides when metal surface is exposed to hydrogen molecules

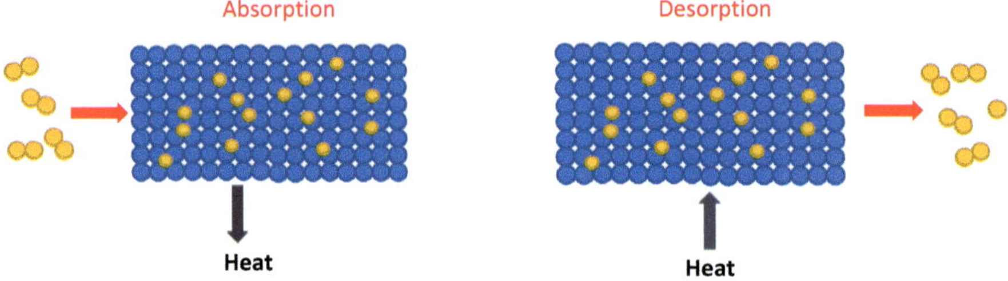

Fig. 16.5 Absorption and desorption processes in metal hydrides

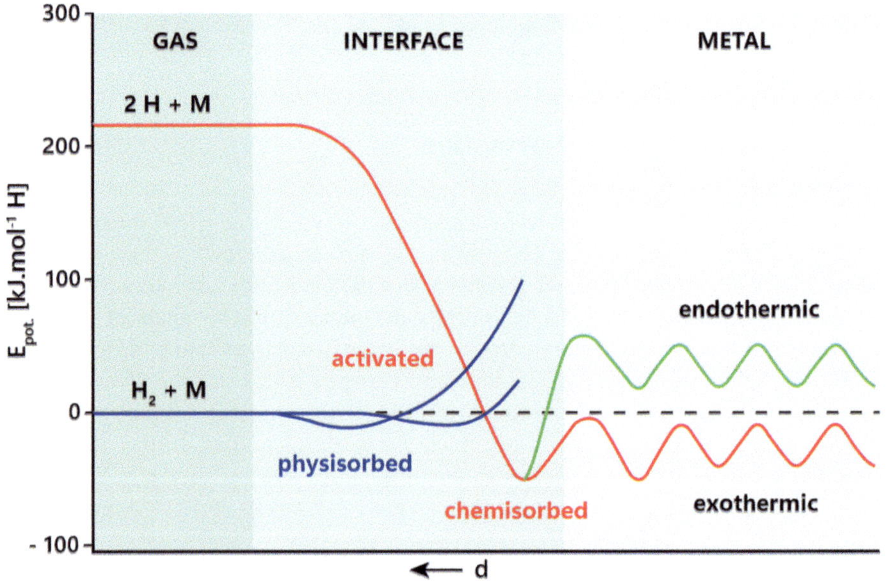

Fig. 16.6 Potential energy diagram showing formation of metal hydrides

To explain the process of hydrogenation or absorption and dehydrogenation or desorption in metal hydrides, it can be represented by a Eq. (16.3) and can be visualized from Fig. 16.5. The "M" which is either a metal or an alloy forms a metal hydride or solid solution or intermetallic on reacting with hydrogen molecule. During the forward reaction of formation of metal hydride (MH) certain amount of heat "Q" is being released as the process of MH formation is exothermic with "Q" being the heat of formation or enthalpy of formation. The metal hydride which is being formed, has a lower energy compared to the metal and the hydrogen molecules, so heat is released in the forward reaction. However, the reverse reaction wherein hydrogen is released from the metal hydrides is an endothermic reaction, i.e., "Q" amount of heat needs to be supplied to desorb the hydrogen, wherein "Q" is termed as enthalpy of decomposition. For the absorption process, the supply pressure should be higher than the equilibrium pressure, the equilibrium pressure changes with change in the temperature and thus depends on temperature, which can be correlated with the van't Hoff equation. The condition for desorption is that the equilibrium pressure of the alloy at any given temperature should be higher than the outlet pressure to get the desired rate of hydrogen. Thus, changing the pressure and temperature will result into either absorption or desorption of hydrogen, in these materials.

Considering in more detail, the formation of metal hydrides can be explained using the potential energy diagram as shown in Fig. 16.6.

Initially, when hydrogen is far away from the metal surface, it is present as a hydrogen molecule. On the metal surface, the initial interaction is through the weak Van der Waals force, and the hydrogen molecule gets physisorbed on to the surface with an adsorption enthalpy of roughly 4–10 kJ/mol, the minimum distance is equal to one molecular hydrogen radius. When it approaches the metal surface, it needs to overcome the energy barrier to dissociate into hydrogen atoms. This dissociation occurs when the required energy for dissociation (depending upon pressure, temperature, and the metal/alloy which is forming metal hydride) is provided and the supplied energy is higher than the activation energy barrier. Thus, on the metal surface hydrogen molecule dissociates into two hydrogen atoms and the required dissociation energy corresponds to 435.9 kJ/mol.

Thereafter, the dissociated hydrogen atom enters the metal surface, and first forms a solid solution, which is generally a bond formation with a metal surface and atom by sharing electron, thus contributing its electron to the band of metal. This bond formation of a hydrogen atom with the surface metal atom results in a chemisorption having stronger bond formation than physisorption and thus this process step has a higher absorption energy in the range of 20–50 kJ/mol. On the surface of metal, the chemisorbed hydrogen due to its high surface mobility interacts with other atoms and forms surface phases, resulting in high surface coverage. Thereafter, the chemisorbed hydrogen atom enters into the subsurface region of the metal, i.e., starts penetrating into the surface ofthe metal. The hydrogen atom then diffuses inside the bulk and occupies the interstitial sites of the metal matrix, forming a metal hydride. When the gas pressure increases, the concentration of hydrogen in the bulk metal matrix will increase until the metal hydride phase starts to form and nucleates, and in that phase the hydrogen will occupy a specific interstitial site, depending upon the crystal

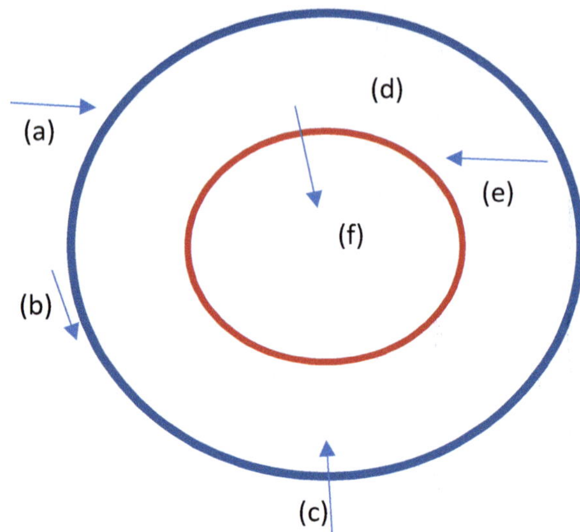

Fig. 16.7 Steps involved in the formation of metal hydrides

structure either the octahedral or tetrahedral site. This hydrogen insertion leads to the metal lattice expansion.

Thus the entire process can be depicted with Fig. 16.7, and the sequence of steps taking place during absorption are as follows:

(a) physisorption;
(b) chemisorption, dissociation;
C) penetration of the metal surface;
(d) formation of the hydride phase;
(e) diffusion;
(f) nucleation and growth of the hydride phase.

The reverse of the sequence of steps mentioned above occurs at the time of desorption.

The hydrogen entering the metal matrix or the alloy occupies the interstitial sites, and gets into the different voids of the lattice. The possibilities are that it can go into either tetrahedral or octahedral voids of the different structures like BCC or HCP or FCC type of lattice.

When a hydrogen atom enters the lattice and the concentration of hydrogen is lower in the metal matrix, it forms a solid solution for a hydrogen atom to the metal atom ratio (H:M) of less than 0.1, it dissolves forming an α-phase. Thereafter, atoms enter into the lattice, leading to expansion of the lattice, also leading to reduction in the symmetry. The typical expansion which is considered to be of around 2–3 Å^3 hydrogen atom, corresponds to 10–20% by volume of the metal lattice. Further, the stability of these hydrides also depends upon the size of these interstitials. If the hydrogen atom enters the BCC lattice, occupying tetrahedral and octahedral voids,

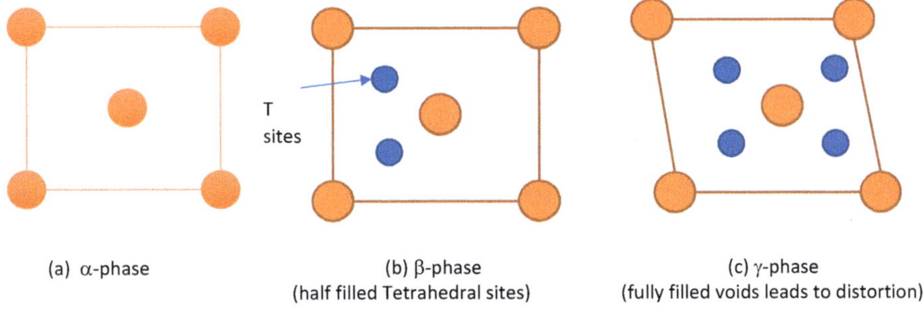

Fig. 16.8 Possible sites occupied by hydrogen atom in the process of forming hydrides

they get distorted largely and in the case of FCC, the hydrogen concentration in the octahedral sites is comparatively lower. During the α-phase, it forms a solid solution, thereafter starts to form hydride and enters into the tetrahedral sites, which are half filled; thereafter, when voids are completely filled leads to significant distortion of the lattice as shown in Fig. 16.8. Once hydrogen concentration exceeds the ratio of 0.1, the formation of hydride occurs and with the resulting deformation or lattice expansion, a lot of stress is created due to expansion at the phase boundaries which is responsible for hydrogen-related decrepitation and later during repeated cycles, reduces the mechanical strength of the materials, making it brittle.

Measurement Techniques for Hydrogen Absorption and Desorption

When the metals or alloys are exposed to hydrogen, they form metal hydrides and the amount of hydrogen uptake is represented in terms of the gravimetric storage capacity. In order to measure the capacity, there are different techniques which can be used. These measurement techniques can be broadly categorized as volumetric, gravimetric, or temperature-programmed desorption as mentioned in Fig. 16.9, which can be carried out for determining the gravimetric storage capacity of the alloy, rate kinetics, pressure-composition-isotherms curves, and the thermodynamic parameters like enthalpy and entropy for the absorption and desorption process. Using these techniques, the hydrogen uptake or release rate, i.e., kinetics, can also be measured, which is important when refueling and controlled or dynamic discharge of the metal hydride-based systems is considered. Also, how long an alloy or material retains its gravimetric storage capacity, i.e., cycle life, can also be quantified (Fig. 16.9).

The schematic showing the principle behind the gravimetric and volumetric techniques are shown in Fig. 16.10a and b respectively

In the volumetric-based measurement technique, a known amount of hydrogen at an initial hydrogen pressure (P_1) is entered into a known calibrated volume (V_1) and the material is allowed to absorb hydrogen. On absorption, the hydrogen pressure

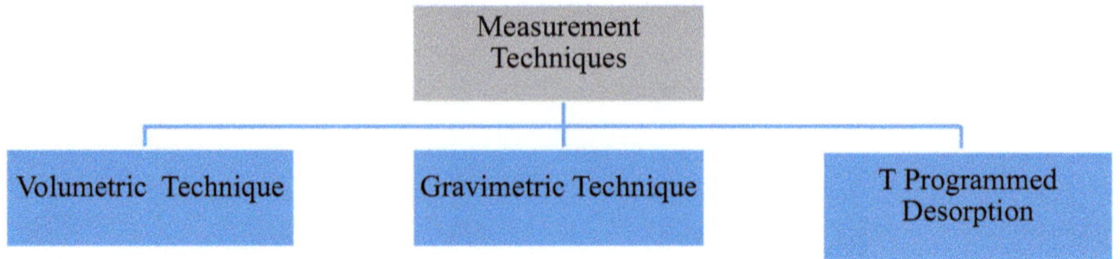

Fig. 16.9 Different techniques to measure hydrogen uptake and release from metal hydrides

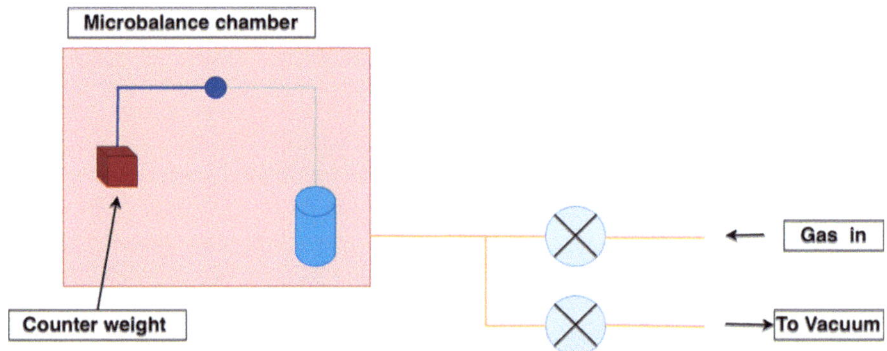

Fig. 16.10 Simple schematic showing the **a** volumetric measurement technique setup and **b** gravimetric measurement technique

decreases (P_2) as the hydrogen is absorbed in to material matrix. Since the pressure difference on absorption can be measured, all the volumes in the systems are calibrated and the initial pressure at which hydrogen was introduced is known, the amount of hydrogen can be calculated using the different equations of state. During desorption, a certain amount of heat is supplied, which is nearly equivalent to the desorption enthalpy. The hydrogen is released from a sample cell characterized as a volume V_2, such that it expands in Volume V_1 initially at lower pressure than V_2, leading to increase in the pressure of the complete calibrated system volume (V_1+V_2), which can be measured, and the amount of hydrogen desorbed can be found using the appropriate equation of state.

In the gravimetric method, instead of recording the change in pressure, the change in mass is measured. A precise, sensitive, and precisely calibrated microbalance is used for measuring the change in mass due to absorption of the material when subjected to different hydrogen gas pressures. This is a direct method but it has issues like the effect of buoyancy leading to measurement error.

The third method is different from the previous two methods: instead of keeping the samples at constant temperature in gravimetric and volumetric-based measurement techniques, in the temperature-programmed desorption method, the desorption process is carried out at varying temperatures.

The variation in the amount of hydrogen uptake at different pressure and at constant temperatures is usually measured using the either volumetric or gravimetric techniques as mentioned above, where pressure is dosed in steps and concentration gradient obtained corresponding to dosed pressure, i.e., aliquot is recorded. In order to find the hydrogen absorption and desorption characteristics, the outcomes of these measurements are plotted in the form of curves which are known as pressure-composition-isotherms (PCI) or pressure-composition-temperature (PCT) curves.

In these isotherm curves for metal hydrides, the concentration of hydrogen, i.e., amount of hydrogen uptake by material in terms of H/M (concentration of hydrogen, where H is number of hydrogen atoms and M is number of metal atoms in the lattice) or wt% is plotted against the pressure.

Considering the PCT isotherm as shown in Fig. 16.11, a typical PCT has three regions as marked in the curve.

If this curve is studied for the absorption characteristics of the alloy, the process starts from 100% metal and 0% hydride phase to 100% metal hydride and 0% metal phase after attaining maximum gravimetric hydrogen capacity for a known mass of alloy. The middle portion of the curve, also referred to as the plateau, the metal and metal hydride portion constitutes equal percentages at the midpoint of width of plateau, i.e., both have 50% phase percentage, which can also be correlated to completion of 50% of the hydrogen storage capacity of an alloy. So, primarily along the plateau, the transformation from the metallic state to the metal hydride state occurs, and after complete conversion of the metal to the metal hydride phase, the metal phase is absent only the hydride phase is present.

According to Gibbs phase rule, along the plateau region, there is a gain of one degree of freedom, and to increase the capacity further beyond the plateau region,

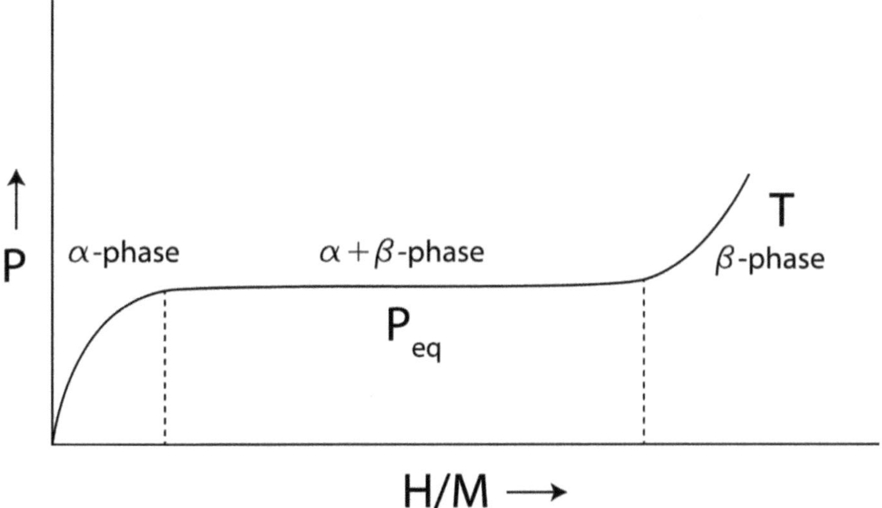

Fig. 16.11 Typical *P–C–T* of a metal hydride

the pressure should be increased, such that hydrogen will enter in solid solution and form the metal hydride. This is a standard phase transition toward the formation of a metal hydride. The plateau pressure changes as temperature changes, wherein plateau pressure increases as temperature increases. Following the PCT study, a van't Hoff plot can be obtained, which represents a linear plot of the logarithm of the plateau pressure as a function of 1/T, wherein the slope and intercept of the van't Hoff curve yields the enthalpy of the reaction and the entropy of the reaction, respectively.

Region I (α-phase)—Initially, for smaller concentration of hydrogen to metal (usually $H/M < 0.1$), the hydrogen gets dissolved in the metal through an exothermic process, wherein the hydrogen molecules loses its entropy and forms solid solution phase. This region is represented as α-phase in Fig. 16.11. Considering the process in this region, there are two components hydrogen and metal, i.e., $C = 2$, there are two phases namely hydrogen and the α-phase, i.e., $F = 2$. Applying this in the Gibbs phase rule, i.e., $F = C + 2 - P$, the number of degrees of freedom for α-phase is 2. Thus, in this region I or α-phase, the concentration varies with the pressure. Since the concentration of hydrogen is low in this phase or region, it is randomly distributed across the metal lattice and its concentration change with temperature is quite slow.

Region II (α + β – phase)—Following region I, when the pressure is increased further, the concentration of hydrogen keeps on increasing and at a particular state when the metal hydride phase starts to nucleate results into formation of region II. Considering the second region, marked as α + β – phase in Fig. 16.11, as the hydrogen concentration increases, the strong hydrogen–hydrogen interaction occurs and the hydride phase starts to nucleate and grow, leading to lattice distortion, such that the H/M is usually one or more. This hydride phase usually has a different crystal structure and a new phase starts to appear.

In region II, the hydrogen is in a solid solution phase, wherein both the α and β phase (the higher concentration phase) are co-existing. In this region, the number of components are two, i.e., $C = 2$, and phases, P = 3 i.e. hydrogen gas, α-phase and β-phase. Applying this in Gibbs phase rule results in degree of freedom, F = 1 for the α + β – phase. That means as compared to region I, a degree of freedom is lost and thus the capacity increases even if the pressure does not increase, so as the supply pressure of gas increases the more hydrogen enters into the metal hydride phase forming the metal hydride. Thus, the number of degrees of freedom for region II is 1 and as seen in Fig. 16.11, the concentration increases at a constant pressure, where in more hydrogen gets in the metal matrix forming hydride, resulting in a plateau region, wherein the corresponding pressure is called plateau pressure or the equilibrium pressure.

Region III (β – phase) —In the region III, the entire phase is β-phase and the α-phase is not present. In region III of Fig. 16.11, there are 2 components, i.e., $C = 2$ and phases, P = 2, i.e. hydrogen and β-phase, using this in Gibbs phase rule gives degree of freedom, F = 2. There are two degrees of freedom and thus an increase in the concentration is observed as the hydrogen gas pressure rises. Since the only hydrogen phase in metal is β-phase in the region III, with increase in pressure the hydrogen enters in the β-phase and the pressure increases with the increase in concentration. Figure 16.11 is a representation of an ideal PCT curve.

However, in the actual case, there are several irreversibilities involved leading to deviations from the ideal scenario of Fig. 16.11.

$$M + \frac{x}{2}\mathrm{H}_2 \leftrightarrow M\mathrm{H}_x + Q$$

Further section briefly discusses the thermodynamics aspects associated with low and high concentration regions in PCI or PCT curve as described in Fig. 16.11.

Low-Concentration Region—As in Fig. 16.11, there are three regions in PCT and in the first region, initially in the low-pressure region or at lower concentration of hydrogen, hydrogen enters into the host metal lattice and gets randomly arranged into the host lattice, with the concentration slowly varies in this case. Considering the thermodynamic equilibrium condition, the chemical potential of hydrogen in gas phase and in α phase is the same and can be written as:

$$\frac{1}{2}\mu_{\mathrm{H}_2}(P, T) = \mu_{\mathrm{H}}(P, T, C_H) \tag{16.4}$$

where the chemical potential for hydrogen molecule is μ_{H_2}, chemical potential of hydrogen atom in metal host is μ_{H}, P is the operating pressure, T is the temperature, and C_{H} is the concentration of hydrogen (i.e., expressed as H/M, where H/M is the number of hydrogen atoms divided by the number of metal atoms).

For low concentration and in the low-pressure region, H_2 gas is considered ideal gas and chemical potential of hydrogen molecule is given by:

$$\frac{1}{2}\mu_{\mathrm{H}_2} = \frac{1}{2}\mu^{\circ}_{\mathrm{H}_2} + RT\ln P^{\frac{1}{2}}_{\mathrm{H}_2} \tag{16.5}$$

$$\frac{1}{2}\mu^{\circ}_{\mathrm{H}_2} = \frac{1}{2}H^{\circ}_{\mathrm{H}_2} - \frac{1}{2}\mathrm{TS}^{\circ}_{\mathrm{H}_2}. \tag{16.6}$$

Chemical potential of H dissolved in a metal matrix is

$$\mu_{\mathrm{H}} = H_{\mathrm{H}} - \mathrm{TS}_{\mathrm{H}} + RT\ln\left(\frac{a}{b-a}\right) \tag{16.7}$$

where

$\mathrm{TS}_H$is the non-configurational (vibration) part of entropy of H in solid solution.
$RT\ln\left(\frac{a}{b-a}\right)$ is the configurational part of entropy.
b = number of interstitial sites per metal atoms.
a = number of sites occupied by hydrogen atoms.
Under the conditions of equilibrium

$$\frac{1}{2}\mu_{H_2} = \mu_H \tag{16.4}$$

$$\frac{1}{2}H^{\circ}_{H_2} - \frac{1}{2}TS^{\circ}_{H_2} + RT\ln P^{\frac{1}{2}}_{H_2} = H_H - TS_H + RT\ln\left(\frac{a}{b-a}\right) \tag{16.8}$$

$$\left(-H_H + \frac{1}{2}H^{\circ}_{H_2}\right) - T\left(-S_H + \frac{S^{\circ}_{H_2}}{2}\right) = -RT\ln\left(\frac{a}{b-a}\right) + RT\ln P^{\frac{1}{2}}_{H_2} \tag{16.9}$$

$$\Delta H_H - T\Delta S_H = RT\ln P^{\frac{1}{2}}_{H_2} + RT\ln\left(\frac{b-a}{a}\right) \tag{16.10}$$

$$\Delta H_H = H_H - \frac{1}{2}H^{\circ}_{H_2} \tag{16.11}$$

$$\Delta S_H = S_H - \frac{1}{2}S^{\circ}_{H_2}. \tag{16.12}$$

For very low concentration range, i.e., region I; $a \ll b$.

The above equation can be rewritten as

$$RT\ln P^{\frac{1}{2}}_{H_2} = \left[\Delta H_H - T\Delta S_H - RT\ln\left(\frac{b-a}{a}\right)\right] \tag{16.13}$$

$$\ln P^{\frac{1}{2}}_{S_2} = \frac{1}{RT}\left[\Delta H_H - T\Delta S_H - RT\ln\left(\frac{b-a}{a}\right)\right] \tag{16.14}$$

$$P^{\frac{1}{2}}_{H_2} = e^{\left\{\frac{\Delta H_H - T\Delta S_H - RT\ln\left(\frac{b-a}{a}\right)}{RT}\right\}} = K_s \tag{16.15}$$

where $K_s$is equilibrium constant and given by

$$K_s = e^{\left\{\frac{\Delta H_H - T\Delta S_H - RT\ln\left(\frac{b-a}{a}\right)}{RT}\right\}}. \tag{16.16}$$

The equation $P^{\frac{1}{2}}_{H_2} = K_s$is called Sievert's law.

Assumptions for validity of Sievert's law are:

1. Gaseous hydrogen can be considered as an ideal gas.
2. H_2 molecules dissociate into Hatoms before dissolving in metals.

Configurational entropy loss of gaseous hydrogen is roughly constant and has the value of $\Delta S_H \sim -65$ kJ/mol.

The Sievert's law states that the amount of the hydrogen being dissolved into the metal phase, i.e., the equilibrium concentration of hydrogen dissolved in the metal is proportional to the square root of the applied pressure.

High-Concentration Region —For higher pressure or the higher concentration region represented by: (a) region II (α + β – phase region) in the PCT curve having metal hydride formation (b) β-phase formation, in that case there exist three phases of hydrogen, the α-phase and β-phase.

The thermodynamic equilibrium suggests that the chemical potential of the hydrogen in the gaseous form is equal to the chemical potential of the alpha phase which is a function of pressure, temperature, and the concentration in the α-phase and that is also equal to the chemical potential of the β-phase. So, the chemical potential of hydrogen, α-phase, and β-phase, all the three coincides.

$$\frac{1}{2}\mu_{H_2} = \mu_\alpha = \mu_\beta \tag{16.17}$$

$$\frac{1}{2}\mu_{H_2} = \frac{1}{2}kT\ln\left(\frac{P}{P_0}\right) + \mu^0_{H_2} \tag{16.18}$$

Now, using the Gibbs phase rule, in this particular high-concentration region, there is only one degree of freedom as discussed earlier. So, among the four variables, i.e., the concentration in the α-phase (c_α), the concentration in the β-phase (c_β), pressure (P), and temperature (T), only one parameter can vary.

For a particular temperature, all the three parameters c_α, c_β and P are considered fixed. The chemical potential or Gibbs free energy for this concentration region can be written as:

$$\Delta G = RT\ln\left(\frac{P}{P_0}\right) = \Delta H - T\Delta S \tag{16.19}$$

$$\ln\left(\frac{P}{P_0}\right) = \frac{\Delta H}{RT} - \frac{\Delta S}{R}. \tag{16.20}$$

This expression is known as van't Hoff equation.

In this expression, ΔS is the entropy change of the hydrogen molecule which from the hydrogen gaseous phase now gets dissolved into the metal to form solid solution or hydride phase. Thus, there is a loss of entropy of that hydrogen molecule and this generally remains constant. The value of ΔS for all the metal hydride system is nearly 130 J/mol K and does not depend upon the nature of the metal. However, the other term in the van't Hoff equation is ΔH or the enthalpy change. This enthalpy change is a vital parameter, that represents the stability of the metal hydride and plays a key role in the design of metal hydride systems. The primary requirement for hydrogen storage materials is high H/M, but at the same time the hydrides formed should not be very stable such that release of hydrogen from the hydrides in case of desorption incurs higher energy expense per unit of hydrogen desorbed and this required energy can be quantified by the ΔH value for the desorption reaction obtained from the van't Hoff plot.

Impact of Hydrogenation on Crystal Structure—When hydrogen enters the host metal lattice, it causes a distortion or expansion of the lattice. In case of some of the metal hydrides, the crystal structure does not change when hydrogen enters in to the lattice, only the expansion is observed. However, in case of the other metal hydrides, even the crystal structure changes as the hydrogen content changes (for example, FeTi or Mg_2Ni, or $CaNi_5$).

As mentioned, the ΔH is a function of hydrogen concentration, and has two contributions, i.e., elastic contribution and electronic contribution. When hydrogen enters into host metal matrix, there is a repulsive force experienced that acts into the lattice, leading to the distortion of the lattice causing local distortion. As the hydrogen atom enters lattice, its own potential energy decreases but the lattice energy increases and there is a local distortion within the lattice. If another hydrogen atom enters in this lattice, its potential energy will be lower than the first hydrogen atom and also the potential energy of lattice is also lower, as H–H interaction dominates (until the two atoms starts to repel, wherein distance between hydrogen atoms decreases such that they start to repel and their distance is limited by repulsive interaction), that means the heat of reaction depends on the concentration. The second contribution is due to electronic contribution and this is due to volume independent terms or effects. Now, this is because of the hydrogen atom that enters into the lattice along with an extra electron. Depending upon the host metal, there is a systematic variation observed for the electronic contribution, this effect can either stabilize or destabilize the hydride formed.

A general observation is that the minimum void size, which hydrogen atom occupies should be greater than 0.4 Å and the minimum H–H separation should be 2.1 Å. When hydrogen enters into the metal lattice, it will occupy the interstitial sites either the tetrahedral sites or the octahedral sites.

The three main crystal lattice, namely FCC, BCC, and HCP and their occupation during hydrogenation will vary. In the FCC lattice, for low-to-medium hydrogen concentration, the hydrogen atom will prefer to occupy the octahedral sites. However, in the HCP lattice, tetrahedral or octahedral sites are occupied enclosed in polyhedral and it gets distorted. This distortion occurs as the lattice parameter ratio c/a deviates from 1.633 and under the low concentration of hydrogen, hydrogen will occupy the tetrahedral sites preferably in the HCP crystal lattice. Similarly, in the BCC crystal lattice the polyhedral will get highly distorted because of the incorporation of hydrogen atom, and hydrogen atom preferably occupies the tetrahedral sites.

In Table 16.2, the tetrahedral "T" and octahedral "O" sites available in the different crystal structures, i.e., in FCC, HCP, and BCC are indicated. During hydrogenation, hydrogen atom will occupy sites in such a way that the lower energy sites will be occupied first and then the higher energy sites, so the order goes from the lowest to the highest energy.

However, some of the sites are left vacant, because of either the mutual repulsion that occurs between the hydrogen–hydrogen atoms or because of the reorganization of the interstitial sublattices as hydrogen is already occupied in these sites.

Table 16.2 Number of sites available in different crystal structures

	O site	T site
FCC	1	2
HCP	1	2
BCC	3	6

Impact of Hydrogenation on Electronic Structure—The hydrogen atom when enters in a metal lattice brings an extra electron. Thus, the electronic structure also changes, the symmetry of the electronic states gets modified and the bandwidth reduces due to the lattice expansion. In the case of alkali or alkaline earth metals, the hydride formed is ionic, and the hydrogen getting in to the metal lattice results in an s-band below the Fermi level. Similarly, s-band in transition metals appears below the fermi level, with the nature of hydrogen being H^- and the electron on insertion moves to Fermi surface.

In general, metal hydride bonding band appears below the metal d-band and these electrons are transferred to the new metal band which is being formed, some of the metal states are also formed below the Fermi level. The hydrides which have more than one hydrogen atom per unit cell, due to the inter hydrogen (H–H) interaction, certain new attributes appear in the lower region of the density of states.

At the same time, a shift in the Fermi level is also observed. DFT studies can be carried out to perform band structure calculations, compute the heat of formation, and impact on the various properties of metal hydrides. Understanding the stability of hydrides is very important when it comes to designing the appropriate material for a specific application, because very stable hydrides formed which require high temperature for release of hydrogen. The stability of hydrides formed depends upon the electronic structure of the metal host.

Overall, the change in the energy of formation can have three major components: (1) the change in the crystal structure from unhydrided to hydride phase, this becomes a deterrent especially in BCC, where this contribution reduces the driving force for hydride formation; (2) introduction of the hydrogen in the metal lattice during hydride formation leading to the expansion or distortion of the lattice, loss of cohesive energy or elastic component, and reduction in the lattice energy. This contribution can be varied and will alter the stability of the metal hydrides, thereby designing of appropriate metal hydrides can be accomplished; (3) bond formation between the metal and hydrogen also helps in hydride formation. The interplay of these three contributions plays a decisive role in the hydrogen storage characteristics of metal hydrides, its stability, and the energy of formation.

Understanding the PCT Diagrams of Metal Hydrides

Typical PCT diagram for a metal hydride has three regions as mentioned in the earlier section, the flat plateau region, wherein both solid solution and hydride phase co-exist is an important region for any metal hydride. The PCT curve shown in Fig. 16.11 is drawn at a particular constant temperature and different curve will be obtained at another temperature, thus the pressure versus composition curves depend on temperature. On changing the temperature, the equilibrium pressure, i.e., the plateau pressure will also change.

Equilibrium or plateau pressure is an important parameter to be considered while designing the metal hydride-based systems along with the selection of an appropriate metal hydride for a particular application. The importance of this plateau pressure is that it is known as the equilibrium dissociation (or formation) pressure of hydride, it's the pressure of hydrogen in the equilibrium with metal–metal hydride phases ($\alpha + \beta$ phases) at any constant temperature. This pressure corresponding to the isotherm, also serves as an important measure of the hydride stability since the thermodynamic parameters (particularly ΔH) obtained through the van't Hoff relation directly give the information about the stability of alloys. However, ΔS mostly remains constant. Thus, while designing a metal hydride system at a fixed initial operating temperature, the consideration of equilibrium plateau pressure is important along with the the overall reaction kinetics of material or alloy.

If the pressure composition curves are drawn at different temperatures, it is observed that as the temperature increases, the equilibrium pressure increases. In Fig. 16.12, the curves are shown for three different temperatures T_1, T_2 and T_3 and critical temperature T_c. It is to be noted that with the increase in temperature, the equilibrium pressure increases and the corresponding plateau width decreases.

The important point to note is that the plateau region tells about the miscibility regime of the 2 phases, i.e., α and β, thus with an increase in temperature the plateau pressure increases, the plateau width decreases, and the miscibility region of α and β phases also decreases. This reduction in plateau width occurs because at higher temperatures, the entropy contribution to the Gibbs free energy becomes increasingly significant, which diminishes the thermodynamic force required for phase-separation. Thus, in the three regions with α, $\alpha + \beta$,and β in a typical PCT with an increase in temperature, the phase region for $\alpha + \beta$ reduces. This trend continues till the critical temperature, i.e., T_C which means the width of the plateau continues to decrease until the plateau disappears. Thus, the plateau width becomes zero and the miscibility also reduces to zero at Tc. Thus instead of α going to $\alpha + \beta$ and then going to β, it is observed that beyond the critical point there is no region II, i.e., $\alpha + \beta$, rather the α phase directly converts to β phase continuously with increase in pressure (Fig. 16.13).

Considering Fig. 16.11, which shows a flat plateau region in a PCT curve, as observed in the case of few hydrides, which is important when a metal hydride-based storage system is designed as a flat plateau ensures the constant release of hydrogen at constant pressure for longer duration at constant temperature. However,

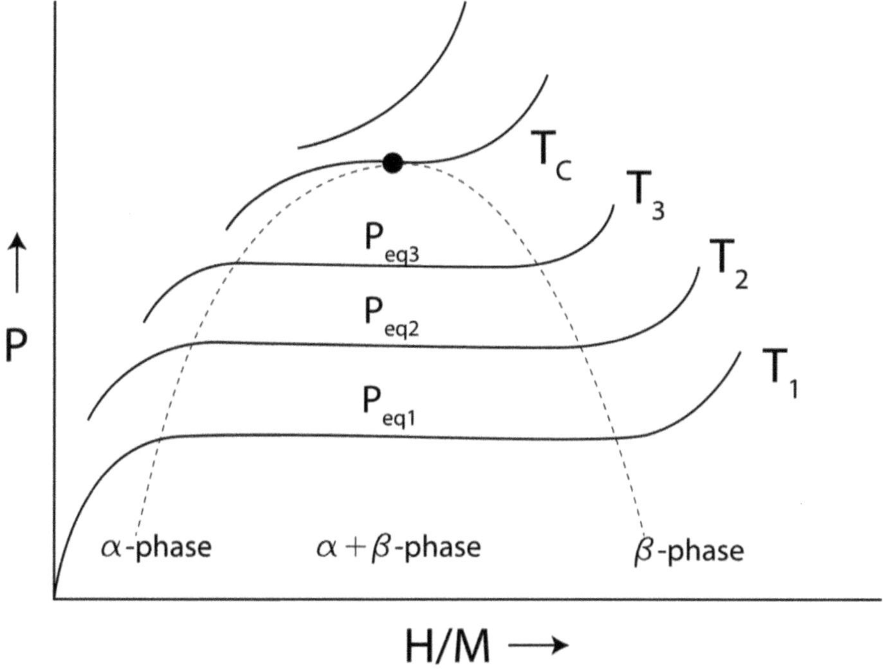

Fig. 16.12 Pressure composition curves drawn at different temperatures

for most of the hydrides, a plateau slope is observed as seen in Fig. 16.13, also called the condition of a sloping plateau. A sloping plateau means that instead of getting a particular or constant value of equilibrium pressure, the equilibrium pressure is observed to vary, and have different values at different locations in the $\alpha+\beta$ region. The significance of the sloping plateau is that it indicates that the different regions of alloys or the host matrix absorb differently at different dosing pressures, resulting in different values of equilibrium pressures in the plateau region.

The reason for the sloping plateau is the presence of different inhomogeneities, impurities or the stresses in the metal matrix due to hydrogen entering the lattice. While designing a hydrogen storage system, the actual plateau in the case of the metal hydride should be selected for that particular application at any given particular temperature. In general, for a sloping plateau the middle point of the plateau region is selected as equilibrium pressure at a given temperature.

Another deviation from the ideal condition is the presence of hysteresis in the PCT curves (Fig. 16.14).

In ideal conditions, during absorption when a PCI is plotted from the measurements after the maximum absorption capacity is reached, then to desorb hydrogen from metal hydrides, the same curve is traced back in the reverse direction. Accordingly, the desortption curve is expected to coincide with absorption curve,

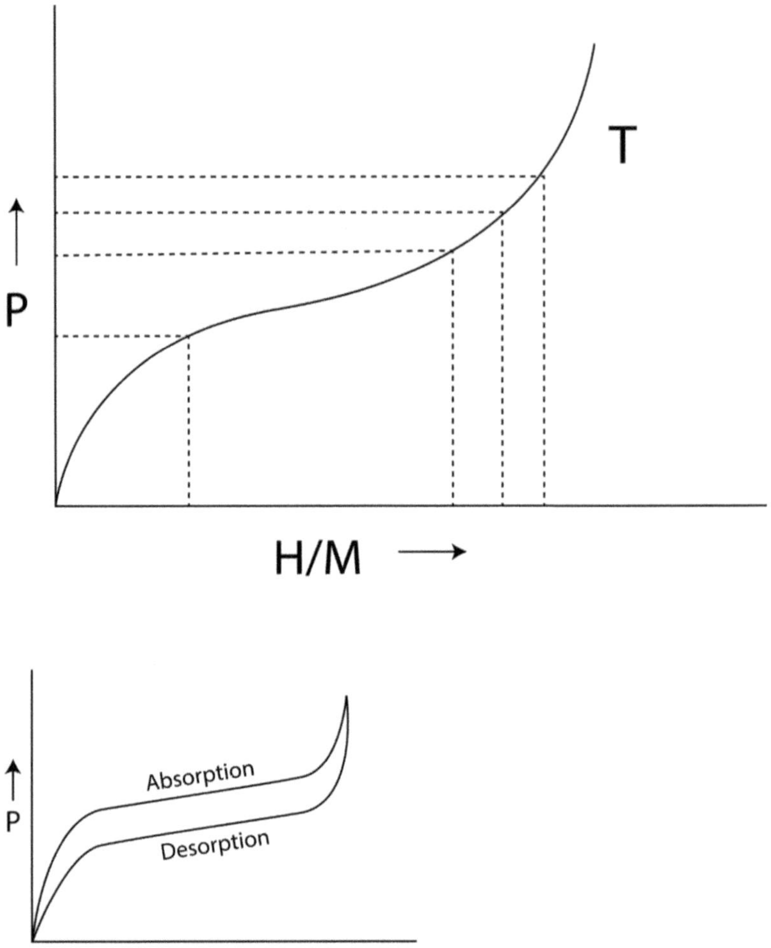

Fig. 16.13 Sloping plateau in the PCT diagram

means desorption curve retracing exactly same path followed during absorption. However, in actual practice for majority of metal hydrides, this trend may not be followed as ideally expected, instead desorption isotherm curve lies below the absorption isotherm curves. Thus, the equilibrium pressure for absorption is not equal to the equilibrium pressure for the desorption, this deviation from the ideal scenario is called hysteresis. There exists a hysteresis gap between the absorption and desorption isotherms and the reason for occurrence of this hysteresis is considered to be the plastic deformations which are not reversible. When hydrogen atoms enters the host metal, lattice expansion is observed resulting in irreversible plastic deformation in the metal matrix. The hysteresis is usually an undesirable phenomenon, but while selecting a metal hydride for any application, consider alloys with lower hysteresis

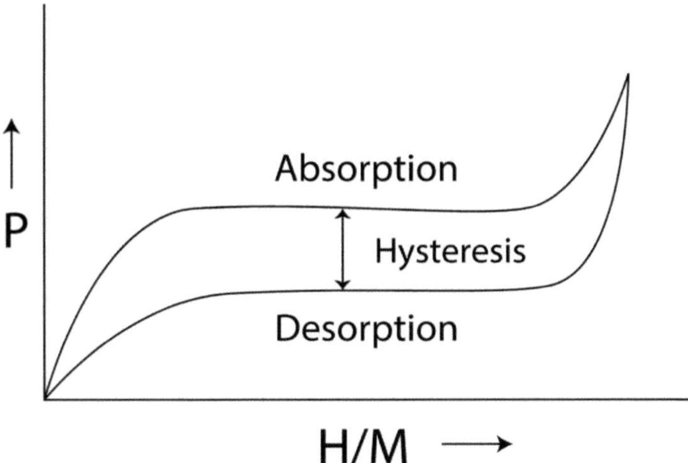

Fig. 16.14 Hysteresis between absorption and desorption isotherm in a PCT diagram

which in turn will have better reversibility. The reason being, this interferes with the effective storage capacity, but here it is important to emphasize that it is not an error, its rather the intrinsic property of the metal hydride, and it appears in every cycle. Hysteresis is must to be considered while designing a hydrogen storage system, because the desorption pressure will be different from the absorption pressure and there might be difference in the reversible storage capacities as well.

Another deviation from the ideal cases discussed above is certain metal hydrides like FeTi, it is observed that instead of getting one plateau in the PCT diagram, there are two or more than two plateaus being observed. There is a plateau which occurs at a lower pressure and another plateau which is at a higher pressure or at a higher concentration. Figure 16.15 shows that for FeTiH there is an α phase which gets hydrogenated to form a β phase, and the β phase further hydrides to give one more phase called the γ phase. Thus, in FeTi, there exists two different plateaus and these are due to two different types of hydrides, i.e., mono hydride and dihydride formation. There is a difference in the stability of these hydride phases and usually the mono hydrides are more stable than the dihydrides. Consideration of the existence of more than one plateau is very important while designing the metal hydride-based system.

Understanding Van't Hoff Plot

The van't Hoff relates the ln P_{eq} to the ΔH value through the expression:

$$\ln\left(\frac{P}{P_0}\right) = \frac{\Delta H}{RT} - \frac{\Delta S}{R}. \tag{16.20}$$

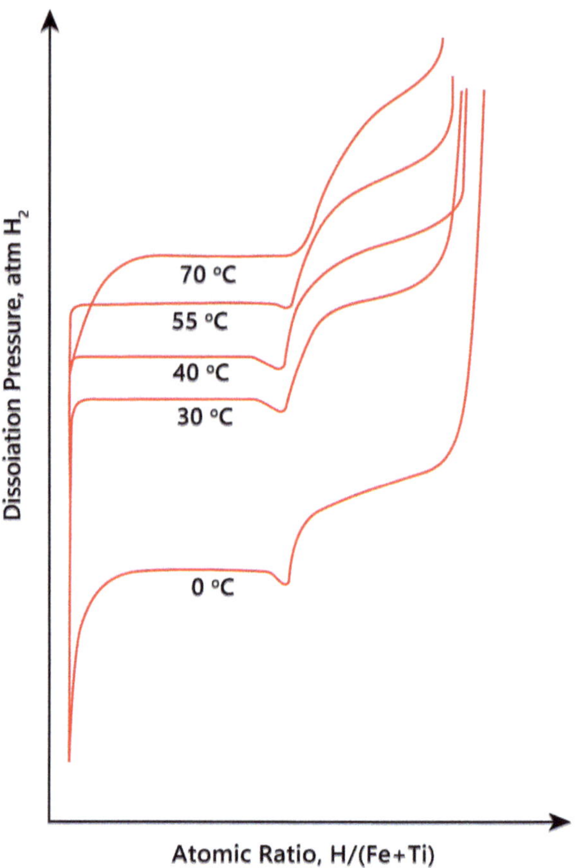

Fig. 16.15 PCT curves for material with multiple plateaus

The value of Peq is obtained on plotting the pressure versus concentration curve at different temperatures in a PCI diagram. On plotting ln P_{eq} with $1/T$ the slope of the straight line is equal to $\Delta H/R$ value, while the intercept is equal to $\Delta S/R$. Thus, the thermodynamic parameters ΔH and ΔS can be obtained from the van't Hoff plot. In case of a flat plateau, the equilibrium pressure remains constant throughout the plateau region and a single value is obtained, as the temperature increases the plateau pressure increases and the plateau width decreases. Usually the middle point of the plateau region can be taken to find the equilibrium pressure.

In case of metal hydrides having a sloping plateau, the pressure also changes with the concentration. In that case, a particular value of *H*/*M is considered*, for sloping plateau condition, and usually the convention is to take the middle point of the plateau region at different temperatures for considering the P_{eq} in the van't Hoff equation.

The third condition could be metal hydrides which exhibit hysteresis in the PCI plots. Thus, in that case there is different equilibrium pressure for absorption and desorption, respectively. Considering the sloping plateau together with hysteresis, the equilibrium pressure to be taken is generally the middle point of the PCT curve of either absorption or desorption, but it has to be consistent, for each tempertaure considered for isotherm study, i.e., T_1, T_2, T_3 while reporting enthalpy change ΔH and entropy change ΔS values $_1$. Thus, from the van't Hoff plot the values of ΔH and ΔS can be obtained. The ΔH value is a key indicator of the stability of the metal hydride, if the ΔH value is high then metal hydride has a higher stability and it will have a low dissociation pressure or high temperature will be required to desorb hydrogen from the metal hydride system.

However, if the value of ΔH is low, then the corresponding metal hydride is less stable and it will exhibit a higher dissociation pressure or low temperature will be required for the desorbing hydrogen. For a metal hydride that operates under optimum set of conditions and from the metal hydride system design perspective, the value of ΔH should neither be very high nor it should be very low.

An optimum or an intermediate value is required for an ease of uptake and release of hydrogen from the metal hydride-based systems, this value is desired to lie in the range of 30–60 kJ/mol.

Terminologies

1. **Volumetric Hydrogen Capacity**—It represents the amount of hydrogen being stored in the system or in the material per unit volume being expressed in g/L. The volume will be of the system or material depending on which one is required to be measured or expressed.
2. **Gravimetric Storage Capacity** — It is the amount of hydrogen stored per unit mass of the material or system. It can be expressed in two ways either as H/M, i.e., the hydrogen to metal atoms ratio or it can also be expressed in terms of wt%.

$$\text{wt\%} = \frac{(\frac{H}{M})M_{\text{H}}}{M_{\text{material}} + \frac{H}{M}M_{\text{H}}} \times 100\% \tag{16.21}$$

where
H/M = hydrogen to metal atom ratio,
M_{H} = molar mass of hydrogen,
M_{material} = molar mass of the host material.

3. **Reversible Storage Capacity of a Material**—It is the quantity of hydrogen absorbed and desorbed between the lower and upper plateau pressures of a hydrogen storage. The reversible storage capacity is lower than the maximum storage capacity and the reversible storage capacity is usually the width of the plateau region and is given by $\Delta(H/M)_r$.

Considering a typical PCT diagram, we can find the reversible and maximum hydrogen storage capacity, hysteresis, and plateau slope as shown in Fig. 16.17

$$\text{Hysteresis} = \frac{P_{\text{eq,abs}}}{P_{\text{eq,des}}} \tag{16.21}$$

$$\text{Plateau slope} = \frac{\mathrm{d}\ln P}{\mathrm{d}\left(\frac{H}{M}\right)}. \tag{16.22}$$

Activation—Surface oxide layers present on the metal hydrides due to their exposure to air acts as a barrier for hydrogen absorption. The top oxide layer can be of different thickness which depends upon the synthesis method. Presence of the surface oxide layer blocks the hydrogen from entering into the metal to form metal hydride; thus, it is crucial to overcome this surface oxide layer barrier, for which activation is being performed and the process is also known as first hydrogenation. Usually, activation is performed at high temperature and pressure, such that hydrogen can penetrate inside this oxide layer and when hydrogen enters into the metal lattice, the expansion occurs while contraction is observed on release. The repeated absorption-desorption cycles result in repeated expansion and contraction resulting in fine alloy particles of few microns to nanometers size, increasing the specific surface area, creating fresh surfaces or active sites that improve both the activation and kinetics of the process. While designing a metal hydride-based system, the activation conditions (high T and P) of the selected metal hydrides are very important.

Kinetics—Hydrogenation is a complex process and involves several steps, any of these can be the rate limiting step. The overall kinetics is limited by the slowest step or the rate limiting step. Plots of reaction fraction with time gives the information on kinetics of a metal hydride and the dependence of this plot on temperature and pressure is helpful in finding out the rate limiting step. Kinetics plays a key role in analyzing the charging and discharging time in the development of a metal hydride-based hydrogen storage system. Factors which affect kinetics includes sample size, particle size, surface properties, purity of gas, and solid.

Cycle Life—While designing a metal hydride-based hydrogen storage system, it is essential that the cycle life of material should be high, which means that the material should retain its gravimetric hydrogen storage capacity over multiple absorption and desorption cycles. This capacity retention over several cycles means capacity kinetics, reversibility, and plateau pressure of alloy or material should remain constant during repeated cycling and ideally during the entire life of use of that material as well as of the system. There are several factors which govern the cycle life of these materials such as disproportionation (the expansion contraction of volume on hydrogenation and dehydrogenation), presence of impurities (metal hydrides get poisoned in presence of impurities), agglomeration of the particles, any structural or phase changes involved, formation of non-hydriding phases, and formation of stable hydrides which are very difficult to desorb under normal conditions. These factors can reduce the cycle life and storage capacity of the metal hydrides. The potential method to improve cycle life can be by means of substitution.

Decrepitation—Self-pulverization of alloys due to change in crystal structure volume during the cycling process makes the metal hydrides brittle in nature. Metal hydride after multiple hydrogenation/dehydrogenation cycles turn into the very fine powder, which could be pyrophoric in nature.

It is an important characteristic that needs to be considered while designing the system, because this decrepitation can affect the alloy packing density, exerting thermal and mechanical stresses on the walls of the container which in turn impacts the heat and mass transport within the metal hydride system. On repeated cycling, the particles break into smaller particles, impacting the porosity, and thereby affecting the gas and heat flow inside the metal hydride bed. The smaller particles obtained due to decrepitation tend to settle down into the bottom of the tank, which can occur with time. Over a longer duration, the decrepitation can lead to bulging of the tank, due to expansion of the lattice in repeated cycling, resulting in rupture of the tank.

Different Types of Hydrides for Hydrogen Storage

Considering the thermodynamic trends discussed in the earlier section through PCI and van't Hoff plots for metal hydrides, many elemental and binary hydrides exhibit lower equilibrium pressures and higher dissociation temperatures as evident from Fig. 16.18, wherein only VH_2 reflects the optimal operational characterstics among all the alloys considered for analysis. This constraints might incur higher energy and economic expense for practical storage systems requirements. Such hydride characteristics might limit their direct applicability from enegineering perspective, where temperature, pressure, thermal management, reversibility, and cyclic efficiency must be optimal for efficient utilization. There exists a narrow subset of hydrides that meets the optimal temperature and pressure requirements for reversible hydrogen storage at near-ambeint temperature and pressure conditions as observed from Fig. 16.18 and Fig. 16.21. In the context of appropriate material selection and storage system development from a practical utilization perspective, hydrides are broadly categorized based on the type of bonding and reaction mechanisms. The different types can be classified as ionic hydrides, covalent hydrides, intermetallic hydrides and complex hydrides. Each class of hydrides exhibit different kinetic characteristics, thermodynamic behavior, storage capacities, and cyclic life, thereby offering different advantages and limitations for their utility in different applications. For example, alkali and alkaline earth metals show high reactivity with hydrogen, but due to highly stable hydride formation, require elevated temperature for hydrogen release, whereas thermodynamic and kinetic characterstics can be tuned with improved cyclic stability in the case of transition metal-based intermetallic hydrides. Complex hydrides (e.g., alanates, borohydrides, amides, imides, etc.) offer higher gravimetric storage capacity in comparison with intermetallic hydrides, but practical operation or utility necessitates the use of catalytic or thermochemical strategies. Herein, a critical examination of these hydride families, understanding their bonding nature, crystal structure, storage capacity, kinetics, thermodynamics, and reusability is essential for the selection of appropriate material for any applications. The following subsections provide an overview of the major hydride categories, discussing their fundamental properties, hydrogen storage behavior, and relevance to practical applications.

Intermetallic Compounds (IMC)

In 1958, Libowitz et al. identified a completely new class of materials and this was a major breakthrough in the domain of hydrogen storage materials. Before 1958, the identified materials as shown in Fig. 16.18 were mostly not suitable for consideration for hydrogen storage, he proposed new class of materials which are known as intermetallic compounds [6]. It was observed that the intermetallic compounds store hydrogen reversibly, with intermediate thermodynamic stability between the two metals. This discovery led to a new horizon of research in the domain of hydrogen storage materials leading to the development of several hundred of intermetallic compounds till date. In the year 1958, they came up with the zirconium nickel (ZrNi) intermetallic compound which forms a ternary hydride. The noteworthy observation was that the ZrNi bond formed in this ternary hydride had a strong destabilizing effect resulting in the thermodynamic stability of ZrNi hydride in between the highly stable zirconium hydride that have an enthalpy of formation, ΔH value of 169 kJ/mol, and the unstable nickel hydride that has an enthalpy of formation, ΔH of 8.8 kJ/mol. This intermediate thermodynamic stability of ternary hydride ZrNi lowers the desorption temperature requirement to 300 °C as opposed to 900 °C for the ZrH_2 at 1 bar. Most of these intermetallic hydrides follows the well-known Miedema rule which is a relationship that correlates the ΔH value of these intermetallic hydrides to their individual hydrides and the intermetallic compounds. As per the Miedema rule,

$$\Delta H\left(A_nB_mH_{x+y}\right) = \Delta H(A_nH_x) + \Delta H\left(B_mH_y\right) - \Delta H(A_nBm). \tag{16.23}$$

This rule provided explanation and method to determine the ΔH values for the majority of intermetallic hydrides. Another breakthrough in the area of metal hydrides was in the year 1974 when the two well-known metal hydrides, $LaNi_5$ and FeTi, were found having low enthalpy of formation [7]. $LaNi_5$ which is an intermetallic compound and forms $LaNi_5H_6$, desorbs hydrogen at 2 bar equilibrium pressure and at a temperature of 20 °C, whereas LaH_2 exhibits 2 bar desorption equilibrium pressure at 1300 degree C.

In the year 1984, Griessen and Drissen introduced a more accurate formula compared to the Miedema rule correlating the enthalpy of reaction with the characteristic energy of the electronic structure of the host metal. This is a semiempirical band structure model with a linear relationship between the ΔH and ΔE.

$$\Delta H = \alpha \Delta E + \beta \tag{16.24}$$

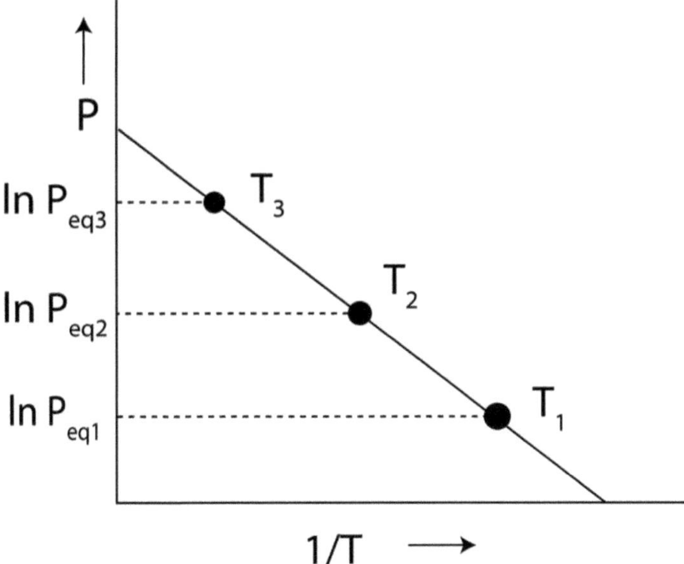

Fig. 16.16 Van't Hoff plot

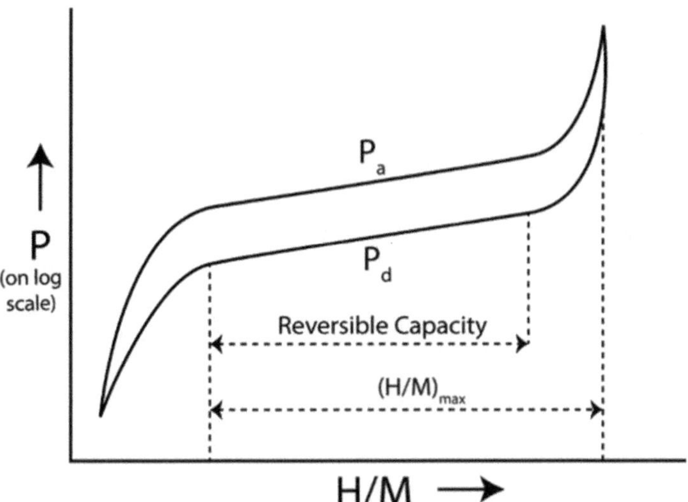

Fig. 16.17 PCT isotherm for an absorption-desorption cycle showing the capacity of hydrogen storage

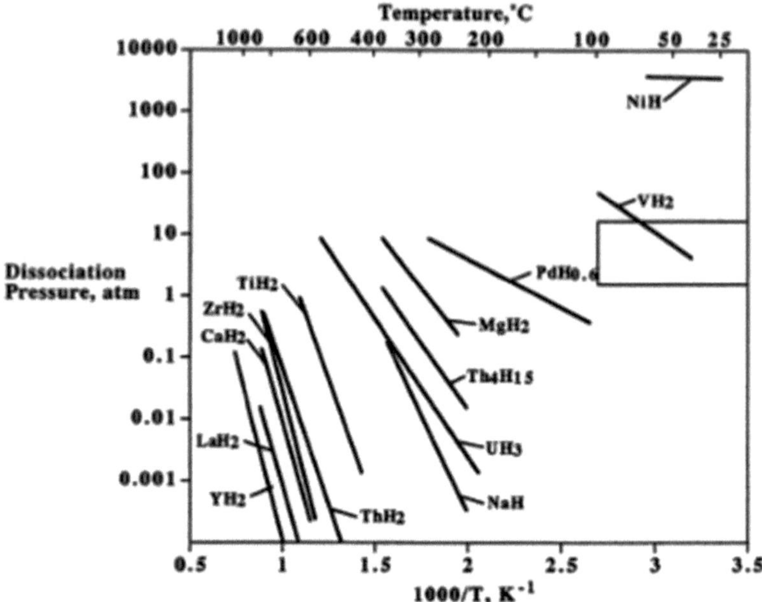

Fig. 16.18 Van't of plot for various binary hydrides (reproduced with permission [5])

$\Delta E = E_F - E_s$ (E_F is the Fermi energy and E_s is center of lowest band of host metal),

$$\alpha = 59.24\,\text{kJ}(\text{eV mol H}_2)^{-1} \quad \text{and} \quad \beta = -270\,\text{kJ}(\text{mol H}_2)^{-1}$$

After this, a wide number of intermetallic compounds were reported. These compounds generally characterized as A_xB_y are formed using two elements "A" and "B." The metal element A is such that it is a transition metal or a rare earth element having a stronger affinity toward hydrogen, forming a stronger bond with hydrogen such that the hydride formed shows high thermal stability. Examples of elements that appear at A site could be Y, La, Ti, Zr, etc. The other element B in these compounds can be a transition metal, which has low affinity toward hydrogen and forms an unstable hydride for example Cr, Mn, Fe, Co, Ni, etc. Thus, a combination of one strong hydride forming element and another element forming an unstable hydride results in an intermetallic compound or alloys which will have intermediate properties between that of the individual A and B elements. Thus, the resulting IMC has different chemical and physical properties compared to both A and B (Fig. 16.20). A typical example could be like at A site La and at B site Ni forming $LaNi_5$ such that the ΔH value of $LaNi_5$ is 32 kJ/mol; however, lanthanum hydride is a stable hydride having a ΔH value of 208 kJ/mol, whereas nickel forms an unstable hydride with a ΔH value of 8.8 kJ/mol. When hydrogen atom enters the interstitial sites of a metal or an alloy, lattice expands resulting in change of lattice parameters or unit cell volume which in turn influences the plateau or equilibrium pressure of hydride. Herein it was observed that the intermetallic compounds which are stoichiometric compounds having ordered phases between two constituent metal elements A and B results in intermediate phases whose structural characterstics determine their hydrogen absorption and desorption behavior.

Now, alloying of these two species A and B increases the degrees of freedom leading to various changes, including the change in the structure, defects, or anti-phase boundaries being created which also helps in improving the hydrogen storage properties of these intermetallic compounds.

The intermetallic compounds can be divided into different classes including AB, AB_2, A_2B, AB_5, A_3B_7, AB_3, etc., as shown in Fig. 16.20. The different compositions exhibit different kinetics and thermodynamic properties. Some examples of these classes are shown in Table 16.3 (Table 16.3; Fig. 16.21).

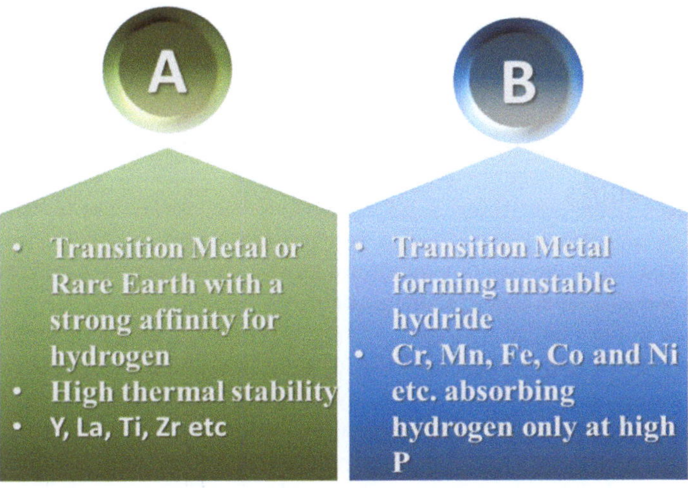

Fig. 16.19 Possible elements in the formation of intermetallic compounds

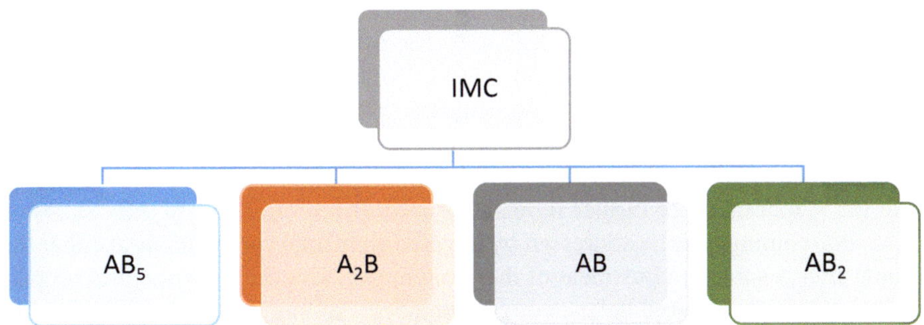

Fig. 16.20 Different classes of metal hydrides

Table 16.3 Different classes of metal hydrides and their hydrogen storage properties

Type of metal hydride	Most studied alloy	Reversible H_2 storage capacity	$\Delta_r H$ kJ mol^{-1} H_2	P_{eq} (bar)	T_{des}	Crystal structure
AB	TiFe	1.5	– 28	5	RT	Cubic Pm3m
AB_2	$ZrMn_2$	0.9	53.2		440	Fd3m
A_2B	Mg_2Ni	3.3/3.6	– 64.5	1	255	Cubic P6222
AB_5	$LaNi_5$	1.28	– 30.8	2	285	Hexagonal P6/mmm
AB_3	$CeNi_3$	0.85	– 35.36			Hexagonal
A_2B_7	Nd_2Ni_7	0.77	– 34	10		
BCC compound	$V_{22}Ti_{35}Cr_{43}$	2.5	– 35	10		Cubic (Im-3m)
Saline hydrides	Mg	< 7.0	– 74.5	1	552	Hexagonal
High entropy alloy	TiVZrHfNb	Maximum = 1.9*H*/*M*	– 59			BCC (Im-3m)

AB_5 Type Alloys

The AB_5-type intermetallic compounds are most widely studied in different aspects such as crystal structure, thermodynamic behavior, and the cyclic ability for its varied applications in the field of hydrogen storage due to the ease of substitution at the A and B sites. AB_5 alloy exhibits a $CaCu_5$-type hexagonal structure, and belongs to the P6/mmm space group. These alloys show good electrochemical properties and are used for electrochemical applications. The origin of AB_5 type of alloys traced to the studies on $SmCo_5$ for their magnetic properties, and further exploration on it revealed that the AB_5 type of alloys having $CaCu_5$-type hexagonal structure exhibits hydrogen storage capabilities. In this, A can be La, Ca, Y, Zr, or Mm (misch metal which is a combination of different rare earths and that is in their naturally occurring composition, may have different compositions depending upon the source, in general the composition is 45% La, 50% Ce, rest Nd, or Pr), whereas potential B elements could be Ni, Sn, Ti, Al, Co, Fe, Cu, Si, etc. (Fig. 16.22). The properties of an AB_5-type alloy can be tailored through the careful choice of the A- and B-site elements. In practice, an alloy optimized for one characteristic, such as hydrogen storage capacity, may exhibit limitations in others, such as cyclic stability, with the synthesis route further influencing overall performance. In AB_5 systems, these shortcomings can be addressed by targeted elemental substitutions at either the A or B sites, enabling adjustment of thermodynamic, kinetic, and structural properties to achieve the desired balance of performance. Among this class of compound, the most widely studied AB_5 alloy is $LaNi_5$, because of its faster kinetics, cyclic life, near-ambient temperature-pressure operation, resistant toward impurities, low

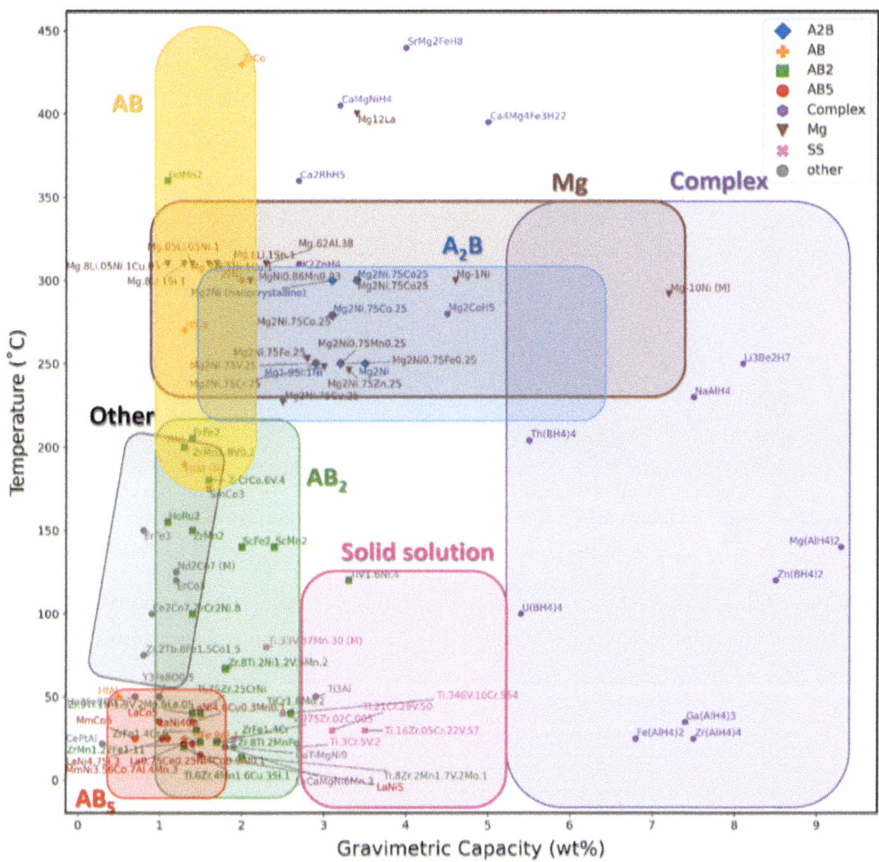

Fig. 16.21 Gravimetric storage capacity as a function of desorption temperature at 1 bar (reproduced with permission from [8])

hysteresis, the desorption equilibrium pressure at room temperature (298 K) is 1.8 bar, and the reversible capacity of 1.25 wt%.

Considering examples with different elements on A side like Mm, La, or Ca with Ni on B side is shown in Table 16.4, the change of element A gives different storage characteristics like $MmNi_5$ has a maximum gravimetric storage capacity of 1.46 wt%, $LaNi_5$ has 1.49 wt%, and $CaNi_5$ has 1.05 wt%.

It can be observed from Table 16.4, that higher the enthalpy values lower is the desorption equilibrium pressure indicating the higher stablity of alloys. It has been found that $LaNi_5$ and $CaNi_5$ on repeated cycling results in significant disproportionation leading to the property changes which can in turn lead to loss in of the reversible capacity. However, this issue of disproportionation can be taken care of by partial substitution on to the Ni side, typical example of this partial substitution is of Sn on Ni side, thereby reducing the disproportionation effect thereby improving cyclic stability of alloy but at an expense of slight reduction in storage capacity. Similarly,

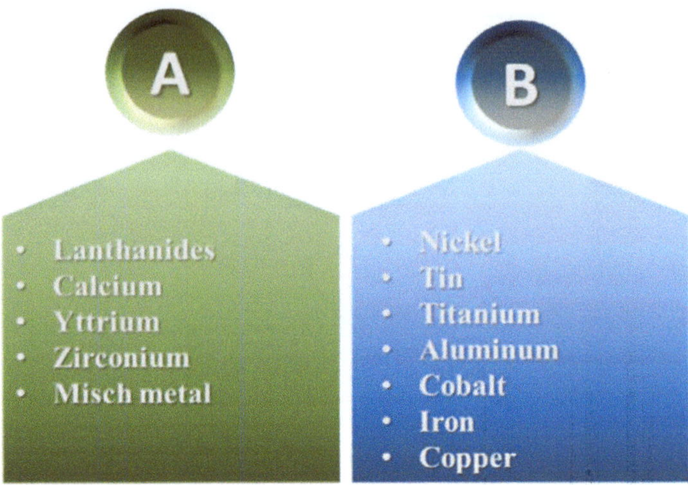

Fig. 16.22 Possible elements at A and B site in AB_5 type of alloys

Table 16.4 AB_5-type metal hydride and effect of substitution to their storage properties

AB_5 type intermetallic compounds	wt%	ΔH (kJ mol^{-1})	P_d at 25 °C (atm)	Hysteresis	Slope
$MmNi_5$	1.46	21.1	23	1.65	0.54
$LaNi_5$	1.49	30.8	1.8	0.13	0.13
$CaNi_5$	1.05	31.9	0.5	0.16	0.19
$MmNi_{4.15}Fe_{0.85}$	1.14	25.3	11.2	0.17	0.36
$LaNi_{4.8}Sn_{0.2}$	1.4	32.8	0.5	0.19	0.22
$MmNi_{3.5}Co_{0.7}Al_{0.8}$	1.24	39.8	0.11	0.2	1.2

substitution in Mm-based alloy on the Ni side with Fe improves the cyclic stability, lowers the hysteresis and desorption equilibrium pressure.

The advantages of AB_5 type of alloys are that they exhibit good kinetics, good cyclic ability, low hysteresis, very good tolerance and toward impurities. However, the disadvantages include low hydrogen storage capacity ($LaNi_5$ at 20 bar shows 1.4 wt%), and high cost when rare earth metals are used. However, the cost can be reduced if in place of expensive rare earth metals, the misch metal is used for large-scale production of alloys, due to its lower cost.

AB_2 Type of Metal Hydrides

AB_2 Laves phase intermetallic compounds are among the most investigated classes of compounds due to their high gravimetric and volumetric energy density as compared to AB_5 and AB intermetallic compounds. AB_2 type of IMC exhibit low activation energy, rapid hydrogenation and dehydrogenation kinetics, long cyclic life and applicability to a wide range of applications by tuning the material's structural properties. The element selection and formation of AB_2 hydrides are such that the A element is approximately 1.225 times larger than the B element. The stability of these type of compounds depends upon their geometry, packing fraction, valence electron concentration, and the electronegativity difference between the A and B elements. Zr, Ti, V are the most widely used A elements in AB_2 compounds, as they have considerable hydrogen storage properties. While, either transition or non-transition metals, such as Cr, Ni, Co, Fe, Mn, etc., which impart strong catalytic properties in sorption kinetics are considered as B elements. AB_2 compounds generally crystallize in three different laves phases, namely hexagonal C14 phase having $MgZn_2$-type structure, C15 cubic phase with $MgCu_2$-type structure, and dihexagonal C36 phase with $MgNi_2$-type structure. These phases are also dependent on electron to atom ratio. The C14 phase is found to be stable at an electron to atom ratio in the range 1.8 to <2.32, whereas the C15 phase is found to be stable with an electron to atom ratio between 1.5 to 1.8 or >2.32. The first compounds with such laves phase was reported in 1960. Zr and Ti-based AB_2 compounds are most widely investigated in the field of solid-state hydrogen storage and related applications. The major limitations associated with binary AB_2 alloys are that, either they are highly stable such as ZrV_2, $ZrCr_2$, $ZrMn_2$, $LaNi_2$ at room temperature or these can be unstable such as $TiCr_2$. Some of the AB_2 alloys exhibit poor kinetics, for example $ZrFe_2$, $ZrCo_2$, $TiMn_2$, these alloys have been studied for hydrogen uptake, wherein they absorb minor fraction of hydrogen at ambient temperature even with the supply pressure of 100 bar, significant hysteresis effect, susceptible to impurities and exhibit plateau slope. The research on AB_2 alloys involves tuning/tailoring the properties of alloys majorly by means of multicomponent or multielement substitution having either stoichiometry or non-stoichiometry to utilize them for different real-time applications such as solid-state hydrogen storage, distributed or hybrid hydrogen storage in combination with compressed state storage, multistage high-pressure hydrogen compression, heat pumps, refrigeration, negative electrodes for Ni-MH batteries, and vehicular applications (Fig. 16.23).

These materials can be slightly sub or superstoichiometric, for example being $ZrMn_{2-x}$, with this change in x the properties can be tuned to get the desirable characteristics like plateau pressure can be either increased, or it can be decreased by changing the composition. We can also get the required properties with such compounds by having multicomponent system, for example, $Zr_{1-x}T_x(Mn,Cr)_{2-y}M_y$, where T could be either Ti, Y, Hf, Sc, Nb, and M can be V, Mo, Mn, Cr, Fe, Co, Ni, Cu, etc. Thus with multicomponent systems, the required hydrogen storage properties can be obtained.

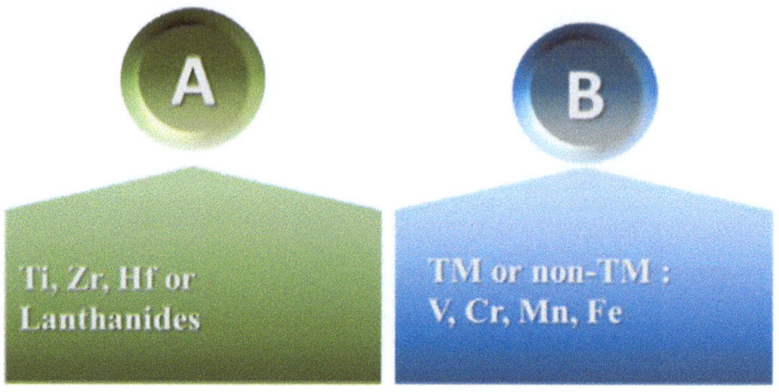

Fig. 16.23 Possible elements at A and B site in AB_2 type of alloys

Table 16.5 AB_2-type metal hydride and effect of substitution to their storage properties

	wt%	ΔH (kJ mol^{-1})	P_d at 25 °C (atm)	Hysteresis	Slope
$TiCr_{1.8}$	2.43	20.2	182	0.11	0.12
$TiMn_{1.5}$	1.86	28.7	8.4	0.93	0.57
$ZrMn_2$	1.77	53.2	0.001	0.99	0.74
$ZrFe_{1.5}Cr_{0.5}$	1.5	25.6	4	0.34	1.26
$TiMn_{1.4}V_{0.62}$	2.15	28.6	3.6		1.4

Some examples of AB_2 type of metal hydrides are shown in Table 16.5. Low hysteresis occurs in $TiCr_{1.8}$, the maximum hydrogen storage capacity of 1.86 wt% in $TiMn_{1.5}$ with reversible capacity of 1.15 wt%.

The advantages with AB_2 type of metal hydrides is a higher capacity compared to the AB_5 type of compounds, good reversibility lower cost, and ability to operate at near-ambeint temperature. However, the limitations associated with these class of compounds are their lower reversible hydrogen storage capacity (~ 1.5 wt%) and pyrophoric nature.

AB-Type Metal Hydrides

AB-type metal hydrides is another class of compounds very widely being studied. However, compared to AB_5 and AB_2 type of compounds, the number of different compositions studied in AB class of hydrides is relatively lesser. The first study was carried out in 1958 on ZrNi, which was found to desorb hydrogen at 300 °C [10]. The structure of compounds is ordered BCC space group *Pm-3m*, and the most widely studied material in AB type of compounds is FeTi, which was studied for the first time by Reilly and Wiswall in 1970 [11]. The biggest advantage of FeTi is that it

Table 16.6 Effect of substitution of Fe on hydrogen storage properties

	wt%	ΔH (kJ mol^{-1})	P_d at 25 °C (atm)	Hysteresis	Slope
TiFe (L)	1.86	28.1	4.1	0.64	0
$TiFe_{0.85}Mn_{0.15}$ (L)	1.9	29.5	2.6	0.62	0.92
$TiFe_{0.8}Ni_{0.2}$ (L)	1.3	41.2	0.1	0.05	0.36

absorbs and desorbs at room temperature, has maximum hydrogen storage capacity of 1.9 wt% at H/M ratio of 0.975, and a reversible capacity of 1.58 wt%. But the major challenge is the activation of FeTi. This alloy activates with repeated cycling at higher temperature of 400°C with pressure in the range 20 to 40 bar. The PCI of FeTi exhibits two plateaus, i.e., two-step absorption and the two plateaus correspond to two different hydrides, β and γ phases, and a large hysteresis is observed.

However, these issues can be addressed by partial substitution of FeTi that changes the hydrogen absorption characteristics.

Table 16.6 shows how the partial substitution of Fe in the FeTi changes the capacity, enthalpy, desorption temperature, hysteresis, and plateau slope. In Table 16.6 the corresponding hydrogen storage characteristics are mentioned for the lower plateau, wherein on substituting Fe with Mn, the desorption pressure reduces and the capacity increases. While substitution with Ni reduces the desorption pressure and hysteresis but also reduces capacity. The alloy $TiFe_{0.85}Mn_{0.15}$ is known to have good cyclic stability [12].

AB type of metal hydrides have advantages that they have very good hydrogen storage capacity, low cost, and operate close to room temperature. However, the major challenge remains is there is a requirement of high temperature and pressure conditions for the activation, these hydrides have low tolerance toward impurities and gets poisoned with the impurities present in the gas, and it is observed that they exhibit more than one plateau with upper plateau instabilities present in such alloys.

Solid Solution Alloys

Another class of hydrides formed are the solid solutions, formed by dissolving one or more minor alloying element in smaller quantities as solute into another major compound which is solvent and constitutes the hydrogen-absorbing phase. The intermetallic compounds are either stoichiometric or nearly stoichiometric compositions. However, solid solutions alloys may not be essentially stoichiometric rather can form non-stoichiometric compositions as well. In general, the solvent could be either Pd or Ti or Zr or V, with Pd the alloy cost is high; with Zr and Ti they form two stable hydrides; and with V being present the capacities are higher. The biggest advantage of these materials is that this material can store large quantities of hydrogen. In constrast to the other type of metal hydrides mentioned above having the maximum capacity $<$

2 wt%, the solid solution alloys may have higher gravimetric capacity of ⩾4wt% with Ti-Fe-V, Ti-V-Cr-Mn. However, in general these alloys have two plateaus and the higher plateau can be considered to provide hydrogen under the required conditions, then the reversible storage capacity reduces to around 2.5 wt%.

Similarly, for Ti-Cr-V the reversible capacity is 2.38 wt%, and they show a very good cycle life exhibting the capacity of 2 wt%, even after more than 1000 cycles. The biggest advantage of these alloys is that the alloy composition can be tuned or tailored to larger extent for getting desired set of operating conditions which might be useful from the storage systems design perspective.

One of the possible variation could be laves phase-related BCC solid solution, e.g., of such alloys includes Ti-V-Mn or Ti-V-Cr or Ti-Cr-V-Mn.

The advantage of Laves phase-related BCC solid solutions is that they show better hydrogen storage performance. These were first identified in the year 1990 and at that time most of the studies were carried out on the vanadium-based BCC solid solutions, which had a maximum hydrogen storage capacity of 3.68 wt% [13]. These alloys have both Laves phase and BCC phase, both of these phases contribute toward hydrogenation and have almost similar equilibrium pressure, example $Zr_{0.5}Ti_{0.5}VMn$ that shows three phases, C14, BCC, and a ZrO_2 oxide phase. Another example is that of Ti-V-Cr, they can give reversible capacity of 2 wt% (forms the dihydride phase), with smaller hysteresis and only one activation cycle is required. But the major disadvantage with these type of solid solution alloys is that they form two different hydride phases corresponding to mono hydride phase which is much more stable and a dihydride phase, which is responsible for giving the required hydrogen storage capacity. The scope for improvement and modification are: (1) destabilizing the monohydride phase so as to get the entire hydrogen desorbed and (2) to enhance the capacity or the reversible capacity of the dihydride phase.

In summary, the advantages of BCC solid solutions are that they operate at room temperature and have very good hydrogen storage capacity. The disadvantages are that they are difficult to activate, some of the compositions have expensive metals like V which increases the cost, high desorption temperature, and sloping plateau.

Mg-Based Hydrides

The basic understanding is that since the gravimetric capacity of the hydrogen storage materials includes the weight of the metal used for alloy formation; thus, if the storage capacity needs to be improved then metals which are lighter in weight need to be used, one of the example is Mg-based hydride. Magnesium hydride is known for its very high storage capacity of 7.6 wt%, but the problem being slow kinetics and higher desorption temperature of 300 to 400 °C as it forms strong stable hydride, specifically an ionic hydride. Another advantage of these hydrides is that they have a lower cost. Thus, the need is to address the problem of sluggish kinetics of absorption and desorption, and the high absorption/desorption temperature. These challenges can be addressed to some extent by adapting different strategies, i.e., by reducing the particle size, by alloying with other elements forming Mg_2Ni, $Mg_{17}La_2$, MgAl, or by forming composites. Ball milling with graphite carbon, forming composites or mixing with different other hydrides like titanium hydride has also proven useful to work with MgH_2 in order to reduce the desorption temperature [14].

However, still the desorption temperature achieved in this class of compounds is much higher than required for the materials-based storage systems development.

Limitations of metal hydrides storage:

- Low gravimetric storage capacity, cost
- Loss of capacity on repeated cycling
- The operating temperature and the pressure not suitable for practical application of alloys having higher gravimetric storage capacity.

The requirement is to modify the metal hydrides to address these challenges. The possible measures are to find new metal hydrides, or else modify the known metal hydrides.

Strategies to Design Better Materials

The requirements with metal hydrides as discussed in the previous sections are to enhance the hydrogen storage capacity, improve kinetics, increase cycle life, reduce cost, and have optimum temperature and pressure for absorption and desorption as per the requirement of various applications. Thus, the need is to consider how the crystal structure, thermodynamics, and kinetics of alloys can be tailored to accomplish desired characterstics useful for real-time applications (Fig. 16.24).

1. **Nanostructuring** —Nanostructuring can increase cycle life, improve kinetics, and result in easier activation of alloys. Nanocrystalline material (< 100 nm) has several defect sites which act as nucleation sites and thereby reduce the activation barrier for hydride formation. With more nucleation sites the kinetics gets faster, with grain boundaries serving as hydrogen pathways to enable hydrogen to reach inside the nanocrystalline material. There are several methods to reduce the particle size of the as-synthesized alloy such as ball milling, cold rolling, high-pressure torsion, etc. For example, in MgH_2, the kinetics improves on ball milling. Mechanical milling can be used to synthesize new alloys, reduce the particle size, and induce metastable phases. It can also create new surfaces, cracks, defects, dislocations, reduce crystallite size, increased number of grain boundaries, high surface area resulting in improved kinetics and diffusion in hydrides.
2. **Alloying**—Intermetallic (AB, AB_2, AB_5, A_2B, stoichiometric or near stoichiometric) and solid solutions (Ti-V-based can be non-stoichiometric) as studied can be modified to tune their hydrogen storage characteristics by means of alloying. Alloying increases degrees of freedom in material formed and creates vacancies, and anti-phase boundaries.

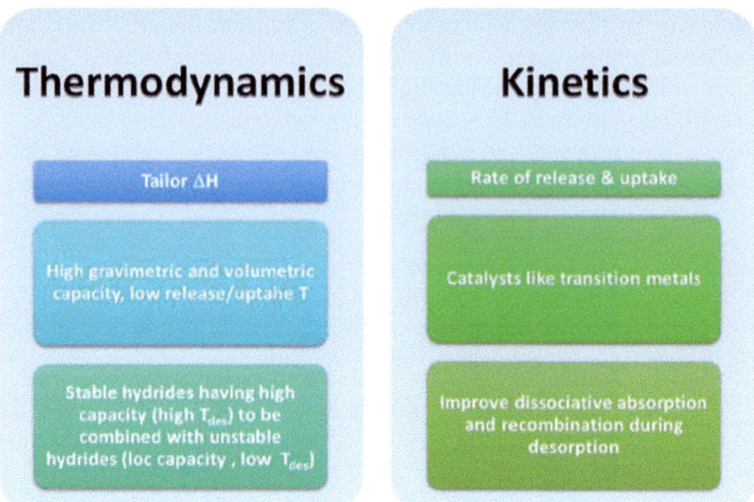

Fig. 16.24 Thermodynamic and kinetics effect on hydrogen storage characteristics

3. **Partial Elemental Substitution** — As discussed in the section on intermetallic compounds, partial substitution of one or more elements with another element can change the thermodynamic properties. This might result in change of storage capacity, altering the equilibrium pressure (cell volume changes on substitution), e.g., $LaNi_5$ (substitution of Ni by Al decreases capacity and substitution by Co results in having two equilibrium pressures). Thus partial elemental substitution can be a means to adjust P_{eq}. However, the major challenge is with the large-scale manufacturing, as small amount of impurities can shift P_{eq} outside desired range, which is defined by the required application. In general, the elemental substitution affects one or more storage characteristics like storage capacity, cyclic life, hysteresis, and plateau slope. Thus, alloy compositions are usually tailored by means of elemental substitution to optimize an alloy for specific application.
4. **Metastable and Amorphous Phases**—Metastable phases (crystalline or amorphous) can be created by melt spinning, ball milling, high-pressure torsion, and atomization. This can lead to creating thermodynamic instability, lowering the desorption temperature. For example, TiFe + 4wt%Ti milled gives $Fe_{46}Ti_{54}$. Amorphous TiCu absorbs 35% more hydrogen than crystalline one as the number of interstitial sites increases which was limiting hydrogen uptake in crystalline counterpart and better kinetics are observed.
5. **Chemical Destabilization**—Destabilization can be achieved in the light hydrides (LiH, MgH_2), complex hydrides which are known for their high capacity but sluggish kinetics. The reason being the high thermodynamic barrier, due to formation of ionic or covalent bond, hydrides become more thermodynamically stable and

have large activation barrier leading to irreversibility. The possibility is to have reactive hydride composites.

In summary, there are large possibilities of tuning the properties of hydrides as there exists wide variety of bonds formed and thus there are many ways to modify existing hydrides toward specific applications. For example:

Alkali/alkaline earth metals—forms ionic hydrides—with hydrogen being in H^- (hydride)—e.g., LiH and NaH,
Covalent hydrides, e.g., $LiNH_2$ and NH_3—with $H^{\delta+}$ (partially positively charged),
$NaAlH_4$ and $LiBH_4$—with hydrogen being $H^{\delta-}$ (partially negatively charged),
Metals—Pd and Ti—hydrogen bonding is metallic—and they can be good catalyst,
Interstitial hydrides—metallic hydrides has been studied in great detail.

Lightweight Hydrides and Other Novel Materials

Complex Hydrides: Another class of materials which can be used for hydrogen storage are the complex hydrides. In complex hydrides, hydrogen is covalently bonded with another metal or non-metal, forming an anion and then this anion is bonded with a metal cation.

The basic difference between complex hydrides and metal hydrides is that the metal hydrides have a very simple operation with absorption and desorption being straightforward (on exposure to hydrogen and under certain set of temperature and pressure conditions) and have good cycle life. The major disadvantage of metal hydride is poor gravimetric capacity, because the elements used like the transition metals or other elements to form the metal hydrides or their alloys or solid solutions, have higher atomic mass which reduces the gravimetric capacity. Complex hydrides are synthesized from lightweight elements, i.e., the elements which are present in the chemical formula of complex hydrides have lower atomic mass as compared to elements of metal hydrides which results in higher gravimetric and volumetric capacity of complex hydrides. There are complex hydrides which can have gravimetric hydrogen storage capacity of more than 13 wt% and a higher volumetric capacity of more than 150 kg/m^3.

In general, their chemical formula can be represented as $M(XH_n)_m$, where M is a metal, X can be a metal or non-metal, and depending upon whether it is a metal or non-metal, the bonding with hydrogen can be covalent or ionic. Complex hydrides have the advantage of high storage capacity but also suffers from challenges in terms of both kinetics and thermodynamic limitations. They have a sluggish kinetics and higher desorption temperature. Further, the reversibility of these complex hydride is another major challenge with most of these hydrides being irreversible or reversible under very stringent set of conditions making them unfavorable for hydrogen storage applications.

The complex hydrides can be categorized into alanates, borohydrides, amides or imides, depending upon whether in the formula the element X present in the complex is Al or B, can also be nitrogen or even have transition metal or carbon.

The alanates, borohydrides, amides, or imides are known for desorbing hydrogen in many steps, through a series of reactions.

1. **Alanates**

- Sodium alanate

$$NaAlH_4 \leftrightarrow 1/3Na_3AlH_6 + 2/3Al + H_2(3.7\,wt\%, 185\,°C) \tag{16.25}$$

$$Na_3AlH_6 \leftrightarrow 3NaH + Al + 3/2H_2\ (1.9\,wt\%, 230\,°C) \tag{16.26}$$

- Lithium alanate

$$3LiAlH_4 \rightarrow Li_3AlH_6 + 2Al + 3H_2\ (5.3\,wt\%, 160\text{–}200\,°C) \tag{16.27}$$

$$Li_3AlH_6 \rightarrow 3LiH + Al + 3/2H_2\ (2.65\,wt\%) \tag{16.28}$$

- Magnesium alanate

$$Mg(AlH_4)_2 \rightarrow MgH_2 + 2Al + 3H_2\ (7\,wt\%,\ 110\text{–}200\,°C) \tag{16.29}$$

$$MgH_2 \rightarrow Mg + H_2\ (2.3\,wt\%,\ 240\text{–}380\,°C) \tag{16.30}$$

$$2Al + Mg \rightarrow 1/2Al_3Mg_2 + 1/2Al\ (400\,°C) \tag{16.31}$$

- Potassium alanate

$$3KAlH_4 \rightarrow K_3AlH_6 + 2Al + 3H_2\ (2.9\,wt\%,\ 300\,°C) \tag{16.32}$$

$$K_3AlH_6 \rightarrow 3KH + Al + 3/2H_2\ (1.4\,wt\%,\ 340\,°C) \tag{16.33}$$

$$3KH \rightarrow 3K + 3/2H_2\ (1.4\,wt\%,\ 430\,°C) \tag{16.34}$$

2. **Borohydrides**

- Sodium borohydride

$$NaBH_4 + (2 + x)H_2O \rightarrow 4H_2 + NaBO_2 \cdot xH_2O \tag{16.35}$$

- Lithium borohydride

$$LiBH_4 \rightarrow LiBH_{4-\varepsilon} + (\varepsilon/2)H_2 : \text{ structural transition at } 108\,^{\circ}\text{C} \quad (16.36)$$

$$LiBH_{4-\varepsilon} \rightarrow LiBH_2 + (1 - \varepsilon/2)H_2 : \text{ first hydrogen peak starting at } 200\,^{\circ}\text{C} \quad (16.37)$$

$$LiBH_2 \rightarrow LiH + B + (1/2)H_2 : \text{ second peak at } 453\,^{\circ}\text{C} \quad (16.38)$$

$$LiBH_4 + 2H_2O \rightarrow LiBO_2 + 4H_2(13.9\,\text{wt\%}) \quad (16.39)$$

- Magnesium borohydride

$$Mg(BH_4)_2 \rightarrow MgH_2 + 2B + 3H_2(11.1\,\text{wt\%}) \quad (16.40)$$

$$MgH_2 + 2B \rightarrow MgB_2 + H_2(4.2\,\text{wt\%}) \quad (16.41)$$

- Calcium borohydride

$$Ca(BH_4)_2 \rightarrow 2/3CaH_2 + 1/3CaB_6 + 10/3H_2\ (9.6\,\text{wt\%}) \quad (16.42)$$

Among the complex hydrides, $NaAlH_4$ received the most interest because, Bogdanovic et al., found that sodium alanate can reversibly store hydrogen and is reversible under milder conditions when it is being reacted with titanium salts [15]. Besides, borohydrides are especially useful due to their high gravimetric capacity and have been of interest for applications. Hydrogen can be released on decomposition of the complex hydrides through either heating, i.e., thermolysis or on reacting with water, i.e., hydrolysis. However, the products formed on hydrolysis are more stable and it gets very difficult to regenerate back the initial material, i.e., very difficult to regenerate.

The stability of the tetrahedral borohydride sites and the presence of metal cations (Li^+, Na^+, Mg^{2+}) introduce an ionic nature in these compounds. Therefore, the borohydrides are observed to be sensitive to moisture and oxygen. However, under inert conditions, thermal decomposition of these borohydrides occurs at higher temperature (250–400 °C) which results in the formation of subsequent metal borides and hydrogen. Various methods have been reported to lower the decomposition temperature of complex borohydride, a few of them are, formation of binary metal borohydrides (e.g., $LiK(BH_4)_2$), increase in the cationic size; application of host material (e.g., activated carbon and carbon nanostructures, etc.); composite formation with another metal hydride (e.g., $LiBH_4$-MgH_2) which was also reported to increase regenerability. Other methods include formation of an ammoniate complex (e.g., $Mg(BH_4)_2$, $2NH_3$) which results in better dehydrogenation behavior. Replacement of the ammonia with metal amides has resulted into a borohydride-amide complex formation of M^1BH_4-M^2NH_2 nature, where M^1, M^2 = Li, Na, Mg, and Ca. The hydrogen generated from the complex has been observed to vary within the range of 7.8–11.3 wt%; however, their decomposition temperature has been observed to be as high as 360–480 °C depending on the chemical composition of the hydride.

Although the metal borohydrides demonstrate high hydrogen storage capacity, the most prominent disadvantages of the metal borohydrides are their extreme susceptibility to atmospheric conditions and high decomposition temperature. Both disadvantages of metal borohydrides can be related to the ionic interaction between the metal and borohydride species. Therefore, incorporating more covalent characteristics in the borohydride material has been observed to make them prone to decompose at lower temperature.

Strategies to tailor the complex hydrides. The different strategies to tailor the properties of the complex hydrides includes:

1. The compositions can be varied, and that can be done either by substitution of the anion or cation, or it can be at both the places at anion and cation site.
2. Formation of composites with different hydrides or metalloids or it could be the coordination of the neutral molecules to the cation in the complex hydrides.
3. Another possibility is tailoring the properties by introducing catalysts or additives, so as to improve the hydrogen absorption and desorption characteristics or kinetics.
4. Nanoconfinement of the complex hydrides by incorporating those into nanostructures, i.e., nanoporous host materials, are found to improve or modify the kinetics or thermodynamics or both for these complex hydrides.

Chemical Hydrides—As mentioned previously, the disadvantages of metal borohydrides being extreme susceptibility to atmospheric conditions and having high decomposition temperature are due to the ionic interaction between the metal and borohydride species. Therefore, incorporating more covalent characteristics in the borohydride material can lower the decomposition temperature.

Interaction between the ammonium cation (NH_4^+) and the borohydride anion (BH_4^-) is found to result in N-B bonds, which shows partial covalent nature. Therefore, analogous to organic compounds (based on C–C skeleton) a wide variety of compounds based on the B-N-based skeletal system have been synthesized and observed to demonstrate prominent hydrogen storage properties. These B-N-based hydrogen storage compounds have been termed as chemical hydrides.

Ammonium Borohydride (NH_4BH_4): The ionic nature of ammonium (NH_4) and borenium (BH_4) groups indicate that it is possible to synthesize ammonium borohydride as solid ionic compound (solid since most ammonium compounds are solid salts). However, the synthesis of ammonium borohydride has been elusive for a long time. The earliest attempt to prepare ammonium borohydride can be traced back to 1950s. In the year 1958, the synthesis of NH_4BH_4 was reported, but it was observed to be stable up to −40 °C [16]. At temperature higher than −40 °C, it slowly decomposed to form diammoniate of diborane (DADB). Nevertheless, despite its apparent instability, ammonium borohydride has been considered a promising hydrogen storage material, with gravimetric hydrogen storage capacity values of 24.7 wt%. The thermal analysis of NH_4BH_4 studies was observed to produce hydrogen with a capacity of 21.7 wt%, with a trace of borazine formation as impurity. Owing to its instability, a stepwise decomposition of ammonium borohydride results in the formation of its partially dehydrogenated more stable product, ammonia borane (AB). AB was found to be a crystalline solid powder, with high hydrogen storage capacity.

Ammonia Borane (AB, NH_3BH_3): Ammonia borane (AB) is a crystalline solid product stable at room temperature. In fact, AB is formed through elimination of one H_2 molecule from ammonium borohydride, and was observed to be stable up to −40 °C. The crystalline white solid (AB, NH_3BH_3) is formed due to covalent interaction between NH_3 and BH_3 groups and demonstrates a gravimetric hydrogen storage capacity of 19.6 wt%, better than $LiBH_4$ (18.4 wt%).

Due to partial covalent interaction between the amine (NH_3) and borane (BH_3) species, the decomposition of chemical hydrides occurs at significantly lower temperature than complex hydrides and with relatively faster kinetics. Among the variants in B-N chemical hydrides, ammonia borane is the most prominent candidate due to its high gravimetric hydrogen storage capacity (GHSC) at room temperature. Unlike other B-N compounds, ammonia borane is not hazardous or explosive, and does not react violently upon its exposure to ambient conditions and moisture. Decomposition of ammonia borane produces significant amount of hydrogen under both the thermolytic and hydrolytic conditions. Despite the moderate reactivity, non-explosive properties and high hydrogen storage capacity, ammonia borane (AB) suffers from various limitations. The structure of NH_3BH_3 molecule indicates that during AB thermolysis, the B-N bond also becomes susceptible to cleavage which results in the formation of harmful gaseous by-products such as diborane (B_2H_6), ammonia (NH_3), and borazine ($B_3N_3H_6$) which contaminates the hydrogen stream. Additionally, the thermal dehydrogenation of AB produces a stable polymeric B-N product, known as polyborazylene (PB). The stability of PB significantly hinders the regenerability of AB. Compared to complex hydride, the regeneration of ammonia borane has been reported to be more plausible; however, a reliable method to regenerate high-purity ammonia borane in significant amount of yield is yet to be discovered.

Hydrazine Borane (HzB, $N_2H_4BH_3$): Hydrazine borane is another candidate for hydrogen storage material from the chemical hydride family and is formed through interaction between hydrazine (N_2H_4) and borane (BH_3). The theoretical gravimetric hydrogen storage capacity of HzB is observed to be 15.4 wt%. The dehydrogenation of hydrazine borane at 140 °C resulted in 6 wt% of hydrogen release within 30 min. However, due to its flammability and shock sensitivity, the hydrazine borane has often observed to be explosive in nature. Thermal decomposition of hydrazine borane through a non-explosive process was also observed to produce pristine hydrogen, but the post-heated end-product after 200 °C was found to be shock-sensitive and explosive.

Hydrazine Bisborane ($N_2H_4(BH_3)_2$): The formation of a hydrazine bisborane also results from the interaction between hydrazine and diborane. The gravimetric storage capacity of hydrazine bisborane (16.8 wt%) was observed to be higher than hydrazine borane (15.4 wt%). Initially, hydrazine bisborane was proposed as an explosive above 100 °C and displayed detonation properties. However, recently there has been a renewed interest in hydrazine bisborane as hydrogen storage material and several works demonstrated that the decomposition of $N_2H_4(BH_3)_2$ is not necessarily explosive under dehydrogenation conditions (< 165 °C). Decomposition studies suggested that up to 150 °C, 10 wt% of hydrogen release is possible, but the hydrogen stream was also observed to be associated with the formation of impurities.

Ammonia Triborane ($NH_3B_3H_7$): As an alternative to prepare high-purity B-N chemical hydrides, ammonia triborane was synthesized from the interaction of $[nBuN_4][B_3H_8]$ and ammonia [17]. The ammonia triborane material has been observed to be more readily soluble in water before dehydrogenation. Hydrolysis of the ammonia triborane at 50 °C in the presence of 5 wt% $Rh\text{-}Al_2O_3$ catalyst was observed to produce 7.6 eqv. of hydrogen. But the post-heated product of the hydrolysis reaction has been observed to be stable meta and poly borate compounds, which have proven to be a challenge for regeneration.

Alongside these compounds, a few other C-substituted amine borane compounds, e.g., guanidium borohydride ($[C(NH_2)_3][BH_4]$) [18], substituted ammonia borane ($MeNH_2BH_3$, Me_2NHBH_3), and sodium bis(borane)dimethylamide have also demonstrated hydrogen storage properties.

High Entropy Alloys (HEA)—HEA involves an entirely new strategy of hydrogen alloying wherein, we can have multiple elements all in high concentration acting as a principal element. This new strategy of alloying has given a multidimensional compositional space to tackle with a wide variety of opportunities, however, still much remains to be studied in this particular class of materials. In the year 2004, it is reported that when multiple elements may be 5 or more are combined in such a way that they have higher concentrations, with the high configurational entropy of mixing, it can overcome the enthalpies of formation for the compounds.

These class of materials differ from the intermetallics. Considering the definition of high-entropy alloys, these can be defined on the basis of composition as well as their configurational entropy. On the basis of composition, the HEAs should have at least 5 elements such that each of these elements has an atomic concentration lying between 5 and 35%.

Another definition is based on the entropy, which states that the multicomponent alloy forms a HEA if the configurational entropy of the alloy at a random state is higher than 1.5R, irrespective of whether it is a single-phase or multiphase system at ambient temperature. The requirement is that the configurational entropy of mixing should be $\geq$ 1.5R, where R is the gas constant for such alloys. However, if this configurational entropy lies in a range of 1.5R–1R, then the alloys are known as medium-entropy alloys and when this value is $\leq$ R, then the alloys are low -entropy alloys or the conventional alloys.

Predicting and understanding the microstructure of these high-entropy alloys are very important and for that the phase stability of these alloys need to be considered. Different approaches, like the computational approach where the phase diagrams are calculated using CALPHAD software, either using density functional theory or ab initio molecular dynamic studies, are carried out to predict the phases present in HEA. The approach is that a complete phase stability map is being drawn considering a series of parameters thermodynamic parameters, such as entropy of mixing, enthalpy of mixing or the ratio of these times temperature, the atomic size difference, valence electron concentration, the electronegativity, and all these parameters are calculated.

The phase stability map is found using the mentioned parameters, with a range of these parameters being defined and when the values lie in the defined range, in that case a particular phase is stable. However, these are the parameters which have been well-reported in literature, but these values are based on the back-tested correlations and there is a little evidence of their predictive capability. There are many other approaches to do the same, but they have their own limitations.

So, there are conflicting literature studies which report dependence on these parameters, at the same time, there are some studies which have also reported that there are little evidences of their predictive capabilities.

(1) ***Thermodynamic Parameters***: Entropy of mixing (ΔS_{mix}), Enthalpy of mixing (ΔH_{mix}) and Ω parameter.

$$\Delta S_{\text{mix}} = -R\sum_{I=1}^{n} c_I \ln c_I \tag{16.43}$$

Entropy of mixing (ΔS_{mix}):
Enthalpy of mixing (ΔH_{mix}):

$$\Delta H_{\text{mix}} = \sum_{I=1, I\neq j}^{n} 4\Delta H_{\text{mix}}^{\text{AB}} c_I c_j \tag{16.44}$$

Ω Parameter:

$$\Omega = \frac{T_m \Delta S_{\text{mix}}}{|\Delta H_{\text{mix}}|}; \quad T_m = \sum_{I=1}^{n} c_I T_{mi} \tag{16.45}$$

($\Omega \geq 1.1$ and $\delta \leq 6.6\%$, for a single-phase solid solution).

(2) ***Atomic Size Difference (δ)***:

$$\delta = \sqrt{\sum_{I=1}^{n} c_I\left(1 - \frac{r_I}{\bar{r}}\right)^2}; \quad \bar{r} = \sum_{I=1}^{n} c_I r_I \tag{16.46}$$

($\delta \geq 9\%$, amorphous; $6.6\% < \delta \leq 9\%$; Multiphase, $\delta \leq 6.6\%$; Single Phase).

(3) ***Valence Electron Concentration and Electronegativity difference***:
Valence electron concentration (VEC):

$$\mathrm{VEC} = \sum_{I=1}^{n} c_I \mathrm{VEC}_I \quad (16.47)$$

(VEC < 6.87 BCC and VEC ≥ 8 FCC).
Electronegativity difference (χ):

$$\Delta\chi = \sqrt{\sum_{I=1}^{n} c_I (\chi_I - \overline{\chi})^2}; \quad \overline{\chi} = \sum_{I=1}^{n} c_I \chi_{\mathrm{i}} \quad (16.48)$$

where n is the number of components, R universal gas constant, $\Delta H_{\mathrm{mix}}^{\mathrm{AB}}$is the enthalpy of mixing for binary equi-atomic AB alloys, T_m is the melting temperature, c_I, $c_j$is the concentration of the element I and j, similar r_I and χ_I atomic radius and Pauling electronegativity for Ith and jth component.

There are different methods to synthesize HEAs:

(a) Solid Mixing: Mechanical alloying.
(b) Liquid Mixing: Arc melting, induction melting, laser engineered net shaping (LENS), laser melting, cladding, etc.
©) Gas Mixing: Atomic layer deposition, pulse laser deposition, sputtering deposition, etc.

Important properties of HEA are:

The diverse characteristics of HEAs can be summarized by four "core effects":

(1) **High Entropy**: Due to the configurational entropy (thermodynamics)
(2) **Severe Lattice Distortion**: Due to large deviation in atomic size of element (structure)
(3) **Sluggish Diffusion**: Kinetic due to lattice distortion (kinetic)
(4) **Cocktail**: Mechanical and physical properties (properties).

Some of the examples of such high-entropy alloys which have been studied for hydrogen storage applications for their hydrogen storage properties includes: the single-phase BCC high-entropy alloy $Ti_{0.3}V_{0.25}Zr_{0.1}Nb_{0.25}Ta_{0.1}$ and this alloy has been studied and reported to have maximum kinetic hydrogen capacity of 2.5 wt% at 33 bar pressure and 373 K along with very good kinetics [19].

Summary of Storage Methods and Economics—With the increasing global hydrogen demand and smooth operation of large-scale hydrogen value chain, the storage is essential. Different methods of hydrogen storage were considered, geological storage to store tens of thousands of tons of hydrogen is already in operation. Selecting an appropriate method of storage can be done based on parameters like amount of hydrogen to be stored, capacity of the system desired, rate of uptake and release desired, how long whether daily or seasonal storage required, the end use applications, cost of storage, and above all safety is very important.

Currently, 85% of the total hydrogen being produced is utilized on site, only 15% of it is being transported, and that transportation is either by means of trucks or pipelines. Most of the storage facilities store hydrogen in the compressed or liquid state in the tanks. Future predictions will depend upon several factors and accordingly the methods will be decided. For example, the requirement of storage can be for a very short-term storage and at the point of export, i.e., at an export port; or the requirement can be to store it for hours at a hydrogen refueling station; or it could be for days to weeks of storage to bridge the mismatch between the supply and demand; and finally it could be a long-term storage and large-scale storage to bridge the seasonal demand which when combined with the electricity or heat demand provides system resilience.

Thus, depending upon capacity, duration of storage, there could be different choices that can be made. For large-scale long-term storage, geological storage is economical; for short-term and small-scale storage compressed, liquid-state or metal hydride storage tanks could be used.

Geological storage is well-proven and provides economies of scale, high-efficiency, and low cost, lower footprint requirement above ground. Salt caverns-based storage have been used since 1970s such as in the UK, and the USA. The cost predicted is 0.6 USD/kg of hydrogen being stored, an efficiency of 98%, higher delivery rate can be obtained. Large salt caverns in the US can store the hydrogen produced from an SMR plant for 30 days with an output of 10–20,000 tons of hydrogen. Salt cavern storages in the UK, can store 1000 tons of hydrogen and some other being in the process of development to store 3500 tons hydrogen being established in Germany. Other options are depleted oil and gas reservoirs, their size is larger as compared to the salt caverns, but there are challenges like purity issues and thus the requirement of purification if need to be used in fuel cells. Aquifers-based storage technologies are still less mature.

However, geological storage is not suitable when the quantities to be stored are small, i.e., small-scale and for short terms, then either storage tanks can be used. With compressed hydrogen tank, high efficiency and high discharge rate can be obtained which can meet the requirement of fuel cells or other applications.

It provides flexibility as it can be made readily available wherever required. The compressed hydrogen tanks are manufactured in the same manner as the compressed natural gas tanks, and type IV tanks are made up of inner liner and with a completely wrapped fiber resin composite. Compared to the CNG tanks, the differences are in terms of size, the pressures at which they operate, store hydrogen, and the production volume. However, unlike the compressed natural gas tanks, the hydrogen storage cost is primarily dominated by the balance of plant and the required composites for fabrication of the tank (Fig. 16.25).

Considering the small number of tanks being produced, i.e., 10,000 units of production per year or the other option considered is production in high volumes like 500,000 units of production per year the cost of BOP could be 50% or 25%, respectively.

Compared to the CNG tanks which store gas at 250 bars, type III or type IV tanks store gas at 350 bar or 700 bar, which means much higher quantity of carbon fibers is required. For example manufacturing of a CNG tanks with operating pressure of 250 bar, and 275 L volume requires around 50 kg of composite, whereas

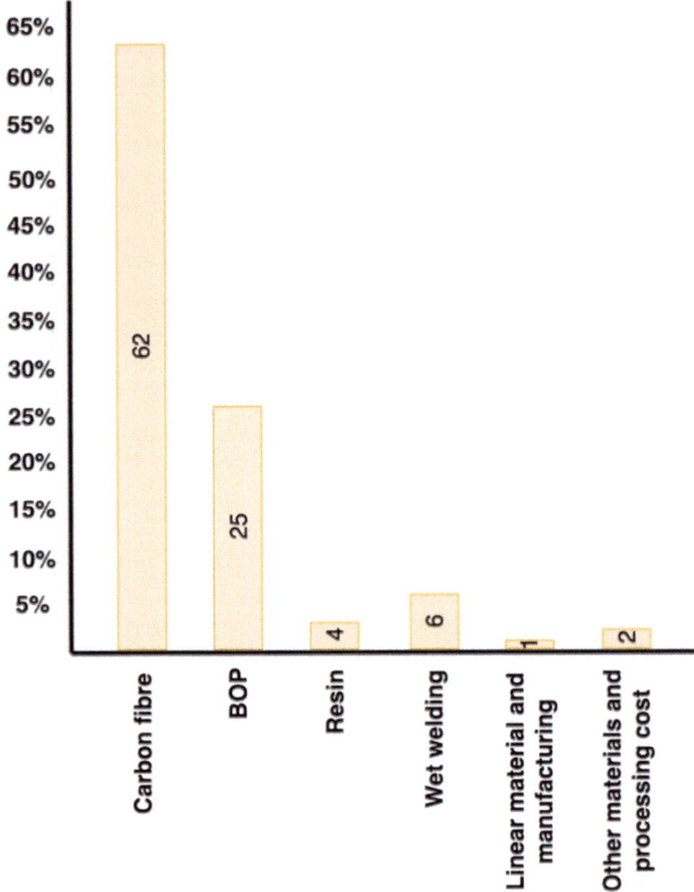

Fig. 16.25 Cost breakup of the components and manufacturing of a compressed hydrogen storage tank

a type IV hydrogen storage tank storing hydrogen at 350 bar pressure, having 245 L volume requires a composite weight of 62 kg. Thus, depending upon the requirement of carbon fiber composite, the price also varies. For type IV compressed hydrogen tank, the typical cost is around 15USD/kWh for a tank which stores 5.6 kg of usable hydrogen for vehicular application, with the major cost component being carbon fiber, roughly 62% of the total cost.

However, as the demand will increase in future and the volume of production will increase, considering lets say with 500,000 units/annum, the balance of plant will amount to 25%, resin 4%, wet welding 6%, linear manufacturing and material cost 1%, and the other cost of the material and processing cost will be 2%. Thus to reduce the cost, the cost of the resin, carbon fiber, and the balance of plant needs to be reduced. Besides, alternatives to carbon fibers also need to be explored.

Another option is storage in the liquid form. The cost of liquefaction depending upon the capacity whether 6000 kg/day or 200,000 kg/day could range from 50 million USD to 800 million USD. There are different cost models including all the production, liquefaction, delivery and dispensing cost being included to get the estimate. With the liquefiers of size 27,000 kg/day, the cost estimate is roughly around 160 million USD, considering production, liquefaction, delivery and dispensing charges, the cost at which finally it is delivered at refueling station is roughly 14.25 USD/kg of hydrogen. Usually, liquefaction as mentioned is a highly energy-intensive process.

The energy required in industrial liquefiers is about 10–20 kWh/kg. There exists variation in capacity of liquefiers, in US small liquefiers have capacity of 6000 kg/day while the larger liquefiers have a capacity of 70,000 kg/day of liquefaction.

Several studies which have been carried out under various DOE projects, such as DOE hydrogen and fuel-cells program. Various models have been used to find out the cost associated with the liquefiers of different capacity like ranging from 6000 to 200,000 kg/day of liquefaction capacity, which is given as [20]:

$$\text{Capital cost} = N * 1,000,000 * 5.6 * C^{0.8} * I \tag{16.49}$$

where, N is number of liquefiers such that each liquefier is having a capacity of 200,000 kg per day of liquefaction, C is liquefier design capacity (MT/day), I is overall chemical engineering plant cost index. Here C is a liquefier design capacity where $5.6 \times C^{0.8}$ is obtained by best fitting the data which is available and is expressed in units of MT/day.

Considering the cost breakup for hydrogen liquefaction for a plant of capacity 27,000 kg/day the major cost is for dispensing, i.e., 51% and another major contributor is in the liquefaction 30% (Fig. 16.26).

The solid-state hydrogen storage method is not yet commercialized on a large scale. Considering the adsorption-based storage for example, MOFs store hydrogen at liquid nitrogen temperature. Considering a tank with storage of 5.6 kg of usable hydrogen capacity, volume165 L, and with hexagonal heat transfer arrangement that has been reported where 32 kg of MOF was stored. Without the metal organic framework, the tank weight was 96 kg. Along with MOF and tank, several other balance of plant components are required for filling the tank, for release of hydrogen, temperature and pressure monitors or regulators, filters, several connections required for usage, heating arrangements to desorb hydrogen at the desired temperature. Metal organic frameworks were fully charged at a pressure of 100 bars and 77 K, the cost of such tanks that was obtained in the work done at SRNL. It is reported that the cost per system at a lower production volume, i.e., 10,000 systems/annum being produced, it is roughly 6300 USD/system while considering higher volume, i.e., 500,000 units/annum being produced, the cost was found to be 3052 USD/system. However, another study from MATI reported slightly different numbers, but at that same time the amount of MOF used, the size of the vessel, the weight of the vessel were different, and they found that the cost was 5777 USD/system and 2945 USD/system, respectively.

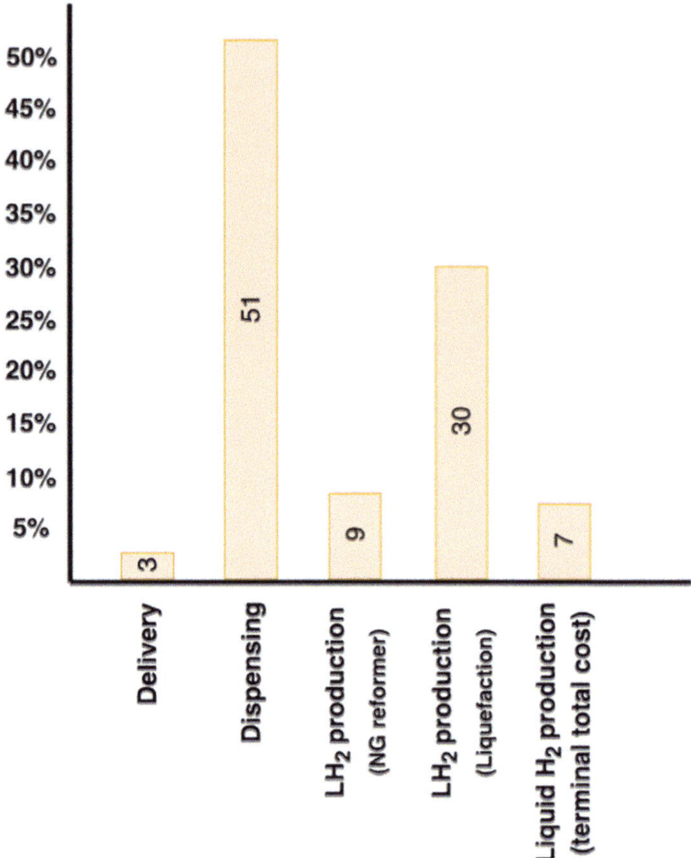

Fig. 16.26 Cost breakup of liquid hydrogen production, storage, transport, and dispensing

The cost of other components other than the tank, in the total cost accounts for about 53% in case of low-volume production and 39% in higher volume.

For absorption-based systems, i.e., the metal hydride-based storage systems, it is reported that these are at par with the compressed hydrogen tank. Studies on integrating an electrolyzer with a metal hydride-based hydrogen storage and fuel-cell system, reported that even without economies of scale the systems are at par and already cost competitive with the 700 bar compressed hydrogen tanks. However, the volumes of manufacturing of compressed hydrogen tanks have already seen the cost benefit due to economies of scale. In the integrated system with electrolyzer and fuel cell together with metal hydride storage, the major cost component in such systems are the electrolyzer and fuel cell.

Thus in metal hydride-based tanks, the economies of scale when being considered, the cost will further come down. The storage system cost, and metal hydride-based storage system cost being minimal.

The metal hydride cost can be calculated using the cost of individual elements and their atomic ratios as follows [20]:

$$C_{MH} = \frac{\left(C_A \frac{x}{x+y} M_A\right) + \left(C_B \frac{y}{x+y} M_B\right)}{\left(\frac{x}{x+y} M_A\right) + \left(\frac{y}{x+y} M_B\right)} \tag{16.50}$$

where the metal hydride considered is of type A_xB_y, A and B are elements in the alloy, M_A and M_B are their molar masses, respectively.

The metal hydride having a gravimetric capacity of 2 wt% if used, then the cost per kg of hydrogen being stored in such a metal hydride-based system is comparable to 700 bar compressed hydrogen tank.

Problems: Q1. Calculate the gravimetric storage capacity of an alloy (in wt%) that stores 667 g of hydrogen in 47.5 kg of alloy.

Solution: Given:

- Mass of hydrogen: $m_{H_2} = 667\,\text{g} = 0.667\,\text{kg}$
- Mass of alloy: $m_{\text{alloy}} = 47.5\,\text{kg}$.

To find: Gravimetric capacity of alloy (in wt%)

$$\text{Gravimetric Capacity (wt\%)} = \frac{m_{\text{H}_2}}{m_{\text{H}_2} + m_{\text{alloy}}} \times 100$$
$$\text{Gravimetric Capacity} = \frac{0.667}{0.667+47.5} \times 100 = \frac{0.667}{48.167} \times 100 \approx 1.38\,\text{wt\%}$$

Q2. Calculate the mass of alloy required to store 750 g of hydrogen, given that the gravimetric storage capacity of the alloy is 1.8 wt%.

Solution: Given:

- Gravimetric capacity: $1.8\,\text{wt\%} = 0.018$
- Mass of hydrogen: $m_{\text{H}_2} = 750\,\text{g} = 0.75\,\text{kg}$

To find: Mass of alloy (m_{alloy}).

Formula:

$$\text{Gravimetric Capacity (wt\%)} = \frac{m_{\text{H}_2}}{m_{\text{H}_2} + m_{\text{alloy}}} \times 100$$

ranging:

$$1.8 = \frac{0.75}{0.75 + m_{\text{alloy}}} \times 100 \Rightarrow 0.018\left(0.75 + m_{\text{alloy}}\right) = 0.75$$

$$0.0135 + 0.018 \cdot m_{\text{alloy}} = 0.75 \Rightarrow 0.018 \cdot m_{\text{alloy}} = 0.7365 \Rightarrow m_{\text{alloy}} = \frac{0.7365}{0.018} = 40.92\,\text{kg}$$

Therefore, storing $750H_2$ it will require 40.92 kg of alloy having 1.8 wt% gravimetric storage capacity.

Q3. Calculate the equilibrium pressure for a metal-hydrogen reaction having:

$$\Delta H = 28.25\,\text{kJ/mol},\ \Delta S = 112\,\text{J/mol K},\ T = 353\,\text{K},\ P_0 = 1\,\text{bar}.$$

Solution:
Given:

- $\Delta H = 28.25\,\text{kJ/mol}$
- $\Delta S = 112\,\text{J/mol K}$
- $T = 353\,\text{K}$
- $P_{\text{ref}} = 1\,\text{bar}$

To find: Equilibrium pressure P_{eq} at 353 K.

Formula:

$$\ln\left(\frac{P_{\text{eq}}}{P_{\text{ref}}}\right) = -\frac{\Delta H}{RT} + \frac{\Delta S}{R} \Rightarrow P_{\text{eq}} = P_{\text{ref}} \cdot \exp\left(-\frac{\Delta H}{RT} + \frac{\Delta S}{R}\right)$$

Substituting values:

$$P_{\text{eq}} = \exp\left(\frac{-28.25 \times 10^3}{8.314 \times 353} + \frac{112}{8.314}\right) \approx \exp(3.847) = 46.78\,\text{bar}$$

∴Equilibrium pressure of an alloy at 353 K is 46.78 bar for the given reaction.

Q4. Calculate the hysteresis for an alloy if the absorption equilibrium pressure is 3 bar and desorption equilibrium pressure is 1.8 bar at a given isotherm temperature.

Solution: Given:

- Absorption equilibrium pressure, $P_{\text{eq,abs}} = 3\,\text{bar}$
- Desorption equilibrium pressure, $P_{\text{eq\,,des}} = 1.8\,\text{bar}$

To find: Hysteresis
Formula:

$$\text{Hysteresis} = \ln\left(\frac{P_{\text{eq,abs}}}{P_{\text{eq,des}}}\right) = \ln\left(\frac{3}{1.8}\right) = \ln(1.6667) = 0.51$$

Hysteresis $= 0.51$.∴.

References

1. Rivard E, Trudeau M, Zaghib K. Hydrogen Storage for Mobility: A Review. Materials. 2019; 12(12):1973. https://doi.org/10.3390/ma12121973
2. Q. Lai et al., Adv. Sustain. Syst. **3**, 1900043 (2019)
3. James, Brian David, Houchins, Cassidy, Huya-Kouadio, Jennie Moton, & DeSantis, Daniel A. (2016). Final Report: Hydrogen Storage System Cost Analysis. https://doi.org/10.2172/1343975
4. G. Sandrock, A panoramic overview of hydrogen storage alloys from a gas reaction point of view. J. Alloys Compd. **293–295**, 877–888 (1999)
5. G. Sandrock, Panoramic overview of hydrogen storage alloys from a gas reaction point of view. J. Alloys Compd. **293**, 877–888 (1999). https://doi.org/10.1016/S0925-8388(99)00384-9
6. A.J. Maeland, G.G. Libowitz, J.P. Lynch, Hydride formation rates of titanium-based b.c.c. solid solution alloys. J. Less-Common Met. **104**, 361–364 (1984). https://doi.org/10.1016/0022-5088(84)90420-X
7. J.J. Reilly, R.H. Wiswall, Formation and properties of iron titanium hydride. Inorg. Chem. **13**, 218–222 (1974). https://doi.org/10.1021/ic50131a042
8. A. Bishnoi, S. Pati, P. Sharma, Architectural design of metal hydrides to improve the hydrogen storage characteristics. J. Power. Sources **608**, 234609 (2024). https://doi.org/10.1016/j.jpowsour.2024.234609
9. G.G. Libowitz, The nature and properties of transition metal hydrides. J. Nucl. Mater. **2**, 1–22 (1960). https://doi.org/10.1016/0022-3115(60)90019-2
10. G.G. Libowitz, H.F. Hayes, T.R.P. Gibb, The system zirconium-nickel and hydrogen. J. Phys. Chem. **62**, 76–79 (1958). https://doi.org/10.1021/j150559a019
11. I.J. Swartzendruber, L.H. Bennett, R.E. Watson, Mossbauer studies of hydrided TiFe [5]. J. Phys. F Met. Phys. **6** (1976). https://doi.org/10.1088/0305-4608/6/12/005
12. P. Modi, K.F. Aguey-Zinsou, Titanium-iron-manganese ($TiFe_{0.85}Mn_{0.15}$) alloy for hydrogen storage: reactivation upon oxidation. Int. J. Hydrogen Energy **44**, 16757–16764 (2019). https://doi.org/10.1016/j.ijhydene.2019.05.005
13. K. Nomura, E. Akiba, H2 absorbing-desorbing characterization of the TiVFe alloy system. J. Alloys Compd. **231**, 513–517 (1995). https://doi.org/10.1016/0925-8388(95)01862-X
14. J. Zhang, Z. Li, Y. Wu, X. Guo, J. Ye, B. Yuan et al., Recent advances on the thermal destabilization of Mg-based hydrogen storage materials. RSC Adv. **9**, 408–428 (2018). https://doi.org/10.1039/C8RA05596C
15. B. Bogdanović, A. Ritter, B. Spliethoff, K. Straßburger, A process steam generator based on the high temperature magnesium hydride/magnesium heat storage system. Int. J. Hydrogen Energy **20**, 811–822 (1995). https://doi.org/10.1016/0360-3199(95)00012-3
16. A. Gutowska, L. Li, Y. Shin, C.M. Wang, X.S. Li, J.C. Linehan et al., Nanoscaffold mediates hydrogen release and the reactivity of ammonia borane. Angew Chemie **117**, 3644–3648 (2005). https://doi.org/10.1002/ange.200462602
17. K.C. Nainan, G.E. Ryschkewitsch, A new synthesis of B3H8− ion. Inorg. Nucl. Chem. Lett. **6**, 765–766 (1970). https://doi.org/10.1016/0020-1650(70)80095-2
18. L.H. Jepsen, M.B. Ley, Y.S. Lee, Y.W. Cho, M. Dornheim, J.O. Jensen et al., Boron-nitrogen based hydrides and reactive composites for hydrogen storage. Mater. Today **17**, 129–135 (2014). https://doi.org/10.1016/j.mattod.2014.02.015
19. C. Zhang, Y. Wu, L. You, X. Cao, Z. Lu, X. Song, Investigation on the activation mechanism of hydrogen absorption in TiZrNbTa high entropy alloy. J. Alloys Compd. **781**, 613–620 (2019). https://doi.org/10.1016/j.jallcom.2018.12.120
20. Current Status of Hydrogen Liquefaction Costs, DOE Report (2019).

Chapter 17
Metal Hydride-Based Systems and Applications

Introduction

The goal toward accomplishing net zero emission targets to limit global temperature rise for the sustainable existence of humankind needs far reaching transition toward exploring and adoption of green/clean alternative energy sources. Hydrogen as an energy carrier is viewed as potential medium to decarbonize energy economy due to its high gravimetric energy density (33 kWh/kg-H_2), greener production routes such as renewable energy integrated water electrolysis, thereby resolving the issue of renewable energy intermittency and cleaner by-products such as water on integration with fuel cells. However, its low volumetric energy density (3 Wh/L) at ambient conditions remains major bottleneck for its long-term safer storage and utilization [1].

Commercial technologies such as compressed hydrogen gas storage (CGH2) and liquefied hydrogen storage (LH2) suffer from major limitations as shown in Fig. 17.1 hampering the mass scale adoption of hydrogen as an energy carrier in the global energy landscape. In CGH2, high-pressure hydrogen compression storage at 350 bar (23 kg/m^3) to 700 bar (40 kg/m^3) in Type 4 tank costs 60–70% of infrastructure cost for compression with potential safety risks. In LH2 storage, hydrogen liquefaction occurs at − 253 °C stored in superinsulated vessels, which have associated boil-off losses, restricting long-term hydrogen storage and have potential risk termed as BLEVE in case of accident. The rapid transition or adoption of hydrogen economy needs a safer medium both for small and large quantities of hydrogen storage. Metal hydride (MH)-based solid-state hydrogen storage technology is viewed as one of the potential mediums of hydrogen storage [2].

Metal hydride-based hydrogen storage systems are characterized to have higher volumetric energy density (100–150 kg/m^3), low operating pressure (< 100 bar), near ambient temperature operation, a safer and viable method for long-term hydrogen storage. Moreover, the tunability of alloy properties enhances the versatility of metal hydride systems, enabling their integration into a wide range of applications including

P. Sharma, *A Textbook on Hydrogen Energy*,
https://doi.org/10.1007/978-981-95-4601-5_17

stationary and back-up power generation, thermal energy storage, hydrogen purification, hydrogen compression, vehicular systems, heat pumps and transformers, as well as refrigeration and air conditioning [3–5] (Fig. 17.1).

When hydrogen storage systems need to be designed, there are several barriers that need to be considered for different applications. It is expected that they should be on par with those for conventional fuel-based performance. For example, in the case of vehicular application, parameters like system weight and volume, cost, fuel quality released from storage, charging and discharging rate, efficiency, durability, and operability should be similar to the current gasoline or diesel-based vehicles in terms of meeting present performance standard.

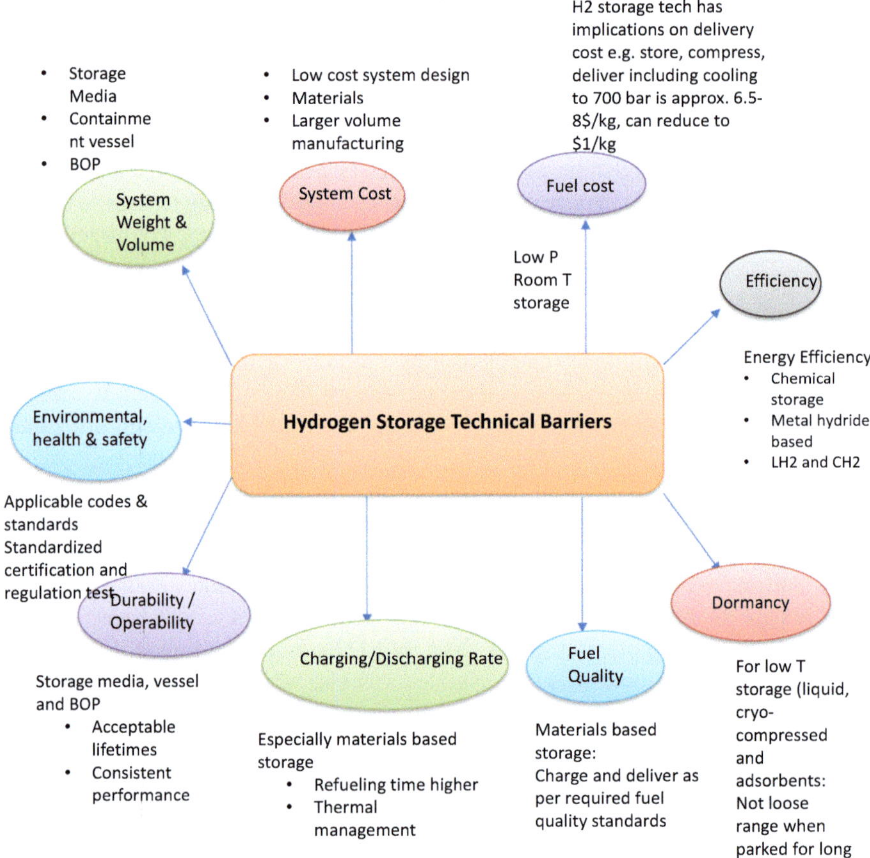

Fig. 17.1 Various technical barriers toward hydrogen storage

Metal Hydride-Based Storage System

A metal hydride is a compound formed by the absorption of hydrogen in atomic form within the interstitials of a metal or an intermetallic compound under different operating conditions of temperature and pressure. The formation of metal hydride is a reversible chemical reaction as shown in Fig. 17.2, wherein the forward reaction is termed as absorption/charging process and reverse reaction is the desorption/discharging reaction. In the absorption reaction, hydrogen reacts with the host metal or alloys forming metal hydride, simultaneously releasing heat, i.e., exothermic reaction and this reaction occurs or favored when hydrogen gas supply pressure is higher than the equilibrium pressure of the alloy at given instantaneous temperature ($P_s > P_{eq,a}$). In the desorption reaction, energy in the form of heat is supplied to the metal hydride so as to convert into metal and molecular hydrogen. Thus, the desorption reaction is endothermic reaction and occurs if equilibrium pressure of alloy is higher than the outlet/end-use delivery pressure ($P_{eq,d} > P_o$). The equilibrium pressure is a temperature-dependent property as represented by van't Hoff equation. The difference between the supply or outlet pressure and the alloy's equilibrium pressure acts as the driving force for these reactions [6].

During absorption, if the released heat is not removed, the alloy's temperature rises, increasing its equilibrium pressure and reducing the driving force which affects rate of hydrogen uptake. Likewise, in desorption, if enough heat is not supplied, the alloy cools down, lowering its equilibrium pressure and weakening the driving force thereby reducing hydrogen release. It can be concluded that the energy or mass transfer to/from the metal hydride-based hydrogen system is limited by heat transfer challenges. Further, the poor thermal conductivity of alloy (0.1–0.3 W/m K) severely impacts the thermodynamic aspects of heat release/supply affecting charging/discharging kinetics [7–9] (Fig. 17.3).

Efficient removal of heat from the metal hydride bed during absorption process and adequate heat input during desorption process becomes crucial for effective

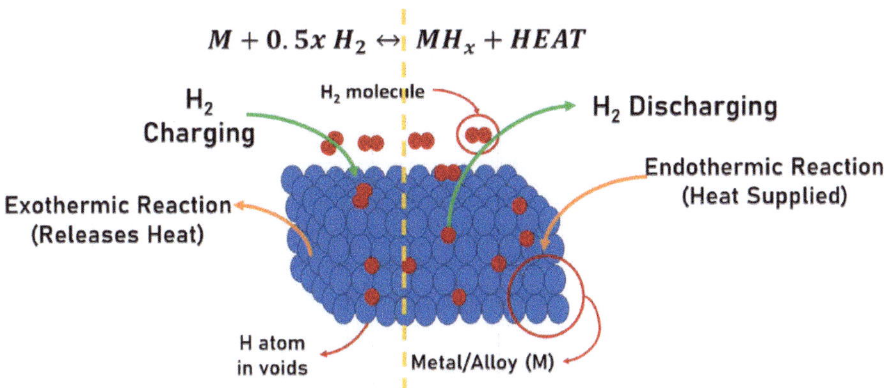

Fig. 17.2 Schematic diagram of the charging and discharging reaction of metal hydrides

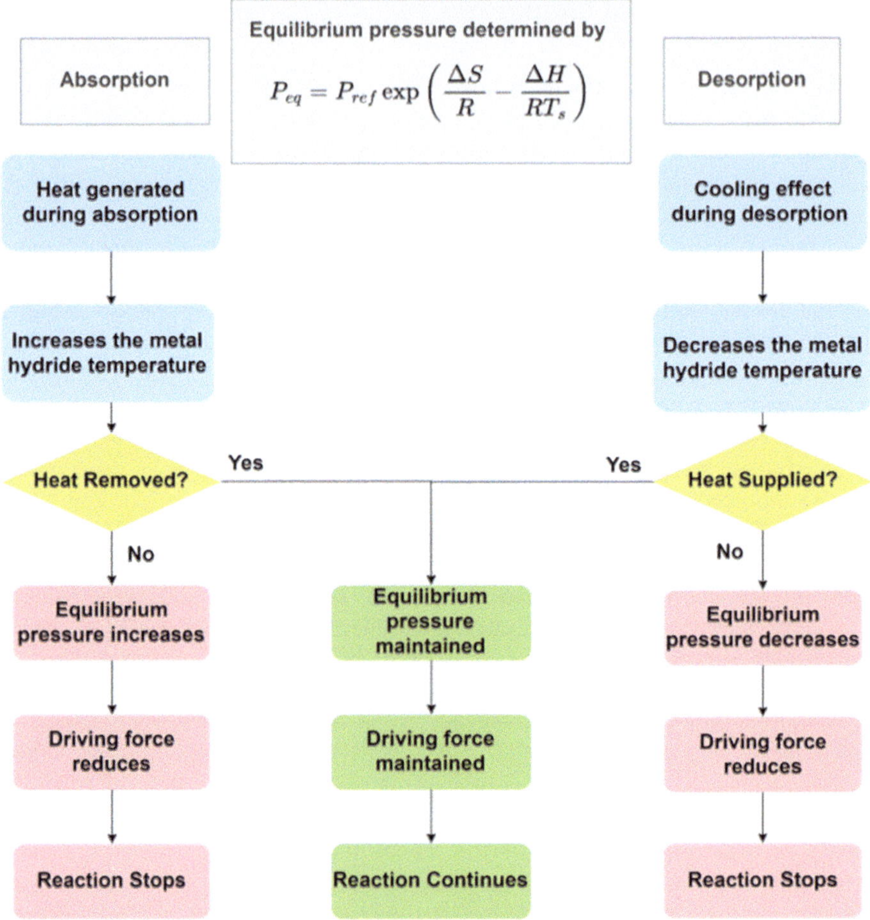

Fig. 17.3 Schematic of the heat transfer associated challenges and its effects on MH system

utilization of metal hydride-based hydrogen storage systems for various end-use applications. This major limitation of heat transfer necessitates the inclusion of an effective heat exchanger mechanism in the development of a metal hydride-based hydrogen storage system with respect to the desired end-use application [10]. Extensive experimental and numerical studies exist in the literature analyzing impact of sensitive parameters influencing the performance of metal hydride systems, along with various heat enhancement techniques to improve their thermal management. A brief overview of these approaches will be presented in the following sections.

Nomenclature	
E_a	Activation energy (J/mol)
Al	Aluminum

(continued)

(continued)

Cu	Copper
P_d	Desorption pressure
k_{eff}	Effective thermal conductivity of metal hydride (W/m K)
k_g	thermal conductivity of hydrogen gas (W/m K)
ΔH	Enthalpy of formation or decomposition (J/mol)
ΔS	Entropy of formation or decomposition(J/mol K)
P_{eq}	Equilibrium pressure
z	Expansion volume (%)
Q_s	Heat source
h_{ex}	Heat transfer coefficient (W/m^2-K)
Q	Heat transfer rate
P_0	Initial pressure
T_{0i}	Initial temperature
D	Inner diameter of MH system
L	Length of reactor (m)
$\dot{m}$	Mass flow rate of hydrogen (kg/m^3 s)
M	Metal
M_g	Molecular mass of hydrogen (kg/kmol)
D_o	Outer diameter of MH tubes
K_p	Permeability (m^2)
P_g	Pressure of hydrogen gas
C	Reaction constant (1/s)
C'	Reaction fraction
P_{ref}	Reference pressure
C_f	Specific heat of HTF (J/kg K)
C_{pg}	Specific heat of hydrogen (J/kg K)
C_{ps}	Specific heat of metal hydride (J/kg K)
C_w	Specific heat of wall (J/kg K)
$V_{storage}$	Storage volume
A_s	Surface area
T_{MH}	Temperature of metal hydride (K)
T_{MHw}	Temperature of metal hydride reactor wall (K)
T_w	Temperature of MH tube wall (K)
k_f	Thermal conductivity of HTF (W/m K)
k_w	Thermal conductivity wall (W/m K)
R	Universal gas constant (J/mol K)
V_f	Velocity of HTF (m/s)
V_g	Velocity of hydrogen gas (m/s)

(continued)

(continued)

*E*ara>	Volume fraction
Greek symbols	
ρ_f	Density of HTF (kg/m^3)
ρ_g	Density of hydrogen gas (kg/m^3)
ρ_{sat}	Bulk Density of metal alloy at saturation state (kg/m^3)
ρ_{emp}	Bulk Density of metal alloy without hydrogen (kg/m^3)
ρ_w	Density of wall (kg/m^3)
ε	Porosity
μ_f	Viscosity of HTF (Pa s)
μ_g	Viscosity of hydrogen (Pa s)
Indices	
A	Absorption
D	Desorption
e	Effective
m	Metal/alloy
w	Reactor wall
Abbreviations	
BOP	Balance of plant
BEV	Battery electric vehicle
COP	Coefficient of performance
CGH2	Compressed gas hydrogen
CT	Cooling tube
ECT	Embedded cooling tube
ENG	Expanded natural graphite
FC	Fuel cell
FCEV	Fuel cell electric vehicle
GRA	Gray relational analysis
HTF	Heat transfer fluid
HTMH	High-temperature metal hydride
ICE	Internal combustion engine
LH2	Liquid hydrogen
LPM	Liters per minute
LTMH	Low-temperature metal hydride
MH	Metal hydride
PCM	Phase change material
PEM	Proton exchange membrane

Design Methodology and Fundamental Governing Equations Associated with Development of Metal Hydride-Based Hydrogen Storage System

Fig. 17.4 represents the fundamental design methodology to be followed for developing any metal hydride-based hydrogen storage systems. As evident from Fig. 17.4, the foremost important step toward the development of any metal hydride-based hydrogen storage system is to understand the end-use application of the MH system (Fig. 17.4).

The fundamental understanding on end-use objective/application of the MH-based hydrogen storage system will necessitate identifying the quantity of hydrogen to be stored in the form of MH in the storage system.

Suppose to meet the end-use objective "m_{H_2}" kg of hydrogen needs to be stored in the form of metal hydride which will require certain mass of alloy having gravimetric storage capacity of "m" wt%.

The mass of alloy "m_{alloy}" required to store "m_{H_2}" mass of hydrogen can be computed as follows:

$$\text{Gravimetric capacity of alloy, } m(\text{wt\%}) = \frac{m_{H_2}}{m_{H_2} + m_{alloy}} \times 100 \tag{17.1}$$

Further, the computed mass of the alloy needs to be stored in a confined space having definite volume. The volume occupied by given mass of alloy ($V_{storage}$) can be computed as follows:

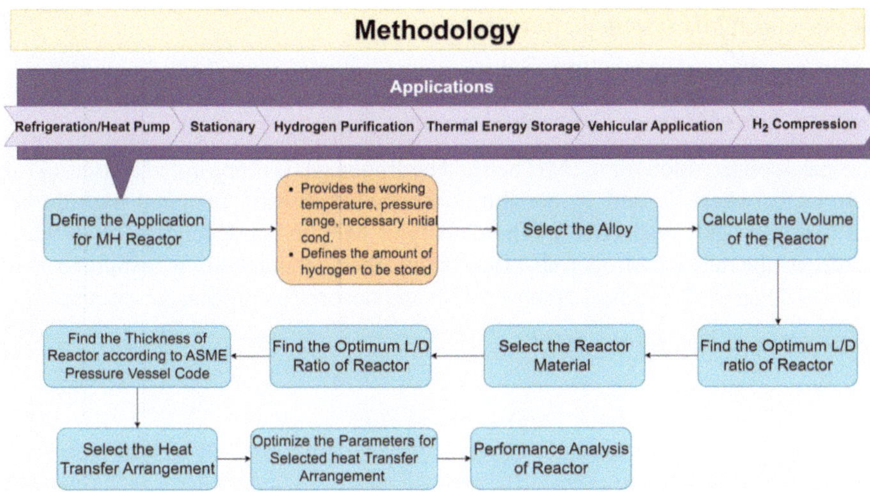

Fig. 17.4 Design methodology for developing a metal hydride-based hydrogen storage system

$$V_{\text{storage}}\left(\text{in m}^3\right) = \frac{m_{\text{alloy}}}{\rho_{\text{e}} \times (1-\varepsilon)} \times \left(1 + \frac{z}{100}\right) \tag{17.2}$$

where ρ_e is the crystalline density of alloy, ε is the porosity of alloy such that $(1 - \varepsilon)$ resembles volume fraction occupied by metal alloy in the vessel and z corresponds to expansion volume. During charging process, the storage of hydrogen in atomic form within interstitials of alloy results into expansion of crystal lattice of the alloy. So, it is essential to account this expansion volume while designing system; else, it may lead to system failure due to repeated thermal and mechanical stress exerted on the portion of storage vessel.

The metal hydride hydrogen storage systems are generally pressure vessels of cylindrical geometry developed using ASME Pressure Vessel Code (Section VIII). So, assuming certain L/D ratio based on the conformability limit, the exact dimension of the system can be calculated.

$$V_{\text{storage}}\left(\text{in m}^3\right) = \frac{\pi}{4} \times D^2 \times L \tag{17.3}$$

Assuming $L/D = y$ ($y = 1,2,3,\ldots n$) in the above equation, L and inner diameter (D) of reactor can be calculated. Further depending upon the working pressure and factor of safety, the wall thickness of the reactor can be calculated using ASME Pressure Vessel Code (Section VIII).

Herein, the basic dimension of cylindrical reactor without any heat exchanger can be calculated. The next step is to understand the fundamental governing equation used to perform mathematical modeling and the behavior of metal hydride-based hydrogen storage system in real-time applications. This will also be helpful in understanding the impact of different process or design parameters and the importance of heat exchangers on MH system performance.

Governing Equations [11, 12]

Conservation of Mass or Continuity Equation

The mass conservation for the MH bed and hydrogen gas is represented by Eqs. (17.4) and (17.5), where ρ_s and ρ_g are the instantaneous density of metal hydride bed and hydrogen gas, respectively. The density of MH bed is dependent on $\dot{m}$ which represents the rate of hydrogen absorbed or desorbed per unit volume computed using Eqs. (17.7) and (17.8) based on absorption and desorption process, respectively.

Continuity equation for the MH bed

$$(1-\varepsilon)\frac{\partial \rho_s}{\partial t} = \dot{m} \tag{17.4}$$

Continuity equation for hydrogen gas

$$\varepsilon\frac{\partial \rho_g}{\partial t} + \nabla.\left(\rho_g \mathbf{V}_g\right) = -\dot{m} \tag{17.5}$$

The density of hydrogen gas is defined by ideal gas equation as given below:

$$\rho_g = \frac{P_g M_g}{RT} \tag{17.6}$$

Reaction Kinetics

For absorption process:

$$\dot{m}_A = C_A \exp\left(-\frac{E_a}{RT_s}\right) \ln\left(\frac{P_g}{P_{eq}}\right) (\rho_{sat} - \rho_s) \tag{17.7}$$

For desorption process:

$$\dot{m}_D = C_D \exp\left(-\frac{E_d}{RT_s}\right) \left(\frac{P_g - P_{eq}}{P_{eq}}\right) (\rho_s - \rho_{emp}) \tag{17.8}$$

The equilibrium pressure of the metal hydride is given by the van't Hoff equation:

$$\frac{P_{eq}}{P_{ref}} = \exp\left[\left(\frac{\Delta S}{R}\right) - \left(\frac{\Delta H}{RT_s}\right)\right] \tag{17.9}$$

Darcy's Law (Conservation of Momentum)

The velocity of hydrogen gas inside MH bed is defined and computed using Darcy's law as follows:

$$\mathbf{V_g} = -\frac{K_p}{\mu_g} \nabla P_g \tag{17.10}$$

The amount of hydrogen absorbed is given by a dimensionless number and is described as:

$$C_A = \frac{\rho_s - \rho_{emp}}{\rho_{sat} - \rho_{emp}} \tag{17.11}$$

The amount of hydrogen desorbed is given by a dimensionless number and is described as:

$$C_D = \frac{\rho_{sat} - \rho_s}{\rho_{sat} - \rho_{emp}} \tag{17.12}$$

Conservation of Energy

Energy Equation for MH Bed

Assuming the local thermal equilibrium between metal hydride bed and hydrogen gas, energy balance can be written as given below:

$$\left(\rho C_{\mathrm{p}}\right)_{\mathrm{e}}\frac{\partial T_{\mathrm{MH}}}{\partial t}+\rho_{\mathrm{g}}C_{\mathrm{pg}}V_{\mathrm{g}}\cdot\nabla T_{\mathrm{MH}}=\nabla\cdot\left(k_{\mathrm{eff}}\nabla T_{\mathrm{MH}}\right)+\frac{\dot{m}\Delta H}{M_{\mathrm{g}}} \tag{17.13}$$

The first term on the left-hand side represents the time rate of change in sensible energy inside the reactor. The second term on the left-hand side quantifies the advection by hydrogen gas through a metal hydride bed. The velocity of hydrogen gas is negligible inside the metal hydride bed and thereby the second term becomes negligible. Despite that in order to make the study comprehensive, the second term on the left hand side is considered in the study. The first term in the right-hand side denotes the heat transfer by conduction inside the MH bed and the second term denotes the heat generation within MH bed. The effect of the radiation term has been neglected in this equation which can be included for studies for high-temperature metal hydride, e.g., MgH_2.

Effective thermal conductivity is defined as:

$$k_{\mathrm{eff}}=\varepsilon k_{\mathrm{g}}+(1-\varepsilon)k_{\mathrm{m}} \tag{17.14}$$

Effective volumetric heat capacity is expressed as follows:

$$\left(\rho C_{\mathrm{p}}\right)_{\mathrm{eff}}=\varepsilon\left(\rho C_{\mathrm{p}}\right)_{\mathrm{g}}+(1-\varepsilon)\left(\rho C_{\mathrm{p}}\right)_{\mathrm{m}} \tag{17.15}$$

Energy equation for MH tube walls

$$\rho_{\mathrm{w}}C_{\mathrm{w}}\frac{\partial T_{\mathrm{w}}}{\partial t}=\nabla\cdot\left(K_{\mathrm{w}}\nabla T_{\mathrm{w}}\right) \tag{17.16}$$

The above equation states that the rate of change of thermal energy within reactor wall is equal to the heat conduction through the reactor wall.

By applying appropriate initial and boundary conditions, above equations can be solved numerically using any CFD software for studying behavior of a simple metal hydride reactor without any heat exchanger in the system. Further, based on applications of MH system, appropriate heat exchanger method can be selected to meet the desired end-use objective.

Methods for Improving Heat Transfer of Metal Hydride-Based Hydrogen Storage Systems

Before proceeding into heat exchanger design towards improving heat transfer capability of metal hydride system, it is important to understand the energy transfer within MH bed by means of conservation of energy equation in the context of MH-based hydrogen storage system as per Eqs. (17.13)–(17.15). The heat transfer process in metal hydride system is characterized by change in sensible energy with respect to time, conduction, convection, and heat generation. The crucial parameters affecting heat transfer are source/sink term, specific heat capacity, thermal conductivity of metal and gas medium.

In this context, the heat transfer improvement methods can be understood in most simplified manner using Fourier Law of Heat Conduction and Newton's Law of Cooling for any given quantity of heat transfer neglecting heat generation for simplification. It can be observed from Eqs. (17.17) and (17.18) that incrementing thermal conductivity, heat transfer surface area, heat transfer coefficient, temperature difference, and reducing alloy bed radius may result in improving heat transfer process from/to the metal hydride-based hydrogen system. Fig. 17.5 provides the brief classification of heat augmentation methods to improve heat transfer in MH system. (Fig. 17.5).

Fourier Law of Heat Conduction

$$Q = kA_S \frac{dT}{dx} \tag{17.17}$$

Newton's Law of Cooling

$$Q = hA_S(T_{MH} - T_O) \tag{17.18}$$

Improving Effective Thermal Conductivity

Majority of the early research studies conclude that thermal conductivity of metal hydride bed is among the most important factors affecting MH system performance. In metal hydride (MH) systems, the thermal conductivity of hydrogen storage alloys can be significantly enhanced through the incorporation of expanded natural graphite (ENG), either by direct mixing or by forming compacted composites with the alloy as shown in Fig. 17.6, as well as through the inclusion of high-conductivity metal foams (such as Al, Cu or Zn) within the metal hydride bed [13–15]. In a comparative study by Sanchez et al. [16], ENG demonstrated a higher effective thermal conductivity (~ 19 W/m K) than aluminum foam (~ 8 W/m K), proven to be an efficient and cost-effective thermal enhancement medium. However, the performance of ENG degrades over repeated cycling, with thermal conductivity declining from an initial 40 W/m K

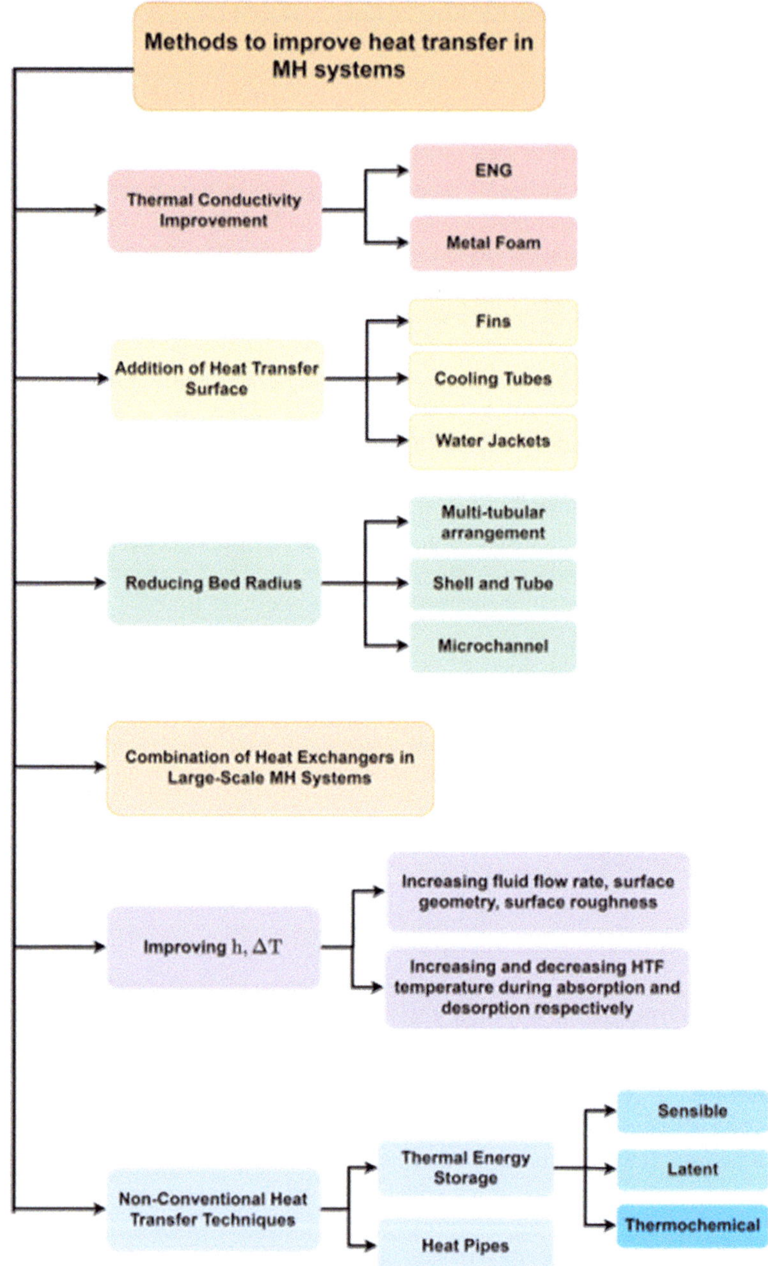

Fig. 17.5 Classification of methods to improve heat transfer in MH systems

Fig. 17.6 Improving thermal conductivity of metal hydride using ENG and metal foams

to a stabilized value of approximately 12.5 W/m K after 250 charge–discharge cycles [17]. The effectiveness of ENG is highly dependent on the metal hydride-to-graphite ratio and the applied compaction pressure. While both ENG and metal foams enhance thermal performance, they introduce trade-offs, including increased system weight, higher costs, and reduced gravimetric hydrogen capacity. Therefore, optimization of additive content and configuration is essential to balance improved heat transfer with overall system efficiency.

Extended Heat Transfer Surface Area

The addition of heat transfer surface area can be accomplished by means of addition of fins, water jackets, and embedded heat transfer fluid tubes. These methods will be explored briefly in this section.

Fins

The addition of heat transfer area in the form of fins has been largely studied in the development of metal hydride systems. The different forms of fin structure have been explored such as internal and external fins, longitudinal and transverse fins, pin fins, plate fins, circular perforated fins, branched fins, hexagonal honeycomb fins as shown in Fig. 17.6 [18–21]. Apart from providing extended heat transfer surface area, fins also contribute in improving effective thermal conductivity depending upon the fin material and quantity of fins. The studies show that transverse fins are prone to non-uniform metal hydride alloy distribution, pressure drop, and structural failure as compared to longitudinal fins due to repeated thermal cycling of metal hydride bed [22]. The inclusion of internal fins must be preferred as compared to external fins for large-scale hydrogen storage reactors as it results in direct removal of heat from the core of MH. However, inclusion of fins in combination with high thermal conductivity enhancers such as foams does not lead to significant improvement in performance. MacDonald and Rowe [23] compared the external fins and annular MH system for meeting fuel cell energy demand wherein it was observed that both show similar performance; however, external fins were not helpful in terms of utility. Askri et al. [24] investigated the impact of internal fin diameter and observed that increasing fin diameter improves absorption performance significantly, but increasing diameter beyond a certain limit has not much effect on performance. Mahvash et al. [25] optimized novel transverse hexagonal honeycomb fin structure for a 50 kg alloy capacity with AB_5 MH reactor; wherein, the major observation remains to be the optimizing number of fins such that trade-off between performance of metal hydride system and addition of parasitic weight due to fins must be optimum.

Heat Transfer Fluid (HTF) Jackets and Tubes

Another means of increasing heat transfer surface area is by adding heat transfer fluid jackets and tubes to the MH system. The inclusion of these mediums also helps in incrementing heat transfer coefficient by means of circulating fluid resulting in better absorption/desorption kinetics as compared to fin only MH systems. The HTF jackets are better solution as compared to system with external fins and for MH system of smaller diameter (< 5 cm) as water jackets only expedites the reaction near surface whereas removal of heat from the core of MH bed remains a concern for system with larger diameter [26]. Hence, inclusion of heat transfer fluid tubes within a system is highly preferred heat exchanger method. There exist numerous studies that have explored various geometries of HTF tubes such as straight, U-shape, spiral tubes as shown in Fig. 17.7 and can be made of different materials of construction such as stainless steel, copper, and aluminum [27–30]. Among all configurations of HTF tubes, spiral/helical tubes offer higher heat transfer area, heat transfer coefficient in limited volume and require higher pumping power as compared to straight tubes for same fluid flow rate.

Muthukumar et al. [27] optimized and investigated the performance of straight HTF tubes within MH reactors. The number of HTF tubes and radial position was optimized and experimentally investigated, wherein it was suggested that this optimization will vary for different scale of hydrogen storage systems. Bao et al. [31] used Taguchi method to optimize the performance of HTF tubes-based MH system. Levesque et al. [30] examined the transient thermal behavior using U-tube as heat exchange mechanism in the presence of Al-foam within MH bed. The circulation of

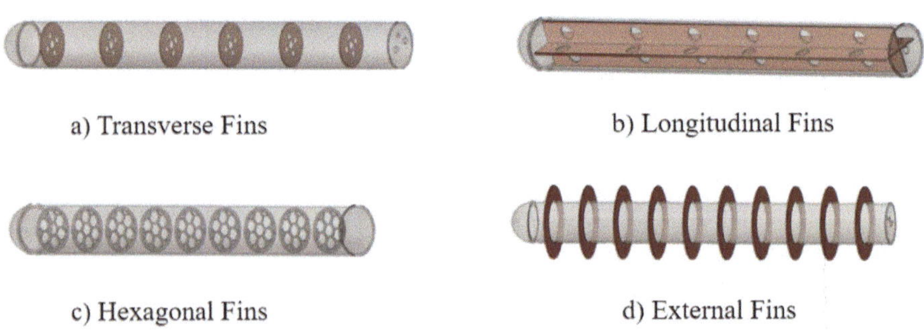

Fig. 17.7 Examples of different types of fins used in MH systems: **a** transverse fins, **b** longitudinal fins, **c** hexagonal fins, **d** external fins

HTF in U-tube enhances the system performance by an order of magnitude over a base case. Visaria and Mudawar [32] investigated the performance of coiled tube heat exchanger and compared with nodular fin-tube heat exchanger. It was concluded that coiled tube heat exchanger occupies only 7% of volume compared to 29% of later but absorption times for 90% reaction were doubled. Tiwari et al. [33] used combination of Taguchi and GRA method to optimize the different geometrical parameters of helical coil with an objective to obtain maximum reaction fraction and high outlet temperature for thermal energy storage application.

The inclusion of heat transfer fluid tubes is a potential heat exchanger medium in metal hydride-based system applications such as thermal energy storage, heat pumps, refrigeration, and air conditioning to extract desired effect. However, there needs strict optimization to optimize the number of HTF tubes, geometrical parameters, heat transfer fluid flow rates, and material of construction for an efficient performance without adding much parasitic weight to system as well as pumping power affecting overall system gravimetric capacity and energy efficiency.

Reducing Bed Radius (or Increasing L/D radius)

In case of large-scale metal hydride reactors storing more than 0.5 kg of hydrogen in single vessel within conformable limits will usually have larger MH bed radius. The larger radius of system will have higher conduction thermal resistance leading to inefficient heat transfer. This limitation can be resolved by reducing bed radius wherein the net amount of metal hydride can be distributed equally in optimum number of metal hydride tubes instead of storing in single vessel. The multitube modular arrangement, shell and tubes with tubes holding metal alloy configuration as shown in Fig. 17.8, microchannel-based MH system are among the most utilized examples in reducing bed thickness or radius of MH system to improve large-scale MH system performance eliminating need of complex heat exchanger arrangement.

Afzal et al. [34] observed that for a large-scale metal hydride reactor having 210 kg alloy, a multitube MH system having 14 MH tubes absorbs 80% hydrogen in half

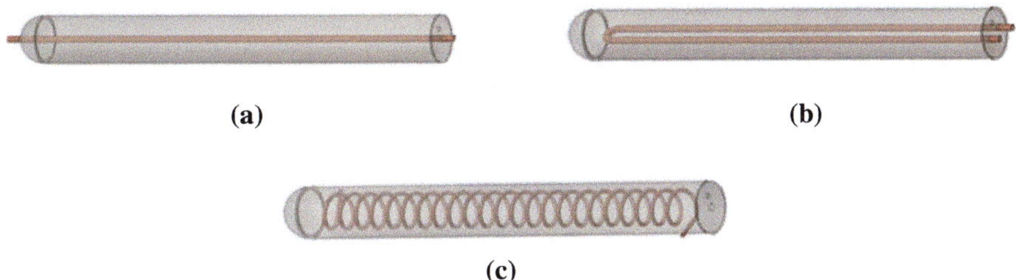

Fig. 17.8 Examples of different types of heat transfer fluid (HTF) tubes used in MH systems: **a** straight HTF tube, **b** U-shape HTF tube, **c** helical tube

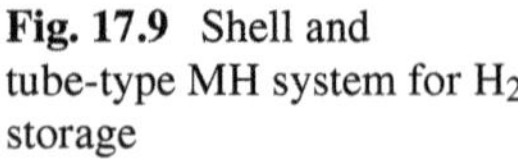

Shell and Tube configuration

Fig. 17.9 Shell and tube-type MH system for H_2 storage

the time as compared to 2 cylindrical reactor MH system having 9 cooling tubes each. Hilali et al. [35] worked on methodology in optimizing the ideal distance between staggered multiple MH tubes for improved performance under the influence of low specific heat fluid such as air. Burger et al. [36] developed and experimentally investigated a performance of 20 MH tube system arranged in aligned arrangement in 4 rows for a fuel cell operation wherein it was observed that developed MH system utilizes 93% of stored hydrogen. In terms of overall conclusion, reducing bed radius methodology for large amount of hydrogen storage will result in performance improvement simultaneously reducing system gravimetric capacity. Herein, it is advised to use optimization technique for selection of optimal number of MH tubes having conformable L/D ratio to overcome the above trade-off (Fig. 17.9).

Non-conventional Methods

Apart from the heat exchanger methods studied above following Fourier law of heat conduction and Newton's law of cooling, there exist many other methods utilized in improving overall system energy efficiency of MH system. These methods are usage of thermal energy storage medium (e.g., PCM, thermochemical medium, etc.) as shown in Fig. 17.10 and heat pipes [37–39]. The use of such methods helps in waste heat recovery, simplification of BOPs for supplying energy during desorption, and improves energy efficiency and economics of MH system.

Herein, storing the energy released from MH bed while charging the metal hydride reactor in appropriate energy storage medium either in sensible or latent form will result in saving waste energy released during absorption as useful energy that can be utilized to desorb hydrogen. The literature studies in terms of thermal energy storage utilize water, thermochemical, and phase change material. Ben Maad et al. [40] numerically studied the heat and mass transfer characteristics for $LaNi_5$-based MH reactor equipped with PCM wherein it was observed that improving thermal conductivity of PCM to 5 W/m K results in significant improvement of sorption

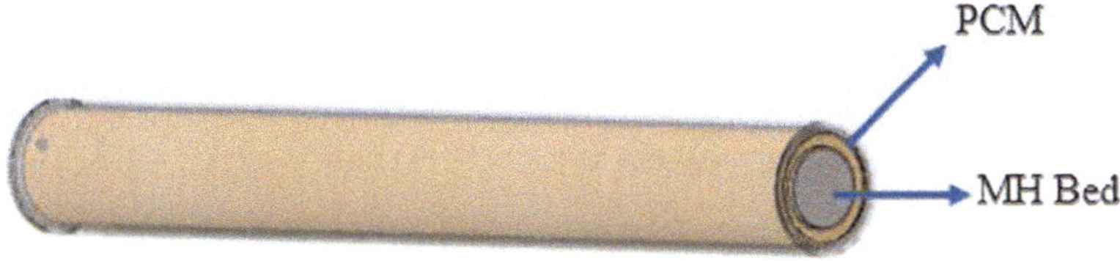

Fig. 17.10 MH system with PCM jacket

performance. Darzi et al. [41] numerically analyzed the absorption and desorption performance of $LaNi_5$-based MH reactor surrounded with PCM jacket with utmost detail on evolution of phase change in PCM due to energy storage or supply. The incorporation of metal foams has been studied for increasing the thermal conductivity of PCM which in turn improves the energy supply to desorb hydrogen. The integration of $LaNi_5$-MH reactor with thermocline-based heat storage unit was investigated by Tiwari and Sharma [42]. It was observed that thermocline unit stored 81.7% of heat released during absorption and delivered 60% of stored thermal energy to desorb hydrogen.

Another most discussed method is the use of heat pipes in recovering or thermal integration of waste heat from fuel cell to desorb hydrogen. Heat pipes having high effective thermal conductivity, transfer heat effectively by liquid–vapor phase change method and offer numerous advantages as compared to other passive thermal management methods [43]. Chung et al. observed 50% improvement in the desorption behavior on using heat pipes [44, 45]. Tetuko et al. [46] modeled thermal integration of fuel cell and MH system using heat pipes to meet the power demand of 500W. It was observed that 5 heat pipes were able to extract 880W of heat from fuel cell at its maximum operating point of which only 20% was required to meet necessary power demand.

Combination of Heat Augmentation Techniques

The transition toward large-scale hydrogen storage (> 0.5 kg-H_2) in metal hydride for safer commercial application having faster charging kinetics and efficient dynamic discharging characteristics; multiple heat exchange methods are combined together to meet the desired objectives. These combinations are discussed but not limited to using high thermal conductivity medium (e.g., ENG compacts, foams) along with fins, fins with embedded cooling tubes, combination of external and internal fins, shell and tube-type MH reactors, water/PCM jackets, and cooling tubes. Among these combinations, few examples are shown in Fig. 17.11.

a) Straight HTF tube with fins

b) U shaped HTF tube with fins

Fig. 17.11 Examples of combination of heat exchanger methods used in MH systems **a** MH system with straight HTF tube with fins, **b** MH system with U-shape HTF with fins

Muthukumar et al. [27] developed 2D mathematical model to optimize 150 kg mischmetal-based hydrogen storage system wherein a reactor with 300 mm diameter, 48 cooling tubes, and a water jacket yields minimum absorption time at operating parameters of 30 bar, 20 °C and 1250 W/m^2. Lototskyy et al. [47] utilized MH system storing 1.7 kg-H_2 for onboard application wherein compacts of ENG and alloy separated by perforated Cu fins were distributed in 40 MH tubes cascaded in 8 cassettes with HTF flowing through each cassette dissipating heat. Gupta et al. [48] experimentally investigated performance of 50 kg-AB_5 MH reactor of shell and tube configuration having 7 MH tubes wherein each tube was embedded with 4 longitudinal Cu fins. There exists a large number of literatures exploring combinations of heat exchanger methods ranging for small-scale hydrogen storage (< 100 g) to large-scale hydrogen storage (5 kg-H_2). However, selection of different combinations of heat exchangers depends on several factors such as end-use applications, mass of hydrogen to be stored, rate of charging/discharging required, energy requirement, space and mass constraints.

Parameters Affecting Charging and Discharging Performance of Metal Hydride-Based Hydrogen Storage System

After discussing the various methods for improving heat transfer within MH system, it is also important to understand the most sensitive factors influencing absorption and desorption performance of metal hydride system. The earlier studies till 90 s in the domain of metal hydride-based hydrogen systems concentrated upon developing correlations, understanding the impact of various design and process parameters on the performance of MH system. The most critical parameters remain to be thermal conductivity of alloy, length-to-diameter ratio (bed radius), supply pressure, heat transfer coefficient, mass of hydrogen to be stored, external temperature, and porosity [34, 49–51]. The sensitivity analysis of these parameters is represented in Fig. 17.12.

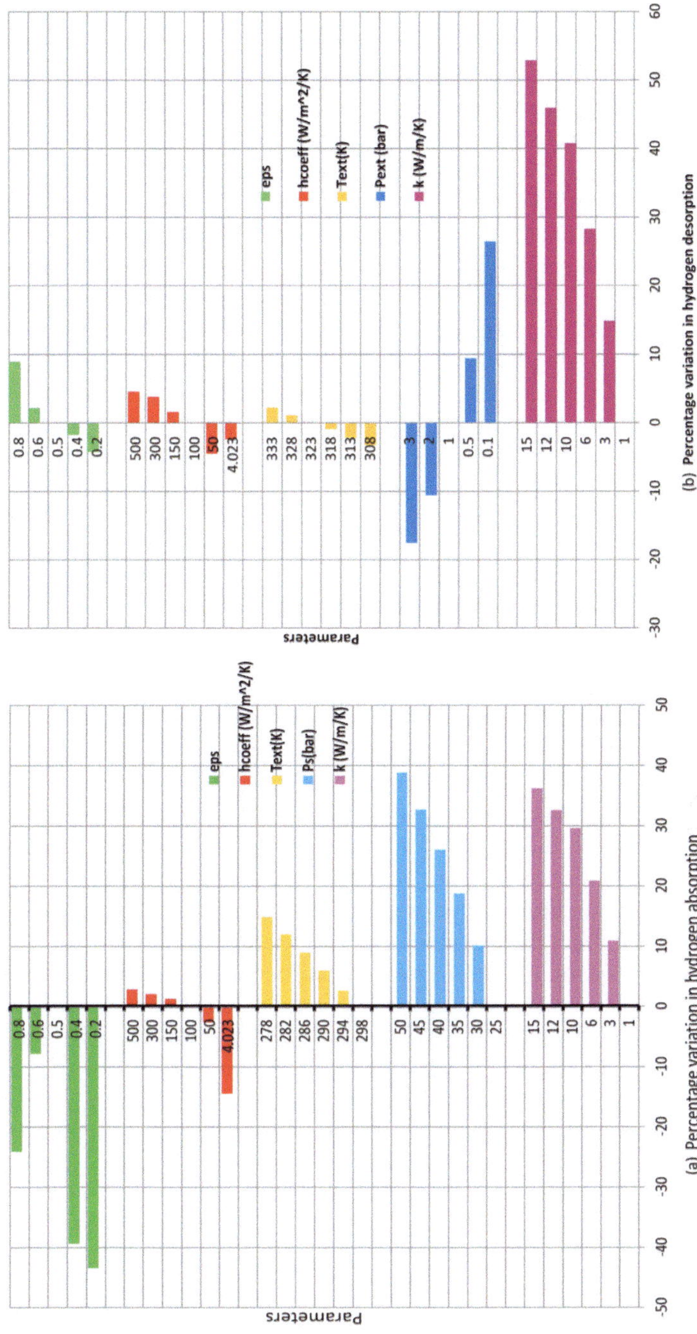

Fig. 17.12 Sensitivity analysis showing the impact of different parameters on the performance (Reproduced from [34] with permission)

Thermal Conductivity of Alloys

The MH system experiences major challenge in terms of utility due to poor thermal conductivity of alloys impacting the hydrogen charging/discharging rates. Increasing the thermal conductivity of metal hydride bed will improve the heat conduction from core of metal alloy during absorption limiting the rise of metal hydride bed temperature thereby maintaining sufficient driving force toward faster hydrogen charging of MH system. Similarly, during desorption improved thermal conductivity of alloy will help in transporting external supplied energy efficiently to desorb hydrogen for end use at desired rate maintaining sufficient equilibrium pressure. Hence, improving thermal conductivity is considered of utmost importance compared to any other parameters for significant improvement in system's performance. In terms of various existing studies on analyzing the sensitivity of thermal conductivity of alloy, it has been observed that increasing effective thermal conductivity of alloy beyond 12 W/m K have least impact in improving system performance. The effective thermal conductivity of alloys can be improved by incorporating ENG, metal foams, fins within MH system.

Supply Pressure (Absorption) and Outlet Pressure (Desorption)

The absorption and desorption process in MH system occurs due to necessary driving force. The difference between hydrogen supply pressure and equilibrium pressure of alloy acts as primary driving force during hydrogen absorption. So, increasing supply pressure during absorption increases driving force for reaction resulting in faster charging. However, the extent of increase in supply pressure will depend on type of alloy, its equilibrium pressure at initial temperature such that increasing supply pressure beyond certain value will not have significant impact on system performance but will affect the overall energy efficiency of the system.

The difference between equilibrium pressure and outlet/end-use required pressure acts as driving force for desorption reaction. The end-use or delivery pressure may be a fixed value, e.g., air cooled PEMFC stacks up to 5 kW requires 0.5 barg-H_2 pressure, PEMFC stacks beyond 5 kW requires 8–12 bar of constant delivery pressure, etc. In such case, to maintain necessary driving force, equilibrium pressure of alloy needs to be sufficiently higher during operation. The equilibrium pressure of alloy can be maintained by appropriate selection of alloy for given application and increasing thermal energy input to alloy.

Length-to-Diameter Ratio (L/D)

The bed radius has been investigated as one of the most influential parameters after thermal conductivity of alloys. Optimizing *L/D* ratio is one of the methods to optimize bed radius and simultaneously increase heat transfer surface area. Increasing *L/D* ratio reduces alloy bed radius which in turn reduces conduction thermal resistance and conduction pathway improving heat transfer from the core of the metal hydride bed. Tiwari et al. [50] performed sensitivity analysis for 0.5 kg hydrogen storage in AB_5 type alloy wherein it was observed that increasing *L/D* ratio beyond 4 yields not much significant performance improvement. Even though increasing *L/D* ratio increases or improves system performance, it has to be optimized such that MH system fits within conformable limit or space for given quantity of hydrogen to be stored.

Heat Transfer Coefficient

Heat transfer coefficient depends on many factors such as surface geometry, surface roughness, surrounding conditions, type of fluid, fluid flow rate, and arrangement of metal hydride tubes or system. Increasing heat transfer coefficient leads to faster heat removal/supply within MH system. Several studies have examined that increasing heat transfer coefficient beyond 1000 W/m^2 K has negligible impact on system's performance [52].

External Heat Transfer Fluid Temperature

It is understood that difference between supply pressure/outlet pressure and equilibrium pressure acts as driving force for absorption and desorption process, respectively. Equilibrium pressure of alloy is temperature dependent property as represented using van't Hoff equation. In the case of an exothermic absorption reaction, bed temperature increases as reaction proceeds which increases the equilibrium pressure and leads to lower driving force impacting reaction rate. So, decreasing external HTF temperature to optimal value is essential during absorption reaction for faster charging kinetics. On the contrary, in the endothermic desorption, increasing external HTF temperature to optimal value is essential to maintain sufficient equilibrium pressure for meeting end-use hydrogen consumption. In this process, the focus should be that changes in the temperature should not incur high energy penalty affecting overall energy efficiency of the system for improving system performance.

1. **Porosity**

Porosity is usually considered as isotropic with value of 0.5 in the majority of metal hydride-based system studies. However, this varies significantly due to manufacturing technique used in synthesis of alloy and repeated thermal cycling of alloy. The variation of the porosity from 0.5 on either side has been found to affect the system adversely. Higher values of porosity of alloy will require larger volume to store net quantity of alloy, whereas lower values may impact the hydrogen diffusion or kinetics of system due to compaction of alloy.

2. **Mass of Hydrogen to be Stored**

The mass of hydrogen to be stored can be considered one of the most sensitive parameters. The increase in amount of hydrogen to be stored is directly proportional to increase in volume of metal hydride reactor. As size of the MH reactor increases, reaction rate reduces which necessitates the inclusion of complex heat exchanger method. Tiwari et al. [50] studied the impact of increasing mass of hydrogen stored from 0.5 to 500 g considering simple cylindrical reactor. It was observed that increasing mass of hydrogen storage for same material properties and process conditions significantly increases reaction completion time.

In conclusion, the majority of sensitivity analysis studies converge stating that MH system's performance is a strong function of thermal conductivity and bed radius or *L/D* ratio whereas weakly dependent on external heat transfer fluid temperature.

Applications of Metal Hydride-Based Hydrogen Storage Systems

The existence or versatility of class of hydrides and ability to tune properties of the alloy using various scientific methods paves way to utilize metal hydride-based hydrogen storage system for various application apart from hydrogen storage as represented in Fig. 17.13. There exists significant research work toward MH system development on a different scale ranging from few grams to kilograms of hydrogen storage. However, majority of these studies are restricted to laboratory-scale reactor development and charging/discharging performance analysis. There is dearth on scaling-up and commercial demonstration of MH system toward its numerous applications (Fig. 17.13).

The major applications of metal hydride systems can be classified as hydrogen purification, hydrogen compression, thermal energy storage, stationary/back-up power generation, vehicular applications, heating and cooling system [2, 53]. These applications will be briefly discussed in this section.

a. Stationary/Back-up Power Generation

Metal hydride-based hydrogen storage system can be integrated with fuel cell to generate electricity along with cleaner by-product as heat and water. MH-H_2 storage

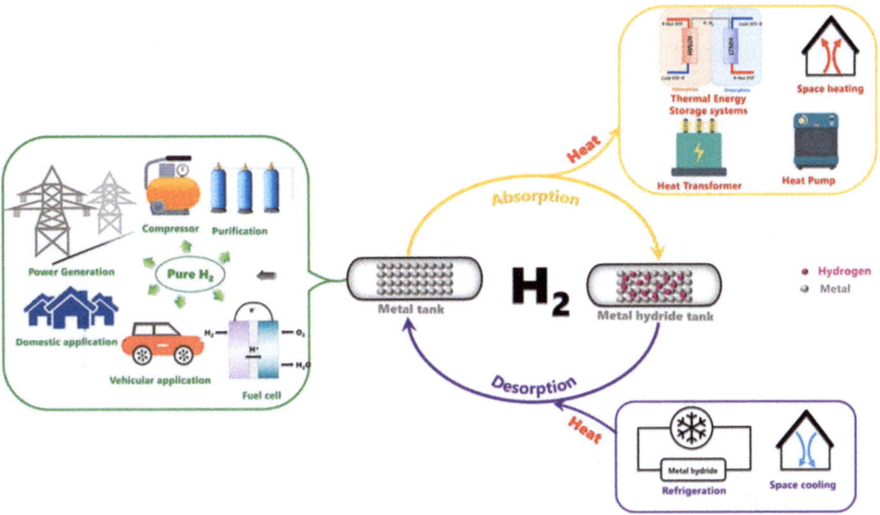

Fig. 17.13 Different applications of metal hydride-based systems (Reproduced from [96] with permission)

system can be a potential medium for resolving renewable energy intermittency wherein surplus energy can be fed to electrolyzer generating green hydrogen at lower pressures of 15–35 bar which can be directly fed to MH-H_2 storage system eliminating compressor as well as hydrogen refueling station requirement for safe and prolonged storage of hydrogen which can be inferred from schematic in Fig. 17.14. This stored hydrogen then further supplied to fuel cell that produces clean electricity as per the demand (Fig. 17.14).

A bench-scale hydrogen energy system comprising PV panels (20-kW), PEM electrolyzer (5 Nm^3/h), PEMFC (3.5 kW), batteries (20 kWh), and 260 kg TiFe-based MH tanks storing 142 kWh energy in the form of H_2 was established and

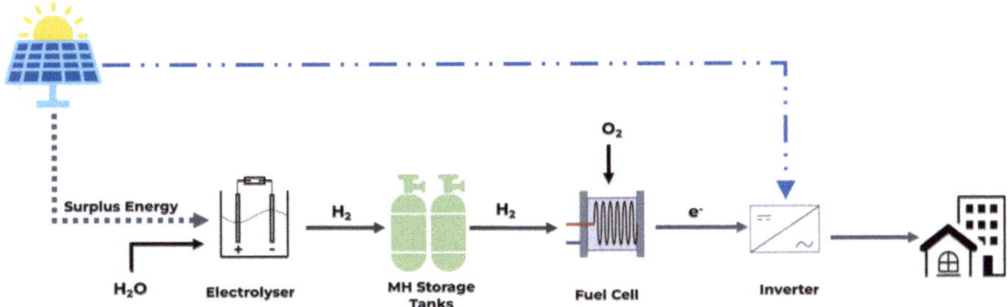

Fig. 17.14 Schematic showing a renewable-based hydrogen production, storage and power generation system

demonstrated for stationary power generation at Fukushima Renewable Energy Institute, AIST, Japan. The hydrogen generation, energy storage in sunshine hours, and power generation in non-sunshine hours/cloudy days were planned and monitored by means of automated building energy management system for 24 h (clear weather)/ 48 h (cloudy days) with an aim to reduce grid power consumption. The demonstration reduces the auxiliary power consumption by 99% to desorb hydrogen by achieving effective utilization of fuel cell heat and achieved > 60% overall efficiency in terms of energy utilization (hydrogen to power). The findings from the demonstration concluded that the developed system can successfully operate throughout the year in all types of weather conditions without requirement of external heat sources for desorbing hydrogen. The research group plans to implement the technology in actual commercial buildings. In terms of scaling the solution, the research group implemented the proposed solution in commercial building with PV panel 64.75-kW capacity and 14 kW PEMFC with remaining component being same. The study assessed that with proposed onsite hydrogen production and Hydro Q-BIC (PV+20 kWh Battery+ 142 kW H_2 energy) solution, annual CO_2 emissions reduced by 53%. Further, the group has scaled this demonstration to Koriyama City Market Building, Japan, with hydrogen storage capacity of 12 kg-H_2 (400kWh). [54–57].

In the major pilot-scale development and demonstration, "Hydrogen Terminal" a joint project between Steinbeis Innovation Centre and TU Braunschweig funded by Federal Ministry of Education and Research of over 20 million euros aimed at storing 500 kg-H_2 (16.67 MWh energy) in the form of hydrogen was inaugurated in December 2023. Hydrogen storage system supplied by GKN hydrogen stores 500 kg-H_2 generated primarily by 1 MW Anion Exchange membrane electrolyzer and 150-kW PV capacity. In the broad context, the project aims to stabilize the grid when conventional fossil power plants are no longer available and also act as hydrogen refueling station for heavy-duty trucks for refueling 350 bar green hydrogen [58].

b. Metal Hydride-Based Hydrogen Purification

Hydrogen purification plays a vital role in hydrogen value chain for utilization of pure hydrogen in various applications such as fuel cell vehicle. There exist various established hydrogen purification techniques such as pressure swing adsorption method (PSA), cryogenic separation, membrane diffusion, metal hydride-based purification system. Conventional industrial methods, such as pressure swing adsorption (PSA), are insufficient in meeting these stringent purity requirements. While PSA typically produces hydrogen with a purity of around 99%, while the fuel cell applications often require a purity level of approximately 99.999%. Metal hydrides and intermetallic compounds are among the most promising candidates for hydrogen purification. These materials selectively absorb hydrogen into their interstitial sites and release it as needed. This selective absorption property enables the production of high-purity hydrogen (Fig. 17.15).

The hydrogen purification using MH-based systems follows three steps, namely absorption, flushing, and desorption, as shown in Fig. 17.15. In the first step, i.e., absorption, the stream of impure gases having CH_4, CO, CO_2, N_2, H_2, etc., from refineries or any other generation source is supplied to metal hydride (MH) alloy, and

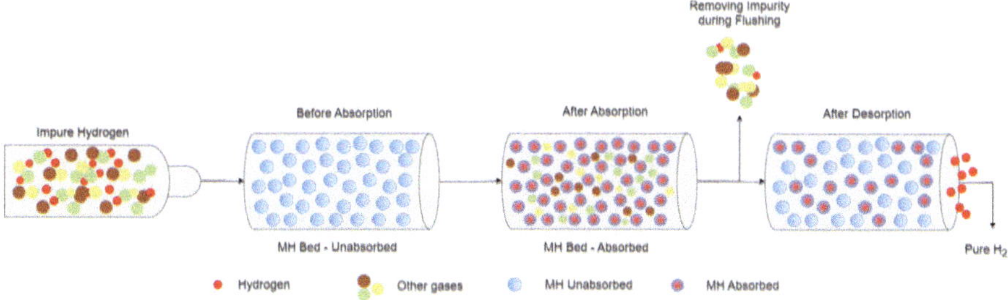

Fig. 17.15 Schematic representing metal hydride-based hydrogen purification process

due to high affinity toward hydrogen, the alloy selectively absorbs only hydrogen in the bulk and rest of the gases will be remain unabsorbed. In the second step, i.e., flushing, the impure gases apart from hydrogen can be purged/flushed to the atmosphere. In the last step, i.e., desorption, the high-purity hydrogen (> 99.99%) is desorbed from the metal hydride by supplying thermal energy input to the alloy equivalent to enthalpy of decomposition. The purification of impure gas stream can be achieved either in single stage or multiple stages depending on the different gaseous impurities and their content in the gaseous stream (Fig. 17.16).

The study on hydrogen purification has been of great interest among the researchers, lately. There are studies that outline purification aspects on small-sample levels, whereas some others have shown noteworthy results on a scaled-up system. Muthukumar and Kumar [59] studied the use of $LaNi_5$-based three-stage system to produce pure hydrogen from an inlet stream of 20% hydrogen and 80% impurities (consisting of CO_2, CO, CH_4, H_2O, etc.). The group reported that the obtained purities of hydrogen were 91%, 99%, and 99.995% in the first, second, and the third stage of their developed MH systems as evident from Fig. 17.16. They noted that while their system produces very pure hydrogen, it is affected by deactivation of alloy and regeneration aspects. Their work provides a strong footing that

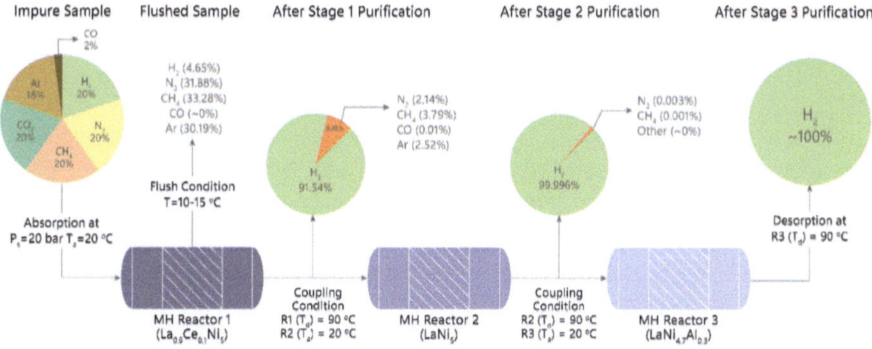

Fig. 17.16 Schematic diagram of hydrogen purification using metal hydrides with three-stage purification system (inspired from [59])

MH systems can be developed for efficient hydrogen purification. Blinov et al. [60] have shown similar studies as that by P Muthukumar, using Sn/Al/Ce-doped $LaNi_5$ derivatives, where a maximum of 83% hydrogen recovery was observed when alloy was subjected to impure hydrogen stream. They reported that a continuous flow purification system based on metal hydrides is effective to produce pure hydrogen in the case when hydrogen percentage in the gas is >10% and that evacuation-based purification is effective for hydrogen being <10%. Ashida et al. [61] studied the use of $LmNi_{4.55}Mn_{0.12}Al_{0.15}Pd_{0.18}$ on hydrogen separation from the products of a Bio-H_2 production plant that majorly contained CH_4 as the contaminant. They reported that with the mentioned metal hydride, the purity of the hydrogen is only 97%.

c. Vehicular Applications

In case of transportation sector, hydrogen powered vehicles are viewed as a potential solution to decarbonize transport sector. Battery electric vehicles (BEVs) are flourishing rapidly but have few major limitations such as lithium reserves and its mining, high charging time, lesser energy density compared to hydrogen (200–300 Wh/kg vs. 33.33 kWh/kg), battery life (~3 to 5 years) and its recycling. These limitations limit BEVs adoption to certain vehicle segments. H_2-FCEVs have potential to be implemented in all form of vehicle segments either low duty or heavy-duty vehicles such as 2/3-wheeler, light commercial vehicles, heavy-duty trucks, material-handling equipment, trains, defense mobility (e.g., submarines). In present scenario, majority of hydrogen powered vehicles are powered using 350–700 bars compressed hydrogen (CGH_2) stored in Type 4 tanks. The major challenges associated with CGH_2 are high pressure, compression energy, infrastructure development (60–70% cost spent on compression facilities), indigenous Type 4 tank development, testing facilities availability and safety restricting mass adoption of hydrogen mobility. Herein, MH-based hydrogen storage has been demonstrated as potential hydrogen storage solution irrespective of its weight or poor gravimetric storage density limitations in vehicular applications.

There exist limited studies in the form of real-time demonstration of large-scale hydrogen storage system based on metal hydride and very few studies have been reported till date related to utilization of MH system in vehicular applications. This is due to the primary fact that the metal hydrides having reversible hydrogen storage capability and favorable thermodynamic properties suffer from lower gravimetric storage capacity. The alloy mass due to low gravimetric capacity combined with self-weight of MH system having higher mass and energy-intensive heat transfer system (e.g., storage and circulation of HTF water or coolants, inclusion of large number of fins) along with other BOP for desorbing H_2 for vehicle propulsion impacts the overall system gravimetric capacity, efficiency, and range of vehicle. MH-based hydrogen storage systems can be indigenously produced/manufactured, directly integrated with electrolyzer for charging thereby eliminating heavy infrastructure requirement. However, MH systems have been demonstrated to be advantageous in vehicular applications such as material-handling mobility device (e.g., forklift, mining locomotives, construction equipment's, submarines) which requires additional counterweight for operation and light-duty vehicles (e.g., 2 and 3 wheelers, golf-cart) having lesser payload with short range of commute [53, 62–65] (Fig. 17.17) Figure 17.17a and b represents an example of dual use of metal hydride systems in heavy-duty vehicular applications wherein hydrogen can be used as fuel to drive vehicle either using fuel cell or as a fuel in internal combustion engine simultaneously extracting cooling effect due to endothermic desorption reaction providing comfort in driver cabin or operating a refrigerated container.

The Sandia National Laboratories (SNL) have been one of the important bodies in the development of solid-state hydrogen storage technology. The 4-ton underground

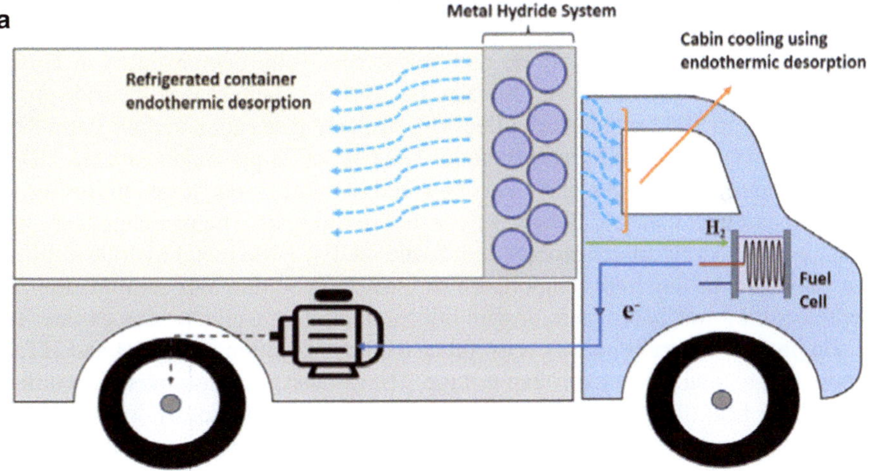

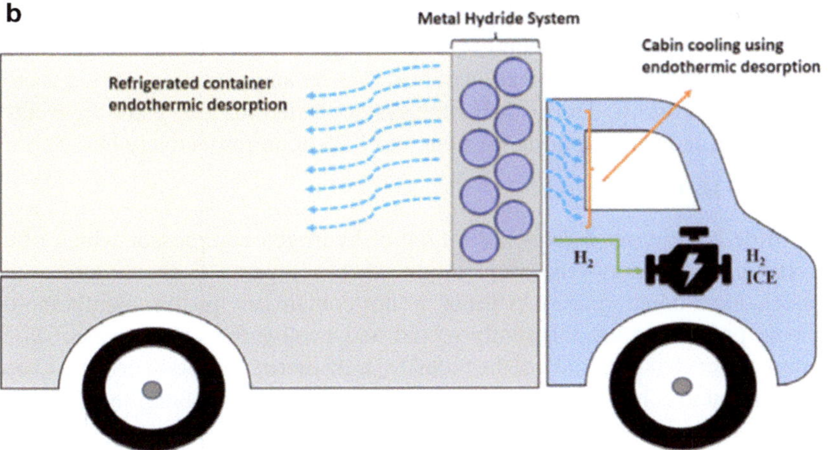

Fig. 17.17 Schematic showing metal hydride-based system on a heavy-duty truck with added cooling effect drawn from the storage system **a** in case of a fuel cell vehicle, **b** in case of an internal combustion engine-based vehicle

mining locomotive was powered using 3 kg hydrogen stored in 213 kg C15 alloy-based MH system developed by SNL on integration with PEMFC. The developed system was comparatively lighter as compared to equivalent energy stored in lead acid battery [66]. Further, SNL also developed a 3 kg $NaAlH_4$-based hydrogen storage system for vehicular applications with General Motors in 2012. This system consists of 4 shell and tube modules having 12 tubes containing $NaAlH_4$. In this case, 10% of stored hydrogen was utilized for catalytic combustion as solution to provide onboard heat to desorb hydrogen [67]. Savannah River National Laboratory has extensively worked on metal hydride-based hydrogen storage technology wherein major projects involves demonstration of H_2 Fuel Bus and GATOR-Industrial Fuel Cell Vehicle powered by hydrogen. The H_2 Fuel Bus was powered by IC Engine using 15 kg hydrogen stored in metal hydride and was successfully demonstrated in City of Augusta, Georgia and Las Vegas, Nevada [68, 69]. The Daimler Benz hydride project during 1979 to 1982 conducted a fleet run of passenger cars and vans completely powered by hydrogen or in hybrid mode. The fleet of 28 passenger cars has 20 cars fully powered by hydrogen and 8 cars powered in hybrid mode tested in Berlin. The other 5 passenger cars and 3 vans were tested in Stuttgart. The passenger cars were fitted with a 5 kg hydrogen storage unit based on Ti-Cr-Mn metal hydride alloy and the vans were having high-temperature and low-temperature MH combination. The demonstration was based on hydrogen injection in internal combustion engine. Two IC Engine vans were primarily retro fitted with combination of LTMH and HTMH amounting to total 5.4 kg hydrogen storage. The exhaust gas from engine was utilized to heat the heat transfer fluid which in turn was utilized to desorb hydrogen from MH reactor for injection in ICE [70]. Toyota Motor Corporation had used BCC alloy-based hybrid metal hydride hydrogen storage system capable of storing 2 kg hydrogen to propel a car up to a range of 250 km with maximum speed of 100 km/h. The vehicle was powered using 20-kW fuel cell developed by Toyota. The vehicle was developed and tested in 1996 [71]. HySA Competence center developed a 1.8 kg-H_2 distributed H_2 storage system wherein 1.7 kg-H_2 stored within Ti-Mn-based MH storage system having combination of heat transfer augmentation techniques for operating electric forklift. The 40 MH tube system tested for heavy and light-duty operation within industry supports 3.1 hour and 7.25 hour, respectively [47, 72].

d. MH-Based H_2 Compression

Metal hydride-based compressors are a type of hydrogen compressor which offer quiet, reliable, and energy-efficient solution when compared to their more traditional mechanical counterparts. As these do not contain any piston, membrane or any moving parts, they are inherently robust and involve low maintenance. MH-based compressors basically utilize the metal hydride thermal sorption process to use energy from heat to achieve hydrogen compression. This also enables the possibility to consume waste industrial heat [73].

The metal hydride hydrogen compressor in simple terms can be understood by means of PCT isotherms of selected alloy as represented in Fig. 17.18, wherein low-pressure hydrogen can be completely absorbed at lower temperature, and as temperature increases, the equilibrium pressure of alloy increases that leads to high-pressure hydrogen desorption from metal hydride system. According to using the van't Hoff equation and plot, P_{eq} increases exponentially with temperature; hence, absorption can be performed at lower temperature and desorption being an endothermic process is carried out at higher temperatures leading to high-pressure hydrogen delivery. The enthalpy and entropy values should be such that the plateau pressure falls within practical range and also has low hysteresis. The selected alloy must also demonstrate high cyclic durability, maintaining its hydrogen storage capacity and thermodynamic behavior over a large number of absorption and desorption cycles. This ensures long-term reliability and performance stability of the compressor under repeated thermal and pressure cycling [74] (Fig. 17.18).

Several notable advancements in metal hydride-based hydrogen compression have been demonstrated globally, showcasing the versatility and scalability of the technology. At Sandia National Laboratories (USA), a two-stage MH compressor prototype utilizing AB_2-type alloys compacted with expanded natural graphite compressed hydrogen from 150 to 875 bar, with a significant temperature swing from 15 to 145 °C. Their low-pressure bed achieved compression from 132 to 634 bar under controlled

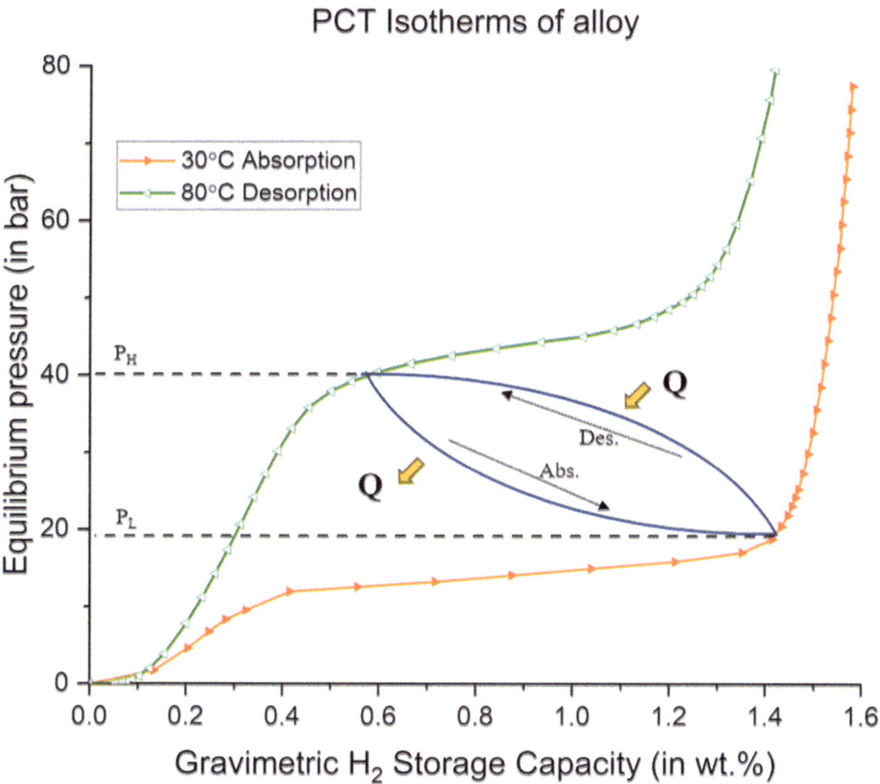

Fig. 17.18 PCT isotherm of a representative metal hydride alloy demonstrating the differences in equilibrium pressure and use for compression applications

thermal cycling [75]. Meanwhile, researchers at Zhejiang University (China) developed a high-capacity double-stage compressor capable of producing hydrogen at pressures exceeding 70 MPa, with a compression capacity of around 2000 L per cycle and a flow rate of 34.6 L/min, using thermal oil at 423 K as the heat source [76]. In Cyprus, HYSTORE Technologies Ltd. engineered a compact MH compressor that raised hydrogen pressure from 7 to 220 bar with only moderate heating at 80 °C, highlighting the potential for low-energy operation [77]. More recently, HySA Systems and TF Design developed a single-stage high-pressure MH compressor prototype capable of compressing hydrogen from 100 bar to 700 bar at working temperatures between 20 and 150 °C, achieving an estimated productivity of approximately 1 Nm^3/h [78]. These examples underscore the adaptability of MH compressors across varying pressure ranges, productivity targets, and thermal conditions, making them promising for both large-scale and decentralized hydrogen infrastructure.

e. Metal Hydride-Based Thermal Energy Storage System

Thermal energy can be stored in different modes such as sensible heat, latent heat, and thermochemical heat. The vital need for thermal energy storage system is to store either excess or waste heat available and later utilize this stored heat to meet heat load demand minimizing the energy supply demand gap along with higher system energy efficiency [79]. The sensible heat storage method primarily works on the principle of change in storage medium temperature without undergoing any phase change. The sensible storage media can be solid (e.g., concrete, sand, stone, etc.), liquid (e.g., water, oil, etc.), and gas (e.g., air, etc.). The latent heat storage system is heat storage due to phase change of material and generally has higher volumetric energy density as compared to sensible storage medium. The medium largely used for latent heat storage are wax, water, metals, etc., depending on applications' temperature range [80]. In thermochemical heat storage method, heat storage is based on reversible chemical reaction and uses heat of decomposition such that process is endothermic. The heat is recovered by a reverse synthesis reaction generally exothermic reaction. This mode of heat storage is generally characterized by high energy storage density, long-term storage, and low heat losses compared to latent and sensible heat storage medium (Fig. 17.19).

Thermochemical heat storage can be accomplished by using metal hydroxides, metal carbonates, and metal hydrides. Among all potential thermochemical storage materials, metal hydrides are of great interest due to high energy storage density, reversibility, low cost, and high exergy efficiency [81]. Studies have found that metal hydride-based thermal storage systems store 5–10 times higher energy as compared to latent and sensible heat storage respectively [82]. Caldwell et al. [83] developed MH-thermal storage system based of Li-H material for supplying continuous heat to the satellite power conversion system. Li-H being expensive material, further studies proposed use of Mg-based materials as cost-effective alternative to be used for metal hydride thermal storage systems due to its high gravimetric hydrogen storage capacity and high operational temperature range [84]. However, MgH_2 suffers from slow kinetics which was later concentrated to be improved by means of doping Mg with

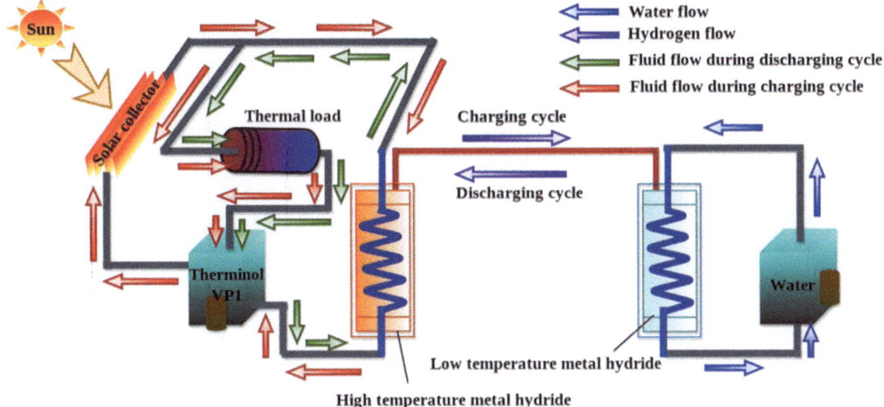

Fig. 17.19 Schematic showing a complete metal hydride-based thermal energy storage system integrated with solar thermal plant

other elements such as Ni and Fe [85]. Max Planck Institute developed metal hydride-based thermal storage system using Ni-doped Mg and Ti-Fe-Cr-Mn alloy pairs for a sterling engine having temperature range of 300–480 °C [86]. Mellouli et al. [87] observed that using Mg_2FeH_6 and Na_3AlH_6 as metal hydride thermal storage systems pairs yields 96% energy recovery of total solar energy stored. D'Entremount et al. [88] used same pair of alloys as used in [87] for thermal energy storage in a working pressure range of 30–70 bar accomplishing energy storage density of 132 kWh/m^3 which was five times higher than the USDOE target. Tiwari et al. [89] numerically investigated performance of coupled Mg_2Ni (HTMH) and $LaNi_5$ (LTMH) thermal energy storage system. It was observed that coupled system has volumetric energy density of 144 kWh/m^3 and recovers 1.22 MJ energy with overall efficiency of 69% within 6 hour.

f. MH-Based Heating and Cooling System

The exothermic and endothermic nature of absorption and desorption is advantageous in terms of extracting the heating and cooling effect of MH bed during thermal cycling respectively as explained graphically in Fig. 17.20. The heat released during absorption can be utilized for developing applications such as heat pumps, whereas heat absorbed during desorption leading to reduction in alloy bed temperature may be utilized for cooling applications. A typical MH heating and cooling system may consist of at least two metal hydride beds connected by a closed-loop hydrogen circuit. One bed, heated by an external thermal source, desorbs hydrogen at high-temperature and equivalent equilibrium pressure. The hydrogen gas then travels to a low-temperature MH bed, where it is absorbed at a lower temperature, releasing heat. This cycling of hydrogen enables the system to extract heat from a low-temperature source and deliver it to a higher-temperature sink—thus functioning as a thermally activated heat pump [90]. The advantages of metal hydride-based heating/cooling systems include silent operation, high reliability due to the absence of moving parts,

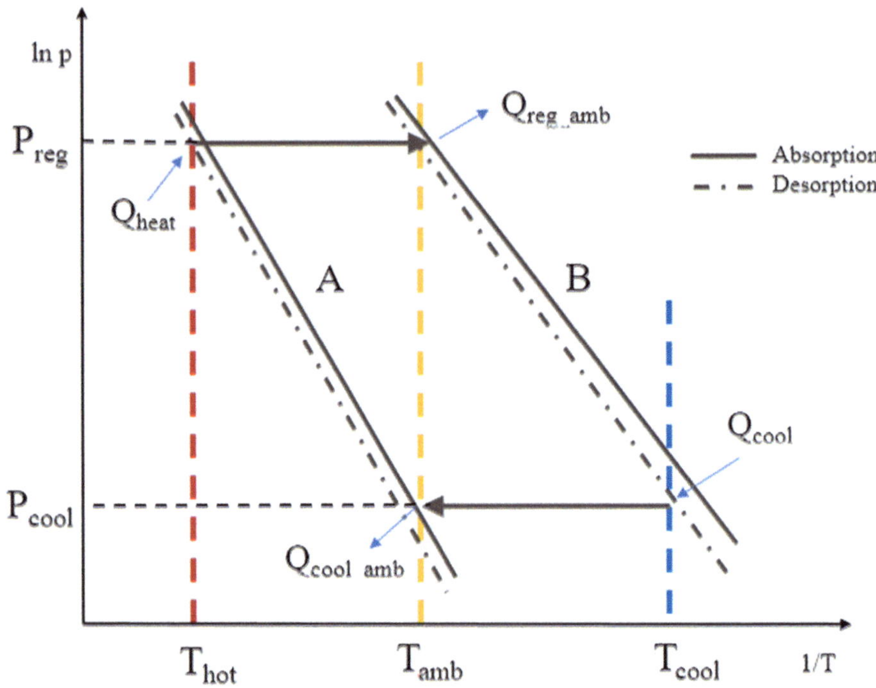

Fig. 17.20 P–T diagram showing the absorption and desorption for a metal hydrides-based systems which can be used for heating and cooling application

compatibility with renewable energy sources, and the use of environmentally benign working fluids (e.g., water). These systems are particularly well-suited for applications in off-grid refrigeration, solar-powered air conditioning, and heat recovery from industrial processes (Fig. 17.20).

Metal hydride (MH) systems are continuously being explored for their use in thermal applications, particularly as heat pumps. Researchers from Kogakuin University, Tokyo, developed a double-stage metal hydride heat pump utilizing three distinct metal hydrides to demonstrate the practical generation of high-temperature heat. This research analyzed experimental demonstration units with capacities of 7.72 and 77 kW, emphasizing the production of high-temperature steam (120–150°C) from a low-grade 80°C heat source [91]. Another example is that of a system specifically designed for cold water production at temperatures below 4°C. This installation notably demonstrated the capability for simultaneous output of cold water and hot water at 50°C, built upon a modular principle to achieve desired productivity and continuous cooling effect generation [92]. Ram Gopal and Murthy [93] evaluated the performance of metal hydride cooling system using Zr-Mn-Fe (HTMH)/ $MmNi_{4.5}Al_{0.5}$ (LTMH) alloy pairs. The MH system with selected alloy pair has a COP of 0.45–0.5 which is similar to conventional systems such as zeolite-water and zeolite-methanol. The low value of COP is attributed to the low enthalpy of formation of low-temperature MH (LTMH) which can be improved by internal heat recovery. University of Stuttgart developed and demonstrated a compact cooling system which was designed to overcome weight and volume constraints, relevant for automotive air conditioning. By employing a capillary tube bundle reaction bed to enhance heat and mass transfer, this system achieved very short half-cycle times, of 100 seconds, and a high specific cooling power of up to 780 W/kg of metal hydride while desorbing[94]. Kolbig et al. [95] investigated the use of Hydralloy C5 in preheating vehicle components and air conditioning within car cabin at − 20 °C. The Hydralloy C5 was coupled with $LaNi_{4.85}Al_{0.15}$ for preheating, wherein the coupled reactions results in the thermal power released by 2 kg $LaNi_{4.85}Al_{0.15}$ was 0.61 kW/kg-MH raising 2 liters lubricant temperature by 20K at − 20 °C and by Hydralloy C5 was 0.76kW/kg-MH at − 20 °C. However, the released peak thermal power by Hydralloy C5 was insufficient against the required conventional 7 kW power for air conditioning primarily affected by endothermic nature of desorption and strong dependency of desorption reaction on ambient temperature.

Problems

Q1. Design a metal hydride reactor considering AB5 as metal hydride to store 1 kg of hydrogen.

Analyze the following:

1. Considering the shape of the metal hydride reactor as a cylinder, calculate the length and diameter of the cylinder if the L/D ratio is 8.
2. Based on this system, calculate the mass of hydrogen absorbed (in kg/m^3 s) at time t at $T = 303$ K and supply pressures of 10, 20, and 30 bar. Finally, comment on the effect of pressure on the system.
3. Based on this system, calculate the mass of hydrogen desorbed (in kg/m^3 s) at time t when temperature $T = 303, 313, 323, 333$ K is varied and at an outlet pressure of 1 bar. Finally, comment on the effect of temperature on the system.

Given: Properties of AB5 metal hydride

Properties	Absorption	Desorption
Gravimetric storage capacity	1.34 wt.%	–
Empty density of alloy (ρ_{emp})	8400 kg/m^3	–
Saturated density of alloy (ρ_{sat})	8513 kg/m^3	–
Instantaneous density of alloy at "t"	8468 kg/m^3	8456 kg/m^3
Porosity (ε)	0.5	0.5
Expansion volume (V_{ex})	20%	20%
Enthalpy change (ΔH)	30,800 J/mol	31,400 J/mol

(continued)

(continued)

Properties	Absorption	Desorption
Entropy change (ΔS)	108 J/mol-K	108 J/mol-K
Activation energy (E_a)	21,179.6 J/mol	16,473 J/mol
Reaction constant (C_r)	59.187[1/s]	9.5[1/s]

Solution

1. **Step I: Mass of alloy required to store 1 kg-H$_2$**

Mass of hydrogen to be stored, $m_{H_2} = 1\ \text{kg} - H_2$

To find: length (L) and diameter (D) of a cylindrical reactor

Gravimetric capacity, wt% $= \frac{m_{H_2}}{m_{\text{alloy}} + m_{H_2}} \times 100$

$$1.34 = \frac{1}{1 + m_{\text{alloy}}} \times 100 \Rightarrow 1 + m_{\text{alloy}} = \frac{100}{1.34} \Rightarrow m_{\text{alloy}} = \frac{100}{1.34} - 1 = 73.63\ \text{kg}$$

Step II: Volume required to store 73.63 kg alloy

$$V = \frac{m_{\text{alloy}}\left(1 + \frac{\%\ \text{expansion}}{100}\right)}{\rho_{\text{emp}} \times (1 - \varepsilon)}$$

Substitute values:

$$V = \frac{73.63 \times 1.2}{8400 \times (1 - 0.5)} = \frac{73.63 \times 1.2}{4200} = 0.02104\ \text{m}^3 = 21.04\ \text{L}$$

So, to store 1 kg of H_2 in alloy with gravimetric capacity 1.34 wt.%, it requires 21.04 L volume.

Step III: Calculation of dimensions of reactor (*L*and *D*)

Assume the vessel is cylindrical in shape with $\frac{L}{D} = 8$

$$V = \frac{\pi D^2 L}{4} = \frac{\pi D^2 \cdot 8D}{4} = 2\pi D^3$$

$$\Rightarrow D = \sqrt[3]{\frac{V}{2\pi}} = \sqrt[3]{\frac{0.02104}{2\pi}} = 0.1496\ \text{m} = 14.96\ \text{cm}$$

$$L = 8 \times D = 8 \times 0.1496 = 1.1968\ \text{m} = 119.68\ \text{cm}$$

Internal diameter (D) and length (L) of the metal hydride reactor for storing 1 kg-H$_2$ are:

$$D = 14.96\ \text{cm},\ L = 119.68\ \text{cm}$$

Mass of hydrogen absorbed $\dot{m}_{abs}$ (in kg/m^3 s) at any time t, with MH system at $T = 303$ K and hydrogen supply pressures $P_s = 10, 20, 30$ bar Absorption Equation

$$\dot{m}_{abs} = C_a \cdot \exp(\tfrac{-E_a}{RT}) \cdot \ln\left(\tfrac{P_s}{P_{eq}}\right)(\rho_{sat} - \rho_s) \tag{17.19}$$

2. **Step I: Calculating P_{eq} at $T = 303$ K**

$$\ln\left(\frac{P_{eq}}{P_{ref}}\right) = \frac{-\Delta H}{RT} + \frac{\Delta S}{R}$$

$$P_{eq} = P_{ref} \times \exp\left(\frac{-30800}{8.314 \times 303} + \frac{108}{8.314}\right) = 2.146\,\text{bar}$$

Step II: Substituting for $\dot{m}_{abs}$

At 10 bar:

$$\dot{m}_{abs} = 59.159 \cdot \exp\left(\frac{-21179.6}{8.314 \times 303}\right) \cdot \ln\left(\frac{10}{2.146}\right) \cdot (8513 - 8468)$$
$$= 0.92\,\text{kg/m}^3\,\text{s}$$

At 20 bar: $\dot{m}_{abs} = 1.33\,\text{kg/m}^3\,\text{s}$.
At 30 bar: $\dot{m}_{abs} = 1.57\,\text{kg/m}^3\,\text{s}$.

Step III: Effect of pressure on absorption

$$\frac{1.33 - 0.92}{0.92} \times 100 = \frac{0.41}{0.92} \times 100 \approx 44.57\%$$
$$\frac{1.57 - 1.33}{1.33} \times 100 = \frac{0.24}{1.33} \times 100 \approx 18.05\%$$

Comment: Increasing the supply pressure from 10 to 20 bar improves the absorption rate by 44.5%. However, increasing it further to 30 bar shows only 18% increase. Beyond 30 bar, it may not significantly enhance performance but reduce energy efficiency.

3. **Step I: Calculating P_{eq} at Different Temperatures**

Mass of hydrogen desorbed (in kg/m^3 s) at different temperatures of MH system:

$$T = 303,\ 313,\ 323,\ 333\ \text{K}$$

at an outlet pressure of 1 bar

Desorption Equation:

$$\dot{m}_{des} = C_D \cdot \exp\left(\frac{-E_a}{RT}\right) \cdot \frac{(P - P_{eq})}{P_{eq}} \cdot (\rho_s - \rho_{eq}) \tag{17.20}$$

$$P_{eq} = P_{ref} \cdot \exp\left(\frac{-\Delta H}{RT} + \frac{\Delta S}{R}\right)$$
$$= \exp\left(\frac{-31400}{8.314 \times 303} + \frac{108}{8.314}\right) \Rightarrow P_{eq,303K} = 1.7\,\text{bar} \quad (17.21)$$

Note: P_{ref} refers to reference pressure $P_{ref} = 1$ bar. Similarly,

$$P_{eq,313K} = 2.52\text{bar},\ P_{eq,323K} = 3.65\,\text{bar},\ P_{eq,333K} = 5.2\,\text{bar}$$

Step II: Substituting P_{eq} and T into desorption equation

$$\dot{m}_{des,303K} = 9.5 \cdot \exp\left(\frac{-16473}{8.314 \times 303}\right) \cdot \left(\frac{1 - 1.7}{1.7}\right) \cdot (8456 - 8400) = -0.32\,\text{kg/m}^3\,\text{s}$$

(Negative sign indicates direction of output hydrogen flow from metal hydride reactor.)

Similarly,

$$\dot{m}_{des,313K} = -0.57\,\text{kg/m}^3\,\text{s}\ \dot{m}_{des,323\,K} = -0.84\,\text{kg/m}^3\,\text{s}\ \dot{m}_{des,333\,K} = -1.12\,\text{kg/m}^3\,\text{s}$$

Comment: Increasing temperature of the metal hydride bed increases rate of desorption. It can be observed that increasing temperature by 10 K (from 303 to 313 K) leads to 78% increase in desorption rate.

Q2. Select an appropriate alloy for a vehicular application from the following such that a light-duty vehicle operating with 5 kW PEM Fuel Cell stack requires constant input pressure of 2 bar.

Assumption: Hydrogen is desorbed without any external energy input at ambient temperature of 303 K.

Observation: During desorption, MH bed temperature drops from 303 to 283 K. Properties of alloys for desorption process are given in the following table.

Properties	Alloy A	Alloy B
Enthalpy change (ΔH)	30,800 j/mol	28,250 j/mol
Entropy change (ΔS)	108 J/mol K	111.92 J/mol K
Gravimetric storage capacity	1.5 wt.%	1.6 wt.%

Solution:

In this problem statement, the vehicle requires constant fixed pressure supply at 2 bar from MH system, i.e., $P = 2$ bar. It is understood that during desorption, difference between equilibrium pressure of hydrogen desorption from MH and outlet/end-use pressure requirement is driving force of desorption such that $P_{eq} > P$. In this case, let us calculate equilibrium pressure at 283 K for both the alloys. The alloy whose equilibrium pressure is higher than the desired pressure at lower temperature can be selected for given applications.

Step I: Calculating P_{eq} for Alloy A & B at 283 K

For Alloy A

$$P_{eq\text{ , alloy A}} = P_{ref} \times \exp\left(\frac{-\Delta H}{RT} + \frac{\Delta S}{R}\right)$$
$$= 1 \times \exp\left(\frac{-30,800}{8.314 \times 283} + \frac{108}{8.314}\right)$$
$$P_{eq\text{, alloy A}} = 0.9\text{ bar}$$

For Alloy B

$$P_{eq\text{, alloy B}} = P_{ref} \times \exp\left(\frac{-\Delta H}{RT} + \frac{\Delta S}{R}\right)$$
$$= 1 \times \exp\left(\frac{-28250}{8.314 \times 283} + \frac{111.92}{8.314}\right)$$
$$P_{eq\text{, alloy B}} = 4.284\text{ bar}$$

Comments: It is evident that p_{eq}, alloy B >p_{eq}, alloy A at 283 K. So alloy B can be chosen for this application. Also, gravimetric capacity is essential but cannot be the only selection criterion for vehicular application.

References

1. The Future of Hydrogen, Futur. Hydrog. (2019). https://doi.org/10.1787/1e0514c4-en
2. M. Hirscher, V.A. Yartys, M. Baricco, J. Bellosta von Colbe, D. Blanchard, R.C. Bowman, D.P. Broom, C.E. Buckley, F. Chang, P. Chen, Y.W. Cho, J.C. Crivello, F. Cuevas, W.I.F. David, P.E. de Jongh, R. V. Denys, M. Dornheim, M. Felderhoff, Y. Filinchuk, G.E. Froudakis, D.M. Grant, E.M.A. Gray, B.C. Hauback, T. He, T.D. Humphries, T.R. Jensen, S. Kim, Y. Kojima, M. Latroche, H.W. Li, M. V. Lototskyy, J.W. Makepeace, K.T. Møller, L. Naheed, P. Ngene, D. Noréus, M.M. Nygård, S. ichi Orimo, M. Paskevicius, L. Pasquini, D.B. Ravnsbæk, M. Veronica Sofianos, T.J. Udovic, T. Vegge, G.S. Walker, C.J. Webb, C. Weidenthaler, C. Zlotea, Materials for hydrogen-based energy storage—past, recent progress and future outlook. J. Alloys Compd. **827** (2020). https://doi.org/10.1016/j.jallcom.2019.153548
3. L. Schlapbach, A. Züttel, Hydrogen storage materials for mobile applications. Nature **414**, 353–358 (2001). https://doi.org/10.1038/35104634
4. A. Züttel, Materials for hydrogen storage. Mater. Today **6**, 24–33 (2003). https://doi.org/10.1016/S1369-7021(03)00922-2

5. P. Muthukumar, A. Kumar, N.N. Raju, K. Malleswararao, M.M. Rahman, A critical review on design aspects and developmental status of metal hydride based thermal machines. Int. J. Hydrogen Energy **43**, 17753–17779 (2018). https://doi.org/10.1016/j.ijhydene.2018.07.157
6. D.P. Broom, The accuracy of hydrogen sorption measurements on potential storage materials. Int. J. Hydrogen Energy **32**, 4871–4888 (2007). https://doi.org/10.1016/j.ijhydene.2007.07.056
7. F. Askri, A. Jemni, S. Ben Nasrallah, Prediction of transient heat and mass transfer in a closed metal-hydrogen reactor. Int. J. Hydrogen Energy **29**, 195–208 (2004). https://doi.org/10.1016/S0360-3199(03)00089-2
8. A. Jemni, S. Ben Nasrallah, Study of two-dimensional heat and mass transfer during absorption in a metal-hydrogen reactor. Int. J. Hydrogen Energy **20**, 43–52 (1995). https://doi.org/10.1016/0360-3199(93)E0007-8.
9. M.R.A.M. Gopal, S. Srinivasa, Studies on heat and mass transfer. Refrig. Air Cond. **20** (1995)
10. M. Afzal, R. Mane, P. Sharma, Heat transfer techniques in metal hydride hydrogen storage: a review. Int. J. Hydrogen Energy **42**, 30661–30682 (2017). https://doi.org/10.1016/j.ijhydene.2017.10.166
11. R. Busqué, R. Torres, J. Grau, V. Roda, A. Husar, Mathematical modeling, numerical simulation and experimental comparison of the desorption process in a metal hydride hydrogen storage system. Int. J. Hydrogen Energy **43**, 16929–16940 (2018). https://doi.org/10.1016/j.ijhydene.2017.12.172
12. M. Afzal, N. Sharma, N. Gupta, P. Sharma, Transient simulation studies on a metal hydride based hydrogen storage reactor with longitudinal fins. J. Energy Storage **51**, 104426 (2022). https://doi.org/10.1016/j.est.2022.104426
13. F. Laurencelle, J. Goyette, Simulation of heat transfer in a metal hydride reactor with aluminium foam. Int. J. Hydrogen Energy **32**, 2957–2964 (2007). https://doi.org/10.1016/j.ijhydene.2006.12.007
14. H.P. Klein, M. Groll, Heat transfer characteristics of expanded graphite matrices in metal hydride beds. Int. J. Hydrogen Energy **29**, 1503–1511 (2004). https://doi.org/10.1016/j.ijhydene.2004.01.017
15. K.J. Kim, B. Montoya, A. Razani, K.H. Lee, Metal hydride compacts of improved thermal conductivity. Int. J. Hydrogen Energy **26**, 609–613 (2001). https://doi.org/10.1016/S0360-3199(00)00115-4
16. A.R. Sánchez, H.P. Klein, M. Groll, Expanded graphite as heat transfer matrix in metal hydride beds. Int. J. Hydrogen Energy **28**, 515–527 (2003). https://doi.org/10.1016/S0360-3199(02)00057-5
17. M. Dieterich, C. Pohlmann, I. Bürger, M. Linder, L. Röntzsch, Long-term cycle stability of metal hydride-graphite composites. Int. J. Hydrogen Energy **40**, 16375–16382 (2015). https://doi.org/10.1016/j.ijhydene.2015.09.013
18. Z. Guo, H.J. Sung, Conjugate heat and mass transfer in metal hydride beds in the hydriding process. Int. J. Heat Mass Transf. **42**, 379–382 (1998). https://doi.org/10.1016/S0017-9310(98)00145-8
19. T. Oi, K. Maki, Y. Sakaki, Heat transfer characteristics of the metal hydride vessel based on the plate-fin type heat exchanger. J. Power. Sources **125**, 52–61 (2004). https://doi.org/10.1016/S0378-7753(03)00822-X
20. B.D. MacDonald, A.M. Rowe, Impacts of external heat transfer enhancements on metal hydride storage tanks. Int. J. Hydrogen Energy **31**, 1721–1731 (2006). https://doi.org/10.1016/j.ijhydene.2006.01.007
21. A. Mallik, P. Sharma, Modeling and numerical simulation of an industrial scale metal hydride reactor based on CFD-Taguchi combined method. Energy Storage **3** (2021). https://doi.org/10.1002/est2.227
22. S.L. Garrison, B.J. Hardy, M.B. Gorbounov, D.A. Tamburello, C. Corgnale, B.A. Vanhassel, D.A. Mosher, D.L. Anton, Optimization of internal heat exchangers for hydrogen storage tanks utilizing metal hydrides. Int. J. Hydrogen Energy **37**, 2850–2861 (2012). https://doi.org/10.1016/j.ijhydene.2011.07.044

23. B.D. MacDonald, A.M. Rowe, A thermally coupled metal hydride hydrogen storage and fuel cell system. J. Power. Sources **161**, 346–355 (2006). https://doi.org/10.1016/j.jpowsour.2006.04.018
24. F. Askri, M. Ben Salah, A. Jemni, S. Ben Nasrallah, Optimization of hydrogen storage in metal-hydride tanks. Int. J. Hydrogen Energy **34,** 897–905 (2009). https://doi.org/10.1016/j.ijhydene.2008.11.021
25. M. Afzal, P. Sharma, Design and computational analysis of a metal hydride hydrogen storage system with hexagonal honeycomb based heat transfer enhancements—part A. Int. J. Hydrogen Energy **46**, 13116–13130 (2021). https://doi.org/10.1016/j.ijhydene.2021.01.135
26. Y. Kaplan, Effect of design parameters on enhancement of hydrogen charging in metal hydride reactors. Int. J. Hydrogen Energy **34**, 2288–2294 (2009). https://doi.org/10.1016/j.ijhydene.2008.12.096
27. P. Muthukumar, A. Singhal, G.K. Bansal, Thermal modeling and performance analysis of industrial-scale metal hydride based hydrogen storage container. Int. J. Hydrogen Energy **37**, 14351–14364 (2012). https://doi.org/10.1016/j.ijhydene.2012.07.010
28. M. Visaria, I. Mudawar, Coiled-tube heat exchanger for high-pressure metal hydride hydrogen storage systems—part 1. Experimental study. Int. J. Heat Mass Transf. **55**, 1782–1795 (2012). https://doi.org/10.1016/j.ijheatmasstransfer.2011.11.035
29. S. Tiwari, N. Gupta, S. Kumar, A. Kumar, P. Sharma, Experimental investigation, development of machine learning model and optimization studies of a metal hydride reactor with embedded helical cooling tube. J. Energy Storage **72**, 108522 (2023). https://doi.org/10.1016/j.est.2023.108522
30. S. Lévesque, M. Ciureanu, R. Roberge, T. Motyka, Hydrogen storage for fuel cell systems with stationary applications—I. Transient measurement technique for packed bed evaluation. Int. J. Hydrogen Energy **25**, 1095–1105 (2000). https://doi.org/10.1016/S0360-3199(00)00023-9
31. Z. Bao, F. Yang, Z. Wu, S. Nyallang Nyamsi, Z. Zhang, Optimal design of metal hydride reactors based on CFD-Taguchi combined method. Energy Convers. Manage. **65**, 322–330 (2013). https://doi.org/10.1016/j.enconman.2012.07.027
32. M. Visaria, I. Mudawar, Experimental investigation and theoretical modeling of dehydriding process in high-pressure metal hydride hydrogen storage systems. Int. J. Hydrogen Energy **37**, 5735–5749 (2012). https://doi.org/10.1016/j.ijhydene.2011.12.140
33. S. Tiwari, P. Sharma, Design of metal hydride reactor with embedded helical coil using optimization methods, Eng. Res. Exp. **3** (2021). https://doi.org/10.1088/2631-8695/abf9e5
34. M. Afzal, P. Sharma, Design of a large-scale metal hydride based hydrogen storage reactor: simulation and heat transfer optimization. Int. J. Hydrogen Energy **43**, 13356–13372 (2018). https://doi.org/10.1016/j.ijhydene.2018.05.084
35. İ Hilali, R. Karadag, H. Bulut, M.A. Aktacir, A study on ideal distance between staggered metal hydride tanks in forced convection. Int. J. Hydrogen Energy **43**, 17970–17977 (2018). https://doi.org/10.1016/j.ijhydene.2018.03.051
36. I. Bürger, M. Dieterich, C. Pohlmann, L. Röntzsch, M. Linder, Standardized hydrogen storage module with high utilization factor based on metal hydride-graphite composites. J. Power. Sources **342**, 970–979 (2017). https://doi.org/10.1016/j.jpowsour.2016.12.108
37. K. Manickam, P. Mistry, G. Walker, D. Grant, C.E. Buckley, T.D. Humphries, M. Paskevicius, T. Jensen, R. Albert, K. Peinecke, M. Felderhoff, Future perspectives of thermal energy storage with metal hydrides. Int. J. Hydrogen Energy **44**, 7738–7745 (2019). https://doi.org/10.1016/j.ijhydene.2018.12.011
38. H. El Mghari, J. Huot, L. Tong, J. Xiao, Selection of phase change materials, metal foams and geometries for improving metal hydride performance. Int. J. Hydrogen Energy **45**, 14922–14939 (2020). https://doi.org/10.1016/j.ijhydene.2020.03.226
39. A.P. Tetuko, B. Shabani, R. Omrani, B. Paul, J. Andrews, Study of a thermal bridging approach using heat pipes for simultaneous fuel cell cooling and metal hydride hydrogen discharge rate enhancement. J. Power. Sources **397**, 177–188 (2018). https://doi.org/10.1016/j.jpowsour.2018.07.030

40. H. Ben Mâad, F. Askri, S. Ben Nasrallah, Heat and mass transfer in a metal hydrogen reactor equipped with a phase-change heat-exchanger. Int. J. Therm. Sci. **99**, 271–278 (2016). https://doi.org/10.1016/j.ijthermalsci.2015.09.003
41. A.A. Rabienataj Darzi, H. Hassanzadeh Afrouzi, A. Moshfegh, M. Farhadi, Absorption and desorption of hydrogen in long metal hydride tank equipped with phase change material jacket. Int. J. Hydrogen Energy **41**, 9595–9610 (2016). https://doi.org/10.1016/j.ijhydene.2016.04.051
42. S. Tiwari, P. Sharma, Integration of metal hydride reactor with thermocline based heat storage system. J. Energy Storage **59**, 106506 (2023). https://doi.org/10.1016/j.est.2022.106506
43. Y. Liu, H. Wang, A.K. Prasad, S.G. Advani, Role of heat pipes in improving the hydrogen charging rate in a metal hydride storage tank. Int. J. Hydrogen Energy **39**, 10552–10563 (2014). https://doi.org/10.1016/j.ijhydene.2014.04.197
44. C.A. Chung, S.W. Yang, C.Y. Yang, C.W. Hsu, P.Y. Chiu, Experimental study on the hydrogen charge and discharge rates of metal hydride tanks using heat pipes to enhance heat transfer. Appl. Energy **103**, 581–587 (2013). https://doi.org/10.1016/j.apenergy.2012.10.024
45. C.A. Chung, Y.Z. Chen, Y.P. Chen, M.S. Chang, CFD investigation on performance enhancement of metal hydride hydrogen storage vessels using heat pipes. Appl. Therm. Eng. **91**, 434–446 (2015). https://doi.org/10.1016/j.applthermaleng.2015.08.039
46. A.P. Tetuko, B. Shabani, J. Andrews, Thermal coupling of PEM fuel cell and metal hydride hydrogen storage using heat pipes. Int. J. Hydrogen Energy **41**, 4264–4277 (2016). https://doi.org/10.1016/j.ijhydene.2015.12.194
47. M. Lototskyy, I. Tolj, Y. Klochko, M.W. Davids, D. Swanepoel, V. Linkov, Metal hydride hydrogen storage tank for fuel cell utility vehicles. Int. J. Hydrogen Energy **45**, 7958–7967 (2020). https://doi.org/10.1016/j.ijhydene.2019.04.124
48. N. Gupta, S. Tiwari, Nitin, R.J. Khandarkar, M. Afzal, P. Sharma, Experimental studies on novel multi tubular reactor with shell having integrated buffer storage. J. Energy Storage **67**, 107491 (2023). https://doi.org/10.1016/j.est.2023.107491
49. M.R. Gopal, S.S. Murthy, Parametric studies on heat and mass transfer in metal hydride beds. Int. J. Hydrogen Energy **20**, 911–917 (1995). https://doi.org/10.1016/0360-3199(95)00026-A
50. S. Tiwari, P. Sharma, Parametric and sensitivity analysis of metal hydride hydrogen storage systems for development of novel design charts, 2022. https://doi.org/10.1002/est2.343
51. A. Kumar, S. Tiwari, N. Gupta, P. Sharma, Machine learning modelling and optimization for metal hydride hydrogen storage systems. Sustain Energy Fuels **8**, 2073–2086 (2024). https://doi.org/10.1039/d4se00031e
52. P. Muthukumar, U. Madhavakrishna, A. Dewan, Parametric studies on a metal hydride based hydrogen storage device. Int. J. Hydrogen Energy **32**, 4988–4997 (2007). https://doi.org/10.1016/j.ijhydene.2007.08.010
53. M.W. Davids, M. Lototskyy, M. Malinowski, D. van Schalkwyk, A. Parsons, S. Pasupathi, D. Swanepoel, T. van Niekerk, Metal hydride hydrogen storage tank for light fuel cell vehicle. Int. J. Hydrogen Energy **44**, 29263–29272 (2019). https://doi.org/10.1016/j.ijhydene.2019.01.227
54. N. Endo, K. Matsumura, Y. Kawakami, M. Ishida, T. Maeda, Operation of metal hydride hydrogen storage systems for hydrogen compression using solar thermal energy. J. Int. Counc. Electr. Eng. **6**, 65–71 (2016). https://doi.org/10.1080/22348972.2016.1173779
55. N. Endo, E. Shimoda, K. Goshome, T. Yamane, T. Nozu, T. Maeda, Operation of a stationary hydrogen energy system using TiFe-based alloy tanks under various weather conditions. Int. J. Hydrogen Energy **45**, 207–215 (2020). https://doi.org/10.1016/j.ijhydene.2019.10.240
56. Y. Segawa, N. Endo, E. Shimoda, T. Maeda, Pilot-scale hydrogen energy utilization system demonstration: a commercial building case study on on-site green hydrogen production and use. Int. J. Hydrogen Energy **47**, 15982–15991 (2022). https://doi.org/10.1016/j.ijhydene.2022.03.073
57. Y. Segawa, N. Endo, E. Shimoda, T. Yamane, T. Maeda, Pilot-scale hydrogen energy utilization system demonstration: a case study of a commercial building with supply and utilization of off-site green hydrogen. Int. J. Hydrogen Energy **50**, 26–36 (2024). https://doi.org/10.1016/j.ijhydene.2023.06.294

58. E. Parliament, M. States, R.A. Podimata, T. Ftt, T. Ftt, H. Kong, A. Ftt, P. Eu-wide, Press release **1**, 32–33 (2011)
59. A. Kumar, P. Muthukumar, Experimental investigation on hydrogen transfer in coupled metal hydride reactors for multistage hydrogen purification application. Appl. Energy **363**, 123076 (2024). https://doi.org/10.1016/j.apenergy.2024.123076
60. D.V. Blinov, V.I. Borzenko, A.V. Bezdudny, A.N. Kazakov, Metal hydride hydrogen storage and purification technologies. J. Phys. Conf. Ser. **2039** (2021). https://doi.org/10.1088/1742-6596/2039/1/012005
61. S. Ashida, N. Katayama, K. Dowaki, M. Kameyama, Study on metal hydride performance for purification and storage of bio-H_2. J. Japan Inst. Energy **96**, 300–306 (2017). https://doi.org/10.3775/jie.96.300
62. M.V. Lototskyy, I. Tolj, L. Pickering, C. Sita, F. Barbir, V. Yartys, The use of metal hydrides in fuel cell applications. Prog. Nat. Sci. Mater. Int. **27**, 3–20 (2017). https://doi.org/10.1016/j.pnsc.2017.01.008
63. C.B. Robledo, L.B. van Leeuwen, A.J.M. van Wijk, Hydrogen fuel cell scooter with plug-out features for combined transport and residential power generation. Int. J. Hydrogen Energy **44**, 29648–29657 (2019). https://doi.org/10.1016/j.ijhydene.2019.04.103
64. A.I. Bevan, A. Züttel, D. Book, I.R. Harris, Performance of a metal hydride store on the "ross Barlow" hydrogen powered canal boat. Faraday Discuss. **151**, 353–367 (2011). https://doi.org/10.1039/c0fd00025f
65. M.V. Lototskyy, I. Tolj, M.W. Davids, Y.V. Klochko, A. Parsons, D. Swanepoel, R. Ehlers, G. Louw, B. van der Westhuizen, F. Smith, B.G. Pollet, C. Sita, V. Linkov, Metal hydride hydrogen storage and supply systems for electric forklift with low-temperature proton exchange membrane fuel cell power module. Int. J. Hydrogen Energy **41**, 13831–13842 (2016). https://doi.org/10.1016/j.ijhydene.2016.01.148
66. A. Miller, D. Barnes, Advanced underground vehicle power and control fuelcell mine locomotive. In: Proceeding 2002 US DOE Hydrogen Program Review, pp. 1–7 (2002). https://www.nrel.gov/docs/fy02osti/32405a30.pdf
67. T.A. Johnson, M.P. Kanouff, D.E. Dedrick, G.H. Evans, S.W. Jorgensen, Model-based design of an automotive-scale, metal hydride hydrogen storage system. Int. J. Hydrogen Energy **37**, 2835–2849 (2012). https://doi.org/10.1016/j.ijhydene.2011.05.030
68. L.K. Heung, T. Motyka, W.A. Summers, hydrogen storage development for utility vehicles. In: 4th International Symposium Hydrogen Power Theoretical and Engineering Solutions, p. 336 (2001)
69. Paper for the Nha Annual U.S.: Hydrogen Conference Savannah River National Laboratory, Development (n.d.)
70. H. Buchner, R. Povel, The daimler-benz hydride vehicle project. Int. J. Hydrogen Energy **7**, 259–266 (1982). https://doi.org/10.1016/0360-3199(82)90089-1
71. Press Release, "Toyota Unveils Fuel Cell Electric Vehicles", (1996)
72. M.V. Lototskyy, I. Tolj, A. Parsons, F. Smith, C. Sita, V. Linkov, Performance of electric forklift with low-temperature polymer exchange membrane fuel cell power module and metal hydride hydrogen storage extension tank. J. Power. Sources **316**, 239–250 (2016). https://doi.org/10.1016/j.jpowsour.2016.03.058
73. M.V. Lototskyy, V.A. Yartys, B.G. Pollet, R.C. Bowman, Metal hydride hydrogen compressors: a review. Int. J. Hydrogen Energy **39**, 5818–5851 (2014). https://doi.org/10.1016/j.ijhydene.2014.01.158
74. M. Lototskyy, V. Linkov, Thermally driven hydrogen compression using metal hydrides. Int. J. Energy Res. 1–21 (2022). https://doi.org/10.1002/er.8189
75. T. Johnson, A. Mallow, R. Bowman, D. Smith, L. Anovitz, C. Jensen, Metal Hydride Compressor for High-Pressure (875 bar) Hydrogen Delivery (2022). https://typeset.io/papers/metal-hydride-compressor-for-high-pressure-875-bar-hydrogen-2abhaf71
76. H. Li, X. Wang, Z. Dong, L. Xu, C. Chen, A study on 70 MPa metal hydride hydrogen compressor. J. Alloys Compd. **502**, 503–507 (2010). https://doi.org/10.1016/j.jallcom.2010.04.206

77. G. Karagiorgis, C.N. Christodoulou, H. von Storch, G. Tzamalis, K. Deligiannis, D. Hadjipetrou, M. Odysseos, M. Roeb, C. Sattler, Design, development, construction and operation of a novel metal hydride compressor. Int. J. Hydrogen Energy **42**, 12364–12374 (2017). https://doi.org/10.1016/j.ijhydene.2017.03.195
78. M.V. Lototskyy, M.W. Davids, D. Swanepoel, R. Ehlers, Y. Klochko, G. Gizer, S. Pasupathi, V. Linkov, V.A. Yartys, Development of a high-pressure 700 bar metal hydride hydrogen compressor. J. Energy Storage **98**, 113072 (2024). https://doi.org/10.1016/j.est.2024.113072
79. J. Pereira da Cunha, P. Eames, Thermal energy storage for low and medium temperature applications using phase change materials—a review. Appl. Energy **177**, 227–238 (2016). https://doi.org/10.1016/j.apenergy.2016.05.097
80. S. Kuravi, J. Trahan, D.Y. Goswami, M.M. Rahman, E.K. Stefanakos, Thermal energy storage technologies and systems for concentrating solar power plants. Prog. Energy Combust. Sci. **39**, 285–319 (2013). https://doi.org/10.1016/j.pecs.2013.02.001
81. C. Corgnale, B. Hardy, T. Motyka, R. Zidan, Metal hydride based thermal energy storage system requirements for high performance concentrating solar power plants. Int. J. Hydrogen Energy **41**, 20217–20230 (2016). https://doi.org/10.1016/j.ijhydene.2016.09.108
82. P. Pardo, A. Deydier, Z. Anxionnaz-Minvielle, S. Rougé, M. Cabassud, P. Cognet, A review on high temperature thermochemical heat energy storage. Renew. Sustain. Energy Rev. **32**, 591–610 (2014). https://doi.org/10.1016/j.rser.2013.12.014
83. R.T. Caldwell, J.W. McDonald, A. Pietsch, Solar-energy receiver with lithium-hydride heat storage. Sol. Energy **9**, 48–60 (1965). https://doi.org/10.1016/0038-092X(65)90161-1
84. D.N. Harries, M. Paskevicius, D.A. Sheppard, T.E.C. Price, C.E. Buckley, Concentrating solar thermal heat storage using metal hydrides. Proc. IEEE **100**, 539–549 (2012). https://doi.org/10.1109/JPROC.2011.2158509
85. A. Reiser, B. Bogdanović, K. Schlichte, Application of Mg-based metal-hydrides as heat energy storage systems. Int. J. Hydrogen Energy **25**, 425–430 (2000). https://doi.org/10.1016/S0360-3199(99)00057-9
86. M. Felderhoff, B. Bogdanović, High temperature metal hydrides as heat storage materials for solar and related applications. Int. J. Mol. Sci. **10**, 335–344 (2009). https://doi.org/10.3390/ijms10010325
87. S. Mellouli, F. Askri, A. Edacherian, T. Alqahtani, S. Algarni, J. Abdelmajid, P. Phelan, Performance analysis of a thermal energy storage system based on paired metal hydrides for concentrating solar power plants. Appl. Therm. Eng. **144**, 1017–1029 (2018). https://doi.org/10.1016/j.applthermaleng.2018.09.014
88. A. d'Entremont, C. Corgnale, M. Sulic, B. Hardy, R. Zidan, T. Motyka, Modeling of a thermal energy storage system based on coupled metal hydrides (magnesium iron–sodium alanate) for concentrating solar power plants. Int. J. Hydrogen Energy **42**, 22518–22529 (2017). https://doi.org/10.1016/j.ijhydene.2017.04.231
89. S. Tiwari, P. Sharma, Numerical and theoretical evaluation of coupled metal-hydride-based thermal storage system with embedded helical tubes. Energy Fuels **36**, 1140–1150 (2022). https://doi.org/10.1021/acs.energyfuels.1c03455
90. M. Bhouri, M. Linder, I. Bürger, Metal hydride reactor for dual use: hydrogen storage and cold production. Int. J. Hydrogen Energy **43**, 23357–23371 (2018). https://doi.org/10.1016/j.ijhydene.2018.10.194
91. S. Suda, Y. Komazaki, H. Narasaki, M. Uchida, Development of a double-stage heat pump: experimental and analytical surveys. J. Less-Common Met. **172–174**, 1092–1110 (1991). https://doi.org/10.1016/S0022-5088(06)80017-2
92. A.S. Chernikov, L.A. Izhvanov, A.I. Solovey, V.P. Frolov, Y.I. Shanin, An installation for water cooling based on a metal hydride heat pump. J. Alloys Compd. **330**, 907–910 (2002). https://doi.org/10.1016/S0925-8388(01)01479-7
93. M. Ram Gopal, S. Srinivasa Murthy, Experiments on a metal hydride cooling system working with ZrMnFe/MmNi4.5Al0.5 pair. Int. J. Refrig. **22**, 137–149 (1999)
94. M. Linder, R. Mertz, E. Laurien, Experimental results of a compact thermally driven cooling system based on metal hydrides. Int. J. Hydrogen Energy **35**, 7623–7632 (2010). https://doi.org/10.1016/j.ijhydene.2010.04.184

95. M. Kölbig, I. Bürger, M. Linder, Thermal applications in vehicles using hydralloy C5 in single and coupled metal hydride systems. Appl. Energy **287** (2021). https://doi.org/10.1016/j.apenergy.2021.116534
96. A. Bishnoi, S. Pati, P. Sharma, Architectural design of metal hydrides to improve the hydrogen storage characteristics. J. Power Sources **608**, 234609 (2024). https://doi.org/10.1016/j.jpowsour.2024.234609

Chapter 18
Hydrogen Transportation

Hydrogen transportation is required, regardless of whether hydrogen production is in a centralized or decentralized manner, and is crucial throughout the entire hydrogen energy supply chain. Transportation from the point of production to the place of utilization is required, and the quantity of hydrogen to be transported, along with the distance over which it will be transported, and the associated economics, determine the method of hydrogen transportation to be used. The major challenge associated with transporting hydrogen is its low density, combined with the complex physical and chemical properties of hydrogen, which makes handling it safely a prime concern and renders it an expensive commodity to transport. Another decisive factor in the economics of moving hydrogen is whether the cost of onsite decentralized hydrogen production offsets or not the cost of centralized production and transporting to another place for utilization. These questions will prevail when the demand for hydrogen, its production and global investments will increase in the future. Although there are proven technologies, when it comes to escalated future demands, the existing technologies need to be tested for those scales or require further development. All this will require a lot of infrastructure building, which will need government support in terms of policies and incentives to meet the demand and make hydrogen accepted and commercially viable globally.

Low density by volume of hydrogen presents challenges in terms of transporting it both domestically and internationally. Due to this low density, the volume required for both storage and transport is significant, and in order to reduce the required space, the following possibilities are:

1. Transport as compressed gaseous hydrogen,
2. Transport as liquid hydrogen,
3. Conversion to high energy density molecules.

A key point is that any of these methods, when employed, involve processes that increase not only the cost but also the energy required for hydrogen transportation. Thus, when economics are considered, then along with transportation costs, the

P. Sharma, *A Textbook on Hydrogen Energy*,
https://doi.org/10.1007/978-981-95-4601-5_18

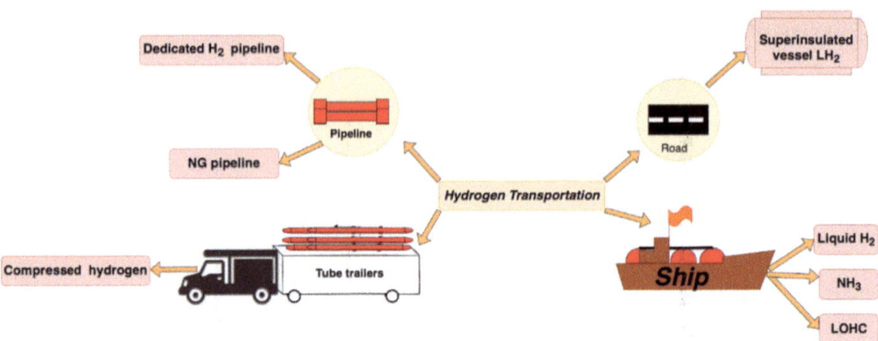

Fig. 18.1 Different ways to transport hydrogen

compression or liquefaction, conversion to other molecules (NH_3 or CH_3OH or LOHCs) and reconversion, all these steps need to be considered. Since there is an energy and cost penalty associated with each of the process or conversion steps, the most economically feasible option is determined based on quantity, distance, end use, and geography.

Thus, the possible options for hydrogen transportation, as shown in Fig. 18.1, include: (1) either in the form of compressed hydrogen through road by means of tube trailers, or can be carried through pipelines. When transported through pipelines, either the existing natural gas pipelines can be used with a certain percentage of blending or have new infrastructure with dedicated hydrogen pipelines; (2) transporting over long distances on a large scale through ships or through road using superinsulated vessels in the form of liquid hydrogen; (3) transporting in the form of liquid organic hydrogen carriers through ships or road or pipelines.

The transport of hydrogen on a large scale over longer distances is attractive for promoting international trade. For example, in the case of countries like Japan and South Korea, as the demand of hydrogen increases, reported studies have shown that green hydrogen production may not be as economical as importing green hydrogen from countries such as Australia and others. The reason is geographical limitations, where there are limits on the available locations for and the cost of producing green hydrogen from renewables may be higher in countries like Japan and South Korea than the cost of importing it from neighboring countries. In such cases, the trade of hydrogen and the large-scale long-distance transport of hydrogen can play a major role. The choice of transportation method will depend on the volume, i.e., the quantities, and the distance, i.e., the length over which they need to be carried, along with the location to be transported and the utilization of the transported hydrogen. Above all, the deciding factor will be economic viability.

Hydrogen Transportation Through Pipelines

Hydrogen dedicated pipelines will be economically feasible when there is consistent demand for large quantities of hydrogen to be transported regularly.

Pipeline-based hydrogen transport offers several advantages, including low operating costs and a longer lifespan, which can range from 40 to 80 years, making it the most economically viable method for transporting large quantities of hydrogen to industries or locations with assured demand for vast quantities of hydrogen. Pipelines also provide a means of storing hydrogen and supplying it as needed. However, the major disadvantage is the requirement for high-capital investment, which increases as the pipeline length increases. Besides, depending upon whether it is intra-country or cross-country, the requirement for right of way makes government support, as well as the certainty of demand, essential in deciding whether this option is suitable for hydrogen transportation. The time periods required for obtaining permissions and laying down pipelines safely and securely for hydrogen transport are also high.

Hydrogen from the place of production can be carried through transmission pipelines and subsequently distributed through the distribution pipelines, as shown in Fig. 18.2. Compared to transmission pipelines, the distribution pipelines have a smaller diameter and hold lower pressure. The requirement for driving force for the flow across any pipeline is the existence of pressure difference between the inlet and outlet of the pipeline. Since the lengths of the pipelines are usually several hundred kilometers, resulting in a substantial pressure drop, the outlet pressure can be very low in this case, making compression essential at specified distances in any pipeline network. Depending on the required conditions at the outlet of the pipeline, such as pressure, flow rate, and diameter, the compression stations may be located every 100–500 km. Large-volume capacity transmission pipelines transport hydrogen and supply it to the distribution pipeline, with the interface occurring at a city gate.

Several key tasks are accomplished at these city gates, including controlling the flow rate, reducing the pressure, and metering the amount of hydrogen being delivered to the utility. At the city gate, various control rooms monitor and regulate the flow of hydrogen The pressure is regulated through automatic controls, and metering and measuring units are utilized. Distribution pipelines are finally connected to the service pipelines to provide the hydrogen delivery service for end-use applications, such as domestic use, as illustrated in Fig. 18.2.

The diameter of transmission and distribution pipelines differ, with transmission pipelines typically having a larger diameter. The natural gas pipeline is usually made of carbon steel or stainless steel, with diameters ranging from 4 to 48 inches. These are high strength steels (X70) with a strength greater than 100 kilo pounds per square inch, and thus a lower thickness required. However, the same material, if exposed to hydrogen, can be more susceptible to hydrogen embrittlement. Thus, specifically for hydrogen transport, low-strength steel is recommended, thus will require high thickness of pipelines thereby increasing the cost. Depending on the pipeline diameter and operating pressure, hydrogen dedicated pipelines can be around 68% more expensive than natural gas pipelines [1]. However, the use of low-strength

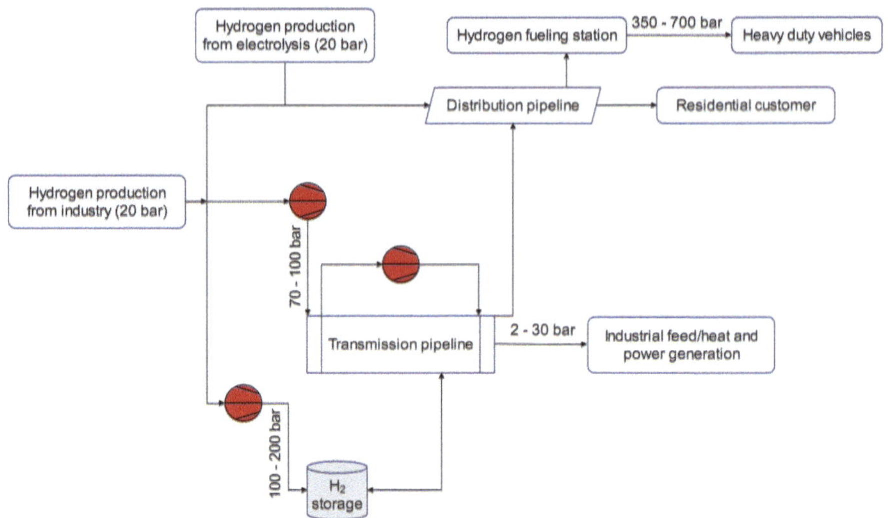

Fig. 18.2 Hydrogen transportation through hydrogen dedicated pipelines

steel will result in a pressure limitation on the flow of hydrogen flow inside these pipelines. If the pressure needs to be increased, the thickness must also increase, resulting in higher cost.

Pipeline Materials

Since hydrogen embrittlement is a primary challenge for metals when exposed to hydrogen under certain operating conditions, regions prone to degradation include cracks, which can lead to embrittlement due to fatigue, as well as heat-affected zones. Thus, the requirement is to use specialized materials that are resistant to hydrogen embrittlement. One possibility is that the high-pressure pipelines in such case can be made up of materials similar to those used for the compressed state storage of hydrogen, i.e., carbon fiber-reinforced composites, which are, in turn, highly cost-intensive. Thus, for laying down hydrogen pipelines, low-grade or strength steel could be used, such as API 5L A, B, X42 and X46, with pressure in these pipelines ranging between 0.34 and 10 MPa, and with diameter in the range of 25–30 cm. However, the diameter varies for transmission or distribution or service pipelines. Transportation pipelines operate at a pressure range of 3.5–10 MPa, while the distribution pipelines with a lower diameter can have pressures in the range of 0.03–1.4 MPa, with a gas velocity of around 10 m/s, as compared to roughly 20 m/s in the case of transmission pipelines.

Pipeline Protection

Since the pipelines, once laid, are designed for continuous use for years and are exposed to various climatic conditions and environmental factors, it is essential to protect them from corrosion. The unprotected steel pipelines, when used for an extended period, will be more susceptible to corrosion and deteriorate rapidly. One of the well-known and widely used methods for protecting pipelines from corrosion is cathodic protection, which has been employed in pipeline networks since the 1930s.

In cathodic protection, a metal rod, typically acting as an anode, is installed in the vicinity of the pipeline. The installed anode takes up the burden of corrosion while preventing the pipeline from corrosion. When that anode gets corroded, it can be easily replaced without incurring a significant cost penalty. The second possible method of pipeline protection is the use of coatings. Coating can be done both internally and externally, protecting the pipelines from direct exposure to harsh environmental conditions and also from internal threats such as hydrogen embrittlement and hydrogen permeation. Similar to natural gas pipelines, the internal coating can be a fusion bonded epoxy or polyethylene. In contrast, the external coating takes the form of a heat sink or sleeves. The third option for pipeline protection is the addition of corrosion inhibitors, i.e., certain additives can be added in certain quantities to the gas flowing through the pipeline, which prevents pipelines from hydrogen embrittlement. Some of the possible examples of different inhibitors studied for addition to hydrogen are carbonyl sulfide, ethylene, chloro-tri-fluoro ethylene, among others However, the major challenge with the inclusion of these additives as corrosion inhibitors is associated with increased cost, toxicity, combustibility. Furthermore, the requirement for additional separation and purification to be integrated at the point of hydrogen usage for specific applications, where high purity hydrogen is required.

Laying Down Pipelines

The various steps involved in laying down pipeline or burying the pipeline below the ground may differ based on geographical locations. In general, the steps involved are discussed in this section. The first step toward laying down any pipeline is preparing the land or surface. This is accomplished by cleaning and grading the working surface appropriately. The usual step involved in burying pipeline is trenching; the trench depth may vary depending upon the pipeline diameter and more importantly, on the applicable legal regulations. Depending on whether the pipeline is a cross-country or an inter-country pipeline, the legal regulations may vary, and accordingly, the trenching depth is determined. Besides trenching, the other options for laying down pipelines are boring or horizontal directional drilling (HDD). Boring and HDD methods are adopted when trenching is not allowed. For example, in cases where, paving road, highway, or river is located along the path of laying down a pipeline, trenching is not feasible; instead, either boring or HDD is used. However, there are certain restrictions or challenges associated with boring or HDD. These two methods may not be compatible with all soil types. At times, it can get difficult to identify any existing pipelines and the use of these methods may cause damage to such pipelines.

The process of laying down pipeline after ground preparation involves having several segments together to form the complete pipeline. However, the pipe sections mostly should be seamless; these seamless pipe sections are made by either hot working or piercing of the billets of solid steel. At times, the pipes formed from plates by welding are also used, which is known as electric resistance welded pipes. Different methods for welding the segments include TIG welding, MIG welding, or laser beam welding. When these pipelines are laid down, the welded segments need to be leveled so that equipment, such as side booms, can be used. Once the entire pipeline is laid down, it must be connected from the initial point, i.e., where hydrogen will be supplied, to the end point. During this process, numerous quality checks are carried out. Since hydrogen has peculiar properties, including high flammability, low ignition energy, and a wide flammability range, it burns with invisible flame. As the smallest molecule, it is much more prone to leakage. Thus, welding and testing of hydrogen pipelines are particularly important and stringent compared to those of natural gas. A significant amount of testing is required for the safe transportation of hydrogen through pipelines. The quality check is conducted through various tests and inspections. The inspections can be radiological inspections, X-ray, or ultrasonic. Especially when it is larger diameter pipeline, which may have higher stresses, multiple levels of these checks are carried out. Once the pipelines are ready after testing and all the relevant standards and protocols of inspections have been followed, they are cleaned of sand, dirt, and the welding debris. Thereafter, pressure testing is carried out, wherein the pipelines are subjected to high pressure for a prolonged duration and must sustain the pressure without leaking. Following the high-pressure testing to ensure leak proof operation, the pipeline is cleaned, dried, and finally, it is ready for operation.

Besides the pipeline being operational, another important aspect of the entire pipeline infrastructure is the monitoring of operations and maintenance to ensure the smooth functioning of these pipelines for several decades. The smooth functioning of the pipeline is ensured and continuously monitored through the establishment of fully equipped control rooms, which utilizes advanced software tools. This enables the entire process of detecting leaks and scheduling repairs and maintenance to be fully automated. The maintenance activities involved includes scheduling maintenance, detecting leaks, shutting down particular sections, and conducting regular inspections, all of which are completely computerized. There are specialized equipment and techniques to monitor and ensure the safe operation of the pipelines, such as the pigging technique, and the instrument used is known as smart pigs. These smart pigs are capable of detecting problems such as the presence of cracks, deformations, or corrosion, and thereafter, relevant preventive measures need to be taken to avoid any accidental scenario. Existing equipment and concepts for natural gas pipelines are already in place, and these can be adapted for hydrogen pipeline transport with minor adjustments to make them compatible with hydrogen-based transport.

The major challenges associated with the hydrogen pipelines are:

1. Due to the properties of hydrogen, all safety regulations and protocols must be strictly followed for safe usage.
2. Gas velocity—The volumetric energy density of hydrogen is 3.9 times lower than that of natural gas, which means that to carry the same energy through a given pipeline at a particular temperature and pressure, the volumetric flow rate required for hydrogen will be significantly higher than that for natural gas. The requirement for a high-volumetric flow rate presents several challenges, including the need for increased compression energy, a higher likelihood of leaks, and hydrogen embrittlement that can become significant.
3. Valves—Valves are required to control, regulate the flow of any gas. For larger diameter pipelines, motor-operated valves are typically used. For hydrogen flow in a pipeline network, even valves and other peripherals should have a higher tolerance to hydrogen and should avoid hydrogen embrittlement by using materials that are not susceptible to this phenomenon, thereby increasing overall costs. Therefore, the valves used in hydrogen pipelines are more expensive than those used in natural gas pipelines. Hydrogen, with its peculiar properties, necessitates continuous or frequent inspection, maintenance, and replacement of these valves over time. This additional requirement further increases the capital cost of these valves, which is already higher in the case of hydrogen pipelines.
4. The most important factor and challenge while selecting the appropriate materials is hydrogen embrittlement. The cost of such specialized materials, which are resistant to hydrogen embrittlement, is higher. Hydrogen embrittlement is a process in which the material loses its ductility and loses its tensile strength. In the process of hydrogen embrittlement, a hydrogen molecule on the metal surface dissociates into a hydrogen atom, which diffuses deep into the steel pipeline, enters the grain boundaries, creates stress, and leads to intergranular cleavage, potentially resulting in rupture or crack propagation and ultimately causing the pipeline to fail.

Gas Flow in Pipelines [2]

Considering gas flow in a pipeline, the flow is possible because of the driving force which is the pressure difference. Let us consider a section of the pipeline such that point 1 represents the inlet and point 2 represents the outlet of the pipe section. The inlet pressure is P_1, outlet pressure is P_2, the elevation of point 1 is H_1 that of point 2 is H_2, the volume flow is Q at a temperature T.

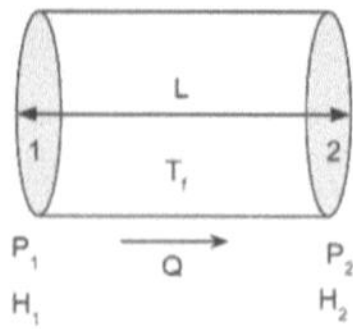

Consider gas flow from points 1 to 2:

- L = Length
- D = Pipe diameter
- T_f = Temperature
- P_1 = Pressure at point 1
- P_2 = Pressure at point 2
- Q = Flow
- h_2, h_1 Elevation at points 2 and 1, respectively

If $P_1 = P_2$, no driving force for gas flow exists. Gas flows primarily due to the pressure difference between points 1 and 2 and partly due to the elevation difference $(h_2 - h_1)$. A gas flow through a pipe encounters pressure drops due to friction between the flow and the pipe. The higher the pipe roughness or length, the greater the pressure drop; additional frictional losses occur due to the presence of elbows, branchings, control valves, and other components. The velocity (V) of gas, which is proportional to the volumetric flow rate Q, also changes depending on the cross-sectional area A of the pipe and the temperature T_f of the gas.

Using Bernoulli's equation, the pressure drop in a gas pipeline can be calculated considering $D, L, h_2 - h_1, Q$, gas specific gravity and gas composition. This basic equation is also called the Fundamental Flow Equation, which also branches into the general flow equation and other equations, such as Panhandle A, Panhandle B, with Panhandle A being the most utilized.

Since Q can vary with T and P, we need to refer to the standard volume flow rate in standard m^3/day (SCMO):

$$Q = 1.1494 \times 10^{-3}\left(\frac{T_b}{P_b}\right)\left[\frac{P_1^2 - e^{\beta} P_2^2}{G T_f L_e Z f}\right]^{0.5} D^{2.5} \tag{18.1}$$

- L is pipe length in km
- β is elevation adjustment parameter, dimensionless parameter
- L_e is equivalent pipe length in km
- D is inside pipe diameter in mm
- T_b is base temperature (288.7 K).
- T_f is average flowing temperature of gas in K.
- G is specific gravity (for H_2, $G = 0.0696$).
- P_1 is inlet pressure in kPa.
- P_2 is outlet pressure in kPa.
- P_b is base pressure in kPa(101.32 kPa).

- Z is the compressibility factor at average temperature and pressure, a dimensionless quantity.
- f is friction factor and is dimensionless.

The elevation correction factor is given by:

$$\beta = 0.06849 \times \left(\frac{h_2 - h_1}{T_f \cdot Z}\right) \tag{18.2}$$

where h_1 and h_2 are inlet and outlet elevations in meters.

The equivalent pipe length is given by:

$$L_e = L\left(\frac{e^{\beta} - 1}{\beta}\right) \tag{18.3}$$

Given P_1 and P_2, we can determine Q or given Q, and we can find pressure drop using Eq. 18.1 considering the pressure drop equation:

- $Q \propto \sqrt{P_1 - P_2}$, meaning the pressure gradient for gas flow is slightly curved compared to a linear drop for liquid flow.
- $Q \propto D^{2.5}$, meaning an increase in D leads to an increase in pipe capacity or possible flow rates.
- $Q \propto \frac{1}{\sqrt{GZLT_f}}$, thus, any increase in G, Z, L or T_f leads to a decrease in Q.

To Calculate Q in Pipelines

1. Z and P_{avg}—Compressibility factor and average flow pressure: Z accounts for the deviation of gas from ideal gas behavior. Typically, Z lies between 0.98 and 1.3 for M_2 in pipelines.

   Average Pressure

$$P_{\text{avg}} = \frac{2}{3} \times \left(\frac{P_1^3 - P_2^3}{P_1^2 - P_2^2}\right) \tag{18.4}$$

2. Friction Factor: Flow through a pipeline can be classified as laminar, turbulent, or transitional (from laminar to turbulent) depending on the Reynolds number (Re).

- Laminar Flow: Re < 2000
- Turbulent Flow: Re > 4000
- Transition Flow: 2000 < Re < 4000

In practice, most gas pipelines operate at flow rates that produce high Reynolds numbers, i.e., in the turbulent regime. The Reynolds number depends on gas properties, pipe diameter, and flow velocity:

$$\text{Re} = \frac{V_{\text{avg}} D \rho_{\text{avg}}}{\mu} \tag{18.5}$$

where

- ρ_{avg} is average gas density
- μ is gas viscosity (in $\mu = f(T, p)$)

The friction factor in general flow equations is referred to as Darcy's friction factor (f). It depends upon:

(a) The internal condition of the pipe wall (roughness or smoothness)
(b) Whether the flow is laminar or turbulent

One way to determine f is graphically using the Moody friction factor diagram.

For turbulent flow, the Colebrook-White equation can be used to calculate the friction factor in a pipeline with roughness (ε) using an iterative method.

$$\frac{1}{\sqrt{f}} = -2\log\left(\frac{\varepsilon}{3.7D} + \frac{2.51}{\mathrm{Re}\sqrt{f}}\right) \tag{18.6}$$

The Colebrook-White equation cannot be solved explicitly. Hence, an iterative solution is required with an initial value of f and Re.

3. Velocity of Gas in Pipeline: Under steady-state conditions, the velocity of gas flow can be calculated with the volumetric flow rate (Q) and pipe cross-sectional area (A). However, due to pressure variation in a gas pipeline (due to frictional losses), the average velocity varies and is a function of the flow rate Q, compressibility factor Z, pipe diameter D, pressure P, and temperature T.

$$V = 14.734\left(\frac{P_b}{T_b}\right)\left(\frac{ZT}{P}\right)\left(\frac{Q}{D^2}\right) \tag{18.7}$$

where V is the gas velocity in m/s.

5. Erosional Velocity: It represents the upper limit of gas velocity in a pipeline. Higher velocities can cause erosion of the pipe wall over a long time. The erosional velocity can be approximated as:

$$V_{\text{max}} = 100\sqrt{0.0513\frac{\mathrm{ZRT}}{\mathrm{GP}}} \tag{18.8}$$

where the gas constant is given by:

$$R = 8.314\frac{\mathrm{kPa{\cdot}m^3}}{\mathrm{kg{\cdot}mol{\cdot}K}} \tag{18.9}$$

5. Pipeline Capacity: Once the flow rate $Q(Sm^3/\text{day})$ is calculated, the design capacity (P_m^3/day) can be computed using:

$$\text{Capacity of pipe } = Q\left(\frac{Sm^3}{day}\right) \times 0.0834\left(\frac{kg_{H_2}}{Sm^3}\right) \tag{18.10}$$

Existing hydrogen dedicated pipelines globally are relatively small and are mostly used for transporting hydrogen to chemical plants and refineries and are operated by key commercial hydrogen gas suppliers. Currently, there are appropriately 5000 km of hydrogen dedicated pipelines, with the USA and Europe leading with 2700 km and 1770 km of pipelines, respectively. Around 1000 km of pipeline construction is underway globally [3]. The longest pipeline is located in the Ruhr region of Germany, spanning from Marl Chemical to Castrop-Rauxel and Leverkusen; the total length of the hydrogen pipeline laid by Air Liquide is 240 km. The other pipelines in Germany, located in Bitterfeld, Schkopau and Leuna by Linde, are 150 km long. Additionally, another pipeline, located in Schleswig-Holstein, is 30 km long. Under the South H_2 corridor project plan, it is intended to have 11,600 km of hydrogen pipeline by 2030 and 40,000 km by 2040 in Europe. Although the cost of hydrogen transported by pipeline varies depending on location, it roughly lies in the range of 11–31 Rs/kg for a 1000 km transport distance.

Hydrogen Flow conditions in Pipelines

Considering that an equal amount of energy is transported in the form of hydrogen in a pipeline will require 4.6 times more energy input to transport hydrogen compared to natural gas. For the same diameter and pressure drop, the flow rate of hydrogen is roughly three times higher than that of methane. Also, the volumetric energy density for hydrogen is 3.29 times lower than that of methane. Considering the operating pressure of hydrogen in the pipeline in the range of 70–100 bar and the erosional velocity to be approximately 2.91 times more than that of methane, in that case, the maximum flow rate for hydrogen could be 2.91 times that of methane. Thus, the maximum energy density of hydrogen in a given pipeline at a certain temperature and pressure is roughly 88.4% of the energy content of methane (i.e., 2.91/3.29 = 0.884). This indicates that to attain higher energy content or higher flow, a higher compression ratio or more compression energy will be required. Roughly, the compression energy required to compress 1 MJ of energy equivalent of hydrogen will be three times more than that for equivalent energy in the case of natural gas.

Blending of Hydrogen in Natural Gas and Transportation

Although hydrogen dedicated pipelines are the most cost-effective method for hydrogen transport, when large volumes and long-distance transmission are required. However, laying down such pipelines is highly cost-intensive; implementing projects requires a significant amount of time due to various approvals, legal regulations, and right-of-way issues, all of which make the process highly time-consuming.

Another option could be blending of hydrogen with natural gas within permissible or safe limits and then transporting it through existing natural gas pipelines. Blending hydrogen with natural gas not only saves time and cost but also serves as a long-term additional means of hydrogen delivery. Thus, when, in the future, dedicated hydrogen pipelines are laid and then the existing natural gas pipelines are not left stranded. However, wherever blended hydrogen could be directly used, this could be a suitable option. At the same time, it can also serve as a means of transporting hydrogen during the initial hydrogen market growth, where the demand of hydrogen is going to increase. If the blending is carried out at a lower concentration, then the total energy equivalent delivered to consumers may require a slightly higher amount; however, the use of blending will lower carbon dioxide emissions.

Currently, various blending standards and regulations exist in different countries, which need to be harmonized and clearly recognized across borders and globally. If this blending is to be done at a lower concentration, it will result in minor modifications to the operation and maintenance of the natural gas pipelines.

If this blending is done at a percentage level of 5–15%, there may be no significant change to the existing infrastructure. However, if blending is to be done at a much higher level, such as 15–50%, then there is a need for more significant changes. This change needs to be implemented by converting end-use appliances or increasing the compression capacity, which will be particularly beneficial for industrial users.

Now, if the blending has to be further exceeded beyond 50%, the challenges will be in multiple areas. The requirement will be in terms of the pipeline material such that it is compatible and safe for transporting such high levels of hydrogen. With such a high level of blending, safety issues needs to be well addressed, and at the same time, modifications will be required at the point of end-use.

There are two possibilities for the use of hydrogen blended with natural gas: i.e., either the blend can be used directly if the end-use appliances or devices, or the industry, can handle the percentage blending being transported, or else the blended hydrogen needs to be separated before it is delivered to the end-use application or utility. However, if the requirement is for pure hydrogen, then separation and purification are added before delivery, which again adds to the cost of the delivered hydrogen.

The biggest advantage of using the existing natural gas pipeline network to transport a safe blend of hydrogen and natural gas is the vast availability of infrastructure. Globally, there are roughly 3 million kilometers of natural gas pipelines in existence, compared to around 5000 kilometers of hydrogen dedicated pipeline. Additionally, globally, there exist substantial underground natural gas storage facilities with a capacity of around 400 billion cubic meters. Besides, there are decades of experience in its safe usage and handling. Already there are well-established safety measures, operational strategies, control and maintenance procedures and protocols in place, which can be further extended. Added to that, there is enormous public acceptance toward the use of natural gas and these infrastructures. The pipeline grids are available geographically over broad locations, and there is an existing interconnectivity.

With all the benefits of the substantial natural gas infrastructure, the requirement is to consider the safe blending level of hydrogen with natural gas, with the least effect on the existing infrastructure in the near term, until hydrogen dedicated infrastructure builds up and takes center stage. The blending percentage or the safe blending limit may vary and may depend upon the existing pipeline design and infrastructure. Important aspects are the design and construction of the existing pipelines, i.e., the integrity of these pipelines, the dimensions of the pipelines, the capacity of these pipelines, materials with which these pipelines are made up of. The design and construction of not only the pipeline but also its peripheral components, as well as the equipment used in conjunction with the pipelines, are important. For example, this includes compressor stations, sensors, monitors, valves, and other related equipment.

Besides the pipeline and peripherals, equally important is the design and construction of the end-use appliances that will ultimately utilize the blended hydrogen. Several literature reports, including studies carried out globally, such as those reported in Germany, have revealed that the existing natural gas pipelines can be converted into dedicated hydrogen pipelines. This conversion will have multiple benefits, a cost reduction by 20–60%, legal regulations and right of ways required, which become the largest hurdles for laying new hydrogen dedicated pipelines, can be resolved faster if existing natural gas pipelines are used, thus a substantial time saving in which such projects can be executed, will be observed

In long run, there can be other benefits like, if the demand of hydrogen grows in future and the renewable integration policies may change to accommodate certain mandates toward use of hydrogen, besides if the demand of natural gas decreases with time in future; in that case, these natural gas pipelines which are huge assets may get stranded, thus blending or this conversion will allow to utilize those assets in future as well.

In addition to the advantages mentioned, there are also challenges in blending hydrogen with the natural gas. Since the volumetric energy density of hydrogen is very low, blending hydrogen with natural gas results in a decrease in the energy content of the delivered gas to the end user. Some industries, such as the chemical industry, also require carbon, which is present in natural gas. With the use of blended hydrogen, these industries will need to use larger gas volumes to meet this requirement. Above all, hydrogen has different properties from natural gas; thus, there will be important safety considerations that need to be adopted, such as the use of new flame detectors. Even though important decisions will be required regarding the end-use appliances, it is essential to identify and avoid any hazardous situations. End-use appliances are designed to handle a specific range of gas compositions. If the gas composition exceeds this range, it may impact their operation. The level of safe blending may vary for different applications, and the upper limit of blending that can be allowed will also depend upon the utility, which could be an appliance or industry, and the utilization of the gas blend. Thus, the level of blending will be decided on case-by-case basis; however, in the entire network of blended hydrogen natural gas pipelines, the determining factor will be the appliance with the lowest tolerance toward the blend composition or percentage of hydrogen, which will set the maximum limit of blending. Thus, these are some of the challenges associated with the blending of hydrogen with the natural gas.

Several studies have been conducted globally to determine the safe blending levels. Some studies conducted in the UK have demonstrated that the existing natural gas pipelines can be utilized for hydrogen transportation, and underground storage, such as salt cavern-based storage, can also be employed for hydrogen storage. When using the blend with certain appliances, the allowable blending percentage can be as high as 23%. Besides appliances, where the high higher blending level could be acceptable are those used for heating or cooking. However, there may be utilities that cannot handle a higher percentage, for example, the chemical industry, where the natural gas is being used as a feedstock. Additionally, for blended gas to be used for existing gas turbines, the seals of gas turbines can handle only 5%, and the gas turbine itself can handle 2%. In gas turbines, it is possible to make minor modifications to gas seals that can increase the allowable blending percentage or enhance the acceptance of higher blending percentages.

When higher blending percentages are to be used in future, there may accordingly be a need to design and deploy modified equipment or appliances to handle that level of blending. Such a change may require a significant investment in infrastructure, manufacturing, and costs globally. However, historically, such a change at the domestic appliance or industry level is not something that has never happened in the past; it was in the 1960s or 1970s several countries including UK, US, Austria, etc. shifted from the use of town gas to natural gas, converted millions of equipment using billions of dollars of funding and replaced those equipment or appliances. Thus, in the future, if a fuel transition occurs, the same can be done when required. It is important that natural gas, being an internationally traded commodity, has uniform standards and regulations, as well as a consistent blending percentage, to facilitate easier transportation and trading globally. Besides, in the future with more knowledge, understanding, and research, it may be possible that a higher blending percentage could be acceptable and utilities may be found to handle that, in that case the developed rules should be able to consider those changes and should not be rigid on the current norms. Thus, the standards should be flexible to address higher blending levels in future; at the same time, policies should be in place to accommodate such possibilities, including the replacement of appliances if required. All these applications, where blended hydrogen with natural gas can be used directly, will not have any other requirements. However, applications that require pure hydrogen will necessitate an additional separation step at the end-use site or before delivery to the utility.

Various methods can be employed for the separation and purification of hydrogen from a blend of hydrogen and natural gas, with the blending percentage and utility requirements determining the method and associated costs. This separation can be achieved through various methods, as previously studied. Hydrogen separation can be achieved through purification, pressure swing adsorption, and membrane-based separation methods, but this again depends on the blending percentage. However, the separation of hydrogen from natural gas increases the cost of hydrogen nominally by around 0.3–0.4 USD/kg of hydrogen. Once hydrogen is separated from the blend, then the separated natural gas and hydrogen need to be recompressed separately to the desired pressure and then transported or used as per the requirement, which further adds up to the cost.

There are several problems associated with natural gas and hydrogen blends; ideally, they should be thoroughly blended. However, due to differences in physical properties, the blend results in stratification, especially at lower speeds and lower temperatures, as well as higher pressures and higher hydrogen volume fractions. This can lead to higher concentrations in certain regions, which may pose safety issues and concerns related to hydrogen embrittlement.

In summary, during the early market growth or in meeting the increasing hydrogen demand, one way of transporting hydrogen could be blending with natural gas. Currently, there are no uniform standards or policies on the level of blending. Given the widespread public acceptance of natural gas and the substantial infrastructure for natural gas transport and storage, blending could be the most feasible option, provided the allowable limits are met, which in turn dependent on several parameters.

Another option for carrying large quantities of hydrogen involves converting hydrogen into high-energy density materials, such as liquid organic hydrogen carriers (LOHC). The LOHCs can be dehydrogenated to release hydrogen, and the carrier molecule must be returned to the recharging location to be hydrogenated. However, the LOHC has physical properties similar to those of oil, is liquid at ambient conditions, and can be safely transported through existing pipelines and other infrastructure, including pumps and storage facilities for oil transport, without requiring much modifications. This method of transportation can significantly reduce costs by saving on infrastructure and maintenance expenses. It will provide flexibility in transportation, offer a cost benefit, and facilitate widespread adoption through early market penetration and growth. For example, ammonia is considered a potential hydrogen carrier, as significant existing infrastructure is available for its transportation. Ammonia liquefies at 10 bar and normal temperature; its liquefaction temperature is 239.7 K, allowing it to be carried through pipelines. The decomposition of ammonia to release hydrogen can be carried out at atmospheric pressure and temperatures ranging from 527 to 627 K, yielding roughly 1 kg of hydrogen from 5.65 kg of ammonia; however, this process is energy intensive.

Hydrogen Transportation Through Ships

Hydrogen dedicated pipelines or blended hydrogen with natural gas, transported through existing natural gas pipelines, is an option for large volumes and long-distance transport of hydrogen. In general, transporting hydrogen via pipeline or road is an option for inland transportation; however, when the transport is required overseas in large quantities and over longer distances, the option being considered is marine transport. Transportation via marine is of interest when it involves intercontinental transport, such as between countries that produce hydrogen from renewables and export it to a country that requires green hydrogen. Thus, liquid hydrogen or other hydrogen dense materials like ammonia or liquid organic hydrogen carriers can be transported via ships in the same manner as oil or liquefied natural gas is transported. Hydrogen imports will enable several countries to diversify their energy sources, thereby enhancing their energy security. Thus, there is a global interest in transporting hydrogen via ships, developing ports, and countries building up infrastructure for this method of hydrogen transportation. Ports are being developed, which will be dedicated for hydrogen or its derivates imports or exports depending on the geographical location or energy dynamics.

Liquid hydrogen can be transported through ships; thus, the shipping cost adds to the cost-intensive liquefaction, making the overall cost of delivered hydrogen high. The inherent disadvantage associated with liquid hydrogen which is boil-off losses, where the lost hydrogen can be used to power the ship. Approximately, 0.2% of the cargo can be consumed per day to power the journey of the ship. The use of the hydrogen gas which gets evaporated due to boil-off loss can be as a fuel either in fuel cells which has been used in Suiso Frontier or using internal combustion engine to propel ship is an alternative way to use to boil-off gas, which improves the energy efficiency, reduces overall costs involved and is and effective way to take care of losses. The challenge with shipping liquid hydrogen is that once transported from the import to export terminal and unloaded, economics work only when during the return journey back to the initial point, a high value liquid product is carried; however, it will not work if the vessel returns empty. When transporting liquid hydrogen via ships, a lot of infrastructure will be required, including liquefaction facilities, storage at the port of import and export, various peripheral equipment, superinsulated vessels, and other necessary components for carrying liquid hydrogen on ship.

The world's first liquid hydrogen tanker, named Suiso Frontier, started its journey on 21st January 2022. It was built by Kawasaki Heavy Industries in Japan and carried liquid hydrogen from the Australian port of Hastings to Kobe. This tanker was 116 m long and carried 75 tons, 1250 m^3 of liquid hydrogen at 20 K temperature over a distance of 9000 km and completed its journey in 16 days. This ship was used to import liquid hydrogen from Australia to Japan port Kobe. In Australia, hydrogen was being produced from brown coal gasification. In the future, the plans are to implement carbon capture and storage to convert it to blue hydrogen. Suiso Frontier was the first to demonstrate the transport of liquid hydrogen via a marine route. This demonstration, conducted through the HySTRA project, provided insight into the understanding of handling, loading, shipping, and unloading of liquid hydrogen, which can be leveraged to advance further in future.

Another possible shipping option for transporting hydrogen could be to incorporate hydrogen into larger molecules and transport these in liquid form via ships. In comparison to laying down hydrogen dedicated pipelines, transportation via ships is economically viable if the transportation distance exceeds 7000 km. Shipping transport also provides flexibility compared to hydrogen pipelines, which are governed and controlled by land laws and relatively fixed to the land. Additionally, with changing geopolitical situations, shipping can offer a more feasible option and may not require long-term commitments of supply and demand between two countries. The Provaris Energy Ltd. received AiP from the American Bureau of Shipping for a shipping concept which could carry compressed hydrogen at 230 bar with a quantity of 430 tons of hydrogen in a volume of 26,000 m^3 and will be powered using dual fuel engines with hydrogen also contributing as fuel.

Hydrogen Energy Carriers for Transporting Hydrogen

Hydrogen can be incorporated into other lager molecules and transported in the high hydrogen density molecules form and transported in liquid form: for example, ammonia, methanol, or liquid organic hydrogen carriers.

These carriers, such as ammonia or liquid organic hydrogen carriers, are much easier to transport compared to hydrogen. However, an important consideration is that these carriers are not always the final products that could be used directly by consumers or the industry. If the demand is for hydrogen and not the carrier molecules, then the requirement is to dehydrogenate or extract hydrogen from these larger molecules and finally use the released hydrogen for the desired application, which further adds to both cost as well as energy requirements.

However, there is a trade-off that will determine the form in which hydrogen will be transported, i.e., the cost savings in transporting liquid hydrogen carrier molecules versus the energy or cost spent in separating hydrogen from the carrier molecules to get pure hydrogen for the desired applications. This important factor will be the determining factor in deciding which particular method of transporting hydrogen is economically and energetically viable. Together with the distance to be transported and the quantity of hydrogen carried will determine the method of transporting hydrogen. Some of the LOHC and their characteristics are mentioned in Tables 18.1 and 18.2.

Studies have shown that transporting hydrogen via pipelines is the most cost-effective option for distances of less than 1500 km. Beyond this distance, it can be transported as ammonia or liquid organic hydrogen carriers, especially when moved overseas. This consideration includes the cost of conversion from hydrogen to ammonia or liquid organic hydrogen carrier and reconversion including all the cost still it remains economical above the distance of 1500 km.

Currently, ammonia can be transported via existing pipelines, and if the demand increases, new pipelines can be laid down for transporting ammonia. However, an interesting finding is that laying down these new pipelines for ammonia transport is cheaper or more economical than transporting via hydrogen dedicated pipelines.

Another option is transporting ammonia via ships; currently, around 20 million tons of ammonia are transported globally via sea routes annually through ships. Besides ammonia, LOHCs can also be easily transported as hydrogen carriers, similar to oil and diesel, and can utilize existing pipelines that are currently in use. However, in carrying LOHC via pipelines, the major challenge is that once hydrogen extracted, the remaining carrier molecule must be returned to the point of origin for recharge. Either a road or a parallel pipeline will be required for this, which adds to the already prevailing high cost. Thus, transporting hydrogen in the form of liquid organic hydrogen carriers will involve both energy and cost associated with conversion (extracting hydrogen from LOHC) and reconversion (converting carrier molecule back to LOHC). Typically, the hydrogenation of the LOHC is an exothermic process, while dehydrogenation is an energy intensive endothermic process that requires roughly 9 kWh/kg of hydrogen.

As per the IEA report, if the cost of hydrogen pipelines including both CAPEX and OPEX for around 1500 km of hydrogen dedicated pipeline is being considered, then the cost of hydrogen transportation by pipeline comes out to be 1 USD/kg of hydrogen being transported. However, transporting in the form of ammonia is cheaper and is expected to be around 0.5 USD/kg, but an additional cost of 1 USD/kg adds up on the top of it for conversion from hydrogen to ammonia, thus making it a total of 1.5 USD/kg required for transporting in the form of ammonia for around 1500 km using pipeline mode. Beyond 1500 km, increasing the distance will increase the cost of hydrogen pipeline. Thus, the relative increase in cost of laying down the hydrogen pipeline is significantly higher than transporting it in the form of ammonia or via ammonia-based pipelines for distances above 1500 km.

Thus, the cost of laying down hydrogen dedicated pipelines beyond a distance of 1500 km is much higher than transporting it as ammonia and that breakeven point comes at 2500 km, wherein at 2500 km transport distance, the cost becomes the same for transporting via hydrogen pipeline or by ammonia-based pipeline, and that includes, in case of ammonia both conversion and reconversion costs. Thus, reports state that at 2500 km transport distance, the cost for both ways is 2 USD/kg. This involves considering the option of transporting hydrogen or ammonia, both of which can be done via a pipeline.

The other possibility is via shipping route if the transportation distance is 1500 km; in that case, the cost when liquid hydrogen is transported via ships is roughly 2 USD/kg, whereas for ammonia it is 1.2 USD/kg, while for LOHC, it is 0.6 USD/kg. The cost of shipping, however, increases as the distance exceeds 1500 km; however, this increase in shipping cost with distance is less compared to the increase in transport cost through hydrogen dedicated pipelines. All the comparisons are when larger quantities and long-distance transport of hydrogen or its derivates is involved.

Hydrogen Transportation via Road

Transporting hydrogen through dedicated pipelines or blending it through existing natural gas pipelines, or transporting it via ships, either in liquid form and its derivatives, are viable options when large quantities are to be transported over longer distances. The other option is transporting via road, either in the form of hydrogen carriers, compressed gaseous hydrogen, or liquid hydrogen. Transportation of hydrogen by road offers multiple advantages, providing versatility, allowing for various routes, and meeting the required demands promptly. The economics and technical feasibility of transporting ammonia and LOHCs via road are assessed for quantities of 5000 kg and 1700 kg, respectively. This also includes considering the carrier molecule's return to the original point via road.

Considering the distance of 500 km, the cost of transporting LOHC is roughly 0.8 USD/kg, and hydrogen extraction and purification incur an additional cost of 2.1 USD/kg, resulting in a total transportation charge of 2.9 USD/kg via road. However,

Table 18.1 LOHC properties and cost comparison

	Price (€/kg)	Density [kg/L]	Liquid temperature range [°C]	Ignition temperature [°C]	Dynamic viscosity 20 °C (mPa-s)	Vapor pressure 40 °C (Pa)	Toxicity (TPI/mg)	Availability (1–10 Rating)
Benzene Cyclohexane	1	0.879	5–80	498	0.648	21,331	–	8–9
Dibenzyltoluene (DBT) Perhydro-dibenzyltoluene	4	1.04 0.91	– 39 to 390 – 45 to 354	450 –	44.1 258	0.07 0.04	13.8	9
N-Ethyl carbozole (NEC) Perhydro-N-ethylcarbazole	40	1.10 0.94	68–270 < 20–280	186 146	121 5.9	0.05 4.4	5.1	1
Carbon dioxide Methanol (MET)	0.03	0.46 0.79	– – 98 to 65	– 440	– 0.6	– 35,400	–	10
Carbon dioxide Formic acid (FA)	0.03	0.47 1.22	– 8–101	– 520	– 1.8	– 11,400	–	10
Toluene (TOL) Methylcyclohexane	0.30	0.88 0.77	– 95 to 111 – 127 to 101	535 260	0.6 0.7	7880 10,900	19.3	10
Naphthalene (NAP) Decalin	0.6	1.14 0.90	80–218 – 43 to 185	540 235	2.0 3.0	114 360	45.8	10
1,2-dihydro-1,2-azaborine (AB) 1,2-BN Cyclohexane	–	– 1.01	– 45 to 87 63–87	22.7 –	0.55 0.57	18,300 18,300	–	1
Phenazine Tetradecahydrophenazine	< 25	1.30 1.00	172–360 –	160 –	3.1 2.2	0.01 0.18	38.0	4
Gasoline							96.4	–

Table 18.2 LOHC assessment on reaction condition and storage capacity

	Hydrogenation condition	Enthalpy of reaction (kJ/mol H_2)	Dehydrogenation condition	Heat of desorption (kJ/mol H_2)	Storage capacity (wt. %)	Energy density (kWh/L)	References
Benzene Cyclohexane		–			7.2		[7, 8]
Dibenzyltoluene (DBT) Perhydro-dibenzyltoluene	150 °C, 50 bar (Ru/Al_2O_3), 240 min	– 65.4	180–270 °C, 90% (Pd /Al_2O_3), 250–25 min	65.4	6.2	1.9	[9]
N- Ethyl carbozole (NEC) Perhydro-N-ethylcarbazole	150 °C, 50 bar (Ru/Al_2O_3),180 min	– 53.2	290–310°C, 97% (Pd /C), 120 min	53.2	5.8	2.5	[9, 10]
Carbon dioxide Methanol (MET)	220–270 °C, 20–80 bar (Cu based)	– 16.5	100–420°C, homogeneous, H_2O	16.5	12.1	3.3	[11–13]
Carbon dioxide Formic acid (FA)	1–100 bar,30–100 °C, homogeneous, H_2O	– 31.2	30–100°C, Ir /$Ce_{0.5}Zr_{0.5}O_2$,H_2O	31.2	4.4	1.8	[14–17]
Toluene (TOL) Methylcyclohexane	120 °C, 30 bar (Pt/CBV-780), 120 min	– 68.3	250–450 °C, (K-Pt/Al_2O_3),95%	68.3	6.2	1.6	[18–20]
Naphthalene (NAP) Decalin	200–300 °C, 69 bar (Al-MCM-41), 480–150 min	– 66.3	210–320 °C, Toluene (Pt-Re/C), 150 min	66.3	7.3	2.2	[21, 23]
1,2-dihydro-1,2-azaborine (AB) 1,2-BN Cyclohexane	80 °C, 10 bar,Pd/C, tetrahydrofuran	– 35.9	80 °C, 99%, 20 min, $FeCl_2$, tetrahydrofuran	35	7.1	0.92.4	[22, 23]
Phenazine Tetradecahydrophenazine	115 °C, 50 bar (Pd_2Ru/SiCN), 1440 min	– 61.3	190 °C, 100% (Pd_2Ru/SiCN),1440 min	61.3	7.2	2.4	[24]

the same charges apply when considering ammonia being transported by road, at 1.5 USD/kg.

Thus, transporting hydrogen adds up high cost to the hydrogen being delivered to the end user or industry. When calculating the final cost of hydrogen delivery, it is important to consider all the possible stages of the supply chain, including conversion, transmission, distribution, storage, reconversion, and other relevant processes. It is possible that one particular option could be cheaper for a specific part of the value chain, but it may prove to be expensive for another part of value chain. At the same time, the technologies mentioned for long-distance transmission are at a different maturity levels. The cost is expected to decrease with economies of scale. However, since the maturity level varies across different technologies, the steps and processes involved may result in varying levels of cost reduction in the future.

Smaller Quantities and Shorter Distance Transport of Hydrogen

If the demand for hydrogen is low and the production plant is located near the transmission or distribution terminal, the preferred option to transport hydrogen is via road. Considering the economics and technical feasibility, as well as the faster deployment and creation of the entire infrastructure for hydrogen transport with greater flexibility, the tube trailer transport of compressed gaseous hydrogen is the simplest, most widely used, and preferred option. The tube trailers can deliver hydrogen to hydrogen refueling stations. The two major factors that significantly impact the energy consumed and cost involved are the payload of hydrogen carried using the tube trailers or trucks, and the energy density of the hydrogen. Thus, through these modes of hydrogen transport, the focus lies on improving the hydrogen storage capacity per unit volume, decreasing the weight of the tank, and reducing costs.

Tube Trailer Delivery of Hydrogen

Hydrogen production can occur either at a central production plant, from which hydrogen is transported via transmission pipelines to the distribution terminal, or at a semicentral production plant where the production plant is co-located with the gas terminal (Fig. 18.3).

Thereafter, hydrogen is compressed from the production pressure (semicentral production plant) or the transmission pipeline pressure to the pressure required for tube trailer or the pressure at which the tube trailers are filled. The tube trailer consists of several high-pressure vessels that are manifolded together and loaded onto a lorry. From the compressed hydrogen loading bays (areas for safely loading and unloading

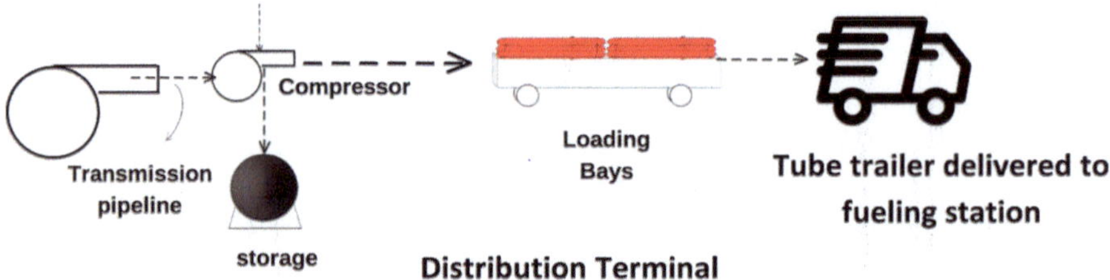

Fig. 18.3 Hydrogen transportation via road using tube trailers

compressed hydrogen), there is interface with the tube trailers, where additional storage can be provided to address demand variability.

At the loading bays, the pressure vessels on the tube trailers are filled with compressed hydrogen, and then the tube trailers deliver hydrogen to the hydrogen refueling station via road.

Typically, on a tube trailer, there are several cylindrical pressure vessels mounted, which are designed to store hydrogen at a certain pressure rating. An important consideration in tube trailer-based compressed hydrogen delivery is the arrangement of these pressure vessels on the trailer. These pressure vessels are packaged together into a container, which is then mounted on a trailer to transport the compressed gaseous hydrogen from the gas distribution terminal to the hydrogen refueling station.

In this tube trailer delivery, several important parameters must be considered. These parameters include the material of construction, dimensions, pressure rating, and packaging of these pressure vessels. Different ISO standards determine or limit the weight of the tube trailer and the dimensions of the vessels to be mounted onto the container (Fig. 18.4).

The pressure vessels are mounted onto the container and are packed in a certain pattern: for example, they can be packed in different configurations as shown in Fig. 18.5.

Depending upon the arrangement of these pressure vessels on the trailer, the total hydrogen payload that the trailer can carry will vary. However, there is a restriction on the total weight of the tube trailer, which includes both the weight of the trailer without hydrogen vessels and including the hydrogen vessels onboard. Thus, how effectively

Fig. 18.4 Picture of a typical tube trailer with compressed hydrogen vessels (Courtesy Ether CT website, https://www.dsaetherct.com/eng/ last accessed on 12-07-2025)

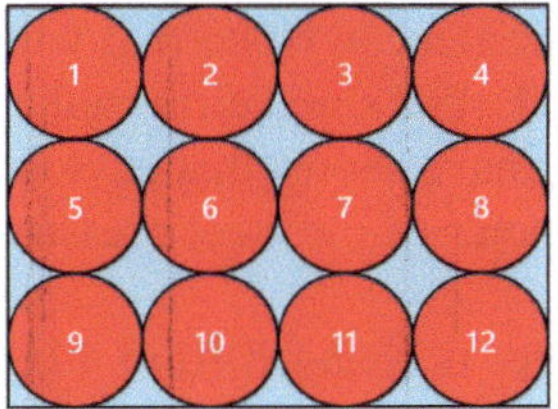

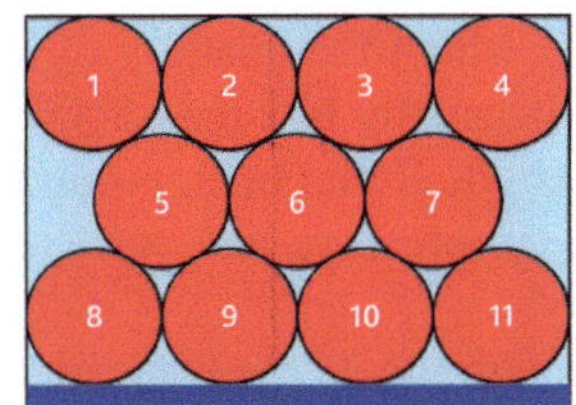

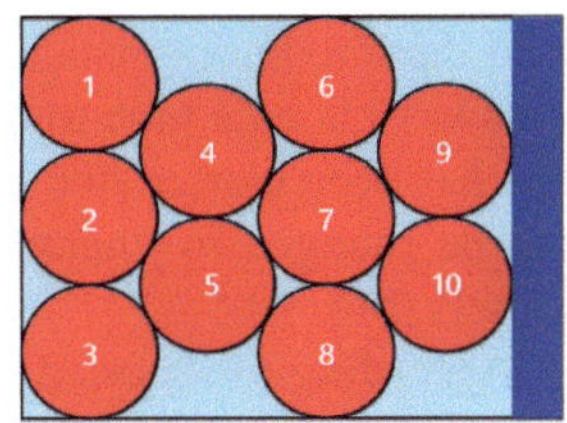

Fig. 18.5 Possible arrangements of compressed vessels on a tube trailer

these pressure vessels are packed together will also determine the effective utilization of available space on the trailer, as there is always a space constraint, and the payload that the trailer can carry is also subjected to a weight constraint. Therefore, the packaging of these pressure vessels should be designed to utilize the available volume effectively. The cost-effective delivery of the compressed hydrogen through a tube trailer will depend on parameters such as the capital cost, which includes both the pressure vessels and their manufacturing, as well as the mounting cost, the payload it can deliver. Thus, optimizations are essential to maximize the payload and are generally expressed in terms of cost per unit payload or the capital cost per kg of the payload being delivered. Several factors determine this cost, including the material from which the pressure vessels are made, the properties of these materials, and the efficiency with which these vessels are packed together.

When designing pressure vessels, the L/D ratio is a very important parameter. If a pressure vessel of a particular length L and diameter D is being considered and if the requirement is to increase the payload or the mass of hydrogen that it is going to carry, then it is essential to increase the pressure. To increase the pressure, the wall thickness needs to be increased, which will, in turn, increase the weight of the vessel and thereby decrease the payload. Thus, increasing the pressure or, in turn, increasing the wall thickness is ineffective in increasing the payload.

Another option is to reduce the diameter, i.e., can decrease the wall thickness. In this case, the volume utilization can be higher, resulting in smaller vessels that can be loaded onto tube trailers. This, in turn, allows for higher volume utilization at a given pressure, leading to a higher payload. However, with smaller vessels, the number of vessels to carry the desired payload will increase, because the diameter has decreased and wall thickness decreased; in that case, there will be more instrumentation requirement like piping or valves, which will in turn again increase the cost. Thus, the design optimization to decrease cost and increase payload is extremely important, and it has been reported that proper optimization can result in a cost decrease of 16%. In addition to the parameters mentioned, another important parameter that governs the economics of tube trailer delivery is the service life, which depends on the type of pressure vessel being used.

For steel pressure vessels, the pressure limit of hydrogen which the vessels can carry is lower due to material properties like tensile strength and there are standards which limit the maximum allowable weight. The maximum allowed payload is usually limited currently to 250–280 kgs and holding pressure of 250 bar along with the maximum distance to which the quantity of hydrogen can be carried with steel vessels on a tube trailer is usually less than 300 km. Lot of simulation studies that has been carried out, like by K Reddi et al. group [4], have shown that payload can be changed by changing the pressure and outer diameter of the steel pressure vessels (Fig. 18.6).

According to reported simulations, if the outer diameter is changed from 40 to 36 inches, the payload could increase from 600 to 690 kg of hydrogen. For a 24 inch diameter, it can store 790 kg of hydrogen. Besides the outer diameter and payload, there is also a link with the cost. As the payload increases, the cost of the trailer and pressure vessel also increases.

The thickness of a pressure vessel is directly proportional to its diameter and inversely proportional to the factor of safety and the strength of the material, which in turn determines the pressure these vessels can hold.

For steel vessels, the factor of safety typically falls within the range of 3–3.5 and is usually considered to be 3. The major challenge mentioned is that to make cost viable, the requirement is to increase the payload by increasing the pressure. However, this will necessitate an increase in wall thickness, which in turn will further increase the weight of the vessel. For steel vessels, there is a limitation on which payload can

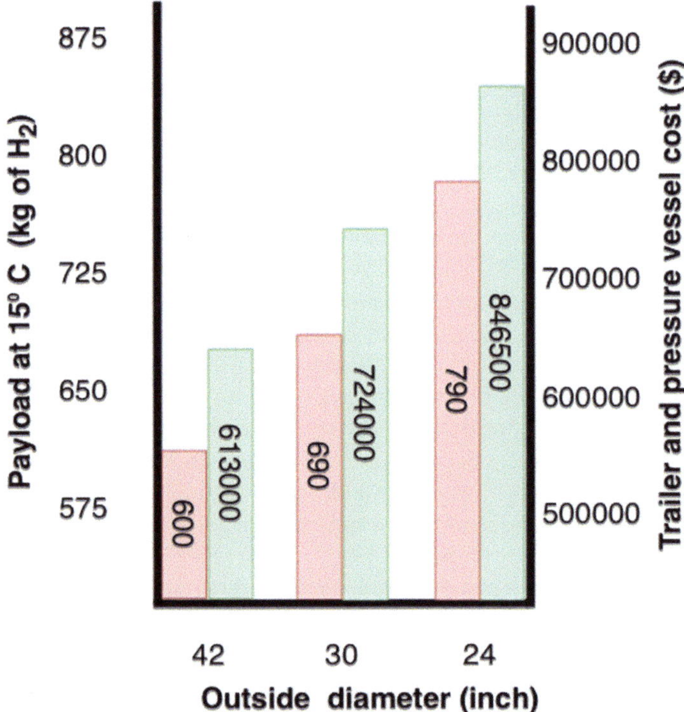

Fig. 18.6 Variation of payload with the change in outer diameter of pressure vessels and impact on associated cost [4]

be increased, and to increase beyond the limit, the material of the pressure vessels should be changed to withstand higher pressure and made of composites.

Studies have revealed that significant cost reduction (10–25%) results when the hydrogen carrying capacity of trailers is increased (from 500 kg to more than 1000 kg). This benefit is based on the distance traveled and number of trips taken. However, with increasing capacity, the impact of operational costs, including manpower, fuel for the trailer, and compression cost, remains lower than the cost benefit of increasing the capacity of hydrogen transported. The cost of compression is the major cost factor in this method of transport.

The other option is the use of composite pressure vessels for transporting a higher payload, utilizing different types of compressed hydrogen tanks, as discussed in the previous chapter, i.e., via Type III and Type IV tanks. With the use of carbon fiber composites, the vessels have high strength, lower weight, can provide higher factor of safety, and thus allowing for higher pressure. Typically, for a composite pressure vessel, the factor of safety to be considered is 2.35. Since composite pressure vessels have higher strength, they require thinner wall to withstand a particular pressure, and as such, the payload that they can deliver is higher. With composite pressure vessels at 500 bar, the payload that can be carried is roughly 1100 kg.

Studies have shown that that the cost varies depending on the dimensions and number of vessels. In Fig. 18.7, the costs for a 30 inch, 9 vessel container and 12 inch, 66-vessel at different pressures are compared.

Due to the limitation on to the empty trailer and truck weight, it has been reported that the weight of an empty trailer and truck could be around 13,600 kg. With the pressure vessels, the weight increases to 22,700 kg. Additionally, the valves used for

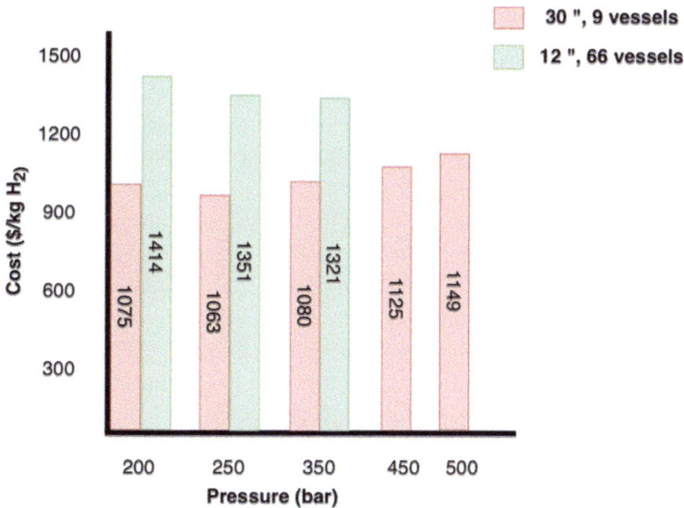

Fig. 18.7 Variation of cost with pressure for different numbers and vessel diameters [4]

these types of vessels weigh approximately 22 kg each. Typically, the shape considered is hemispherical, and the cost associated with the material varies accordingly. When the overall cost of the different pressure vessels is considered, the total cost comprises two major components: the cost of materials and the cost of manufacturing. For steel pressure vessels, the cost breakdown is 80% corresponding to the material cost and 20% to the manufacturing cost. For composite material pressure vessels, the material cost accounts for 40% and the manufacturing cost for 60%, as reported. The cost of the empty trailer is roughly around 15.05 lakh; however, there may be variations to this amount as well.

Type III or Type IV can be used to carry hydrogen at pressures between 200 and 500 bar, transporting quantities ranging from 200 to 1000 kg. With Type IV tanks holding a pressure of 500 bar and a length in the range of 3–12 m, they can carry 240–1115 kg of hydrogen in different pressure tanks, which can number from 20 to 100. These tanks could be fixed in protective frames for safe transportation in a truck. With tube trailer delivery, usually the pressure with type III tank is 200–250 bar and carrying roughly 380 kg in a trailer, while with Type IV, the quantity of hydrogen transported can be roughly 560–900 kg and over a distance of up to 200 km to applications where relatively smaller quantities are required or to the hydrogen refueling stations [5]. Hexagon Purus GmBH has manufactured such trailer vessels. The largest reported capacity is 1300 kg at a pressure of 517 bar on a trailer being developed by Calvera Hydrogen. The Type IV vessels are more expensive than Type I, but the increased cost is offset by the higher payload, i.e., the ability to carry larger quantities of hydrogen per trip.

However, the major challenges with transporting hydrogen via tube trailer are that the increase in payload requirement necessitates an increase in pressure because the density of hydrogen is very low. Nevertheless, if the pressure is increased, there is not much effective improvement in the payload because of the higher weight of the vessels. With the use of better materials, hydrogen is carried in pressure vessels at higher pressures. Therefore, it is essential to consider the effect of repeated cycling on the vessels, and for that, frequent testing will become necessary. The improvement in payload could be achieved by storing cold and compressed gas, thereby achieving higher energy density at a lower pressure.

When the tube trailer is emptied, the complete gas pressure cannot be delivered, and a certain amount of gas pressure remains in the vessels. Studies are required to determine the minimum pressure at which the trailer needs to be returned back for refueling, so that the entire economics becomes more viable. For example, in the case of a 20 ton vehicle transporting 400 kg of hydrogen, around 85% of this can be used or delivered, and the remaining 15% must remain in the trailer after delivery. Besides, there are other important parameters to be considered when the interfacing of the tube trailer with the refueling infrastructure is to be considered like the time required to fill the empty tube trailer, the requirements for operation at the distribution terminal, the compressor output pressure and how it gets affected by the different parameters, the required supply pressure, what is the maximum pressure available after emptying the tube trailer. During the loading of tube trailer, the rate at which filling is carried out without a significant temperature rise is crucial, as there is a direct relationship between the loading time and the resulting temperature rise. The pressure available at the tube trailer and its effect on the terminal, as well as the hydrogen refueling station, is another important parameter to consider. The impact of all these parameters on the delivery cost of hydrogen needs to be studied. The compressors used at the gas stations at the distribution terminal should have high throughput, deliver high pressure should be reliable, and their availability and the associated cost also plays a key role. Delivery of gaseous hydrogen by tube trailer to the market for industrial use is currently the most mature method of hydrogen distribution, suitable for small quantities of hydrogen and over shorter distances. However, the major limitation in tube trailer-based distribution is the lower storage efficiency, due to weight, the requirement for high delivery pressure, and the quantity of hydrogen needed.

For a hydrogen fuel cell vehicle with an onboard Type III tank storing at a pressure of 350 bar, a pressure of 450 bar will be required at the refueling station. Thus, 450 bar will be the storage pressure on the tube trailer, which will provide the required pressure for filling the tank on vehicle. For a Type IV pressure tank on FCEV having onboard storage of 700 bar, the delivery pressure from the tube trailer needs to be still at a higher pressure, roughly 880 bar, which will add to the cost of tube trailer-based delivery; if, however, the tube trailer delivers at a pressure of 200 bar, then there will be a need for onsite compression. Most industrial gas companies are developing delivery trailers that can operate at higher pressures using carbon fiber-reinforced composite cylinders; however, that increases the cost. There have been recent advancements in which these delivery tube trailers also feature integrated hydrogen dispensing systems, allowing the onboard storage tank of a fuel cell vehicle to be refueled without the need for a hydrogen refueling station.

Liquid Hydrogen Superinsulated Tanker Trucks

The distribution of hydrogen using compressed gas tube trailers is a viable option up to a distance of 300 km, with a capacity 400 kg of hydrogen. Beyond this, the economics of hydrogen transport will work more in favor of liquid hydrogen-based transport via road. Since the density of liquid hydrogen is much higher than that of compressed gaseous hydrogen, much higher quantities can be carried over relatively longer distances in this form.

In case of compressed hydrogen transport, the pressure required for transport via tube trailers increases. The larger tanks with compressed hydrogen become highly expensive. When the cost versus distance of transport is being considered, initially till a certain distance compressed tube trailer-based transport is economical; however, as the distance increases, the compressed hydrogen transport becomes expensive and the liquid hydrogen transport starts to get cheaper for higher distances and larger volumes; there lies a breakeven point wherein the compressed hydrogen beyond a certain distance becomes more expensive than the liquid hydrogen transport.

Thus, another method of hydrogen transport is in the form of liquid hydrogen, transported via tanker trucks on roads. This method is being considered when there is a reliable and long-term demand, which is expected to continue into the long-term future; it is a mature technology.

The cost of liquefaction is extremely high; but, liquid hydrogen has a significantly higher volumetric energy density. When hydrogen is transported in the form of liquid hydrogen, there is a breakeven point at which the liquefaction cost can be offset by savings on hydrogen transport costs, as more quantities of hydrogen can be carried in liquid form compared to compressed gas form. Thus, if the savings can offset the higher liquefaction cost in term of hydrogen transportation cost, then highly insulated liquid hydrogen tankers or trucks could be more economical and viable.

Liquid hydrogen tankers can carry 4000 kg of liquid hydrogen over 4000 km. This is roughly about 5–6 times that of the composite tube trailers pressure vessels and 15–20 times that of the steel tube trailers. However, above a distance of 4000 km, this option may not be economically viable. For distances beyond this, when LH_2 being carried it get heated up, it evaporates and undergoes boil-off losses, resulting in a rise in pressure. As such, beyond this distance, it would not be a viable option.

As shown in Fig. 18.8, hydrogen can be transported from the point of production to the point of end use via liquid hydrogen tanker trucks. Either the hydrogen production plant could be centralized or a semicentralized plant. From a centralized hydrogen production plant, hydrogen can be carried through transmission pipelines, liquefied using a liquefier and stored in the cryogenic storage vessels. Liquid hydrogen is pumped using cryogenic pumps to load onto the tankers.

However, if the hydrogen production facility is a semicentral production plant, wherein production and liquefaction are done onsite at the same location as the distribution terminal, the gaseous hydrogen being transported is liquefied using a liquefier and then stored in a cryogenic storage container and finally, pumped using

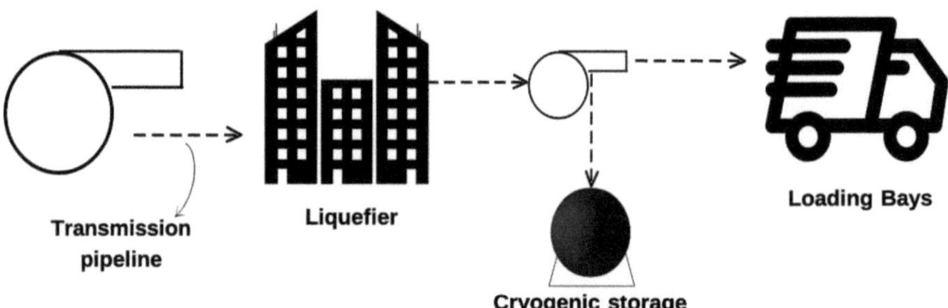

Fig. 18.8 Schematic of a typical liquid hydrogen transportation facility

Fig. 18.9 Picture of a liquid hydrogen transportation truck. *Source of image* Liquid hydrogen technology for vehicles, Joachim Wolf, MRS Bulleting (2002) 684–687

a cryogenic pump into the tanker at the loading bay. These tankers carry liquid hydrogen to the hydrogen refueling stations or to utilities.

The liquefiers used for hydrogen liquefaction usually have a daily capacity that is either equal to or exceeds the average daily demand for liquid hydrogen to meet the transportation needs. The cryogenic storage facility at the distribution terminal is designed to store 5–7 days of output obtained from the liquefier (Fig. 18.9).

Liquefaction, being an energy and cost-intensive process, suffers from boil-off losses; thus, liquefaction requires more than 1 USD/kg of hydrogen. The boil-off losses depends upon the shape and size of the storage tank: for example, for a 50 m^3 container the boil-off rate is 0.4% per day while for a 20,000 m^3 of tank the boil of rate is 0.06% per day. The tankers carrying liquid hydrogen from the distribution terminal to the hydrogen refueling are unloaded to the cryogenic storage facility at the hydrogen refueling station. Besides the boil-off losses during the transport of hydrogen, substantial losses are also observed during unloading of liquid hydrogen (approx. 5% losses by volume) at the hydrogen refueling stations. Thus, the number of such deliveries needs to be limited, the route needs to be well planned in order to optimize the losses, and requirement is to minimize these losses. Various measures are being researched and implemented to reduce these boil-off losses.

The well-known global liquid hydrogen companies are Air Liquide, Linde, and Praxair, among others. Linde has a liquid hydrogen container that is 50 ft in length that can carry 3000 kg of liquid hydrogen at a pressure of 12 bar. Similarly, another vessel that is 13.7 m in length can carry 4000 kg of liquid hydrogen at 12 bar pressure. Linde reported that with 3853 kg of liquid hydrogen in a tank, the pressure buildup is to 3.4 bar in 158 hours, with the maximum holding capacity of tank being 12 bar. This implies that the pressure inside the tank is much lower and will not be released into the environment.

However, the entire amount of liquid hydrogen stored is not delivered to the facility or utility, around 10% of it is required to maintain the temperature in the tank.

With liquid hydrogen tanker-based transportation, the major challenge is boil-off losses, which must be minimized through appropriate design of the tankers. Otherwise, any hydrogen lost through boil-off needs to be captured by specific means and then utilized. The high energy requirement for liquefaction, roughly 12–15 kWh/kg of hydrogen, or about 30% of the energy content of hydrogen that goes in liquefaction, is another biggest challenge.

To summarize, the transport of smaller volumes over shorter distances can be achieved using compressed gas tube trailers. If the distance and volume increase to medium distances and intermediate capacities i.e., 4000 kg and 4000 km, then liquid hydrogen tanker trucks will be more economically viable. For transport over long distances, boil-off losses are higher and liquid hydrogen will not be cost-effective. For larger quantities over long distances and when it is necessary to transfer directly to industries, compressed hydrogen transport via pipeline is the cheapest way. However, in cases where connectivity through pipeline is not possible, such as for intercontinental hydrogen transport, ships will be the most economical choice. Transporting liquid hydrogen using ships can be economical, as large quantities can be shipped between an exporting country rich in renewables-based hydrogen production and an energy importing country. However, the boil-off losses and high cost of liquid hydrogen production, shipping, and storage tanks are the major disadvantages. Carrying energy dense material like the liquid organic hydrogen carriers or ammonia, or liquid hydrogen via ships has its own advantages and disadvantages in terms of conversion and reconversion costs; hazardous nature of some of the LOHCs, loss of hydrogen in the process, and taking back the carrier molecule to the point of hydrogenation adds up to the cost of shipping.

Economics of Hydrogen Transport

The life cycle cost assessment for transporting hydrogen can be estimated using the following equation:

$$C = C_o + C_c + C_s + C_t + C_d \tag{18.11}$$

where C is the total annual cost, C_o is the other cost, C_c is the compression cost, C_s is the storage cost, C_t is the transportation cost and C_d is the distribution cost.

$$\mathrm{LCC} = \sum_{1}^{T} \frac{C + \mathrm{MC}}{(1 + r)^n} \tag{18.12}$$

where, LCC is the life cycle cost, M is the ratio of annual maintenance cost to the annual cost, r is the discount rate and T is the project life in years. The calculation of levelized cost of hydrogen (LCOH) requires knowledge of both capital expenditure (CAPEX) and operational expenditure (OPEX). The major contributors to the CAPEX, while considering LCOH, are the cost of purchasing equipment, the cost of infrastructure, and the other indirect costs involved. In contrast, OPEX primarily involves energy consumption in the process, the maintenance costs, and any other recurring costs. LCOH can be determined by considering the capital cost and annual operating expenses, along with the quantity of hydrogen being transported.

Hydrogen Refueling Stations

Hydrogen refueling station is a complex system and consists of several components like, compressors, storage vessels, dispenser, high pressure filling lines, chiller or heat exchanger or the pre-cooling unit. The hydrogen refueling stations can be of two types, either it can be an onsite hydrogen refueling station or off-site refueling stations.

Onsite refueling stations are considered in cases where either the hydrogen production site is far from the hydrogen refueling station or is not connected to the distribution terminal. In those cases, hydrogen is being produced onsite using either a reformer or an electrolyzer. The other possible option is when hydrogen is carried through transmission pipelines and distribution pipelines to the service pipelines, then it is considered as an off-site hydrogen refueling station. Hydrogen produced in a central production plant or a semicentral production plant is transported using tube trailers or insulated trucks to the refueling station at an off-site hydrogen refueling station.

A refueling station can be designed for dispensing either compressed gaseous hydrogen or liquid hydrogen, and depending on this, the components of the hydrogen refueling stations will differ.

Compressed Gas Hydrogen Refueling Station

A typical hydrogen refueling station will have components as shown in Fig. 18.10 where hydrogen is carried from the distribution terminal through delivery tube trailers at a certain pressure. Thereafter is compressed using a compressor from the tube trailer pressure to the high-pressure buffer storage pressure. Before filling the vehicle tank with hydrogen, it needs to be cooled to a certain temperature and then dispensed

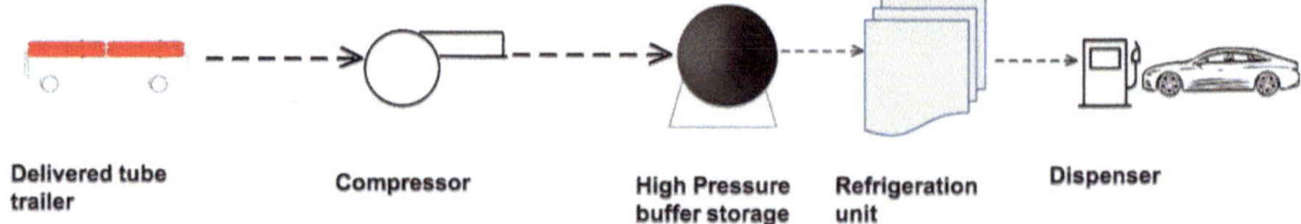

Fig. 18.10 Schematic of a typical hydrogen refueling station

into the onboard tank in a fuel cell vehicle. Another important unit in the refueling station is the high-pressure buffer storage, which determines the capacity of a hydrogen refueling station. Depending upon the application, this storage can be either a high-pressure storage facility, which could be composite pressure tanks, i.e., Type 3 or Type 4 having volume typically of around 1–20 m^3 depending upon the size of the fuelling station or it could be a low-pressure tank made of a seamless steel vessel, important is that all these should be as per the standard like the design should be as per the ASME standard and TSG 21-2016 (Fig. 18.11).

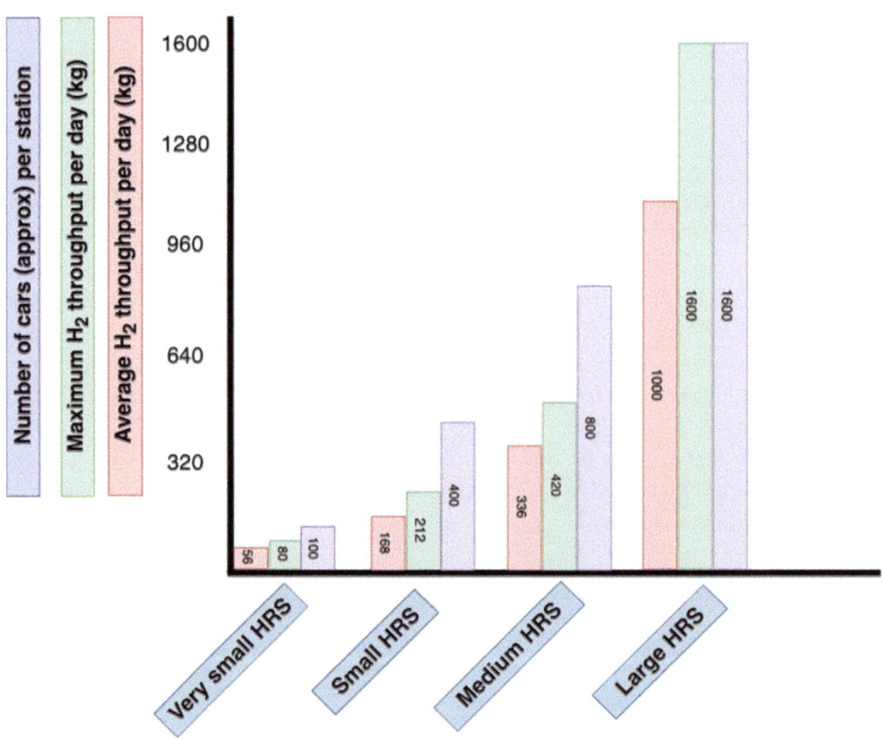

Fig. 18.11 Comparison of different sized hydrogen refueling stations based on the number of cars arriving for refueling, hydrogen throughput, and average hydrogen delivery per day (Data taken from [4])

Depending on the capacity of the buffer storage, there can be different types of hydrogen refueling stations, i.e., very small, small, medium, or large hydrogen refueling stations. The factors determining this classification are the maximum hydrogen throughput per day in kg, the average hydrogen throughput per day and the approximate number of cars that arrive at the station for refueling per day. For a very small and small refueling station, only one refueling point is typically present. For a medium sized refueling station, there can be two refueling points, and for large hydrogen refueling station, four refueling points can be included. The average hydrogen throughput for a very small, small, medium and large refueling station can be roughly 56, 168, 336 and 1000 kg, respectively. The average number of cars that come for refueling at very small, small, medium, and large hydrogen refueling stations is roughly 10, 30, 60, and 120, respectively.

The major component of a compressed gas hydrogen refueling station is the dispenser. The dispenser consists of several components, including high-pressure piping, filling lines, a shutoff pneumatic valve, and a supporting explosion proof solenoid valve, etc. There are several temperature and pressure sensors, temperature and pressure regulators, various monitors, and different control mechanisms including the gas guns, a chip controller, and a complete control panel within the dispenser. All these components are housed within the dispenser. The components have several functions, including preventing over pressurization, ambient temperature compensation, preventing immediate protective actions like shutoffs in case of any emergency, hose break prevention, and providing a form of sequential gas extraction.

All components and operations are designed in accordance with standards, such as typical standard followed, SAE J2601. The fuelling nozzle, as per SAE 2799 standard, and the refueling connector must be according to the SAE J2600 standard.

Besides the components mentioned, a refrigeration unit is also important to maintain temperature as per the SAE J2601 standard. Since the requirement from any refueling station is for fast refueling, providing a minimum filling time and maximum allowable filling pressure, such that when the car arrives at the station, the passenger should not have to wait for longer durations. A short filling time, along with a high final state of charge, will require removal of heat generated during the filling process. As per the SAE J2601 standard, the hydrogen before filling should be pre-cooled to a temperature of − 33 to − 40 °C. This standard is to ensure that the maximum temperature rise during the filling of hydrogen on the onboard vehicle tank should not go beyond 85 °C; for that, refrigeration is being achieved through thermal refrigeration cycle. The refrigeration unit comprises of a chiller, heat exchanger, and associated piping. Hydrogen, when extracted from the high-pressure buffer storage before being dispensed into the vehicle, has to go through chiller. The coolant circulates through a heat exchanger, extracting heat from the hydrogen, and then returns back to the chiller.

In the whole process, the temperature of hydrogen reduces in the range of − 33 to − 40°C. The entire pre-cooling unit is cost-intensive; the total equipment cost of the unit is about 10% of the total cost of the entire dispensing systems. An essential requirement for designing a hydrogen refueling station is that the pre-cooling unit

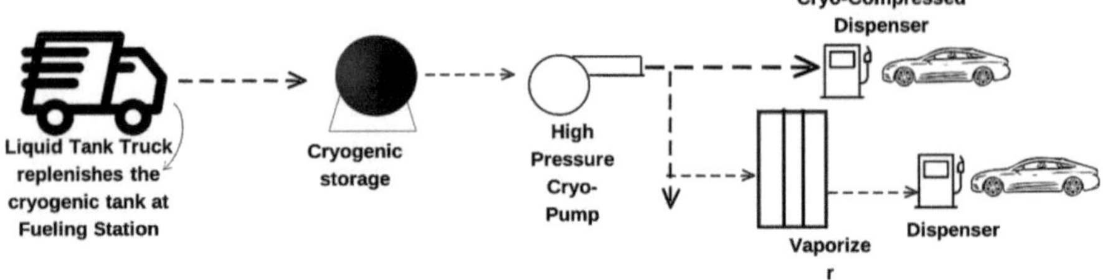

Fig. 18.12 Schematic of a typical liquid hydrogen refueling station

should be in close proximity to the dispenser to minimize energy losses. However, the pre-cooling unit is not required if the filling station is a liquid hydrogen filling station with a cryo-pump being used.

Liquid Hydrogen Refueling Station

The other option is refueling hydrogen onboard on a vehicle in liquid form through a liquid hydrogen refueling station. Such refueling stations are similar to the compressed gas hydrogen refueling stations. Hydrogen can be transported from the distribution terminal by a means of liquid hydrogen trucks which are used to replenish the storage tank located onsite at the refueling station. Thus, at the location of liquid hydrogen refueling station, a cryogenic superinsulated vessel storage can be either underground or above the ground is co-located. At the time of filling, either a high-pressure cryo-pump is used to refill a cryo-compressed pressure tank on the vehicle or a vaporizer is used to convert the liquid hydrogen into compressed hydrogen and refill onboard tank of the vehicle. For the second option, the liquid hydrogen has to be evaporated and converted into gaseous hydrogen and then filled into the vehicle tank through a dispenser. The limitation with the liquid hydrogen refueling station is that the maximum hydrogen boil-off rate should be limited to 1.4% of the capacity per day, in 24 h considering that the ambient temperature is at 15 °C (Fig. 18.12).

Dispensing Concepts

There are two different dispensing concepts at the hydrogen refueling station one is multibank dispensing and another being booster dispensing.

Multibank Dispensing

In the case of multibank dispensing, there are multiple storage tanks: for example, three tanks or banks are shown in Fig. 18.13 which corresponds to a low-pressure bank, a medium pressure bank, and a high-pressure bank. Initially, all the banks have the same pressure, i.e., during the filling cycle, all three pressure banks are filled to the same pressure level; however, while filling the onboard tank, the pressure drops in a particular vessel. The biggest advantage of this method of dispensing is that the reduced pressure can still be utilized to fill the onboard tank.

Considering the mechanism of dispensing in this concept of multibank dispensing, there are three steps in the filling of the car tank. Initially, the hydrogen from the lower pressure bank flows into the onboard tank; usually, the pressure filling with the low-pressure bank is in the range of 10–15 MPa. To fill the car tank to 15 MPa, the pressure in the low-pressure bank should be at least 16 MPa. Once the required pressure is achieved in the car tank, the flow immediately switches automatically to the medium pressure bank. The medium pressure bank fills the car tank to a pressure of 20–25 MPa, against the pressure in the medium pressure bank of 35 MPa, which is required for filling. Once the achievable pressure with the medium pressure bank

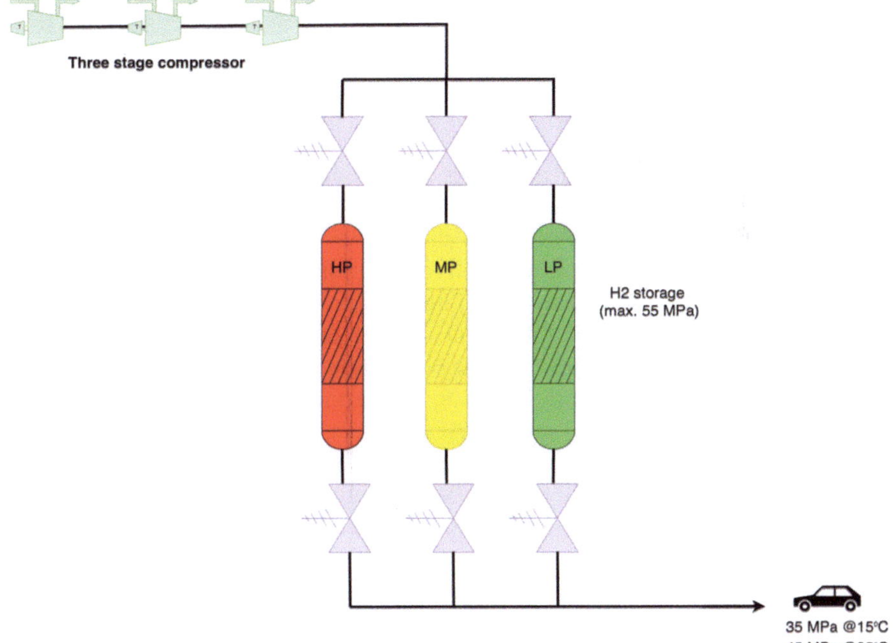

Fig. 18.13 Multibank dispensing concept for a hydrogen refueling station

is obtained, the flow switches automatically to the high-pressure bank. The high-pressure bank finally delivers the required pressure of 35 MPa and filling is completed of the Type III tank onboard to the required pressure of 35 MPA.

An important point to note is that the maximum pressure in the car tank is 35 MPa, considering that filling has been carried out at 15 °C. However, while filling hydrogen, the temperature increases; thus, a pre-cooling unit is present to restrict the maximum temperature to 85 °C as per the standard. During the filling of hydrogen into the car tank, the maximum allowable temperature that it can reach is 85 °C. However, at 85 °C, the pressure of 35 MPa corresponds to 43.8 MPa, which means when the temperature comes back to normal, the filling level will not remain the same; thus, the filling pressure should be essentially higher than 35 MPa such that when the temperature reaches 15 °C, the final pressure in the tank is 35 MPa. Thus, in order to achieve a filling pressure of 35 MPa at 15 °C, when the temperature reaches maximum of 85 °C, the corresponding filling pressure should be 43.8 MPa. Now, to fill the car tank to 43.8 MPa, the pressure in the high-pressure bank must be still higher, for the gas to flow across pressure difference. The high-pressure bank should be at a pressure higher than 43.8 MPa, and usually, it should be at 44.8 MPa or around 45 MPa.

In the case of Type IV tank in the vehicle, the desired filling pressure is 70 MPa; with the same logic as above, the high-pressure bank should be at 88 MPa.

Booster Dispensing

Another concept of dispensing is booster dispensing, which can be of two types (Fig. 18.14).

Onsite hydrogen produced using an electrolyzer is compressed and stored in a hydrogen storage vessel at the hydrogen refueling station, as shown in Fig. 18.14. The pressure at the hydrogen storage unit lies in the range of 15–30 MPa, i.e., lower than the pressure required for filling into the car tank. Thus, there is a primary compressor to raise the pressure from the production pressure, i.e., the electrolyzer output pressure of approximately 2 MPa, to an intermediate pressure of 15–30 MPa for storage in the storage facility at HRS. As and when the car arrives at HRS for refueling, the pressure level can be increased to the desired filling pressure, depending on whether the car has a Type III or Type IV tank onboard, i.e., 35 or 70 MPa, respectively. Thus, the second stage compressor, called the booster compressor, increases

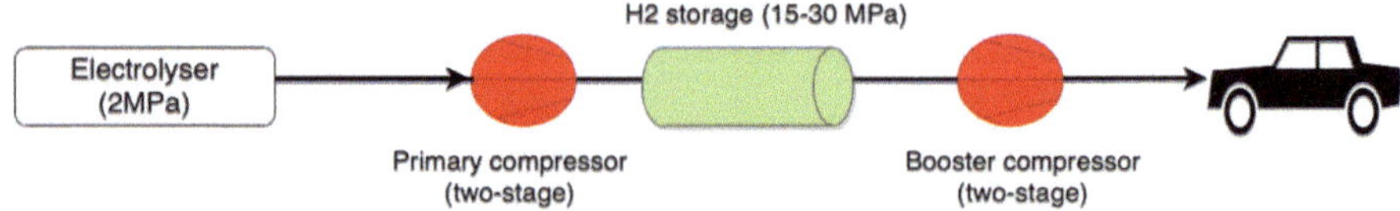

Fig. 18.14 Booster dispensing concept for hydrogen refueling station

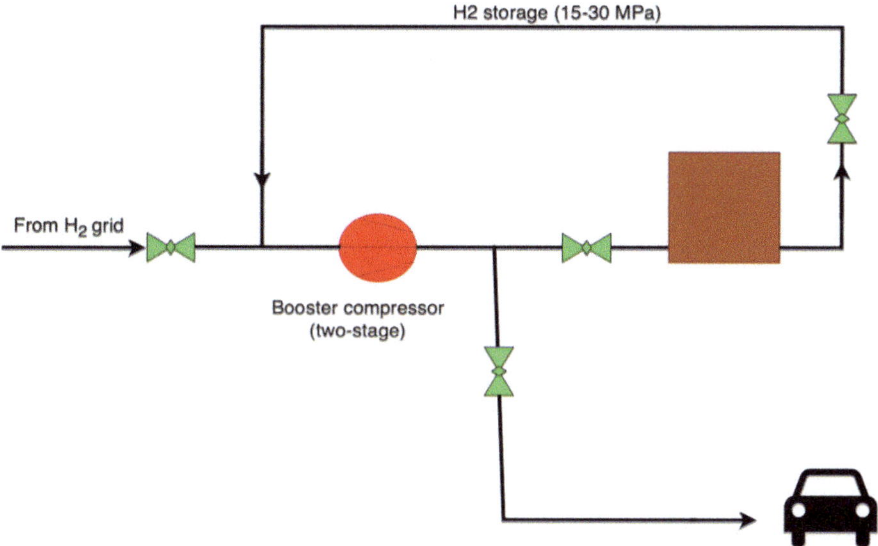

Fig. 18.15 Another possibility of the booster dispensing concept for hydrogen refueling station

the pressure from the storage facility pressure to the desired filling pressure of the car tank, and the filling is carried out through the dispenser.

The another possibility is shown in Fig. 18.15; the use of a single compressor, called the booster compressor, boosts the pressure to the desired filling pressure of the car tank. The same booster compressor is used to compress and fill the storage vessel present onsite at the HRS. Thus, in this concept, a single compressor is used for both for pressurizing the hydrogen storage unit, which stores at pressure in the range of 15–30 MPa, and for filling the car tank to the desired pressure, either 35 MPa or 70 MPa, depending on the car tank type.

Figure 18.16 shows a picture of a hydrogen CNG filling station a blend, which is at Faridabad in Haryana and is Indian Oil Corporation Limited (IOCL) and the reference is from the India country status report on hydrogen and fuel cells [6].

To establish the hydrogen infrastructure for fuel cell electric vehicles, hydrogen refueling stations will be required. It is crucial to establish such infrastructure in urban areas with the required standards of space availability. Obtaining that space in a populated city will be critical; therefore, it is vital that the required footprint area be as small as possible and that all safety features and protocols for establishing such dispensing units are followed.

Currently, globally, there are about 540 hydrogen refueling stations. Among these, Japan has the highest number, with 154 filling stations, followed by South Korea with 112, Germany with 91, and China with 85. These numbers may vary depending upon the references and the date of reporting. The number of hydrogen refueling station globally is still relatively low, and they are custom designed. To commission an HRS, a high capital and operational costs are involved. Typically, the cost is around

Fig. 18.16 Picture of the IOCL HCNG refueling station located in Faridabad, Haryana, India

1.9 million USD for a hydrogen refueling station with a capacity of 1500 kg per day. Definitely, as the number of such installations increases over time, the economies of scale will bring down the cost.

The quality of hydrogen being dispensed at these stations should meet the standards, such as ISO 14687. A crucial aspect is the social or public acceptance for such an infrastructure development, and in that, reliability plays a significant role. The typical downtimes of such hydrogen refueling station should be as low as possible, a single trip not being able to fill the fuel may result in the customer getting annoyed and accompanied by reduction in acceptability. Thus, the crucial aspect will be the availability of the trained manpower to keep these HRS operational all the time and attend to maintenance issues promptly, so that the downtime can be as low as possible. In order to improve the public acceptance in the initial phase, mobile hydrogen refueling stations can also cater to the initial demand.

Solved Problem 1

Calculate the gas flow through a 36-inch pipeline operating at maximum capacity with a pipe length of 500 km.

Given Data:

$$P_1 = 70\,\text{bar}$$

$$P_2 = 28\,\text{bar}$$
$$Z = 1.031$$
$$T = 288.15\,\text{K}$$
$$G = 0.0696$$
$$V_{avg} = 35\,\text{m/s}$$
$$\rho_{avg} = 4.58\,\text{kg/m}^3$$
$$\mu = 8.7 \times 10^{-6}\,\text{kg/m} \cdot \text{s}$$
$$h_2 - h_1 = 100$$

Solution:

Average Pressure

$$P_{\text{avg}} = \tfrac{2}{3} \times \left(\tfrac{P_1^3 - P_2^3}{P_1^2 - P_2^2}\right) = \tfrac{2}{3} \times \tfrac{70^3 - 28^3}{70^2 - 28^2} = 52\,\text{bar}$$

Erosional Velocity

$$V_{\text{max}} = 100\sqrt{0.05131 \times \frac{(1.031 \times 8.314 \times 288.15)}{(0.0696 \times 7000)}}$$
$$V_{\text{max}} = 51.01\,\text{m/s}$$

Reynolds Number

$$\text{Re} = \frac{V_{\text{av}} D \rho_{\text{av}}}{\mu} = \frac{35 \times 0.895 \times 4.38}{0.0000087}$$
$$\text{Re} = 1.65 \times 10^7$$

Pipeline Dimensions:

- Outer diameter = 36.0 inches
- Inner diameter = 35.2 inches
- Pipe thickness = 0.8 inches

Colebrook Equation

$$\frac{1}{\sqrt{f}} = -2\log\left(\frac{\varepsilon}{3.7D} + \frac{2.51}{\mathrm{Re}\sqrt{f}}\right)$$

$f = 0.0194$ (calculated iteratively using $f = 1$ as the starting value) Pipeline Diameter:

$$D = 0.06849\left(\frac{h_2 - h_1}{T_f Z}\right)$$
$$= 0.06849 \times 0.0696\left(\frac{100}{288.15 \times 1.031}\right) = 0.0017$$

Equivalent Length:

$$L_e = L\left(\frac{e^{\lambda} - 1}{\lambda}\right) = 500\left(\frac{e^{0.0017} - 1}{0.0017}\right) = 500.43\ \text{km}$$

Flow Rate:

$$Q = 1.1494 \times 10^3\left(\frac{T_b}{P_b}\right)\left(\frac{P_1^2 - e^{\lambda}P_2^2}{GT_f L_e Z f}\right)^{0.5} \times D^{2.5}$$

(D is the inner diameter in mm)

$$Q = 1.1494 \times 10^3\left(\frac{288.15}{101}\right)\left(\frac{7000^2 - e^0.0017 \times 2800^2}{0.0696 \times 288.15 \times 500.43 \times 1.031 \times 0.009}\right)^{0.5}$$
$$\times (895.3)^{2.5} = 51,255,602.23\ \text{Sm}^3/\text{day}$$
$$= 51.25\ \text{million Sm}^3/\text{day}$$

Inlet Velocity:

$$V_{\text{inlet}} = 14.734\left(\frac{P_b}{T_b}\right)\left(\frac{ZT}{P}\right)\left(\frac{Q_b}{D^2}\right)$$
$$= 14.734\left(\frac{101}{288.15}\right)\left(\frac{1.031 \times 288.15}{7000}\right)\left(\frac{51,255,602.23}{895.3^2}\right) = 13.99\ \text{m/s}$$

Capacity (kgH_2/day) $= Q \times 0.0834$.

Solved Problem 2

Problem Statement: Using the data provided below, calculate the transportation cost per kg of hydrogen for the compressed hydrogen scenario. Determine the following:

- Number of trips per year,
- total kilometers driven per year,
- total drive time per year,
- total load and unload time per year,
- number of trucks required,
- number of drivers required,
- total capital cost,
- variable non-fuel O&M cost per year,
- fixed operating cost per year,
- annual fuel usage (liters),
- annual fuel cost,
- annual labour cost,
- cost of capital charges per year,
- total annual cost,
- transportation cost per kg of hydrogen.

Given data		
Description	Unit	Value
Actual hydrogen production	kg/day	135,000
Truck tube/hydrogen capacity	kg	181
Truck tube unit cost	Rs./module	4,500,000
Truck undercarriage cost	Rs./trailer	2,700,000
Operating days per year	days/year	350
Truck load and unload time	hours/trip	2
Truck availability	hr/day	24
Driver availability	hr/day	12
Driver wages	Rs./hr	50
General facilities	% of Capital Cost	15
Engineering, permitting & start-up	% of Capital Cost	8
Working capital, land & misc	% of Capital Cost	4
diesel price	Rs./litre	25
Delivery distance	km (one-way)	200
Non-fuel variable O&M cost	% of Capital Cost per year	1
Fixed O&M costs	% of Capital Cost per year	3
Capital charges	% per year	12
Contingencies	% of Capital Cost	6
Per km fuel consumption	L/km	0.38
Time per Trip	hr./trip	8

Solution

$$\text{Annual hydrogen production} = 135{,}000\ \text{kg/day} \times 350\ \text{days/year}$$
$$= 47{,}250{,}000\ \text{kg/year}$$

$$\text{Number of trips per year} = \text{Annual hydrogen production/Truck capacity}$$
$$= 47{,}250{,}000\,\text{kg}/181\ \text{kg/trip}$$
$$= 261{,}049.72\,\text{trips/year} \approx 261{,}050\,\text{trips/year}$$

$$\text{Total km per trip} = 200\,\text{km(one} - \text{way)} \times 2$$
$$= 400\,\text{km/trip}$$

$$\text{Total km driven per year} = 261{,}050\,\text{trips/year} \times 400\,\text{km/trip}$$
$$= 104{,}420{,}000\,\text{km/year}$$

$$\text{Total drive time per year} = 261{,}050\,\text{trips/year} \times 8\,\text{hr/trip}$$
$$= 2{,}088{,}400\,\text{hr/year}$$

$$\text{Total load and unload time per year} = 261{,}050\,\text{trips/year} \times 2\,\text{hr/trip}$$
$$= 522{,}100\,\text{hr/year}$$

$$\text{Total time per trip} = \text{Drive time} + \text{Load/Unload time} = 8\,\text{hr/trip} + 2\,\text{hr/trip}$$
$$= 10\,\text{hr/trip}$$

$$\text{Trips per truck per day} = \text{Truck availability/Total time per trip}$$
$$= 24\,\text{hr/day}/10\,\text{hr/trip}$$
$$= 2.4\,\text{trips/day}$$

$$\text{Trips per truck per year} = 2.4\text{trips/day} \times 350\,\text{days/year}$$
$$= 840\,\text{trips/year}$$

$$\text{Trucks required} = \text{Number of trips per year/Trips per truck per year}$$
$$= 261{,}050\,\text{trips/year}/840\,\text{trips/year} = 310.77 \approx 311\,\text{trucks}$$

$$\text{Drive time per truck per day} = 2.4\,\text{trips/day} \times 8\,\text{hr/trip} = 19.2\,\text{hr/day}$$

$$\begin{aligned}\text{Drivers per truck} &= \text{Drive time per truck per day/Driver availability} \\ &= 19.2\,\text{hr/day}/12\,\text{hr/day} = 1.6\,\text{drivers/truck}\end{aligned}$$

$$\text{Total drivers required} = 311\ \text{trucks} \times 1.6\,\text{drivers/truck} = 497.6 \approx 498\,\text{drivers}$$

$$\begin{aligned}\text{Cost per truck} &= \text{Truck tube unit cost} + \text{Truck undercarriage cost} \\ &= 4{,}500{,}000\,\text{Rs./module} + 2{,}700{,}000\,\text{Rs./trailer} \\ &= 7{,}200{,}000\,\text{Rs./truck}\end{aligned}$$

$$\text{Base capital cost} = 311\ \text{trucks} \times 7{,}200{,}000\,\text{Rs./truck} = 2{,}239{,}200{,}000\,\text{Rs.}$$

Additional costs:

- General Facilities = 15% × 2,239,200,000 Rs. = 335,880,000 Rs.
- Engineering, Permitting & Start-up = 8% × 2,239,200,000 Rs. = 179,136,000 Rs.
- Working Capital, Land & Misc. = 4% × 2,239,200,000 Rs. = 89,568,000 Rs.
- Contingencies = 6% × 2,239,200,000 Rs. = 134,352,000 Rs.

Total capital cost = 2,239,200,000 + 335,880,000 + 179,136,000 + 89,568,000 + 134,352,000 = 2,978,136,000 Rs.

$$\text{Variable non - fuel O\&M cost per year} = 1\% \times \text{Total capital cost}$$

$$= 0.01 \times 2{,}978{,}136{,}000\text{Rs.} = 29{,}781{,}360\,\text{Rs./year}$$

$$\text{Fixed operating cost per year} = 3\% \times \text{Total capital cost}$$

$$= 0.03 \times 2{,}978{,}136{,}000\,\text{Rs.} = 89{,}344{,}080\,\text{Rs./year}$$

$$\begin{aligned}\text{Annual fuel usage} &= \text{Total km driven per year} \times \text{Per km fuel consumption} \\ &= 104{,}420{,}000\,\text{km/year} \times 0.38\,\text{L/km} \\ &= 39{,}679{,}600\,\text{L/year}\end{aligned}$$

$$\begin{aligned}\text{Annual fuel cost} &= \text{Annual fuel usage} \times \text{Diesel price} \\ &= 39{,}679{,}600\,\text{L/year} \times 25\,\text{Rs./litre} \\ &= 991{,}990{,}000\,\text{Rs./year}\end{aligned}$$

$$\begin{aligned}\text{Annual labour cost} &= \text{Total drivers} \times \text{Driver availability} \times \text{Driver wages} \\ &\quad \times \text{Operating days} \\ &= 498\,\text{drivers} \times 12\,\text{hr/day} \times 50\,\text{Rs./hr} \times 350\,\text{days/year} \\ &= 104{,}790{,}000\,\text{Rs./year Cost of capital charges per year} \\ &= 12\% \times \text{Total capital cost } = 0.12 \times 2{,}978{,}136{,}000\,\text{Rs.} \\ &= 357{,}376{,}320\,\text{Rs./year}\end{aligned}$$

$$\begin{aligned}&\text{Total annual cost} = \text{Variable non-fuel O\&M cost} + \text{Fixed operating cost}\\&\quad + \text{Annual fuel cost} + \text{Annual labour cost} + \text{Capital charges} = 29{,}781{,}360\\&\quad + 89,344{,}080 + 991{,}990{,}000 + 104{,}790{,}000 + 357{,}376{,}320\\&= 1{,}573{,}281{,}760\,\text{Rs./year}\end{aligned}$$

$$\begin{aligned}&\text{Transportation cost per kg of hydrogen}\\&\quad = \text{Total annual cost/Annual hydrogen production}\\&\quad = 1{,}573{,}281{,}760\,\text{Rs./year}/47{,}250{,}000\,\text{kg/year} = 33.29\,\text{Rs./kg}\end{aligned}$$

Final Answers

Metric	Value
Number of trips per year	261,050 trips
Total km driven per year	104,420,000 km
Total drive time per year	2,088,400 h
Total load and unload time per year	522,100 h
Trucks required	311 trucks
Drivers required	498 drivers
Total capital cost	2,978,136,000 Rs
Variable non-fuel O&M cost per year	29,781,360 Rs
Fixed operating cost per year	89,344,080 Rs
Annual fuel usage	39,679,600 L
Annual fuel cost	991,990,000 Rs
Annual labour cost	104,790,000 Rs
Cost of capital charges per year	357,376,320 Rs
Total annual cost	1,573,281,760 Rs
Transportation cost per kg of hydrogen	33.29 Rs./kg

Solved Problem 3: Hydrogen Transportation Cost Analysis

Exam Question: Hydrogen Transportation Cost Analysis

A transportation company is evaluating the cost of delivering hydrogen through a pipeline. Using the data below, calculate the hydrogen transportation cost per kg, considering all relevant capital, operating, and energy costs. Show all intermediate calculations, including:

- Compressor power to overcome friction losses
- Annual energy consumption
- Compressor capital cost

- Total direct capital cost (compressor + pipeline)
- General facilities cost
- Engineering, permitting and start-up costs
- Contingencies
- Working capital, land and miscellaneous costs
- Total capital cost
- Variable non-fuel O&M cost
- Fixed operating cost
- Annual energy cost
- Cost of capital charges
- Total annual cost
- Cost per kg of hydrogen

Given Data

Parameter	Unit	Value
General facilities	% of capital cost	15
Engineering, permitting and start-up	% of capital cost	8
Working capital, land and misc	% of capital cost	4
Diesel price	Rs./litre	25
Delivery distance	km (one-way)	200
Contingencies	% of capital cost	6
Non-fuel variable O&M cost	% of capital cost per year	1
Fixed O&M costs	% of capital cost per year	3
Capital charges	% per year	12
Energy cost rate	Rs./kWh	3
Compressor capital cost	Rs./kW	45,000
Compressor size	kW	4000
Pipeline cost	Rs./km	27,960,000
Pipe diameter	m	0.25
Delivery pressure	MPa	2
Temperature	K	283
Operating days	Days	350
Operating hours	Hr	24
Total hydrogen transported	TPD	150
Efficiency of the compressor		55%
Friction factor	f	≈ 0.015
Pipe roughness for steel	ε	≈ 0.000045 m
Dynamic viscosity of hydrogen at 283 K	μ	$\approx 8.8 \times 10^{-6}$ Pa·s

Solution: Hydrogen Transportation Cost Analysis

Step 1: Constants and Unit Conversions

- Total hydrogen transported: 150 TPD = 150,000 kg/day
- Annual hydrogen transported: 150,000 kg/day × 350 days = 52,500,000 kg/year
- Delivery distance: 200 km
- Operating hours per year: $350\,\text{days} \times 24\,\text{hr/day} = 8400\,\text{hr/year}$
- Mass flow rate: $150{,}000\,\text{kg/day} \div (24 \times 3600)\text{s/day} = 1.736\,\text{kg/s}$
- Delivery pressure: 2 MPa = 2×10^6Pa
- Gas constant for hydrogen, R = 4124 J/kg·K

Step 2: Velocity of Hydrogen in Pipeline

Using the ideal gas law, $P = \rho RT$, density is:

$$\rho = P/(\text{RT}) = (2 \times 10^6)/(4124 \times 283) = 1.714\,\text{kg/m}^3$$

Cross-sectional area of pipe, $A = \pi r^2 = \pi(0.25/2)^2 = 0.0491\ \text{m}^2$.
Mass flow rate = ρAV

$$V = \text{Mass flow rate}/(\rho A) = 1.736/(1.714 \times 0.0491) = 20.63\,\text{m/s}$$

Step 3: Compressor Power to Overcome Friction Losses

Friction pressure drop using the Darcy-Weisbach equation:

$$\Delta P = f(L/D)(\rho v^2/2)$$
$$\Delta P = 0.015 \times (200 \times 1000/0.25) \times (1.714 \times 20.63^2/2)$$
$$= 4{,}383{,}315\ \text{Pa}$$

Power required (before efficiency):

$$\text{Power} = (\text{Mass flow rate} \times \Delta P)/\rho$$

$$\text{Power} = (1.736 \times 4{,}383{,}315)/1.714 = 4{,}441{,}586\,\text{W} = 4441.6\,\text{kW}$$

Adjust for compressor efficiency (55%):

$$\text{Actual Power} = 4441.6/0.55 = 8075.6\ \text{kW}.$$

Note: Given the compressor size is 4000 kW, so we assume multiple compressors or capacity adjustment for this calculation.

Step 4: Annual Energy Consumption

$$\begin{aligned}\text{Energy} &= \text{Power} \times \text{Operating hours} = 8075.6\,\text{kW} \times 8400\,\text{hr}\\ &= 67{,}835{,}040\,\text{kWh/year}\end{aligned}$$

Step 5: Compressor Capital Cost

Using the actual power required:

$$\text{Cost} = 8075.6\,\text{kW} \times 45{,}000\,\text{Rs./kW} = 363{,}402{,}000\,\text{Rs.}$$

Step 6: Total Direct Capital Cost (Compressor + Pipeline)

Pipeline cost:

$$\text{Pipeline Cost} = 200\,\text{km} \times 27{,}960{,}000\,\text{Rs./km} = 5{,}592{,}000{,}000\,\text{Rs.}$$

Total direct capital cost:

$$\text{Total} = 363{,}402{,}000 + 5{,}592{,}000{,}000 = 5{,}955{,}402{,}000\,\text{Rs.}$$

Step 7: General Facilities Cost

$$\text{Cost} = 5\% \times 5{,}955{,}402{,}000 = 0.15 \times 5{,}955{,}402{,}000 = 893{,}310{,}300\,\text{Rs.}$$

Step 8: Engineering, Permitting and Start-up Cost

$$\text{Cost} = 8\% \times 5{,}955{,}402{,}000 = 0.08 \times 5{,}955{,}402{,}000 = 476{,}432{,}160\,\text{Rs.}$$

Step 9: Contingencies

$$\text{Cost} = 6\% \times 5{,}955{,}402{,}000 = 0.06 \times 5{,}955{,}402{,}000 = 357{,}324{,}120\,\text{Rs.}$$

Step 10: Working Capital, Land and Misc. costs

$$\text{Cost} = 4\% \times 5{,}955{,}402{,}000 = 0.04 \times 5{,}955{,}402{,}000 = 238{,}216{,}080\,\text{Rs.}$$

Step 11: Total Capital Cost

$$\begin{aligned}\text{Total} &= 5{,}955{,}402{,}000 + 893{,}310{,}300 + 476{,}432{,}160\\ &\quad + 357{,}324{,}120 + 238{,}216{,}080 = 7{,}920{,}684{,}660\,\text{Rs.}\end{aligned}$$

Step 12: Variable Non-Fuel O&M Cost

$$\begin{aligned}\text{Cost} &= 1\% \times 7{,}920{,}684{,}660 = 0.01 \times 7{,}920{,}684{,}660 \\ &= 79{,}206{,}847\,\text{Rs./year}\end{aligned}$$

Step 13: Fixed Operating Cost

$$\begin{aligned}\text{Cost} &= 3\% \times 7{,}920{,}684{,}660 = 0.03 \times 7{,}920{,}684{,}660 \\ &= 237{,}620{,}540\,\text{Rs./year}\end{aligned}$$

Step 14: Annual Energy Cost

$$\begin{aligned}\text{Cost} &= 67{,}835{,}040\,\text{kWh/year} \times 3\,\text{Rs./kWh} \\ &= 203{,}505{,}120\,\text{Rs./year}\end{aligned}$$

Step 15: Cost of Capital Charges

$$\begin{aligned}\text{Cost} &= 12\% \times 7{,}920{,}684{,}660 = 0.12 \times 7{,}920{,}684{,}660 \\ &= 950{,}482{,}159\,\text{Rs./year}\end{aligned}$$

Step 16: Total Annual Cost

$$\begin{aligned}\text{Total} &= 79{,}206{,}847 + 237{,}620{,}540 + 203{,}505{,}120 + 950{,}482{,}159 \\ &= 1{,}470{,}814{,}666\ \text{Rs./year}\end{aligned}$$

Step 17: Cost per kg of Hydrogen

$$\begin{aligned}\text{Cost per kg} &= \text{Total Annual Cost/Annual Hydrogen Transported} \\ &= 1{,}470{,}814{,}666/52{,}500{,}000 = 28.02\,\text{Rs./kg}\end{aligned}$$

Final Results

Component	Value (Rs. or Rs./year)
Compressor power	8075.6 kW
Annual energy consumption	67,835,040 kWh/year
Compressor capital cost	363,402,000
Total direct capital cost	5,955,402,000
General facilities	893,310,300
Engineering, permitting & start-up	476,432,160

(continued)

(continued)

Component	Value (Rs. or Rs./year)
Contingencies	357,324,120
Working capital, land & misc	238,216,080
Total capital cost	7,920,684,660
Variable non-fuel O&M cost	79,206,847
Fixed operating cost	237,620,540
Annual energy cost	203,505,120
Cost of capital charges	950,482,159
Total annual cost	1,470,814,666
Cost per kg of hydrogen	28.02 Rs./kg

References

1. Hydrogen Transport and storage cost report, Report by Department of Energy Security & Net Zero by UK Government, last accessed on 12-07-2023. Link to view: https://www.gov.uk/government/publications/hydrogen-transport-and-storage-cost
2. M.A. Khan, C. Young, D.B. Layzell, The techno-economics of hydrogen pipelines. Transit. Accel. Tech. Briefs **1**(2), 1–40 (2021). ISSN 2564-1379
3. Z. Xie, Q. Jin, G. Su, W. Lu, A review of hydrogen storage and transportation: progresses and challenges. Energies **17**, 4070 (2024)
4. K. Reddi, A. Elgowainy, N. Rustagi, E. Gupta, Techno-economic analysis of conventional and advanced high-pressure tube trailer configurations for compressed hydrogen gas transportation and refueling. IJHE **43**, 4428–4438 (2018)
5. M. Yang, R. Hunger, S. Berrettoni, B. Sprecher, B. Wang, A review of hydrogen storage and transport technologies. Clean Energy 7(2023), 190–216 (2023)
6. India Country status report on hydrogen and fuel cells, link to access the same is: https://dst.gov.in/sites/default/files/Country%20status%20report%20final%20Hydrogen.pdf. Last accessed on 12 July 2025
7. Z. Xia, H. Liu, H. Lu, Z. Zhang, Y. Chen, Study on catalytic properties and carbon deposition of Ni-Cu/SBA-15 for cyclohexane dehydrogenation. Appl. Surf. Sci. **422**, 905–912 (2017). https://doi.org/10.1016/j.apsusc.2017.04.245
8. Y. Ni, Z. Han, Y. Chai, G. Wu, L. Li, Catalytic hydrogen storage in liquid hydrogen carriers. EES Catalysis **1**, 459–494 (2023). https://doi.org/10.1039/D3EY00076A
9. N. Brückner, Evaluation of industrially applied heat-transfer fluids as liquid organic hydrogen carrier systems. ChemSusChem (n.d.). https://doi.org/10.1002/cssc.201300426. Accessed 12 July 2025
10. M. Yang, Y. Dong, S. Fei, H. Ke, H. Cheng, A comparative study of catalytic dehydrogenation of perhydro-N-ethylcarbazole over noble metal catalysts. Int. J. Hydrogen Energy **39**, 18976–18983 (2014). https://doi.org/10.1016/j.ijhydene.2014.09.123
11. M. Nielsen, E. Alberico, W. Baumann, H.J. Drexler, H. Junge, S. Gladiali et al., Low-temperature aqueous-phase methanol dehydrogenation to hydrogen and carbon dioxide. Nature **495**, 85–89 (2013). https://doi.org/10.1038/nature11891
12. A. Monney, E. Barsch, P. Sponholz, H. Junge, R. Ludwig, M. Beller, Base-free hydrogen generation from methanol using a bi-catalytic system. Chem. Commun. **50**, 707–709 (2014). https://doi.org/10.1039/C3CC47306F

13. S.G. Jadhav, P.D. Vaidya, B.M. Bhanage, J.B. Joshi, Catalytic carbon dioxide hydrogenation to methanol: a review of recent studies. Chem. Eng. Res. Des. **92**, 2557–2567 (2014). https://doi.org/10.1016/j.cherd.2014.03.005
14. Y. Maenaka, T. Suenobu, S. Fukuzumi, Catalytic interconversion between hydrogen and formic acid at ambient temperature and pressure. Energy Environ. Sci. **5**, 7360–7367 (2012). https://doi.org/10.1039/C2EE03315A
15. J.F. Hull, Reversible hydrogen storage using CO_2 and a proton-switchable iridium catalyst in aqueous media under mild temperatures and pressures. Nat. Chem. (n.d.). https://www.nature.com/articles/nchem.1295. Accessed 12 July 2025
16. S.F. Hsu, A rechargeable hydrogen battery based on Ru catalysis. Angew. Chem. Int. Ed. (n.d.). https://doi.org/10.1002/anie.201310972. Accessed 12 July 2025
17. K.C. Park, D.J. Yim, S.K. Ihm, Characteristics of Al-MCM-41 supported Pt catalysts: effect of Al distribution in Al-MCM-41 on its catalytic activity in naphthalene hydrogenation. Catal. Today **74**, 281–290 (2002). https://doi.org/10.1016/S0920-5861(02)00024-X
18. K. Thomas, C. Binet, T. Chevreau, D. Cornet, J.-P. Gilson, Hydrogenation of toluene over supported Pt and Pd catalysts: influence of structural factors on the sulfur tolerance. J. Catal. **212**, 63–75 (2002). https://doi.org/10.1006/jcat.2002.3780
19. M.D. Irfan Hatim, M.A. Umi Fazara, A. Muhammad Syarhabil, F. Riduwan, Catalytic dehydrogenation of methylcyclohexane (MCH) to toluene in a palladium/alumina hollow fibre membrane reactor. Procedia Eng. **53**, 71–80 (2013). https://doi.org/10.1016/j.proeng.2013.02.012
20. S. Yolcular, Ö. Olgun, Ni/Al_2O_3 catalysts and their activity in dehydrogenation of methylcyclohexane for hydrogen production. Catal. Today **138**, 198–202 (2008). https://doi.org/10.1016/j.cattod.2008.07.020
21. S. Hodoshima, S. Takaiwa, A. Shono, K. Satoh, Y. Saito, Hydrogen storage by decalin/naphthalene pair and hydrogen supply to fuel cells by use of superheated liquid-film-type catalysis. Appl. Catal. A **283**, 235–242 (2005). https://doi.org/10.1016/j.apcata.2005.01.010
22. P.G. Campbell, L.N. Zakharov, D.J. Grant, D.A. Dixon, S.-Y. Liu, Hydrogen storage by boron−nitrogen heterocycles: a simple route for spent fuel regeneration. J. Am. Chem. Soc. **132**, 3289–3291 (2010). https://doi.org/10.1021/ja9106622
23. S.-Y. Liu, Hydrogen storage by novel CBN heterocycle materials final report DOE Award No. DE-FG36-08GO18143 Project Period : September 1, 2008–September 30 2012 Recipient: University of Oregon Department of Chemistry 1253 University of Oregon Eugene, OR 97403-2012
24. D. Forberg, T. Schwob, M. Zaheer, M. Friedrich, N. Miyajima, R. Kempe, Single-catalyst high-weight% hydrogen storage in an N-heterocycle synthesized from lignin hydrogenolysis products and ammonia. Nat. Commun. **7**, 1–6 (2016). https://doi.org/10.1038/ncomms13201

Chapter 19
Hydrogen Sensing

Hydrogen has properties such as high flammability, a wide flammability limit, a low ignition energy, and being the smallest molecule makes hydrogen much more prone to leak. It can leak from various joints, valves, fittings, storage systems, and from various equipment. Thus, the early identification of leaks, their detection, and subsequent sensing are extremely important. This detection can occur in small amounts, i.e., traces like ppm or ppb levels, when the application is identifying impurities of hydrogen in other gaseous mixture, or it could be a detection at low concentration as low as 4 vol% of hydrogen, to avoid the formation of a flammable mixture with air. Hydrogen detection in large concentrations close to even 100 vol% is also required for applications where the concentration needs to be monitored or controlled according to the process or application requirements. Thus, detection and monitoring make sensing essential to avoid hazardous situations, to make the product or process useful, and, above all, with the increasing hydrogen market, it will become a must in the entire hydrogen energy value chain, including production, storage, transportation, and utilization.

The use of sensor is to identify the presence of hydrogen or measure its concentration based on the application. The hydrogen sensors are required for a variety of applications, as mentioned below:

- Metallurgical processes—like galvanization, etc.
- Synthesis of ammonia and methanol—to identify any leaks,
- hydrogenation of hydrocarbons—hydrogen is used as a feedstock, so any form of leakage needs to be identified, and the content of hydrogen must also be determined,
- desulfurization of hydrocarbons—hydrogen is used as feedstock, so any form of leakage needs to be identified, and the content of hydrogen must also be determined,
- production of rocket fuels,
- space applications—being used for several decades,
- use as a coolant in turbines,

P. Sharma, *A Textbook on Hydrogen Energy*,
https://doi.org/10.1007/978-981-95-4601-5_19

- in transformers—oil degradation may result in the formation of hydrogen that needs to be detected for early fault detection,
- lightening industry—hydrogen acts as a contaminant,
- semiconductor industry—hydrogen is used for passivation,
- nuclear reactor safety—to identify hydrogen present and take control measures,
- coal mines—to avoid hazardous situations on release of hydrogen from methane or from coal dust explosions.

Requirements for sensors

There are several requirements that these sensors should meet, as mentioned below:

- Detection in a wide concentration range, from as low as 0.01–10% (to identify leaks and avoid hazardous situations) and as high as 100% (where the only gas required is hydrogen, e.g., fuel cells).
- Accuracy should be high, and the detection should be precise; i.e., sensitivity should be high.
- No interference with any other gases other than hydrogen, cross-sensitivity with other gases, as well as humidity should be low.
- Operable under a stringent set of conditions, even at high temperature (in processes like thermochemical, etc.) or high pressure (Type III or Type IV tanks) or in the presence of gas chlorides.
- The output signal from the sensors should have low noise and should give a stable signal.
- Response time should be very low, preferably less than a second, and the recovery time should be as low as possible.
- Should be small, compact, robust.
- It should have a long life.
- Have low power consumption.
- The design and size should be such that they can be easily integrated for a variety of applications.
- Easy to calibrate, operate, and maintain.
- Low in cost.

Hydrogen Sensing Technologies

Certain analytical equipment, such as a gas chromatograph, mass spectrometer, IR spectroscope, UV spectroscope, and chemiluminescence spectrometer, can be used to identify even trace amounts of hydrogen. These techniques have a good response and detection range. However, the biggest drawback of such techniques is that they are large in size and heavy in weight, have high capital costs, require trained manpower to operate and maintain, and are also not portable. Thus, the analytical equipment mentioned is not very useful when continuous monitoring is required. A hydrogen

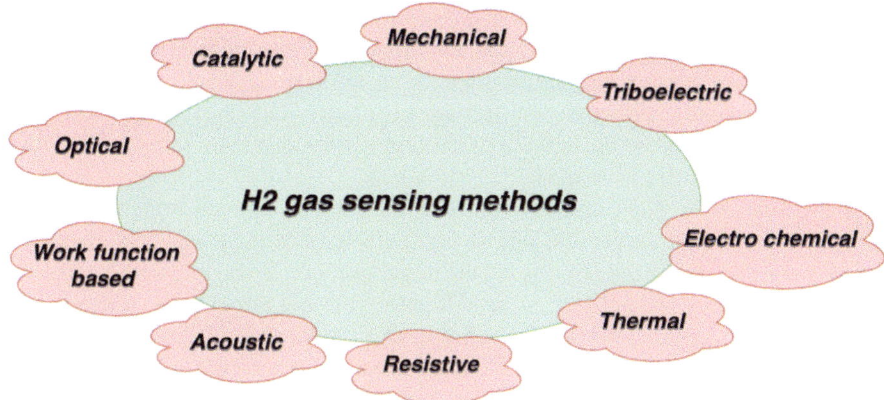

Fig. 19.1 Different hydrogen sensing technologies

sensing device needs to be simple to operate, small and compact in size, which can be used anywhere, and should be cost-effective.

Principle Behind Gas Sensing

Any material or sensing device when exposed to hydrogen undergoes a change in property, and this quantifiable change in property can be calibrated against the hydrogen concentration to sense hydrogen presence; this is the basic principle of operation for hydrogen or any gas sensors. Depending upon which property of the material changes on being exposed to hydrogen, there are different types of sensors as indicated in Fig. 19.1. Besides the sensing material or device, the change in the property needs to be converted into a measurable signal, i.e., electrical signal through a transducer, which could be measured and then analyzed to determine the quantity of hydrogen.

Based on the methods of hydrogen sensing, the sensors can be classified into mechanical sensors, triboelectric sensors, electrochemical sensors, thermal sensors, resistive sensors, acoustic sensors, work function-based sensors, optical sensors, and catalytic sensors.

Thermal Sensors

The principle behind the operation of a thermal sensor is the change in heat transfer in the presence of hydrogen, and the property considered in these sensors is thermal conductivity. The thermal conductivity of hydrogen is higher than that of air. At 23 °C, the thermal conductivity values for hydrogen and air are 0.174 W/mK and 0.026 W/

mK, respectively. The extent of heat loss from a sensing element will depend on the thermal conductivity of the surrounding environment. For example, if there is more hydrogen in the surrounding environment, then the heat dissipation will increase because the thermal conductivity of hydrogen is higher. That means higher the heat dissipation, higher will be the amount of hydrogen present, and this can be calibrated to identify and quantify hydrogen.

The operation of such sensors is based on the Wheatstone bridge principle, wherein there are two resistors: one is for the reference gas, which is air and is sealed into a cell, while the other is for the target gas, i.e., hydrogen, and is exposed. These two resistors are housed in the reference cell and target cells, respectively. They are connected in a Wheatstone bridge, as shown in Fig. 19.2. When there is only air and no hydrogen in the surrounding area, since both the reference resistor and the target resistor are exposed to air, both resistors dissipate an equal amount of heat, and the Wheatstone bridge remains balanced. However, when hydrogen is present in the surrounding environment, the heat loss of the target cell is higher in the presence of hydrogen than the heat loss from reference resistor housed in reference cell. A difference in heat loss will result in a higher temperature of the reference bead compared to the target bead, leading to a difference in the resistance between the two, thereby causing imbalance in the Wheatstone bridge. This difference can be measured and be calibrated to detect or sense the hydrogen leaks.

Another variant of the thermal sensor is one where, instead of a reference cell, there are two resistive elements, with one at a higher temperature and another at a lower temperature, such that the difference between the two is kept constant. Since

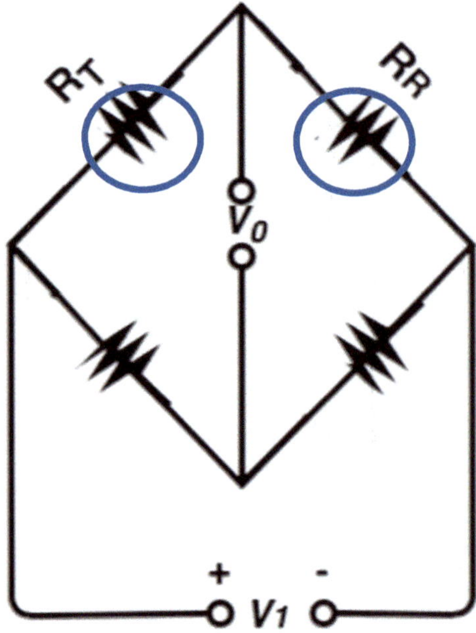

Fig. 19.2 Schematic of a thermal sensor

heat will be transferred from the hot to the cold element, and in the presence of hydrogen, the heat loss will be greater. Thus, to maintain the temperature of the hot element, some extra power must be supplied, which can be used as a measure and calibrated to determine the hydrogen gas concentration.

Thermal sensors have a wide detection limit, which can lie in the range of < 1 to 100% hydrogen; the 100% hydrogen means no air or oxygen present. However, the challenge lies in accurate detection at the lower limit (< 0.2%). The cross-sensitivity from other gases is very low and has been shown to have a low poisoning effect (due to absence of metals that act as catalysts) in the presence of other gases or hydrogen. The life of such sensors is high, i.e., more than five years of operation. However, the disadvantage is a slow response time of around 5 seconds; besides, if there is any change in the ambient temperature, such sensors are impacted by it and a temperature correction is absolutely necessary. Another disadvantage is the possibility of interference from any other gas with a similar thermal conductivity to that of hydrogen, such as helium, argon, methane, or carbon monoxide. Thermal sensors have a long history of use, dating back to as early as 1913 for airships filling [1]. Recent research is focused on reducing the size of these sensors through the use of micromechanical fabrication techniques. Such devices require lower power, have lower costs, and offer faster response times.

Electrochemical Sensors

Electrochemical sensors work on the principle of charge transfer phenomena. The charge transfer takes place between the electrodes with an electrolytic medium in between. Another possibility is sensing the change in electrical properties that occur due to electrochemical reactions. Thus, based on these, the electrochemical sensors are of two types: amperometric or potentiometric type sensors, which depend on the type of signal output from these sensors. In the case of an amperometric sensor, a constant voltage is applied, and the diffusion-limited current is measured. The process is called diffusion-limited when the rate of diffusion of the gas onto the sensing electrode is slower than the rate of the reaction. Another type is a potentiometric sensor in which the electromotive force or potential difference is measured. Thus, such sensors are typically electrochemical devices with two electrodes: one of which serves as a sensing electrode and the other one as the counterelectrode; a third electrode acts as a reference electrode, as shown in Fig. 19.3.

In the amperometric sensor, the voltage is kept constant, and a potentiostat is connected to measure the diffusion-limited current. The electrodes are typically made up of noble metals, such as platinum, which facilitates the oxidation of hydrogen gas. The liquid electrolyte facilitates the charge transfer process. An electrolyte can be a solid, such as an acidic electrolyte, the most common being Nafion. A gas permeable layer, also known as a diffusion layer, serves the purpose of limiting diffusion. The diffusion layer serves multiple purposes, permeating the analyte while selectively allowing hydrogen gas through it, while simultaneously limiting other gases to reduce

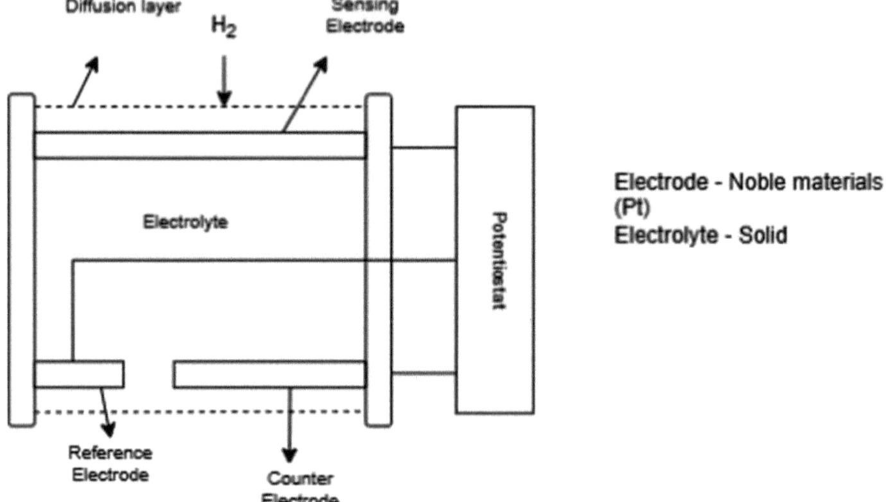

Fig. 19.3 Schematic of an amperometric sensor (Inspired from [1])

cross-interference from unwanted gases. It also prevents the electrolyte from drying and any leakage from the cell. This gas permeable layer selectively allows hydrogen and could be of Teflon, a perfluorinated polymer. Hydrogen gas gets oxidized at the sensing electrode and is represented as:

$$H_2 \rightarrow 2H^+ + 2e^- \quad (19.1)$$

The reduction of oxygen on the counterelectrode is given by the equation:

$$\frac{1}{2}O_2 + 2H^+ + 2e^- \rightarrow H_2O \quad (19.2)$$

According to Eq. (19.1), the flow of electrons occurs from the anode to the cathode, generating an electric current in the process. The value of the electrical current is directly proportional to the concentration of the hydrogen gas and is given by Faraday's law as:

$$I = zFQ \quad (19.3)$$

where Z is the number of electrons transferred per molecule, F is Faraday's constant (value is 96486.7 A-s/mol), and Q is the rate of conversion of hydrogen given in mol/s.

Thus, the current can be measured, which can be used to calibrate and find out the hydrogen concentration present in the environment. The temperature range in which amperometric sensors can be used is extensive, from – 20 °C to + 80 °C. At the lower temperatures, the important measure is that the electrolyte should not freeze. In this case, these sensors can be used even beyond 80 °C; in that case, ceramic electrodes should be used. However, the impact of temperature on the diffusion process and the electrochemical reaction needs to be considered. The impact of pressure is in terms of availability of hydrogen gas for the electrochemical reaction. The impact of humidity is on the proton conduction due to the water content present.

The advantages of amperometric sensors include a wide range of detection, from 5 ppm in presence of argon to 100% detection, low power consumption, and a long life of 5 years or more. However, the disadvantage is the longer response time (20–50 s); the major challenge is the requirement of low humidity levels. The cross-sensitivity from other gases, including CO or CO_2, can be prevented by an appropriate choice of gas permeable layer. Ideally, there should be no interference from other gases. The reported membranes include those made of platinum, palladium, fluorinated ethylene propylene, and tungsten carbide, among others.

Another type of electrochemical sensor is a potentiometric sensor. In this, the potential difference, or electromotive force, between the two electrodes (the sensing and the reference electrodes) is measured ideally under the conditions of zero current. Thus, the measured electrode potential in such sensors can be calibrated and can be used for finding the hydrogen gas concentration using the Nernst equation given below:

$$E = E_0 + \left[\frac{RT}{zF}\right] \ln\left(\frac{a}{a_0}\right) \tag{19.4}$$

where E_0 is the standard electrode potential, E is the electrode potential, a is the chemical activity of the analyte, and a_0 is the chemical activity of the reference electrode; the rest are same as described above. The chemical activity of the analyte is proportional to the hydrogen gas concentration. The structure/configuration of both the amperometric and potentiometric sensors remains the same; however, the significant difference is that in the case of potentiometric type of sensors, the signal is independent of the sensor size and geometry, which is not the case with amperometric sensors. The biggest advantage of obtaining a signal independent of sensor size and geometry is that the size can be reduced without compromising the sensitivity, thus making miniaturization possible. However, the major disadvantage of potentiometric sensors is the logarithmic variation of the electrode potential, which can lead to lower accuracy at higher hydrogen gas concentrations, unlike the linear response exhibited by amperometric sensors. Amperometric sensors are more widely used than potentiometric sensors. The electrodes are usually made up of noble metals like Pt, Pd, Au, Ag, and a solid electrolyte is used like the In-doped $CaZrO_3$, $SrYb_{0.05}Ce_{0.95}O_{3-x}$, α-alumina, polybenzimidazole-type polymer or even oxide-conducting yttria stabilized zirconia are being reported in literature.

The advantages of electrochemical sensors include that they consume lower power, usually operate at lower temperatures, and can have a wide range of detection, ranging from 10 ppm levels in nitrogen to 100% detection. The palladium used on ITO is reported to have a limit of detection of 0.1%. Tin oxide-based scaffolds have been used and reported to operate at a higher temperature of 450 °C while also having a lower limit of detection of 40 ppm. The disadvantage is a high response time in the range of 10–100 s, however, few literature reports mentioned a faster response time of 2 s. The advantage of such sensors is their very low cross-sensitivity, and they can be used in a variety of mediums, even in aqueous solutions or at very high temperatures, such as in molten metals.

Electrical Conductivity or Resistivity-Based Sensors

The hydrogen sensing property used for measuring hydrogen concentration in this class of sensors is the change in electrical conductivity, resistivity, or resistance of the material. The concept behind the operation of these types of sensors is that the electrical properties of the metal oxides change when exposed to reducing gases like hydrogen, and this change in properties can be linked to the hydrogen gas concentration. Usually, these sensors consist of a metal–oxide–semiconductor-type structure, wherein a substrate is present in the structure on which various other films are being deposited. Above the substrate, there can be a heating element which is used in case the metal oxides operate at higher temperatures, as shown in Fig. 19.4. The structure thus has an insulating layer and on the top of it is the metal oxide layer to which electrodes are deposited. The electrodes are usually made of noble metals, such as Au, Ag, and Pt. The insulating layer can be of Al_2O_3 or SiO_2, etc., while the metal oxide used can be ZnO, VO_2, WO_3, TiO_2, SnO_2, In_2O_3, FeO, Fe_2O_3, NiO, V_2O_5, Sb_2O_5, $SrCe_{0.95}Yb_{0.05}O_3$, etc.

The principle of operation of metal oxide-based sensors is that when the environmental oxygen gets adsorbed onto the surface of the metal oxides, then the following reaction is observed in the absence of hydrogen:

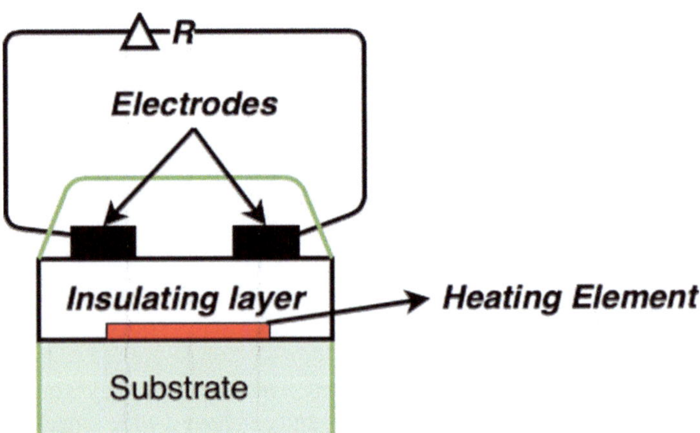

Fig. 19.4 Schematic of an electrical conductivity-based sensor

$$O_2 + 2e^- \rightarrow 2O^- \tag{19.5}$$

The surface oxygen takes up free electrons, and thus, the conductivity reduces, or the resistivity increases.

However, when hydrogen or the target gas is available in the surroundings, the adsorbed oxygen on the surface of metal oxides reacts with the adsorbed hydrogen molecule through the following reactions:

$$H_2 + O^- \rightarrow H_2O + e^- \tag{19.6}$$

This reaction of adsorbed oxygen with a hydrogen molecule results in the formation of water and releases the free electron, which increases the conductivity or reduces the resistance of the film. The resistance measured between the two electrodes or the current between these electrodes is calibrated such that the change in resistance or current provides a measure of the hydrogen gas concentration.

When metallic nanoparticles or films, such as those of Pd, form metal hydrides upon exposure to hydrogen, then the resistance of the film changes which is calibrated to the hydrogen gas concentration.

A digitized electrode is present above the metal oxide layer; the other variation of resistivity-based sensors is the printed circuit board, which uses a copper electrode. The future gas sensing electrodes could be on a flexible substrate, which can be inkjet printed to detect hydrogen gas. Micro-electro-mechanical system (MEMS) technology has been used recently to fabricate In or F-doped SnO_2 devices, which have been reported to be small in size, consume lower power, and are highly sensitive to hydrogen with very low cross-interference from other gases. Thus, the hydrogen concentration detected ranges from 10 ppm–2%.

The advantage of these electrical conductivity or resistivity-type sensors is that they are simple in design and easy to operate. The sensitivity of the sensors can be increased, and cross-inference from the other gases like CO, CH_4, C_3H_8, H_2S, etc., can be reduced by doping metal oxides with catalytic metals or by depositing a thin gas filtering layer to prevent the diffusion of gases other than hydrogen and also optimizing the operating temperature of the gas sensor.

The advantage of electrochemical sensors is that they have low power consumption and can be used at both room temperature and high temperatures. Electrochemical sensors are the most widely available type of sensor in the market, and most of these are based on tin oxide. The response time is higher and can range from 4–20 s; these sensors are affected by the temperature and humidity present.

Work Function-Based Sensors

Work function is the minimum energy required to remove an electron from the metal surface to infinity. In such sensors, a metal that is sensitive to hydrogen is deposited onto an oxide layer. The principle behind the operation of such sensors is the change in the work function of the metal when exposed to hydrogen. The hydrogen gets diffused and adsorbed between the metal and oxide layers, which leads to the polarization of the hydrogen atoms and a change in the work function of the metal. This change can be measured as a change in voltage and used to determine the hydrogen concentration. The structure of these sensors consists of three layers: a metal, an oxide layer, and a semiconductor layer, forming MOS structure. Based on the structure, there can be different types of such devices as mentioned below:

1. Metal–semiconductor Schottky diode device (MS or MIS),
2. metal–insulator–semiconductor transistor device (MOSFET or MISFET),
3. metal–insulator–semiconductor capacitor device.

1. The **Schottky type** sensor has a metal layer deposited over a semiconductor substrate or an oxide layer. The fermi level of a semiconductor aligns with the Fermi level of the metal, by the difference in work functions between the two materials is called the Schottky barrier height. The alignment also occurs when there is thin insulating layer present between the metal and the semiconductor forming a metal–insulator–semiconductor type of structure as shown in Fig. 19.5. The metals which are known to be most sensitive toward hydrogen are Pd or Pt, when these are exposed to hydrogen than the hydrogen molecule dissociates into hydrogen atom, diffuse to the interface between metal and insulator/semiconductor, get polarized to form a sort of dipole layer which results into the change in the Schottky barrier height or the work function of the metal. The change in work function of the metal changes the I-V characteristics curve, resulting in a change in voltage at constant current; these effects can be measured as a function of the hydrogen gas concentration. Thus, the voltage change created adds to the externally applied voltage, and this ΔV can be calibrated or used as a measure of gas concentration. The semiconductors used in such Schottky type of sensors are CdS, Si, SiC, InP, GaN, InAlP, among others. The choice of semiconductor depends on the operating temperature. In addition to Pt and Pd, other possible metals include Ru, Au, Ag, PdAg or Ni, etc. Literature reports indicate that MIS/MOS structures have higher sensitivity than MS, as the Fermi level pinning effect is absent in these structures. The most widely used insulator material is SiO_2, while the other materials reported include WO_3, HfO_2,, etc. The sensitivity of such sensors is high, allowing them to detect hydrogen in very low concentrations, even up to 10 ppm to 30%, for concentration monitoring purposes. The other possible advantage is a faster response time of roughly 1 s and a longer lifespan.
2. MISFET/MOSFET devices are based on FETs to convert the change in work function of metal into an electrical signal, which can be calibrated to measure hydrogen gas concentration. The device has three layers: a metal layer, typically a hydrogen sensitive metal like Pt or Pd or their alloys; an oxide layer, usually SiO_2; and a substrate, usually silicon. The metal layer acts as gate, and the source and drain regions are located on the silicon substrate as shown in Fig. 19.5. The source is grounded for reference, and positive bias is applied to the gate, which

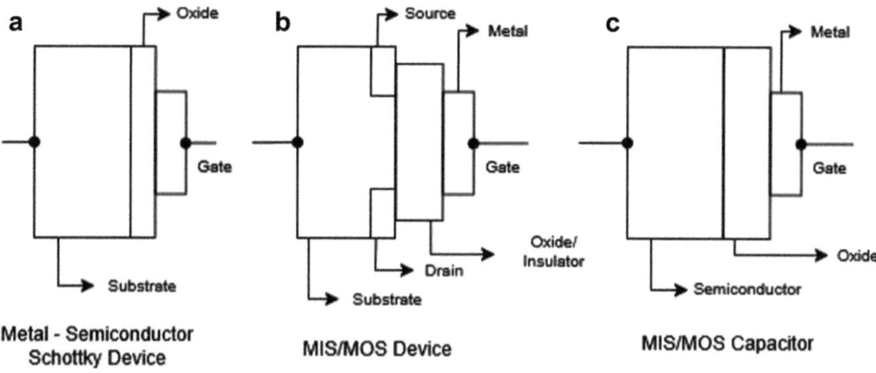

Fig. 19.5 Schematic of work function-based sensors **a** M-S type, **b** MIS/MOS type, **c** MIS/MOS capacitor type sensor (Inspired from [1])

creates a channel between source and drain through which current flows, which can be varied by changing the potential on the gate.

The principle of operation is the same as mentioned for MS device. A hydrogen molecule dissociates into hydrogen atom, which diffuses to the interface and gets polarized, thereby changing the work function of the metal giving a measurable voltage signal which corresponds to the hydrogen gas concentration. In order to avoid humidity effect and variations due to changes in ambient temperature, these sensors can be operated at higher temperatures, providing a better response. However, this heating is undesirable for many applications, and at the same time, power consumption also increases, which can be reduced by heating only selective regions in the presence of hydrogen. Besides a gap between metal and insulator that allows easy access of gas is called suspended gate technology, allowing better sensitivity, measurements possible at room temperature, and the use of other metals other than Pd, like Ir or Pt, is possible. However, for applications demanding high-temperature operations or those under corrosive environments, SiC or GaN which are wide bandgap semiconductors are used. These materials provide a wide range of detection, from 10 ppb in nitrogen to even as high as 50%.

There are numerous literature reports on further enhancing performance and mitigating the influence of variation of ambient temperature, by using two different metals for two-gate device fabrication. Overall, MOSFET hydrogen sensors have a fast response time, reported to be as low as 1 ms to few seconds, a wide range of detection limits, and a long life of up to 10 years.

3. MIS/MOS capacitor

The MOS/MIS sensors are based on change in capacitance to measure hydrogen gas concentration. The structure of such a device is similar to the above-mentioned devices, as shown in Fig. 19.5. The only difference is the presence of thicker oxide or insulating layer to prevent current conduction and allow the buildup of charges. A semiconductor substrate has a layer of oxide, and on the top of it is the hydrogen sensitive metal layer. When hydrogen is present in the environment, it gets absorbed, and the corresponding reaction at the metal oxide interface results in a change in capacitance. This observed change is recorded in the C-V plot with the change in flat band voltage observed to be proportional to the hydrogen gas concentration.

According to literature reports, the gate in such devices can be of Pt or Pd, where Pd is used for detecting hydrogen at lower concentrations, while Pt is used for detecting hydrogen at higher concentrations. Another low-cost option and substitute for Pd or Pt is Ni, as per the reported literature studies. Ni-based MOS capacitors can detect hydrogen in the range of 50–1000 ppm with a response time of 20–50 s. With a Pd/Ni thin film MOS capacitor, the detection range is about 10ppm – 1% with a few seconds response time.

Different insulator materials studied include Al_2O_3, Ta_2O_5, Si_3N_4, etc. The other possible devices reported include structures like metal/ferroelectric/metal configuration with the ferroelectric being $(Ba,Sr)TiO_3$, amorphous or lead zirconate titanate.

Alternatives to Pt metal in such sensors could be Ru, Ni, Au, Ag, Ir, Pd–Ag; however, Pt is known to have better sensitivity and a faster response. The semiconductor used depends on the sensor operating temperature; for example, at a temperature of around 250 °C, Si is preferred. However, for higher temperatures, wider band gap semiconductors like SiC, InP, GaN or InAlP, etc., are preferred.

Mechanical Sensors

The change in the mechanical properties of metal when exposed to hydrogen is the principle behind the working of mechanical sensors. The metal like Pd, which is sensitive to hydrogen, is coated onto a microcantilever. When the hydrogen sensitive metal is exposed to hydrogen, the hydrogen molecule dissociates into hydrogen atoms, and those hydrogen atoms sit in the interstitial positions in the metal lattice, causing lattice expansion or change in the mechanical properties of the metal. This expansion causes the microcantilever to bend, which can be measured as a parameter to estimate hydrogen gas concentration. Pd film is coated only on one side of the cantilever, so that when hydrogen is absorbed, it creates stresses, as the substrate on which Pd is coated prevents the volume expansion. These stresses result in mechanical bending of the cantilever, causing curvature that can be calibrated to detect the hydrogen concentration. The Pd film, if it is porous, will increase the diffusivity, leading to increased gas interaction and thereby increasing the sensitivity while reducing response time. However, the biggest challenge with such sensors is that multiple cycling with absorption or desorption, i.e., expansion and contraction in the film, leads to delamination or removal of the Pd film due to loss of adhesion. There are different ways to improve the adhesion by fabricating trenches on the surface of

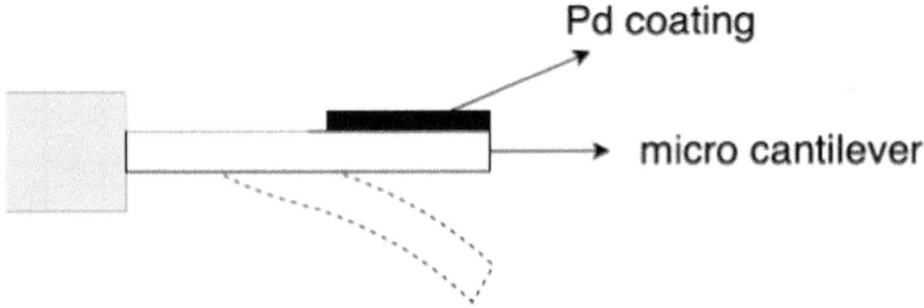

Fig. 19.6 Schematic of a mechanical sensor

cantilever, thus better designing, and the fabrication route effects the adhesion of the film.

The other possibility is that the cleaved side of the optical fiber is used to fabricate a micromachined cantilever. An intermediate thin layer (of Cr or Zr) can be deposited to increase the adhesion between the hydrogen sensitive metal film and the substrate used in the cantilever. Besides instead of using pure Pd, the use of an alloy of Pd, for example, Pd–Ag or Pd–Ni, can be deposited and is known to reduce delamination. The bending of the cantilever can be measured using optical methods.

Another technique used for hydrogen leak detection is explicitly required when no other gas is present in the system, which involves an uncoated silicon microcantilever (USMC), capable of sensing at a concentration as low as 0.02%. For the USMC, the working principle is that when there is a hydrogen leakage, an additional mass density is present in the surroundings, which causes a decrease in the resonance frequency of the silicon microcantilever. This can be calibrated to measure the hydrogen gas concentration (Fig. 19.6).

Optical Sensors

The optical properties of hydrogen sensitive materials change when exposed to hydrogen, and this change is measured to detect the concentration of hydrogen gas. Materials like metals or oxides are coated along the length of an optical fiber or onto the tip of the optical fiber. When these surfaces are exposed to hydrogen, it gets adsorbed and an expansion occurs radially and axially, which can be monitored to identify the gas concentration. The optical fibers used in these optical sensors are known as optodes or optrodes. Pd is the hydrogen sensitive material used, and a thin layer of Ti can be deposited between the Pd and the optical fiber to enhance the adhesion.

Different measurement techniques can be used to monitor or measure changes in optical properties when exposed to hydrogen.

1. Measurement of reflectivity—In this method, a hydrogen sensing material, usually Pd, is deposited on the cleaved portion of the optical fiber, wherein the

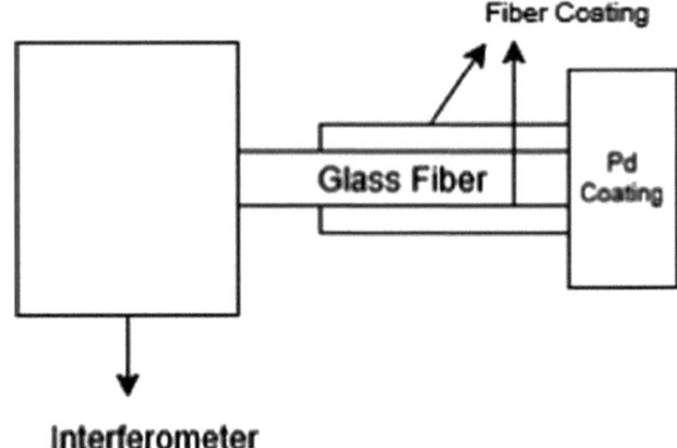

Fig. 19.7 Schematic of an optical sensor (Inspired from [2])]

glass fiber is centrally located. When the sensing element is exposed to hydrogen, its reflectivity changes and this change is used to detect hydrogen, which can then be finally correlated to the hydrogen gas concentration.

2. Interferometric measurement—The structure of the device remains same as with the case of reflectivity measurement; the cleaved region of the central optical fiber is coated with hydrogen sensitive material like Pd. When the device is exposed to hydrogen, the dimension or the refractive index of the Pd layer changes which causes a phase change in the light beam, and this can be measured using an interferometer. The change is correlated to the hydrogen gas concentration. The interferometers which can be used can be either Fabry–Perot or Michelson or Mach–Zehnder apparatus (Fig. 19.7).

3. Surface plasmon resonance—The surface plasmons are defined as the surface electromagnetic wave that propagates parallel to the interface between the metal and dielectric and is highly sensitive toward any structural changes of the metal surface. There are several ways to generate surface plasmons, including the use of a prism or grating coupler, optical fibers, or wave guides. However, few materials support surface plasmon resonance and are sensitive to hydrogen, such as Pd, Au, Ag, etc. When these materials are exposed to hydrogen, a change in the refractive index occurs; this change can be correlated with changes in the resonant wavelength, resonant angle of incident, or intensity of reflected light. The most commonly used hydrogen sensing material is Pd. However, the major issue with the hydrogen sensitive material films is blistering or delamination, which occurs due to repeated cycles leading to expansion and compression during hydrogen insertion and disinsertion into the interstitial sites on absorption and desorption of hydrogen in palladium. The delamination can be reduced by alloying Pd with other metals like Ni, where alloying with Ni not only increases stability of the film but also shifts the phase transformation toward higher concentrations. Advanced devices have been reported with use of chemochromic oxides like WO_3, Au, Ag as surface plasmon support materials and Pd as the sensitive material. Such sensors have a lower detection limit of even 0.04% of hydrogen in nitrogen.

4. Measurement of evanescent field interaction—Evanescent field refers to the electromagnetic field at the boundary of certain mediums or interfaces, like at the core of optical fiber and decays exponentially with distance from the core. The hydrogen sensitive material is deposited on the core after removing the cladding of optical fiber. When hydrogen is present in the surroundings, the interaction with the layer of hydrogen sensitive material leads to a change in the refractive index, resulting in a change in the evanescent field, consequently, a change in transmittance. The hydrogen sensitive material used includes Pd or Pd/Au.
5. Fiber Bragg grating—The cladding is removed, and the core of the optical fiber is etched to make a grating. The grating on the core modulates the light, resulting in the periodic modulation of the refractive index. The grating is then coated with hydrogen sensitive material. When exposed to hydrogen, the resulting expansion of the palladium and reflection of light from a grating of a specific wavelength result in a change in the Bragg wavelength. These sensors are affected by temperature. The other option is a long-period grating, which has a much more extended period of around 0.1–1 mm, compared to the fiber Bragg grating (which has same period as the optical wavelength). The long-period grating has higher sensitivity toward hydrogen and a lower impact of temperature compared to the fiber Bragg grating. Such sensors have faster response and better hydrogen gas sensitivity over the range of 0–4%.
6. Optical time domain reflectometry—Another type of optical sensor is the optical time domain reflectometry. In this method, the reflectance is measured along the entire length of the fiber optic cable, rather than at a particular point. This method allows for the use of a single device to sense over a larger area. The reflectance measurement is carried out time-resolved, and, considering the speed of light, the spatially resolved signal is utilized to measure hydrogen gas concentration, which can be used for sensing over a longer distance.
7. Use of chemochromic oxides—Chemochromic oxides are known to change color when their oxidation state changes, and this change occurs when exposed to hydrogen, which can be calibrated to measure the hydrogen gas concentration. The most commonly known chemochromic oxide is WO_3. The coating of WO_3 is done at an end of the optical fiber, WO_3 upon reacting with hydrogen, forms tungsten bronze and changes color due to a change in oxidation state.

In summary, the optical sensors have advantages such as faster response times and can be used without being exposed to certain oxidizers or flowing an electric current through them, which can initiate any hazardous situations and act as a source of accidents. The detection range of the optical sensors is broad, ranging from 0.1 to 100%. Wide area detection is possible with optical sensors.

A significant disadvantage of optical sensors is their cross-sensitivity to other gases, as well as interference from ambient light, which affects the detection of hydrogen. Additionally, they are also affected by humidity. However, interference from electromagnetic noise is not a concern with this type of sensor.

Acoustic Sensors

The acoustic sensors are based on the principle that the surface property of the piezoelectric material changes when exposed to hydrogen, generating a surface acoustic wave. These changes are monitored to measure the hydrogen gas concentration. Various devices can be used for detecting surface acoustic waves; the simplest and oldest technology is the quartz crystal microbalance.

Quartz crystal microbalance—A small disk of quartz with electrodes on both sides is printed and coated with hydrogen sensitive material (Pt, Pt-Ir, etc.). When a deformation is induced in the structure of the quartz crystal, it resonates. The resonance frequency changes and is related to the amount of gas adsorbed on the surface of quartz crystal and can be used to detect the hydrogen gas concentration by measuring the change in the resonance frequency (Fig. 19.8).

The biggest disadvantage of using QCM to detect hydrogen is the cross-sensitivity to other gases present in the environment, as well as the impact on precision due to changes in ambient temperature. The detection or response time is very high in the case of QCM. Several ways to reduce the cross-sensitivity have been reported, for example, by including materials that are only sensitive toward hydrogen, such as functionalized graphene oxide. An increase in temperature due to any reaction, like catalytic oxidation of hydrogen, can also change (increase) the resonance frequency and be used for detection.

Peizoelectric-Based Sensors—The change in hydrogen concentration in air alters the acoustic properties of materials. A piezoelectric sound resonance cavity can be used to monitor the resonance state. Such type of sensors have been reported in literature to have a low detection limit of even 8 ppm.

Another option is to use a surface acoustic wave on a piezoelectric wave for hydrogen sensing. The structure of such sensors consists of a piezoelectric material substrate with a hydrogen sensitive film and two interdigitized sets of electrodes or

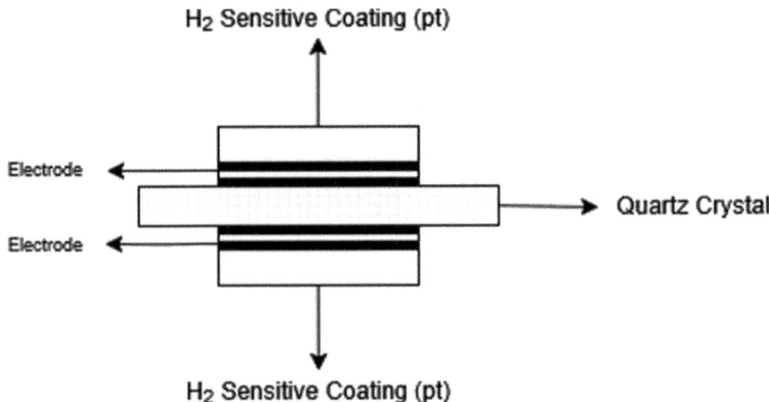

Fig. 19.8 Schematic of quartz crystal microbalance-based sensor (Inspired from [2])

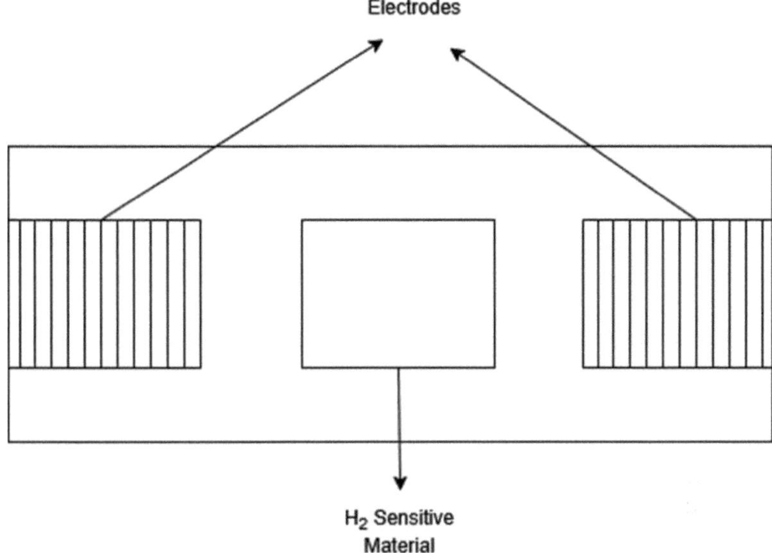

Fig. 19.9 Schematic of an acoustic type sensor (Inspired from [2])

transducers deposited on its surface. The role of one transducer is to convert the electrical signal applied to an acoustic wave, while another does the reverse, i.e., converts the wave to an electrical signal. When a hydrogen sensitive film is exposed to hydrogen in the environment, a change in surface acoustic wave causes change in the electrical signal. The surface acoustic wave is very sensitive to the surface properties. When the hydrogen sensitive material is exposed to hydrogen, leading to absorption of hydrogen, a change in output signal with respect to the input signal occurs during the surface acoustic wave propagation. This change is monitored to measure the hydrogen concentration (Fig. 19.9).

Different materials can be used for the construction of acoustic sensors. $LiNbO_3$ is the most widely used piezoelectric material, either Pd or Pd-doped SnO_2, which can be coated over $LiNbO_3$, as well as Pd-doped graphene or carbon nanotubes, Pd on WO_3, and Pd coated ZnO, can be used in SAW sensors. When hydrogen is adsorbed by Pd, its mass density and electrical conductivity change, which are considered for measurement to calibrate the hydrogen concentration in the environment. However, the biggest challenge is the repeatability. Between different SAW sensors, it is difficult to obtain consistent results due to the requirement for completely identical hydrogen sensitive material films on the surface, which is difficult to achieve.

Sound velocity measurements

The principle behind sound velocity-based hydrogen sensors is that the velocity of sound in hydrogen is much higher than the velocity of sound in air, which can be used as a measure of hydrogen detection in the environment. The velocity of sound in hydrogen is 1314 m/s while in air it is 346 m/s; this difference can be used as a means to measure hydrogen concentration. A sound generator is used to create acoustic waves by irradiating a sensitizer with a laser, and then a receiver is used to detect how much time the acoustic wave takes to propagate in the presence of different gas concentrations. Thus, the propagation time is calibrated against the gas concentration and can be used to detect hydrogen gas. When Pd is exposed to hydrogen, its mass density and mechanical properties change, and thereby, the propagation of sound wave in Pd also changes. The change in the speed of light while traveling through Pd can be measured to detect hydrogen. Temperature variation also affects the acoustic wave velocity and can lead to erroneous results, but that can be addressed by using a reference.

The acoustic sensors have several advantages, including high sensitivity, reliability, lower power consumption, fast response times, and a wide range of detection from ppm levels to 100%. But at the same time the disadvantage is that these sensors are sensitive toward the other sound waves or the vibrations. Additionally, there can be cross-sensitivity toward other gases and a temperature-dependent impact. So, the noise could be an interference in these sensors. They are unable to operate at high temperatures, and long-term stability is the biggest challenge; their use is therefore limited.

Catalytic Sensors

The working principle of the catalytic sensors is based on measuring the heat released during the oxidation of a certain combustible gas, such as hydrogen, on the surface of the catalyst. Hydrogen reacts with an oxidant exothermically, releasing a heat of combustion of 142 MJ/kg. The release of heat or the temperature change can be used as a means to sense the hydrogen gas concentration.

The most widely used catalytic gas sensors are the pellistor type. The construction of these sensors involves two Pt coils, which are covered by ceramic beads and connected in a Wheatstone bridge. Among the two beads, one is an active bead, and the other is an inactive bead. The active bead is coated with hydrogen sensitive material and is used for sensing hydrogen or temperature change, while the inactive bead acts a reference. When an electric current is passed through the elements inside the bead, the resistive losses cause heating of the ceramic beads to a temperature higher than RT. Initially, when both the ceramic beads are exposed to air, the Wheatstone bridge is balanced. If hydrogen is present in the environment, hydrogen gas molecules chemisorb on the hydrogen sensitive material, usually Pd, and are oxidized by the adsorbed surface oxygen, forming water. The heat released causes a temperature change and, in turn, the resistance of the active bead, which creates an imbalance in the Wheatstone bridge that can be measured to detect the hydrogen gas concentration. The pellistor type of sensors are the most widely used hydrogen sensors and has been conventionally used even long back in coal mines for detecting gases (Fig. 19.10).

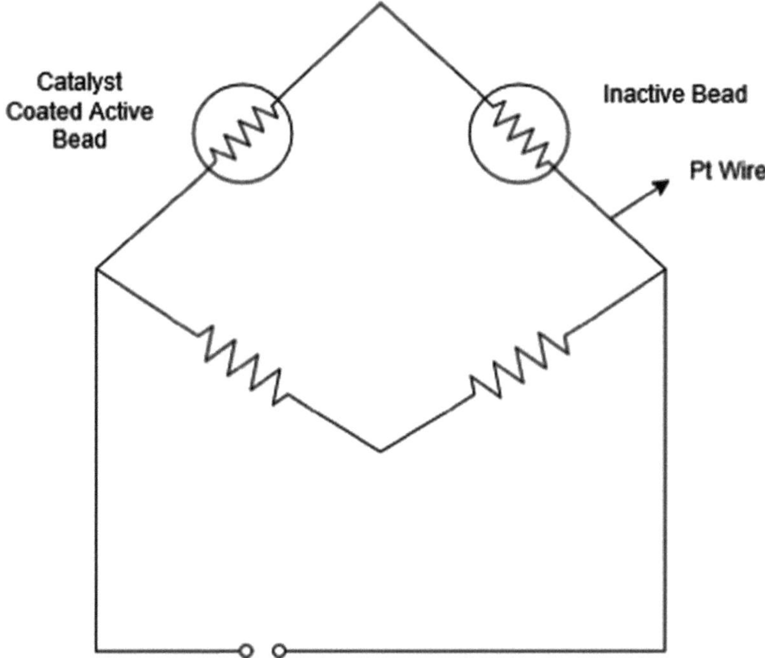

Fig. 19.10 Schematic of a catalytic sensor

Platinum- and Pt-based materials, such as platinum nanoparticles loaded graphene aerogel, have been studied and have shown a good sensitivity of 1.6% at 10,000 ppm of hydrogen gas, with a lower detection limit of 65 ppm, with a fast response time of 0.97 s and a reported power consumption of 2.2 mW.

The biggest advantage is since both beads are subjected to the same set of external conditions, the interference or variation due to ambient temperature changes or humidity gets mitigated. Such sensors are suitable for operating under a wide range of temperatures, pressures, and relative humidity. The lifetime of such sensors is approximately 3–5 years of operation. The major disadvantage of pellistor type of sensors is their cross-sensitivity to other hydrocarbons, and if any impurity is present in hydrogen, it can get strongly adsorbed on the surface of Pd, leading to a decrease in the surface active sites for oxidation to take place. Impurities in general, which are typically present in the hydrogen produced from hydrocarbons, include S, halogen, P or Si containing compounds which can poison these sensor materials. To increase the active sites, either the surface area of the active bead needs to be increased by increasing its size or making it porous, or else include filters which could prevent poisoning from other gases. Thus, the requirement for a larger size of active bead is another challenge that hampers its miniaturization. A high power requirement to heat the beads again is a disadvantage with such sensors. However, research is underway to address these issues, with studies on planar or microcatalytic sensors. Reports have shown a reduction in power consumption from around 0.5 W to as low as 30 mW. The detection limit of such sensors ranges from 0–4% and response time is around 1 s. The power consumption by sensors has been reported to further reduce with the use of conducting oxides in place of hydrogen sensitive materials. With a temperature rise, the conductivity change is monitored to detect hydrogen; the conducting oxide reportedly used is TiO_2.

Another class of catalytic sensors is the thermoelectric sensors, where the working principle is similar to pellistor-type sensors, with the difference that the thermoelectric effect generated due to the exothermic oxidation of hydrogen on the surface of a hydrogen sensitive material is used to detect hydrogen. The Seebeck or thermoelectric effect is the change in temperature between the two points of a metal or semiconductor leading to a voltage difference between these two points that can be measured. In Seebeck effect-based sensors, when the oxidation of a combustible gas occurs on the surface bead, it occurs exothermically, increasing the temperature thereby inducing thermoelectric voltage that can be measured and calibrated against the amount of hydrogen gas present in the surroundings. The change in voltage on temperature change between the active and reference points can be correlated to the hydrogen gas concentration as follows:

$$V = s\Delta T \tag{19.7}$$

where s is the Seebeck coefficient.

This can be related to the rate of combustion as follows:

$$V = s\Delta H.\left[k.\mathrm{e}^{\left(\frac{-E_a}{\mathrm{RT}}\right)}\right] \tag{19.8}$$

where E_a is the activation energy for the combustion reaction and ΔH is the heat of combustion, k is a constant.

The structure of such a thermoelectric sensor includes an insulating substrate, such as glass or alumina. A thick or thin film of thermoelectric materials like Si, SiGe, P-doped SiGe, alkali-doped NiO, $Si_3N_4/SiO_2/Si$, Bi_2Te_3/Al_2O_3, etc., is deposited on the reference film; half of this thermoelectric film is coated with hydrogen sensitive material, usually Pt or Pd. However, in order to increase the operating temperature (> 100 °C) to facilitate oxidation, the power requirement of such sensors is typically higher. With lot of research, different structures and devices have been proposed, like microheaters and micromachined devices for which the power consumption has been reported to be reduced to even 0.34 W. A wide range of detection has been reported in literature, with different materials being used for sensors ranging from 10 ppm to 3%. However, the major disadvantage is the slow response, with response time reported in the range of 2.5–55 s.

With the use of Pt nanoparticles, the response time is reported to reduce to < 150 ms and power consumption is as low as 0.22 mW and detection level of 10 ppm.

Triboelectric Sensors

The triboelectric effect is the process of charge separation that occurs when two dielectric surfaces come into contact and are separated periodically. The structure of such sensors includes dielectric materials with metallic electrodes deposited on the surfaces. The principle behind the operation of such sensors is that when the two such surfaces are brought together and then separated, one of the surfaces loses electrons and acquires positive charge, while the other acquires a negative charge. With such charges developed on the dielectric material's surface, the electrodes in contact acquire opposite charges, and thereby a potential difference is created that can be measured as current when connected to an external load. In the presence of air, the environmental oxygen adsorbs on the surface, capturing electrons and acquiring a negative charge. Consequently, the current is not observed, as the electrons are not available to produce the triboelectric effect. However, in the presence of hydrogen, the reaction of hydrogen with chemisorbed surface oxygen, after electrons are taken up, reacts to form water and the electrons on the oxygen ions are available to participate in the triboelectric effect. Thus, the electrons that are freed result in an increased surface charge density, giving rise to a voltage drop, and that output voltage can be measured and calibrated to the hydrogen gas concentration.

Pd-decorated ZnO nanorods with micron-sized pyramids of polydimethylsiloxane (PDMS) are reported to be used in a triboelectric sensor, where Pd is used as the hydrogen sensitive material. With Pd having lower work function, it allows electron transfer from ZnO to Pd, and micropyramids enable smooth contacting and breaking contact between the two surfaces. Similarly, Pd on ITO with polyethylene terephthalate films (PET) has also been shown to exhibit triboelectric properties and act as hydrogen sensor. The biggest advantage of triboelectric sensors is that they are self-powered devices with fast response; however, the major challenge that remains is the complex fabrication process and design.

In summary, different hydrogen sensors operate on different working principles, and based on these working principles, the sensitivity, the level of detection, operating parameters, power requirements, and complexity may vary. Thus, the choice of sensor to be used depends on the application or is highly application-specific. For example, metal oxide sensors are suitable for detecting low gas concentrations, while thermal conductivity sensors are more effective for higher gas concentrations. One hydrogen sensor cannot be used for all sets of applications. The electrochemical or pellistor type of sensors is suitable for operation under harsh environmental conditions and can be integrated with filling stations (SMART). A semiconductor gas sensor or a thermal conductivity-based sensor can provide a wide range of detection and a fast response, making them suitable for use with fuel cells (such as H_2-Semicon sensor or Xensor's sensors). A table of comparison of the different types of sensors described in this chapter is included (Table 19.1)

Table 1 .

S. No.	Type	Property	Measure	Detection Limit	Response Time	Materials	Challenges	Advantages	Ref
1	Thermal Sensors	Thermal Conductivity (K) difference	Heat loss from a body to surroundings	< 1 to even 100 % H2, difficult to detect at low range (0.2%)	<4s	Pt	Interfere with gases, e.g., He, Ar, CH4, CO and similar K,Impacted by varying T	Life> 5years, low power consumption ~5 mW, wide detection range, low response time,	3
2	Electrochemical	Charge transfer or charge transport or electrical properties - Amperometric sensor	diffusion-limited current at constant applied voltage	5ppm in Ar to 100%	20–50 sec	Electrode -Pt,Electrolyte – Solid (Nafion) or Liquid,Gas permeable layer - PTFE	Electrode -Pt,Electrolyte – Solid (Nafion) or Liquid,Gas permeable layer - PTFE	Wide detection range, low P consumption, wide T operation (-20^{0}C-80^{0}C), most widely used	4, 5
		Potential difference or EMF – Potentiometric sensors	Electrode potential or activity is related to H_2 gas concentration	Pd on ITO (0.1%), SnO2 based scaffold can operate at (40 ppm) high T	<2 sec	Electrodes -Pd, Pt, Au, Ag Electrolyte – αAl2O3, In-doped CaZrO3, phosphoro-silica glass, YSZ	Long response time	Signal independent miniaturization possible, low power consumption, wide temperature operation, wide detection range	6, 7

(continued)

Table 1 (continued)

S. No.	Type	Property	Measure	Detection Limit	Response Time	Materials	Challenges	Advantages	Ref
3	Electrical conductivity (metal oxide semiconducting type of structural)	Electrical conductivity or resistivity change	Change in Resistance or current	900 ppm,0.1% – 100% also reported	4–20s Even 75ms reported	Metal - Pt, Au, Ag, Cu Oxide - ZnO nanorods, SnO2, ZnO, V2O5, In2O3, TiO2, VO2, WO3, NiO, etc.	Low sensitivity, high response time, output depends on T & humidity, adsorbed surface oxygen impacts detection	Wide T detection from RT to higher T possible, with MEMS small size possible, lower power consumption, low cost	8, 9
4	Work function-based sensors – Metal semiconductor Schottky device	Change in the work function of metal and shifting of energy levels at M-I interface	measurable voltage change ΔV	10 ppm – 30%	1s	Metal -Pt, Pd, Au, Ag, IrPt Semiconductor- Si, CdS, SiC, InP, GaN, InAlP oxide- SiO2, WO3, HfO2 or Ga2O3	Fermi level pinning effect observed reducing detection of hydrogen	Better sensitivity, wide T and detection range possible, fast response, long life	10-14
	Work function-based sensors – Metal insulator semiconductor transistor device			10 ppb – 50%	1ms – <3s	Metal – Pt, Ir Insulator- SiO2 Semiconductor – Si, SiC, GaN	Influence of T and humidity observed on output signal, higher power consumption	Long life up to 10 years, wide T operation RT to 150^0C, low detection limit possible, fast response time,	15-18

(continued)

Table 1 (continued)

S. No.	Type	Property	Measure	Detection Limit	Response Time	Materials	Challenges	Advantages	Ref
	Work function-based sensors – Metal insulator semiconductor capacitor device			10–1000ppm in N2, 10ppm-100%	20–50s	Metal – Pt, Pd, Ni Insulator- SiO2, Al2O3, Si3N4, Ta2O5 Semiconductor – Si	T of operation 175 - 250^{0}C, output depends on adsorbed surface oxygen	High stability so longer life, high sensitivity	19-21
5	Mechanical Sensors	Physical Property change, bending on expansion of the hydrogen sensitive material	Volume expansion in Pd results in bending of cantilever, change in response frequency	50ppm–4%	30s	Pd, Cr-Pd,Pd Ag, Pd Ni Delamination issue can be addressed by alloying Pd and having an intermediate layer of another metal (Cr)	Complex fabrication, delamination of metal, high response time, output depends on humidity	Sensitivity improves, can be used under stringent operating conditions, very low power consumption	22, 23

(continued)

Table 1 (continued)

S. No.	Type	Property	Measure	Detection Limit	Response Time	Materials	Challenges	Advantages	Ref
6	Optical Sensors	Change in optical properties	1. Reflectivity change 2. Interferometric-change in refractive 3. Surface plasma resonance – change in refractive index 4. Evanescent field change 5. Chemochromic oxide - change color 6. Fiber brags grating - change in brags wavelength 7. Optical domain reflectometry	As low as 0.04% to 100%		Pd on optical fibers; Pd, Pt, Au, Ag on optical fibers; WO_3 MoO_3 as sensing material	Delamination of metal layer, Cross-sensitivity with other gases, impact of ambient temperature and outside light	Faster response, wide T operation -15 to 50^0C, can be used in harsh operating conditions	24-32

(continued)

Table 1 (continued)

S. No.	Type	Property	Measure	Detection Limit	Response Time	Materials	Challenges	Advantages	Ref
7	Acoustics sensors	Surface properties of piezoelectric material, surface acoustic wave	Resonance frequency – QCM; electrical signal-SAW; sound velocity measurement	10ppm–100% 0.1–100%	<2s <30s	QCM – Pt-Ir on quartz; SAW and sound velocity sensor:- Piezoelectric material - LiNbO3, LiTaO3, WO3 Hydrogen sensitive material – Pd, Pd-CNT, PD-ZnO nanorods	Sensitive to other sound waves or Vibration, cannot operate at high temperature, stability poor, cross-interface from other gases	High sensitivity & reliable,Low power consumption,Wide range of detection	33-36
8	Catalytic sensors	Heat released due to combustion can be measured - pellistor type	Change of temperature	1.6% sensitivity at 10,000ppm of H_2 gas, 5–95%	0.97s	Pt	High-temperature operation (300^0C)	low power consumption (2.2 mW), low Level of detection (65ppm) and negligible cross-sensitivity, wide T range of operation (-20 to 70^0C)	37-42
		Seebeck type T different Voltage generated	Change in voltage	10 ppm – 3%	<50s	Pt nanoparticles, SiGe, alkali-doped NiO	Higher power requirement, interference with other gases, higher response time	Wide Temperature operation (RT to 100^0C), high stability and long life, low power consumption (0.34W)	

(continued)

Table 1 (continued)

S. No.	Type	Property	Measure	Detection Limit	Response Time	Materials	Challenges	Advantages	Ref
9	Triboelectric	Charge separation	potential difference			PDMS pyramids on Pd/ZnO nanorods, Pd coated ITO on PET Films	Complex design and complex fabrication	Self powered devices, fast response	43, 44

References

1. T. Hübert, L. Boon-Brett, G. Black, U. Banach, Hydrogen sensors—a review. Sens. Actuators B **157**, 329–352 (2011)
2. P.S. Chauhan, S. Bhattacharya, Hydrogen gas sensing methods, materials, and approach to achieve parts per billion level detection: a review. Int. J. Hydrog. Energy **44**(47), 26076–26099 (2019)
3. L. Boon-Brett, G. Black, P. Moretto, J. Bousek, A comparison of test methods for the measurement of hydrogen sensor response and recovery times, Int. J. Hydrog. Energy 357652–7663 (2010)
4. Y. Chao, S. Yao, W. J. Buttner, J. R. Stetter, Amperometric sensor for selective and stable hydrogen measurements, Sensors and Actuators B. Chem. **106** 784–790 (2005)
5. C. Ramesh, N.Murugan, M.V.Krishnaiah, V. Ganesan, G. Periaswami, Improved nafion-based amperometric sensor for hydrogen in argon, J. solid. state. electrochem. **12**1109-1116 (2008)
6. Y. Okuyama, et al., A new type of hydrogen sensor for molten metals usable up to 1600 K, Electrochim. Acta **55** 470–474 (2009)
7. M. Nogami, M. Matsumura, Y. Daiko, Hydrogen sensor prepared using fast proton-conducting glass films, Sens. Actuators B **120** 266–269 (2006)
8. B. Hoffheins, et al., Development of low cost sensors for hydrogen safety applications, Proc. of the 1999 U.S. DOE Hydrogen Program Rev. NREL/CP-570–269 (1999)
9. R.K. Joshi, S. Krishnan, M. Yoshimura, A. Kumar, Pd nanoparticles and thin films for room temperature hydrogen sensor, Nanoscale Res. Lett. **4** 1191–1196 (2009)
10. C.T. Lee, J.T. Yan, Sensing mechanisms of Pt/beta- Ga2O3/GaN hydrogen sensor diodes, Sens. Actuators B: Chem. **147** 723–729 (2010)
11. S. Kandasamy, A. Trinchi, W. Wlodarski, E. Comini, G. Sberveglieri, Hydrogen and hydrocarbon gas sensing performance of Pt/WO3/SiC MROSiC devices, Sens. Actuators B: Chem. **111** 111–116 (2005)
12. W.M. Tang, C.H. Leung, P.T. Lai, Improved sensing characteristics of MISiC Schottky-diode hydrogen sensor by using HfO2 as gate insulator, Microelectron. Reliab. **48** 1780–1785 (2008)
13. T.J. Anderson, H.T. Wang, B.S. Kang, F. Ren, S.J. Pearton, A. Osinsky, A. Dabiran, P.P. Chow, Effect of bias voltage polarity on hydrogen sensing with AlGaN/GaN Schottky diodes, Appl. Surf. Sci. **255** 2524–2526 (2008)
14. X. Yu, C. Li, Z.N. Low, J. Lin, T.J. Anderson, H.T. Wang, F. Ren, Y.L. Wang, C.Y. Chang, S.J. Pearton, C.H. Hsu, A. Osinsky, A. Dabiran, P. Chow, C. Balaban, J. Painter, Wireless hydrogen sensor network using AlGaN/GaN high electron mobility transistor differential diode sensors, Sens. Actuators B: Chem. **135** 188–194 (2008)
15. K. Yokosawa, K. Saitoh, S. Nakano, Y. Goto, K. Tsukada, FET hydrogen-gas sensor with direct heating of catalytic metal, Sens. Actuators B: Chem. **130** 90–99 (2008)
16. K. Scharnagl, et al., Low temperature hydrogen detection at high concentrations: comparison of platinum and iridium, Sens. Actuators B: Chem. **80** 163–168 (2001)
17. T.J. Anderson, H.T. Wang, B.S. Kang, F. Ren, S.J. Pearton, A. Osinsky, A. Dabiran, P.P. Chow, Effect of bias voltage polarity on hydrogen sensing with AlGaN/GaN Schottky diodes, Appl. Surf. Sci. **255** 2524–2526 (2008)
18. R. Loloee, B. Chorpening, S. Beer, R.N. Ghosh, Hydrogen monitoring for power plant applications using SiC sensors, Sens. Actuators B: Chem. **129** 200–210 (2008)
19. J.H. Song, W. Lu, Operation of Pt/AlGaN/GaN-heterojunction field-effect transistor hydrogen sensors with low detection limit and high sensitivity, IEEE Electron Device Lett. **29** 1193–1195 (2008)
20. C. Lu, Z. Chen, K. Saito, Hydrogen sensors based on Ni/SiO2/Si MOS capacitors, Sens. Actuators B: Chem. **122** 556–559 (2007)
21. W. Zhu, O.K. Tan, X. Yao, Amorphous ferroelectric (Ba0.67Sr0.33)Ti1.02O3 thin films with enhanced H2 induced interfacial polarization potential, J. Appl. Phys. **84** 5134–5139 (1998)
22. J. Deng, W. Zhu, O.K. Tan, X. Yao, Amorphous Pb(Zr, Ti)O3 Thin film hydrogen gas sensor, Sens. Actuators B: Chem. **77** 416–420 (2001)

23. D.R. Baselt, et al., Design and performance of a microcantilever-based hydrogen sensor, Sens. Actuators B: Chem. **88** 120–131 (2003)
24. D.J. Kirby, et al., A differential capacitive thin film hydrogen sensor, Sens. Actuators B: Chem. **141** 424–430 (2009)
25. W. Zhu, O.K. Tan, X. Yao, Amorphous ferroelectric (Ba0.67Sr0.33)Ti1.02O3 thin films with enhanced H2 induced interfacial polarization potential, J. Appl. Phys. **84** 5134–5139 (1998)
26. X. Bevenot, A. Trouillet, C. Veillas, H. Gagnaire, M. Clement, Hydrogen leak detection using an optical fibre sensor for aerospace applications, Sens. Actuators B: Chem. **67** 57–67 (2000)
27. A. Trouillet, E. Marin, C. Veillas, Fibre gratings for hydrogen sensing, Meas. Sci. Technol. **17** 1124–1128 (2006)
28. E. Maciak, Z. Opilsk, Transition metal oxides covered Pd film for optical H2 gas detection, Thin Solid Films **515** 8351–8355 (2007)
29. F. Farahi, P. Akhavan Leilabady, J.D.C. Jones, D.A. Jackson, Interferometric fibreoptic hydrogen sensor, J. Phys. E: Sci. Instrum. **20** 432–434 (1987)
30. B. Chadwick, M. Gal, Enhanced optical detection of hydrogen using the excitation of surface plasmons in palladium, Appl. Surface Sci. **68** 135–138 (1993)
31. D.K. Benson, C.E. Tracy, G.S Hishmeh, P.A. Ciszek, D.P. Haberman, Low-Cost, Fiber-Optic Hydrogen Gas Detector Using Guided-Wave, Surface-Plasmon Resonance in Chemochromic Thin Films, Report NREL/CP-590-25611 (1998). M. Morjan, H. Zuchner, K. Cammann, Contributions to a reliable hydrogen sensor based on surface plasmon surface resonance spectroscopy, Surface Sci. **603** 1353–1359 (2009)
32. M. Morjan, H. Zuchner, K. Cammann, Contributions to a reliable hydrogen sensor based on surface plasmon surface resonance spectroscopy, Surface Sci. **603** 1353–1359 (2009)
33. S. Sekimoto, et al., A fibre-optic evanescent-wave hydrogen gas sensor using palladium supported tungsten oxide, Sens. Actuators B: Chem. **66** 142–145 (2000)
34. D. Luna-Moreno, D. Monzon-Hernandez, J. Villatoro, G. Badenes, Optical fiber hydrogen sensor based on core diameter mismatch and annealed Pd–Au thin films, Sens. Actuators B: Chem. **125** 66–71 (2007)
35. S. Dong, F. Bai, J.F. Li, D. Viehland, Sound-resonance hydrogen sensor, Appl. Phys. Lett. **82** 4590–4592 (2003)
36. A. D'Amico, A. Palma, E. Verona, Surface acoustic wave hydrogen sensor, Sens. Actuators **3** 31–39 (1982)
37. W.P. Jakubik, Investigations of thin film structures of WO3 and WO3 with Pd for hydrogen detection in a surface acoustic wave sensor system, Thin Solid Films **515** 8345–8350 (2007)
38. S. Lomperski, M. Anselmi, I. Huhtiniemi, Ultrasonic and resistive hydrogen sensors for inert gas–water vapour atmospheres, Meas. Sci. Technol. **11**1–8 (2000)
39. E. Lee, I. Hwang, J. Cha, H. Lee, W. Lee, J. Pak, J. Lee, B. Ju, Micromachined catalytic combustible hydrogen gas sensors, Sens. Actuators B: Chem. **153** 392–397 (2011)
40. W. Shin, et al., Thermoelectric thick-film hydrogen gas sensor operating at room temperature, Jpn. J. Appl. Phys. **40** 1232–1234 (2001)
41. F. Qiu, W. Shin, M. Matsumiya, N. Izu, I. Matsubara, N. Murayama, Miniaturization of thermoelectric hydrogen sensor prepared on glass substrate with low-temperature crystallized SiGe film, Sens. Actuators B: Chem. **103** 252–259 (2004)
42. K. Tajima, F. Qiu, W. Shin, N. Izu, I. Matsubara, N. Murayama, Micromechanical fabrication of low-power thermoelectric hydrogen sensor, Sens. Actuators B: Chem. **108** 973–978 (2005)
43. W. Shin, K. Tajima, Y. Choi, N. Izu, I. Matsubara, N. Murayama, Planar catalytic combustor film for thermoelectric hydrogen sensor, Sens. Actuators B: Chem. **108** 455–460 (2005)
44. K. Tajima, F. Qiu, W. Shin, N. Sawaguchi, N. Izu, I. Matsubara, N. Murayama, Thermoelectric properties of RF-sputtered SiGe thin film for hydrogen gas sensor, Jpn. J. Appl. Phys. **43** 5978–5983 (2004)
45. L. Boon-Brett, J. Bousek, P. Moretto, Reliability of commercial available hydrogen sensors for detection of hydrogen at critical concentrations: part II –selected sensor test results, Int. J. Hydrogen Energy **34** 562–571 (2009)
46. B. Hoffheins, et al., Development of low cost sensors for hydrogen safety applications, Proc. of the 1999 U.S. DOE Hydrogen Program Rev. NREL/CP-570-269

Chapter 20
Hydrogen Safety

Hydrogen safety is crucial throughout the entire hydrogen energy value chain. Hydrogen has been used as a chemical feedstock and as an industrial gas for overa century. The industry has an excellent safety record when using it in the controlled industrial environment and premises. However, there has been a very limited historical evidence of its use and the risk assessment associated with it, when used by the general public.

For a very long time, the perspective of the general public was perceived by certain incidents like the Hindenburg disaster in 1937, hydrogen bomb, or the Challenger space shuttle accident—the explosion that occurred in 1986. However, over time, there has been a tremendous technological advancement that has been built up, along with the confidence gained from the long industrial usage of hydrogen, which is changing the perspective. Importantly, as the use and demand of hydrogen align with the global policies on hydrogen, climate change commitments, and the urge to decarbonize hard-to-abate sectors, the use of hydrogen will increase as an energy carrier. As technologies enter the marketplace, their use will increase even among laymen who have no knowledge of the characteristics of hydrogen. Thus, primary consideration when using any hydrogen based technology is to have adequate safety measures and protocols in place, as well as sufficiently trained personnel to safely operate to the hydrogen facilities. In addition to following safety protocols, it is equally important to educate the end users on how to avoid unwanted hazardous situations and how to respond if such situations arises. The properties of hydrogen that make it different from conventional fuels and the necessary safe practices, as well as possible hazards along the entire hydrogen value chain and existing regulations and standards will be the focus of this chapter.

The acceptable level of safety in hydrogen can be determined by calculating the associated risk. The risk is a product of the probability of an accident or hazard occurring and the cost or the consequences associated with it, which could include loss in terms of property, infrastructure, and at times, even lives.

P. Sharma, *A Textbook on Hydrogen Energy*,
https://doi.org/10.1007/978-981-95-4601-5_20

However, in case of hydrogen, there is very limited knowledge about the probability of occurrence as well as the consequences associated with it. Therefore, various techniques can be employed to assess the risk associated while hydrogen usage in the different energy systems. These techniques are primarily predictive, employing several methods to identify risks in the hydrogen energy systems, such as quantitative risk assessment (QRA), also commonly known as probabilistic risk assessment. However, these techniques are used for large-scale and complex systems and can still be applied to small, simple systems, but are generally more expensive. Several softwares have been developed, and various existing programs enable a detailed study of the possible hazards, their consequences, and the risk associated with the use of hydrogen. In fact, risk management is about identifying, assessing, and prioritizing the risk. When a detailed study of possible risk is conducted, it is important to identify the potential implications and which risks could be more devastating. Thus, the study of risk assessment involves identifying the risk, assessing it in terms of consequences, and understanding how resource management can be done economically to monitor, minimize chances, and control the probability and impact of catastrophic events.

Historical Accidents

The past incidents related to hydrogen had resulted in great setback, leading to reduced interest and inducing fear in the public towards the use of hydrogen and hydrogen-based technologies for a long time. The well-known Hindenburg disaster, which occurred in New Jersey on May 6, 1937, involved hydrogen, where hydrogen was used to provide buoyancy to the giant airship. Fire started from the tail end and resulted in an explosion, and the fire engulfed the 240 tons of airship, resulting in a huge loss of lives, with 36 people on board lost their lives; it was considered that the accident occurred because of hydrogen. However, several studies were subsequently carried out to understand the reasons behind this disaster. It was found that the outer shell of the aircraft and the paint of the airship were having flammable material and the weather conditions on the day of accident were turbulent, and during that electrostatic charges led to electrical sparks which initiated explosion in the airship leading to the well-known Hindenburg disaster (Fig. 20.1).

Another incident that occurred in Houston on October 23, 1989, was a hydrogen vapor cloud explosion. The incident occurred in a very large polyethylene plant, resulting in extensive damage to the plant, causing loss of 22 lives and 100 injuries. Upon investigation, it was found that the release of processed gas during maintenance was the cause of the accident. Studies have shown that several flaws were present in the plant's installation, no hazard assessment was conducted; and there was an inadequate safe distance between the control room and the reactor. Further, there

Fig. 20.1 Picture of the Hindenburg disaster. *Source* https://en.wikipedia.org/wiki/Hindenburg_disaster

was not even permit for the maintenance, which was carried by the not very experienced individuals in handling such situations; the design errors in the construction of the building and its ventilation system ultimately led to the catastrophic event.

In Frankfurt in the year 1991, a compressed hydrogen tank storing hydrogen at 45 bar pressure, having a volume of 100 m^3 burst. The explosion was so powerful that the shock waves and missiles originating from the compressed hydrogen tank caused extensive damage to the industrial plant. The reason for the catastrophe, on investigation was found to be, the welding points, where metallic cracks were present, and these cracks propagated from inside the tank to the outside. Those welding point became the stress centers, where hydrogen, being the smallest molecule, diffused quickly and penetrated through these cracks, joints, welds, thus resulting in further increasing the crack size and ultimately leading to the failure of the tank or its rupture. However, following that incident in the country, several measures were taken, leading to the improvements in standards and safety protocols, which ultimately resulted in the enhanced hydrogen-related safety features.

Hydrogen tank explosion at alkaline electrolysis plant in South Korea on May 23, 2019, resulted in six injuries and two loss of lives. The accident was found to have been caused by oxygen spillover at a low load condition and human error, resulting in a rupture of the hydrogen separator. The ignition started due to static electricity in the absence of proper earthing [1].

In 2003, the fasteners on a hydrogen tube trailer securing the cylinders together failed, resulting in the cylinders falling on the road, causing a hydrogen leak. A spark from nearby vehicle ignited the hydrogen released, triggering a deflagration and subsequent explosion. This incident caused damage to nearby vehicle and houses.

These are some of the incidents being reported widely in literature. Besides, there is a large respository including reporting of incidents that have occurred in past and the details are available on the website of "The Center for Hydrogen Safety" at the Pacific Northwest National Laboratory (PNNL). The Centre initiated the reporting of all incidents on the website https://h2tools.org/ in 2006. The purpose was to understand the incident and share the lessons learned from each incident related to hydrogen safety. Several other organizations joined this initiative; and academia, researchers, and industry are conducting self-reporting of incidents globally. The website reports more than 200 such incidents.

Thus, as per the literature reports and studies carried out, the list of accidents being analyzed and observations shows that out of the total accidents that have occurred and been reported in past, a significant portion of those have occurred in the laboratories as shown in Fig. 20.2 [3]. Thereafter follow the refueling stations, then commercial facilities, power plant, tube trailer or delivery system, the chemical plant, the nuclear fuel processing, refineries, storage and the production facility. Thus, a significant number of such incidents have been reported to have occurred in laboratories, the two primary reasons were identified as the cause of these accidents, human error and the equipment failure [4].

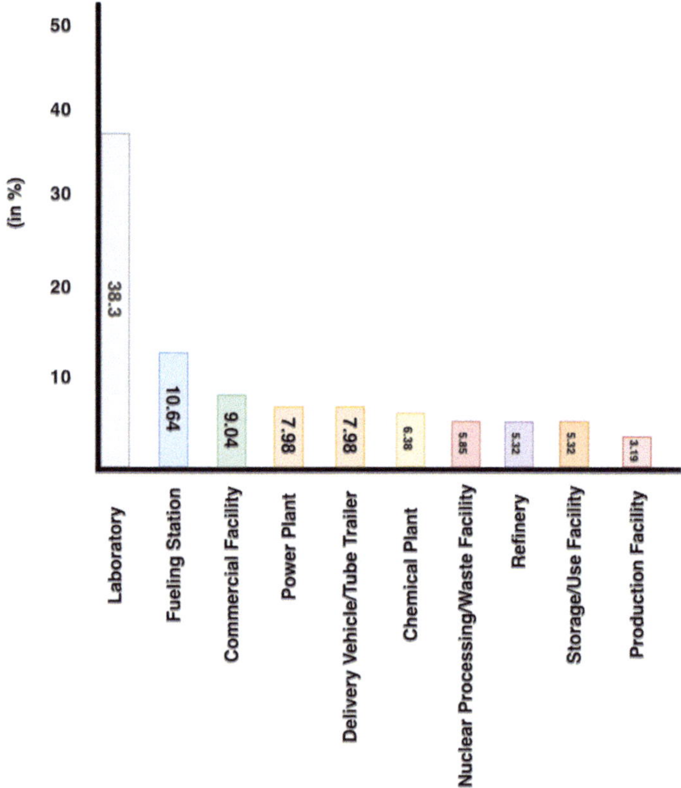

Fig. 20.2 Different incidents reported, along with their location of occurrence. Data taken from [2]

The major reason for equipment failure was associated with the breakdown of piping, fitting, and valves and their subsequent rupture. The other reasons were associated with the failure of hydrogen storage equipment due to damage of the outlet of the cylinders at the points of valves or regulators. Further failure at the fueling system, safety equipment, ventilation system, pressure relief devices are mentioned in Fig. 20.3.

Among the major reasons for the accidents, the equipment failure was the dominant factor; other reasons were human error and design flaws, as mentioned in Fig. 20.4. The reasons reported for equipment failure were hydrogen embrittlement and the facilities fatigue, which were the most significant factors. Human error reported in the incidents was mostly due to lack of training and awareness of how to handle accidental situations, the safety protocols were not followed, and the improper maintenance led to such accidents. In these reported incidents, the losses have been observed to be in terms of property, injuries and sometimes human life, as well as facility shutdowns for repair, and in some cases there were no losses, as mentioned in Fig. 20.5.

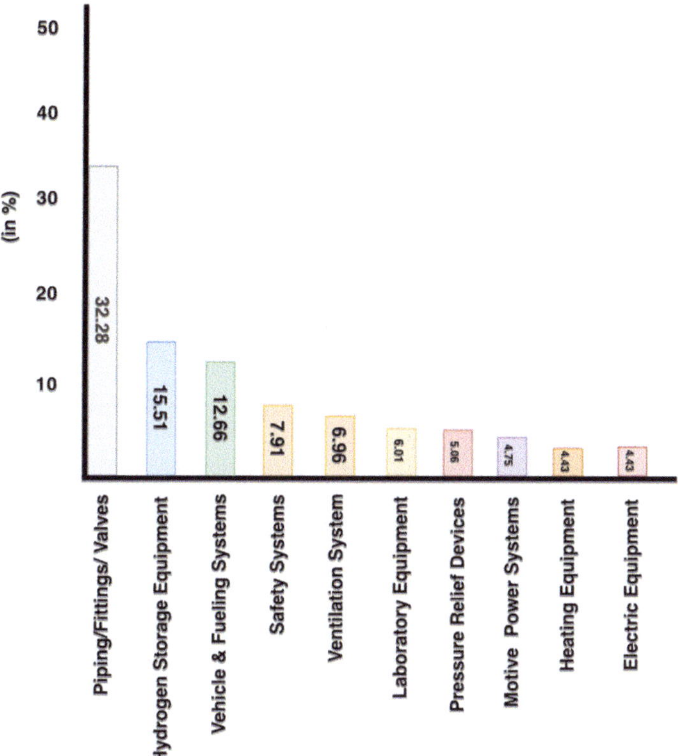

Fig. 20.3 Percentage of accidents occurred at different hydrogen facilities and components involved. Data taken from [2]

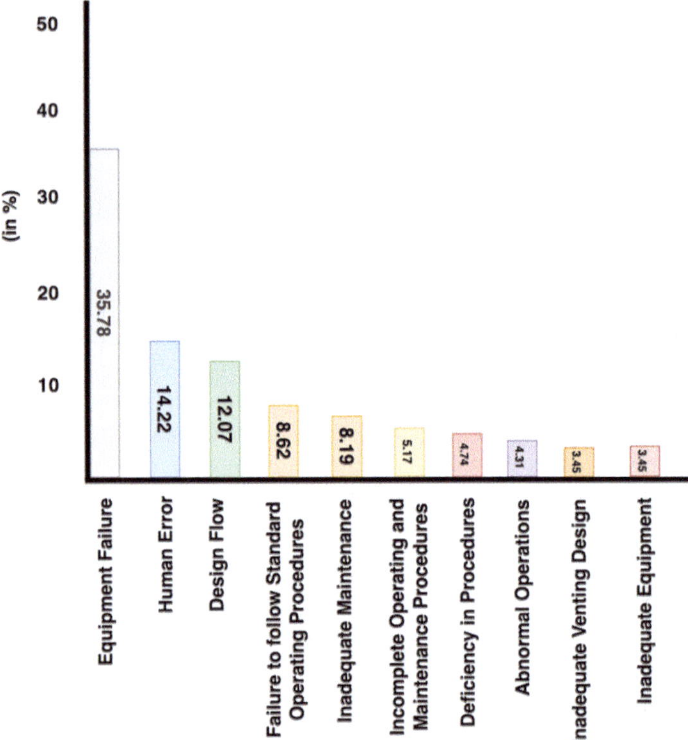

Fig. 20.4 Reasons for the hydrogen-related accidents. Data taken from [2]

The most important thing after understanding such incidents and learning the causes, risks, and impacts is to prevent such catastrophic events from happening in the future. Several projects have been undertaken globally to study the risks associated with these incidents, to determine their origin, and to explore how they can be avoided. With those studies, improved technologies, infrastructure, and strategies were put in place. The required safety measures associated with the hydrogen energy value chain will be discussed in the following sections. Understanding hazardous situations and measures to avoid them requires knowledge of the properties and characteristics of hydrogen, which makes it distinct from the other conventional fuels and challenges associated with its use.

Most of the accidents reported where fire or explosion has resulted, occured due to four ignition sources: (1) negative Joule–Thomson effect can be a contributing factor along with others, (2) diffusion ignition, (3) hot surface, and (4) electrostatic ignition.

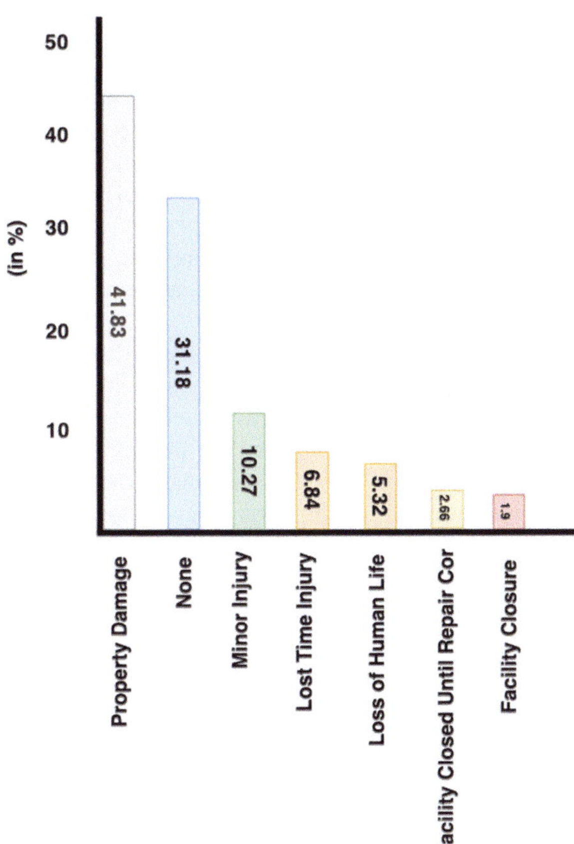

Fig. 20.5 Outcomes of the hydrogen-related accidents. Data taken from [2]

Important Characteristics and Properties of Gaseous Hydrogen Associated with Accidents

Hydrogen is a colorless, odorless, tasteless gas, not detectable by human senses. Thus, in case of any hydrogen leakage, it's not possible to detect that leak by human senses.

Hydrogen is a potential asphyxiant, i.e., hydrogen, when released in a confined environment, it displaces oxygen. If the oxygen level in the surroundings reduces below 20%, it can pose a health hazard to human beings. If oxygen level lies in the range of 15–19% by volume, it leads to decreased capability to perform tasks, with level between 12 and 15% causes deeper respiration, faster pulse and poor coordination; between 10 and 12% giddiness; between 8 and 10% leads to nausea and unconsciousness; for oxygen level less than 8% by volume can lead to loss of life.

The buoyancy of hydrogen is highest on earth; it is 14 times lighter than air. In the event of leak, it moves upward very quickly, which is significant safety advantage of hydrogen. The tendency of hydrogen to flow away from the site of leakage is and advantage while considering its safe usage. Besides, the diffusivity of hydrogen is very high, providing the advantage that in event of leak, the concentration reduces to below the lower flammability limit, which is again a safety asset. Thus, the important characteristic is that it does not get accumulate in unconfined space, which is the biggest advantage in comparison to the heavier hydrocarbons. Hydrocarbons in turn, form a huge combustible cloud that could lead to much more dangerous accidents than hydrogen. One such accident happened in the past in Flixborough in 1974 and in Brucefield in 2005, where the hydrocarbon cloud explosion resulted in a significant loss (Table 20.1).

The impact of buoyancy in the safe dispersion of hydrogen is more significant compared to the diffusivity. Although diffusivity also plays an important role, the diffusivity of hydrogen is approximately three times higher than that of methane and an order of magnitude higher than that of gasoline vapors. Although this makes hydrogen diffusion very fast and makes it safe in case of leakage compared to other hydrocarbons, at the same time, it poses a challenge when it comes to storing hydrogen.

Since natural gas is closest to hydrogen, a comparison of fuel properties is shown in Fig. 20.1. The density of hydrogen is lower than that of natural gas by an order of magnitude. Hydrogen has a low viscosity, and, together with its small size, it becomes very easy for hydrogen to achieve very high flow rates. In the event of a leak occurring at the piping, fittings, seals, or valves, the small size and low viscosity result in a speedy escape of hydrogen. Another important characteristic of hydrogen is that it burns in clean environment with an invisible flame. In a clean environment where there is no material in vicinity that could burn, the flame is invisible. However, in the presence of other material, the volatiles will be released and make the flame visible. The average adiabatic flame temperature for hydrogen is very high, i.e., 2403 K, and can cause very severe burns as well as very severe incidents.

The flammability limit of hydrogen in air, at NTP, ranges between 4% and 74vol%, as compared to that for natural gas, which is 5–15%. This property is crucial, as leaner mixtures of hydrogen with air or oxidizer can form a combustible mixture. At the same time, a concentrated mixture of hydrogen in air can also form a combustible

Table 20.1 Properties of hydrogen as compared to other conventional fuels in relation to the possible hazards

Parameter	Hydrogen	NG (85% CH_4)
Density (STP, g/L)	0.089	0.717
MIE (in air, mJ)	0.017	0.31
Flammability limits (in air, %, NTP)	4–75	5–15
Energy density (LHV, MJ/kg)	119.96	50.07
Boiling point (°C)	– 253	– 162
Ignition temperature (°C)	574	650
Burning velocity (m/s)	2.65–3.25	0.38

mixture; provided there is an ignition source, which can lead to fire. The flammability limit depends upon temperature and pressure, can further expand at temperatures other than NTP, for example, at 100 °C, LFL is 3%.

The burning velocity of hydrogen is 2.65–3.25 m/s, compared to natural gas, which is 0.38 m/s, i.e., an order of magnitude higher for hydrogen. Thus, the hydrogen fire is more intense than a natural gas fire. The burning velocity depends on temperature, pressure, and composition; the important consequence is that it is challenging to arrest, quench, or confine a hydrogen flame. Thus, hydrogen fires deplete oxygen, causing asphyxiation, flame travels faster, and explosions can result. In the case of an explosion, it is very difficult to confine. Besides, with detonation, the velocity involved is three orders of magnitude higher than that of deflagration, and the pressures are also significantly higher.

In the event of a hydrogen leak, the escaped hydrogen ignites easily in the presence of any ignition source. Whether the ignition source is a strong or a very weak source, the minimum ignition energy of hydrogen in air is 0.017 mJ, compared to natural gas which is 0.31 mJ (5% of the natural gas). The ignition source could be an open flame, an electrical short circuit, a hot surface, a spark, a person smoking, or even it could be a static charge that could initiate the ignition. Thus, it is essential when considering hydrogen safety that, no ignition source should be in the proximity of the place where hydrogen is located or used. The hydrogen–air mixtures pose a significantly higher safety challenge than the natural gas-air mixture.

Hydrogen, by itself, is an electrical insulator; however, when hydrogen flows, or in the case of any agitation during transfer, static charges may be developed, which can be catastrophic. Therefore, all storage vessels are kept grounded to avoid such hazardous situation.

The quenching distance for hydrogen is the lowest with a value of 0.64 mm, as against 2 mm for gasoline. Hydrogen flames travel very close to the cylinder, making it very difficult to quench and also have increased likelihood of backfire.

Hydrogen, being the smallest molecule, easily infiltrates various materials, such as steel, leading to changes into the mechanical properties of the materials, thereby reducing strength and increasing chances of failure due to hydrogen embrittlement.

Properties of Liquid Hydrogen Associated with Accidents

The properties of liquid hydrogen are important and need to be considered while using, storing, and transporting it. The properties of compressed and liquid hydrogen differ; the major difference is that the density of liquid hydrogen is substantially high, surpassing that of air. Due to the high density of liquid hydrogen in the event of spill, the cloud of liquid hydrogen moves close to the ground. Thus, liquid hydrogen cloud continues to stay close to the ground, and over time, it gains heat from the environment, and liquid hydrogen starts to evaporate until it completely converts into gaseous hydrogen. However, during this process, if there is an ignition source in the vicinity, it can lead to a fire or explosion.

The peripherals such as pipes, orifices, vents, and valves attached to the hydrogen facility can also get affected by air inflow. If the moisture in the air liquefies or solidifies, it can block, plug, or damage these components. A pipe between two valves if allows air to enter, then it condenses, forming ice. When the air regasifies, the pressure increases, which can lead to mechanical failure of the pipe. Failure of the pipe can inturn lead to the release of hydrogen in the form of jet. Thus, the mechanical failure could even lead to potentially boiling liquid expanding vapor cloud explosion BLEVE. While liquefying hydrogen or condensing any gas, the volume of the gas reduces. As the volume reduces, it creates a vacuum; which can draw air inside along with the moisture; air will solidify and get stored along with the liquid hydrogen. During repeated filling, i.e., repeated pressurization, the concentration of air that enters the liquid hydrogen will increase, and at times, it may lead to concentration lying in the range of formation of an explosive mixture with hydrogen.

When the liquid hydrogen is warmed for maintenance, the liquid in the tank will regasify, which can result in a very high pressure. Increased pressure together with if any condensed air was present which also gets evaporated, this situation can also lead to the formation of an explosive mixture. To avoid such situation, these storage vessels are kept under positive pressure, preventing any air inflow into the vessel and avoiding the formation of flammable mixture. The liquid hydrogen vessels are warmed periodically; purging is performed to remove condensed air and reduce the oxygen level to below 2%, achieved through repeated warming and purging.

In case of a liquid hydrogen spill, it gains heat from the surroundings, evaporates and converts into gaseous hydrogen. The evaporated gaseous hydrogen which is at normal boiling point will further get warmed up with heat from the surroundings and converts into gaseous hydrogen at normal temperature and pressure condition. This conversion from liquid hydrogen to gaseous hydrogen will result in a tremendous volume increase. The volume expansion ratio of liquid hydrogen when it evaporates to gaseous hydrogen at normal boiling point during the transition phase and then finally to gaseous hydrogen at its normal temperature and pressure is 847, and the corresponding increase in pressure from an initial pressure of 1 bar is to a pressure of 177 MPa. Since superinsulated liquid hydrogen storage tanks are not designed to handle this excessively high pressure, a pressure relief valv is, an integral part of liquid hydrogen storage facilities, which releases the pressure or gaseous hydrogen.

Similarly, the expansion of a compressed gaseous hydrogen, in the event of a hydrogen leak, the expansion ratio for the gaseous hydrogen stored at 25 MPa and released to atmospheric pressure is 240.

Hydrogen Embrittlement

The major challenge in the hydrogen energy value chain is to select the appropriate materials for hydrogen storage and associated piping, fittings, tubing, valves, and all peripheral components. When materials are subjected to hydrogen or come into contact with it for a longer duration at a specific temperature and pressure, their properties change.

A deterioration in the mechanical properties of metals is observed; i.e., the metals usually change their characteristics from ductile and becomes brittle. The choice of material is very important and the selection should be such that, the material should not undergo embrittlement, should be metallurgically stable and do not undergo phase change over time and with repeated thermal cycling.

The properties of gaseous and liquid hydrogen have been considered, which are crucial regarding hydrogen safety. The possible accidents or hazards possible due to the properties mentioned in the section are described in the later sections.

Types of Hydrogen Hazards

Hydrogen hazards can be classified into three types: physiological hazards, physical hazards, and chemical hazards (Fig. 20.6).

Physiological Hazards—In case of a hydrogen leak, when the released hydrogen forms a flammable mixture with air or oxygen, and if an ignition source is located

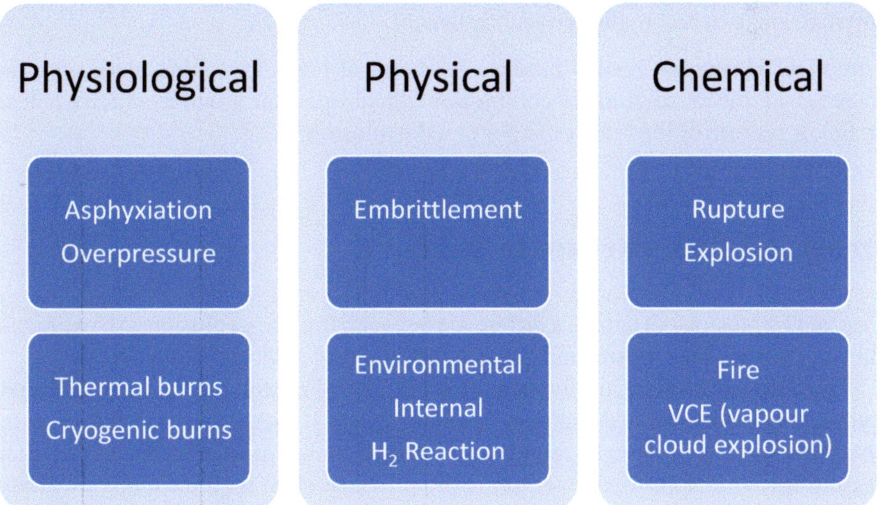

Fig. 20.6 Classification of hydrogen hazards

nearby, it can result in fire or explosion. A person, if present in the vicinity of the affected area can have severe injuries or burns. At the same time, the blast waves from explosions can cause over-pressurization injuries, depending upon the duration for which the person is exposed, and can lead to several different types of injuries or organ failures.

Thermal burns occur due to the heat released, i.e., the immediate heat radiation from a fire, which can cause injuries to the person, property damage, and other losses. Besides, during a hydrogen fire, the surrounding air temperature also increases, which can affect the persons in contact.

Thermal burns which depend upon several parameters, such as the duration of exposure, the burning rate, environmental temperature and pressure conditions, and the burning surface area, can lead to damage.

Other than the direct injuries and in the absence of the ignition source, the leakage of hydrogen can displace oxygen. If the oxygen reduces below 20% by volume, it can cause asphyxiation or severe health hazards. Especially in confined spaces, the hydrogen release can lead to oxygen deficiency. In the range of 15–19 vol% oxygen reduced human activity, a 12–15% faster pulse, deep respiration and poor coordination can occur. Additionally 10–12% dizziness, 8–10% nausea, and unconsciousness, below 6% person can go in coma and is life threatening.

Acoustic hazard could be due to blast sound and can lead to acoustic trauma and hearing impairment.

With liquid hydrogen, contact can lead to cryogenic burns, frostbite, or hypothermia.

Physical Hazards—It primarily occurs due to embrittlement, which causes change in the mechanical properties of materials leading to failure, rupture or leakage from various systems. Hydrogen embrittlement is of different types like environmental, internal, and external hydrogen embrittlement.

Chemical Hazards—results because of chemical reactions taking place with the material, or due to oxidation or combustion of hydrogen–air mixture, which can lead to fire or rupture of the tank, or vapor cloud explosion.

Important Terms Related to Safety

Fire and Explosion—There is a difference between fire and explosion. An explosion may not always lead to fire, and fire may not always lead to an explosion.

The term explosion is used when there is a loss of confinement; i.e., initially the hydrogen gas is contained within a confined space. If there is an unusual pressure rise that causes the fuel to be released and loss of containment from the vessel, then it constitutes to explosion. However, in the presence of an ignition source, with the loss of confinement a fire can also result, if the released hydrogen forms a flammable mixture.

Deflagration—Deflagration refers to the process when the flame front moves through a flammable mixture in the form of subsonic wave with respect to the unburned mixture of hydrogen and the oxidant. Deflagration in open areas may cause over pressure of 10 kPa; however, in closed places or enclosures, it will be much higher and can be mitigated by venting.

Detonation—Detonation refers to the process in which the flame front coupled with the shock wave propagates through a flammable mixture in the form of supersonic wave with respect to the unburnt mixture. Detonation propagates 2–3 orders of magnitude faster (speed 1600–2000 m/s) than deflagration, along with a pressure that can be 15–20 times higher than the initial pressure (ranging from 1000 to1500 kPa). However, detonation is known to cause much more devastation than deflagration, and it propagates thousand times faster than deflagration.

BLEVE—Boiling liquid expanding vapor explosion is a physical phenomenon where a sudden release of a large mass of pressurized liquid into the atmosphere takes place, i.e., loss of confinement. This expansion can lead to two phenomena: either a fireball or a flash fire. Thus, if the flammable mixture ignites, forming a spherical burning cloud, it is known as fireball; it could be in a confined space or it could be in an unconfined space; accordingly it is known as unconfined vapor cloud explosion (UVCE), or it could be in a confined space so-called confined vapor cloud explosion (CVCE). The resulting event can be a flash fire, which occurs when a very dilute or very concentrated fuel–air mixture, within the flammability limit, ignites. In flash fire, the flame travels through a cloud at a low speed and with lower pressure.

Pool Fire—Liquid hydrogen spills on the ground and gets ignited, it is referred to as a pool fire. In the pool fire, the liquid, upon ignition, burns with a stationary diffusion flame, and the evaporated gaseous hydrogen also undergoes combustion.

Jet Fire—Also called spray fire, occurs due to the ignition of a flammable mixture with fuel being continuously released from a facility with significant momentum and in a particular direction, resulting in a turbulent diffusion flame. The heat flux arising from jet fire can engulf objects in the vicinity, leading to the failure of storage vessels and increased devastation.

Flash Fire—also known as vapor cloud fire—occurs when a flammable mixture ignites and the flame front propagates at subsonic speed and negligible overpressure. In the case of flash fire, the flammable mixture burns, but an explosion does not occur; the hazard could be thermal burn-related injuries or asphyxiation due to oxygen depletion.

Ignition Kernel—Iginition kernel refers to the very initial shape of the flame when the air–fuel mixture ignition just starts, after only a few milliseconds of combustion. The term refers to the very early stage of flame initiation.

Fireball—When the ignition occurs immediately, and the sufficient mixing of fuel or oxidant has not occurred, such that the surface only forms a flammable mixture while the core is still rich in fuel and deprived of oxidant, then the ignition results in a fireball. Thus, a fireball is a spherical cloud of flammable vapor with a core rich in fuel and an outer envelope forming a flammable mixture, which burns from the outer surface. Upon ignition, the fireball rises due to the buoyancy of the combustion gases, forming a mushroom-shaped cloud. Fireball has a very high-temperature and high-radiant heat. Generally sudden, rapid release of liquid hydrogen can lead to a fireball.

Possible Causes of Hydrogen-Related Hazards—The causes of hydrogen-related hazards can be broadly classified into two types: one related to material properties, and the second arising from the handling of systems. These reasons can lead to hydrogen hazards, and accidents can be evaluated through either experimental techniques or computational techniques.

Experimental techniques, although they provide valuable insights, can prove to be very expensive and sometimes become impossible to carry those, especially when they involve large-scale incidents that need to be replicated multiple times.

However, CFD techniques are widely used as predictive tools to understand the influencing parameters and identify possible physical phenomena that could result in hazard risk components. Several tools are used to understand the hazards, the risks associated with them, and the possible consequences resulting from these hazards (Fig. 20.7).

Hydrogen leak can be caused due to various reasons like permeation through materials due to small size of hydrogen; due to corrosion or fatigue of facility; leak may develop over a period of time because of wear of materials, instrumental or component leaks from valves, joints, fittings, etc.; rupture of the equipment or vessel due to internal or external aggression either caused due to impact or chemical reaction or can be thermally induced or due to over pressure. Considering the different

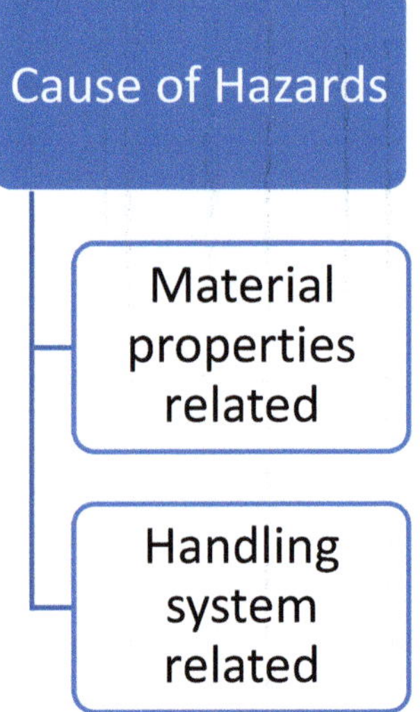

Fig. 20.7 Causes of hydrogen-related hazards

risk components related to hydrogen hazards at different hydrogen facilities and at different hydrogen applications.

1. **Materials-Related Hazards**

a. **Hydrogen Embrittlement (HE)**—It is the change in the mechanical properties of metals or non-metals when exposed to hydrogen. Material characteristic changes from ductile to brittle. Hydrogen embrittlement is a complex phenomenon that depends on several parameters, including exposure duration, metal purity, gas concentration, temperature and pressure conditions, material surface properties, mechanical properties, alloy composition, and material microstructure.

   Hydrogen embrittlement is of three types: environmental hydrogen embrittlement wherein the crack propagation starts from the outer surface; internal hydrogen embrittlement in which the cracks start to propagate from the inner surface; and hydrogen reaction embrittlement in which hydrogen gets absorbed into the metal then reacts with the metal forming metal hydride or it can react with the carbon present in material thereby resulting in decarbonization of the steel and leading to failure. The environmental hydrogen embrittlement and internal hydrogen embrittlement are observed in the temperature range of 200–300 K, while the hydrogen reaction embrittlement occurs above room temperature. Hydrogen embrittlement is known to increase with the increase in the strength of the steel, increasing exposure time, and with the increase in the hydrogen pressure.

   There are various ways to address the challenges associated with hydrogen embrittlement or mitigate its impact. The possible ways include having an oxide layer coating, reducing stress concentration inside the material of the confinement vessel or piping, etc., with the help of different additives added to the hydrogen, by proper selection of material and grain structure.

b. **Permeation of Hydrogen**—Hydrogen can migrate through the permeable material. Hydrogen permeation is defined as the flux of hydrogen also defined as the rate of hydrogen migrating into the permeable material leading to hydrogen embrittlement over a period of time. Permeation of hydrogen is insignificant in case of Type I, Type II, and Type III tanks; however, in Type IV tanks, this is also restricted. Hydrogen permeation depends on various parameters, such as temperature, pressure, and aging of the tank. In case of Type IV tanks, as the tank ages, the permeation increases.

c. **Change in the Properties at Low Temperature or at the Cryogenic Temperature**—When materials are subjected to liquid hydrogen temperature, like the storage tanks or pipelines or other peripheral components, in that case a change in their mechanical properties is observed. At low temperatures, change from ductile to brittle behavior an occur. In addition certain unprecedented plastic deformation or a phase transformation resulting in changes to the elastic and mechanical properties of the material can also occur. At liwuid hydrogen temperature, thermal contraction of the materials can occur, the severity of it, varies from material to material. Thus, when materials are exposed to liquid hydrogen at such low temperatures, both the embrittlement and thermal contractions play a dominant role. The thermal contraction is very low in case of metals, it is < 1% while for structured plastics it ranges between 1 and 2.5%. In past, an accident related to the embrittlement at extremely low temperature has been reported to have occurred in Cleveland, USA, where an LNG vessel ruptured and spilled its contents. The process occurred due to the domino effect, where a nearby vessel caught fire. The missiles from that vessel initiated rupture of this tank, resulting in the spilling of its contents, which then exploded and ignited, causing a lot of damage to the property. It lead to 128 loss of lives, along with 200–400 injuries. Thus, it is very important that the impact of low temperature on to the properties of the materials in contact should be monitored to avoid any hazardous situation.

d. **Liner Blistering in Type IV Tanks**—Type IV tanks feature an inner polymer liner and a carbon fiber-reinforced composite. Hydrogen is filled in these tanks at higher pressure; a certain amount of hydrogen gets absorbed into the inner polymer. During the release or utilization of hydrogen, if the depressurization rate exceeds the rate at which hydrogen escapes by diffusion during depressurization, delamination or blistering of the polymer liner results. The blistering of the liner depends on parameters such as the initial filling pressure, discharge rate, and dwell time, all of which could be responsible for failures in the Type IV tank.

e. **Damage to the Carbon Fibers**—The Type IV tanks have a highly complicated design, structure, and construction, and all safety requirements must be met. The operation should be such that any damage to the tank, including fiber matrix cracking/breakage or delamination, must be avoided. The dome structure should be designed to withstand overpressure and any potential impact. The design of such tanks is intended to withstand any external impact and should not be affected by environmental conditions. The hydrogen filling characteristics of the vessel are determined by the fiber density, stacking sequence, and orientation phenomena of the carbon fibers; any damage to the carbon fibers can lead to catastrophic incidents.

2. **Handling of Hydrogen-Related Hazards**—Handling of hydrogen can also give rise to accidents at hydrogen facilities and at the place of its utilization. Usually these hazards are associated with the loss of containment or confinement, in which either gaseous hydrogen or liquid hydrogen is suddenly released due to system handling-related issues and can lead to several types of hazards.

a. **Gaseous Hydrogen Leakage**—Hydrogen is the smallest molecule, and it is much more prone to leakages. Hydrogen can leak through joints, connects, valves or even through vessels from cracks, etc. The crack formation or deformations may occur because of the embrittlement, leading to the rapid release of gaseous hydrogen. The rapid release of gaseous hydrogen can be followed by either immediate ignition or delayed ignition, and each of these scenarios will have different impacts. Thus, gaseous hydrogen leakage can lead to hazards after release, which could be hydrogen jet fire or delayed ignition which can even result in explosion.

b. **Hydrogen Jet Fire**—It is phenomenon that results from a gaseous hydrogen leak. If the ignition occurs immediately after the release of gaseous hydrogen, it can lead to hydrogen jet fire. The source of ignition can be an open flame, an electrical short circuit, a spark from an electrical equipment, a hot instrument or surface, a heating device or element, a person smoking nearby the facility. The hydrogen jet fire is a high-velocity, turbulent flame that arises from the combustion of a flammable hydrogen–air mixture. The hydrogen jet fire occurs in the event of a release in a certain direction with extremely high momentum, which can lead to property damage and result in severe injuries.

c. **Delayed Ignition or Explosion**—In the event of a delayed ignition following gaseous hydrogen leakage or release, it can result in an explosion. Due to the ignition delay, the hydrogen and air gets sufficient time to mix well, forming a flammable mixture. In the presence of an ignition source with the required ignition energy, the uniform mixture of fuel and oxidant ignites, causing an explosion.
d. **Thermal Radiation**—Due to the presence of a high-temperature source, such as a flame or fire, any nearby facilities where hydrogen is being stored or utilized can result in catastrophic events. The presence of a high-temperature source in the vicinity can cause an increase in temperature and pressure inside the gaseous or liquid hydrogen storage vessels, as well as changes in the material properties of the tank, leading to failure. This can lead to damage to the facility or equipment, storage tank, and damage to the structure of the different hydrogen facilities. The surrounding fire, flame, thermal radiation, or high temperature can also lead to loss of confinement in a compressed gaseous hydrogen vessel or liquid hydrogen vessel (Fig. 20.8).

In the compressed gaseous hydrogen vessel, a temperature rise can lead to expansion of the gas, increasing the pressure inside the tank, and ultimately, vessel's rupture.

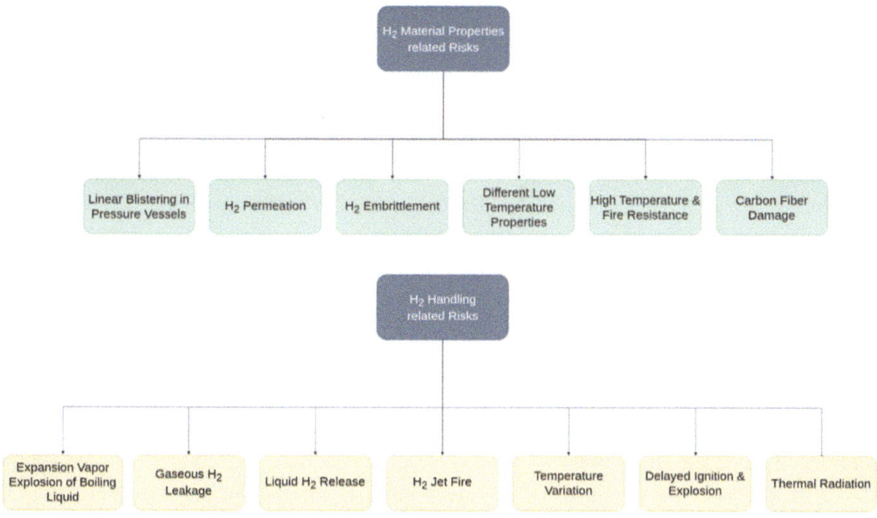

Fig. 20.8 Materials-based and systems handling-related possible hydrogen hazards. Redrawn from [5]

Additionally, it supplies the necessary heat for the evaporation of liquid hydrogen. The change in the mechanical properties of the vessel and the evaporation of liquid hydrogen can cause increase in the pressure, ultimately leading to rupture of the liquid hydrogen tank.

The thermal radiation could be because of a nearby incident, wherein a primary incident could be a fire, or a rise in temperature can lead to a secondary incident or, in fact, at times can lead to a series of accidents, such that the final outcome is much more devastating than the primary accident; this is known as the domino effect.

e. **Temperature Variation**—The filling of hydrogen to high pressure in an onboard hydrogen storage tank on a fuel cell vehicle is done through a pressure difference. The requirement is that the filling should be carried out as quickly as possible, so that the customer doesn't have to wait in long queues, and the experience should be same as that for conventional gasoline-based vehicles refueling. During the filling process, the gas is compressed into the tank, and the faster filling can lead to increase in temperature. During the vehicle utilization, the reverse of is observed; i.e., cooling effect is observed. Thus, the resulting heating and cooling effect causes temperature variations, which can lead to changes in the mechanical properties of the liner material and of the carbon fiber resin composite. There could be stress concentrations in the regions between the liner and the composite, and over time, these can lead to catastrophic events or result in reduced life of the storage tank.
f. **Liquid Hydrogen Release**—The liquid hydrogen release can occur from the liquid hydrogen tank in the form of a flow jet, where the liquid hydrogen is released and forms a pool onto the ground. After release, due to the high ambient temperature, the spilled liquid hydrogen vaporizes, forming a cloud of liquid hydrogen. This cloud, once formed, will continue to move either horizontally or move downward toward ground. In this case, there are several phenomena possible; it can result in flash evaporation, where the vaporization of the liquid hydrogen occurs causing a flash evaporation, or a partial vaporization leading to the formation of a pool of liquid hydrogen.

This pool of liquid hydrogen forms a boiling vapor film due to the vaporization of the liquid hydrogen. When it is close to the ground, it will freeze the ground. Over time, the liquid hydrogen condenses the surrounding air and humidity in the environment. This condensation fo surrounding air and humidity contributes increases the moisture content and solid particles within the liquid hydrogen cloud, making it more dense.

Thus, two competing effects occur at that instant: One is due to the condensation of the atmospheric humidity makes the cloud denser, resulting in a negative buoyancy effect. At the same time, the another opposing effect is the phase transition taking place from the liquid hydrogen to gaseous hydrogen due to ambient conditions, and it has a positive buoyancy effect. Thus, there will be two competing effects; however, the outcome will depend on several environmental parameters, such as wind speed, environmental conditions, and the external temperature, among others

g. **Boiling Liquid Expanding Vapor Explosion**—BLEVE is a physical phenomenon where rapid release of hydrogen occurs from a containment vessel, releasing the hydrogen at atmospheric pressure. However, it is at a temperature higher than the boiling point. The rapid release of hydrogen is actually an explosion phenomenon that occurs because of rupture of the vessel. In a BLEVE, the internal pressure of hydrogen increases beyond the holding strength of the liquid hydrogen vessel resulting in an explosion. It is a physical explosion, not a chemical reaction explosion, and can occur in either confined or unconfined space, accordingly; it is called a confined vapor cloud explosion or an unconfined vapor cloud explosion. An important point to note is that the vapor cloud explosions are much more hazardous compared to jet fire or flash fire.

A review of liquid hydrogen-related accidents has revealed that roughly two-thirds of the accidents were associated with liquefaction and liquid hydrogen storage. In contrast, the remaining accidents occurred during the transportation of liquid hydrogen. The major reasons identified for such accidents were: equipment failure, incorrect operation, improper maintenance, environmental or weather conditions, design and construction flaws, domino effect, on road accidents or impact. These reasons have an lead to either fire, explosion, BLEVE, no major hazard, or the accumulation and diffusion of released liquid hydrogen [6].

Possible Risks, Components, or Accidents with Gaseous Hydrogen

The release of gaseous hydrogen from the containment vessel can result in several hazards, caused by an immediate or a delayed ignition. When the gaseous hydrogen leakage or release occurs into the environment, it forms a flammable mixture with air. Depending upon the time taken to form the flammable mixture in composition, the consequences can be different. In case of delayed ignition it gets sufficient time to form a uniform flammable mixture. In the event of an immediate ignition, the hydrogen leaked does not get sufficient time to form a uniform flammable mixture. Thus, depending on whether the ignition is immediate or delayed, there will be different consequences. The third possibility is the rupture of the vessel; if no ignition source is present, hydrogen will diffuse and disperse without igniting. Different factors thereafter control the level of catastrophe, the possible consequences, and impacts on the nearby neighborhood or property, as well as the lives due to such accidents. Outcomes of the incident depend upon parameters like the release pressure, depends upon the leak size resulted into the release, the environment (closed or open), the type of ignition source, immediate or delayed ignition or no ignition. The size of the release will also determine whether it is buoyancy driven; if the release size is small, hydrogen has a very high buoyancy and diffusivity, it will travel very fast upward. However, if the size of release is considerable and occurs very quickly, it may not have sufficient time to diffuse and may not undergo much buoyancy

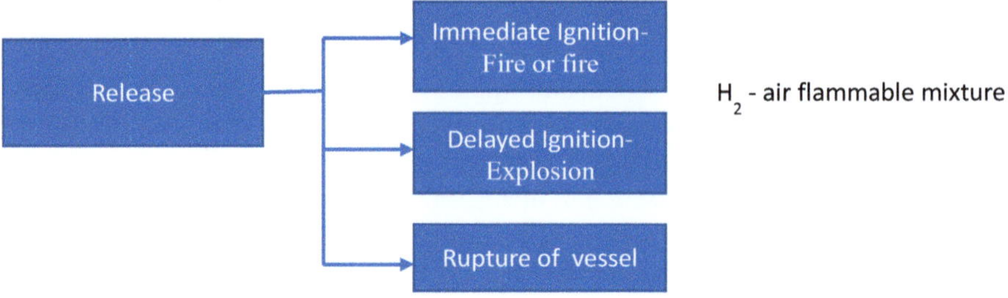

Fig. 20.9 Possible outcomes of gaseous hydrogen release

driven diffusion, and may not diffuse quickly to get diluted with the air. Another important parameter which governs the consequences is whether the release occurred in a confined space or an unconfined space. There will be different challenges and hazards depending upon whether an ignition source was present or not in the vicinity of the release. In the presence of an ignition source, whether the ignition occurred immediately or was delayed. In the absence of an ignition source, if the pressure vessel ruptures due to external aggression or a mechanical impact, there will be a loss of confinement, and different parameters will decide the consequences of the gaseous hydrogen release (Fig. 20.9).

Gaseous Hydrogen Release in an Unconfined Space—Hydrogen has a high buoyancy and high diffusivity, being the smallest molecule, it diffuses very fast. Thus, in the case of gaseous hydrogen release in a unconfined space, it rises rapidly and becomes diluted, leading to a concentration laying outside the flammability limit. Depending on whether the ignition occurs immediately or is delayed, the consequences will vary.

If the release is immediately followed by ignition from an ignition source, it will burn like a jet fire or flash fire. Jet fire is a turbulent diffusion flame that arises from the combustion of a flammable mixture, characterized by a continuous leak or release with significant momentum and in a particular direction. Usually, a jet fire is triggered by the release of high-pressure hydrogen into the atmosphere, resulting in a high velocity jet. The pressure ratio at which the gas is released and that of the atmospheric pressure decides on whether the jet formed is (1) a subsonic jet, (2) a critical jet, or (3) an under expanded jet. In the case of a subsonic jet, the gas release is fully expanded at the leak point, such as nozzle or outlet, and is usually driven by buoyancy or momentum. The critical jet (required pressure of release is 2 bar) gas velocity at the outlet reaches the local sonic speed. An unexpanded jet, where the gas released continues to expand and also accelerate into the environment, is usually momentum driven. The associated hazard with jet fire will depend on parameters such as flame length, radiative heat fluxes, and the thermal effects of the immediate ignition will determine the extent of the catastrophe. Factors such as the dispersion characteristic of the jet formed, the length of the jet until it forms a combustible mixture, flame length, among others, govern the intensity of the hazards.

In fact, the release will burn from outside, but the core remains rich in hydrogen. Since not enough oxidizer is present, it appears as a shape of a fireball; the thermal effects will be experienced by persons nearby or the structure, building, or enclosures in the immediate neighborhood.

a delayed ignition, the released hydrogen gets enough time to mix with the air in the atmosphere, and composition lied in the flammable limit, such that it will burn completely, like a flash fire.

The flash fire resulting in deflagration causes overpressure effects that impact the people present in the surroundings, as well as various installations and facilities. The effect is aggravated if there is a congestion. Presence of several nearby facilities, or obstacles on the site can result in different consequences and severity. The presence of obstacles or congestion can lead to different catastrophic events, which can even accelerate or escalate the hazards. Deflagration which initiated due to the release alongwith with delayed ignition, can even lead to detonation, which is much more hazardous. If certain obstacles are present or if it occurs in a confined space, the consequences are different.

Gaseous Hydrogen Release in a Confined Space—If the gaseous hydrogen release occurs in a confined space, like a parking space, in a tunnel, or a garage, in that case different types of catastrophic incidents can happen depending upon ventilation. In a confined space, accumulation of hydrogen will occur, which will depend upon whether there is a ventilation or not in that space.

Within the confined space, the released hydrogen mixes with air and may form a flammable mixture, which can burn in presence of an ignition source. In the event of ignition, various thermal effects occur, accompanied by a pressure rise leading to destruction of the building, enclosure, or structure. The process that causes such destructions is known as the pressure peaking phenomenon, which occurs when the rate of hydrogen release is very high in a confined space. When the pressure developed inside the confined space exceeds the strength of that building or enclosure or structure, that can lead to two possibile occurences: an ignited or unignited release of hydrogen depending on whether the ignition source is present or not. The pressure peaking phenomenon is complicated, and it depends on various factors: hydrogen release rate, the size of the enclosure, and the size of the vent.

If gaseous hydrogen released in a confined space and in the absence of an ignition source, hydrogen displaces air, reduces the oxygen concentration and can lead to asphyxiation to persons present in that enclosure or confined space. This is usually observed when the hydrogen release rate is high and the size of vent and enclosure is small.

Hydrogen Buildup—When hydrogen is released in a confined space, it starts accumulating, and the hydrogen concentration builds up in that confined space; the process is called hydrogen buildup. Depending on the ventilation rate, the concentration level of hydrogen, as well as its distribution in the confined space, may vary.

There are three possibilities; if there is **no ventilation** in that space, in that case hydrogen will continue to build up. Depending on how long it has been released before being shut-off, the hydrogen concentration level can even reach up to 100%.

In the case of **natural ventilation** present in the confined space, the buoyancy effect drives hydrogen out, thereby mitigating the catastrophe related to hydrogen buildup.

The third possibility is that there is no ventilation possible, or if it is insufficient to reach the safety targets, then **mechanical ventilation may be considered**. This mechanical ventilation can be achieved through fans or specialized mechanical devices, which vent off, i.e., remove the hydrogen released, thereby aiding to limit the hydrogen concentration in the confined space. Mechanical ventilation can be provided in such confined spaces through venting panels, which are usually present on the roof of the container, building, or enclosure. With hydrogen buildup in a confined space in the presence of an ignition source, it can lead to an explosion, resulting in overpressure. This overpressure caused can be reduced or released with the help of venting panels. During the explosion, when the flames are directed upwards, that aerial overpressure at ground level is limited by use of venting panels. This pressure can be reduced by diverting it out through these venting panels, thereby limiting container deformation or destruction. This is a highly efficient mitigation method that can limit the aftereffects of an explosion in a confined space.

The release of gaseous hydrogen in the absence of an ignition source can lead to loss of containment or confinement due to overpressure, ultimately resulting in a burst or rupture. The process is a physical explosion, and the tank releases all its contents. During the rupture of the pressure vessel, debris, projectiles, or missiles, or fragments of the pressure vessels will fly out in all directions. The rupture of the tank can even occur at the normal pressure, i.e., the operating pressure, if the tank gets degraded due to corrosion attack, fatigue, or embrittlement. Any external mechanical impact on the pressurized vessel can also lead to the rupture. The rupture can also occur at a certain bursting pressure, and this overpressure could have resulted from the overfiling of the tank. Compressed hydrogen tanks are equipped with a pressure relief valve; the failure of pressure relief valve can also lead to rupture. Any outside fire can lead to temperature rise, resulting in expansion and an increase in pressure, changes in the mechanical properties of the tank, leading to the rupture of the pressurized vessel.

Possible Risks, Components, or Accidents with Liquid Hydrogen Release

Liquid hydrogen release can lead to different types of hazards, similar to those associated with gaseous hydrogen, if a liquid hydrogen spill occurs. In such cases, it can undergo either immediate or delayed ignition. The possibilities are immediate ignition, delayed ignition, flash evaporation, partial evaporation, cryogenic boiling pool formation, or BLEVE (Fig. 20.10).

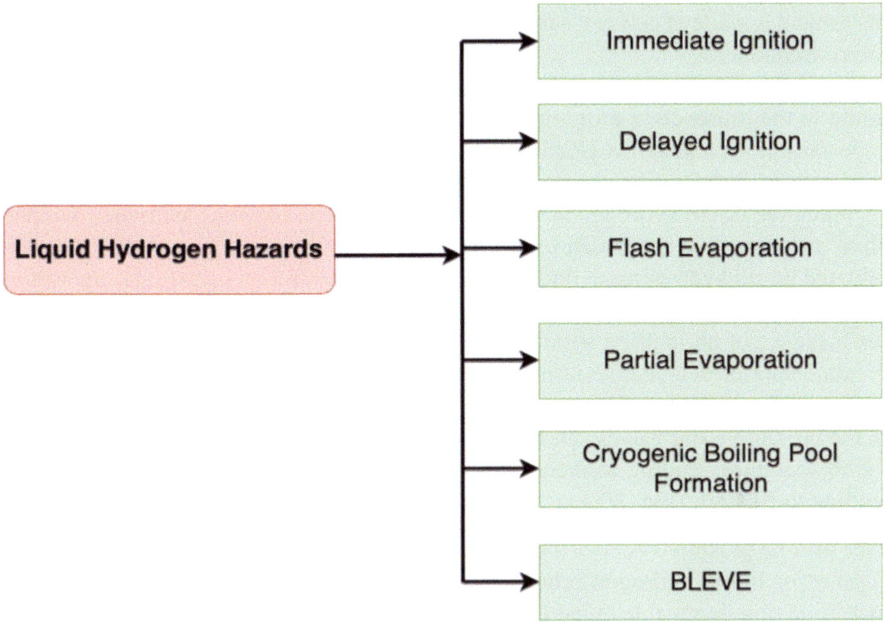

Fig. 20.10 Possible outcomes of liquid hydrogen release

The consequences of liquid hydrogen release or spill, followed by immediate ignition, are the same as those in the case of gaseous hydrogen release. A high-pressure jet is formed following the overpressure effects, and ignition can result in the same hazards as in the case of gaseous hydrogen.

In case of delayed ignition of pressurized liquid hydrogen release, it forms a cloud due to relatively high density. The cloud moves either parallel to the ground horizontally or it will move downward immediately after the release of liquid hydrogen, as a result of condensation of the air and atmospheric humidity, makes it even denser. If the pressure of release is low or the release diameter is large, other phenomena can result such as jet or a rain-out phenomenon, either any one of these or both. During this rain-out phenomenon, hydrogen droplets fall onto the ground, inducing a hydrogen pool.

The spilled hydrogen can form a pool of liquid hydrogen resulting in cryogenic hydrogen pool vaporization. The liquid hydrogen vaporizes due to the temperature difference between liquid hydrogen and the ambient, forming a flammable cloud, a high expansion ratio, results in significant volume increase.

During the evaporation of the liquid hydrogen cloud, the top layers will be boiling films; while close to ground the low temperature results in the ground to freeze. Thus, depending on the environmental conditions, both propagation and dispersion of the cloud will occur, the wind conditions will have a significant impact on this process. Additionally, the ground conditions also impact the phenomenon, such as whether the ground is made of cement, sand, or gravel, it is observed that the gravel aggravates the consequences.

unconfined vapor cloud explosion, the cold and reactive hydrogen–air cloud can ignite in the presence of an ignition source. The presence of obstacles, such as the vaporizer, pipe rack, and vegetation, can lead to flame acceleration. In the worst-case scenario, the deflagration can convert into detonation.

When the liquid hydrogen is stored at a temperature above its saturation temperature and atmospheric pressure, the tank can rupture, leading to a BLEVE. The released liquid hydrogen can undergo either flash evaporation or partial evaporation, forming a pool. It will then condense the surrounding air, forming a visible cloud. In the presence of an ignition source, it can ignite, depending on whether the ignition is immediate or delayed, resulting in a fireball or a flash fire. The liquid hydrogen vessels are equipped with pressure relief valves; however, any failure or blockage of the pressure relief valves can lead to such incidents. Any nearby fire can increase the temperature, leading to increase in pressure inside the liquid hydrogen vessel, leading to BLEVE (Fig. 20.11).

Fire and Explosion—As per the earlier sections, the most damaging risk with any gaseous or liquid hydrogen release is the occurrence of fire or explosion. Fire or explosion of a flammable cloud can occur either in the form of vapor cloud fire, also called flash fire or vapor cloud explosion, and this depends upon whether the flame front, which travels through the flammable mixture, undergoes sufficient acceleration for the deflagration-to-detonation transition with overpressure or blast wave being created as it travels. The initiation of such events can be attributed to the presence of

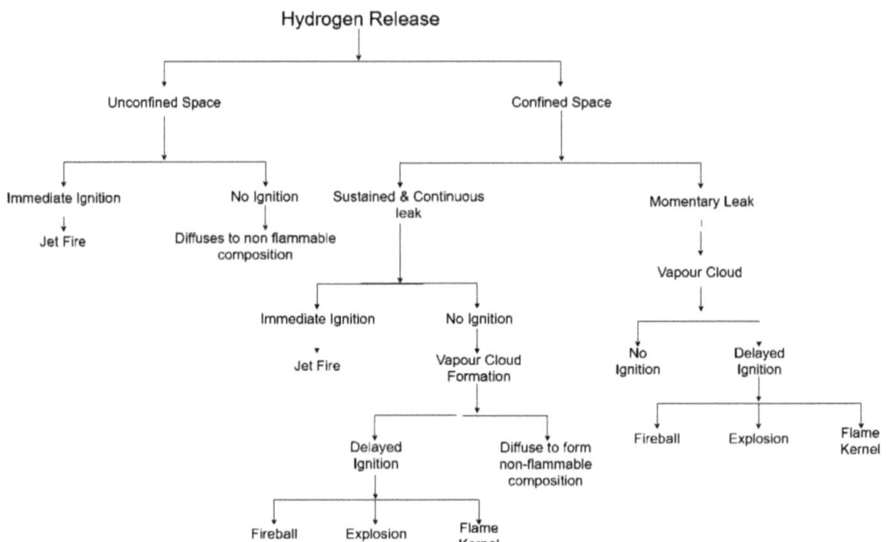

Fig. 20.11 Schematic showing the risk analysis of hydrogen release, including initiating factors and possible outcomes

ignition source, which may include any hot surface, electrostatic charge, the negative Joule–Thomson effect, or diffusion ignition. Since the ignition energy for hydrogen is very small 0.31 mJ (static charge which a human body can detect is 1 mJ) any static charge arising when electrical equipment which is not properly grounded, charge presence in gas flow due to coating on inner surface of piping, static discharge from person, coronary discharge arising between flanges, improper grounding of storage vessels or piping can lead to electrostatic ignition. Diffusion ignition, where the shock wave is formed and propagates through a flammable mixture, with the heating effect of the wave traveling along with the return reflected wave, can create a heating effect, such that the temperature in front of a shock wave can exceed autoignition temperature. Hot surface ignition can be much more pronounced when a flammable mixture is partially heated, providing the required heat to raise the temperature above the autoignition temperature. The negative Joule–Thomson effect, although it does not increase the temperature beyond the autoignition temperature, can become a contributing factor alongside other factors to fire or explosion.

A vapor cloud fire (slow deflagration) or explosion (detonation) can initiate with deflagration and eventually lead to detonation. DDT depends on the environmental conditions, the presence or absence of obstacles or congestion, and the intensity of turbulence in the deflagration flame. During this transition from deflagration to detonation, various intermediate steps are involved. The process begins with a slow deflagration, which evolves into a fast deflagration, then progresses to a pre-detonation stage leading to an overdriven detonation and finally a stable detonation. The biggest culprit seen in case of hydrogen facilities is the hydrogen embrittlement which can cause catastrophic incidents.

Hydrogen Embrittlement

Hydrogen can enter the lattice structure of metals, leading to hydrogen embrittlement. Hydrogen ingression into metal or corrosion attack in the presence of moisture reduces mechanical strength through environmental hydrogen embrittlement, which can be caused by stress corrosion-induced cracking due to the presence of hydrogen, hydrogen-induced cracking, sulfide-induced stress cracking, stress corrosion-related cracking, and debonding of elements induced by hydrogen. Exposure to pure hydrogen for longer durations under high pressure can cause a change in mechanical properties from ductile to brittle, leading to failures.

Hydrogen embrittlement can occur in facilities that deal with liquid hydrogen, particularly when the temperature is below the ductile–brittle phase transition temperature, such as those used for storing or transporting liquid hydrogen. Any instances of thermal stresses should be avoided, like a decrease in temperature from RT to 20 K, and materials that can sustain such a sudden temperature variation should be selected.

At high pressures, hydrogen can permeate and diffuse into mild steel, causing decarburization and leading to embrittlement; thus, an appropriate choice of materials is essential. The regions of concentration of such effects are grain boundaries, welding points, dislocations, and interfaces, among others. Austenitic stainless steel, aluminum alloys, and copper alloys are considered better for hydrogen handling. Both high-strength steel beyond 100 ksi and low-strength steel are more prone to hydrogen embrittlement. Thus, the selection of material, along with protection from corrosion, is important, as any facility fatigue or crack becomes pronounced due to hydrogen embrittlement, which also reduces the toughness of the material.

Hydrogen embrittlement starts with adsorption on the surface of the material, followed by the dissociation of the hydrogen molecule into atomic hydrogen. Hydrogen then diffuses into the material, sits in the crystal structure, causing stresses at dislocations, defects, and voids, leading to microcrack formation and thereby reducing the strength and ductility of the metal or alloy. At high temperature above 220 °C and pressures above 7 bar, hydrogen molecule adsorbs and dissociates into hydrogen atoms, which diffuse into steel and react with non-metallic phases, such as carbides (Fe_3C). This reaction can form methane, which, due to its larger size, cannot diffuse out rather builds up pressure, leading to embrittlement. Beyond 540 °C, in the presence of steam, hydrogen can lead to decarbonization, resulting in the formation of cracks. The presence of intermetallic elements can result in the formation of brittle hydrides when exposed to hydrogen, leading to hydrogen embrittlement. The presence of H_2S and moisture can accelerate sulfide-induced stress cracking.

Hydrogen Risk Assessment

Hydrogen risk analysis provides a reliable connection between the scientific knowledge gained from numerical analysis, theoretical models, experimental data, and the best industrial practices. Industrial practices are crucial when considering risk analysis, as hydrogen has been very safely used in industries for more than a century, establishing a good safety record (Fig. 20.12).

The key steps involved in the hydrogen risk analysis are, first, to identify the key risk drivers that can lead to certain hazardous situations and the influencing parameters that can further aggravate the situation. Then, the secondary measure is to establish mitigation strategies to prevent potential accidents. The outcomes of the hydrogen risk assessment forms the basis for the regulations, codes, and standards. Both the preventive and the protective measures identified form, the basis for the regulation, codes and standards.

The objective of having those at place is to reduce the chances of accidental scenarios, reduce the impact of any hydrogen-related accident at the same time reduce the frequency of such accidents, i.e., occurrence of such accidents.

The various outcomes of hydrogen risk assessment can be utilized to operate different hydrogen-based systems and facilities, as well as to conduct a failure analysis of related hydrogen technologies or systems (Fig. 20.13).

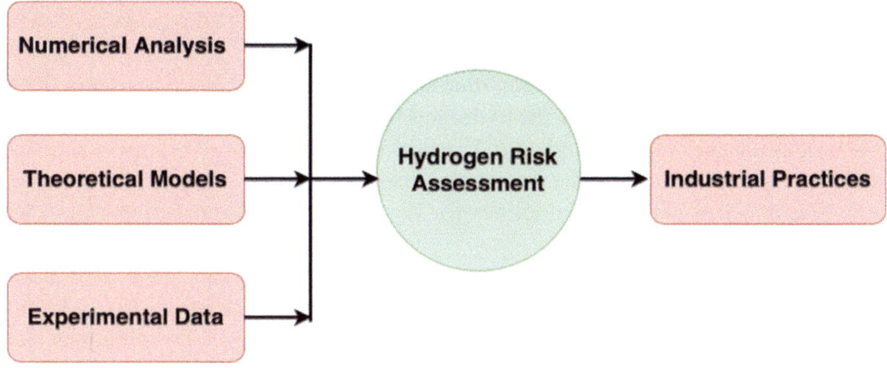

Fig. 20.12 Components of hydrogen risk assessment

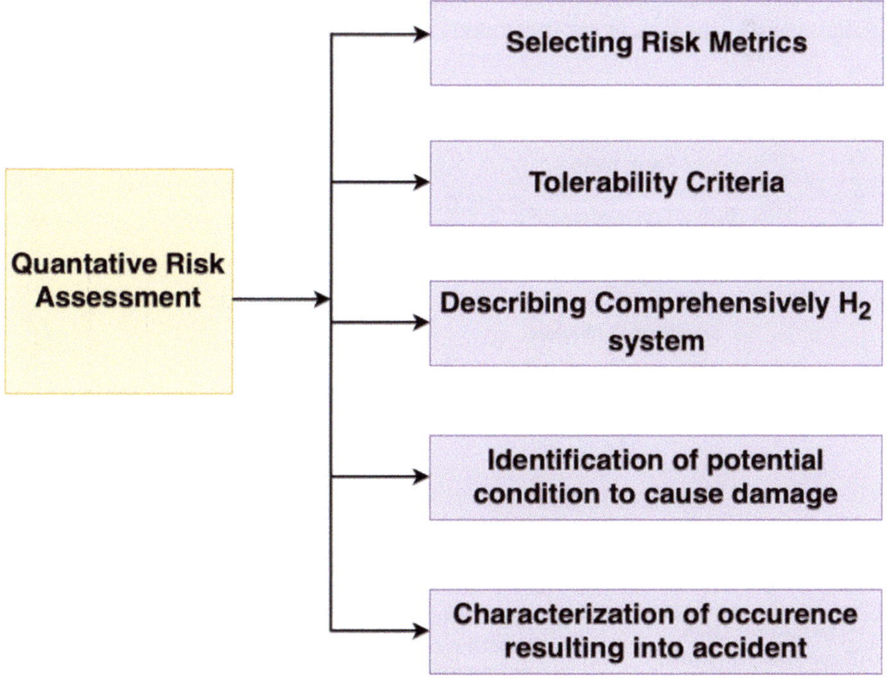

Fig. 20.13 Steps involved in quantitative risk assessment (QRA)

The quantitative risk assessment (QRA) can be conducted using various procedures and approaches, with multiple implementation options being available.

In general, the process involves selecting the risk matrices, identifying the tolerability criteria, describing comprehensively what the hydrogen systems involved are that are being targeted, then identification of the conditions that have resulted in the catastrophe, the potential causes of that damage, and, ultimately, the characterization of the resulting incident.

Thus, the decision whether to continue with the design, build up a system, keep operating the system, or modify the hydrogen-based system or technologies or products becomes easier if the complete risk involved in each of these is known and also what is the acceptable risk to the user is identified through the risk acceptance procedure. The analysis provides guidelines for designing, operating, and maintenance conditions and parameters.

There are various data-driven approaches available that can serve as a baseline for predicting different risks or providing guidelines associated with hydrogen energy systems. Various computational techniques, including CFD techniques, can be employed to understand potential hazardous situations, accidental scenarios, and the physics that could lead to catastrophic incidents.

When performing modeling for the quantitative risk assessment, it can be a single step in QRA or an integration of different steps in QRA. The different models used for qualitative risk assessment are illustrated in Fig. 20.14.

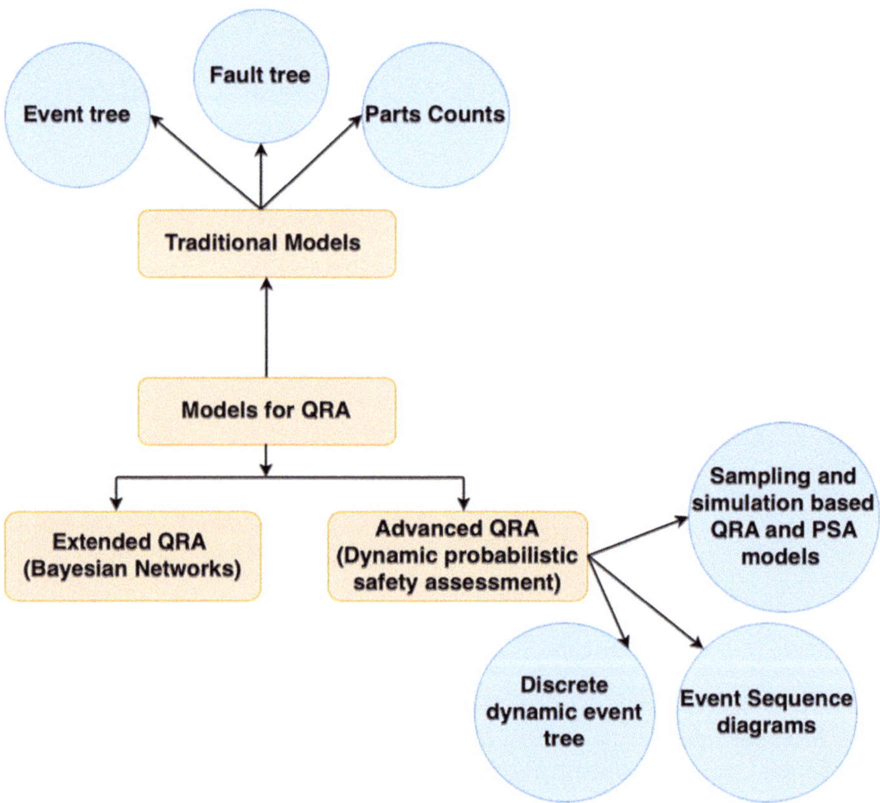

Fig. 20.14 Different models for quantitative risk assessment

These models could be either traditional models, such as the well-known event tree analysis, wherein the first step involves examining the initiating step that has resulted in a catastrophic incident. This could be incidents like a nearby fire that has occurred, over-pressurization or a certain leak that has developed due to embrittlement, thus identifying the initiating event or cause, thereafter looking at what has been the influencing parameters which have led to aggravating the catastrophic incident, and then how these influencing parameters have interacted with the initiating event leading to the consequences. Finally, it is about categorizing the possible outcomes or consequences of the catastrophic incident or initiating event.

The different traditional models including the event tree analysis, fault tree analysis, and parts counts are preferred. Another possibility is the extended quantitative risk assessment, which utilizes Bayesian networks. Finally, the advanced QRA models, which utilize dynamic probabilistic safety assessment tools, include discrete dynamic event trees, event sequence diagrams, sampling and simulation-based quantitative risk assessment, and probabilistic safety assessment models.

Different techniques have been used to model the different hazardous situations; for e.g., CFD techniques are used to understand that in case of a hydrogen release or leak how the released hydrogen will diffuse; how the jet fires can cause the loss of property or the damage to the structures; how an explosion can be hazardous; and what could be the consequences of each of these events. Several advanced CFD tools are available to analyze the consequences of various scenarios and predict potential hazards, as well as to inform the safe design of infrastructure, equipment, or products. It helps in planning, mitigate and take measures, as well as understanding the possible risks. With advanced CFD tools, it is possible to simulate complex geometries that are very difficult to analyze; these tools can accurately replicate the real-world boundary conditions. Although the experimental techniques provide a better insight, they have limitations. They can be very expensive, and at times, large-scale accidents cannot be mimicked or replicated for risk analysis; in such cases, the computational technique plays a decisive role.

Some examples of the quantitative risk assessment are as follows:

(a) It can be used to determine the minimum safe distance between the target, such as a facility, structure, or people, at a hydrogen refueling station. This is particularly helpful when a hydrogen refueling station is located in a congested urban area. In such cases, the design of the station can be worked out with the help of understanding the risks involved.
(b) The risk associated with the hydrogen production process to the neighborhood or surroundings of that plant.
(c) If onboard hydrogen storage system is subjected to fire, what could be the resulting hazards.

These are some of the examples where QRA has been carried out to understand the risks and hazards associated with these situations. Some of these possibilities are also illustrated in Fig. 20.15.

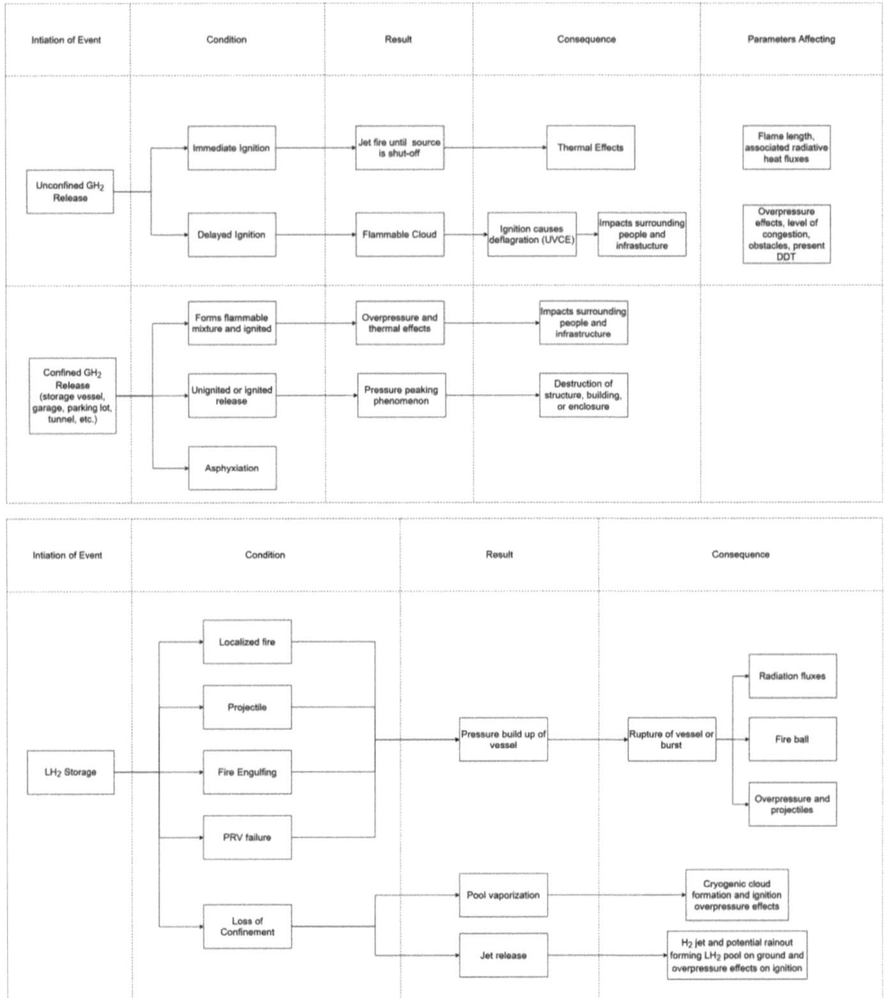

Fig. 20.15 Understanding the possibilities with gaseous and liquid hydrogen release and possible hazards

Ways to Mitigate Hydrogen-Related Hazards

The reasons associated with hydrogen-related hazards and their consequences have been discussed in the earlier sections. The requirement is to avoid such scenarios and to identify and mitigate risk factors very early, thereby preventing the occurrence of accidental situations. There are in general two different ways to mitigate hydrogen-related hazards—(1) active methods and (2) passive methods. These can be implemented both at the (1) equipment or component level or storage facility level and (2) infrastructure level, like enclosure or building.

1. **Mitigation Strategy for Infrastructure**—In order to avoid deflagration or detonation in a building's premises, enclosure, or structure, either active or passive systems can be employed. Passive systems can be used to mitigate accidental situations with infrastructure in two ways: either by reducing the hydrogen gas concentration below the flammability limit with the help of ventilation and explosion diverters, or by post-ignition venting to protect people from injuries or infrastructure from destruction. However, the active method involves use of a hydrogen sensor to detect hydrogen release and vent it out, thereby avoiding the formation of a flammable mixture. Additionally, spray systems can be activated after sensing, preventing ignition.

   Ignition of a flammable mixture leading to deflagration in a confined space can be more dangerous, resulting in overpressure (6–10 times the initial pressure). Furthermore, if obstacles are present, reflected shock can exacerbate the situation. The best way to mitigate deflagration in a closed space is to vent or have explosion vents. The requirement for venting is that the gas or explosion is directed to a safe area to avoid injuries or destruction. There are various standards available for venting design, NFPA 68, BS EN 14994:2007, BS EN 14797:2006, etc.

   Detonation is the worst-case scenario where a complex combination of combustion and shock wave propagates faster than the speed of sound. The detonability limit depends on the geometry and size of the system, the concentration of hydrogen, the nature of the oxidant (for a hydrogen–air mixture, it lies in the range of 18.2–58.9 vol%, for a hydrogen–oxygen mixture, 15–90 vol%), and also on temperature.

   In order to prevent such situations, hydrogen facilities should be preferably in open spaces with no obstacles or congestion, and no ignition sources should be present at the facility. Natural ventilation should be available, and sensors should be installed to facilitate early identification of leaks or releases. Flame arrestors and diluents can be used in order to avoid DDT.
2. **Mitigation at the Storage System or Component Level**—According to fire safety norms, the hazardous situations in the storage systems and piping can be mitigated using passive systems, such as flame arrestors, explosion diverters, and venting to relieve pressure and prevent deflagration. However, these may not be adequate in the event of detonation, and reignition is likely to occur due to the presence of hot, compressed flammable gas downstream. Thus, active systems are required where both the flame and shock need to be mitigated in the piping system. For this, sensors are used to detect the exact location and velocity of the flame front. A detonation arresting system, consisting of a flame arresting spray, is used to extinguish the flame. During the propagation of shock wave, further propagation can be stopped by closing the actuating valve, if it occurs in a piping system.

Safety Features Available at Hydrogen Facilities to Avoid Accidents

The equipment and systems for hydrogen storage or utilization are all well equipped with various safety features. These safety devices are employed to avoid any unprecedented situations or hazards, or catastrophic accidents. Electrolyzers have various process monitoring devices to track pressure and temperature, detect leaks, and identify any malfunctions in the electrolyzer. Leak and flame detectors are installed, which activate to give warning signals, and the corresponding valves shut off the supply automatically in the event of any accidental leakage.

In compressed gaseous hydrogen pressure vessels, the primary risk is the formation of a flammable mixture and its subsequent release, which can lead to fire and thermal aggressions. Thus, to avoid these accidental situations, pressure vessels are well equipped with thermally activated pressure relief devices, which in the case of unprecedented situations could release the excess pressure being created inside the vessel.

In the compressed gaseous hydrogen trailers, which carry the pressure vessels, all the vessels are isolated, and an isolation valve is present with each vessel. The tube trailers also feature thermally activated pressure relief devices (TPRDs), which are especially important when the loaded tanks are Type IV. TPRDs are used to prevent over-pressurization or burst in the event of a fire, and they are located on top and point upward. Therefore, any leak or release should be ideally diffused away promptly. Before the tube trailer begins its journey to deliver hydrogen, various leak-tightness tests are conducted to ensure the safety of the operation.

Pipelines are always well equipped with pressure monitoring devices to detect any leaks into the network. Periodic inspections are performed to detect any leaks in the coating and to prevent any major releases. Cathodic protection is provided to prevent corrosion in pipelines.

If the storage or transport is in the form of liquid organic hydrogen carrier, it is essential to avoid the contact with the of toxic materials of the persons working in the vicinity. Any leakage from the liquid hydrogen carrier containers must be avoided to prevent exposure to dangerous chemicals for both personnel and the environment. Various safety features are included in such containers, including the use of double-skinned tanks and adsorbent materials in case of any leakage, along with spill pools. At the same time, regular monitoring of leaks is being conducted in such tanks. Other safety features and procedures include tightness testing, use of collection trays to prevent the spread of the chemical, overfill protection to limit the risk of chemical spreading due to overfilling.

Gaseous hydrogen refueling stations (HRSs) are highly complex in construction, and a large number of safety measures are taken to prevent any accidents. The hoses and fittings used are qualified and validated to the relevant standards. At the same time, these hoses and fittings are periodically tested, checked, monitored, and replaced whenever necessary. Hydrogen detectors are used at HRS to activate warnings, leading to the automatic shut-off of the valves. Flame detectors are installed at HRS. There are automatic shut-off valves that limit the hydrogen inventory in case of an accidental release. Pressure monitors detect any abnormal pressure drop that could arise due to a leak or piping rupture. Locations of HRS are usually in well-ventilated space to avoid any flammable conditions in case of a release. Hoses are grounded to avoid any static electricity that could be generated during the refueling. Automatic leak testing and flow restrictions are included to automatically close and limit the flow rate upon warning and by closing the hydrogen feeding valves. Hose break-away devices are installed, and emergency punch stops are provided at a distance from the facility to close the hydrogen supply valves in case of an emergency. Shock protection is provided to safeguard the dispenser against mechanical impact. The slabs are grounded to avoid any sparks caused by static electricity.

Fuel cell electric vehicles are equipped with various safety devices, including thermally activated pressure relief devices, leak detectors, and automatic shut-off valves. Proper thermal insulation is provided, the high-pressure line is kept very short, and an excess flow valve is included. Shock detectors, an absorbing shield, and earthing connections are also provided in the fuel cell electric vehicle.

In liquid hydrogen trailers, two safety valves are provided, at least one of which is pneumatic, and all storage vessels are isolated by means of valves. Additionally, a road safety valve is included to evacuate in the event of any overpressure. A rupture disk bursts if the pressure in the trailer rises, and pressure relief devices are included to limit the risk associated with boil-off.

With liquid hydrogen storage tanks, different monitors like the pressure and temperature monitors, as well as level monitors are provided as safety devices. Further pressure relief devices and rupture disk are provided to prevent occurrence of any catastropic incident. The various possible hazards and ways to prevent and mitigate are enlisted in Table 20.2.

Regulations, Codes, and Standards

After understanding the hydrogen-related properties, possible hazards, and past accidents, it is essential to comprehend the various regulations, codes, and standards in place to prevent such catastrophic incidents. Not only are all these requirements necessary, but there is also a dire need to train manpower, not only the people operating the hydrogen-based technologies or plants, but also the personnel responsible for maintaining these facilities. Nevertheless, also the first responders who will attend any unprecedented situation, and the general public who will use these technologies when they reach the marketplace need to be trained. The need to follow best practices, use safety equipment at the facility, train workers, adhere to protocols for operation and maintenance, design and build facilities and infrastructure in accordance with standards, and conduct regular inspections and maintenance is essential for the safe use of hydrogen.

Large quantities of hydrogen have been used safely as a chemical feedstock and industrial gas for decades. Codes, regulations, and standards governing hydrogen storage, distribution, and use at industrial sites are well established. However, the use of hydrogen and hydrogen-based technologies in the consumer market by larger

Table 20.2 Possible hydrogen hazards, the preventive and mitigation measure, and the safety feature in the entire hydrogen value chain

H_2 value chain	Possible hazards	Preventive measures	Mitigation	Safety features	Some standards
Production via electrolysis	Possible Leak; ignition; corrosion; malfunction; static discharge; improper earthing	Monitor P & T Avoid ignition sources. Sensors/detectors for flame, ignition & leak	Activate warning system. Shut-off valves	H_2 sensors. H_2 detection system. Flame detector. P & T monitoring. ATEX-certified equipment	
Gaseous H_2 storage & transport	Leaks; pipeline corrosion; jet fire; ignition/explosion; thermal burns; asphyxiation; fire; vessel rupture; fireball; over-pressurization; projectiles; flash fire; thermal aggression	Isolation of all storage by valves avoid leaks, over-pressurization Natural & forced ventilation at the storage facility Avoid any ignition sources. Avoid electric static charge or short circuit	Venting of pressure Avoiding deflagration-to-detonation transition. Flame & shock mitigation Determine exact location & speed of flame front in pipes. Flame-arresting system. Use actuating valve	Thermally activated pressure relief device. Isolation valves. Leak-tightness test in trailers. H_2 sensors in pipelines. P monitors, cathodic protection. Flame arrestors. Explosion diverter	NFPA 55
LOHC	Chemical spilling; toxicity, leakage; overfilling, low flammability, corrosive; inhalation ingestion dangerous	Avoid spillage, leakage & overfilling	Check filling level, Avoid direct contact with people & facility. Collect any spillage or leakage	Collection tray, Level indicator, Double-walled tanks. Protective use of adsorbent materials	

(continued)

Table 20.2 (continued)

H_2 value chain	Possible hazards	Preventive measures	Mitigation	Safety features	Some standards
Hydrogen Refueling station	Leakage	Qualified as per standards like fitting hoses, dispenser Avoid accidental leakage, Periodic replacement of hoses. Leak-detection flame arrestors' detectors should be present P & T monitors present. Detect any abnormal pressure drop due to leaks. Space should be well ventilated and have provisions for forced ventilation. Grounded structures hose grounded	Avoid formation of flammable mixtures. Avoid ignition sources. Limit flow rate. Avoid mechanical impact Prevent sparks or electrical short circuit. Leak test before filling	Active warning, shut-off valves in case of leakage, Curtail H_2 inventory. Use of flame detectors & arrestors. Detect abnormal pressure drop due to leak and act accordingly. Limit flow rate, close Hydrogen feeding valves, Maintain safe distancing	Hydrogen detector, flame detectors, automatic shut-off valves, P & T Monitors Qualified & validated as per standard hose. ATEX-certified equipment. Grounded hose, flow restrictors, Leak testers, Hose breakaway device on dispenser. Emergency stop device with push button option away from dispenser

(continued)

Table 20.2 (continued)

H_2 value chain	Possible hazards	Preventive measures	Mitigation	Safety features	Some standards
Fuel Cell Electric Vehicles	Rupture of pressure vessel. Case of accident or collision. Over turning release of Hydrogen	Thermal insulation around gas storage. Avoid over-pressurization, high-pressure line should be as short as possible. Earthing connection to the car refueling nozzle	Prevent any ignition source pressure monitoring. Hydrogen leak detection limit hydrogen inventory in case of leak. Limit flow rate in case of release or rupture of piping. Activate warning and shutdown valves	TPRD Hydrogen leak detector. Automatic shut-off valve. Excess flow valve shock detector shock absorbing shield. Hydrogen feeding valve, proper earthing connection	
Liquid Hydrogen storage & delivery	Ignition, cryogenic boiling pool formation, BLEVE, vapor cloud explosion, flash fire, fireball, over pressure creation, projectiles, jet fire, fire pool ignition	All storage should be isolated by valve. Avoid over pressure, limit risk of boil-off. Temperature and pressure monitoring, level monitoring, insulation detection, periodic inspection. Avoid ignition source nearby. Maintain safe distance	Difficult to mitigate one aftereffects of spillage or rupture of liquid hydrogen vessels	Pneumatic safety valves, road safety valve for transport, rupture disk, PRD	ISO13984:1999(en), ISO13985:2006(en), NFPA50B, NFPA2, CGA P-12–2017, CGA H-3–2019

public will be new and is expected to grow as hydrogen market demand rises in several sectors. Thus, the development of codes and standards for this market segment is crucial to create an environment that accepts commercial hydrogen products and systems, fosters confidence in them, and promotes public acceptance. Before considering the existing codes, standards, and regulations, the definitions of each of these terms are as follows:

Regulations—Regulations are the legal requirements or laws. These need to be followed and respected by all citizens. These are treated as the highest level of rules, which not only describe the physical and operational features of a particular product or technology, but also indicate the performance indices. Not only performance indices, but also the limiting values such as tolerances, efficiencies are indicated. At the same time, regulations impose an implicit restriction on the use of non-standard or non-compliant items or products.

An important point to note is that standards are sufficient for business use. However, regulations are required to save lives, ensure public safety, and maintain environmental compatibility. Any given product or system can not unduly compromise those, as regulations are laws.

Codes—Codes are defined as legal binding. These can be a collection of laws, ordinances, or regulations, or certain statutory requirements that are being adopted by the government's legislative authority. Codes are involved in ensuring the adequate physical properties or health conditions of a product, equipment, or technology.

For code, it must be predictable, acceptable, practical, and consistent, with a minimum set of harmonized requirements applied to protect life and ensure the safety and well-being of the public. Thus, codes are adopted by government bodies to regulate construction or manufacturing within a particular jurisdiction, i.e., a city or country. The primary purpose of having code is to regulate a new or proposed product, equipment, or construction.

Codes of Practice—Codes of practice explain the basic functions for handling safely and problem preventive maintenance, and they are intended to guarantee trouble-free operation. Thus, these are basic rules that may not be well documented written in text, but they are being followed in industrial practices. For example, in the automotive industry, we know that there is a sequence in which the paddles are included in a car, that the steering wheel is turned in a particular direction. These are not well documented, but they are widely accepted and are referred to as a code of practice. These can be different for each equipment or product, but they share some basic elements that are based on generic features and functionalities of the technology. These arebased on the common understanding of the relevant people and how they use a particular product or system. Codes of practice can also be referred to as industry standards. They may not be derived from formal agreements or written documents. However, they exist and are followed as the best, most efficient, and safest practice.

Standards—The context of standards is that if there is a particular technology which has to interface with other systems or if there are specific standard components which need to be used with different platforms, then there should be a compatibility which needs to be documented and acceptable by not only the manufacturer of the component or technology but also by the user and to ensure that the document is standard, which links all the stakeholders together. There has to be a written document which is agreed upon in terms of its contents and scope by the developers, designers, and manufacturers of the product and systems in that given field of technology. Such document is called as standard and is developed and maintained by special organizations or interest groups called as SDO, i.e., Standards Development Organization. These are used to build a particular technology around the case and a branch of technology.

For global markets, standards should be developed internationally to improve the compatibility with the market and minimize duplication or confusion. International standards are developed by the International Organization for Standards (ISO) and the International Electrotechnical Commission (IEC). The standards developed by ISO and IEC are well recognized, accepted, and adopted at both national, regional, and international levels by all stakeholders, including manufacturers, trade organizations, purchasers, consumers, certification bodies, testing laboratories, and authorities. The standards are developed considering the best practices and experience of industry, researchers, consumers, regulators globally, and cover the common requirements in various countries. Standards constitute the basis to avoid the unnecessary technical trade barriers. Thus, the standards are meant to harmonize the requirements for interfacing technical systems and are agreed upon by different organizations, not as a legal requirement, but as the state of the art. Standards require a legal status if they are referenced in regulations, laws, or other legally binding documents. The inclusion of standards in law through referencing is achieved by means of a detailed statement of technical requirements, which is included in the code or regulation, either in its entirety or in its relevant parts. The use of ISO or IEC standards in regulations is intended to support regional, national, or global policies.

Standards can be referenced directly, by quoting them within the legal text using their title or identification number. The inclusion of identification number avoids reproducing the full standard in legal documents and eliminates the need for permission from regulators to use copyrighted material. The entire standard or a part of it can be included to support the regulation. The direct referencing can be either dated or undated. In dated any revision or amendment to standards also needs to be considered by the authority, and required modifications in legal writing should include the amendment or revision. At the same time, undated referencing - with quoting only the title or identification number of a standard, without specifying a date - provides much more flexibility. In the event of any amendment or revision to the standard, the latest version should be adopted as the reference.

Indirect referencing involves recognizing and registering the standard as an official document outside the regulation text. In this case, the list of standards deemed suitable is compiled and published through an official process, with control remaining with the regulator. In the case of amendment or revision to the standard, only the list needs to be updated. The date of standards publication needs to be included to ensure its validity and certainty, and it is required to be kept up to date and accessible to users through website and other means.

Absence of globally harmonized standards can lead to technical trade barriers (TBTs). Thus, common references and harmonized standards recognized between regions and countries are essential for international trade and competition. International standards developed and accepted by trade partners serve as a common trade language and the key recipe for the World Trade Organization's agreement on TBT.

Technical Specification—A technical specification is a pre-standardization document that has not yet reached to the level of an international standard. Technical specification can be a document that is still under preparation, whose acceptance as a standard is uncertain, or whose approval is pending, or that requires further discussion before it can be considered for an international standard. Standard organizations, such as ISO or IEC committees or subcommittees, review the technical specification document submitted within three years of its publication. The committee can also re-examine the groundwork behind technical specification during this period to determine whether it can be converted into an international standard.

Another document, similar to the technical specification, is called the publicly available specification (PAS). A PAS is a document that has not yet reached the stage of meeting the requirements of an international standard. Thus, PAS can be considered an intermediate document before being converted to a full-fledged standard, and a period of three years, extendable to another three years, is given to complete or withdraw.

Similarly, a technical report is a document that does not meet the level of international standard; it only provides information about a subject through surveys and is meant for providing information purposes only, with no standing, and can be withdrawn if required.

The society of Automotive Engineers is responsible for preparing documents related to automotive engineering and consists of members from the automotive industry and other engineering experts. SAE is responsible for developing automotive standards, documents, reports, and best practices information. These outcomes are very well accepted to guide the procedure, specifications, and practices.

The primary distinction between codes, standards, and regulations is that regulations are procedures or rules that must be followed and enforced by law. Codes are typically provided regarding the environment, construction, product, or system involved. Standards are meant to ensure that the business, developed for the components and systems, includes the testing protocols that need to be followed.

Governing Bodies

The organizations that prepare the codes are the International Code Council (ICC) and the National Fire Protection Association (NFPA) which develop different codes for hydrogen and fuel cells. For developing the different standards and in order to have uniform standards, some of these could be accepted by countries and are the national standards, but some of the standards are globally accepted and become International Standards. The international standards are being developed by various organizations like International Standards Organization (ISO), International Electrotechnical Commission IEC or the American Society of Mechanical Engineers (ASME), American Petroleum Institute (API), American National Standards Institute (ANSI), The European Committee for Standardization (CEN), Institute of Electrical and Electronics Engineers (IEEE). Besides these key bodies the organizations involved in developed International standards include: Compressed gas association (CGA), International Society of Automation (ISA), Occupational Safety and Health Administration (OSHA), Society of Automotive Engineers (SAE), National Fire Protection Association (NFPA), Underwriters Laboratory (UL), American Gas Association (AGA), Institution of Gas Engineers and Managers (IGEM), Manufacturers Standardization Society of the Valve and Fittings Industry (MSS), Association of Materials Protection and Performance (AMPP), European Industrial Gas Association (EIGA), United States Department of Energy (US DOE), GuoBiao Standards (GB), and CSA Group.

The committees working toward developing standards comprises of members from industry, government, fire safety specialist, officials who are involved in providing the permits, people from the different organizations like standard development organization or code development organizations. Different bodies or members which form such committees, to come up with different standards, codes, and regulations are mentioned in Fig. 20.16.

Fig. 20.16 Different bodies or members involved in the establishing of regulations, codes, and standards

Fig. 20.17 Entities in India responsible for hydrogen-related standards

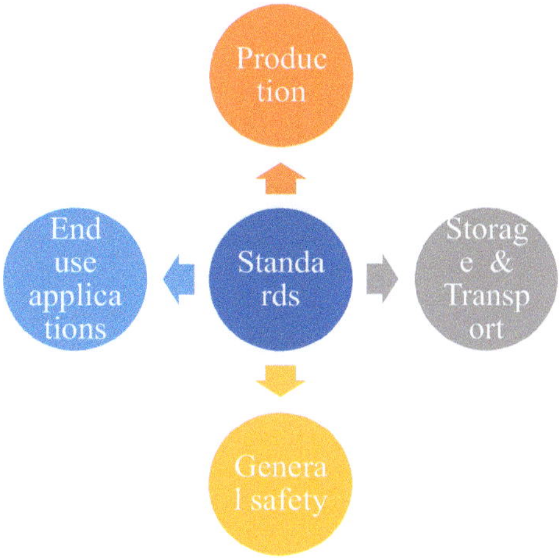

BIS
OISD
PESO
PNGRB
MORTH

Fig. 20.18 Standards in the field of hydrogen

Production
End use applications
Standards
Storage & Transport
General safety

In India, the launch of National Green Hydrogen Mission has accelerated the developments in the entire hydrogen energy value chain, aiming to make India as the global hub in Green Hydrogen. The target is to meet domestic demand and become an exporter of hydrogen and its derivatives. Standards are being developed, adopted, or under development in the fields of: hydrogen production, hydrogen storage and transportation, general hydrogen safety, and end-use. Ministry of New and Renewable Energy is the key ministry coordinating and implementing the mission. Entities in India responsible for developing or adopting the standards in the area of hydrogen energy are Bureau of Indian Standards (BIS), Oil Industry Safety Directorate (OISD), Ministry of Road Transport and Highways (MoRTH), Petroleum and Explosives Safety Organizations (PESO), Petroleum and Natural Gas Regulatory Board (PNGRB), as mentioned in Figs. 20.17 and 20.18.

In India, 87 standards have been either adopted or developed, while 59 are under development as on 2024. Additionally, 52 standards that exist globally can be considered for adoption in India. Thus, in all, there are 8 standards in production, 102 in storage and transport, 52 in end-use applications, and 39 in general safety. The distribution is as follows:

While 52 are being to be considered for adoption, there are 3 areas in terms of components where there are no global standards (Tables 20.3 and 20.4).

Table 20.3 Current status of hydrogen-related standards in India

	Adopted from international standards or developed in India	Under development
Hydrogen production	3	0
Hydrogen storage and transport	87	36
End-use application	33	7
General safety	14	16
Total	87	59

Table 20.4 A table summarizing some of the standards, both Indian and international, is as follows:

	Hydrogen value chain domain	International standard	Indian standard
1	Hydrogen production from hydrocarbons and biomass	ISO 16110-1:2007	IS 16512 (Part 1):2016
2	Hydrogen Production from Electrolysis	ISO 22734:2019	IS 16509:2020
3	Quality of hydrogen from electrolyzers for fuel cell vehicles	ISO 14687:2019	
4	Pressure swing adsorption	SO/TS 19883:2017	
5	Test protocols for performance for H2 production from hydrocarbon reforming	ISO 16110-2:2010	
6	Measure the performance of electrolyzers	(ISO/AWI 22734-1)	
7	Compressed gaseous hydrogen storage	ISO 11114-4:2017104 and ASME STP/PT-003-2005 and NFPA 55	IS 7285:Part 1:2018
8	Liquid hydrogen storage	ISO 13985:2006 and EIGA (EIGA Doc 06/19) and NFPA 55	
9	Onboard storage on FCEV	ISO 19881:2018	IS 16735:2018
10	Pipeline Transport	ISO 19880-1:2020 and ASME B31.12-2019	
11	Pure hydrogen and blended hydrogen pipelines	CGA G-5.6 and EIGA Doc 121/14	
12	Transporting compressed gaseous hydrogen by road	ISO 19881:2018	IS 8198:2004
13	Fuel cell powered vehicles	ISO 12619-1 and ISO 23828:2022 and SAE J2579_201806	IS 23828:2013 and IS 12619-1

(continued)

Table 20.4 (continued)

	Hydrogen value chain domain	International standard	Indian standard
14	TPRD	ISO 19882:2018	
15	Heavy duty industrial vehicles	CSA HPIT 1-2015	
16	BOS of Fuel cell systems	EC 62,282 code	
17	Complete hydrogen dispensing system	ISO 19880-1:2020	
18	Handling of methanol as hydrogen carrier		IS 7444:197
19	Safe handling of ammonia	AS/NZS 2022:2003 and ANSI/CGA G-2.1-2014	IS 4IS 517544:2000 and S 8198:Part 7:1988
20	Hydrogen safety across value chain	ISO/TR 15916:2015	IS 15201:2002

References

1. M. Calabrese, M. Portarapillo, A. Di Nardo, V. Venezia, M. Turco, G. Luciani, A. Di Benedetto, Hydrogen safety challenges: a comprehensive review on production, storage, transport, utilisation, and CFD based consequence and risk assessment. Energies **17**, 1350 (2024)
2. F. Yang, T. Wang, X. Deng, J. Dang, Z. Huang, S. Hu, Y. Li, M. Ouyang, Review on hydrogen safety issues: incident statistics, hydrogen diffusion, and detonation process. Int. J. Hydrogen Energy **46**, 31467–31488 (2021)
3. S. C. Weiner, L.L. Fassbender, Lessons learned from safety events. Int. J. Hydrog. Energy **37**, 17358–17363 (2012)
4. F. Yang, T. Wang, X. Deng, J. Dang, Z. Huang, H. Song, Y. Li, M. Ouyang, Review on hydrogen safety issues: incident statistics, hydrogen diffusion, and detonation process. Int. J. Hydrogen Energy **46**, 31467–31488 (2021)
5. E. Abohamzeh, F. Khan, F. Salehi, M. Sheikholeslami, R. Abbassi, Review of hydrogen safety during storage, transmission, and applications processes. J. Loss Prev. Process. Ind. **72**, 104569 (2021)
6. B. J. Lowesmith, G. Hankinson, S. Chynowyth, Safety issues of the liquefaction, storage and transportation of liquid hydrogen: an analysis of incidents and HAZIDS. Int. J. Hydrog. Energy **39**, 20516–20521 (2014)
7. Accelerating the Implementation of the National Green Hydrogen Mission Assessment of Standards to Enable the Ecosystem, report by CEEW (2023)
8. Green hydrogen standards and approval systems in India, Report by MNRE and CEEW (2024)

Chapter 21
Hydrogen Utilization and Its Global Status

A crucial part of the hydrogen energy value chain is its utilization for various applications and in different sectors. Hydrogen has conventionally been used in sectors such as petroleum refining and chemical industry; however, increasing concerns over climate change and growing awareness of the need to reduce GHG emissions have driven strong global efforts to shift energy usage and lower emissions. Hydrogen is an excellent energy carrier, that can offer solutions to many energy-related challenges, including providing energy security, replacing the use of fossil fuels in the major sectors, and thus playing a key role in decarbonization. Considering the current scenario, a clear global energy crisis could be forecasted, as different geopolitical situations are changing rapidly with the chances of disruptions becoming quite prominent. The overdependence of some of fossil fuels and subsequent supply cuts can lead to market vulnerabilities. Thus, in the current conditions, the self-reliance of a country in fuel sector, with its own strengths in renewables and the various energy storage solutions, is of prime importance. Together with the clearly visible, drastic global climate change implications in terms of flash floods, drought, unseasonal rainfalls, poor air quality indices, shift in seasons, and the emergence of new diseases are impacting every individual's life, making it essential to find substitutes for the current fossil fuel usage.

Hydrogen received interest in earlier years, as well in the 1970s, when the interest was its use for FCEV, in the 1990s and early 2000s, again a lot of global interest was seen, when the climate change implications were the reason. Still in all these times, the major utilization was expected in the transportation segment, and the interest did not sustain much. However, the global interest in the last few years has not been confined only to transportation applications; rather the interest has been multidimensional, covering a wide range of sectors. The interest is quite evident across the various stakeholders including industry, academia, researchers, policymakers, government, and investors. Hydrogen has emerged as a strong contender not only for transportation, but also for sectors and fields where there is no other possible solution to decarbonization other than hydrogen.

P. Sharma, *A Textbook on Hydrogen Energy*,
https://doi.org/10.1007/978-981-95-4601-5_21

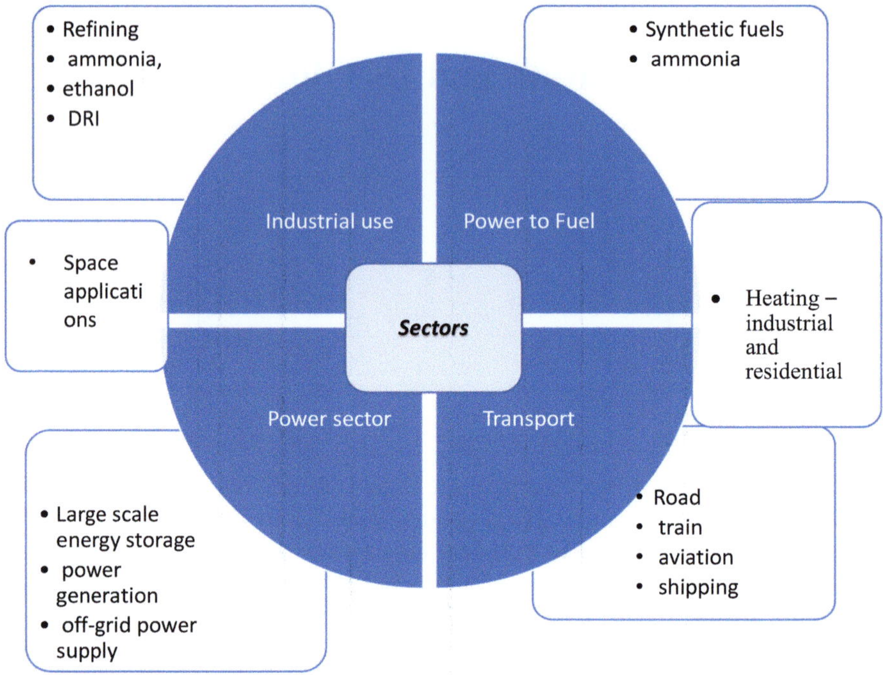

Fig. 21.1 Different sectors of hydrogen utilization

A. Hydrogen Utilization in various sectors - Hydrogen has been used in refineries and for fertilizer manufacturing, where ammonia is the major requirement, for several decades. Another major conventional sector is the chemical industry where hydrogen is used for manufacturing various chemicals, such as ammonia, methanol, and dimethyl ether. Hydrogen is also used for direct reduction of iron in the iron and steel industry. The new and future sectors, where hydrogen demand will increase, include transport, power generation and storage, as well as domestic applications such as heating and conversion into liquid fuels, i.e., Fischer–Tropsch fuels.

The different sectors where hydrogen will find its use are shown in Fig. 21.1.

Global hydrogen demand as, reported in the IEA report, is 97Mt, with China having the highest demand, followed by the USA and the Middle East in the year 2023. The demand was primarily in the conventional sectors, including refineries and chemical industry. With most of the demand being met by hydrogen produced from fossil fuels. The demand increased from previous years, but it is being observed that the demand was driven by industrial growth or economic development, rather than policy support.

1. Hydrogen Use in Refineries

- The refineries that used to be major producers of hydrogen, with the fuel norms getting stricter and stricter, are now major consumers of hydrogen. In refineries, hydrogen is required for various processes including hydrocracking, hydrotreatment, hydrodenitrogenation, hydrodesulfurization processes, isomerization. The hydrogen produced in the refineries is being consumed internally, i.e., captive hydrogen for converting the crude oil into useful products like gasoline, diesel, jet fuel, kerosene, liquefied petroleum gas, etc., which drive the economy of any country. In refineries, the use of hydrogen is highly optimized and its management is crucial. The increasing use of hydrogen in refineries over time is due to the growing demand for clean fuels, stricter regulations, declining heavy fuel oil demand, and the rising supply of heavy and sour crude. Therefore, optimizing the hydrogen network in refineries is crucial. Hydrogen is required in various hydroprocessing, like hydrotreating used for removing the various contaminants and upgrading the oil fraction; this is required due to environmental standards getting stricter with the BS-VI being implemented; besides, the impurities like sulfur, nitrogen, aromatics, oxygenates, etc. need to be taken care of to protect the downstream process catalysts. In hydrotreating, the sulfur or nitrogen, etc., containing impurities are hydrogenated to convert into H_2S and NH_3 which are later scrubbed to remove from the stream. Besides, hydrocracking, which involves cracking and the hydrogenation of heavier hydrocarbons like gas oils to get the lighter fractions (naptha, distillates, etc.), i.e., the refined fuels which have smaller molecules and a higher H/C ratio, as the lighter the fraction more is value associated with the fuel, all this requires hydrogen in the process. Thus, multiple requirements have increased the demand for hydrogen in refineries including the change in demand for products, changes in in feedstock supply, environmental regulations, and technology development.

$$C_{12}H_8S + 5H_2 \rightarrow C_{12}H_{16} + H_2S \tag{21.1}$$

$$C_{12}H_9N + 8H_2 \rightarrow C_{12}H_{22} + NH_3 \tag{21.2}$$

$$R - SH + H_2 \rightarrow R - H + H_2S \tag{21.3}$$

The other processes that require hydrogen in refineries include hydrogenation of olefins, hydrodearomization, hydrodeoxygenation, hydrodemetallization, etc. (Fig. 21.2).

Most of these processes use gray hydrogen where the hydrogen is produced via processes like steam methane reforming of natural gas, naptha, liquefied petroleum gas, or partial oxidation of oil in a hydrogen generation unit. A certain amount of hydrogen is also obtained through various refinery processes, which is

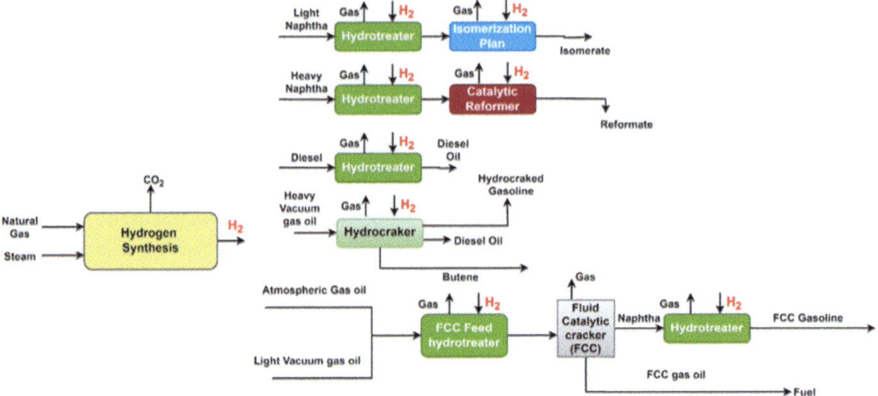

Fig. 21.2 Schematic showing hydrogen use in various refinery processes

used internally, and also as merchant hydrogen for these processes. The requirement of hydrogen in a refinery depends on both the feedstock and the required product characteristics. The need is to replace the gray hydrogen currently used with green hydrogen in the future, where the hydrogen will be primarily produced from renewables like solar or wind, or from biomass, thereby reducing the greenhouse gas emissions.

Global and Indian Status

According to the IEA report 2024, the global demand for hydrogen in refineries is 43 million tons in 2023. Out of the 43 Mt, the gray hydrogen component was 45%, in contrast 35% of the demand was met from the by-product hydrogen (naptha reforming and steam cracking) obtained as a part of different refinery processes. Still, this amount is insufficient for meeting the demand in the refinery, while the remaining 20% is obtained from merchant hydrogen. Most of the hydrogen used in refineries is unabated hydrogen, where the emissions have been released into the environment. The impact of using green hydrogen from either electrolysis or biomass gasification, or having carbon capture and sequestration, will add to the cost. In the near term, integrating with CCS could be a possibility, and in the long term, both CCS, together with the use of green hydrogen from electrolyzers, could be the solution. However, the cost of green hydrogen will depend on whether the electrolyzers installed are solely meant for green hydrogen production of large capacity or based on curtailed electricity, where the cost of electricity remains the major factor in deciding the LCOH in large electrolyzers with 24 $\times$ 7 operations. Similarly, in biomass gasification, the cost and availability of feedstock remain a challenge.

The low-emission hydrogen demand was 250 ktpa in 2023 as per the IEA report and is expected to grow in the years to come. Some of the large capacity projects as per the IEA includes GALP in Portugal Sines refinery and Shell's refinery in Rhineland, Germany, with two electrolyzers of capacity 100 MW, BP's HyVal project at Castellon (Spain) with 25 MW capacity extendable to 2GW in future, two projects with one from Air Liquide and another from Air Products for steam methane reforming integrated with CCUS (both retrofitting existing hydrogen production units) linked to the Porthos carbon capture and storage (CCS) project in Rotterdam, Shell's Polaris project, in the Scotford refinery and chemicals complex (Canada) integrated with CCS. The fully functional projects along with those which are in the early stage of development or FID can increase the capacity of low-emission hydrogen in refineries to 1.5 Mtpa. A lot of progress is seen in Europe, North America, and India, with large projects announcement. Along with traditional refineries, interest is also growing toward biorefineries for upgrading biofuels. Two such projects are being developed in Canada and France. Additionally, at the same time there is a need to explore how pure hydrogen can be directly used in furnace. Projects like Essar in Stanlow refinery in the UK and, Idemitsu in Japan using ammonia in furnace.

The REFHYNE Project in Germany, funded by the joint undertaking of the European Commission and supported by ITM and Shell, is targeted utilizing green hydrogen in an existing refinery for the refinery processes.

In India, there are currently 23 refineries with a total refining capacity of 258,116 MT capacity as of 01.04.2025 [1].

Thus, there are options for using green hydrogen in refineries by retrofitting the existing refineries for operations where gray hydrogen was to be replaced with green hydrogen, or including carbon capture and sequestration. Besides, whatever percentage of merchant hydrogen is used in refineries also need to be replaced with green hydrogen. The possible challenges in this makeover include cost increase, which will reduce margins of refineries. It is difficult to implement these, due to already established operations in place. Associated CO_2 emissions with hydrogen production are 920 Mt. However, the current demand of hydrogen is met by various fossil fuels. To address the GHG emissions issues and reduced emissions, two possibilities are either to integrate carbon capture, use, and sequestration or use low-carbon electrolytic hydrogen. The two aspects considered are that capturing hydrogen from SMR plants is the lowest cost option and integrating CCUS with it will also require a policy support.

The demand for hydrogen will increase due to growing refining activity and rising requirements for treatment until 2030, after which there are studies which show that there will be a decline in demand. There are different scenarios, such as net zero emissions scenario which predicts that demand will decrease to 25 million tons by 2030 and drop to 10 million tons by 2050. According to the announced pledged scenarios and the net zero emissions scenario, which is more optimistic, findings differ. The demand is expected to drop after 2030; other factors will drive such as improvements in the efficiency. There will also be increased electrification which will contribute to slowing down the demand after 2030.

2. Hydrogen Use in Chemical Industry

Hydrogen is an ingredient in the chemical formula of most of the chemicals; however, the major requirement in the chemical industry is for the production of ammonia for fertilizers and for synthesis of methanol, with 60% of the total demand in ammonia while 30% for methanol synthesis and rest 10% is for iron and steel industry in the process of direct reduction of iron (DRI). The total demand for hydrogen in chemical industry in the 2023 was 54 Mt. Ammonia is primarily used in fertilizer industry for urea and ammonium nitrate synthesis; rest of its use is in the area of speciality chemicals, synthetic fibers, etc. Methanol is required for the synthesis of several chemicals, including methyl methacrylate and formaldehyde. Methanol is also used for the conversion to fuels, producing gasoline. Besides methanol to gasoline, the other uses of methanol include methanol to olefins, methanol to plastics, methanol to aromatics as well as for other high-value products. Some of the chemical processes produce hydrogen as a by-product. One of the most important sources of this by-product hydrogen is the Chlor-alkali process, where electrolysis of brine gives hydrogen. Additionally, steam cracking of oil products also results in by-product hydrogen. China is the major consumer of hydrogen for industrial use followed by Middle East, North America, India, and Europe.

Besides, it is expected that the demand of hydrogen in the industrial sector will increase; however to decarbonize this sector requirement will be use of low-emission hydrogen. This will require having new renewables-based hydrogen generation plants together with retrofitting exists chemical plants for carbon capture and sequestration. Besides the conventional chemical processes, there will be increasing demand for new processes, such as industrial heating and DRI, where the demand of hydrogen is expected to rise. The projects that will be using low-emission hydrogen are at FID and will amount to around 750 ktpa capacity by 2030. On the top of the hydrogen being produced, a significant quantity merchant hydrogen is required in the chemical production plants and that demand will grow with increase in demand of chemicals.

A significant portion of energy demand in industries stems from processes that require heating, such as industrial boilers and for refining, where fossil fuels are currently used. Several projects have been initiated globally to use hydrogen in industrial boilers, substituting the fossil fuels with ammonia where low-temperature heat is required in industrial boilers, in glass manufacturing, blended hydrogen use in cement kilns and asphalt production, as well as for heat treatment in the aluminum industry.

Chemical industry is a major sector that utilizes about 40% of hydrogen in pure and or as mixed gases of the total global hydrogen demand. The sector is 2nd and 3rd largest hydrogen consumer used for producing ammonia, fertilizers and methanol.

The two major industries where hydrogen is obtained as are the chlor-alkali processes, and another is upgrading or cracking the high-value chemical.

The demand of hydrogen in the chemical industry, is expected to grow and will reach 57 million tons per year by 2030.

On the other hand, the rate of increase of demand may slow down, with an increase in energy efficiency and the material efficiency strategies when being followed in this sector alongwith mandatory norms associated with the reducing emissions in future.

Emissions could be reduced by using cleaner paths, which including use of low carbon hydrogen and incorporating carbon capture use and sequestration. However, use of these methods will add to the energy input as well as cost. With the use of low cost renewable electricity in countries with high renewable energy potential,, the green hydrogen production and use in this sector can become cost competitive.

However, the additional requirement with the renewables-based electricity will be need for buffer storage, due to the associated intermittency of the renewables. Further, if the renewable power plants are not co-located with the chemical industry, there will be an additional cost associated with the transportation of hydrogen.

3. Hydrogen Use in Iron and Steel Industry

According to the data from World Steel Association, the global crude steel production in 2024 was reported to be 1884.6 Mt, and it is further expected to increase at a rate of 1.2% per year. China is the world's largest crude steel producer, followed by India, Japan, and the USA.

The iron and steel industry is one of the most carbon-intensive industries which is contributing roughly 9% of energy related. This industry currently employs two major steel production pathways: blast furnace–basic oxygen furnace (BF–BOF) and the electric arc furnace (EAF). The dominant (71% of total production) BF–BOF route is economical. The emissions from BF-BOF are very high, exceeding 2 ton CO_2/ton steel produced, is a major contributor of GHG emissions. There are several emerging low-carbon steel production processes ranging from blast furnace with carbon capture and sequestration-top gas recycling (CCS-TGR), natural gas-based direct reduction of iron ore (DRI), hydrogen-based direct reduction of iron ore (HDRI), etc. The emphasis of the government globally is on alternate technologies and processes that could reduce GHG emissions in steel manufacturing, increase efficiency of the processes, and improve and waste utilization.

It is expected that the energy consumption of the steel production will cut down by around 14% by 2050, led by technological advancements and fuel switching. Processes with a low-carbon footprint that have the potential to decarbonize the iron and steel industry are CCS integrated with BF–BOF, CCS integrated DRI, and green hydrogen-based DRI. The most feasible and practical alternative to BF–BOF process is the green hydrogen-based direct reduction (HDRI) integrated with EAF. The use of green hydrogen to decarbonize the iron and steel industry can reduce CO_2 emissions by 2.3 Gt per year globally (IEA report).

3.1 Hydrogen-Based Direct Reduction of Iron Ore

In direct reduction, the iron ore is directly reduced to sponge iron without melting it to liquid iron. This process currently uses gaseous reducing agents, which are produced by coal gasification or the steam reforming of natural gas. The penetration of DRI plants in the USA is growing due to the availability of cheap shale gas. The emissions are low from the process, but are still not enough to reach the decarbonization targets; and hence, the global penetration of green hydrogen-based DRI has to be increased.

The advantages of the process are

- Relatively simple design,
- low space requirement,
- lower capital expenditure.

Small pellets are formed by pelletizing iron ore, which are used as feed for the shaft furnace. The green hydrogen produced by electrolyzers can be used to reduce the pellets and form sponge iron. The sponge iron produced in the shaft furnace is fed to the EAF which can be powered by electricity generated by renewables (Fig. 21.3).

CO_2 emissions are unavoidable as long as coke, coal, and carbonaceous fuels are used in the blast furnace but they can be reduced by measures such as the use of biomass-based fuels and hydrogen, use of CCS technology, and top gas recycling (TGR).

3.2 BF-BOF integrated with Carbon Capture and Storage (CCS)

The most carbon-intensive process in the BF–BOF steelmaking is the blast furnace. The blast furnace exhaust gas contains around 20% CO_2, and hence, blast furnace is the section where the majority of the research is being carried out. To reduce the GHG emissions, post-combustion, the BF emissions are transferred to the amine capture unit for removing CO_2 from the gas. A joint exhaust gas stream from all the major emitting sources of a BF–BOF unit, when subjected to amine capture, can potentially reduce CO_2 emissions by 50–75%. Another process that can be used for capturing carbon is with the use of Selexol solvent; it can capture CO_2 from the BF, MEA solvent when used in BF can capture CO_2.

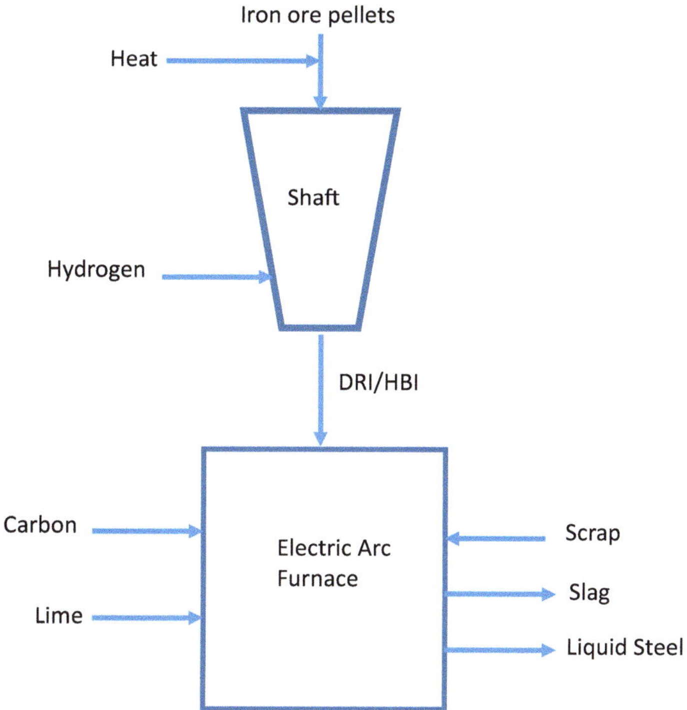

Fig. 21.3 HDRI–EAF process

Similarly, the CCS can also be used in DRI process to absorb the CO_2 produced; adding CCS to processes such as BF, DRI is expected to add extra costs compared to the conventional processes (cost varies regionally). These extra costs exceed the current profit margins of the iron and steel industry and therefore are not cost-competitive; the costs are likely to decrease in the years to come.

3.4 Top Gas Recycling (TGR)

In the TGR blast furnace, the top gas stream, which is rich in reducing gases such as H_2 and CO, is re-injected into the furnace. This has a CO_2 reduction potential of around 5–10%, which can be increased to 50–60% by integrating it with CCS. In this pathway, CO_2 is removed from the top gas stream, and the reducing gas-rich stream is heated and injected into the BF; the CO_2 captured in the process is piped for storage or further use. The CO_2 reduction potential can be further increased to over 80% by the use of biomass-based BF and carbon storage (TGRBF charcoal + CCS).

3.5 Current Status of Low-Carbon Steel Production Technologies

The iron and steel industry has tremendous potential for carbon reduction. Considering the increasing demand, emission reduction by cutting the steel production is not a feasible plan. Hence, the implementation of low-carbon steel manufacturing programs and technologies is necessary.

Many countries are already working on pilot or lab scale projects implementing these technologies; some key projects are

1. ULCOS program in EU,
2. HYBRIT program in Sweden and Finland,
3. COURSE 50 program in Japan,
4. POSCO program in South Korea,
5. AISI program in USA.

Out of all these, the main projects which are under focus worldwide are UCLOS and HYBRIT [2].

4. Non-conventional Areas of Hydrogen Utilization

4.1 Buildings or for Domestic Applications

Currently, there is not any significant utilization of hydrogen for domestic applications. The energy requirement in the building sector is for cooking, space heating, and to meet the hot water requirement. However, this demand varies from region to region and is very difficult to be replaced with the use of hydrogen fuel due to multiple hurdles. The choice of energy use is specific, to the customer and their requirements, type of building (whether commercial or residential), cost of equipment, location, convenience of the user among the other factors. There are multiple competitors for the use of hydrogen in this segment involving electrical heating, natural gas heating, solar thermal heating. In the near term, the option could be the use of blended hydrogen, as decarbonization of the sector is necessary due to the sector being a major energy consumer, accounting for roughly 30% of global energy needs. There are over 37 projects in the UK, Europe, and Asia, which aim to provide heating solutions, micro cogeneration plants and fuel cell demonstrations respectively. The major projects are heating solution in H21 north of England and ENE-FARM project in Japan for installation of fuel cells with combined heat and power units in commercial and residential buildings. Similar demonstration had been in Korea for a building complex, where the installation of fuel cells is demonstarted. There are few demonstration projects started in the Netherlands and Scotland for residential use.

Among the two options, blending natural gas for heating requires rigorous prior testing of blending in low concentrations to ensuring safety, efficiency, and environmental impact, which can be tested in the existing equipment. However, the use of pure hydrogen in buildings will involve several infrastructural changes, and at the same time, the economics will play a dominating role, along with safety concerns.

4. 2 **Electricity Generation**

Currently, there is negligible use of hydrogen in the power generation sector, with less than even 0.2% of power plants using mixed gases, rather than pure hydrogen. The possible options includes co-firing of ammonia in coal fired power plants, hydrogen gas turbines, and combined cycle gas turbines. Besides, power generation, hydrogen serves as a medium for energy storage from renewables, i.e., for long-term and large-scale energy storage. Some of the examples of use of hydrogen in this sector includes, a hydrogen-fired combined cycle gas turbine providing heat and electricity to a community in Kobe, Japan; a gas turbine that uses 95% hydrogen has been in use for more than two decades in Korea. In addition to this, both internal combustion engines and fuel cells are used to produce both power and heat. As per the IEA 2024 report, hydrogen and ammonia used for power generation had led to power capacity of around 330 MW in 2023. However, out of the total installed capacity, 60% is from fuel cells and the remaining 40% is from gas turbines and combined cycle gas turbine.

4.3 **Industrial Heat**

Another major segment where hydrogen can be used in the future is providing the high temperature industrial heat, which is used in various industries for different processes and operations such as melting, gasifying, drying, and for various chemical reactions. This heat is either directly supplied in a furnace or supplied indirectly, such as through the production of steam, which is then used to meet the heating demand. The requirement for heat can be categorized into three categories; a requirement for a low-temperature heat, i.e., less than 100 °C, medium temperature 100–400 °C, or high temperature greater than 400 °C. The major global demand for high temperature industrial heat comes from the cement, iron and steel, and other industries. This demand for industrial heat is expected to increase. However, minor reductions are expected which will be due to the increase in energy and material efficiencies in future. Currently, hydrogen is not being used in this sector, but it is expected that in future, there will be a demand for hydrogen or hydrogen-rich fuels to reduce the emissions associated with industrial heat supply. The demand could be met with either hydrogen, but the usage of pure hydrogen has several challenges. Therefore, the option will be hydrogen-rich fuels which can reduce the emissions associated with industrial heat production.

4.4 **Hydrogen Use in Transport Sector**

The major application that may find its way into the future of hydrogen is in the transportation sector. The transport sector accounts for a significant portion of total global greenhouse gas emissions, and 90% of the energy used in the transportation sector comes from oil products. Either electrification using battery electric vehicle or use of hydrogen in fuel cell vehicles or ICE-based vehicles will result in reduction of the emissions related to the transportation sector.

With the use of fuel cell electric vehicles, there is a strong potential to reduce local air pollution, as they have no tailpipe emission like the battery electric vehicles. It is also possible to use synthetic fuels other than hydrogen such as synthetic methane, methanol, ammonia, etc. which can find their application in a wide range of transportation applications.

If hydrogen is being produced through an electrolytic route and then used to produce synthetic liquid fuels, it is known as power to liquid. The production of these synthetic fuels is expected to increase, and this demand for hydrogen in synthetic fuel production is projected to grow to 7 million tons or higher by 2030. This use of synthetic fuels has several advantages, such as the usage of existing conventional infrastructure which will contribute ot limited changes in the energy value chain.

However, there will be significant efficiency losses involved due to several conversion steps involved in the whole process: first to hydrogen and then into synthetic fuel. For the aviation sector, synthetic jet fuel will have potential; for the shipping segment, ammonia could be used. The two major polluting, aviation and shipping sectors have great difficulty in changing to an entirely new fuel, these sectors are very difficult to directly electrify or use hydrogen. Thus, the use of synthetic fuels or hydrogen-rich fuels will have a potential in future.

Under road transport, in June 2021 there were approximately 40,000 fuel cell electric vehicles on road, with a 70% of the annual growth was observed between 2017 and 2021(as per IEA report). However, the rate of increase in hydrogen refueling stations was not significant, with only 20% annual increase observed during this period. The use of hydrogen is in the different types of forklifts, with approximately 25,000 operating globally, and this is one of the key applications of fuel cell-based operations.

Among the bus transport, China has the largest deployment of 400 buses, which was in 2018. There are 11 companies globally that manufacture fuel cell buses. The fuel cell buses could find its way for intercity transport. Advantage of fuel cell buses is that they can provide longer range, and shorter refueling time. Additionally, the use of hydrogen infrastructure can provide flexibility in terms of routing as well as operational flexibility. At times of the year, when there is a change in the route or an extended route, fuel cell-based bus can provide that flexibility. For delivery fleets that require long-range operations, it will be cost-effective to have hydrogen stations along the route

Among trucks, China is the leading country having 412 registered fuel cell trucks and 100 vans. Outside China, FedeX and UPS, are also expanding their range in the US and the French postal service, they are also acquiring hydrogen fuel cell trucks [Ref. IEA 2023 report].

Among rail, two Alstrom trains in Germany have traveled more than 180,000 km of distance. Austria will have a regular passenger service; UK, Netherland, and India will also have pilot fuel cell rail. France, Italy, and UK have placed orders for trains.

Among the shipping sector, Green Ammonia is viewed as a potential alternative to power ships. For aviation, Airbus is exploring a hydrogen powered aircraft concept, which is expected to be used by 2035. According to the announced pledge scenario, the hydrogen demand in this particular sector is expected to be 520 petajoules, and it is expected to rise to 15 times by 2050.

The positive aspect of this utilization in transport is the cost reduction over the past two decades. Further cost reduction is required to implement and widespread deployment. The expected cost reduction should be in fuel cell stack cost, hydrogen storage cost, refueling cost, among others. Besides the availability of infrastructure, such as hydrogen refueling stations will also contribute significantly towards its widespread usage.

B. Energy Conversion Devices Used for Various Applications

Use in Internal Combustion Engines

The use of hydrogen in internal combustion engines, ways back to 1807 when DE. Rivaz patented the technology to use hydrogen in IC engines. Recent global interest is focused on replacing fossil fuels, through retrofitting the existing engines, or by using pure hydrogen engines to reduce the emissions. In the ICE, the emissions are CO_2 and also NO_X, SO_X, unburnt hydrocarbons, and particulate matter.

Considering a typical spark ignition engine and the combustion process in it, when an ignition source, i.e., a spark, initiates ignition, then some fuel which is in the closest vicinity of the ignition source burns; during this burning, it generates heat which is transferred to the other nearby molecules of the fuel-oxidant mixture thereby results in flame front propagation as shown in Fig. 21.4. Flame front propagation depends on the calorific value, which in turn determines how the heat propagates from the ignition front to the subsequent layers of combustible mixture, ultimately reaching the ignition temperature. In the flame propagation, the factors responsible for flame propagation include flammability limits, fuel reactivity, ignition energy provided, thermal conductivity of fuel, calorific value of fuel, air-to-fuel ratio. For hydrogen, these properties are exceptionally good and results in very high flame speed as compared to any other hydrocarbon. Besides, due to high thermal conductivity of hydrogen, the heat is very quickly transferred from the point of ignition to the surroundings; this is very essential because it may be possible that the temperature may be high but the energy may be in a localized zone and needs to be transferred to the rest of the fuel to break the bonds and for the reaction to take place. Due to low ignition energy even though the autoignition temperature is high, and the hydrogen reactivity is high, the energy requirement to achieve a sustained combustion is very low for hydrogen.

The wide flammability limit of hydrogen allows for lean combustion, which helps limit the NOx emissions at low concentration of hydrogen, in the flammable mixture,

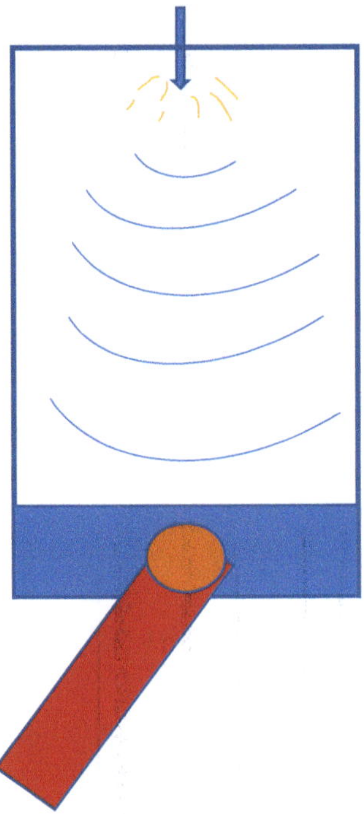

Fig. 21.4 Schematic showing injection of fuel, spark and corresponding flame propagation in an internal combustion engine

thereby enabling for cold start. Since the autoignition temperature of hydrogen is very high, high compression ratio can be achieved.

In the process of combustion, when the inlet valve opens, and the air–fuel mixture is taken in the combustion chamber, the spark is ignited from spark plug, wleading to flame. Flame propagates through the air–fuel mixture and throughout the cylinder as shown in schematic in Fig. 12.4. The time taken to consume the entire fuel in the cylinder depends on the flame speed which acts as a limit to the speed that the engine can achieve.

The amount of fuel injected burns completely in a fraction of time (1–10 ms), from the position of spark plug to the bottom dead center. The time taken in a power stroke in a four-stroke engine is 10 ms. With an engine speed of 2000 rpm, 4000 rpm, or 10,000 rpm, the time taken for the power stroke reduces, making engine operation very fast, and all the four strokes happen very quickly. In an internal combustion engine, an increase in acceleration leads to an increase in vehicle speed and, consequently, engine speed.. Thus, as the engine speed increases during acceleration, the time available for complete combustion during a power stroke reduces; in this case, the flame speed plays a crucial role. A fuel that burns and travels fast, alongwith high reactivity, and having very high flame speed is desired, which is the case for hydrogen. Compared to gasoline, roughly one-fifth of time is required for combustion under similar conditions; thus, a very high rpm can be achieved in case of hydrogen.

Thus, higher power output can be achieved at the highest possible rpm with hydrogen, with minimal fuel input, the number of cycles per unit time increases, thereby yielding more power per unit time. Due to high autoignition temperature, hydrogen is not suitable for compression ignition engine; rather it is meant for spark ignition engines.

For hydrogen, the density is very low, i.e., 10 MJ/L as compared to 34.6 MJ/L in case of petrol. Gasoline vapor with stoichiometric fuel–air mixture in a 1000 cc engine, correspond to 3.5 kJ of energy. However, in case of hydrogen due to very low density under the similar conditions, the energy in a power stroke is 3.0 kJ and has a lower mixture calorific value (deliver 85% that of petrol under stoichiometric conditions). Thus, the requirement is for more of hydrogen to provide the required power in the power stroke;. Thus, with hydrogen as a fuel in ICE, there is a requirement for turbo charging, i.e., the delivery of pressurized hydrogen. The turbo charging involves injecting more fuel per cycle. However considering the emissions, it is not advisable to increase fuel–air mixture and a lean fuel mixture is desirable for a given stoichiometric mixture. At stoichiometric conditions, hydrogen occupies 30% of the volume, compared to 1–2% in case of gasoline or other conventional fuels.

Hydrogen has a high flame velocity which is around 10 times that of gasoline and it burns very fast, which helps in attaining higher rpm. The chances of knocking are higher with hydrogen, because of its high flame velocity, i.e., high heat release rate. Hydrogen burns very quickly and completely, fuel combusts in a very short time, resulting into higher peak pressure leading to higher impact on the engine piston and cylinder causing the knocking phenomenon. Knocking is an undesirable phenomenon; in order to avoid it, leaner mixture is helpful, which means less of fuel being injected per cycle leading to less energy per cycle, i.e., moderate pressure. Thus, to reduce chances of knocking, with lean mixture and high flammability limits, the equivalence ratio can be 0.2. For hydrogen the flammability limit is wide, lean mixtures can burn together with high diffusivity ensures good premixing resulting in complete combustion of the fuel.

In a spark ignition (SI) engine, as the spark ignites and the flame propagates, there is less fuel behind the spark plug, and the fuel combusts from the piston to the bottom dead center. Close to the wall of the cylinder, where heat loss from the cylinder wall occurs, and thus, wall quenching phenomenon is observed, in which the temperature near the wall drops below the ignition temperature, and fuel near wall do not burn in petrol engine. For hydrogen ICE, if the fuel does not burn completely and if the unburnt hydrogen is ejected with the exhaust, that will pose a challenge. However, the low quenching distance of hydrogen, low ignition energy, and high thermal conductivity of hydrogen make the situation different, resulting in less fuel wastage, better combustion, and low emissions due to unburnt fuel.

NO_x emissions occurs usually at high temperatures. The combustion rate is high in case of hydrogen as a fuel in ICE, it reaches to higher temperature in a shorter duration. For leaner mixture and lower air–fuel ratio, it is advantageous for hydrogen and the lower temperature operation leads to lower NO_x emissions.

In 2005–2007, BMW introduced BMW 7 series hydrogen car, which had higher efficiency than the diesel engine. The model was hydrogen petrol bi-fuel vehicle, with spark ignition and had hydrogen in liquid form. The injection was direct, at 300 bar, and it achieved a mileage of 50 km/kg, with an average fuel consumption of 3.7 kg per 100 km. Several OEMs are actively engaged in research toward developing hydrogen fuelled internal combustion engines, where different options are considered, such as direct and port injections. The other option is fuel cell-based vehicles such as Toyota Mirai or Honda Clarity, which provide high efficiency compared to IC engine-based cars.

Use of Hydrogen in Fuel Cells

Fuel cell is an electrochemical device used to convert the chemical energy in hydrogen fuel in to electrical energy. Fuel cell has several commonalities with the other energy conversion devices like internal combustion engines, battery, and electrolyzer. Fuel cell like internal combustion engine is an energy conversion device. Fuel cell is like an electrolyzer as it is an electrochemical device. The principle of operation is very similar to those for electrolyzer and the major difference will be outlined in this section.

The difference between an electrolyzer and fuel cell is that the electrolyzer takes electricity and water as input and produces hydrogen and oxygen; while fuel cell does reverse of it, takes oxygen and hydrogen as input and produces water and electricity as the output.

The demonstration of a first gas voltage battery was done by William Grove in 1839, when he proved that an electrochemical reaction between hydrogen and oxygen can produce electricity. Later Langer and Monde introduced the term fuel cell, and at that time, they used coal as a fuel and obtained a current density of 20 A/m^2 at a voltage of 0.73 V [3].

This cell was modified for the first time to make it into alkaline fuel cell. It was in 1950 that NASA first started developing fuel cells and they used the device for space missions. After the 1970s, there was a lot of interest in fuel cell, especially for the vehicular application.

Fundamentals

The fundamentals of the fuel cell are much similar to those described for electrolyzers in chapter 7. However, a brief overview of these two technologies and advancements in the field of fuel cells is included under this section. Fuel cell is an electrochemical device, takes hydrogen and oxygen as input, and gives electricity and water as output. As long as we provide the fuel and oxidant, i.e., hydrogen and oxygen/air, it will ideally supply electricity. In that analogy it works like an internal combustion engine.

However, fuel cells being an electrochemical device convert the chemical energy of fuel into electrical energy; as such they have a similarity with the batteries, both of which are electrochemical devices. However, the major difference between fuel cell and battery is that, in fuel cell,, ideally we can get electricity continuously as long as we supply fuel. In contrast, a battery stores chemical energy and uses it up. There is a similarity between fuel cell and an IC engine; i.e., the engine converts the chemical energy of the fuel into energy/heat via reaction. In the hydrogen IC engine, hydrogen and oxygen combine to give water which can be converted later into mechanical energy and then into electrical energy. In an IC engine, hydrogen oxygen reacts to produce water and heat; on a molecular scale, there occurs collision between hydrogen and oxygen and producing water. However, on an atomic scale, the hydrogen–hydrogen bond breaks, oxygen–oxygen bond breaks, and the hydrogen–oxygen bond is formed; this bond reconfiguration occurs through electron transfer (Fig. 21.5).

The bond configuration energy of the products formed is higher than that of reactants, and the difference in energy is released as heat during the process. During the bonding reconfiguration, electron transfer takes place from one bonding state to another bonding state. This electron transfer in the combustion reaction occurs in the internal combustion engine in a matter of picoseconds and at sub-atomic scale, with the energy difference between the bonding configurations being released in the form of heat. This heat evolved in IC engine is converted to mechanical energy and thereafter to electrical energy, if the required output is electricity. The number of steps involved in converting chemical energy into electrical energy in the case of an internal combustion engine are higher than in a fuel cell. As such, the process becomes complex and inefficient due to losses at every step in hydrogen ICE. This electron transfer which occurs on a time scale of picoseconds, and at sub-atomic scale, is difficult to harness.

$$\text{Fuel} + \text{Oxidant} \rightarrow \text{Eletric Power} + \text{Waste Heat} + \text{Water}$$

However, if that electron transfer during bonding reconfiguration could be harnessed, then we can directly convert this chemical energy into electrical energy, and this the phenomenon that happens in a fuel cell. The electrons that are transferred at the sub-atomic scale and in picoseconds can be harnessed by partially separating the reactants, hydrogen and oxygen in case of a fuel cell. In that case, the

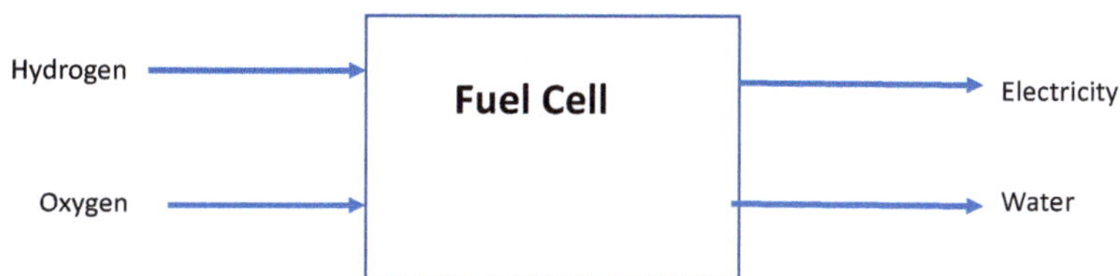

Fig. 21.5 A black box showing inputs and outputs forms a typical fuel cell

electron transfer occurs over a partially extended length.

$$H_2 \rightarrow 2H^+ + 2e^- \quad (21.1)$$

$$2H^+ + \frac{1}{2}O_2 + 2e^- \rightarrow H_2O \quad (21.2)$$

The above two reactions are separated partially by means of an electrolyte, and the two half-cell reactions (H_2-O_2 fuel cell) occur at the electrodes surface. The fuel cell is an electrochemical device that directly converts the chemical energy into electrical energy, the reaction between the fuel with oxidant, combining to produce electricity, results in a certain amount of waste heat being liberated, along with water as the output.

Thus, in a fuel cell there are certain advantages of both batteries and internal combustion engines. In fuel cells, there is a direct conversion of chemical energy into electrical energy, involving less number of steps, making it more efficient. As against battery the advantage of fuel cell is that there is an independent scaling up possible of power and capacity, by having higher capacity fuel cell and fuel reservoir so as to increase the power and energy delievered.

Advantages of fuel cells

- More efficient than combustion engines as there is direct conversion of chemical energy into electrical energy.
- Independent scaling between power and capacity,
- Can be easily scaled up,
- Reliable and long lasting, as they are no moving parts, they offer silent operation,
- No undesirable products and particulate emissions,

Disadvantages of fuel cells

- High cost,
- Limited power density,
- Fuel availability and storage issues,
- Use of alternate fuels, the auxiliary equipment adds to the cost and complexity,
- Poisoning issues of catalyst due to impurities present in the fuel,
- Start-stop cycling or transient response can lead to possible degradation.

Basic Principle of Operation of Fuel Cells

In fuel cell, the current produced directly depends upon the reaction surface area or the interfacial area which is the area where electrode, electrolyte, and reactants meet. Thus, to scale it up and increase the current produced by a fuel cell, the requirement is to increase reaction area or the interfacial area, which is achieved by making the structure thin and planar.

At the same time, porous electrodes not only increase the surface area but also the surface to volume ratio and making the access of gas easier in the fuel cell. In the basic operation of a fuel cell, there are primarily four steps involved:

(1) **Reactant Delivery to the Fuel Cell**

- Continuous efficient delivery of reactants is required.

A continuous supply of reactants is required to maintain continuous electrical output. Operation under higher current conditions at a higher rating, needs the supply to be even higher. This is provided by the use of flow field.

- Flow field plates having channels to transport gases.

For the continuous supply of the gases on both sides of the fuel cell, the flow field plates and porous electrodes are required, which will help in easier transport of the gases. These flow field plates are the essential component of a fuel cell and have channels or grooves such that it transport gases throughout the surface of the fuel cell. The size, shape, design of these patterns, grooves, channels are important, and these determine the performance of the fuel cell.

(2) **Electrochemical Reaction**

- Current generation directly related to rate of electrochemical reaction

During the electrochemical reactions and the two half-cell reactions, the current generated is directly proportional to the rate of the electrochemical reactions occurring on the anode and cathode sides.

- High current output is desired, so catalysts are used

To achieve higher current output, the rate of electrochemical reaction should be faster. As such, there is a requirement of catalyst to accelerate the reaction and the catalyst plays a dominating role. The appropriate choice of catalyst is essential, along with the design of the fuel cell, design of the reaction surface area are important towards achieving a better performance of the fuel cell.

(3) **Ion Conduction Through the Electrolyte and Electron Conduction Through External Circuit**

- Ion conduction is more difficult, as ions are larger and more massive than electrons.

During the third step, conduction of ions through the electrolyte and electron conduction through the external circuit is required. The construction of fuel cell is such that the electrolyte separates the two electrochemical half-cell reactions and the two electrodes and the two reactants, in such a manner that ions which are formed on one side are transported through the electrolyte and consumed on the other side of the fuel cell. Similarly, the electrons are produced at the electrode are conducted through the external circuit to the other electrode, where they are consumed. The electrolyte should be such that it should allow the conduction of ions, but not of electrons. The ion conduction is more difficult compared to the electron conduction because the ions are more massive and larger in size compared to that of electron.

- Ion transport is less efficient, contributing to resistance losses.

Ion transport is comparatively less efficient than electronic conduction, and contributes to the resistance losses.

- Thin electrolyte should be used for ion transport.

To address the issue of losses, the electrolyte is made thin and that reduces the path for ion transport across the electrolyte.

(4) **Product Removal from Fuel Cell**

- Product removal is important, else it prevents new fuel and oxidant reactions to take place.

The last step involved is removal of the product as they are formed from the fuel cell.

This is an important step, the requirements are similar to those for reactant delivery and the product removal, in order to facilitate the new fuel and oxidant reaching the reaction surface.

- Water flooding may be an issue in many fuel cells.

Thermodynamic Fundamentals

Electrical work done by charge = Potential difference*Charge

$$W = EQ \tag{21.3}$$

$$Q = nF \tag{21.4}$$

(n denotes moles of electrons transferred and F is Faraday's constant.)

Max electrical work is the negative of Gibbs free energy (for constant P, T)

$$W = -\Delta G \tag{21.5}$$

$$\Delta G = -nFE \tag{21.6}$$

In a hydrogen–oxygen fuel cell,

$$H_2 + 0.5O_2 \leftrightarrow H_2O \tag{21.7}$$

For liquid water as product, ΔG = -237 kJ/mol.

Standard state reversible voltage generated by fuel cell, $E = -\Delta G/nF = 1.23$ V

$$\text{Thermal efficiency} = \frac{\text{Useful energy}}{\Delta H} \tag{21.8}$$

$$\text{Ideal efficiency} = \frac{\Delta G}{\Delta H} = \frac{237.1}{285.8} = 0.83 \tag{21.9}$$

$$\text{Thermal efficiency} = \frac{\text{Useful Power}}{\Delta G/0.83} = \frac{0.83 * V_{\text{actual}}}{V_{\text{ideal}}} \tag{21.10}$$

Ideal fuel cell voltage at 1 atm atmosphere, 25 °C is 1.229 V

$$\frac{0.83 * V_{\text{actual}}}{1.229} = 0.675 * V_{\text{actual}} = \text{Voltage Efficiency} \tag{21.11}$$

Net cell efficiency = Voltage efficiency ∗ Fuel Utilization Factor

Different Types of Fuel Cells

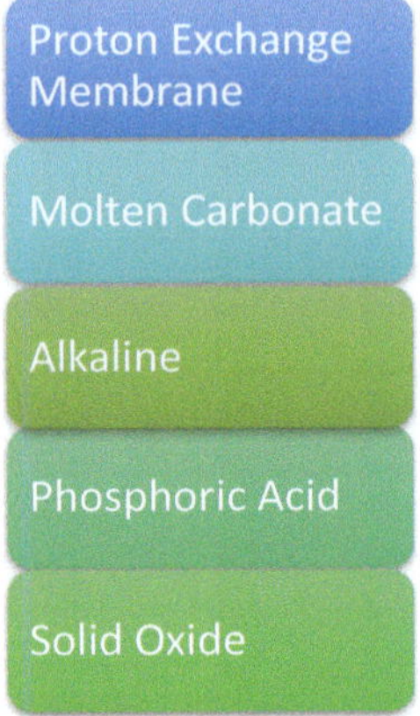

Polymer Electrolyte Membrane Fuel Cell (PEMFC)

The polymer electrolyte membrane fuel cell has the highest power density as compared to the other types of fuel cells. The electrolyte is a polymer electrolyte membrane or a proton-conducting membrane. The most commonly used membrane is Nafion, and the membrane must be operated at a certain temperature. To maintain conductivity, the membrane must remain hydrated; however, it cannot operate beyond 90 °C.

In a PEMFC, a platinum-based electrocatalyst is coated on both sides of the membrane and are supported on a porous carbon electrode material. The combination of electrodes, catalyst, and membrane is known as membrane electrode assembly. PEMFC operates at lower temperature, as such can be used for various portable applications. The reactions are:

$$\text{Anode}: \ 2H_2 \rightarrow 4H^+ + 4e^- \tag{21.12}$$

$$\text{Cathode}: \ O_2 + 4H^+ + 4e^- \rightarrow 2H_2O \tag{21.13}$$

$$\text{Overall}: \ 2H_2 + O_2 \rightarrow 2H_2O + \text{electric energy} + \text{heat} \tag{21.14}$$

Alkaline Fuel Cell (AFC)

Alkaline fuel cell is the most mature and widely used technology, which was first used in Apollo mission by NASA. AFC can be operated in a wide temperature range. The electrolyte used is low cost KOH, and the biggest advantage of alkaline fuel cell is the use of non-noble metal catalyst, which reduces the cost; have lower crossover problems compared to the polymer electrolyte membrane fuel cell.

The current density for AFC is similar to PEM fuel cell, and the reactions are:

$$\text{Anode}: \ 2H_2 + 4OH^- \rightarrow 4H_2O + 4e^- \quad (21.15)$$

$$\text{Cathode}: \ O_2 + 2H_2O + 4e^- \rightarrow 4OH^- \quad (21.16)$$

$$\text{Overall}: \ 2H_2 + O_2 \rightarrow 2H_2O + \text{electric energy} + \text{heat} \quad (21.17)$$

Phosphoric Acid Fuel Cell (PAFC)—In a phosphoric acid fuel cell, either dilute or concentrated H_3PO_4 is used as the electrolyte. This was the first commercial fuel cell which was used for stationary applications. The components must be acid resistant as an acidic electrolyte is used. The components used should be tolerant to carbon monoxide, with the maximum allowable limit is 1.5%. PAFC provides the flexibility, in the choice of fuel, however, sulfur needs to be removed. The major disadvantage associated with phosphoric acid fuel cells, includelower efficiency and high weight. The reactions involved are:

$$\text{Anode}: \ 2H_2 \rightarrow 4H^+ + 4e^- \quad (21.18)$$

$$\text{Cathode}: \ O_2 + 4H^+ + 4e^- \rightarrow 2H_2O \quad (21.19)$$

$$\text{Overall}: \ 2H_2 + O_2 \rightarrow 2H_2O + \text{electric energy} + \text{heat} \quad (21.20)$$

Molten Carbonate Fuel Cell—These fuel cells are high temperature fuel cell; the electrolyte is made of molten carbonate salts; and they require carbon dioxide at cathode, which is produced and consumed within the fuel cell. Since MFC is a high temperature fuel cell, there are issues associated with the stability of the material. This fuel cell is more resistant to impurities compared to the other types of fuel cell, but the major challenge is short life and high cost. The operating reactions are:

$$\text{Anode}: \ 2H_2 + 2CO_3^{2-} \rightarrow 2H_2O + 2CO_2 + 4e^- \quad (21.21)$$

$$\text{Cathode}: \ O_2 + 2CO_2 + 4e^- \rightarrow 2CO_3^{2-} \quad (21.22)$$

$$\text{Overall}:\ 2H_2 + O_2 \rightarrow 2H_2O + \text{electric energy} + \text{heat} \tag{21.23}$$

Solid Oxide Fuel Cell—SOFC is high operating temperature fuel cell. A wide variety of fuels can be used in case of solid oxide fuel cell. Since the electrolyte used is solid, the issues related to managing the liquids absent, unlike other types of fuel cell. Further, the use of solid electrolyte requires reduced electrolyte thickness, and it produces high-grade by-product heat that can be utilized for cogeneration. The major disadvantage is the material degradation at high temperatures. The requirement is to have a lower cost ceramic materials which are stable under these stringent operating conditions. The reactions are as follows:

$$\text{Anode}:\ 2H_2 + 2O^- \rightarrow 2H_2O + 4e^- \text{ and } 2CO + 2O^- \rightarrow 2CO_2 + 4e^- \tag{21.24}$$

$$\text{Cathode}:\ O_2 + 4e^- \rightarrow 2O^- \tag{21.25}$$

$$\text{Overall}:\ 2H_2 + O_2 \rightarrow 2H_2O \quad \text{and} \quad 2CO + O_2 \rightarrow 2CO_2 \tag{21.26}$$

A comparison of the different fuel cell technologies is presented in Table 21.1.

Table 21.1 Comparison the different fuel cell technologies [6]

Characteristic	Polymer electrolyte (PEMFC)	Alkaline (AFC)	Phosphoric acid (PAFC)	Molten carbonate (MCFC)	Solid oxide (SOFC)
Operating temperature (°C)	40–80	65–220	205	650	600–1000
Electrolyte	Hydrated polymeric ion exchange membrane	Mobilized or immobilized potassium hydroxide in asbestos matrix	Immobilized liquid phosphoric acid in SiC	Immobilized liquid molten carbonate in $LiAlO_2$	Perovskites
Electrode	Carbon	Platinum	Carbon	Nickel and nickel oxide	Perovskite and perovskite/ metal cermet
Catalyst	Platinum			Electrode material	
Charge carrier	H^+	OH^-	H^+	CO_3^{2-}	O^{2-}
Fuel compatibility	Hydrogen, methanol	Hydrogen	Hydrogen	Hydrogen, methane	Hydrogen, methane, CO

Components of a PEM Fuel Cell

(a) **Membrane Electrode Assembly**

- **Polymer Electrolyte Membrane**

Permits necessary ions to pass between the anode and cathode.

- **Catalyst Layer**

On anode, the catalyst enables hydrogen molecules to split into protons and electrons. On the cathode, the catalyst facilitates oxygen reduction by reacting with the protons generated at the anode, thereby producing water.

- **Gas Diffusion Layers**

Facilitates the transport of reactants into the catalyst layer and the removal of the water formed.

(b) **Hardware**

- **Bipolar Plates**

Provide electrical conduction between cells and strength to the cell stack.

- **Gasket**

Added around the edges of the MEA to make a gas-tight seal.

Current–Voltage Characteristics of a Fuel Cell

In a typical fuel cell, the theoretical value of the standard reversible voltage is 1.23 V; however, several irreversibilities are involved in the fuel cell. The power produced by a fuel cell stack is given by:

$$\begin{aligned}\text{Power produced by a fuel cell stack} &= \text{Current density} \\ &\quad * \text{Geometric area} * \text{Voltage from fuel cell stack}\end{aligned} \tag{21.27}$$

$$\begin{aligned}\text{Voltage from a fuel cell stack} &= \text{Voltage from each fuel cell} \\ &\quad * \text{No. of cells in the stack}\end{aligned} \tag{21.28}$$

$$\begin{aligned}\text{Output voltage of a fuel cell} &= \text{Reversible voltage of cell} \\ &\quad - \text{Irreversible losses}\end{aligned} \tag{21.29}$$

(Reversible voltage of cell is the voltage under thermodynamic reversible conditions, while the irreversible losses in the fuel cell is overpotential.)

The irreversible losses can arise for several reasons; Table 21.2 summarizes the different losses or overvoltages associate with the fuel cell and their origin.

The performance of a PEM fuel cell can be represented by plotting the current-voltage characteristics, i.e., the plot between cell voltage and the current density. Thus, the actual output voltage is given by the ideal voltage minus the sum of the different over voltages involved in the fuel cell

$$\begin{aligned}\text{Real output voltage} &= \text{Ideal voltage (i.e. thermodynamically predicted)} \\ &\quad - (\text{Activation} + \text{Ohmic} + \text{Conc. Losses}) \qquad (21.30)\end{aligned}$$

In the lower current density regime activation losses are predominant. In the real operational region, ohmic losses are predominant, i.e., in the medium current densities. While in the higher current density regime, mass transport losses are predominant (Fig. 21.6).

Several parameters affect the performance of a fuel cell.

The current density and the effect of temperature on the fuel cell are:

Current Density

- Losses directly dependent on current density.
- Ohmic losses dominate in normal operation.
- High current density gives high power density, but results in lower efficiency.

Temperature

Table 21.2 Different losses in a typical fuel cell and their origin

Loss	Cause
Activation overpotential	Slowness of reaction on electrode surface
Ohmic overpotential	Resistance to flow of ions and electrons
Concentration overpotentials	Mass transfer limitations at high current

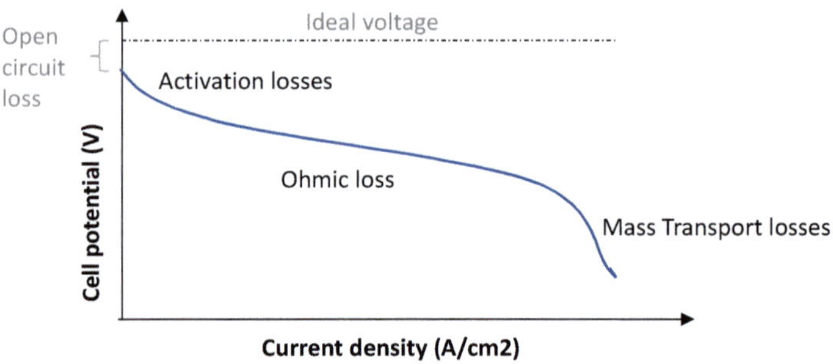

Fig. 21.6 *I–V* characteristic curve for a fuel cell

- Losses decline exponentially with increasing temperature.
- The higher the activation energy, greater the impact of temperature.
- The impact of temperature on cell resistance varies with the material.
- For aqueous electrolyte, high temperature leads to membrane dehydration and loss of conductivity.
- Mass transport losses not affected by temperature.

Pressure

- Increase in operating pressure has benefits.
- Gas solubility and mass transfer rates increase with pressure.
- Electrolyte loss by evaporation, reduced at high pressure.
- Disadvantages: thicker piping required, additional cost for pressurization.

Gas Composition

- Electrochemical reactions cause changes in gas composition, the variation are due to changes at the inlet and outlet of a fuel cell.
- Change in gas composition causes reduction in cell voltage.
- Cell voltage adjusts to the lowest electrode potential, given by the Nernst equation.

Developments

There have been several recent developments in the domain of fuel cell that are being considered as major reason for improvements and are helpful toward the green energy transition. Some of the key fuel cell manufacturing companies include Ballard, fuel cell energy, Mitsubishi plug power and Cummins.

Fuel cells are being widely explored for their use in distributed power as well as vehicular application, and they have a wide scope of application in automobiles that could provide a longer driving range along with shorter fuelling time. The Asia–Pacific has emerged as a major market for fuel cell because of the government policies in countries such as China, Japan, and South Korea where interest in fuel cell has increased, with the primary applications includes heavy-duty transport and energy storage.

In 2020, South Korea surpassed USA and China, becoming a leading market for fuel cell electric vehicles with over 10,000 vehicles. In December 2021, Toyota unveiled a new hydrogen production facility and a gas refueling station, at its factory in Melbourne, Australia. Hyzon Motors announced the delivery of 29 fuel cell electric trucks to be used in China by a major steam conglomerate.

In 2021, Edison Motors signed an agreement with Plug Power to develop and market a hydrogen fuel powered electric bus, as per IEA's global EV Outlook. China accounts for 94% of the fuel cell buses and 99% of the fuel cell trucks globally. The higher the deployment, the more will be the impact of economies of scale, towardsbringing down the cost of the fuel cell.

Global Hydrogen Strategies

Japan—Japan was the first country to release the strategy document in the year 2017, and a revised document was published in 2023. In the revised document, the key features include a midterm target of production of 12 million tons per annum of hydrogen, which was a sixfold increase from the existing level. Pathway to the low-carbon hydrogen to reduce the emissions to 3.4 kg of CO_2 per kg hydrogen produced and a target of 10% share of Japanese companies in the global electrolyzer manufacturing, were other key feature. Various support schemes were highlighted, including cluster development, support to establish internationally competitive industrial clusters, and a producer support scheme aimed to expand supply volume and reducing cost.

Canada—In the year 2020, Canada published a national strategy with the aim of achieving net zero by 2050. The major targets included encouraging early deployment of hydrogen hubs in sectors that are already mature, utilizing hydrogen and demonstrations in upcoming areas for emerging applications, employing regulations to drive near-term investments, and framing new policy and regulatory measures needed to reach net zero by 2050. Activities were planned for short term and the progress was reported in the year 2024. Further aim was that the growth and diversification of the sector will take place from 2025 to 2030, followed by rapid expansion until 2050 to completely realize the full benefits of the hydrogen strategy. In the year 2022, a Government initiative: the Clean Hydrogen Investment Tax Credit was announced where a refundable investment tax credit for investments in electrolysis and reforming with CCUS was considered. Key findings toward developing hydrogen hubs and growth in the sectoral utilization of hydrogen have been witnessed.

Germany—Germany introduced its hydrogen strategy in 2020 and revised it in 2023. Additionally an import strategy for hydrogen and its derivatives was published in 2024. The key points in the updated strategy focused on accelerating market ramp up, along with ensuring sufficient availability of hydrogen. Increasing the domestic electrolyser capacity in 2030 to 10GW, while rest to be met through imports. Developing the high-performance hydrogen infrastructure with 1800 km of hydrogen pipeline and additional 4500 km in Europe connecting all large generation and storage facilities by 2032. Diversified use of hydrogen in various sectors including power generation, buildings, air as well as naval and for decentralized heating applications. To establish Germany as a lead provider of hydrogen technology in the world. All initiatives were supported with efficient planning, approved procedures, ensured standard and certification in a globally coordinated manner.

China—China's Hydrogen Energy Development strategy was laid for 2021–2035, with a major focus on integrating hydrogen into the energy transition. The targets set were to establish 80 GW of electrolyzer capacity by 2030; and to have 50,000 hydrogen fuel cell vehicles by 2025. Consideration was given to the hydrogen's potential in industries such as steelmaking, chemicals, and refining to reduce fossil fuel dependence. In 2023, China introduced the "Guidelines for the Construction of Hydrogen Energy Industry Standard System (2023 Version)" to establish a comprehensive standard system for hydrogen production, storage, transportation, and utilization. An initiative to build up international collaboration is already underway.

South Korea—South Korea developed a hydrogen Economy Roadmap in 2019. With the government's efforts, the world's first legislation on the hydrogen industry came into force in 2021, the "Hydrogen Economy Promotion and Safety Management Act" (Hydrogen Act), as a step towards the hydrogen economy. This had a two-track approach, supporting a two-track approach that combined hydrogen economy and measures for safety management. The target for annual green hydrogen production is 3 MMTPA by 2040, 6.2 million FCEV by 2040, 15 GW of fuel cell power generation by 2040, 700 km of hydrogen pipelines by 2030, and huge investment in hydrogen R&D is planned for future. Implemented clean hydrogen portfolio standard in 2022 for use of green hydrogen in power generation. Development of specialized zones to foster innovation and collaboration among hydrogen-related industries are the other key features.

Australia—Australia launched the hydrogen strategy in 2019 and revised in 2024. The targets were considered in two phases up to 2025 and after 2025. Target is of 15 MMTPA of green hydrogen by 2050. Australia has announced huge investment in hydrogen pipeline and for the development of green hydrogen industry. Some of the announced incentives include a 2$/kg of renewable hydrogen incentive for up to 10 years project as well as $4 billion in the Hydrogen Headstart program, administered by ARENA, to provide revenue support for large-scale renewable hydrogen projects. A $3 billion of fund has been set aside for supporting and finance renewables, and low-emissions technologies. This initiative also includes funding for hydrogen electrolyzers and fuel switching.

South Africa—In 2022, South Africa unveiled its Hydrogen Society Roadmap & Commercialization, which was approved as a blueprint to support the development of hydrogen-based economy, leveraging South Africa's platinum group metals (PGMs) for electrolyzers and fuel cells. The strategy was announced with the targets of green hydrogen production of 500 kTPA by 2030, 10 GW of electrolyzer capacity by 2030 in Northern Cape and 15 GW nationwide. The focus of the roadmap was on decarbonizing heavy-duty transport, both grid and off-grid clean power generation, decarbonizing hard to abate sectors, creating export market, escalating the local manufacturing capabilities, and scaling up the supply chain of green hydrogen.

European Union—The European Union introduced its hydrogen strategy in 2020. A three phased roadmap was announced, where in phase I (2020–24) was targeted towards manufacturing more than 6 GW electrolyzer capacity and 1 MMTPA of green hydrogen. In Phase 2 (2025–2030), to scale up electrolyzer manufacturing to 40 GW and green hydrogen to 10 MMTPA alongwith integration in industries, transport and other sectors. In Phase 3 to extend the use of green hydrogen to the hard to abate sectors.

USA—USA announced the U.S. National Clean Hydrogen Strategy and Roadmap in 2023. Targets announced were hydrogen production of 10, 20 and 50 MMTPA by

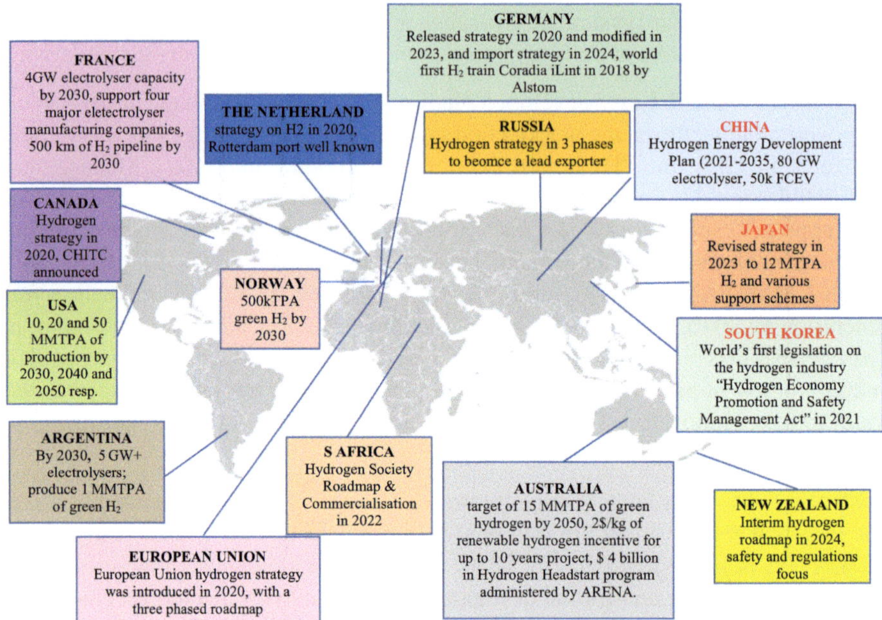

Fig. 21.7 World map showing the major countries who announce their hydrogen strategies

2030, 2040 and 2050 resp. With DOE's "Hydrogen Shot" initiative of 1-1-1, to bring down the cost of hydrogen to 1 USD for 1 kg in 1 decade. Seven hubs were selected for federal funding. The key sectors considered in the strategy includes heavy-duty transport, industrial processes and supporting local manufacturing with innovation and R&D support. Fig. 21.7 is a representation on world map showing the different announcements globally.

National Green Hydrogen Mission of India

India launched the National Green Hydrogen Mission in January 2023. The key aim of NGHM is to make India a global leader in the green hydrogen value chain, encompassing production, storage, utilization, and export. The mission aligns with the overall goals of achieving energy independence by 2047 and net zero by 2070.

Figures 21.8 and 21.9 show the major aims of the NGHM and how these will be achieved, respectively. According to the NGHM document, the mission target is to achieve 5 MMTPA of green hydrogen production by 2030, which will be supported by 120 GW of renewable energy. Achieving these targets will reduce India's dependence on imports, provide energy security, create jobs and reduce greenhouse gas emissions. The goals of NGHM will be strategically achieved through the scaling up of green hydrogen production leading to economies of scale, cost reduction in electrolyzer manufacturing, and renewable energy pricing; this will be achieved by providing suitable policy support and deploying various infrastructure required for enhancing the green hydrogen ecosystem.

The mission has two phases. The first phase focuses on enhancing the demand of green hydrogen, increasing the electrolyzer manufacturing capacity and transitioning from fossil fuel usage to renewable energy sources. In the second phase, cost competitiveness with fossil fuels is expected; various pilot projects will be initiated in new sectors, and the R&D projects supported in phase I will be scaled up; This initiative will lead to enhanced penetration into all the sectors, facilitating deep decarbonization of economy, which is shown in Fig. 21.10 (Fig. 21.11).

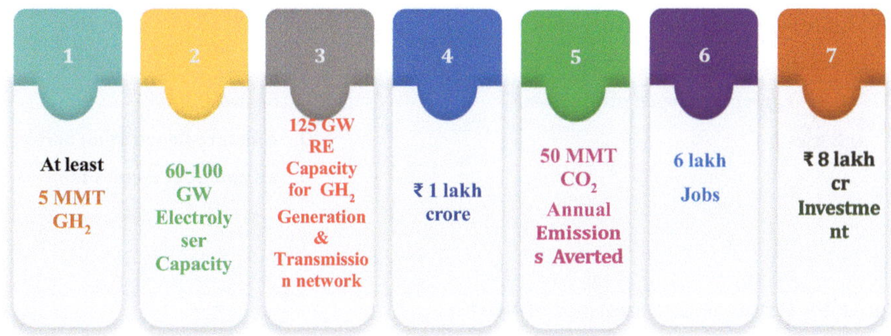

Fig. 21.8 The major goals of the NGHM

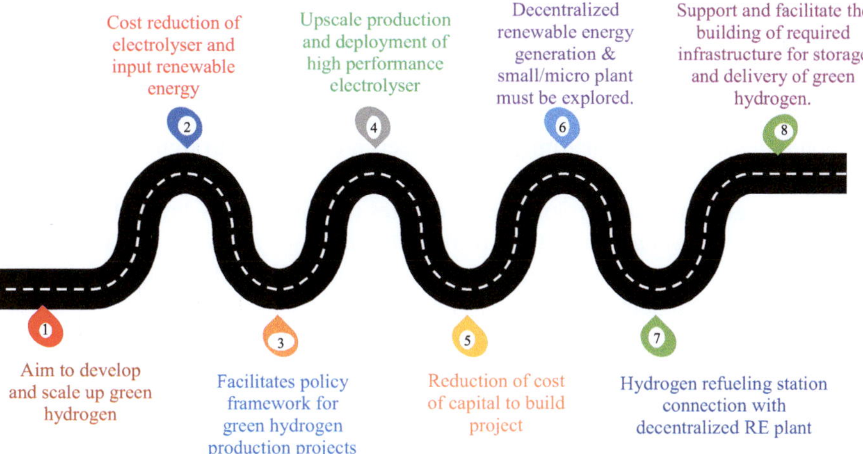

Fig. 21.9 Pathway to reach the goal of NGHM

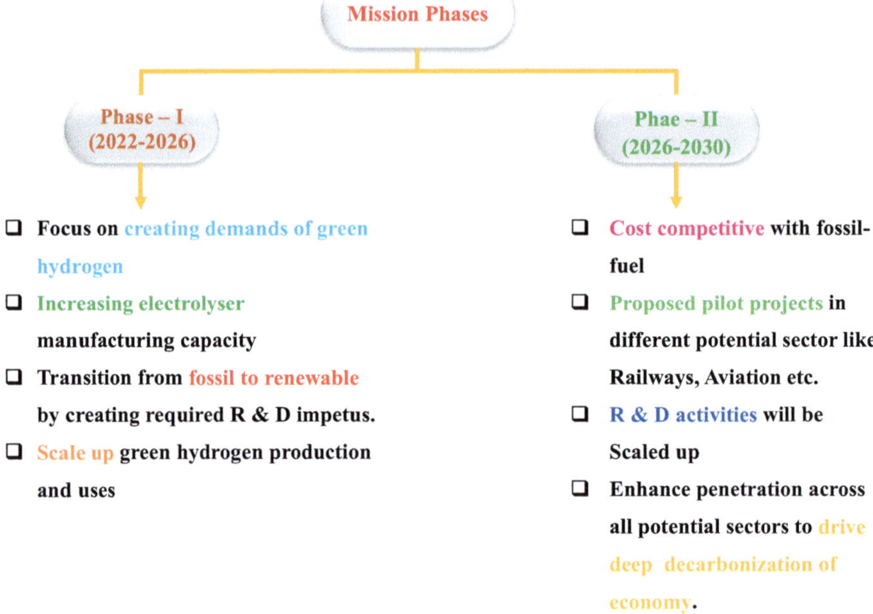

Fig. 21.10 Phases of implementation of NGHM

Toward the end of this book and chapter, the major barriers to the global implementation of the hydrogen economy, the key takeaways from the book, and the way forward are being summarized:

- **Cost**: One of the major barriers to the implementation and widespread deployment of hydrogen-based technologies is cost. In particular, the high cost of green hydrogen compared to conventional fuels remains a key challenge. However, the economies of scale and learning curve effects are expected to significantly reduce the cost over time.
- **Technological maturity**: Some technologies remain at a low level of technological readiness level and need to be proven at scale, e.g., gas turbines that operating exclusively on H_2 and maritime hydrogen transport; currently, only one prototype vessel have been demonstrated to transport LH_2.
- **Efficiency**: The complete hydrogen energy value chain involves several steps, each of which incur - whether in production, transport, conversion or end-use - resulting in a relatively low overall efficiency.
- **Sufficient renewable electricity**: Electrolytic H_2 will consume close to 21,000 TWh by 2050—almost as much electricity as is produced globally today; when all the end-use sectors are electrified, a lack of sufficient renewable electricity may become a bottleneck
- **Policy and regulatory uncertainty**: over 140 countries have pledged to achieve NZE within the coming decades; the speed with which these goals will be achieved

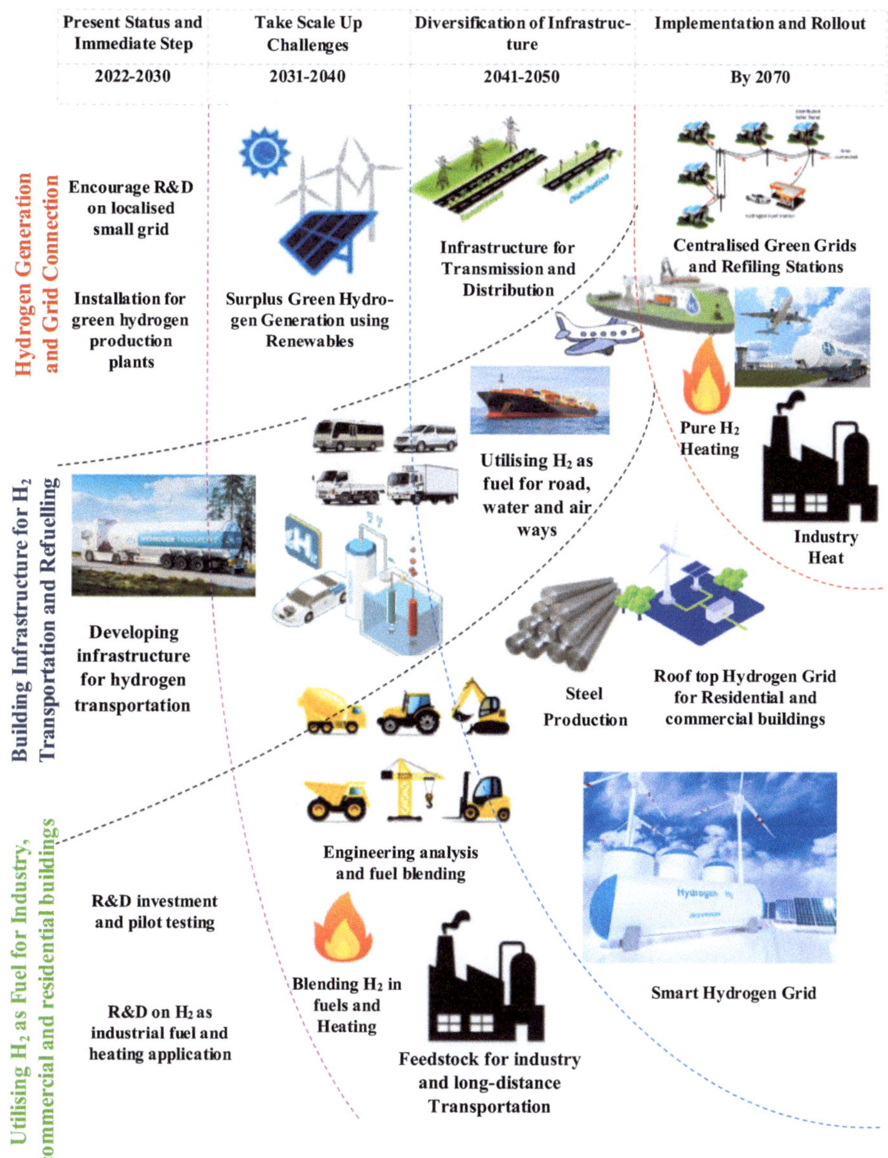

Fig. 21.11 Projected road map for building national hydrogen infrastructure (reproduced from [7] with permission)

remains uncertain; thus there is a need of stable, long-term policy framework to support development and deployment at scale.
- **Standards and certification**: Harmonized standards required for international trade. Safety measures and standards to be at place for domestic utilization as well. The standards, regulations, and regulatory certainty should be in place to increase the international trade and, at the same time, accelerate the goals of the net zero energy scenario or the announced pledges scenario.
- **Chicken-and-egg problem**: without demand, investments remain too risky and this will grow when cost comes down with the economies of scale; but without economies of scale, the technology remains too costly.

Key Takeaways

- Renewables together with energy efficiency will be responsible for energy transition.
- Hydrogen will play a key role toward climate change mitigation strategies, 12% of final energy consumption by 2050will be by green H_2.
- H_2 will allow transport of renewable energy over long distances, promoting energy trade.
- Countries with ample renewables deployment will become major producers of green H_2.
- In terms of hydrogen trade some countries will become major exporters and the business will become more competitive.
- Cross-border trade of hydrogen will increase; new bilateral energy relations will be established.
- Many countries will become exporters, some of those countries are currently energy importers (Chile, Morocco, Namibia).
- Demand is expected to increase from 2035 and is anticipated to drive significant growth and economic competitiveness.
- The use of hydrogen will reduce import dependence, address price volatility, and increase the resilience of the energy systems, providing flexibility and diversification.
- Requirement will be harmonized rules, standards, and norms, to increase deployment of H_2 technologies and trade.
- Initial market growth could come from applications where hydrogen is the only possible solution.

Some sectors are already using hydrogen in the current scenario, such as industrial processes and refineries. Other sectors require preparation before the technology can be rolled out on a larger scale, such as hydrogen blending and light- and medium-duty vehicles. Finally, certain segments - including marine, rail, mining, aviation, heat and power and steel - still require demonstration projects before wider deployment.

The major driving factors that will enable enhanced hydrogen deployment are cost reduction and the creation of demand. Policy support will be required including carbon pricing, and other measures that mandate the use of green

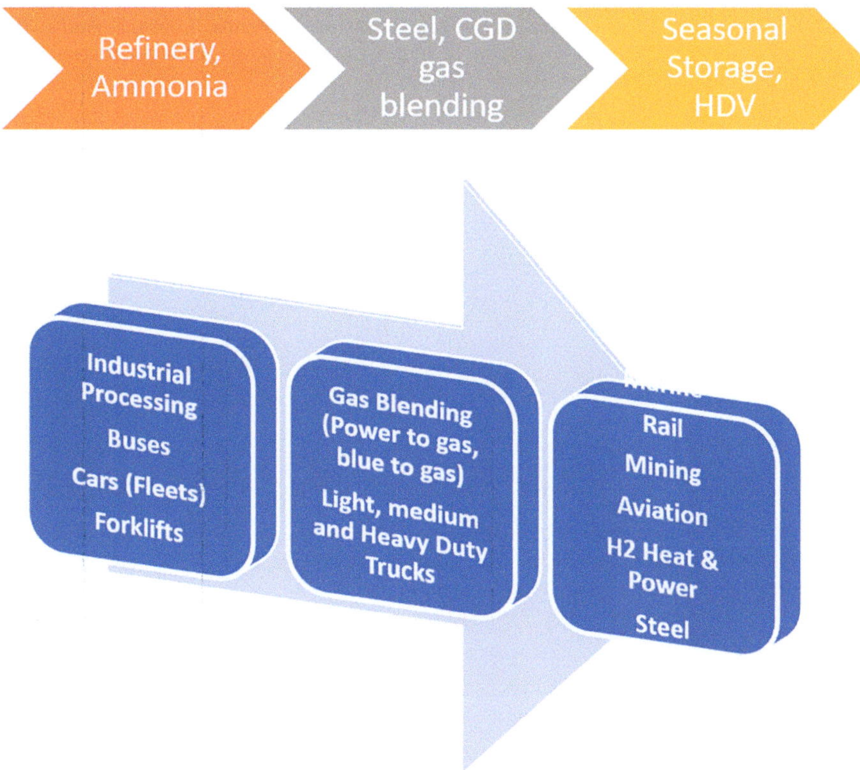

Fig. 21.12 Sectors where hydrogen is used currently and the projected growth will be in new additional sectors

hydrogen. An appropriate regulatory framework is essential to facilitate a transition from gry to green hydrogen, which is increasingly being prioritized to meet the environmental and energy needs. (Fig. 21.12).

References

1. Ministry of Planning and Infrastructure Cell website, http://ppac.gov.in/infrastructure/installed-refinery-capacity. Accessed 12 Jul 2025
2. Z. Fan, S. Julio Friedmann, Low-carbon production of iron and steel: Technology options, economic assessment, and policy, Joule, 5(4), 829–862, 2021. ISSN 2542–4351.
3. Global hydrogen review 2023, IEA report, https://www.iea.org/reports/global-hydrogen-review. Accessed 12 Jul 2025
4. Global hydrogen review 2024, IEA report 2024, https://www.iea.org/reports/global-hydrogen-review-2024. Accessed 12 Jul 2025
5. U. Lucia, Overview on fuel cells. Renew. Sustain. Energy Rev. **30**, 164–169 (2014)
6. A. Arshad et al., Energy and exergy analysis of fuel cells: A review. Thermal Science and Engineering Progress **9**, 308–321 (2019)

7. P. Muthukumar, A. Kumar, M. Afzal, S. Bhogilla, P. Sharma, A. Parida, S. Jana, E. Anil kumar, R.K. Pai, I.P. Jain, Review on large scale hydrogen storage systems for better sustainability. IJHE **48**, 33223–33259 (2023)

MIX
Papier aus verantwortungsvollen Quellen
Paper from responsible sources
FSC® C105338

If you have any concerns about our products,
you can contact us on
ProductSafety@springernature.com

In case Publisher is established outside the EU,
the EU authorized representative is:
Springer Nature Customer Service Center GmbH
Europaplatz 3, 69115 Heidelberg, Germany

Printed by Libri Plureos GmbH
in Hamburg, Germany